Calculus: Early Transcendentals Vol I

Includes Enhanced Web Assign Access

MATH 171 - Community College of Philadelphia

Stewart

8th Edition

CENGAGE
Learning·

Australia • Brazil • Japan • Korea • Mexico • Singapore • Spain • United Kingdom • United States

Calculus: Early Transcendentals: Includes Enhanced Web Assign Access, MATH 171 - Community College of Philadelphia, 8th Edition

Calculus: Early Transcendentals, Eighth Edition
James Stewart

For product information and technology assistance, contact us at
Cengage Learning Customer & Sales Support, 1-800-354-9706

For permission to use material from this text or product,
submit all requests online at **cengage.com/permissions**
Further permissions questions can be emailed to
permissionrequest@cengage.com

Compilation © 2015 Cengage Learning

ISBN: 978-1-337-05924-4

WCN: 01-100-101

Cengage Learning
20 Channel Center Street
Boston, MA 02210
USA

Cengage Learning is a leading provider of customized learning solutions with office locations around the globe, including Singapore, the United Kingdom, Australia, Mexico, Brazil, and Japan. Locate your local office at:
www.international.cengage.com/region.

Cengage Learning products are represented in Canada by Nelson Education, Ltd.

For your lifelong learning solutions, visit **www.cengage.com/custom.**

Visit our corporate website at **www.cengage.com.**

To the Student

Reading a calculus textbook is different from reading a newspaper or a novel, or even a physics book. Don't be discouraged if you have to read a passage more than once in order to understand it. You should have pencil and paper and calculator at hand to sketch a diagram or make a calculation.

Some students start by trying their homework problems and read the text only if they get stuck on an exercise. I suggest that a far better plan is to read and understand a section of the text before attempting the exercises. In particular, you should look at the definitions to see the exact meanings of the terms. And before you read each example, I suggest that you cover up the solution and try solving the problem yourself. You'll get a lot more from looking at the solution if you do so.

Part of the aim of this course is to train you to think logically. Learn to write the solutions of the exercises in a connected, step-by-step fashion with explanatory sentences—not just a string of disconnected equations or formulas.

The answers to the odd-numbered exercises appear at the back of the book, in Appendix I. Some exercises ask for a verbal explanation or interpretation or description. In such cases there is no single correct way of expressing the answer, so don't worry that you haven't found the definitive answer. In addition, there are often several different forms in which to express a numerical or algebraic answer, so if your answer differs from mine, don't immediately assume you're wrong. For example, if the answer given in the back of the book is $\sqrt{2} - 1$ and you obtain $1/(1 + \sqrt{2})$, then you're right and rationalizing the denominator will show that the answers are equivalent.

The icon ⌂ indicates an exercise that definitely requires the use of either a graphing calculator or a computer with graphing software. But that doesn't mean that graphing devices can't be used to check your work on the other exercises as well. The symbol CAS is reserved for problems in which the full resources of a computer algebra system (like Maple, Mathematica, or the TI-89) are required.

You will also encounter the symbol ⊘, which warns you against committing an error. I have placed this symbol in the margin in situations where I have observed that a large proportion of my students tend to make the same mistake.

Tools for Enriching Calculus, which is a companion to this text, is referred to by means of the symbol TEC and can be accessed in the eBook via Enhanced WebAssign and CourseMate (selected Visuals and Modules are available at www.stewartcalculus.com). It directs you to modules in which you can explore aspects of calculus for which the computer is particularly useful.

You will notice that some exercise numbers are printed in red: 5. This indicates that *Homework Hints* are available for the exercise. These hints can be found on stewartcalculus.com as well as Enhanced WebAssign and CourseMate. The homework hints ask you questions that allow you to make progress toward a solution without actually giving you the answer. You need to pursue each hint in an active manner with pencil and paper to work out the details. If a particular hint doesn't enable you to solve the problem, you can click to reveal the next hint.

I recommend that you keep this book for reference purposes after you finish the course. Because you will likely forget some of the specific details of calculus, the book will serve as a useful reminder when you need to use calculus in subsequent courses. And, because this book contains more material than can be covered in any one course, it can also serve as a valuable resource for a working scientist or engineer.

Calculus is an exciting subject, justly considered to be one of the greatest achievements of the human intellect. I hope you will discover that it is not only useful but also intrinsically beautiful.

JAMES STEWART

Calculators, Computers, and Other Graphing Devices

Advances in technology continue to bring a wider variety of tools for doing mathematics. Handheld calculators are becoming more powerful, as are software programs and Internet resources. In addition, many mathematical applications have been released for smartphones and tablets such as the iPad.

Some exercises in this text are marked with a graphing icon ⌂, which indicates that the use of some technology is required. Often this means that we intend for a graphing device to be used in drawing the graph of a function or equation. You might also need technology to find the zeros of a graph or the points of intersection of two graphs. In some cases we will use a calculating device to solve an equation or evaluate a definite integral numerically. Many scientific and graphing calculators have these features built in, such as the Texas Instruments TI-84 or TI-Nspire CX. Similar calculators are made by Hewlett Packard, Casio, and Sharp.

You can also use computer software such as *Graphing Calculator* by Pacific Tech (www.pacifict.com) to perform many of these functions, as well as apps for phones and tablets, like Quick Graph (Colombiamug) or Math-Studio (Pomegranate Apps). Similar functionality is available using a web interface at WolframAlpha.com.

In general, when we use the term "calculator" in this book, we mean the use of any of the resources we have mentioned.

The CAS icon is reserved for problems in which the full resources of a *computer algebra system* (CAS) are required. A CAS is capable of doing mathematics (like solving equations, computing derivatives or integrals) *symbolically* rather than just numerically.

Examples of well-established computer algebra systems are the computer software packages Maple and Mathematica. The WolframAlpha website uses the Mathematica engine to provide CAS functionality via the Web.

Many handheld graphing calculators have CAS capabilities, such as the TI-89 and TI-Nspire CX CAS from Texas Instruments. Some tablet and smartphone apps also provide these capabilities, such as the previously mentioned MathStudio.

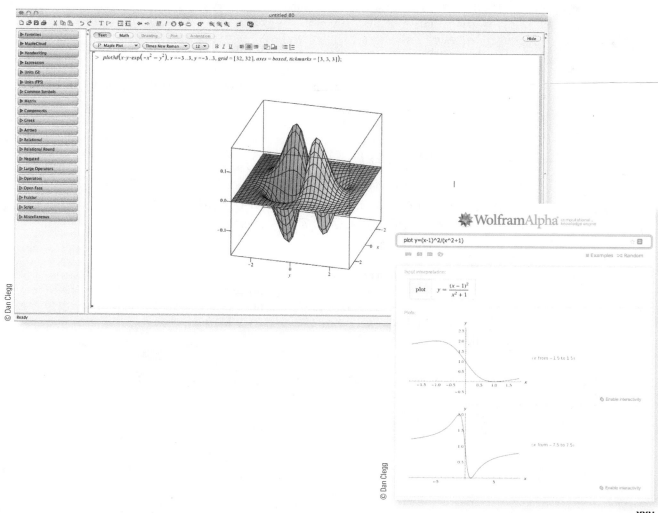

Diagnostic Tests

Success in calculus depends to a large extent on knowledge of the mathematics that precedes calculus: algebra, analytic geometry, functions, and trigonometry. The following tests are intended to diagnose weaknesses that you might have in these areas. After taking each test you can check your answers against the given answers and, if necessary, refresh your skills by referring to the review materials that are provided.

A Diagnostic Test: Algebra

1. Evaluate each expression without using a calculator.
   (a) $(-3)^4$ (b) -3^4 (c) 3^{-4}

   (d) $\dfrac{5^{23}}{5^{21}}$ (e) $\left(\dfrac{2}{3}\right)^{-2}$ (f) $16^{-3/4}$

2. Simplify each expression. Write your answer without negative exponents.
   (a) $\sqrt{200} - \sqrt{32}$
   (b) $(3a^3b^3)(4ab^2)^2$
   (c) $\left(\dfrac{3x^{3/2}y^3}{x^2y^{-1/2}}\right)^{-2}$

3. Expand and simplify.
   (a) $3(x + 6) + 4(2x - 5)$ (b) $(x + 3)(4x - 5)$
   (c) $(\sqrt{a} + \sqrt{b})(\sqrt{a} - \sqrt{b})$ (d) $(2x + 3)^2$
   (e) $(x + 2)^3$

4. Factor each expression.
   (a) $4x^2 - 25$ (b) $2x^2 + 5x - 12$
   (c) $x^3 - 3x^2 - 4x + 12$ (d) $x^4 + 27x$
   (e) $3x^{3/2} - 9x^{1/2} + 6x^{-1/2}$ (f) $x^3y - 4xy$

5. Simplify the rational expression.
   (a) $\dfrac{x^2 + 3x + 2}{x^2 - x - 2}$ (b) $\dfrac{2x^2 - x - 1}{x^2 - 9} \cdot \dfrac{x + 3}{2x + 1}$

   (c) $\dfrac{x^2}{x^2 - 4} - \dfrac{x + 1}{x + 2}$ (d) $\dfrac{\dfrac{y}{x} - \dfrac{x}{y}}{\dfrac{1}{y} - \dfrac{1}{x}}$

6. Rationalize the expression and simplify.

(a) $\dfrac{\sqrt{10}}{\sqrt{5} - 2}$

(b) $\dfrac{\sqrt{4 + h} - 2}{h}$

7. Rewrite by completing the square.

(a) $x^2 + x + 1$

(b) $2x^2 - 12x + 11$

8. Solve the equation. (Find only the real solutions.)

(a) $x + 5 = 14 - \frac{1}{2}x$

(b) $\dfrac{2x}{x + 1} = \dfrac{2x - 1}{x}$

(c) $x^2 - x - 12 = 0$

(d) $2x^2 + 4x + 1 = 0$

(e) $x^4 - 3x^2 + 2 = 0$

(f) $3|x - 4| = 10$

(g) $2x(4 - x)^{-1/2} - 3\sqrt{4 - x} = 0$

9. Solve each inequality. Write your answer using interval notation.

(a) $-4 < 5 - 3x \leq 17$

(b) $x^2 < 2x + 8$

(c) $x(x - 1)(x + 2) > 0$

(d) $|x - 4| < 3$

(e) $\dfrac{2x - 3}{x + 1} \leq 1$

10. State whether each equation is true or false.

(a) $(p + q)^2 = p^2 + q^2$

(b) $\sqrt{ab} = \sqrt{a}\,\sqrt{b}$

(c) $\sqrt{a^2 + b^2} = a + b$

(d) $\dfrac{1 + TC}{C} = 1 + T$

(e) $\dfrac{1}{x - y} = \dfrac{1}{x} - \dfrac{1}{y}$

(f) $\dfrac{1/x}{a/x - b/x} = \dfrac{1}{a - b}$

ANSWERS TO DIAGNOSTIC TEST A: ALGEBRA

1. (a) 81 (b) −81 (c) $\frac{1}{81}$
(d) 25 (e) $\frac{9}{4}$ (f) $\frac{1}{8}$

2. (a) $6\sqrt{2}$ (b) $48a^5b^7$ (c) $\dfrac{x}{9y^7}$

3. (a) $11x - 2$ (b) $4x^2 + 7x - 15$
(c) $a - b$ (d) $4x^2 + 12x + 9$
(e) $x^3 + 6x^2 + 12x + 8$

4. (a) $(2x - 5)(2x + 5)$ (b) $(2x - 3)(x + 4)$
(c) $(x - 3)(x - 2)(x + 2)$ (d) $x(x + 3)(x^2 - 3x + 9)$
(e) $3x^{-1/2}(x - 1)(x - 2)$ (f) $xy(x - 2)(x + 2)$

5. (a) $\dfrac{x + 2}{x - 2}$ (b) $\dfrac{x - 1}{x - 3}$
(c) $\dfrac{1}{x - 2}$ (d) $-(x + y)$

6. (a) $5\sqrt{2} + 2\sqrt{10}$ (b) $\dfrac{1}{\sqrt{4 + h} + 2}$

7. (a) $\left(x + \frac{1}{2}\right)^2 + \frac{3}{4}$ (b) $2(x - 3)^2 - 7$

8. (a) 6 (b) 1 (c) −3, 4
(d) $-1 \pm \frac{1}{2}\sqrt{2}$ (e) $\pm 1, \pm\sqrt{2}$ (f) $\frac{2}{3}, \frac{22}{3}$
(g) $\frac{12}{5}$

9. (a) $[-4, 3)$ (b) $(-2, 4)$
(c) $(-2, 0) \cup (1, \infty)$ (d) $(1, 7)$
(e) $(-1, 4]$

10. (a) False (b) True (c) False
(d) False (e) False (f) True

If you had difficulty with these problems, you may wish to consult the Review of Algebra on the website **www.stewartcalculus.com**.

B Diagnostic Test: Analytic Geometry

1. Find an equation for the line that passes through the point $(2, -5)$ and
    (a) has slope -3
    (b) is parallel to the x-axis
    (c) is parallel to the y-axis
    (d) is parallel to the line $2x - 4y = 3$

2. Find an equation for the circle that has center $(-1, 4)$ and passes through the point $(3, -2)$.

3. Find the center and radius of the circle with equation $x^2 + y^2 - 6x + 10y + 9 = 0$.

4. Let $A(-7, 4)$ and $B(5, -12)$ be points in the plane.
    (a) Find the slope of the line that contains A and B.
    (b) Find an equation of the line that passes through A and B. What are the intercepts?
    (c) Find the midpoint of the segment AB.
    (d) Find the length of the segment AB.
    (e) Find an equation of the perpendicular bisector of AB.
    (f) Find an equation of the circle for which AB is a diameter.

5. Sketch the region in the xy-plane defined by the equation or inequalities.
    (a) $-1 \le y \le 3$ (b) $|x| < 4$ and $|y| < 2$
    (c) $y < 1 - \frac{1}{2}x$ (d) $y \ge x^2 - 1$
    (e) $x^2 + y^2 < 4$ (f) $9x^2 + 16y^2 = 144$

ANSWERS TO DIAGNOSTIC TEST B: ANALYTIC GEOMETRY

1. (a) $y = -3x + 1$ (b) $y = -5$
    (c) $x = 2$ (d) $y = \frac{1}{2}x - 6$

2. $(x + 1)^2 + (y - 4)^2 = 52$

3. Center $(3, -5)$, radius 5

4. (a) $-\frac{4}{3}$
    (b) $4x + 3y + 16 = 0$; x-intercept -4, y-intercept $-\frac{16}{3}$
    (c) $(-1, -4)$
    (d) 20
    (e) $3x - 4y = 13$
    (f) $(x + 1)^2 + (y + 4)^2 = 100$

5.

(a)

(b)

(c)

(d)

(e)

(f)

If you had difficulty with these problems, you may wish to consult the review of analytic geometry in Appendixes B and C.

C Diagnostic Test: Functions

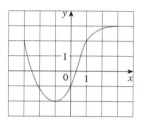

FIGURE FOR PROBLEM 1

1. The graph of a function f is given at the left.
   (a) State the value of $f(-1)$.
   (b) Estimate the value of $f(2)$.
   (c) For what values of x is $f(x) = 2$?
   (d) Estimate the values of x such that $f(x) = 0$.
   (e) State the domain and range of f.

2. If $f(x) = x^3$, evaluate the difference quotient $\dfrac{f(2 + h) - f(2)}{h}$ and simplify your answer.

3. Find the domain of the function.

   (a) $f(x) = \dfrac{2x + 1}{x^2 + x - 2}$ (b) $g(x) = \dfrac{\sqrt[3]{x}}{x^2 + 1}$ (c) $h(x) = \sqrt{4 - x} + \sqrt{x^2 - 1}$

4. How are graphs of the functions obtained from the graph of f?

   (a) $y = -f(x)$ (b) $y = 2f(x) - 1$ (c) $y = f(x - 3) + 2$

5. Without using a calculator, make a rough sketch of the graph.

   (a) $y = x^3$ (b) $y = (x + 1)^3$ (c) $y = (x - 2)^3 + 3$
   (d) $y = 4 - x^2$ (e) $y = \sqrt{x}$ (f) $y = 2\sqrt{x}$
   (g) $y = -2^x$ (h) $y = 1 + x^{-1}$

6. Let $f(x) = \begin{cases} 1 - x^2 & \text{if } x \leqslant 0 \\ 2x + 1 & \text{if } x > 0 \end{cases}$

   (a) Evaluate $f(-2)$ and $f(1)$.   (b) Sketch the graph of f.

7. If $f(x) = x^2 + 2x - 1$ and $g(x) = 2x - 3$, find each of the following functions.
   (a) $f \circ g$ (b) $g \circ f$ (c) $g \circ g \circ g$

ANSWERS TO DIAGNOSTIC TEST C: FUNCTIONS

1. (a) -2 (b) 2.8
   (c) $-3, 1$ (d) $-2.5, 0.3$
   (e) $[-3, 3], [-2, 3]$

2. $12 + 6h + h^2$

3. (a) $(-\infty, -2) \cup (-2, 1) \cup (1, \infty)$
   (b) $(-\infty, \infty)$
   (c) $(-\infty, -1] \cup [1, 4]$

4. (a) Reflect about the x-axis
   (b) Stretch vertically by a factor of 2, then shift 1 unit
   downward
   (c) Shift 3 units to the right and 2 units upward

5. (a)   (b)   (c)

   (d)   (e)   (f)

   (g)   (h)

6. (a) $-3, 3$ (b)

7. (a) $(f \circ g)(x) = 4x^2 - 8x + 2$
(b) $(g \circ f)(x) = 2x^2 + 4x - 5$
(c) $(g \circ g \circ g)(x) = 8x - 21$

> If you had difficulty with these problems, you should look at sections 1.1–1.3 of this book.

D   Diagnostic Test: Trigonometry

1. Convert from degrees to radians.
(a) $300°$ (b) $-18°$

2. Convert from radians to degrees.
(a) $5\pi/6$ (b) 2

3. Find the length of an arc of a circle with radius 12 cm if the arc subtends a central angle of $30°$.

4. Find the exact values.
(a) $\tan(\pi/3)$ (b) $\sin(7\pi/6)$ (c) $\sec(5\pi/3)$

5. Express the lengths a and b in the figure in terms of θ.

FIGURE FOR PROBLEM 5

6. If $\sin x = \frac{1}{3}$ and $\sec y = \frac{5}{4}$, where x and y lie between 0 and $\pi/2$, evaluate $\sin(x + y)$.

7. Prove the identities.
(a) $\tan\theta \sin\theta + \cos\theta = \sec\theta$ (b) $\dfrac{2\tan x}{1 + \tan^2 x} = \sin 2x$

8. Find all values of x such that $\sin 2x = \sin x$ and $0 \leqslant x \leqslant 2\pi$.

9. Sketch the graph of the function $y = 1 + \sin 2x$ without using a calculator.

ANSWERS TO DIAGNOSTIC TEST D: TRIGONOMETRY

1. (a) $5\pi/3$ (b) $-\pi/10$

2. (a) $150°$ (b) $360°/\pi \approx 114.6°$

3. 2π cm

4. (a) $\sqrt{3}$ (b) $-\frac{1}{2}$ (c) 2

5. (a) $24\sin\theta$ (b) $24\cos\theta$

6. $\frac{1}{15}\left(4 + 6\sqrt{2}\right)$

8. $0, \pi/3, \pi, 5\pi/3, 2\pi$

9.

> If you had difficulty with these problems, you should look at Appendix D of this book.

A Preview of Calculus

By the time you finish this course, you will be able to calculate the length of the curve used to design the Gateway Arch in St. Louis, determine where a pilot should start descent for a smooth landing, compute the force on a baseball bat when it strikes the ball, and measure the amount of light sensed by the human eye as the pupil changes size.

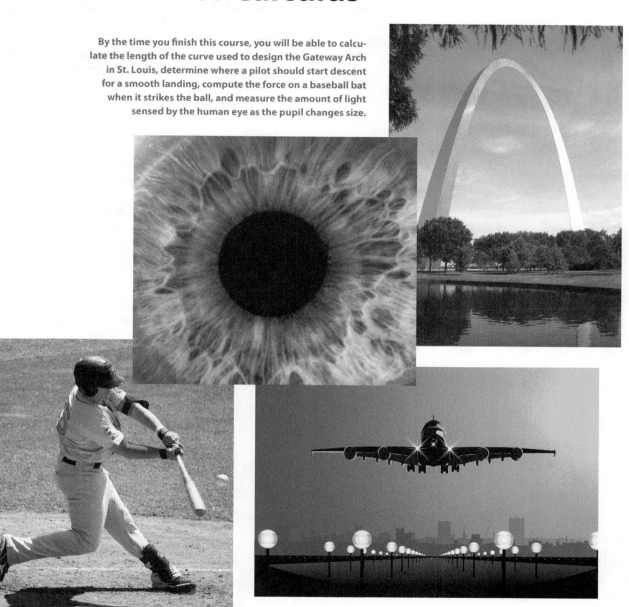

CALCULUS IS FUNDAMENTALLY DIFFERENT FROM the mathematics that you have studied previously: calculus is less static and more dynamic. It is concerned with change and motion; it deals with quantities that approach other quantities. For that reason it may be useful to have an overview of the subject before beginning its intensive study. Here we give a glimpse of some of the main ideas of calculus by showing how the concept of a limit arises when we attempt to solve a variety of problems.

$$A = A_1 + A_2 + A_3 + A_4 + A_5$$

FIGURE 1

The Area Problem

The origins of calculus go back at least 2500 years to the ancient Greeks, who found areas using the "method of exhaustion." They knew how to find the area A of any polygon by dividing it into triangles as in Figure 1 and adding the areas of these triangles.

It is a much more difficult problem to find the area of a curved figure. The Greek method of exhaustion was to inscribe polygons in the figure and circumscribe polygons about the figure and then let the number of sides of the polygons increase. Figure 2 illustrates this process for the special case of a circle with inscribed regular polygons.

FIGURE 2

Let A_n be the area of the inscribed polygon with n sides. As n increases, it appears that A_n becomes closer and closer to the area of the circle. We say that the area of the circle is the limit of the areas of the inscribed polygons, and we write

$$A = \lim_{n \to \infty} A_n$$

TEC In the Preview Visual, you can see how areas of inscribed and circumscribed polygons approximate the area of a circle.

The Greeks themselves did not use limits explicitly. However, by indirect reasoning, Eudoxus (fifth century BC) used exhaustion to prove the familiar formula for the area of a circle: $A = \pi r^2$.

We will use a similar idea in Chapter 5 to find areas of regions of the type shown in Figure 3. We will approximate the desired area A by areas of rectangles (as in Figure 4), let the width of the rectangles decrease, and then calculate A as the limit of these sums of areas of rectangles.

FIGURE 3   **FIGURE 4**

The area problem is the central problem in the branch of calculus called *integral calculus*. The techniques that we will develop in Chapter 5 for finding areas will also enable us to compute the volume of a solid, the length of a curve, the force of water against a dam, the mass and center of gravity of a rod, and the work done in pumping water out of a tank.

The Tangent Problem

Consider the problem of trying to find an equation of the tangent line t to a curve with equation $y = f(x)$ at a given point P. (We will give a precise definition of a tangent line in

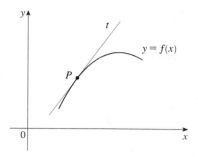

FIGURE 5
The tangent line at P

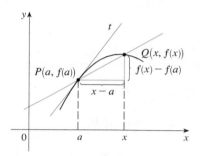

FIGURE 6
The secant line at PQ

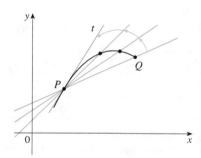

FIGURE 7
Secant lines approaching the
tangent line

Chapter 2. For now you can think of it as a line that touches the curve at P as in Figure 5.) Since we know that the point P lies on the tangent line, we can find the equation of t if we know its slope m. The problem is that we need two points to compute the slope and we know only one point, P, on t. To get around the problem we first find an approximation to m by taking a nearby point Q on the curve and computing the slope m_{PQ} of the secant line PQ. From Figure 6 we see that

$$\boxed{1} \qquad m_{PQ} = \frac{f(x) - f(a)}{x - a}$$

Now imagine that Q moves along the curve toward P as in Figure 7. You can see that the secant line rotates and approaches the tangent line as its limiting position. This means that the slope m_{PQ} of the secant line becomes closer and closer to the slope m of the tangent line. We write

$$m = \lim_{Q \to P} m_{PQ}$$

and we say that m is the limit of m_{PQ} as Q approaches P along the curve. Because x approaches a as Q approaches P, we could also use Equation 1 to write

$$\boxed{2} \qquad m = \lim_{x \to a} \frac{f(x) - f(a)}{x - a}$$

Specific examples of this procedure will be given in Chapter 2.

The tangent problem has given rise to the branch of calculus called *differential calculus,* which was not invented until more than 2000 years after integral calculus. The main ideas behind differential calculus are due to the French mathematician Pierre Fermat (1601–1665) and were developed by the English mathematicians John Wallis (1616–1703), Isaac Barrow (1630–1677), and Isaac Newton (1642–1727) and the German mathematician Gottfried Leibniz (1646–1716).

The two branches of calculus and their chief problems, the area problem and the tangent problem, appear to be very different, but it turns out that there is a very close connection between them. The tangent problem and the area problem are inverse problems in a sense that will be described in Chapter 5.

■ Velocity

When we look at the speedometer of a car and read that the car is traveling at 48 mi/h, what does that information indicate to us? We know that if the velocity remains constant, then after an hour we will have traveled 48 mi. But if the velocity of the car varies, what does it mean to say that the velocity at a given instant is 48 mi/h?

In order to analyze this question, let's examine the motion of a car that travels along a straight road and assume that we can measure the distance traveled by the car (in feet) at 1-second intervals as in the following chart:

t = Time elapsed (s)	0	1	2	3	4	5
d = Distance (ft)	0	2	9	24	42	71

As a first step toward finding the velocity after 2 seconds have elapsed, we find the average velocity during the time interval $2 \leqslant t \leqslant 4$:

$$
\text{average velocity} = \frac{\text{change in position}}{\text{time elapsed}}
$$

$$
= \frac{42 - 9}{4 - 2}
$$

$$
= 16.5 \text{ ft/s}
$$

Similarly, the average velocity in the time interval $2 \leqslant t \leqslant 3$ is

$$
\text{average velocity} = \frac{24 - 9}{3 - 2} = 15 \text{ ft/s}
$$

We have the feeling that the velocity at the instant $t = 2$ can't be much different from the average velocity during a short time interval starting at $t = 2$. So let's imagine that the distance traveled has been measured at 0.1-second time intervals as in the following chart:

t	2.0	2.1	2.2	2.3	2.4	2.5
d	9.00	10.02	11.16	12.45	13.96	15.80

Then we can compute, for instance, the average velocity over the time interval $[2, 2.5]$:

$$
\text{average velocity} = \frac{15.80 - 9.00}{2.5 - 2} = 13.6 \text{ ft/s}
$$

The results of such calculations are shown in the following chart:

Time interval	[2, 3]	[2, 2.5]	[2, 2.4]	[2, 2.3]	[2, 2.2]	[2, 2.1]
Average velocity (ft/s)	15.0	13.6	12.4	11.5	10.8	10.2

The average velocities over successively smaller intervals appear to be getting closer to a number near 10, and so we expect that the velocity at exactly $t = 2$ is about 10 ft/s. In Chapter 2 we will define the instantaneous velocity of a moving object as the limiting value of the average velocities over smaller and smaller time intervals.

In Figure 8 we show a graphical representation of the motion of the car by plotting the distance traveled as a function of time. If we write $d = f(t)$, then $f(t)$ is the number of feet traveled after t seconds. The average velocity in the time interval $[2, t]$ is

$$
\text{average velocity} = \frac{\text{change in position}}{\text{time elapsed}} = \frac{f(t) - f(2)}{t - 2}
$$

which is the same as the slope of the secant line PQ in Figure 8. The velocity v when $t = 2$ is the limiting value of this average velocity as t approaches 2; that is,

$$
v = \lim_{t \to 2} \frac{f(t) - f(2)}{t - 2}
$$

and we recognize from Equation 2 that this is the same as the slope of the tangent line to the curve at P.

FIGURE 8

Thus, when we solve the tangent problem in differential calculus, we are also solving problems concerning velocities. The same techniques also enable us to solve problems involving rates of change in all of the natural and social sciences.

■ The Limit of a Sequence

In the fifth century BC the Greek philosopher Zeno of Elea posed four problems, now known as *Zeno's paradoxes*, that were intended to challenge some of the ideas concerning space and time that were held in his day. Zeno's second paradox concerns a race between the Greek hero Achilles and a tortoise that has been given a head start. Zeno argued, as follows, that Achilles could never pass the tortoise: Suppose that Achilles starts at position a_1 and the tortoise starts at position t_1. (See Figure 9.) When Achilles reaches the point $a_2 = t_1$, the tortoise is farther ahead at position t_2. When Achilles reaches $a_3 = t_2$, the tortoise is at t_3. This process continues indefinitely and so it appears that the tortoise will always be ahead! But this defies common sense.

FIGURE 9

One way of explaining this paradox is with the idea of a *sequence*. The successive positions of Achilles ($a_1, a_2, a_3, \ldots$) or the successive positions of the tortoise ($t_1, t_2, t_3, \ldots$) form what is known as a sequence.

In general, a sequence $\{a_n\}$ is a set of numbers written in a definite order. For instance, the sequence

$$\left\{1, \tfrac{1}{2}, \tfrac{1}{3}, \tfrac{1}{4}, \tfrac{1}{5}, \ldots\right\}$$

can be described by giving the following formula for the nth term:

$$a_n = \frac{1}{n}$$

FIGURE 10

We can visualize this sequence by plotting its terms on a number line as in Figure 10(a) or by drawing its graph as in Figure 10(b). Observe from either picture that the terms of the sequence $a_n = 1/n$ are becoming closer and closer to 0 as n increases. In fact, we can find terms as small as we please by making n large enough. We say that the limit of the sequence is 0, and we indicate this by writing

$$\lim_{n \to \infty} \frac{1}{n} = 0$$

In general, the notation

$$\lim_{n \to \infty} a_n = L$$

is used if the terms a_n approach the number L as n becomes large. This means that the numbers a_n can be made as close as we like to the number L by taking n sufficiently large.

The concept of the limit of a sequence occurs whenever we use the decimal representation of a real number. For instance, if

$$a_1 = 3.1$$

$$a_2 = 3.14$$

$$a_3 = 3.141$$

$$a_4 = 3.1415$$

$$a_5 = 3.14159$$

$$a_6 = 3.141592$$

$$a_7 = 3.1415926$$

$$\vdots$$

then

$$\lim_{n \to \infty} a_n = \pi$$

The terms in this sequence are rational approximations to π.

Let's return to Zeno's paradox. The successive positions of Achilles and the tortoise form sequences $\{a_n\}$ and $\{t_n\}$, where $a_n < t_n$ for all n. It can be shown that both sequences have the same limit:

$$\lim_{n \to \infty} a_n = p = \lim_{n \to \infty} t_n$$

It is precisely at this point p that Achilles overtakes the tortoise.

The Sum of a Series

Another of Zeno's paradoxes, as passed on to us by Aristotle, is the following: "A man standing in a room cannot walk to the wall. In order to do so, he would first have to go half the distance, then half the remaining distance, and then again half of what still remains. This process can always be continued and can never be ended." (See Figure 11.)

FIGURE 11

Of course, we know that the man can actually reach the wall, so this suggests that perhaps the total distance can be expressed as the sum of infinitely many smaller distances as follows:

3

$$1 = \frac{1}{2} + \frac{1}{4} + \frac{1}{8} + \frac{1}{16} + \cdots + \frac{1}{2^n} + \cdots$$

Zeno was arguing that it doesn't make sense to add infinitely many numbers together. But there are other situations in which we implicitly use infinite sums. For instance, in decimal notation, the symbol $0.\overline{3} = 0.3333\ldots$ means

$$\frac{3}{10} + \frac{3}{100} + \frac{3}{1000} + \frac{3}{10,000} + \cdots$$

and so, in some sense, it must be true that

$$\frac{3}{10} + \frac{3}{100} + \frac{3}{1000} + \frac{3}{10,000} + \cdots = \frac{1}{3}$$

More generally, if d_n denotes the nth digit in the decimal representation of a number, then

$$0.d_1 d_2 d_3 d_4 \ldots = \frac{d_1}{10} + \frac{d_2}{10^2} + \frac{d_3}{10^3} + \cdots + \frac{d_n}{10^n} + \cdots$$

Therefore some infinite sums, or infinite series as they are called, have a meaning. But we must define carefully what the sum of an infinite series is.

Returning to the series in Equation 3, we denote by s_n the sum of the first n terms of the series. Thus

$$s_1 = \tfrac{1}{2} = 0.5$$

$$s_2 = \tfrac{1}{2} + \tfrac{1}{4} = 0.75$$

$$s_3 = \tfrac{1}{2} + \tfrac{1}{4} + \tfrac{1}{8} = 0.875$$

$$s_4 = \tfrac{1}{2} + \tfrac{1}{4} + \tfrac{1}{8} + \tfrac{1}{16} = 0.9375$$

$$s_5 = \tfrac{1}{2} + \tfrac{1}{4} + \tfrac{1}{8} + \tfrac{1}{16} + \tfrac{1}{32} = 0.96875$$

$$s_6 = \tfrac{1}{2} + \tfrac{1}{4} + \tfrac{1}{8} + \tfrac{1}{16} + \tfrac{1}{32} + \tfrac{1}{64} = 0.984375$$

$$s_7 = \tfrac{1}{2} + \tfrac{1}{4} + \tfrac{1}{8} + \tfrac{1}{16} + \tfrac{1}{32} + \tfrac{1}{64} + \tfrac{1}{128} = 0.9921875$$

$$\vdots$$

$$s_{10} = \tfrac{1}{2} + \tfrac{1}{4} + \cdots + \tfrac{1}{1024} \approx 0.99902344$$

$$\vdots$$

$$s_{16} = \frac{1}{2} + \frac{1}{4} + \cdots + \frac{1}{2^{16}} \approx 0.99998474$$

Observe that as we add more and more terms, the partial sums become closer and closer to 1. In fact, it can be shown that by taking n large enough (that is, by adding sufficiently many terms of the series), we can make the partial sum s_n as close as we please to the number 1. It therefore seems reasonable to say that the sum of the infinite series is 1 and to write

$$\frac{1}{2} + \frac{1}{4} + \frac{1}{8} + \cdots + \frac{1}{2^n} + \cdots = 1$$

In other words, the reason the sum of the series is 1 is that

$$\lim_{n \to \infty} s_n = 1$$

In Chapter 11 we will discuss these ideas further. We will then use Newton's idea of combining infinite series with differential and integral calculus.

Summary

We have seen that the concept of a limit arises in trying to find the area of a region, the slope of a tangent to a curve, the velocity of a car, or the sum of an infinite series. In each case the common theme is the calculation of a quantity as the limit of other, easily calculated quantities. It is this basic idea of a limit that sets calculus apart from other areas of mathematics. In fact, we could define calculus as the part of mathematics that deals with limits.

After Sir Isaac Newton invented his version of calculus, he used it to explain the motion of the planets around the sun. Today calculus is used in calculating the orbits of satellites and spacecraft, in predicting population sizes, in estimating how fast oil prices rise or fall, in forecasting weather, in measuring the cardiac output of the heart, in calculating life insurance premiums, and in a great variety of other areas. We will explore some of these uses of calculus in this book.

In order to convey a sense of the power of the subject, we end this preview with a list of some of the questions that you will be able to answer using calculus:

1. How can we explain the fact, illustrated in Figure 12, that the angle of elevation from an observer up to the highest point in a rainbow is 42°? (See page 285.)

2. How can we explain the shapes of cans on supermarket shelves? (See page 343.)

3. Where is the best place to sit in a movie theater? (See page 465.)

4. How can we design a roller coaster for a smooth ride? (See page 182.)

5. How far away from an airport should a pilot start descent? (See page 208.)

6. How can we fit curves together to design shapes to represent letters on a laser printer? (See page 657.)

7. How can we estimate the number of workers that were needed to build the Great Pyramid of Khufu in ancient Egypt? (See page 460.)

8. Where should an infielder position himself to catch a baseball thrown by an outfielder and relay it to home plate? (See page 465.)

9. Does a ball thrown upward take longer to reach its maximum height or to fall back to its original height? (See page 609.)

10. How can we explain the fact that planets and satellites move in elliptical orbits? (See page 868.)

11. How can we distribute water flow among turbines at a hydroelectric station so as to maximize the total energy production? (See page 980.)

12. If a marble, a squash ball, a steel bar, and a lead pipe roll down a slope, which of them reaches the bottom first? (See page 1052.)

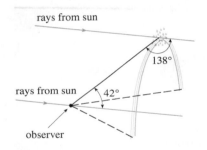

FIGURE 12

1 Functions and Models

Pictura Collectus/Alamy

Often a graph is the best way to represent a function because it conveys so much information at a glance. Shown is a graph of the vertical ground acceleration created by the 2011 earthquake near Tohoku, Japan. The earthquake had a magnitude of 9.0 on the Richter scale and was so powerful that it moved northern Japan 8 feet closer to North America.

Seismological Society of America

THE FUNDAMENTAL OBJECTS THAT WE deal with in calculus are functions. This chapter prepares the way for calculus by discussing the basic ideas concerning functions, their graphs, and ways of transforming and combining them. We stress that a function can be represented in different ways: by an equation, in a table, by a graph, or in words. We look at the main types of functions that occur in calculus and describe the process of using these functions as mathematical models of real-world phenomena.

1.1   Four Ways to Represent a Function

Functions arise whenever one quantity depends on another. Consider the following four situations.

A. The area A of a circle depends on the radius r of the circle. The rule that connects r and A is given by the equation $A = \pi r^2$. With each positive number r there is associated one value of A, and we say that A is a *function* of r.

B. The human population of the world P depends on the time t. The table gives estimates of the world population $P(t)$ at time t, for certain years. For instance,

$$P(1950) \approx 2{,}560{,}000{,}000$$

But for each value of the time t there is a corresponding value of P, and we say that P is a function of t.

C. The cost C of mailing an envelope depends on its weight w. Although there is no simple formula that connects w and C, the post office has a rule for determining C when w is known.

D. The vertical acceleration a of the ground as measured by a seismograph during an earthquake is a function of the elapsed time t. Figure 1 shows a graph generated by seismic activity during the Northridge earthquake that shook Los Angeles in 1994. For a given value of t, the graph provides a corresponding value of a.

Year	Population (millions)
1900	1650
1910	1750
1920	1860
1930	2070
1940	2300
1950	2560
1960	3040
1970	3710
1980	4450
1990	5280
2000	6080
2010	6870

FIGURE 1
Vertical ground acceleration
during the Northridge earthquake

Calif. Dept. of Mines and Geology

Each of these examples describes a rule whereby, given a number (r, t, w, or t), another number (A, P, C, or a) is assigned. In each case we say that the second number is a function of the first number.

> A **function** f is a rule that assigns to each element x in a set D exactly one element, called $f(x)$, in a set E.

We usually consider functions for which the sets D and E are sets of real numbers. The set D is called the **domain** of the function. The number $f(x)$ is the **value of f at x** and is read "f of x." The **range** of f is the set of all possible values of $f(x)$ as x varies throughout the domain. A symbol that represents an arbitrary number in the *domain* of a function f is called an **independent variable**. A symbol that represents a number in the *range* of f is called a **dependent variable**. In Example A, for instance, r is the independent variable and A is the dependent variable.

FIGURE 2

Machine diagram for a function f

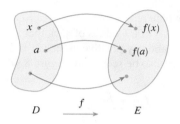

FIGURE 3

Arrow diagram for f

It's helpful to think of a function as a **machine** (see Figure 2). If x is in the domain of the function f, then when x enters the machine, it's accepted as an input and the machine produces an output $f(x)$ according to the rule of the function. Thus we can think of the domain as the set of all possible inputs and the range as the set of all possible outputs.

The preprogrammed functions in a calculator are good examples of a function as a machine. For example, the square root key on your calculator computes such a function. You press the key labeled $\sqrt{\ }$ (or $\sqrt{x}$) and enter the input x. If $x < 0$, then x is not in the domain of this function; that is, x is not an acceptable input, and the calculator will indicate an error. If $x \geqslant 0$, then an *approximation* to $\sqrt{x}$ will appear in the display. Thus the $\sqrt{x}$ key on your calculator is not quite the same as the exact mathematical function f defined by $f(x) = \sqrt{x}$.

Another way to picture a function is by an **arrow diagram** as in Figure 3. Each arrow connects an element of D to an element of E. The arrow indicates that $f(x)$ is associated with x, $f(a)$ is associated with a, and so on.

The most common method for visualizing a function is its graph. If f is a function with domain D, then its **graph** is the set of ordered pairs

$$\{(x, f(x)) \mid x \in D\}$$

(Notice that these are input-output pairs.) In other words, the graph of f consists of all points (x, y) in the coordinate plane such that $y = f(x)$ and x is in the domain of f.

The graph of a function f gives us a useful picture of the behavior or "life history" of a function. Since the y-coordinate of any point (x, y) on the graph is $y = f(x)$, we can read the value of $f(x)$ from the graph as being the height of the graph above the point x (see Figure 4). The graph of f also allows us to picture the domain of f on the x-axis and its range on the y-axis as in Figure 5.

FIGURE 4

FIGURE 5

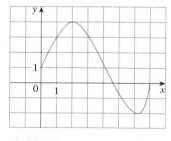

FIGURE 6

The notation for intervals is given in Appendix A.

EXAMPLE 1   The graph of a function f is shown in Figure 6.
(a) Find the values of $f(1)$ and $f(5)$.
(b) What are the domain and range of f?

SOLUTION
(a) We see from Figure 6 that the point $(1, 3)$ lies on the graph of f, so the value of f at 1 is $f(1) = 3$. (In other words, the point on the graph that lies above $x = 1$ is 3 units above the x-axis.)

When $x = 5$, the graph lies about 0.7 units below the x-axis, so we estimate that $f(5) \approx -0.7$.

(b) We see that $f(x)$ is defined when $0 \leqslant x \leqslant 7$, so the domain of f is the closed interval $[0, 7]$. Notice that f takes on all values from -2 to 4, so the range of f is

$$\{y \mid -2 \leqslant y \leqslant 4\} = [-2, 4] \qquad \blacksquare$$

FIGURE 7

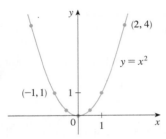

FIGURE 8

The expression

$$\frac{f(a + h) - f(a)}{h}$$

in Example 3 is called a **difference quotient** and occurs frequently in calculus. As we will see in Chapter 2, it represents the average rate of change of $f(x)$ between $x = a$ and $x = a + h$.

EXAMPLE 2 Sketch the graph and find the domain and range of each function.

(a) $f(x) = 2x - 1$ (b) $g(x) = x^2$

SOLUTION

(a) The equation of the graph is $y = 2x - 1$, and we recognize this as being the equation of a line with slope 2 and y-intercept -1. (Recall the slope-intercept form of the equation of a line: $y = mx + b$. See Appendix B.) This enables us to sketch a portion of the graph of f in Figure 7. The expression $2x - 1$ is defined for all real numbers, so the domain of f is the set of all real numbers, which we denote by $\mathbb{R}$. The graph shows that the range is also $\mathbb{R}$.

(b) Since $g(2) = 2^2 = 4$ and $g(-1) = (-1)^2 = 1$, we could plot the points $(2, 4)$ and $(-1, 1)$, together with a few other points on the graph, and join them to produce the graph (Figure 8). The equation of the graph is $y = x^2$, which represents a parabola (see Appendix C). The domain of g is $\mathbb{R}$. The range of g consists of all values of $g(x)$, that is, all numbers of the form x^2. But $x^2 \geq 0$ for all numbers x and any positive number y is a square. So the range of g is $\{y \mid y \geq 0\} = [0, \infty)$. This can also be seen from Figure 8. ■

EXAMPLE 3 If $f(x) = 2x^2 - 5x + 1$ and $h \neq 0$, evaluate $\dfrac{f(a + h) - f(a)}{h}$.

SOLUTION We first evaluate $f(a + h)$ by replacing x by $a + h$ in the expression for $f(x)$:

$$f(a + h) = 2(a + h)^2 - 5(a + h) + 1$$

$$= 2(a^2 + 2ah + h^2) - 5(a + h) + 1$$

$$= 2a^2 + 4ah + 2h^2 - 5a - 5h + 1$$

Then we substitute into the given expression and simplify:

$$\frac{f(a + h) - f(a)}{h} = \frac{(2a^2 + 4ah + 2h^2 - 5a - 5h + 1) - (2a^2 - 5a + 1)}{h}$$

$$= \frac{2a^2 + 4ah + 2h^2 - 5a - 5h + 1 - 2a^2 + 5a - 1}{h}$$

$$= \frac{4ah + 2h^2 - 5h}{h} = 4a + 2h - 5$$

■

■ Representations of Functions

There are four possible ways to represent a function:

- verbally   (by a description in words)
- numerically   (by a table of values)
- visually   (by a graph)
- algebraically (by an explicit formula)

If a single function can be represented in all four ways, it's often useful to go from one representation to another to gain additional insight into the function. (In Example 2, for instance, we started with algebraic formulas and then obtained the graphs.) But certain functions are described more naturally by one method than by another. With this in mind, let's reexamine the four situations that we considered at the beginning of this section.

A. The most useful representation of the area of a circle as a function of its radius is probably the algebraic formula $A(r) = \pi r^2$, though it is possible to compile a table of values or to sketch a graph (half a parabola). Because a circle has to have a positive radius, the domain is $\{r \mid r > 0\} = (0, \infty)$, and the range is also $(0, \infty)$.

B. We are given a description of the function in words: $P(t)$ is the human population of the world at time t. Let's measure t so that $t = 0$ corresponds to the year 1900. The table of values of world population provides a convenient representation of this function. If we plot these values, we get the graph (called a *scatter plot*) in Figure 9. It too is a useful representation; the graph allows us to absorb all the data at once. What about a formula? Of course, it's impossible to devise an explicit formula that gives the exact human population $P(t)$ at any time t. But it is possible to find an expression for a function that *approximates* $P(t)$. In fact, using methods explained in Section 1.2, we obtain the approximation

$$P(t) \approx f(t) = (1.43653 \times 10^9) \cdot (1.01395)^t$$

Figure 10 shows that it is a reasonably good "fit." The function f is called a *mathematical model* for population growth. In other words, it is a function with an explicit formula that approximates the behavior of our given function. We will see, however, that the ideas of calculus can be applied to a table of values; an explicit formula is not necessary.

t (years since 1900)	Population (millions)
0	1650
10	1750
20	1860
30	2070
40	2300
50	2560
60	3040
70	3710
80	4450
90	5280
100	6080
110	6870

FIGURE 9

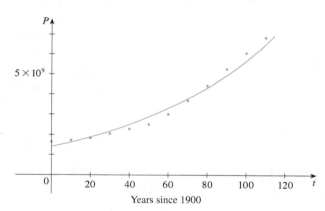

FIGURE 10

The function P is typical of the functions that arise whenever we attempt to apply calculus to the real world. We start with a verbal description of a function. Then we may be able to construct a table of values of the function, perhaps from instrument readings in a scientific experiment. Even though we don't have complete knowledge of the values of the function, we will see throughout the book that it is still possible to perform the operations of calculus on such a function.

A function defined by a table of values is called a *tabular* function.

w (ounces)	$C(w)$ (dollars)
$0 < w \le 1$	0.98
$1 < w \le 2$	1.19
$2 < w \le 3$	1.40
$3 < w \le 4$	1.61
$4 < w \le 5$	1.82
.	.
.	.
.	.

C. Again the function is described in words: Let $C(w)$ be the cost of mailing a large envelope with weight w. The rule that the US Postal Service used as of 2015 is as follows: The cost is 98 cents for up to 1 oz, plus 21 cents for each additional ounce (or less) up to 13 oz. The table of values shown in the margin is the most convenient representation for this function, though it is possible to sketch a graph (see Example 10).

D. The graph shown in Figure 1 is the most natural representation of the vertical acceleration function $a(t)$. It's true that a table of values could be compiled, and it is even possible to devise an approximate formula. But everything a geologist needs to know—amplitudes and patterns—can be seen easily from the graph. (The same is true for the patterns seen in electrocardiograms of heart patients and polygraphs for lie-detection.)

In the next example we sketch the graph of a function that is defined verbally.

EXAMPLE 4 When you turn on a hot-water faucet, the temperature T of the water depends on how long the water has been running. Draw a rough graph of T as a function of the time t that has elapsed since the faucet was turned on.

SOLUTION The initial temperature of the running water is close to room temperature because the water has been sitting in the pipes. When the water from the hot-water tank starts flowing from the faucet, T increases quickly. In the next phase, T is constant at the temperature of the heated water in the tank. When the tank is drained, T decreases to the temperature of the water supply. This enables us to make the rough sketch of T as a function of t in Figure 11. ∎

FIGURE 11

In the following example we start with a verbal description of a function in a physical situation and obtain an explicit algebraic formula. The ability to do this is a useful skill in solving calculus problems that ask for the maximum or minimum values of quantities.

EXAMPLE 5 A rectangular storage container with an open top has a volume of 10 m³. The length of its base is twice its width. Material for the base costs $10 per square meter; material for the sides costs $6 per square meter. Express the cost of materials as a function of the width of the base.

SOLUTION We draw a diagram as in Figure 12 and introduce notation by letting w and $2w$ be the width and length of the base, respectively, and h be the height.

The area of the base is $(2w)w = 2w^2$, so the cost, in dollars, of the material for the base is $10(2w^2)$. Two of the sides have area wh and the other two have area $2wh$, so the cost of the material for the sides is $6[2(wh) + 2(2wh)]$. The total cost is therefore

$$C = 10(2w^2) + 6[2(wh) + 2(2wh)] = 20w^2 + 36wh$$

FIGURE 12

To express C as a function of w alone, we need to eliminate h and we do so by using the fact that the volume is 10 m³. Thus

$$w(2w)h = 10$$

which gives

$$h = \frac{10}{2w^2} = \frac{5}{w^2}$$

Substituting this into the expression for C, we have

$$C = 20w^2 + 36w\left(\frac{5}{w^2}\right) = 20w^2 + \frac{180}{w}$$

Therefore the equation

$$C(w) = 20w^2 + \frac{180}{w} \qquad w > 0$$

expresses C as a function of w. ∎

PS In setting up applied functions as in Example 5, it may be useful to review the principles of problem solving as discussed on page 71, particularly *Step 1: Understand the Problem.*

EXAMPLE 6 Find the domain of each function.

(a) $f(x) = \sqrt{x + 2}$

(b) $g(x) = \dfrac{1}{x^2 - x}$

Domain Convention

If a function is given by a formula and the domain is not stated explicitly, the convention is that the domain is the set of all numbers for which the formula makes sense and defines a real number.

SOLUTION

(a) Because the square root of a negative number is not defined (as a real number), the domain of f consists of all values of x such that $x + 2 \geqslant 0$. This is equivalent to $x \geqslant -2$, so the domain is the interval $[-2, \infty)$.

(b) Since

$$g(x) = \frac{1}{x^2 - x} = \frac{1}{x(x - 1)}$$

and division by 0 is not allowed, we see that $g(x)$ is not defined when $x = 0$ or $x = 1$. Thus the domain of g is

$$\{x \mid x \neq 0, x \neq 1\}$$

which could also be written in interval notation as

$$(-\infty, 0) \cup (0, 1) \cup (1, \infty)$$ ∎

The graph of a function is a curve in the xy-plane. But the question arises: Which curves in the xy-plane are graphs of functions? This is answered by the following test.

> **The Vertical Line Test** A curve in the xy-plane is the graph of a function of x if and only if no vertical line intersects the curve more than once.

The reason for the truth of the Vertical Line Test can be seen in Figure 13. If each vertical line $x = a$ intersects a curve only once, at (a, b), then exactly one function value is defined by $f(a) = b$. But if a line $x = a$ intersects the curve twice, at (a, b) and (a, c), then the curve can't represent a function because a function can't assign two different values to a.

For example, the parabola $x = y^2 - 2$ shown in Figure 14(a) is not the graph of a function of x because, as you can see, there are vertical lines that intersect the parabola twice. The parabola, however, does contain the graphs of *two* functions of x. Notice that the equation $x = y^2 - 2$ implies $y^2 = x + 2$, so $y = \pm\sqrt{x + 2}$. Thus the upper and lower halves of the parabola are the graphs of the functions $f(x) = \sqrt{x + 2}$ [from Example 6(a)] and $g(x) = -\sqrt{x + 2}$. [See Figures 14(b) and (c).]

We observe that if we reverse the roles of x and y, then the equation $x = h(y) = y^2 - 2$ *does* define x as a function of y (with y as the independent variable and x as the dependent variable) and the parabola now appears as the graph of the function h.

(a) This curve represents a function.

(b) This curve doesn't represent a function.

FIGURE 13

(a) $x = y^2 - 2$

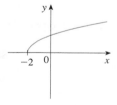

(b) $y = \sqrt{x + 2}$

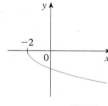

(c) $y = -\sqrt{x + 2}$

FIGURE 14

■ Piecewise Defined Functions

The functions in the following four examples are defined by different formulas in different parts of their domains. Such functions are called **piecewise defined functions**.

EXAMPLE 7 A function f is defined by

$$f(x) = \begin{cases} 1 - x & \text{if } x \leqslant -1 \\ x^2 & \text{if } x > -1 \end{cases}$$

Evaluate $f(-2)$, $f(-1)$, and $f(0)$ and sketch the graph.

SOLUTION Remember that a function is a rule. For this particular function the rule is the following: First look at the value of the input x. If it happens that $x \leqslant -1$, then the value of $f(x)$ is $1 - x$. On the other hand, if $x > -1$, then the value of $f(x)$ is x^2.

Since $-2 \leqslant -1$, we have $f(-2) = 1 - (-2) = 3$.

Since $-1 \leqslant -1$, we have $f(-1) = 1 - (-1) = 2$.

Since $0 > -1$, we have $f(0) = 0^2 = 0$.

How do we draw the graph of f? We observe that if $x \leqslant -1$, then $f(x) = 1 - x$, so the part of the graph of f that lies to the left of the vertical line $x = -1$ must coincide with the line $y = 1 - x$, which has slope -1 and y-intercept 1. If $x > -1$, then $f(x) = x^2$, so the part of the graph of f that lies to the right of the line $x = -1$ must coincide with the graph of $y = x^2$, which is a parabola. This enables us to sketch the graph in Figure 15. The solid dot indicates that the point $(-1, 2)$ is included on the graph; the open dot indicates that the point $(-1, 1)$ is excluded from the graph. ∎

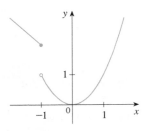

FIGURE 15

The next example of a piecewise defined function is the absolute value function. Recall that the **absolute value** of a number a, denoted by $|a|$, is the distance from a to 0 on the real number line. Distances are always positive or 0, so we have

$$|a| \geqslant 0 \qquad \text{for every number } a$$

For a more extensive review of absolute values, see Appendix A.

For example,

$$|3| = 3 \quad |-3| = 3 \quad |0| = 0 \quad |\sqrt{2} - 1| = \sqrt{2} - 1 \quad |3 - \pi| = \pi - 3$$

In general, we have

$$\boxed{\begin{aligned} |a| &= a \quad \text{if } a \geqslant 0 \\ |a| &= -a \quad \text{if } a < 0 \end{aligned}}$$

(Remember that if a is negative, then $-a$ is positive.)

EXAMPLE 8 Sketch the graph of the absolute value function $f(x) = |x|$.

SOLUTION From the preceding discussion we know that

$$|x| = \begin{cases} x & \text{if } x \geqslant 0 \\ -x & \text{if } x < 0 \end{cases}$$

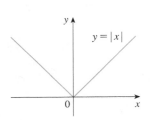

FIGURE 16

Using the same method as in Example 7, we see that the graph of f coincides with the line $y = x$ to the right of the y-axis and coincides with the line $y = -x$ to the left of the y-axis (see Figure 16). ∎

FIGURE 17

Point-slope form of the equation of a line:
$$y - y_1 = m(x - x_1)$$
See Appendix B.

EXAMPLE 9 Find a formula for the function f graphed in Figure 17.

SOLUTION The line through $(0, 0)$ and $(1, 1)$ has slope $m = 1$ and y-intercept $b = 0$, so its equation is $y = x$. Thus, for the part of the graph of f that joins $(0, 0)$ to $(1, 1)$, we have

$$f(x) = x \qquad \text{if } 0 \leqslant x \leqslant 1$$

The line through $(1, 1)$ and $(2, 0)$ has slope $m = -1$, so its point-slope form is

$$y - 0 = (-1)(x - 2) \qquad \text{or} \qquad y = 2 - x$$

So we have $\qquad\qquad\qquad f(x) = 2 - x \qquad \text{if } 1 < x \leqslant 2$

We also see that the graph of f coincides with the x-axis for $x > 2$. Putting this information together, we have the following three-piece formula for f:

$$f(x) = \begin{cases} x & \text{if } 0 \leqslant x \leqslant 1 \\ 2 - x & \text{if } 1 < x \leqslant 2 \\ 0 & \text{if } x > 2 \end{cases}$$

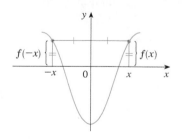

FIGURE 18

EXAMPLE 10 In Example C at the beginning of this section we considered the cost $C(w)$ of mailing a large envelope with weight w. In effect, this is a piecewise defined function because, from the table of values on page 13, we have

$$C(w) = \begin{cases} 0.98 & \text{if } 0 < w \leqslant 1 \\ 1.19 & \text{if } 1 < w \leqslant 2 \\ 1.40 & \text{if } 2 < w \leqslant 3 \\ 1.61 & \text{if } 3 < w \leqslant 4 \\ \vdots \end{cases}$$

The graph is shown in Figure 18. You can see why functions similar to this one are called **step functions**—they jump from one value to the next. Such functions will be studied in Chapter 2.

◼ Symmetry

If a function f satisfies $f(-x) = f(x)$ for every number x in its domain, then f is called an **even function**. For instance, the function $f(x) = x^2$ is even because

$$f(-x) = (-x)^2 = x^2 = f(x)$$

The geometric significance of an even function is that its graph is symmetric with respect to the y-axis (see Figure 19). This means that if we have plotted the graph of f for $x \geqslant 0$, we obtain the entire graph simply by reflecting this portion about the y-axis.

If f satisfies $f(-x) = -f(x)$ for every number x in its domain, then f is called an **odd function**. For example, the function $f(x) = x^3$ is odd because

$$f(-x) = (-x)^3 = -x^3 = -f(x)$$

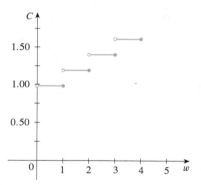

FIGURE 19

An even function

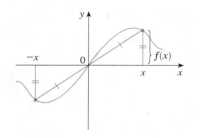

FIGURE 20

An odd function

The graph of an odd function is symmetric about the origin (see Figure 20). If we already have the graph of f for $x \geq 0$, we can obtain the entire graph by rotating this portion through $180°$ about the origin.

EXAMPLE 11 Determine whether each of the following functions is even, odd, or neither even nor odd.

(a) $f(x) = x^5 + x$ (b) $g(x) = 1 - x^4$ (c) $h(x) = 2x - x^2$

SOLUTION

(a)
$$f(-x) = (-x)^5 + (-x) = (-1)^5 x^5 + (-x)$$
$$= -x^5 - x = -(x^5 + x)$$
$$= -f(x)$$

Therefore f is an odd function.

(b)
$$g(-x) = 1 - (-x)^4 = 1 - x^4 = g(x)$$

So g is even.

(c)
$$h(-x) = 2(-x) - (-x)^2 = -2x - x^2$$

Since $h(-x) \neq h(x)$ and $h(-x) \neq -h(x)$, we conclude that h is neither even nor odd. ∎

The graphs of the functions in Example 11 are shown in Figure 21. Notice that the graph of h is symmetric neither about the y-axis nor about the origin.

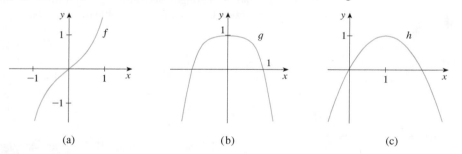

FIGURE 21   (a)   (b)   (c)

■ Increasing and Decreasing Functions

The graph shown in Figure 22 rises from A to B, falls from B to C, and rises again from C to D. The function f is said to be increasing on the interval $[a, b]$, decreasing on $[b, c]$, and increasing again on $[c, d]$. Notice that if x_1 and x_2 are any two numbers between a and b with $x_1 < x_2$, then $f(x_1) < f(x_2)$. We use this as the defining property of an increasing function.

FIGURE 22

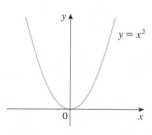

FIGURE 23

A function f is called **increasing** on an interval I if

$$f(x_1) < f(x_2) \qquad \text{whenever } x_1 < x_2 \text{ in } I$$

It is called **decreasing** on I if

$$f(x_1) > f(x_2) \qquad \text{whenever } x_1 < x_2 \text{ in } I$$

In the definition of an increasing function it is important to realize that the inequality $f(x_1) < f(x_2)$ must be satisfied for *every* pair of numbers x_1 and x_2 in I with $x_1 < x_2$.

You can see from Figure 23 that the function $f(x) = x^2$ is decreasing on the interval $(-\infty, 0]$ and increasing on the interval $[0, \infty)$.

1.1   EXERCISES

1. If $f(x) = x + \sqrt{2 - x}$ and $g(u) = u + \sqrt{2 - u}$, is it true that $f = g$?

2. If

$$f(x) = \frac{x^2 - x}{x - 1} \qquad \text{and} \qquad g(x) = x$$

is it true that $f = g$?

3. The graph of a function f is given.
  (a) State the value of $f(1)$.
  (b) Estimate the value of $f(-1)$.
  (c) For what values of x is $f(x) = 1$?
  (d) Estimate the value of x such that $f(x) = 0$.
  (e) State the domain and range of f.
  (f) On what interval is f increasing?

4. The graphs of f and g are given.

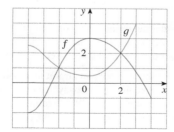

  (a) State the values of $f(-4)$ and $g(3)$.
  (b) For what values of x is $f(x) = g(x)$?

  (c) Estimate the solution of the equation $f(x) = -1$.
  (d) On what interval is f decreasing?
  (e) State the domain and range of f.
  (f) State the domain and range of g.

5. Figure 1 was recorded by an instrument operated by the California Department of Mines and Geology at the University Hospital of the University of Southern California in Los Angeles. Use it to estimate the range of the vertical ground acceleration function at USC during the Northridge earthquake.

6. In this section we discussed examples of ordinary, everyday functions: Population is a function of time, postage cost is a function of weight, water temperature is a function of time. Give three other examples of functions from everyday life that are described verbally. What can you say about the domain and range of each of your functions? If possible, sketch a rough graph of each function.

7–10 Determine whether the curve is the graph of a function of x. If it is, state the domain and range of the function.

7.

8.

9.

10.

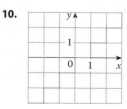

11. Shown is a graph of the global average temperature T during the 20th century. Estimate the following.
    (a) The global average temperature in 1950
    (b) The year when the average temperature was 14.2°C
    (c) The year when the temperature was smallest? Largest?
    (d) The range of T

Source: Adapted from *Globe and Mail* [Toronto], 5 Dec. 2009. Print.

12. Trees grow faster and form wider rings in warm years and grow more slowly and form narrower rings in cooler years. The figure shows ring widths of a Siberian pine from 1500 to 2000.
    (a) What is the range of the ring width function?
    (b) What does the graph tend to say about the temperature of the earth? Does the graph reflect the volcanic eruptions of the mid-19th century?

Source: Adapted from G. Jacoby et al., "Mongolian Tree Rings and 20th-Century Warming," *Science* 273 (1996): 771–73.

13. You put some ice cubes in a glass, fill the glass with cold water, and then let the glass sit on a table. Describe how the temperature of the water changes as time passes. Then sketch a rough graph of the temperature of the water as a function of the elapsed time.

14. Three runners compete in a 100-meter race. The graph depicts the distance run as a function of time for each runner. Describe in words what the graph tells you about this race. Who won the race? Did each runner finish the race?

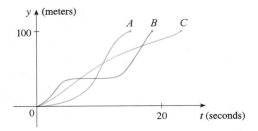

15. The graph shows the power consumption for a day in September in San Francisco. (P is measured in megawatts; t is measured in hours starting at midnight.)
    (a) What was the power consumption at 6 AM? At 6 PM?
    (b) When was the power consumption the lowest? When was it the highest? Do these times seem reasonable?

Pacific Gas & Electric

16. Sketch a rough graph of the number of hours of daylight as a function of the time of year.

17. Sketch a rough graph of the outdoor temperature as a function of time during a typical spring day.

18. Sketch a rough graph of the market value of a new car as a function of time for a period of 20 years. Assume the car is well maintained.

19. Sketch the graph of the amount of a particular brand of coffee sold by a store as a function of the price of the coffee.

20. You place a frozen pie in an oven and bake it for an hour. Then you take it out and let it cool before eating it. Describe how the temperature of the pie changes as time passes. Then sketch a rough graph of the temperature of the pie as a function of time.

21. A homeowner mows the lawn every Wednesday afternoon. Sketch a rough graph of the height of the grass as a function of time over the course of a four-week period.

22. An airplane takes off from an airport and lands an hour later at another airport, 400 miles away. If t represents the time in minutes since the plane has left the terminal building, let $x(t)$ be the horizontal distance traveled and $y(t)$ be the altitude of the plane.
    (a) Sketch a possible graph of $x(t)$.
    (b) Sketch a possible graph of $y(t)$.

(c) Sketch a possible graph of the ground speed.

(d) Sketch a possible graph of the vertical velocity.

23. Temperature readings T (in °F) were recorded every two hours from midnight to 2:00 PM in Atlanta on June 4, 2013. The time t was measured in hours from midnight.

t	0	2	4	6	8	10	12	14
T	74	69	68	66	70	78	82	86

(a) Use the readings to sketch a rough graph of T as a function of t.

(b) Use your graph to estimate the temperature at 9:00 AM.

24. Researchers measured the blood alcohol concentration (BAC) of eight adult male subjects after rapid consumption of 30 mL of ethanol (corresponding to two standard alcoholic drinks). The table shows the data they obtained by averaging the BAC (in mg/mL) of the eight men.

(a) Use the readings to sketch the graph of the BAC as a function of t.

(b) Use your graph to describe how the effect of alcohol varies with time.

t (hours)	BAC	t (hours)	BAC
0	0	1.75	0.22
0.2	0.25	2.0	0.18
0.5	0.41	2.25	0.15
0.75	0.40	2.5	0.12
1.0	0.33	3.0	0.07
1.25	0.29	3.5	0.03
1.5	0.24	4.0	0.01

Source: Adapted from P. Wilkinson et al., "Pharmacokinetics of Ethanol after Oral Administration in the Fasting State," *Journal of Pharmacokinetics and Biopharmaceutics* 5 (1977): 207–24.

25. If $f(x) = 3x^2 - x + 2$, find $f(2)$, $f(-2)$, $f(a)$, $f(-a)$, $f(a + 1)$, $2f(a)$, $f(2a)$, $f(a^2)$, $[f(a)]^2$, and $f(a + h)$.

26. A spherical balloon with radius r inches has volume $V(r) = \frac{4}{3}\pi r^3$. Find a function that represents the amount of air required to inflate the balloon from a radius of r inches to a radius of $r + 1$ inches.

27–30 Evaluate the difference quotient for the given function. Simplify your answer.

27. $f(x) = 4 + 3x - x^2$, $\quad \dfrac{f(3 + h) - f(3)}{h}$

28. $f(x) = x^3$, $\quad \dfrac{f(a + h) - f(a)}{h}$

29. $f(x) = \dfrac{1}{x}$, $\quad \dfrac{f(x) - f(a)}{x - a}$

30. $f(x) = \dfrac{x + 3}{x + 1}$, $\quad \dfrac{f(x) - f(1)}{x - 1}$

31–37 Find the domain of the function.

31. $f(x) = \dfrac{x + 4}{x^2 - 9}$

32. $f(x) = \dfrac{2x^3 - 5}{x^2 + x - 6}$

33. $f(t) = \sqrt[3]{2t - 1}$

34. $g(t) = \sqrt{3 - t} - \sqrt{2 + t}$

35. $h(x) = \dfrac{1}{\sqrt[4]{x^2 - 5x}}$

36. $f(u) = \dfrac{u + 1}{1 + \dfrac{1}{u + 1}}$

37. $F(p) = \sqrt{2 - \sqrt{p}}$

38. Find the domain and range and sketch the graph of the function $h(x) = \sqrt{4 - x^2}$.

39–40 Find the domain and sketch the graph of the function.

39. $f(x) = 1.6x - 2.4$

40. $g(t) = \dfrac{t^2 - 1}{t + 1}$

41–44 Evaluate $f(-3)$, $f(0)$, and $f(2)$ for the piecewise defined function. Then sketch the graph of the function.

41. $f(x) = \begin{cases} x + 2 & \text{if } x < 0 \\ 1 - x & \text{if } x \geq 0 \end{cases}$

42. $f(x) = \begin{cases} 3 - \frac{1}{2}x & \text{if } x < 2 \\ 2x - 5 & \text{if } x \geq 2 \end{cases}$

43. $f(x) = \begin{cases} x + 1 & \text{if } x \leq -1 \\ x^2 & \text{if } x > -1 \end{cases}$

44. $f(x) = \begin{cases} -1 & \text{if } x \leq 1 \\ 7 - 2x & \text{if } x > 1 \end{cases}$

45–50 Sketch the graph of the function.

45. $f(x) = x + |x|$

46. $f(x) = |x + 2|$

47. $g(t) = |1 - 3t|$

48. $h(t) = |t| + |t + 1|$

49. $f(x) = \begin{cases} |x| & \text{if } |x| \leq 1 \\ 1 & \text{if } |x| > 1 \end{cases}$

50. $g(x) = ||x| - 1|$

51–56 Find an expression for the function whose graph is the given curve.

51. The line segment joining the points $(1, -3)$ and $(5, 7)$

52. The line segment joining the points $(-5, 10)$ and $(7, -10)$

53. The bottom half of the parabola $x + (y - 1)^2 = 0$

54. The top half of the circle $x^2 + (y - 2)^2 = 4$

55.

56.

57–61 Find a formula for the described function and state its domain.

57. A rectangle has perimeter 20 m. Express the area of the rectangle as a function of the length of one of its sides.

58. A rectangle has area 16 m². Express the perimeter of the rectangle as a function of the length of one of its sides.

59. Express the area of an equilateral triangle as a function of the length of a side.

60. A closed rectangular box with volume 8 ft³ has length twice the width. Express the height of the box as a function of the width.

61. An open rectangular box with volume 2 m³ has a square base. Express the surface area of the box as a function of the length of a side of the base.

62. A Norman window has the shape of a rectangle surmounted by a semicircle. If the perimeter of the window is 30 ft, express the area A of the window as a function of the width x of the window.

63. A box with an open top is to be constructed from a rectangular piece of cardboard with dimensions 12 in. by 20 in. by cutting out equal squares of side x at each corner and then folding up the sides as in the figure. Express the volume V of the box as a function of x.

64. A cell phone plan has a basic charge of $35 a month. The plan includes 400 free minutes and charges 10 cents for each additional minute of usage. Write the monthly cost C as a function of the number x of minutes used and graph C as a function of x for $0 \leqslant x \leqslant 600$.

65. In a certain state the maximum speed permitted on freeways is 65 mi/h and the minimum speed is 40 mi/h. The fine for violating these limits is $15 for every mile per hour above the maximum speed or below the minimum speed. Express the amount of the fine F as a function of the driving speed x and graph $F(x)$ for $0 \leqslant x \leqslant 100$.

66. An electricity company charges its customers a base rate of $10 a month, plus 6 cents per kilowatt-hour (kWh) for the first 1200 kWh and 7 cents per kWh for all usage over 1200 kWh. Express the monthly cost E as a function of the amount x of electricity used. Then graph the function E for $0 \leqslant x \leqslant 2000$.

67. In a certain country, income tax is assessed as follows. There is no tax on income up to $10,000. Any income over $10,000 is taxed at a rate of 10%, up to an income of $20,000. Any income over $20,000 is taxed at 15%.
(a) Sketch the graph of the tax rate R as a function of the income I.
(b) How much tax is assessed on an income of $14,000? On $26,000?
(c) Sketch the graph of the total assessed tax T as a function of the income I.

68. The functions in Example 10 and Exercise 67 are called *step functions* because their graphs look like stairs. Give two other examples of step functions that arise in everyday life.

69–70 Graphs of f and g are shown. Decide whether each function is even, odd, or neither. Explain your reasoning.

69.

70.
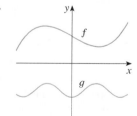

71. (a) If the point $(5, 3)$ is on the graph of an even function, what other point must also be on the graph?
(b) If the point $(5, 3)$ is on the graph of an odd function, what other point must also be on the graph?

72. A function f has domain $[-5, 5]$ and a portion of its graph is shown.
(a) Complete the graph of f if it is known that f is even.
(b) Complete the graph of f if it is known that f is odd.

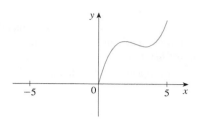

73–78 Determine whether f is even, odd, or neither. If you have a graphing calculator, use it to check your answer visually.

73. $f(x) = \dfrac{x}{x^2 + 1}$

74. $f(x) = \dfrac{x^2}{x^4 + 1}$

75. $f(x) = \dfrac{x}{x + 1}$

76. $f(x) = x \, |x|$

77. $f(x) = 1 + 3x^2 - x^4$

78. $f(x) = 1 + 3x^3 - x^5$

79. If f and g are both even functions, is $f + g$ even? If f and g are both odd functions, is $f + g$ odd? What if f is even and g is odd? Justify your answers.

80. If f and g are both even functions, is the product fg even? If f and g are both odd functions, is fg odd? What if f is even and g is odd? Justify your answers.

1.2 Mathematical Models: A Catalog of Essential Functions

A **mathematical model** is a mathematical description (often by means of a function or an equation) of a real-world phenomenon such as the size of a population, the demand for a product, the speed of a falling object, the concentration of a product in a chemical reaction, the life expectancy of a person at birth, or the cost of emission reductions. The purpose of the model is to understand the phenomenon and perhaps to make predictions about future behavior.

Figure 1 illustrates the process of mathematical modeling. Given a real-world problem, our first task is to formulate a mathematical model by identifying and naming the independent and dependent variables and making assumptions that simplify the phenomenon enough to make it mathematically tractable. We use our knowledge of the physical situation and our mathematical skills to obtain equations that relate the variables. In situations where there is no physical law to guide us, we may need to collect data (either from a library or the Internet or by conducting our own experiments) and examine the data in the form of a table in order to discern patterns. From this numerical representation of a function we may wish to obtain a graphical representation by plotting the data. The graph might even suggest a suitable algebraic formula in some cases.

FIGURE 1
The modeling process

The second stage is to apply the mathematics that we know (such as the calculus that will be developed throughout this book) to the mathematical model that we have formulated in order to derive mathematical conclusions. Then, in the third stage, we take those mathematical conclusions and interpret them as information about the original real-world phenomenon by way of offering explanations or making predictions. The final step is to test our predictions by checking against new real data. If the predictions don't compare well with reality, we need to refine our model or to formulate a new model and start the cycle again.

A mathematical model is never a completely accurate representation of a physical situation—it is an *idealization*. A good model simplifies reality enough to permit mathematical calculations but is accurate enough to provide valuable conclusions. It is important to realize the limitations of the model. In the end, Mother Nature has the final say.

There are many different types of functions that can be used to model relationships observed in the real world. In what follows, we discuss the behavior and graphs of these functions and give examples of situations appropriately modeled by such functions.

■ Linear Models

The coordinate geometry of lines is reviewed in Appendix B.

When we say that y is a **linear function** of x, we mean that the graph of the function is a line, so we can use the slope-intercept form of the equation of a line to write a formula for the function as

$$y = f(x) = mx + b$$

where m is the slope of the line and b is the y-intercept.

A characteristic feature of linear functions is that they grow at a constant rate. For instance, Figure 2 shows a graph of the linear function $f(x) = 3x - 2$ and a table of sample values. Notice that whenever x increases by 0.1, the value of $f(x)$ increases by 0.3. So $f(x)$ increases three times as fast as x. Thus the slope of the graph $y = 3x - 2$, namely 3, can be interpreted as the rate of change of y with respect to x.

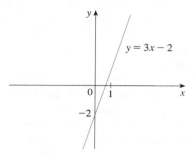

x	$f(x) = 3x - 2$
1.0	1.0
1.1	1.3
1.2	1.6
1.3	1.9
1.4	2.2
1.5	2.5

FIGURE 2

EXAMPLE 1

(a) As dry air moves upward, it expands and cools. If the ground temperature is 20°C and the temperature at a height of 1 km is 10°C, express the temperature T (in °C) as a function of the height h (in kilometers), assuming that a linear model is appropriate.
(b) Draw the graph of the function in part (a). What does the slope represent?
(c) What is the temperature at a height of 2.5 km?

SOLUTION

(a) Because we are assuming that T is a linear function of h, we can write

$$T = mh + b$$

We are given that $T = 20$ when $h = 0$, so

$$20 = m \cdot 0 + b = b$$

In other words, the y-intercept is $b = 20$.

We are also given that $T = 10$ when $h = 1$, so

$$10 = m \cdot 1 + 20$$

The slope of the line is therefore $m = 10 - 20 = -10$ and the required linear function is

$$T = -10h + 20$$

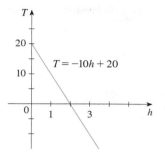

FIGURE 3

(b) The graph is sketched in Figure 3. The slope is $m = -10°C/km$, and this represents the rate of change of temperature with respect to height.

(c) At a height of $h = 2.5$ km, the temperature is

$$T = -10(2.5) + 20 = -5°C \qquad \blacksquare$$

If there is no physical law or principle to help us formulate a model, we construct an **empirical model**, which is based entirely on collected data. We seek a curve that "fits" the data in the sense that it captures the basic trend of the data points.

EXAMPLE 2 Table 1 lists the average carbon dioxide level in the atmosphere, measured in parts per million at Mauna Loa Observatory from 1980 to 2012. Use the data in Table 1 to find a model for the carbon dioxide level.

SOLUTION We use the data in Table 1 to make the scatter plot in Figure 4, where t represents time (in years) and C represents the CO_2 level (in parts per million, ppm).

Table 1

Year	CO_2 level (in ppm)	Year	CO_2 level (in ppm)
1980	338.7	1998	366.5
1982	341.2	2000	369.4
1984	344.4	2002	373.2
1986	347.2	2004	377.5
1988	351.5	2006	381.9
1990	354.2	2008	385.6
1992	356.3	2010	389.9
1994	358.6	2012	393.8
1996	362.4		

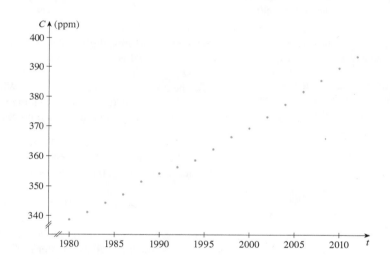

FIGURE 4 Scatter plot for the average CO_2 level

Notice that the data points appear to lie close to a straight line, so it's natural to choose a linear model in this case. But there are many possible lines that approximate these data points, so which one should we use? One possibility is the line that passes through the first and last data points. The slope of this line is

$$\frac{393.8 - 338.7}{2012 - 1980} = \frac{55.1}{32} = 1.721875 \approx 1.722$$

We write its equation as

$$C - 338.7 = 1.722(t - 1980)$$

or

1
$$C = 1.722t - 3070.86$$

Equation 1 gives one possible linear model for the carbon dioxide level; it is graphed in Figure 5.

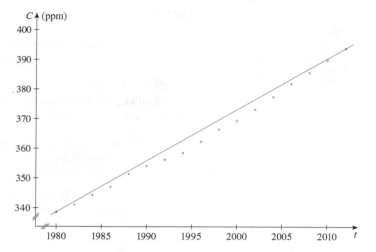

FIGURE 5
Linear model through first
and last data points

A computer or graphing calculator finds the regression line by the method of **least squares**, which is to minimize the sum of the squares of the vertical distances between the data points and the line. The details are explained in Section 14.7.

Notice that our model gives values higher than most of the actual CO_2 levels. A better linear model is obtained by a procedure from statistics called *linear regression*. If we use a graphing calculator, we enter the data from Table 1 into the data editor and choose the linear regression command. (With Maple we use the fit[leastsquare] command in the stats package; with Mathematica we use the Fit command.) The machine gives the slope and y-intercept of the regression line as

$$m = 1.71262 \qquad b = -3054.14$$

So our least squares model for the CO_2 level is

$$\boxed{2} \qquad C = 1.71262t - 3054.14$$

In Figure 6 we graph the regression line as well as the data points. Comparing with Figure 5, we see that it gives a better fit than our previous linear model.

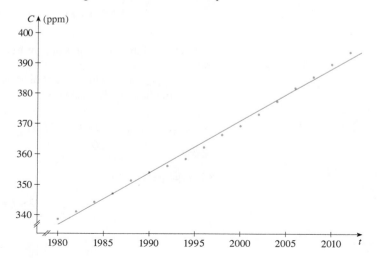

FIGURE 6
The regression line

EXAMPLE 3 Use the linear model given by Equation 2 to estimate the average CO_2 level for 1987 and to predict the level for the year 2020. According to this model, when will the CO_2 level exceed 420 parts per million?

SOLUTION Using Equation 2 with $t = 1987$, we estimate that the average CO_2 level in 1987 was

$$C(1987) = (1.71262)(1987) - 3054.14 \approx 348.84$$

This is an example of *interpolation* because we have estimated a value *between* observed values. (In fact, the Mauna Loa Observatory reported that the average CO_2 level in 1987 was 348.93 ppm, so our estimate is quite accurate.)

With $t = 2020$, we get

$$C(2020) = (1.71262)(2020) - 3054.14 \approx 405.35$$

So we predict that the average CO_2 level in the year 2020 will be 405.4 ppm. This is an example of *extrapolation* because we have predicted a value *outside* the time frame of observations. Consequently, we are far less certain about the accuracy of our prediction.

Using Equation 2, we see that the CO_2 level exceeds 420 ppm when

$$1.71262t - 3054.14 > 420$$

Solving this inequality, we get

$$t > \frac{3474.14}{1.71262} \approx 2028.55$$

We therefore predict that the CO_2 level will exceed 420 ppm by the year 2029. This prediction is risky because it involves a time quite remote from our observations. In fact, we see from Figure 6 that the trend has been for CO_2 levels to increase rather more rapidly in recent years, so the level might exceed 420 ppm well before 2029. ∎

■ Polynomials

A function P is called a **polynomial** if

$$P(x) = a_n x^n + a_{n-1} x^{n-1} + \cdots + a_2 x^2 + a_1 x + a_0$$

where n is a nonnegative integer and the numbers $a_0, a_1, a_2, \ldots, a_n$ are constants called the **coefficients** of the polynomial. The domain of any polynomial is $\mathbb{R} = (-\infty, \infty)$. If the leading coefficient $a_n \neq 0$, then the **degree** of the polynomial is n. For example, the function

$$P(x) = 2x^6 - x^4 + \tfrac{2}{5}x^3 + \sqrt{2}$$

is a polynomial of degree 6.

A polynomial of degree 1 is of the form $P(x) = mx + b$ and so it is a linear function. A polynomial of degree 2 is of the form $P(x) = ax^2 + bx + c$ and is called a **quadratic function**. Its graph is always a parabola obtained by shifting the parabola $y = ax^2$, as we will see in the next section. The parabola opens upward if $a > 0$ and downward if $a < 0$. (See Figure 7.)

A polynomial of degree 3 is of the form

$$P(x) = ax^3 + bx^2 + cx + d \qquad a \neq 0$$

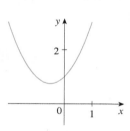

(a) $y = x^2 + x + 1$

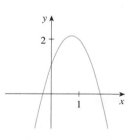

(b) $y = -2x^2 + 3x + 1$

FIGURE 7
The graphs of quadratic functions are parabolas.

and is called a **cubic function**. Figure 8 shows the graph of a cubic function in part (a) and graphs of polynomials of degrees 4 and 5 in parts (b) and (c). We will see later why the graphs have these shapes.

FIGURE 8

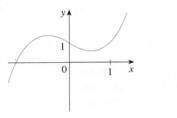

(a) $y = x^3 - x + 1$

(b) $y = x^4 - 3x^2 + x$

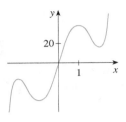

(c) $y = 3x^5 - 25x^3 + 60x$

Polynomials are commonly used to model various quantities that occur in the natural and social sciences. For instance, in Section 3.7 we will explain why economists often use a polynomial $P(x)$ to represent the cost of producing x units of a commodity. In the following example we use a quadratic function to model the fall of a ball.

Table 2

Time (seconds)	Height (meters)
0	450
1	445
2	431
3	408
4	375
5	332
6	279
7	216
8	143
9	61

EXAMPLE 4 A ball is dropped from the upper observation deck of the CN Tower, 450 m above the ground, and its height h above the ground is recorded at 1-second intervals in Table 2. Find a model to fit the data and use the model to predict the time at which the ball hits the ground.

SOLUTION We draw a scatter plot of the data in Figure 9 and observe that a linear model is inappropriate. But it looks as if the data points might lie on a parabola, so we try a quadratic model instead. Using a graphing calculator or computer algebra system (which uses the least squares method), we obtain the following quadratic model:

3    $$h = 449.36 + 0.96t - 4.90t^2$$

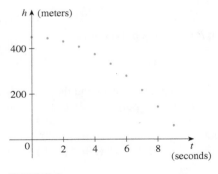

FIGURE 9
Scatter plot for a falling ball

FIGURE 10
Quadratic model for a falling ball

In Figure 10 we plot the graph of Equation 3 together with the data points and see that the quadratic model gives a very good fit.

The ball hits the ground when $h = 0$, so we solve the quadratic equation

$$-4.90t^2 + 0.96t + 449.36 = 0$$

The quadratic formula gives

$$t = \frac{-0.96 \pm \sqrt{(0.96)^2 - 4(-4.90)(449.36)}}{2(-4.90)}$$

The positive root is $t \approx 9.67$, so we predict that the ball will hit the ground after about 9.7 seconds. ■

■ Power Functions

A function of the form $f(x) = x^a$, where a is a constant, is called a **power function**. We consider several cases.

(i) $a = n$, where n is a positive integer

The graphs of $f(x) = x^n$ for $n = 1, 2, 3, 4$, and 5 are shown in Figure 11. (These are polynomials with only one term.) We already know the shape of the graphs of $y = x$ (a line through the origin with slope 1) and $y = x^2$ [a parabola, see Example 1.1.2(b)].

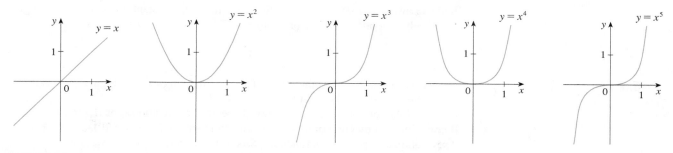

FIGURE 11 Graphs of $f(x) = x^n$ for $n = 1, 2, 3, 4, 5$

The general shape of the graph of $f(x) = x^n$ depends on whether n is even or odd. If n is even, then $f(x) = x^n$ is an even function and its graph is similar to the parabola $y = x^2$. If n is odd, then $f(x) = x^n$ is an odd function and its graph is similar to that of $y = x^3$. Notice from Figure 12, however, that as n increases, the graph of $y = x^n$ becomes flatter near 0 and steeper when $|x| \geq 1$. (If x is small, then x^2 is smaller, x^3 is even smaller, x^4 is smaller still, and so on.)

A **family of functions** is a collection of functions whose equations are related. Figure 12 shows two families of power functions, one with even powers and one with odd powers.

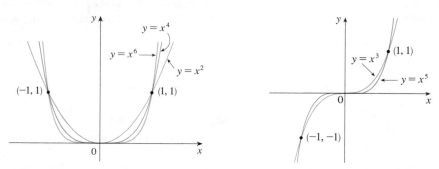

FIGURE 12

(ii) $a = 1/n$, where n is a positive integer

The function $f(x) = x^{1/n} = \sqrt[n]{x}$ is a **root function**. For $n = 2$ it is the square root function $f(x) = \sqrt{x}$, whose domain is $[0, \infty)$ and whose graph is the upper half of the parabola $x = y^2$. [See Figure 13(a).] For other even values of n, the graph of $y = \sqrt[n]{x}$ is similar to that of $y = \sqrt{x}$. For $n = 3$ we have the cube root function $f(x) = \sqrt[3]{x}$ whose domain is $\mathbb{R}$ (recall that every real number has a cube root) and whose graph is shown in Figure 13(b). The graph of $y = \sqrt[n]{x}$ for n odd ($n > 3$) is similar to that of $y = \sqrt[3]{x}$.

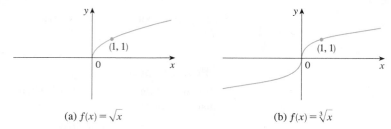

FIGURE 13
Graphs of root functions

(a) $f(x) = \sqrt{x}$ (b) $f(x) = \sqrt[3]{x}$

(iii) $a = -1$

The graph of the **reciprocal function** $f(x) = x^{-1} = 1/x$ is shown in Figure 14. Its graph has the equation $y = 1/x$, or $xy = 1$, and is a hyperbola with the coordinate axes as its asymptotes. This function arises in physics and chemistry in connection with Boyle's Law, which says that, when the temperature is constant, the volume V of a gas is inversely proportional to the pressure P:

$$V = \frac{C}{P}$$

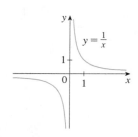

FIGURE 14
The reciprocal function

where C is a constant. Thus the graph of V as a function of P (see Figure 15) has the same general shape as the right half of Figure 14.

Power functions are also used to model species-area relationships (Exercises 30–31), illumination as a function of distance from a light source (Exercise 29), and the period of revolution of a planet as a function of its distance from the sun (Exercise 32).

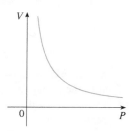

FIGURE 15
Volume as a function of pressure at constant temperature

■ Rational Functions

A **rational function** f is a ratio of two polynomials:

$$f(x) = \frac{P(x)}{Q(x)}$$

where P and Q are polynomials. The domain consists of all values of x such that $Q(x) \neq 0$. A simple example of a rational function is the function $f(x) = 1/x$, whose domain is $\{x \mid x \neq 0\}$; this is the reciprocal function graphed in Figure 14. The function

$$f(x) = \frac{2x^4 - x^2 + 1}{x^2 - 4}$$

is a rational function with domain $\{x \mid x \neq \pm 2\}$. Its graph is shown in Figure 16.

■ Algebraic Functions

A function f is called an **algebraic function** if it can be constructed using algebraic operations (such as addition, subtraction, multiplication, division, and taking roots) starting with polynomials. Any rational function is automatically an algebraic function. Here are two more examples:

FIGURE 16
$f(x) = \dfrac{2x^4 - x^2 + 1}{x^2 - 4}$

$$f(x) = \sqrt{x^2 + 1} \qquad g(x) = \frac{x^4 - 16x^2}{x + \sqrt{x}} + (x - 2)\sqrt[3]{x + 1}$$

When we sketch algebraic functions in Chapter 4, we will see that their graphs can assume a variety of shapes. Figure 17 illustrates some of the possibilities.

FIGURE 17   (a) $f(x) = x\sqrt{x+3}$ (b) $g(x) = \sqrt[4]{x^2 - 25}$ (c) $h(x) = x^{2/3}(x-2)^2$

An example of an algebraic function occurs in the theory of relativity. The mass of a particle with velocity v is

$$m = f(v) = \frac{m_0}{\sqrt{1 - v^2/c^2}}$$

where m_0 is the rest mass of the particle and $c = 3.0 \times 10^5$ km/s is the speed of light in a vacuum.

◼ Trigonometric Functions

The Reference Pages are located at the back of the book.

Trigonometry and the trigonometric functions are reviewed on Reference Page 2 and also in Appendix D. In calculus the convention is that radian measure is always used (except when otherwise indicated). For example, when we use the function $f(x) = \sin x$, it is understood that $\sin x$ means the sine of the angle whose radian measure is x. Thus the graphs of the sine and cosine functions are as shown in Figure 18.

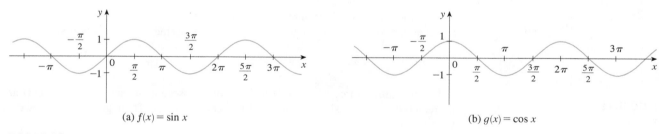

(a) $f(x) = \sin x$ (b) $g(x) = \cos x$

FIGURE 18

Notice that for both the sine and cosine functions the domain is $(-\infty, \infty)$ and the range is the closed interval $[-1, 1]$. Thus, for all values of x, we have

$$-1 \leq \sin x \leq 1 \qquad -1 \leq \cos x \leq 1$$

or, in terms of absolute values,

$$|\sin x| \leq 1 \qquad |\cos x| \leq 1$$

Also, the zeros of the sine function occur at the integer multiples of π; that is,

$$\sin x = 0 \qquad \text{when} \qquad x = n\pi \quad n \text{ an integer}$$

An important property of the sine and cosine functions is that they are periodic functions and have period 2π. This means that, for all values of x,

$$\sin(x + 2\pi) = \sin x \qquad \cos(x + 2\pi) = \cos x$$

The periodic nature of these functions makes them suitable for modeling repetitive phenomena such as tides, vibrating springs, and sound waves. For instance, in Example 1.3.4 we will see that a reasonable model for the number of hours of daylight in Philadelphia t days after January 1 is given by the function

$$L(t) = 12 + 2.8 \sin\left[\frac{2\pi}{365}(t - 80)\right]$$

EXAMPLE 5 What is the domain of the function $f(x) = \dfrac{1}{1 - 2 \cos x}$?

SOLUTION This function is defined for all values of x except for those that make the denominator 0. But

$$1 - 2 \cos x = 0 \iff \cos x = \frac{1}{2} \iff x = \frac{\pi}{3} + 2n\pi \quad \text{or} \quad x = \frac{5\pi}{3} + 2n\pi$$

where n is any integer (because the cosine function has period 2π). So the domain of f is the set of all real numbers except for the ones noted above. ■

The tangent function is related to the sine and cosine functions by the equation

$$\tan x = \frac{\sin x}{\cos x}$$

and its graph is shown in Figure 19. It is undefined whenever $\cos x = 0$, that is, when $x = \pm\pi/2, \pm 3\pi/2, \dots$. Its range is $(-\infty, \infty)$. Notice that the tangent function has period π:

$$\tan(x + \pi) = \tan x \qquad \text{for all } x$$

The remaining three trigonometric functions (cosecant, secant, and cotangent) are the reciprocals of the sine, cosine, and tangent functions. Their graphs are shown in Appendix D.

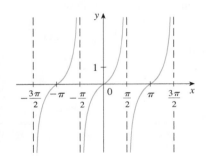

FIGURE 19
$y = \tan x$

■ Exponential Functions

The **exponential functions** are the functions of the form $f(x) = b^x$, where the base b is a positive constant. The graphs of $y = 2^x$ and $y = (0.5)^x$ are shown in Figure 20. In both cases the domain is $(-\infty, \infty)$ and the range is $(0, \infty)$.

Exponential functions will be studied in detail in Section 1.4, and we will see that they are useful for modeling many natural phenomena, such as population growth (if $b > 1$) and radioactive decay (if $b < 1$).

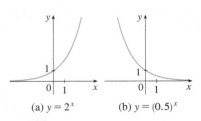

(a) $y = 2^x$ (b) $y = (0.5)^x$

FIGURE 20

■ Logarithmic Functions

The **logarithmic functions** $f(x) = \log_b x$, where the base b is a positive constant, are the inverse functions of the exponential functions. They will be studied in Section 1.5. Figure

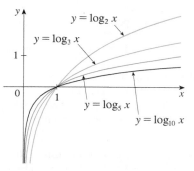

FIGURE 21

21 shows the graphs of four logarithmic functions with various bases. In each case the domain is $(0, \infty)$, the range is $(-\infty, \infty)$, and the function increases slowly when $x > 1$.

EXAMPLE 6 Classify the following functions as one of the types of functions that we have discussed.

(a) $f(x) = 5^x$

(b) $g(x) = x^5$

(c) $h(x) = \dfrac{1 + x}{1 - \sqrt{x}}$

(d) $u(t) = 1 - t + 5t^4$

SOLUTION

(a) $f(x) = 5^x$ is an exponential function. (The x is the exponent.)

(b) $g(x) = x^5$ is a power function. (The x is the base.) We could also consider it to be a polynomial of degree 5.

(c) $h(x) = \dfrac{1 + x}{1 - \sqrt{x}}$ is an algebraic function.

(d) $u(t) = 1 - t + 5t^4$ is a polynomial of degree 4.   ∎

1.2   EXERCISES

1–2 Classify each function as a power function, root function, polynomial (state its degree), rational function, algebraic function, trigonometric function, exponential function, or logarithmic function.

1. (a) $f(x) = \log_2 x$

(b) $g(x) = \sqrt[4]{x}$

(c) $h(x) = \dfrac{2x^3}{1 - x^2}$

(d) $u(t) = 1 - 1.1t + 2.54t^2$

(e) $v(t) = 5^t$

(f) $w(\theta) = \sin\theta\,\cos^2\theta$

2. (a) $y = \pi^x$

(b) $y = x^\pi$

(c) $y = x^2(2 - x^3)$

(d) $y = \tan t - \cos t$

(e) $y = \dfrac{s}{1 + s}$

(f) $y = \dfrac{\sqrt{x^3 - 1}}{1 + \sqrt[3]{x}}$

3–4 Match each equation with its graph. Explain your choices. (Don't use a computer or graphing calculator.)

3. (a) $y = x^2$ (b) $y = x^5$ (c) $y = x^8$

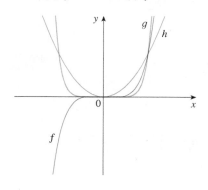

4. (a) $y = 3x$ (b) $y = 3^x$ (c) $y = x^3$ (d) $y = \sqrt[3]{x}$

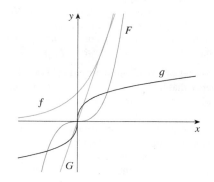

5–6 Find the domain of the function.

5. $f(x) = \dfrac{\cos x}{1 - \sin x}$

6. $g(x) = \dfrac{1}{1 - \tan x}$

7. (a) Find an equation for the family of linear functions with slope 2 and sketch several members of the family.

(b) Find an equation for the family of linear functions such that $f(2) = 1$ and sketch several members of the family.

(c) Which function belongs to both families?

8. What do all members of the family of linear functions $f(x) = 1 + m(x + 3)$ have in common? Sketch several members of the family.

9. What do all members of the family of linear functions $f(x) = c - x$ have in common? Sketch several members of the family.

10. Find expressions for the quadratic functions whose graphs are shown.

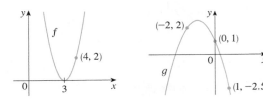

11. Find an expression for a cubic function f if $f(1) = 6$ and $f(-1) = f(0) = f(2) = 0$.

12. Recent studies indicate that the average surface temperature of the earth has been rising steadily. Some scientists have modeled the temperature by the linear function $T = 0.02t + 8.50$, where T is temperature in °C and t represents years since 1900.
   (a) What do the slope and T-intercept represent?
   (b) Use the equation to predict the average global surface temperature in 2100.

13. If the recommended adult dosage for a drug is D (in mg), then to determine the appropriate dosage c for a child of age a, pharmacists use the equation $c = 0.0417D(a + 1)$. Suppose the dosage for an adult is 200 mg.
   (a) Find the slope of the graph of c. What does it represent?
   (b) What is the dosage for a newborn?

14. The manager of a weekend flea market knows from past experience that if he charges x dollars for a rental space at the market, then the number y of spaces he can rent is given by the equation $y = 200 - 4x$.
   (a) Sketch a graph of this linear function. (Remember that the rental charge per space and the number of spaces rented can't be negative quantities.)
   (b) What do the slope, the y-intercept, and the x-intercept of the graph represent?

15. The relationship between the Fahrenheit (F) and Celsius (C) temperature scales is given by the linear function $F = \frac{9}{5}C + 32$.
   (a) Sketch a graph of this function.
   (b) What is the slope of the graph and what does it represent? What is the F-intercept and what does it represent?

16. Jason leaves Detroit at 2:00 PM and drives at a constant speed west along I-94. He passes Ann Arbor, 40 mi from Detroit, at 2:50 PM.
   (a) Express the distance traveled in terms of the time elapsed.
   (b) Draw the graph of the equation in part (a).
   (c) What is the slope of this line? What does it represent?

17. Biologists have noticed that the chirping rate of crickets of a certain species is related to temperature, and the relationship appears to be very nearly linear. A cricket produces 113 chirps per minute at 70°F and 173 chirps per minute at 80°F.
   (a) Find a linear equation that models the temperature T as a function of the number of chirps per minute N.
   (b) What is the slope of the graph? What does it represent?
   (c) If the crickets are chirping at 150 chirps per minute, estimate the temperature.

18. The manager of a furniture factory finds that it costs $2200 to manufacture 100 chairs in one day and $4800 to produce 300 chairs in one day.
   (a) Express the cost as a function of the number of chairs produced, assuming that it is linear. Then sketch the graph.
   (b) What is the slope of the graph and what does it represent?
   (c) What is the y-intercept of the graph and what does it represent?

19. At the surface of the ocean, the water pressure is the same as the air pressure above the water, 15 lb/in². Below the surface, the water pressure increases by 4.34 lb/in² for every 10 ft of descent.
   (a) Express the water pressure as a function of the depth below the ocean surface.
   (b) At what depth is the pressure 100 lb/in²?

20. The monthly cost of driving a car depends on the number of miles driven. Lynn found that in May it cost her $380 to drive 480 mi and in June it cost her $460 to drive 800 mi.
   (a) Express the monthly cost C as a function of the distance driven d, assuming that a linear relationship gives a suitable model.
   (b) Use part (a) to predict the cost of driving 1500 miles per month.
   (c) Draw the graph of the linear function. What does the slope represent?
   (d) What does the C-intercept represent?
   (e) Why does a linear function give a suitable model in this situation?

21–22 For each scatter plot, decide what type of function you might choose as a model for the data. Explain your choices.

21.

(a)   (b)

22.

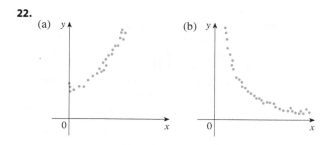

(a)

(b)

23. The table shows (lifetime) peptic ulcer rates (per 100 population) for various family incomes as reported by the National Health Interview Survey.

Income	Ulcer rate (per 100 population)
$4,000	14.1
$6,000	13.0
$8,000	13.4
$12,000	12.5
$16,000	12.0
$20,000	12.4
$30,000	10.5
$45,000	9.4
$60,000	8.2

(a) Make a scatter plot of these data and decide whether a linear model is appropriate.
(b) Find and graph a linear model using the first and last data points.
(c) Find and graph the least squares regression line.
(d) Use the linear model in part (c) to estimate the ulcer rate for an income of $25,000.
(e) According to the model, how likely is someone with an income of $80,000 to suffer from peptic ulcers?
(f) Do you think it would be reasonable to apply the model to someone with an income of $200,000?

24. Biologists have observed that the chirping rate of crickets of a certain species appears to be related to temperature. The table shows the chirping rates for various temperatures.
(a) Make a scatter plot of the data.
(b) Find and graph the regression line.
(c) Use the linear model in part (b) to estimate the chirping rate at 100°F.

Temperature (°F)	Chirping rate (chirps/min)	Temperature (°F)	Chirping rate (chirps/min)
50	20	75	140
55	46	80	173
60	79	85	198
65	91	90	211
70	113		

25. Anthropologists use a linear model that relates human femur (thighbone) length to height. The model allows an anthropologist to determine the height of an individual when only a partial skeleton (including the femur) is found. Here we find the model by analyzing the data on femur length and height for the eight males given in the following table.
(a) Make a scatter plot of the data.
(b) Find and graph the regression line that models the data.
(c) An anthropologist finds a human femur of length 53 cm. How tall was the person?

Femur length (cm)	Height (cm)	Femur length (cm)	Height (cm)
50.1	178.5	44.5	168.3
48.3	173.6	42.7	165.0
45.2	164.8	39.5	155.4
44.7	163.7	38.0	155.8

26. When laboratory rats are exposed to asbestos fibers, some of them develop lung tumors. The table lists the results of several experiments by different scientists.
(a) Find the regression line for the data.
(b) Make a scatter plot and graph the regression line. Does the regression line appear to be a suitable model for the data?
(c) What does the y-intercept of the regression line represent?

Asbestos exposure (fibers/mL)	Percent of mice that develop lung tumors	Asbestos exposure (fibers/mL)	Percent of mice that develop lung tumors
50	2	1600	42
400	6	1800	37
500	5	2000	38
900	10	3000	50
1100	26		

27. The table shows world average daily oil consumption from 1985 to 2010 measured in thousands of barrels per day.
(a) Make a scatter plot and decide whether a linear model is appropriate.
(b) Find and graph the regression line.
(c) Use the linear model to estimate the oil consumption in 2002 and 2012.

Years since 1985	Thousands of barrels of oil per day
0	60,083
5	66,533
10	70,099
15	76,784
20	84,077
25	87,302

Source: US Energy Information Administration

28. The table shows average US retail residential prices of electricity from 2000 to 2012, measured in cents per kilowatt hour.
(a) Make a scatter plot. Is a linear model appropriate?
(b) Find and graph the regression line.
(c) Use your linear model from part (b) to estimate the average retail price of electricity in 2005 and 2013.

Years since 2000	Cents/kWh
0	8.24
2	8.44
4	8.95
6	10.40
8	11.26
10	11.54
12	11.58

Source: US Energy Information Administration

29. Many physical quantities are connected by *inverse square laws*, that is, by power functions of the form $f(x) = kx^{-2}$. In particular, the illumination of an object by a light source is inversely proportional to the square of the distance from the source. Suppose that after dark you are in a room with just one lamp and you are trying to read a book. The light is too dim and so you move halfway to the lamp. How much brighter is the light?

30. It makes sense that the larger the area of a region, the larger the number of species that inhabit the region. Many ecologists have modeled the species-area relation with a power function and, in particular, the number of species S of bats living in caves in central Mexico has been related to the surface area A of the caves by the equation $S = 0.7A^{0.3}$.
(a) The cave called *Misión Imposible* near Puebla, Mexico, has a surface area of $A = 60 \text{ m}^2$. How many species of bats would you expect to find in that cave?
(b) If you discover that four species of bats live in a cave, estimate the area of the cave.

31. The table shows the number N of species of reptiles and amphibians inhabiting Caribbean islands and the area A of the island in square miles.
(a) Use a power function to model N as a function of A.
(b) The Caribbean island of Dominica has area 291 mi². How many species of reptiles and amphibians would you expect to find on Dominica?

Island	A	N
Saba	4	5
Monserrat	40	9
Puerto Rico	3,459	40
Jamaica	4,411	39
Hispaniola	29,418	84
Cuba	44,218	76

32. The table shows the mean (average) distances d of the planets from the sun (taking the unit of measurement to be the distance from the earth to the sun) and their periods T (time of revolution in years).
(a) Fit a power model to the data.
(b) Kepler's Third Law of Planetary Motion states that "The square of the period of revolution of a planet is proportional to the cube of its mean distance from the sun."
Does your model corroborate Kepler's Third Law?

Planet	d	T
Mercury	0.387	0.241
Venus	0.723	0.615
Earth	1.000	1.000
Mars	1.523	1.881
Jupiter	5.203	11.861
Saturn	9.541	29.457
Uranus	19.190	84.008
Neptune	30.086	164.784

1.3   New Functions from Old Functions

In this section we start with the basic functions we discussed in Section 1.2 and obtain new functions by shifting, stretching, and reflecting their graphs. We also show how to combine pairs of functions by the standard arithmetic operations and by composition.

Transformations of Functions

By applying certain transformations to the graph of a given function we can obtain the graphs of related functions. This will give us the ability to sketch the graphs of many functions quickly by hand. It will also enable us to write equations for given graphs.

Let's first consider **translations**. If c is a positive number, then the graph of $y = f(x) + c$ is just the graph of $y = f(x)$ shifted upward a distance of c units (because each y-coordinate is increased by the same number c). Likewise, if $g(x) = f(x - c)$, where $c > 0$, then the value of g at x is the same as the value of f at $x - c$ (c units to the left of x). Therefore the graph of $y = f(x - c)$ is just the graph of $y = f(x)$ shifted c units to the right (see Figure 1).

> **Vertical and Horizontal Shifts** Suppose $c > 0$. To obtain the graph of
>
> $y = f(x) + c$, shift the graph of $y = f(x)$ a distance c units upward
> $y = f(x) - c$, shift the graph of $y = f(x)$ a distance c units downward
> $y = f(x - c)$, shift the graph of $y = f(x)$ a distance c units to the right
> $y = f(x + c)$, shift the graph of $y = f(x)$ a distance c units to the left

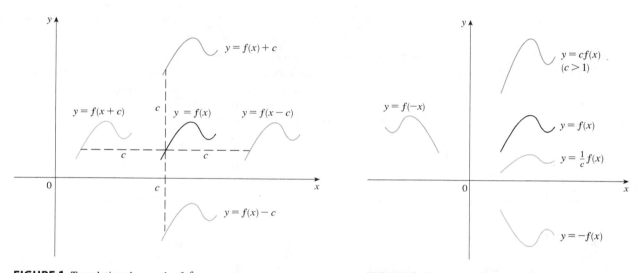

FIGURE 1 Translating the graph of f

FIGURE 2 Stretching and reflecting the graph of f

Now let's consider the **stretching** and **reflecting** transformations. If $c > 1$, then the graph of $y = cf(x)$ is the graph of $y = f(x)$ stretched by a factor of c in the vertical direction (because each y-coordinate is multiplied by the same number c). The graph of $y = -f(x)$ is the graph of $y = f(x)$ reflected about the x-axis because the point (x, y) is replaced by the point $(x, -y)$. (See Figure 2 and the following chart, where the results of other stretching, shrinking, and reflecting transformations are also given.)

> **Vertical and Horizontal Stretching and Reflecting** Suppose $c > 1$. To obtain the graph of
>
> $y = cf(x)$, stretch the graph of $y = f(x)$ vertically by a factor of c
> $y = (1/c)f(x)$, shrink the graph of $y = f(x)$ vertically by a factor of c
> $y = f(cx)$, shrink the graph of $y = f(x)$ horizontally by a factor of c
> $y = f(x/c)$, stretch the graph of $y = f(x)$ horizontally by a factor of c
> $y = -f(x)$, reflect the graph of $y = f(x)$ about the x-axis
> $y = f(-x)$, reflect the graph of $y = f(x)$ about the y-axis

Figure 3 illustrates these stretching transformations when applied to the cosine function with $c = 2$. For instance, in order to get the graph of $y = 2 \cos x$ we multiply the y-coordinate of each point on the graph of $y = \cos x$ by 2. This means that the graph of $y = \cos x$ gets stretched vertically by a factor of 2.

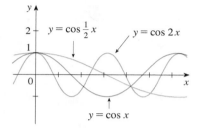

FIGURE 3

EXAMPLE 1 Given the graph of $y = \sqrt{x}$, use transformations to graph $y = \sqrt{x} - 2$, $y = \sqrt{x - 2}$, $y = -\sqrt{x}$, $y = 2\sqrt{x}$, and $y = \sqrt{-x}$.

SOLUTION The graph of the square root function $y = \sqrt{x}$, obtained from Figure 1.2.13(a), is shown in Figure 4(a). In the other parts of the figure we sketch $y = \sqrt{x} - 2$ by shifting 2 units downward, $y = \sqrt{x - 2}$ by shifting 2 units to the right, $y = -\sqrt{x}$ by reflecting about the x-axis, $y = 2\sqrt{x}$ by stretching vertically by a factor of 2, and $y = \sqrt{-x}$ by reflecting about the y-axis.

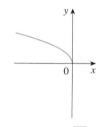

(a) $y = \sqrt{x}$ (b) $y = \sqrt{x} - 2$ (c) $y = \sqrt{x - 2}$ (d) $y = -\sqrt{x}$ (e) $y = 2\sqrt{x}$ (f) $y = \sqrt{-x}$

FIGURE 4

EXAMPLE 2 Sketch the graph of the function $f(x) = x^2 + 6x + 10$.

SOLUTION Completing the square, we write the equation of the graph as

$$y = x^2 + 6x + 10 = (x + 3)^2 + 1$$

This means we obtain the desired graph by starting with the parabola $y = x^2$ and shifting 3 units to the left and then 1 unit upward (see Figure 5).

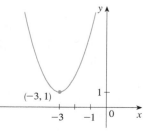

FIGURE 5

(a) $y = x^2$ (b) $y = (x + 3)^2 + 1$

EXAMPLE 3 Sketch the graphs of the following functions.

(a) $y = \sin 2x$ (b) $y = 1 - \sin x$

SOLUTION

(a) We obtain the graph of $y = \sin 2x$ from that of $y = \sin x$ by compressing horizontally by a factor of 2. (See Figures 6 and 7.) Thus, whereas the period of $y = \sin x$ is 2π, the period of $y = \sin 2x$ is $2\pi/2 = \pi$.

FIGURE 6

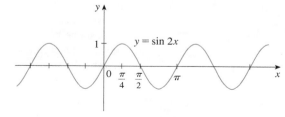

FIGURE 7

(b) To obtain the graph of $y = 1 - \sin x$, we again start with $y = \sin x$. We reflect about the x-axis to get the graph of $y = -\sin x$ and then we shift 1 unit upward to get $y = 1 - \sin x$. (See Figure 8.)

FIGURE 8

EXAMPLE 4 Figure 9 shows graphs of the number of hours of daylight as functions of the time of the year at several latitudes. Given that Philadelphia is located at approximately 40°N latitude, find a function that models the length of daylight at Philadelphia.

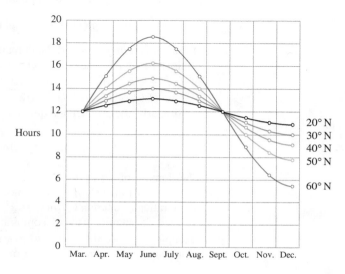

FIGURE 9
Graph of the length of daylight from
March 21 through December 21
at various latitudes

Source: Adapted from L. Harrison,
Daylight, Twilight, Darkness and Time
(New York: Silver, Burdett, 1935), 40.

SOLUTION Notice that each curve resembles a shifted and stretched sine function. By looking at the blue curve we see that, at the latitude of Philadelphia, daylight lasts about 14.8 hours on June 21 and 9.2 hours on December 21, so the amplitude of the curve (the factor by which we have to stretch the sine curve vertically) is $\frac{1}{2}(14.8 - 9.2) = 2.8$.

By what factor do we need to stretch the sine curve horizontally if we measure the time t in days? Because there are about 365 days in a year, the period of our model should be 365. But the period of $y = \sin t$ is 2π, so the horizontal stretching factor is $2\pi/365$.

We also notice that the curve begins its cycle on March 21, the 80th day of the year, so we have to shift the curve 80 units to the right. In addition, we shift it 12 units upward. Therefore we model the length of daylight in Philadelphia on the tth day of the year by the function

$$L(t) = 12 + 2.8 \sin\left[\frac{2\pi}{365}(t - 80)\right]$$

■

Another transformation of some interest is taking the *absolute value* of a function. If $y = |f(x)|$, then according to the definition of absolute value, $y = f(x)$ when $f(x) \geq 0$ and $y = -f(x)$ when $f(x) < 0$. This tells us how to get the graph of $y = |f(x)|$ from the graph of $y = f(x)$: The part of the graph that lies above the x-axis remains the same; the part that lies below the x-axis is reflected about the x-axis.

EXAMPLE 5 Sketch the graph of the function $y = |x^2 - 1|$.

SOLUTION We first graph the parabola $y = x^2 - 1$ in Figure 10(a) by shifting the parabola $y = x^2$ downward 1 unit. We see that the graph lies below the x-axis when $-1 < x < 1$, so we reflect that part of the graph about the x-axis to obtain the graph of $y = |x^2 - 1|$ in Figure 10(b). ■

(a) $y = x^2 - 1$

(b) $y = |x^2 - 1|$

FIGURE 10

■ Combinations of Functions

Two functions f and g can be combined to form new functions $f + g$, $f - g$, fg, and f/g in a manner similar to the way we add, subtract, multiply, and divide real numbers. The sum and difference functions are defined by

$$(f + g)(x) = f(x) + g(x) \qquad (f - g)(x) = f(x) - g(x)$$

If the domain of f is A and the domain of g is B, then the domain of $f + g$ is the intersection $A \cap B$ because both $f(x)$ and $g(x)$ have to be defined. For example, the domain of $f(x) = \sqrt{x}$ is $A = [0, \infty)$ and the domain of $g(x) = \sqrt{2 - x}$ is $B = (-\infty, 2]$, so the domain of $(f + g)(x) = \sqrt{x} + \sqrt{2 - x}$ is $A \cap B = [0, 2]$.

Similarly, the product and quotient functions are defined by

$$(fg)(x) = f(x)g(x) \qquad \left(\frac{f}{g}\right)(x) = \frac{f(x)}{g(x)}$$

The domain of fg is $A \cap B$, but we can't divide by 0 and so the domain of f/g is $\{x \in A \cap B \mid g(x) \neq 0\}$. For instance, if $f(x) = x^2$ and $g(x) = x - 1$, then the domain of the rational function $(f/g)(x) = x^2/(x - 1)$ is $\{x \mid x \neq 1\}$, or $(-\infty, 1) \cup (1, \infty)$.

There is another way of combining two functions to obtain a new function. For example, suppose that $y = f(u) = \sqrt{u}$ and $u = g(x) = x^2 + 1$. Since y is a function of u and u is, in turn, a function of x, it follows that y is ultimately a function of x.

We compute this by substitution:

$$y = f(u) = f(g(x)) = f(x^2 + 1) = \sqrt{x^2 + 1}$$

The procedure is called *composition* because the new function is *composed* of the two given functions f and g.

In general, given any two functions f and g, we start with a number x in the domain of g and calculate $g(x)$. If this number $g(x)$ is in the domain of f, then we can calculate the value of $f(g(x))$. Notice that the output of one function is used as the input to the next function. The result is a new function $h(x) = f(g(x))$ obtained by substituting g into f. It is called the *composition* (or *composite*) of f and g and is denoted by $f \circ g$ ("f circle g").

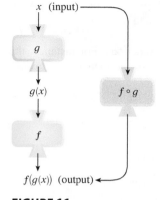

FIGURE 11

The $f \circ g$ machine is composed of the g machine (first) and then the f machine.

> **Definition** Given two functions f and g, the **composite function** $f \circ g$ (also called the **composition** of f and g) is defined by
>
> $$(f \circ g)(x) = f(g(x))$$

The domain of $f \circ g$ is the set of all x in the domain of g such that $g(x)$ is in the domain of f. In other words, $(f \circ g)(x)$ is defined whenever both $g(x)$ and $f(g(x))$ are defined. Figure 11 shows how to picture $f \circ g$ in terms of machines.

EXAMPLE 6 If $f(x) = x^2$ and $g(x) = x - 3$, find the composite functions $f \circ g$ and $g \circ f$.

SOLUTION We have

$$(f \circ g)(x) = f(g(x)) = f(x - 3) = (x - 3)^2$$

$$(g \circ f)(x) = g(f(x)) = g(x^2) = x^2 - 3 \qquad \blacksquare$$

NOTE You can see from Example 6 that, in general, $f \circ g \neq g \circ f$. Remember, the notation $f \circ g$ means that the function g is applied first and then f is applied second. In Example 6, $f \circ g$ is the function that *first* subtracts 3 and *then* squares; $g \circ f$ is the function that *first* squares and *then* subtracts 3.

EXAMPLE 7 If $f(x) = \sqrt{x}$ and $g(x) = \sqrt{2 - x}$, find each of the following functions and their domains.
(a) $f \circ g$ (b) $g \circ f$ (c) $f \circ f$ (d) $g \circ g$

SOLUTION

(a) $$(f \circ g)(x) = f(g(x)) = f\left(\sqrt{2 - x}\right) = \sqrt{\sqrt{2 - x}} = \sqrt[4]{2 - x}$$

The domain of $f \circ g$ is $\{x \mid 2 - x \geqslant 0\} = \{x \mid x \leqslant 2\} = (-\infty, 2]$.

(b) $$(g \circ f)(x) = g(f(x)) = g\left(\sqrt{x}\right) = \sqrt{2 - \sqrt{x}}$$

For $\sqrt{x}$ to be defined we must have $x \geqslant 0$. For $\sqrt{2 - \sqrt{x}}$ to be defined we must have $2 - \sqrt{x} \geqslant 0$, that is, $\sqrt{x} \leqslant 2$, or $x \leqslant 4$. Thus we have $0 \leqslant x \leqslant 4$, so the domain of $g \circ f$ is the closed interval $[0, 4]$.

If $0 \leqslant a \leqslant b$, then $a^2 \leqslant b^2$.

(c) $$(f \circ f)(x) = f(f(x)) = f\left(\sqrt{x}\right) = \sqrt{\sqrt{x}} = \sqrt[4]{x}$$

The domain of $f \circ f$ is $[0, \infty)$.

(d) $$(g \circ g)(x) = g(g(x)) = g(\sqrt{2-x}) = \sqrt{2 - \sqrt{2-x}}$$

This expression is defined when both $2 - x \geqslant 0$ and $2 - \sqrt{2-x} \geqslant 0$. The first inequality means $x \leqslant 2$, and the second is equivalent to $\sqrt{2-x} \leqslant 2$, or $2 - x \leqslant 4$, or $x \geqslant -2$. Thus $-2 \leqslant x \leqslant 2$, so the domain of $g \circ g$ is the closed interval $[-2, 2]$. ■

It is possible to take the composition of three or more functions. For instance, the composite function $f \circ g \circ h$ is found by first applying h, then g, and then f as follows:

$$(f \circ g \circ h)(x) = f(g(h(x)))$$

EXAMPLE 8 Find $f \circ g \circ h$ if $f(x) = x/(x+1)$, $g(x) = x^{10}$, and $h(x) = x + 3$.

SOLUTION

$$(f \circ g \circ h)(x) = f(g(h(x))) = f(g(x+3))$$

$$= f((x+3)^{10}) = \frac{(x+3)^{10}}{(x+3)^{10} + 1}$$ ■

So far we have used composition to build complicated functions from simpler ones. But in calculus it is often useful to be able to *decompose* a complicated function into simpler ones, as in the following example.

EXAMPLE 9 Given $F(x) = \cos^2(x + 9)$, find functions f, g, and h such that $F = f \circ g \circ h$.

SOLUTION Since $F(x) = [\cos(x + 9)]^2$, the formula for F says: First add 9, then take the cosine of the result, and finally square. So we let

$$h(x) = x + 9 \qquad g(x) = \cos x \qquad f(x) = x^2$$

Then

$$(f \circ g \circ h)(x) = f(g(h(x))) = f(g(x + 9)) = f(\cos(x + 9))$$

$$= [\cos(x + 9)]^2 = F(x)$$ ■

1.3   EXERCISES

1. Suppose the graph of f is given. Write equations for the graphs that are obtained from the graph of f as follows.
   (a) Shift 3 units upward.
   (b) Shift 3 units downward.
   (c) Shift 3 units to the right.
   (d) Shift 3 units to the left.
   (e) Reflect about the x-axis.
   (f) Reflect about the y-axis.
   (g) Stretch vertically by a factor of 3.
   (h) Shrink vertically by a factor of 3.

2. Explain how each graph is obtained from the graph of $y = f(x)$.
   (a) $y = f(x) + 8$ (b) $y = f(x + 8)$
   (c) $y = 8f(x)$ (d) $y = f(8x)$
   (e) $y = -f(x) - 1$ (f) $y = 8f\left(\frac{1}{8}x\right)$

3. The graph of $y = f(x)$ is given. Match each equation with its graph and give reasons for your choices.

(a) $y = f(x - 4)$ (b) $y = f(x) + 3$
(c) $y = \frac{1}{3}f(x)$ (d) $y = -f(x + 4)$
(e) $y = 2f(x + 6)$

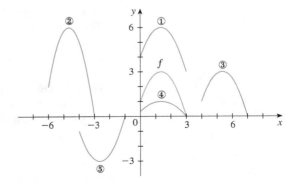

4. The graph of f is given. Draw the graphs of the following functions.

(a) $y = f(x) - 3$
(b) $y = f(x + 1)$
(c) $y = \frac{1}{2}f(x)$
(d) $y = -f(x)$

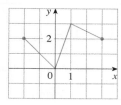

5. The graph of f is given. Use it to graph the following functions.

(a) $y = f(2x)$
(b) $y = f\left(\frac{1}{2}x\right)$
(c) $y = f(-x)$
(d) $y = -f(-x)$

6–7 The graph of $y = \sqrt{3x - x^2}$ is given. Use transformations to create a function whose graph is as shown.

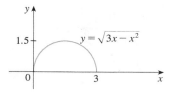

6.

7.

8. (a) How is the graph of $y = 2 \sin x$ related to the graph of $y = \sin x$? Use your answer and Figure 6 to sketch the graph of $y = 2 \sin x$.
(b) How is the graph of $y = 1 + \sqrt{x}$ related to the graph of $y = \sqrt{x}$? Use your answer and Figure 4(a) to sketch the graph of $y = 1 + \sqrt{x}$.

9–24 Graph the function by hand, not by plotting points, but by starting with the graph of one of the standard functions given in Section 1.2, and then applying the appropriate transformations.

9. $y = -x^2$

10. $y = (x - 3)^2$

11. $y = x^3 + 1$

12. $y = 1 - \dfrac{1}{x}$

13. $y = 2 \cos 3x$

14. $y = 2\sqrt{x + 1}$

15. $y = x^2 - 4x + 5$

16. $y = 1 + \sin \pi x$

17. $y = 2 - \sqrt{x}$

18. $y = 3 - 2 \cos x$

19. $y = \sin\left(\frac{1}{2}x\right)$

20. $y = |x| - 2$

21. $y = |x - 2|$

22. $y = \dfrac{1}{4} \tan\left(x - \dfrac{\pi}{4}\right)$

23. $y = \left|\sqrt{x} - 1\right|$

24. $y = |\cos \pi x|$

25. The city of New Orleans is located at latitude 30°N. Use Figure 9 to find a function that models the number of hours of daylight at New Orleans as a function of the time of year. To check the accuracy of your model, use the fact that on March 31 the sun rises at 5:51 AM and sets at 6:18 PM in New Orleans.

26. A variable star is one whose brightness alternately increases and decreases. For the most visible variable star, Delta Cephei, the time between periods of maximum brightness is 5.4 days, the average brightness (or magnitude) of the star is 4.0, and its brightness varies by ±0.35 magnitude. Find a function that models the brightness of Delta Cephei as a function of time.

27. Some of the highest tides in the world occur in the Bay of Fundy on the Atlantic Coast of Canada. At Hopewell Cape the water depth at low tide is about 2.0 m and at high tide it is about 12.0 m. The natural period of oscillation is about 12 hours and on June 30, 2009, high tide occurred at 6:45 AM. Find a function involving the cosine function that models the water depth $D(t)$ (in meters) as a function of time t (in hours after midnight) on that day.

28. In a normal respiratory cycle the volume of air that moves into and out of the lungs is about 500 mL. The reserve and residue volumes of air that remain in the lungs occupy about 2000 mL and a single respiratory cycle for an average human takes about 4 seconds. Find a model for the total volume of air $V(t)$ in the lungs as a function of time.

29. (a) How is the graph of $y = f(|x|)$ related to the graph of f?
(b) Sketch the graph of $y = \sin |x|$.
(c) Sketch the graph of $y = \sqrt{|x|}$.

30. Use the given graph of f to sketch the graph of $y = 1/f(x)$. Which features of f are the most important in sketching $y = 1/f(x)$? Explain how they are used.

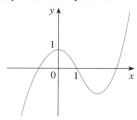

31–32 Find (a) $f + g$, (b) $f - g$, (c) fg, and (d) f/g and state their domains.

31. $f(x) = x^3 + 2x^2$, $g(x) = 3x^2 - 1$

32. $f(x) = \sqrt{3 - x}$, $g(x) = \sqrt{x^2 - 1}$

33–38 Find the functions (a) $f \circ g$, (b) $g \circ f$, (c) $f \circ f$, and (d) $g \circ g$ and their domains.

33. $f(x) = 3x + 5$, $g(x) = x^2 + x$

34. $f(x) = x^3 - 2$, $g(x) = 1 - 4x$

35. $f(x) = \sqrt{x + 1}$, $g(x) = 4x - 3$

36. $f(x) = \sin x$, $g(x) = x^2 + 1$

37. $f(x) = x + \dfrac{1}{x}$, $g(x) = \dfrac{x + 1}{x + 2}$

38. $f(x) = \dfrac{x}{1 + x}$, $g(x) = \sin 2x$

39–42 Find $f \circ g \circ h$.

39. $f(x) = 3x - 2$, $g(x) = \sin x$, $h(x) = x^2$

40. $f(x) = |x - 4|$, $g(x) = 2^x$, $h(x) = \sqrt{x}$

41. $f(x) = \sqrt{x - 3}$, $g(x) = x^2$, $h(x) = x^3 + 2$

42. $f(x) = \tan x$, $g(x) = \dfrac{x}{x - 1}$, $h(x) = \sqrt[3]{x}$

43–48 Express the function in the form $f \circ g$.

43. $F(x) = (2x + x^2)^4$ **44.** $F(x) = \cos^2 x$

45. $F(x) = \dfrac{\sqrt[3]{x}}{1 + \sqrt[3]{x}}$ **46.** $G(x) = \sqrt[3]{\dfrac{x}{1 + x}}$

47. $v(t) = \sec(t^2) \tan(t^2)$ **48.** $u(t) = \dfrac{\tan t}{1 + \tan t}$

49–51 Express the function in the form $f \circ g \circ h$.

49. $R(x) = \sqrt{\sqrt{x} - 1}$ **50.** $H(x) = \sqrt[8]{2 + |x|}$

51. $S(t) = \sin^2(\cos t)$

52. Use the table to evaluate each expression.
   (a) $f(g(1))$ (b) $g(f(1))$ (c) $f(f(1))$
   (d) $g(g(1))$ (e) $(g \circ f)(3)$ (f) $(f \circ g)(6)$

x	1	2	3	4	5	6
$f(x)$	3	1	4	2	2	5
$g(x)$	6	3	2	1	2	3

53. Use the given graphs of f and g to evaluate each expression, or explain why it is undefined.
   (a) $f(g(2))$ (b) $g(f(0))$ (c) $(f \circ g)(0)$
   (d) $(g \circ f)(6)$ (e) $(g \circ g)(-2)$ (f) $(f \circ f)(4)$

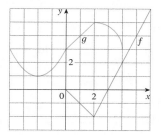

54. Use the given graphs of f and g to estimate the value of $f(g(x))$ for $x = -5, -4, -3, \ldots, 5$. Use these estimates to sketch a rough graph of $f \circ g$.

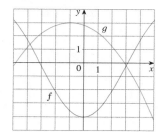

55. A stone is dropped into a lake, creating a circular ripple that travels outward at a speed of 60 cm/s.
   (a) Express the radius r of this circle as a function of the time t (in seconds).
   (b) If A is the area of this circle as a function of the radius, find $A \circ r$ and interpret it.

56. A spherical balloon is being inflated and the radius of the balloon is increasing at a rate of 2 cm/s.
   (a) Express the radius r of the balloon as a function of the time t (in seconds).
   (b) If V is the volume of the balloon as a function of the radius, find $V \circ r$ and interpret it.

57. A ship is moving at a speed of 30 km/h parallel to a straight shoreline. The ship is 6 km from shore and it passes a lighthouse at noon.
   (a) Express the distance s between the lighthouse and the ship

as a function of d, the distance the ship has traveled since noon; that is, find f so that $s = f(d)$.
(b) Express d as a function of t, the time elapsed since noon; that is, find g so that $d = g(t)$.
(c) Find $f \circ g$. What does this function represent?

58. An airplane is flying at a speed of 350 mi/h at an altitude of one mile and passes directly over a radar station at time $t = 0$.
(a) Express the horizontal distance d (in miles) that the plane has flown as a function of t.
(b) Express the distance s between the plane and the radar station as a function of d.
(c) Use composition to express s as a function of t.

59. The **Heaviside function** H is defined by

$$H(t) = \begin{cases} 0 & \text{if } t < 0 \\ 1 & \text{if } t \geq 0 \end{cases}$$

It is used in the study of electric circuits to represent the sudden surge of electric current, or voltage, when a switch is instantaneously turned on.
(a) Sketch the graph of the Heaviside function.
(b) Sketch the graph of the voltage $V(t)$ in a circuit if the switch is turned on at time $t = 0$ and 120 volts are applied instantaneously to the circuit. Write a formula for $V(t)$ in terms of $H(t)$.
(c) Sketch the graph of the voltage $V(t)$ in a circuit if the switch is turned on at time $t = 5$ seconds and 240 volts are applied instantaneously to the circuit. Write a formula for $V(t)$ in terms of $H(t)$. (Note that starting at $t = 5$ corresponds to a translation.)

60. The Heaviside function defined in Exercise 59 can also be used to define the **ramp function** $y = ctH(t)$, which

represents a gradual increase in voltage or current in a circuit.
(a) Sketch the graph of the ramp function $y = tH(t)$.
(b) Sketch the graph of the voltage $V(t)$ in a circuit if the switch is turned on at time $t = 0$ and the voltage is gradually increased to 120 volts over a 60-second time interval. Write a formula for $V(t)$ in terms of $H(t)$ for $t \leq 60$.
(c) Sketch the graph of the voltage $V(t)$ in a circuit if the switch is turned on at time $t = 7$ seconds and the voltage is gradually increased to 100 volts over a period of 25 seconds. Write a formula for $V(t)$ in terms of $H(t)$ for $t \leq 32$.

61. Let f and g be linear functions with equations $f(x) = m_1 x + b_1$ and $g(x) = m_2 x + b_2$. Is $f \circ g$ also a linear function? If so, what is the slope of its graph?

62. If you invest x dollars at 4% interest compounded annually, then the amount $A(x)$ of the investment after one year is $A(x) = 1.04x$. Find $A \circ A$, $A \circ A \circ A$, and $A \circ A \circ A \circ A$. What do these compositions represent? Find a formula for the composition of n copies of A.

63. (a) If $g(x) = 2x + 1$ and $h(x) = 4x^2 + 4x + 7$, find a function f such that $f \circ g = h$. (Think about what operations you would have to perform on the formula for g to end up with the formula for h.)
(b) If $f(x) = 3x + 5$ and $h(x) = 3x^2 + 3x + 2$, find a function g such that $f \circ g = h$.

64. If $f(x) = x + 4$ and $h(x) = 4x - 1$, find a function g such that $g \circ f = h$.

65. Suppose g is an even function and let $h = f \circ g$. Is h always an even function?

66. Suppose g is an odd function and let $h = f \circ g$. Is h always an odd function? What if f is odd? What if f is even?

1.4 Exponential Functions

The function $f(x) = 2^x$ is called an *exponential function* because the variable, x, is the exponent. It should not be confused with the power function $g(x) = x^2$, in which the variable is the base.

In general, an **exponential function** is a function of the form

$$f(x) = b^x$$

In Appendix G we present an alternative approach to the exponential and logarithmic functions using integral calculus.

where b is a positive constant. Let's recall what this means.
If $x = n$, a positive integer, then

$$b^n = \underbrace{b \cdot b \cdot \cdots \cdot b}_{n \text{ factors}}$$

If $x = 0$, then $b^0 = 1$, and if $x = -n$, where n is a positive integer, then

$$b^{-n} = \frac{1}{b^n}$$

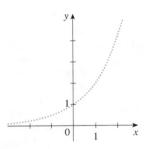

FIGURE 1
Representation of $y = 2^x$, x rational

If x is a rational number, $x = p/q$, where p and q are integers and $q > 0$, then

$$b^x = b^{p/q} = \sqrt[q]{b^p} = \left(\sqrt[q]{b}\right)^p$$

But what is the meaning of b^x if x is an irrational number? For instance, what is meant by $2^{\sqrt{3}}$ or 5^π?

To help us answer this question we first look at the graph of the function $y = 2^x$, where x is rational. A representation of this graph is shown in Figure 1. We want to enlarge the domain of $y = 2^x$ to include both rational and irrational numbers.

There are holes in the graph in Figure 1 corresponding to irrational values of x. We want to fill in the holes by defining $f(x) = 2^x$, where $x \in \mathbb{R}$, so that f is an increasing function. In particular, since the irrational number $\sqrt{3}$ satisfies

$$1.7 < \sqrt{3} < 1.8$$

we must have

$$2^{1.7} < 2^{\sqrt{3}} < 2^{1.8}$$

and we know what $2^{1.7}$ and $2^{1.8}$ mean because 1.7 and 1.8 are rational numbers. Similarly, if we use better approximations for $\sqrt{3}$, we obtain better approximations for $2^{\sqrt{3}}$:

$$1.73 < \sqrt{3} < 1.74 \quad \Rightarrow \quad 2^{1.73} < 2^{\sqrt{3}} < 2^{1.74}$$

$$1.732 < \sqrt{3} < 1.733 \quad \Rightarrow \quad 2^{1.732} < 2^{\sqrt{3}} < 2^{1.733}$$

$$1.7320 < \sqrt{3} < 1.7321 \quad \Rightarrow \quad 2^{1.7320} < 2^{\sqrt{3}} < 2^{1.7321}$$

$$1.73205 < \sqrt{3} < 1.73206 \quad \Rightarrow \quad 2^{1.73205} < 2^{\sqrt{3}} < 2^{1.73206}$$

$$\vdots \qquad\qquad \vdots \qquad\qquad \vdots \qquad\qquad \vdots$$

A proof of this fact is given in
J. Marsden and A. Weinstein,
Calculus Unlimited (Menlo Park, CA:
Benjamin/Cummings, 1981).

It can be shown that there is exactly one number that is greater than all of the numbers

$$2^{1.7}, \quad 2^{1.73}, \quad 2^{1.732}, \quad 2^{1.7320}, \quad 2^{1.73205}, \quad \ldots$$

and less than all of the numbers

$$2^{1.8}, \quad 2^{1.74}, \quad 2^{1.733}, \quad 2^{1.7321}, \quad 2^{1.73206}, \quad \ldots$$

We define $2^{\sqrt{3}}$ to be this number. Using the preceding approximation process we can compute it correct to six decimal places:

$$2^{\sqrt{3}} \approx 3.321997$$

Similarly, we can define 2^x (or b^x, if $b > 0$) where x is any irrational number. Figure 2 shows how all the holes in Figure 1 have been filled to complete the graph of the function $f(x) = 2^x$, $x \in \mathbb{R}$.

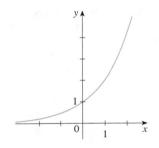

FIGURE 2
$y = 2^x$, x real

The graphs of members of the family of functions $y = b^x$ are shown in Figure 3 for various values of the base b. Notice that all of these graphs pass through the same point $(0, 1)$ because $b^0 = 1$ for $b \neq 0$. Notice also that as the base b gets larger, the exponential function grows more rapidly (for $x > 0$).

If $0 < b < 1$, then b^x approaches 0 as x becomes large. If $b > 1$, then b^x approaches 0 as x decreases through negative values. In both cases the x-axis is a horizontal asymptote. These matters are discussed in Section 2.6.

FIGURE 3

You can see from Figure 3 that there are basically three kinds of exponential functions $y = b^x$. If $0 < b < 1$, the exponential function decreases; if $b = 1$, it is a constant; and if $b > 1$, it increases. These three cases are illustrated in Figure 4. Observe that if $b \neq 1$, then the exponential function $y = b^x$ has domain $\mathbb{R}$ and range $(0, \infty)$. Notice also that, since $(1/b)^x = 1/b^x = b^{-x}$, the graph of $y = (1/b)^x$ is just the reflection of the graph of $y = b^x$ about the y-axis.

FIGURE 4   (a) $y = b^x$, $0 < b < 1$ (b) $y = 1^x$ (c) $y = b^x$, $b > 1$

One reason for the importance of the exponential function lies in the following properties. If x and y are rational numbers, then these laws are well known from elementary algebra. It can be proved that they remain true for arbitrary real numbers x and y.

www.stewartcalculus.com
For review and practice using the Laws of Exponents, click on *Review of Algebra*.

Laws of Exponents If a and b are positive numbers and x and y are any real numbers, then

1. $b^{x+y} = b^x b^y$ **2.** $b^{x-y} = \dfrac{b^x}{b^y}$ **3.** $(b^x)^y = b^{xy}$ **4.** $(ab)^x = a^x b^x$

EXAMPLE 1 Sketch the graph of the function $y = 3 - 2^x$ and determine its domain and range.

For a review of reflecting and shifting graphs, see Section 1.3.

SOLUTION First we reflect the graph of $y = 2^x$ [shown in Figures 2 and 5(a)] about the x-axis to get the graph of $y = -2^x$ in Figure 5(b). Then we shift the graph of $y = -2^x$

upward 3 units to obtain the graph of $y = 3 - 2^x$ in Figure 5(c). The domain is $\mathbb{R}$ and the range is $(-\infty, 3)$.

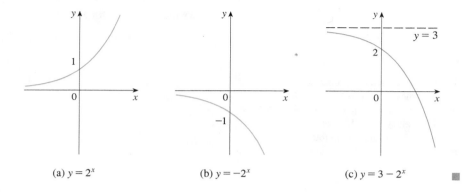

FIGURE 5   (a) $y = 2^x$ (b) $y = -2^x$ (c) $y = 3 - 2^x$ ∎

EXAMPLE 2 Use a graphing device to compare the exponential function $f(x) = 2^x$ and the power function $g(x) = x^2$. Which function grows more quickly when x is large?

SOLUTION Figure 6 shows both functions graphed in the viewing rectangle $[-2, 6]$ by $[0, 40]$. We see that the graphs intersect three times, but for $x > 4$ the graph of $f(x) = 2^x$ stays above the graph of $g(x) = x^2$. Figure 7 gives a more global view and shows that for large values of x, the exponential function $y = 2^x$ grows far more rapidly than the power function $y = x^2$.

Example 2 shows that $y = 2^x$ increases more quickly than $y = x^2$. To demonstrate just how quickly $f(x) = 2^x$ increases, let's perform the following thought experiment. Suppose we start with a piece of paper a thousandth of an inch thick and we fold it in half 50 times. Each time we fold the paper in half, the thickness of the paper doubles, so the thickness of the resulting paper would be $2^{50}/1000$ inches. How thick do you think that is? It works out to be more than 17 million miles!

FIGURE 6   **FIGURE 7**   ∎

■ Applications of Exponential Functions

The exponential function occurs very frequently in mathematical models of nature and society. Here we indicate briefly how it arises in the description of population growth and radioactive decay. In later chapters we will pursue these and other applications in greater detail.

   First we consider a population of bacteria in a homogeneous nutrient medium. Suppose that by sampling the population at certain intervals it is determined that the population doubles every hour. If the number of bacteria at time t is $p(t)$, where t is measured in hours, and the initial population is $p(0) = 1000$, then we have

$$p(1) = 2p(0) = 2 \times 1000$$

$$p(2) = 2p(1) = 2^2 \times 1000$$

$$p(3) = 2p(2) = 2^3 \times 1000$$

It seems from this pattern that, in general,

$$p(t) = 2^t \times 1000 = (1000)2^t$$

This population function is a constant multiple of the exponential function $y = 2^t$, so it exhibits the rapid growth that we observed in Figures 2 and 7. Under ideal conditions (unlimited space and nutrition and absence of disease) this exponential growth is typical of what actually occurs in nature.

What about the human population? Table 1 shows data for the population of the world in the 20th century and Figure 8 shows the corresponding scatter plot.

Table 1

t (years since 1900)	Population (millions)
0	1650
10	1750
20	1860
30	2070
40	2300
50	2560
60	3040
70	3710
80	4450
90	5280
100	6080
110	6870

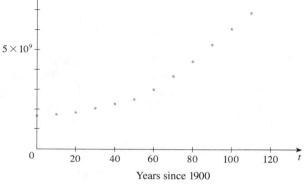

FIGURE 8
Scatter plot for world
population growth

The pattern of the data points in Figure 8 suggests exponential growth, so we use a graphing calculator with exponential regression capability to apply the method of least squares and obtain the exponential model

$$P = (1436.53) \cdot (1.01395)^t$$

where $t = 0$ corresponds to 1900. Figure 9 shows the graph of this exponential function together with the original data points. We see that the exponential curve fits the data reasonably well. The period of relatively slow population growth is explained by the two world wars and the Great Depression of the 1930s.

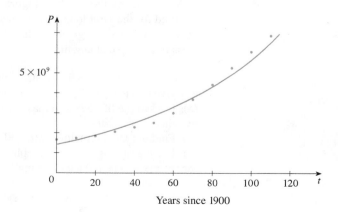

FIGURE 9
Exponential model for
population growth

In 1995 a paper appeared detailing the effect of the protease inhibitor ABT-538 on the human immunodeficiency virus HIV-1.[1] Table 2 shows values of the plasma viral load $V(t)$ of patient 303, measured in RNA copies per mL, t days after ABT-538 treatment was begun. The corresponding scatter plot is shown in Figure 10.

Table 2

t (days)	$V(t)$
1	76.0
4	53.0
8	18.0
11	9.4
15	5.2
22	3.6

FIGURE 10 Plasma viral load in patient 303

The rather dramatic decline of the viral load that we see in Figure 10 reminds us of the graphs of the exponential function $y = b^x$ in Figures 3 and 4(a) for the case where the base b is less than 1. So let's model the function $V(t)$ by an exponential function. Using a graphing calculator or computer to fit the data in Table 2 with an exponential function of the form $y = a \cdot b^t$, we obtain the model

$$V = 96.39785 \cdot (0.818656)^t$$

In Figure 11 we graph this exponential function with the data points and see that the model represents the viral load reasonably well for the first month of treatment.

FIGURE 11
Exponential model for viral load

We could use the graph in Figure 11 to estimate the **half-life** of V, that is, the time required for the viral load to be reduced to half its initial value (see Exercise 33). In the next example we are given the half-life of a radioactive element and asked to find the mass of a sample at any time.

EXAMPLE 3 The half-life of strontium-90, ^{90}Sr, is 25 years. This means that half of any given quantity of ^{90}Sr will disintegrate in 25 years.
(a) If a sample of ^{90}Sr has a mass of 24 mg, find an expression for the mass $m(t)$ that remains after t years.
(b) Find the mass remaining after 40 years, correct to the nearest milligram.
(c) Use a graphing device to graph $m(t)$ and use the graph to estimate the time required for the mass to be reduced to 5 mg.

1. D. Ho et al., "Rapid Turnover of Plasma Virions and CD4 Lymphocytes in HIV-1 Infection," *Nature* 373 (1995): 123–26.

SOLUTION

(a) The mass is initially 24 mg and is halved during each 25-year period, so

$$m(0) = 24$$

$$m(25) = \frac{1}{2}(24)$$

$$m(50) = \frac{1}{2} \cdot \frac{1}{2}(24) = \frac{1}{2^2}(24)$$

$$m(75) = \frac{1}{2} \cdot \frac{1}{2^2}(24) = \frac{1}{2^3}(24)$$

$$m(100) = \frac{1}{2} \cdot \frac{1}{2^3}(24) = \frac{1}{2^4}(24)$$

From this pattern, it appears that the mass remaining after t years is

$$m(t) = \frac{1}{2^{t/25}}(24) = 24 \cdot 2^{-t/25} = 24 \cdot (2^{-1/25})^t$$

This is an exponential function with base $b = 2^{-1/25} = 1/2^{1/25}$.

(b) The mass that remains after 40 years is

$$m(40) = 24 \cdot 2^{-40/25} \approx 7.9 \text{ mg}$$

(c) We use a graphing calculator or computer to graph the function $m(t) = 24 \cdot 2^{-t/25}$ in Figure 12. We also graph the line $m = 5$ and use the cursor to estimate that $m(t) = 5$ when $t \approx 57$. So the mass of the sample will be reduced to 5 mg after about 57 years. ■

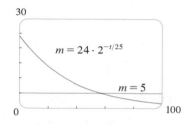

FIGURE 12

■ The Number e

Of all possible bases for an exponential function, there is one that is most convenient for the purposes of calculus. The choice of a base b is influenced by the way the graph of $y = b^x$ crosses the y-axis. Figures 13 and 14 show the tangent lines to the graphs of $y = 2^x$ and $y = 3^x$ at the point $(0, 1)$. (Tangent lines will be defined precisely in Section 2.7. For present purposes, you can think of the tangent line to an exponential graph at a point as the line that touches the graph only at that point.) If we measure the slopes of these tangent lines at $(0, 1)$, we find that $m \approx 0.7$ for $y = 2^x$ and $m \approx 1.1$ for $y = 3^x$.

FIGURE 13

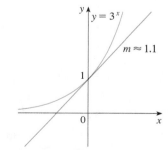

FIGURE 14

It turns out, as we will see in Chapter 3, that some of the formulas of calculus will be greatly simplified if we choose the base b so that the slope of the tangent line to $y = b^x$

FIGURE 15
The natural exponential function
crosses the y-axis with a slope of 1.

TEC Module 1.4 enables you to graph
exponential functions with various
bases and their tangent lines in order
to estimate more closely the value of b
for which the tangent has slope 1.

FIGURE 16

at $(0, 1)$ is *exactly* 1. (See Figure 15.) In fact, there *is* such a number and it is denoted by the letter e. (This notation was chosen by the Swiss mathematician Leonhard Euler in 1727, probably because it is the first letter of the word *exponential*.) In view of Figures 13 and 14, it comes as no surprise that the number e lies between 2 and 3 and the graph of $y = e^x$ lies between the graphs of $y = 2^x$ and $y = 3^x$. (See Figure 16.) In Chapter 3 we will see that the value of e, correct to five decimal places, is

$$e \approx 2.71828$$

We call the function $f(x) = e^x$ the **natural exponential function**.

EXAMPLE 4 Graph the function $y = \frac{1}{2}e^{-x} - 1$ and state the domain and range.

SOLUTION We start with the graph of $y = e^x$ from Figures 15 and 17(a) and reflect about the y-axis to get the graph of $y = e^{-x}$ in Figure 17(b). (Notice that the graph crosses the y-axis with a slope of -1). Then we compress the graph vertically by a factor of 2 to obtain the graph of $y = \frac{1}{2}e^{-x}$ in Figure 17(c). Finally, we shift the graph downward one unit to get the desired graph in Figure 17(d). The domain is $\mathbb{R}$ and the range is $(-1, \infty)$.

(a) $y = e^x$

(b) $y = e^{-x}$

(c) $y = \frac{1}{2}e^{-x}$

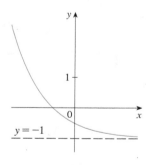

(d) $y = \frac{1}{2}e^{-x} - 1$

FIGURE 17

How far to the right do you think we would have to go for the height of the graph of $y = e^x$ to exceed a million? The next example demonstrates the rapid growth of this function by providing an answer that might surprise you.

EXAMPLE 5 Use a graphing device to find the values of x for which $e^x > 1,000,000$.

SOLUTION In Figure 18 we graph both the function $y = e^x$ and the horizontal line $y = 1{,}000{,}000$. We see that these curves intersect when $x \approx 13.8$. Thus $e^x > 10^6$ when $x > 13.8$. It is perhaps surprising that the values of the exponential function have already surpassed a million when x is only 14.

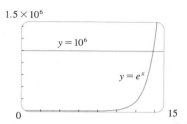

1.5×10^6

$y = 10^6$

$y = e^x$

0   15

FIGURE 18

1.4 EXERCISES

1–4 Use the Law of Exponents to rewrite and simplify the expression.

1. (a) $\dfrac{4^{-3}}{2^{-8}}$ (b) $\dfrac{1}{\sqrt[3]{x^4}}$

2. (a) $8^{4/3}$ (b) $x(3x^2)^3$

3. (a) $b^8(2b)^4$ (b) $\dfrac{(6y^3)^4}{2y^5}$

4. (a) $\dfrac{x^{2n} \cdot x^{3n-1}}{x^{n+2}}$ (b) $\dfrac{\sqrt{a}\sqrt{b}}{\sqrt[3]{ab}}$

5. (a) Write an equation that defines the exponential function with base $b > 0$.
   (b) What is the domain of this function?
   (c) If $b \neq 1$, what is the range of this function?
   (d) Sketch the general shape of the graph of the exponential function for each of the following cases.
       (i) $b > 1$
       (ii) $b = 1$
       (iii) $0 < b < 1$

6. (a) How is the number e defined?
   (b) What is an approximate value for e?
   (c) What is the natural exponential function?

7–10 Graph the given functions on a common screen. How are these graphs related?

7. $y = 2^x$, $y = e^x$, $y = 5^x$, $y = 20^x$

8. $y = e^x$, $y = e^{-x}$, $y = 8^x$, $y = 8^{-x}$

9. $y = 3^x$, $y = 10^x$, $y = \left(\frac{1}{3}\right)^x$, $y = \left(\frac{1}{10}\right)^x$

10. $y = 0.9^x$, $y = 0.6^x$, $y = 0.3^x$, $y = 0.1^x$

11–16 Make a rough sketch of the graph of the function. Do not use a calculator. Just use the graphs given in Figures 3 and 13 and, if necessary, the transformations of Section 1.3.

11. $y = 4^x - 1$ **12.** $y = (0.5)^{x-1}$

13. $y = -2^{-x}$ **14.** $y = e^{|x|}$

15. $y = 1 - \frac{1}{2}e^{-x}$ **16.** $y = 2(1 - e^x)$

17. Starting with the graph of $y = e^x$, write the equation of the graph that results from
   (a) shifting 2 units downward.
   (b) shifting 2 units to the right.
   (c) reflecting about the x-axis.
   (d) reflecting about the y-axis.
   (e) reflecting about the x-axis and then about the y-axis.

18. Starting with the graph of $y = e^x$, find the equation of the graph that results from
   (a) reflecting about the line $y = 4$.
   (b) reflecting about the line $x = 2$.

19–20 Find the domain of each function.

19. (a) $f(x) = \dfrac{1 - e^{x^2}}{1 - e^{1-x^2}}$ (b) $f(x) = \dfrac{1 + x}{e^{\cos x}}$

20. (a) $g(t) = \sqrt{10^t - 100}$ (b) $g(t) = \sin(e^t - 1)$

21–22 Find the exponential function $f(x) = Cb^x$ whose graph is given.

21.   **22.**

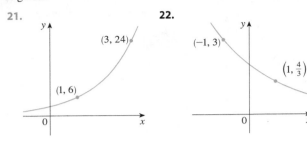

23. If $f(x) = 5^x$, show that

$$\frac{f(x + h) - f(x)}{h} = 5^x \left(\frac{5^h - 1}{h} \right)$$

24. Suppose you are offered a job that lasts one month. Which of the following methods of payment do you prefer?
  I. One million dollars at the end of the month.
  II. One cent on the first day of the month, two cents on the second day, four cents on the third day, and, in general, 2^{n-1} cents on the nth day.

25. Suppose the graphs of $f(x) = x^2$ and $g(x) = 2^x$ are drawn on a coordinate grid where the unit of measurement is 1 inch. Show that, at a distance 2 ft to the right of the origin, the height of the graph of f is 48 ft but the height of the graph of g is about 265 mi.

26. Compare the functions $f(x) = x^5$ and $g(x) = 5^x$ by graphing both functions in several viewing rectangles. Find all points of intersection of the graphs correct to one decimal place. Which function grows more rapidly when x is large?

27. Compare the functions $f(x) = x^{10}$ and $g(x) = e^x$ by graphing both f and g in several viewing rectangles. When does the graph of g finally surpass the graph of f?

28. Use a graph to estimate the values of x such that $e^x > 1{,}000{,}000{,}000$.

29. A researcher is trying to determine the doubling time for a population of the bacterium *Giardia lamblia*. He starts a culture in a nutrient solution and estimates the bacteria count every four hours. His data are shown in the table.

Time (hours)	0	4	8	12	16	20	24
Bacteria count (CFU/mL)	37	47	63	78	105	130	173

(a) Make a scatter plot of the data.
(b) Use a graphing calculator to find an exponential curve $f(t) = a \cdot b^t$ that models the bacteria population t hours later.

(c) Graph the model from part (b) together with the scatter plot in part (a). Use the TRACE feature to determine how long it takes for the bacteria count to double.

G. lamblia

30. A bacteria culture starts with 500 bacteria and doubles in size every half hour.
(a) How many bacteria are there after 3 hours?
(b) How many bacteria are there after t hours?
(c) How many bacteria are there after 40 minutes?
(d) Graph the population function and estimate the time for the population to reach 100,000.

31. The half-life of bismuth-210, ^{210}Bi, is 5 days.
(a) If a sample has a mass of 200 mg, find the amount remaining after 15 days.
(b) Find the amount remaining after t days.
(c) Estimate the amount remaining after 3 weeks.
(d) Use a graph to estimate the time required for the mass to be reduced to 1 mg.

32. An isotope of sodium, ^{24}Na, has a half-life of 15 hours. A sample of this isotope has mass 2 g.
(a) Find the amount remaining after 60 hours.
(b) Find the amount remaining after t hours.
(c) Estimate the amount remaining after 4 days.
(d) Use a graph to estimate the time required for the mass to be reduced to 0.01 g.

33. Use the graph of V in Figure 11 to estimate the half-life of the viral load of patient 303 during the first month of treatment.

34. After alcohol is fully absorbed into the body, it is metabolized with a half-life of about 1.5 hours. Suppose you have had three alcoholic drinks and an hour later, at midnight, your blood alcohol concentration (BAC) is 0.6 mg/mL.
(a) Find an exponential decay model for your BAC t hours after midnight.
(b) Graph your BAC and use the graph to determine when you can drive home if the legal limit is 0.08 mg/mL.

Source: Adapted from P. Wilkinson et al., "Pharmacokinetics of Ethanol after Oral Administration in the Fasting State," *Journal of Pharmacokinetics and Biopharmaceutics* 5 (1977): 207–24.

35. Use a graphing calculator with exponential regression capability to model the population of the world with the data from 1950 to 2010 in Table 1 on page 49. Use the model to estimate the population in 1993 and to predict the population in the year 2020.

36. The table gives the population of the United States, in millions, for the years 1900–2010. Use a graphing calculator

Year	Population	Year	Population
1900	76	1960	179
1910	92	1970	203
1920	106	1980	227
1930	123	1990	250
1940	131	2000	281
1950	150	2010	310

with exponential regression capability to model the US population since 1900. Use the model to estimate the population in 1925 and to predict the population in the year 2020.

37. If you graph the function

$$f(x) = \frac{1 - e^{1/x}}{1 + e^{1/x}}$$

you'll see that f appears to be an odd function. Prove it.

38. Graph several members of the family of functions

$$f(x) = \frac{1}{1 + ae^{bx}}$$

where $a > 0$. How does the graph change when b changes? How does it change when a changes?

1.5   Inverse Functions and Logarithms

Table 1 gives data from an experiment in which a bacteria culture started with 100 bacteria in a limited nutrient medium; the size of the bacteria population was recorded at hourly intervals. The number of bacteria N is a function of the time t: $N = f(t)$.

Suppose, however, that the biologist changes her point of view and becomes interested in the time required for the population to reach various levels. In other words, she is thinking of t as a function of N. This function is called the *inverse function* of f, denoted by f^{-1}, and read "f inverse." Thus $t = f^{-1}(N)$ is the time required for the population level to reach N. The values of f^{-1} can be found by reading Table 1 from right to left or by consulting Table 2. For instance, $f^{-1}(550) = 6$ because $f(6) = 550$.

Table 1 N as a function of t

t (hours)	$N = f(t)$ = population at time t
0	100
1	168
2	259
3	358
4	445
5	509
6	550
7	573
8	586

Table 2 t as a function of N

N	$t = f^{-1}(N)$ = time to reach N bacteria
100	0
168	1
259	2
358	3
445	4
509	5
550	6
573	7
586	8

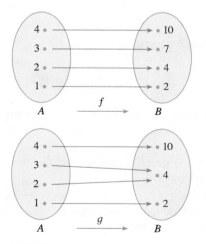

FIGURE 1
f is one-to-one; g is not.

Not all functions possess inverses. Let's compare the functions f and g whose arrow diagrams are shown in Figure 1. Note that f never takes on the same value twice (any two inputs in A have different outputs), whereas g does take on the same value twice (both 2 and 3 have the same output, 4). In symbols,

$$g(2) = g(3)$$

but $$f(x_1) \neq f(x_2)   \text{ whenever } x_1 \neq x_2$$

Functions that share this property with f are called *one-to-one functions*.

In the language of inputs and outputs, this definition says that f is one-to-one if each output corresponds to only one input.

> **1** **Definition** A function f is called a **one-to-one function** if it never takes on the same value twice; that is,
>
> $$f(x_1) \neq f(x_2) \qquad \text{whenever } x_1 \neq x_2$$

If a horizontal line intersects the graph of f in more than one point, then we see from Figure 2 that there are numbers x_1 and x_2 such that $f(x_1) = f(x_2)$. This means that f is not one-to-one.

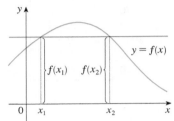

FIGURE 2
This function is not one-to-one because $f(x_1) = f(x_2)$.

Therefore we have the following geometric method for determining whether a function is one-to-one.

> **Horizontal Line Test** A function is one-to-one if and only if no horizontal line intersects its graph more than once.

EXAMPLE 1 Is the function $f(x) = x^3$ one-to-one?

SOLUTION 1 If $x_1 \neq x_2$, then $x_1^3 \neq x_2^3$ (two different numbers can't have the same cube). Therefore, by Definition 1, $f(x) = x^3$ is one-to-one.

SOLUTION 2 From Figure 3 we see that no horizontal line intersects the graph of $f(x) = x^3$ more than once. Therefore, by the Horizontal Line Test, f is one-to-one. ∎

FIGURE 3
$f(x) = x^3$ is one-to-one.

EXAMPLE 2 Is the function $g(x) = x^2$ one-to-one?

SOLUTION 1 This function is not one-to-one because, for instance,

$$g(1) = 1 = g(-1)$$

and so 1 and -1 have the same output.

SOLUTION 2 From Figure 4 we see that there are horizontal lines that intersect the graph of g more than once. Therefore, by the Horizontal Line Test, g is not one-to-one. ∎

FIGURE 4
$g(x) = x^2$ is not one-to-one.

One-to-one functions are important because they are precisely the functions that possess inverse functions according to the following definition.

> **2** **Definition** Let f be a one-to-one function with domain A and range B. Then its **inverse function** f^{-1} has domain B and range A and is defined by
>
> $$f^{-1}(y) = x \iff f(x) = y$$
>
> for any y in B.

FIGURE 5

This definition says that if f maps x into y, then f^{-1} maps y back into x. (If f were not one-to-one, then f^{-1} would not be uniquely defined.) The arrow diagram in Figure 5 indicates that f^{-1} reverses the effect of f. Note that

$$\text{domain of } f^{-1} = \text{range of } f$$

$$\text{range of } f^{-1} = \text{domain of } f$$

For example, the inverse function of $f(x) = x^3$ is $f^{-1}(x) = x^{1/3}$ because if $y = x^3$, then

$$f^{-1}(y) = f^{-1}(x^3) = (x^3)^{1/3} = x$$

CAUTION Do not mistake the -1 in f^{-1} for an exponent. Thus

$$f^{-1}(x) \quad \text{does } not \text{ mean} \quad \frac{1}{f(x)}$$

The reciprocal $1/f(x)$ could, however, be written as $[f(x)]^{-1}$.

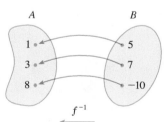

FIGURE 6
The inverse function reverses inputs and outputs.

EXAMPLE 3 If $f(1) = 5$, $f(3) = 7$, and $f(8) = -10$, find $f^{-1}(7)$, $f^{-1}(5)$, and $f^{-1}(-10)$.

SOLUTION From the definition of f^{-1} we have

$$f^{-1}(7) = 3 \qquad \text{because} \qquad f(3) = 7$$

$$f^{-1}(5) = 1 \qquad \text{because} \qquad f(1) = 5$$

$$f^{-1}(-10) = 8 \qquad \text{because} \qquad f(8) = -10$$

The diagram in Figure 6 makes it clear how f^{-1} reverses the effect of f in this case. ■

The letter x is traditionally used as the independent variable, so when we concentrate on f^{-1} rather than on f, we usually reverse the roles of x and y in Definition 2 and write

$$\boxed{3} \qquad f^{-1}(x) = y \quad \Longleftrightarrow \quad f(y) = x$$

By substituting for y in Definition 2 and substituting for x in (3), we get the following **cancellation equations**:

$$\boxed{4} \qquad \begin{aligned} f^{-1}(f(x)) &= x \qquad \text{for every } x \text{ in } A \\ f(f^{-1}(x)) &= x \qquad \text{for every } x \text{ in } B \end{aligned}$$

The first cancellation equation says that if we start with x, apply f, and then apply f^{-1}, we arrive back at x, where we started (see the machine diagram in Figure 7). Thus f^{-1} undoes what f does. The second equation says that f undoes what f^{-1} does.

FIGURE 7

$$x \longrightarrow \boxed{f} \longrightarrow f(x) \longrightarrow \boxed{f^{-1}} \longrightarrow x .$$

For example, if $f(x) = x^3$, then $f^{-1}(x) = x^{1/3}$ and so the cancellation equations become

$$f^{-1}(f(x)) = (x^3)^{1/3} = x$$

$$f(f^{-1}(x)) = (x^{1/3})^3 = x$$

These equations simply say that the cube function and the cube root function cancel each other when applied in succession.

Now let's see how to compute inverse functions. If we have a function $y = f(x)$ and are able to solve this equation for x in terms of y, then according to Definition 2 we must have $x = f^{-1}(y)$. If we want to call the independent variable x, we then interchange x and y and arrive at the equation $y = f^{-1}(x)$.

---

5   **How to Find the Inverse Function of a One-to-One Function f**

STEP 1   Write $y = f(x)$.

STEP 2   Solve this equation for x in terms of y (if possible).

STEP 3   To express f^{-1} as a function of x, interchange x and y. The resulting equation is $y = f^{-1}(x)$.

EXAMPLE 4   Find the inverse function of $f(x) = x^3 + 2$.

SOLUTION According to (5) we first write

$$y = x^3 + 2$$

Then we solve this equation for x:

$$x^3 = y - 2$$

$$x = \sqrt[3]{y - 2}$$

Finally, we interchange x and y:

$$y = \sqrt[3]{x - 2}$$

In Example 4, notice how f^{-1} reverses the effect of f. The function f is the rule "Cube, then add 2"; f^{-1} is the rule "Subtract 2, then take the cube root."

Therefore the inverse function is $f^{-1}(x) = \sqrt[3]{x - 2}$.   ■

The principle of interchanging x and y to find the inverse function also gives us the method for obtaining the graph of f^{-1} from the graph of f. Since $f(a) = b$ if and only if $f^{-1}(b) = a$, the point (a, b) is on the graph of f if and only if the point (b, a) is on the graph of f^{-1}. But we get the point (b, a) from (a, b) by reflecting about the line $y = x$. (See Figure 8.)

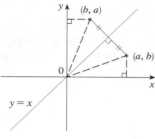

FIGURE 8

FIGURE 9

Therefore, as illustrated by Figure 9:

> The graph of f^{-1} is obtained by reflecting the graph of f about the line $y = x$.

FIGURE 10

EXAMPLE 5 Sketch the graphs of $f(x) = \sqrt{-1 - x}$ and its inverse function using the same coordinate axes.

SOLUTION First we sketch the curve $y = \sqrt{-1 - x}$ (the top half of the parabola $y^2 = -1 - x$, or $x = -y^2 - 1$) and then we reflect about the line $y = x$ to get the graph of f^{-1}. (See Figure 10.) As a check on our graph, notice that the expression for f^{-1} is $f^{-1}(x) = -x^2 - 1$, $x \geq 0$. So the graph of f^{-1} is the right half of the parabola $y = -x^2 - 1$ and this seems reasonable from Figure 10. ∎

Logarithmic Functions

If $b > 0$ and $b \neq 1$, the exponential function $f(x) = b^x$ is either increasing or decreasing and so it is one-to-one by the Horizontal Line Test. It therefore has an inverse function f^{-1}, which is called the **logarithmic function with base b** and is denoted by $\log_b$. If we use the formulation of an inverse function given by (3),

$$f^{-1}(x) = y \iff f(y) = x$$

then we have

> [6] $$\log_b x = y \iff b^y = x$$

Thus, if $x > 0$, then $\log_b x$ is the exponent to which the base b must be raised to give x. For example, $\log_{10} 0.001 = -3$ because $10^{-3} = 0.001$.

The cancellation equations (4), when applied to the functions $f(x) = b^x$ and $f^{-1}(x) = \log_b x$, become

> [7] $$\log_b(b^x) = x \quad \text{for every } x \in \mathbb{R}$$
> $$b^{\log_b x} = x \quad \text{for every } x > 0$$

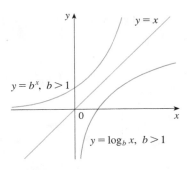

FIGURE 11

The logarithmic function $\log_b$ has domain $(0, \infty)$ and range $\mathbb{R}$. Its graph is the reflection of the graph of $y = b^x$ about the line $y = x$.

Figure 11 shows the case where $b > 1$. (The most important logarithmic functions have base $b > 1$.) The fact that $y = b^x$ is a very rapidly increasing function for $x > 0$ is reflected in the fact that $y = \log_b x$ is a very slowly increasing function for $x > 1$.

Figure 12 shows the graphs of $y = \log_b x$ with various values of the base $b > 1$. Since $\log_b 1 = 0$, the graphs of all logarithmic functions pass through the point $(1, 0)$.

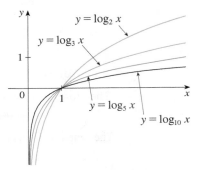

FIGURE 12

The following properties of logarithmic functions follow from the corresponding properties of exponential functions given in Section 1.4.

Laws of Logarithms   If x and y are positive numbers, then

1. $\log_b(xy) = \log_b x + \log_b y$

2. $\log_b\left(\dfrac{x}{y}\right) = \log_b x - \log_b y$

3. $\log_b(x^r) = r \log_b x$ (where r is any real number)

EXAMPLE 6 Use the laws of logarithms to evaluate $\log_2 80 - \log_2 5$.

SOLUTION Using Law 2, we have

$$\log_2 80 - \log_2 5 = \log_2\left(\frac{80}{5}\right) = \log_2 16 = 4$$

because $2^4 = 16$.   ∎

■ Natural Logarithms

Notation for Logarithms

Most textbooks in calculus and the sciences, as well as calculators, use the notation $\ln x$ for the natural logarithm and $\log x$ for the "common logarithm," $\log_{10} x$. In the more advanced mathematical and scientific literature and in computer languages, however, the notation $\log x$ usually denotes the natural logarithm.

Of all possible bases b for logarithms, we will see in Chapter 3 that the most convenient choice of a base is the number e, which was defined in Section 1.4. The logarithm with base e is called the **natural logarithm** and has a special notation:

$$\log_e x = \ln x$$

If we put $b = e$ and replace $\log_e$ with "ln" in (6) and (7), then the defining properties of the natural logarithm function become

$$\boxed{8} \qquad \ln x = y \iff e^y = x$$

$$\boxed{9} \qquad \begin{aligned} \ln(e^x) &= x & x \in \mathbb{R} \\ e^{\ln x} &= x & x > 0 \end{aligned}$$

In particular, if we set $x = 1$, we get

$$\ln e = 1$$

EXAMPLE 7 Find x if $\ln x = 5$.

SOLUTION 1 From (8) we see that

$$\ln x = 5 \qquad \text{means} \qquad e^5 = x$$

Therefore $x = e^5$.

(If you have trouble working with the "ln" notation, just replace it by $\log_e$. Then the equation becomes $\log_e x = 5$; so, by the definition of logarithm, $e^5 = x$.)

SOLUTION 2 Start with the equation

$$\ln x = 5$$

and apply the exponential function to both sides of the equation:

$$e^{\ln x} = e^5$$

But the second cancellation equation in (9) says that $e^{\ln x} = x$. Therefore $x = e^5$. ■

EXAMPLE 8 Solve the equation $e^{5-3x} = 10$.

SOLUTION We take natural logarithms of both sides of the equation and use (9):

$$\ln(e^{5-3x}) = \ln 10$$
$$5 - 3x = \ln 10$$
$$3x = 5 - \ln 10$$
$$x = \tfrac{1}{3}(5 - \ln 10)$$

Since the natural logarithm is found on scientific calculators, we can approximate the solution: to four decimal places, $x \approx 0.8991$. ■

EXAMPLE 9 Express $\ln a + \tfrac{1}{2}\ln b$ as a single logarithm.

SOLUTION Using Laws 3 and 1 of logarithms, we have

$$\ln a + \tfrac{1}{2}\ln b = \ln a + \ln b^{1/2}$$
$$= \ln a + \ln \sqrt{b}$$
$$= \ln(a\sqrt{b})$$
■

The following formula shows that logarithms with any base can be expressed in terms of the natural logarithm.

> **10 Change of Base Formula** For any positive number b ($b \neq 1$), we have
>
> $$\log_b x = \frac{\ln x}{\ln b}$$

PROOF Let $y = \log_b x$. Then, from (6), we have $b^y = x$. Taking natural logarithms of both sides of this equation, we get $y \ln b = \ln x$. Therefore

$$y = \frac{\ln x}{\ln b}$$ ∎

Scientific calculators have a key for natural logarithms, so Formula 10 enables us to use a calculator to compute a logarithm with any base (as shown in the following example). Similarly, Formula 10 allows us to graph any logarithmic function on a graphing calculator or computer (see Exercises 43 and 44).

EXAMPLE 10 Evaluate $\log_8 5$ correct to six decimal places.

SOLUTION Formula 10 gives

$$\log_8 5 = \frac{\ln 5}{\ln 8} \approx 0.773976$$ ∎

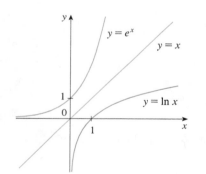

FIGURE 13
The graph of $y = \ln x$ is the reflection of the graph of $y = e^x$ about the line $y = x$.

Graph and Growth of the Natural Logarithm

The graphs of the exponential function $y = e^x$ and its inverse function, the natural logarithm function, are shown in Figure 13. Because the curve $y = e^x$ crosses the y-axis with a slope of 1, it follows that the reflected curve $y = \ln x$ crosses the x-axis with a slope of 1.

In common with all other logarithmic functions with base greater than 1, the natural logarithm is an increasing function defined on $(0, \infty)$ and the y-axis is a vertical asymptote. (This means that the values of $\ln x$ become very large negative as x approaches 0.)

EXAMPLE 11 Sketch the graph of the function $y = \ln(x - 2) - 1$.

SOLUTION We start with the graph of $y = \ln x$ as given in Figure 13. Using the transformations of Section 1.3, we shift it 2 units to the right to get the graph of $y = \ln(x - 2)$ and then we shift it 1 unit downward to get the graph of $y = \ln(x - 2) - 1$. (See Figure 14.)

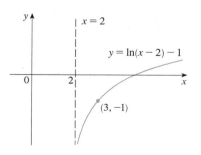

FIGURE 14

Although $\ln x$ is an increasing function, it grows *very* slowly when $x > 1$. In fact, $\ln x$ grows more slowly than any positive power of x. To illustrate this fact, we compare approximate values of the functions $y = \ln x$ and $y = x^{1/2} = \sqrt{x}$ in the following table and we graph them in Figures 15 and 16. You can see that initially the graphs of $y = \sqrt{x}$ and $y = \ln x$ grow at comparable rates, but eventually the root function far surpasses the logarithm.

x	1	2	5	10	50	100	500	1000	10,000	100,000
$\ln x$	0	0.69	1.61	2.30	3.91	4.6	6.2	6.9	9.2	11.5
$\sqrt{x}$	1	1.41	2.24	3.16	7.07	10.0	22.4	31.6	100	316
$\dfrac{\ln x}{\sqrt{x}}$	0	0.49	0.72	0.73	0.55	0.46	0.28	0.22	0.09	0.04

FIGURE 15

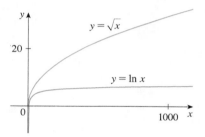

FIGURE 16

■ Inverse Trigonometric Functions

When we try to find the inverse trigonometric functions, we have a slight difficulty: Because the trigonometric functions are not one-to-one, they don't have inverse functions. The difficulty is overcome by restricting the domains of these functions so that they become one-to-one.

You can see from Figure 17 that the sine function $y = \sin x$ is not one-to-one (use the Horizontal Line Test). But the function $f(x) = \sin x$, $-\pi/2 \leqslant x \leqslant \pi/2$, *is* one-to-one (see Figure 18). The inverse function of this restricted sine function f exists and is denoted by $\sin^{-1}$ or arcsin. It is called the **inverse sine function** or the **arcsine function**.

FIGURE 17

FIGURE 18
$y = \sin x$, $-\frac{\pi}{2} \leqslant x \leqslant \frac{\pi}{2}$

Since the definition of an inverse function says that

$$f^{-1}(x) = y \iff f(y) = x$$

we have

$$\sin^{-1}x = y \iff \sin y = x \quad \text{and} \quad -\frac{\pi}{2} \leqslant y \leqslant \frac{\pi}{2}$$

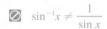

Thus, if $-1 \leqslant x \leqslant 1$, $\sin^{-1}x$ is the number between $-\pi/2$ and $\pi/2$ whose sine is x.

EXAMPLE 12 Evaluate (a) $\sin^{-1}\left(\frac{1}{2}\right)$ and (b) $\tan\left(\arcsin\frac{1}{3}\right)$.

SOLUTION

(a) We have

$$\sin^{-1}\left(\tfrac{1}{2}\right) = \frac{\pi}{6}$$

because $\sin(\pi/6) = \frac{1}{2}$ and $\pi/6$ lies between $-\pi/2$ and $\pi/2$.

(b) Let $\theta = \arcsin\frac{1}{3}$, so $\sin\theta = \frac{1}{3}$. Then we can draw a right triangle with angle θ as in Figure 19 and deduce from the Pythagorean Theorem that the third side has length $\sqrt{9-1} = 2\sqrt{2}$. This enables us to read from the triangle that

$$\tan\left(\arcsin\tfrac{1}{3}\right) = \tan\theta = \frac{1}{2\sqrt{2}}$$

FIGURE 19

The cancellation equations for inverse functions become, in this case,

$$\sin^{-1}(\sin x) = x \quad \text{for } -\frac{\pi}{2} \leqslant x \leqslant \frac{\pi}{2}$$

$$\sin(\sin^{-1}x) = x \quad \text{for } -1 \leqslant x \leqslant 1$$

FIGURE 20
$y = \sin^{-1}x = \arcsin x$

The inverse sine function, $\sin^{-1}$, has domain $[-1, 1]$ and range $[-\pi/2, \pi/2]$, and its graph, shown in Figure 20, is obtained from that of the restricted sine function (Figure 18) by reflection about the line $y = x$.

The **inverse cosine function** is handled similarly. The restricted cosine function $f(x) = \cos x$, $0 \leqslant x \leqslant \pi$, is one-to-one (see Figure 21) and so it has an inverse function denoted by $\cos^{-1}$ or arccos.

$$\cos^{-1}x = y \iff \cos y = x \quad \text{and} \quad 0 \leqslant y \leqslant \pi$$

The cancellation equations are

$$\cos^{-1}(\cos x) = x \quad \text{for } 0 \leqslant x \leqslant \pi$$

$$\cos(\cos^{-1}x) = x \quad \text{for } -1 \leqslant x \leqslant 1$$

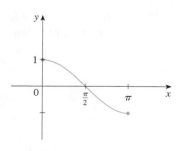

FIGURE 21
$y = \cos x, \, 0 \leqslant x \leqslant \pi$

FIGURE 22
$y = \cos^{-1}x = \arccos x$

The inverse cosine function, $\cos^{-1}$, has domain $[-1, 1]$ and range $[0, \pi]$. Its graph is shown in Figure 22.

The tangent function can be made one-to-one by restricting it to the interval $(-\pi/2, \pi/2)$. Thus the **inverse tangent function** is defined as the inverse of the function $f(x) = \tan x$, $-\pi/2 < x < \pi/2$. (See Figure 23.) It is denoted by $\tan^{-1}$ or arctan.

$$\tan^{-1}x = y \iff \tan y = x \quad \text{and} \quad -\frac{\pi}{2} < y < \frac{\pi}{2}$$

EXAMPLE 13 Simplify the expression $\cos(\tan^{-1}x)$.

SOLUTION 1 Let $y = \tan^{-1}x$. Then $\tan y = x$ and $-\pi/2 < y < \pi/2$. We want to find $\cos y$ but, since $\tan y$ is known, it is easier to find $\sec y$ first:

$$\sec^2 y = 1 + \tan^2 y = 1 + x^2$$

$$\sec y = \sqrt{1 + x^2} \qquad \text{(since } \sec y > 0 \text{ for } -\pi/2 < y < \pi/2\text{)}$$

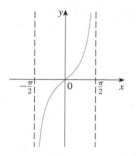

FIGURE 23
$y = \tan x, \ -\frac{\pi}{2} < x < \frac{\pi}{2}$

Thus $\qquad \cos(\tan^{-1}x) = \cos y = \dfrac{1}{\sec y} = \dfrac{1}{\sqrt{1 + x^2}}$

SOLUTION 2 Instead of using trigonometric identities as in Solution 1, it is perhaps easier to use a diagram. If $y = \tan^{-1}x$, then $\tan y = x$, and we can read from Figure 24 (which illustrates the case $y > 0$) that

$$\cos(\tan^{-1}x) = \cos y = \frac{1}{\sqrt{1 + x^2}}$$

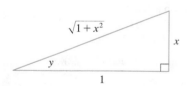

FIGURE 24

The inverse tangent function, $\tan^{-1} = \arctan$, has domain $\mathbb{R}$ and range $(-\pi/2, \pi/2)$. Its graph is shown in Figure 25.

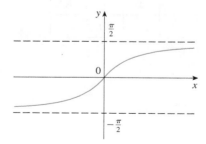

FIGURE 25
$y = \tan^{-1}x = \arctan x$

We know that the lines $x = \pm\pi/2$ are vertical asymptotes of the graph of tan. Since the graph of $\tan^{-1}$ is obtained by reflecting the graph of the restricted tangent function about the line $y = x$, it follows that the lines $y = \pi/2$ and $y = -\pi/2$ are horizontal asymptotes of the graph of $\tan^{-1}$.

The remaining inverse trigonometric functions are not used as frequently and are summarized here.

$$\boxed{11} \quad y = \csc^{-1}x \ (|x| \geqslant 1) \quad \Leftrightarrow \quad \csc y = x \quad \text{and} \quad y \in (0, \pi/2] \cup (\pi, 3\pi/2]$$

$$y = \sec^{-1}x \ (|x| \geqslant 1) \quad \Leftrightarrow \quad \sec y = x \quad \text{and} \quad y \in [0, \pi/2) \cup [\pi, 3\pi/2)$$

$$y = \cot^{-1}x \ (x \in \mathbb{R}) \quad \Leftrightarrow \quad \cot y = x \quad \text{and} \quad y \in (0, \pi)$$

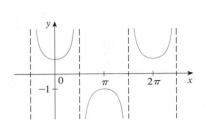

FIGURE 26
$y = \sec x$

The choice of intervals for y in the definitions of $\csc^{-1}$ and $\sec^{-1}$ is not universally agreed upon. For instance, some authors use $y \in [0, \pi/2) \cup (\pi/2, \pi]$ in the definition of $\sec^{-1}$. [You can see from the graph of the secant function in Figure 26 that both this choice and the one in (11) will work.]

1.5   EXERCISES

1. (a) What is a one-to-one function?
   (b) How can you tell from the graph of a function whether it is one-to-one?

2. (a) Suppose f is a one-to-one function with domain A and range B. How is the inverse function f^{-1} defined? What is the domain of f^{-1}? What is the range of f^{-1}?
   (b) If you are given a formula for f, how do you find a formula for f^{-1}?
   (c) If you are given the graph of f, how do you find the graph of f^{-1}?

3–14 A function is given by a table of values, a graph, a formula, or a verbal description. Determine whether it is one-to-one.

3.

x	1	2	3	4	5	6
$f(x)$	1.5	2.0	3.6	5.3	2.8	2.0

4.

x	1	2	3	4	5	6
$f(x)$	1.0	1.9	2.8	3.5	3.1	2.9

5.

6.

7.

8.

9. $f(x) = 2x - 3$

10. $f(x) = x^4 - 16$

11. $g(x) = 1 - \sin x$

12. $g(x) = \sqrt[3]{x}$

13. $f(t)$ is the height of a football t seconds after kickoff.

14. $f(t)$ is your height at age t.

15. Assume that f is a one-to-one function.
   (a) If $f(6) = 17$, what is $f^{-1}(17)$?
   (b) If $f^{-1}(3) = 2$, what is $f(2)$?

16. If $f(x) = x^5 + x^3 + x$, find $f^{-1}(3)$ and $f(f^{-1}(2))$.

17. If $g(x) = 3 + x + e^x$, find $g^{-1}(4)$.

18. The graph of f is given.
   (a) Why is f one-to-one?
   (b) What are the domain and range of f^{-1}?
   (c) What is the value of $f^{-1}(2)$?
   (d) Estimate the value of $f^{-1}(0)$.

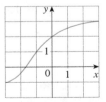

19. The formula $C = \frac{5}{9}(F - 32)$, where $F \geqslant -459.67$, expresses the Celsius temperature C as a function of the Fahrenheit temperature F. Find a formula for the inverse function and interpret it. What is the domain of the inverse function?

20. In the theory of relativity, the mass of a particle with speed v is

$$m = f(v) = \frac{m_0}{\sqrt{1 - v^2/c^2}}$$

where m_0 is the rest mass of the particle and c is the speed of light in a vacuum. Find the inverse function of f and explain its meaning.

21–26 Find a formula for the inverse of the function.

21. $f(x) = 1 + \sqrt{2 + 3x}$

22. $f(x) = \dfrac{4x - 1}{2x + 3}$

23. $f(x) = e^{2x-1}$

24. $y = x^2 - x, \quad x \geqslant \frac{1}{2}$

25. $y = \ln(x + 3)$

26. $y = \dfrac{1 - e^{-x}}{1 + e^{-x}}$

 27–28 Find an explicit formula for f^{-1} and use it to graph f^{-1}, f, and the line $y = x$ on the same screen. To check your work, see whether the graphs of f and f^{-1} are reflections about the line.

27. $f(x) = \sqrt{4x + 3}$

28. $f(x) = 1 + e^{-x}$

29–30 Use the given graph of f to sketch the graph of f^{-1}.

29.

30.

31. Let $f(x) = \sqrt{1 - x^2}, \ 0 \leqslant x \leqslant 1$.
(a) Find f^{-1}. How is it related to f?
(b) Identify the graph of f and explain your answer to part (a).

32. Let $g(x) = \sqrt[3]{1 - x^3}$.
(a) Find g^{-1}. How is it related to g?
(b) Graph g. How do you explain your answer to part (a)?

33. (a) How is the logarithmic function $y = \log_b x$ defined?
(b) What is the domain of this function?
(c) What is the range of this function?
(d) Sketch the general shape of the graph of the function $y = \log_b x$ if $b > 1$.

34. (a) What is the natural logarithm?
(b) What is the common logarithm?
(c) Sketch the graphs of the natural logarithm function and the natural exponential function with a common set of axes.

35–38 Find the exact value of each expression.

35. (a) $\log_2 32$ (b) $\log_8 2$

36. (a) $\log_5 \frac{1}{125}$ (b) $\ln(1/e^2)$

37. (a) $\log_{10} 40 + \log_{10} 2.5$
(b) $\log_8 60 - \log_8 3 - \log_8 5$

38. (a) $e^{-\ln 2}$ (b) $e^{\ln(\ln e^3)}$

39–41 Express the given quantity as a single logarithm.

39. $\ln 10 + 2 \ln 5$ **40.** $\ln b + 2 \ln c - 3 \ln d$

41. $\frac{1}{3} \ln(x + 2)^3 + \frac{1}{2} [\ln x - \ln(x^2 + 3x + 2)^2]$

42. Use Formula 10 to evaluate each logarithm correct to six decimal places.
(a) $\log_5 10$ (b) $\log_3 57$

43–44 Use Formula 10 to graph the given functions on a common screen. How are these graphs related?

43. $y = \log_{1.5} x, \quad y = \ln x, \quad y = \log_{10} x, \quad y = \log_{50} x$

44. $y = \ln x, \quad y = \log_{10} x, \quad y = e^x, \quad y = 10^x$

45. Suppose that the graph of $y = \log_2 x$ is drawn on a coordinate grid where the unit of measurement is an inch. How many miles to the right of the origin do we have to move before the height of the curve reaches 3 ft?

46. Compare the functions $f(x) = x^{0.1}$ and $g(x) = \ln x$ by graphing both f and g in several viewing rectangles. When does the graph of f finally surpass the graph of g?

47–48 Make a rough sketch of the graph of each function. Do not use a calculator. Just use the graphs given in Figures 12 and 13 and, if necessary, the transformations of Section 1.3.

47. (a) $y = \log_{10}(x + 5)$ (b) $y = -\ln x$

48. (a) $y = \ln(-x)$ (b) $y = \ln|x|$

49–50 (a) What are the domain and range of f?
(b) What is the x-intercept of the graph of f?
(c) Sketch the graph of f.

49. $f(x) = \ln x + 2$ **50.** $f(x) = \ln(x - 1) - 1$

51–54 Solve each equation for x.

51. (a) $e^{7-4x} = 6$ (b) $\ln(3x - 10) = 2$

52. (a) $\ln(x^2 - 1) = 3$ (b) $e^{2x} - 3e^x + 2 = 0$

53. (a) $2^{x-5} = 3$ (b) $\ln x + \ln(x - 1) = 1$

54. (a) $\ln(\ln x) = 1$ (b) $e^{ax} = Ce^{bx}$, where $a \neq b$

55–56 Solve each inequality for x.

55. (a) $\ln x < 0$ (b) $e^x > 5$

56. (a) $1 < e^{3x-1} < 2$ (b) $1 - 2 \ln x < 3$

57. (a) Find the domain of $f(x) = \ln(e^x - 3)$.
(b) Find f^{-1} and its domain.

58. (a) What are the values of $e^{\ln 300}$ and $\ln(e^{300})$?
(b) Use your calculator to evaluate $e^{\ln 300}$ and $\ln(e^{300})$. What do you notice? Can you explain why the calculator has trouble?

CAS **59.** Graph the function $f(x) = \sqrt{x^3 + x^2 + x + 1}$ and explain why it is one-to-one. Then use a computer algebra system to find an explicit expression for $f^{-1}(x)$. (Your CAS will produce three possible expressions. Explain why two of them are irrelevant in this context.)

CAS **60.** (a) If $g(x) = x^6 + x^4$, $x \geqslant 0$, use a computer algebra system to find an expression for $g^{-1}(x)$.
(b) Use the expression in part (a) to graph $y = g(x)$, $y = x$, and $y = g^{-1}(x)$ on the same screen.

61. If a bacteria population starts with 100 bacteria and doubles every three hours, then the number of bacteria after t hours is $n = f(t) = 100 \cdot 2^{t/3}$.
(a) Find the inverse of this function and explain its meaning.
(b) When will the population reach 50,000?

62. When a camera flash goes off, the batteries immediately begin to recharge the flash's capacitor, which stores electric charge given by
$$Q(t) = Q_0(1 - e^{-t/a})$$
(The maximum charge capacity is Q_0 and t is measured in seconds.)
(a) Find the inverse of this function and explain its meaning.
(b) How long does it take to recharge the capacitor to 90% of capacity if $a = 2$?

63–68 Find the exact value of each expression.

63. (a) $\cos^{-1}(-1)$ (b) $\sin^{-1}(0.5)$

64. (a) $\tan^{-1}\sqrt{3}$ (b) $\arctan(-1)$

65. (a) $\csc^{-1}\sqrt{2}$ (b) $\arcsin 1$

66. (a) $\sin^{-1}(-1/\sqrt{2})$ (b) $\cos^{-1}(\sqrt{3}/2)$

67. (a) $\cot^{-1}(-\sqrt{3})$ (b) $\sec^{-1} 2$

68. (a) $\arcsin(\sin(5\pi/4))$ (b) $\cos(2\sin^{-1}(\frac{5}{13}))$

69. Prove that $\cos(\sin^{-1} x) = \sqrt{1 - x^2}$.

70–72 Simplify the expression.

70. $\tan(\sin^{-1}x)$ **71.** $\sin(\tan^{-1}x)$ **72.** $\sin(2\arccos x)$

73-74 Graph the given functions on the same screen. How are these graphs related?

73. $y = \sin x$, $-\pi/2 \leqslant x \leqslant \pi/2$; $y = \sin^{-1}x$; $y = x$

74. $y = \tan x$, $-\pi/2 < x < \pi/2$; $y = \tan^{-1}x$; $y = x$

75. Find the domain and range of the function
$$g(x) = \sin^{-1}(3x + 1)$$

76. (a) Graph the function $f(x) = \sin(\sin^{-1}x)$ and explain the appearance of the graph.
(b) Graph the function $g(x) = \sin^{-1}(\sin x)$. How do you explain the appearance of this graph?

77. (a) If we shift a curve to the left, what happens to its reflection about the line $y = x$? In view of this geometric principle, find an expression for the inverse of $g(x) = f(x + c)$, where f is a one-to-one function.
(b) Find an expression for the inverse of $h(x) = f(cx)$, where $c \neq 0$.

---

1   **REVIEW**

CONCEPT CHECK   Answers to the Concept Check can be found on the back endpapers.

1. (a) What is a function? What are its domain and range?
(b) What is the graph of a function?
(c) How can you tell whether a given curve is the graph of a function?

2. Discuss four ways of representing a function. Illustrate your discussion with examples.

3. (a) What is an even function? How can you tell if a function is even by looking at its graph? Give three examples of an even function.
(b) What is an odd function? How can you tell if a function is odd by looking at its graph? Give three examples of an odd function.

4. What is an increasing function?

5. What is a mathematical model?

6. Give an example of each type of function.
(a) Linear function   (b) Power function
(c) Exponential function   (d) Quadratic function
(e) Polynomial of degree 5   (f) Rational function

7. Sketch by hand, on the same axes, the graphs of the following functions.
(a) $f(x) = x$ (b) $g(x) = x^2$
(c) $h(x) = x^3$ (d) $j(x) = x^4$

8. Draw, by hand, a rough sketch of the graph of each function.

(a) $y = \sin x$ (b) $y = \tan x$ (c) $y = e^x$
(d) $y = \ln x$ (e) $y = 1/x$ (f) $y = |x|$
(g) $y = \sqrt{x}$ (h) $y = \tan^{-1}x$

9. Suppose that f has domain A and g has domain B.

(a) What is the domain of $f + g$?
(b) What is the domain of fg?
(c) What is the domain of f/g?

10. How is the composite function $f \circ g$ defined? What is its domain?

11. Suppose the graph of f is given. Write an equation for each of the graphs that are obtained from the graph of f as follows.

(a) Shift 2 units upward.   (b) Shift 2 units downward.
(c) Shift 2 units to the right.   (d) Shift 2 units to the left.
(e) Reflect about the x-axis.

(f) Reflect about the y-axis.
(g) Stretch vertically by a factor of 2.
(h) Shrink vertically by a factor of 2.
(i) Stretch horizontally by a factor of 2.
(j) Shrink horizontally by a factor of 2.

12. (a) What is a one-to-one function? How can you tell if a function is one-to-one by looking at its graph?
(b) If f is a one-to-one function, how is its inverse function f^{-1} defined? How do you obtain the graph of f^{-1} from the graph of f?

13. (a) How is the inverse sine function $f(x) = \sin^{-1}x$ defined? What are its domain and range?
(b) How is the inverse cosine function $f(x) = \cos^{-1}x$ defined? What are its domain and range?
(c) How is the inverse tangent function $f(x) = \tan^{-1}x$ defined? What are its domain and range?

TRUE-FALSE QUIZ

Determine whether the statement is true or false. If it is true, explain why. If it is false, explain why or give an example that disproves the statement.

1. If f is a function, then $f(s + t) = f(s) + f(t)$.

2. If $f(s) = f(t)$, then $s = t$.

3. If f is a function, then $f(3x) = 3f(x)$.

4. If $x_1 < x_2$ and f is a decreasing function, then $f(x_1) > f(x_2)$.

5. A vertical line intersects the graph of a function at most once.

6. If f and g are functions, then $f \circ g = g \circ f$.

7. If f is one-to-one, then $f^{-1}(x) = \dfrac{1}{f(x)}$.

8. You can always divide by e^x.

9. If $0 < a < b$, then $\ln a < \ln b$.

10. If $x > 0$, then $(\ln x)^6 = 6 \ln x$.

11. If $x > 0$ and $a > 1$, then $\dfrac{\ln x}{\ln a} = \ln \dfrac{x}{a}$.

12. $\tan^{-1}(-1) = 3\pi/4$

13. $\tan^{-1}x = \dfrac{\sin^{-1}x}{\cos^{-1}x}$

14. If x is any real number, then $\sqrt{x^2} = x$.

EXERCISES

1. Let f be the function whose graph is given.

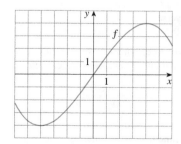

(a) Estimate the value of $f(2)$.
(b) Estimate the values of x such that $f(x) = 3$.
(c) State the domain of f.
(d) State the range of f.
(e) On what interval is f increasing?

(f) Is f one-to-one? Explain.
(g) Is f even, odd, or neither even nor odd? Explain.

2. The graph of g is given.

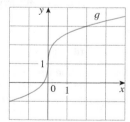

(a) State the value of $g(2)$.
(b) Why is g one-to-one?
(c) Estimate the value of $g^{-1}(2)$.
(d) Estimate the domain of g^{-1}.
(e) Sketch the graph of g^{-1}.

3. If $f(x) = x^2 - 2x + 3$, evaluate the difference quotient

$$\frac{f(a + h) - f(a)}{h}$$

4. Sketch a rough graph of the yield of a crop as a function of the amount of fertilizer used.

5–8 Find the domain and range of the function. Write your answer in interval notation.

5. $f(x) = 2/(3x - 1)$ **6.** $g(x) = \sqrt{16 - x^4}$

7. $h(x) = \ln(x + 6)$ **8.** $F(t) = 3 + \cos 2t$

9. Suppose that the graph of f is given. Describe how the graphs of the following functions can be obtained from the graph of f.
(a) $y = f(x) + 8$ (b) $y = f(x + 8)$
(c) $y = 1 + 2f(x)$ (d) $y = f(x - 2) - 2$
(e) $y = -f(x)$ (f) $y = f^{-1}(x)$

10. The graph of f is given. Draw the graphs of the following functions.
(a) $y = f(x - 8)$ (b) $y = -f(x)$
(c) $y = 2 - f(x)$ (d) $y = \frac{1}{2}f(x) - 1$
(e) $y = f^{-1}(x)$ (f) $y = f^{-1}(x + 3)$

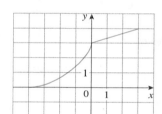

11–16 Use transformations to sketch the graph of the function.

11. $y = (x - 2)^3$ **12.** $y = 2\sqrt{x}$

13. $y = x^2 - 2x + 2$ **14.** $y = \ln(x + 1)$

15. $f(x) = -\cos 2x$ **16.** $f(x) = \begin{cases} -x & \text{if } x < 0 \\ e^x - 1 & \text{if } x \geqslant 0 \end{cases}$

17. Determine whether f is even, odd, or neither even nor odd.
(a) $f(x) = 2x^5 - 3x^2 + 2$
(b) $f(x) = x^3 - x^7$
(c) $f(x) = e^{-x^2}$ (d) $f(x) = 1 + \sin x$

18. Find an expression for the function whose graph consists of the line segment from the point $(-2, 2)$ to the point $(-1, 0)$ together with the top half of the circle with center the origin and radius 1.

19. If $f(x) = \ln x$ and $g(x) = x^2 - 9$, find the functions
(a) $f \circ g$, (b) $g \circ f$, (c) $f \circ f$, (d) $g \circ g$, and their domains.

20. Express the function $F(x) = 1/\sqrt{x + \sqrt{x}}$ as a composition of three functions.

21. Life expectancy improved dramatically in the 20th century. The table gives the life expectancy at birth (in years) of males born in the United States. Use a scatter plot to choose an appropriate type of model. Use your model to predict the life span of a male born in the year 2010.

Birth year	Life expectancy	Birth year	Life expectancy
1900	48.3	1960	66.6
1910	51.1	1970	67.1
1920	55.2	1980	70.0
1930	57.4	1990	71.8
1940	62.5	2000	73.0
1950	65.6		

22. A small-appliance manufacturer finds that it costs $9000 to produce 1000 toaster ovens a week and $12,000 to produce 1500 toaster ovens a week.
(a) Express the cost as a function of the number of toaster ovens produced, assuming that it is linear. Then sketch the graph.
(b) What is the slope of the graph and what does it represent?
(c) What is the y-intercept of the graph and what does it represent?

23. If $f(x) = 2x + \ln x$, find $f^{-1}(2)$.

24. Find the inverse function of $f(x) = \dfrac{x + 1}{2x + 1}$.

25. Find the exact value of each expression.
(a) $e^{2\ln 3}$ (b) $\log_{10} 25 + \log_{10} 4$
(c) $\tan\left(\arcsin \frac{1}{2}\right)$ (d) $\sin\left(\cos^{-1}\left(\frac{4}{5}\right)\right)$

26. Solve each equation for x.
(a) $e^x = 5$ (b) $\ln x = 2$
(c) $e^{e^x} = 2$ (d) $\tan^{-1}x = 1$

27. The half-life of palladium-100, ^{100}Pd, is four days. (So half of any given quantity of ^{100}Pd will disintegrate in four days.) The initial mass of a sample is one gram.
(a) Find the mass that remains after 16 days.
(b) Find the mass $m(t)$ that remains after t days.
(c) Find the inverse of this function and explain its meaning.
(d) When will the mass be reduced to 0.01g?

28. The population of a certain species in a limited environment with initial population 100 and carrying capacity 1000 is

$$P(t) = \frac{100,000}{100 + 900e^{-t}}$$

where t is measured in years.
(a) Graph this function and estimate how long it takes for the population to reach 900.
(b) Find the inverse of this function and explain its meaning.
(c) Use the inverse function to find the time required for the population to reach 900. Compare with the result of part (a).

Principles of Problem Solving

There are no hard and fast rules that will ensure success in solving problems. However, it is possible to outline some general steps in the problem-solving process and to give some principles that may be useful in the solution of certain problems. These steps and principles are just common sense made explicit. They have been adapted from George Polya's book *How To Solve It*.

1 UNDERSTAND THE PROBLEM

The first step is to read the problem and make sure that you understand it clearly. Ask yourself the following questions:

> *What is the unknown?*
>
> *What are the given quantities?*
>
> *What are the given conditions?*

For many problems it is useful to

> *draw a diagram*

and identify the given and required quantities on the diagram.

Usually it is necessary to

> *introduce suitable notation*

In choosing symbols for the unknown quantities we often use letters such as a, b, c, m, n, x, and y, but in some cases it helps to use initials as suggestive symbols; for instance, V for volume or t for time.

2 THINK OF A PLAN

Find a connection between the given information and the unknown that will enable you to calculate the unknown. It often helps to ask yourself explicitly: "How can I relate the given to the unknown?" If you don't see a connection immediately, the following ideas may be helpful in devising a plan.

Try to Recognize Something Familiar Relate the given situation to previous knowledge. Look at the unknown and try to recall a more familiar problem that has a similar unknown.

Try to Recognize Patterns Some problems are solved by recognizing that some kind of pattern is occurring. The pattern could be geometric, or numerical, or algebraic. If you can see regularity or repetition in a problem, you might be able to guess what the continuing pattern is and then prove it.

Use Analogy Try to think of an analogous problem, that is, a similar problem, a related problem, but one that is easier than the original problem. If you can solve the similar, simpler problem, then it might give you the clues you need to solve the original, more difficult problem. For instance, if a problem involves very large numbers, you could first try a similar problem with smaller numbers. Or if the problem involves three-dimensional geometry, you could look for a similar problem in two-dimensional geometry. Or if the problem you start with is a general one, you could first try a special case.

Introduce Something Extra It may sometimes be necessary to introduce something new, an auxiliary aid, to help make the connection between the given and the unknown. For instance, in a problem where a diagram is useful the auxiliary aid could be a new line drawn in a diagram. In a more algebraic problem it could be a new unknown that is related to the original unknown.

Take Cases We may sometimes have to split a problem into several cases and give a different argument for each of the cases. For instance, we often have to use this strategy in dealing with absolute value.

Work Backward Sometimes it is useful to imagine that your problem is solved and work backward, step by step, until you arrive at the given data. Then you may be able to reverse your steps and thereby construct a solution to the original problem. This procedure is commonly used in solving equations. For instance, in solving the equation $3x - 5 = 7$, we suppose that x is a number that satisfies $3x - 5 = 7$ and work backward. We add 5 to each side of the equation and then divide each side by 3 to get $x = 4$. Since each of these steps can be reversed, we have solved the problem.

Establish Subgoals In a complex problem it is often useful to set subgoals (in which the desired situation is only partially fulfilled). If we can first reach these subgoals, then we may be able to build on them to reach our final goal.

Indirect Reasoning Sometimes it is appropriate to attack a problem indirectly. In using proof by contradiction to prove that P implies Q, we assume that P is true and Q is false and try to see why this can't happen. Somehow we have to use this information and arrive at a contradiction to what we absolutely know is true.

Mathematical Induction In proving statements that involve a positive integer n, it is frequently helpful to use the following principle.

Principle of Mathematical Induction Let S_n be a statement about the positive integer n. Suppose that

1. S_1 is true.

2. S_{k+1} is true whenever S_k is true.

Then S_n is true for all positive integers n.

This is reasonable because, since S_1 is true, it follows from condition 2 (with $k = 1$) that S_2 is true. Then, using condition 2 with $k = 2$, we see that S_3 is true. Again using condition 2, this time with $k = 3$, we have that S_4 is true. This procedure can be followed indefinitely.

3 CARRY OUT THE PLAN

In Step 2 a plan was devised. In carrying out that plan we have to check each stage of the plan and write the details that prove that each stage is correct.

4 LOOK BACK

Having completed our solution, it is wise to look back over it, partly to see if we have made errors in the solution and partly to see if we can think of an easier way to solve the problem. Another reason for looking back is that it will familiarize us with the method of solution and this may be useful for solving a future problem. Descartes said, "Every problem that I solved became a rule which served afterwards to solve other problems."

These principles of problem solving are illustrated in the following examples. Before you look at the solutions, try to solve these problems yourself, referring to these Principles of Problem Solving if you get stuck. You may find it useful to refer to this section from time to time as you solve the exercises in the remaining chapters of this book.

EXAMPLE 1 Express the hypotenuse h of a right triangle with area 25 m² as a function of its perimeter P.

PS Understand the problem

SOLUTION Let's first sort out the information by identifying the unknown quantity and the data:

Unknown: hypotenuse h

Given quantities: perimeter P, area 25 m²

It helps to draw a diagram and we do so in Figure 1.

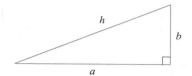

FIGURE 1

In order to connect the given quantities to the unknown, we introduce two extra variables a and b, which are the lengths of the other two sides of the triangle. This enables us to express the given condition, which is that the triangle is right-angled, by the Pythagorean Theorem:

$$h^2 = a^2 + b^2$$

The other connections among the variables come by writing expressions for the area and perimeter:

$$25 = \tfrac{1}{2}ab \qquad P = a + b + h$$

Since P is given, notice that we now have three equations in the three unknowns a, b, and h:

1	$h^2 = a^2 + b^2$
2	$25 = \tfrac{1}{2}ab$
3	$P = a + b + h$

Although we have the correct number of equations, they are not easy to solve in a straightforward fashion. But if we use the problem-solving strategy of trying to recognize something familiar, then we can solve these equations by an easier method. Look at the right sides of Equations 1, 2, and 3. Do these expressions remind you of anything familiar? Notice that they contain the ingredients of a familiar formula:

$$(a + b)^2 = a^2 + 2ab + b^2$$

Using this idea, we express $(a + b)^2$ in two ways. From Equations 1 and 2 we have

$$(a + b)^2 = (a^2 + b^2) + 2ab = h^2 + 4(25)$$

From Equation 3 we have

$$(a + b)^2 = (P - h)^2 = P^2 - 2Ph + h^2$$

Thus
$$h^2 + 100 = P^2 - 2Ph + h^2$$

$$2Ph = P^2 - 100$$

$$h = \frac{P^2 - 100}{2P}$$

This is the required expression for h as a function of P. ∎

As the next example illustrates, it is often necessary to use the problem-solving principle of *taking cases* when dealing with absolute values.

EXAMPLE 2 Solve the inequality $|x - 3| + |x + 2| < 11$.

SOLUTION Recall the definition of absolute value:

$$|x| = \begin{cases} x & \text{if } x \geq 0 \\ -x & \text{if } x < 0 \end{cases}$$

73

It follows that

$$|x - 3| = \begin{cases} x - 3 & \text{if } x - 3 \geq 0 \\ -(x - 3) & \text{if } x - 3 < 0 \end{cases}$$

$$= \begin{cases} x - 3 & \text{if } x \geq 3 \\ -x + 3 & \text{if } x < 3 \end{cases}$$

Similarly

$$|x + 2| = \begin{cases} x + 2 & \text{if } x + 2 \geq 0 \\ -(x + 2) & \text{if } x + 2 < 0 \end{cases}$$

$$= \begin{cases} x + 2 & \text{if } x \geq -2 \\ -x - 2 & \text{if } x < -2 \end{cases}$$

PS Take cases

These expressions show that we must consider three cases:

$$x < -2 \qquad -2 \leq x < 3 \qquad x \geq 3$$

CASE I If $x < -2$, we have

$$|x - 3| + |x + 2| < 11$$

$$-x + 3 - x - 2 < 11$$

$$-2x < 10$$

$$x > -5$$

CASE II If $-2 \leq x < 3$, the given inequality becomes

$$-x + 3 + x + 2 < 11$$

$$5 < 11 \qquad \text{(always true)}$$

CASE III If $x \geq 3$, the inequality becomes

$$x - 3 + x + 2 < 11$$

$$2x < 12$$

$$x < 6$$

Combining cases I, II, and III, we see that the inequality is satisfied when $-5 < x < 6$. So the solution is the interval $(-5, 6)$. ∎

In the following example we first guess the answer by looking at special cases and recognizing a pattern. Then we prove our conjecture by mathematical induction.

In using the Principle of Mathematical Induction, we follow three steps:

Step 1 Prove that S_n is true when $n = 1$.

Step 2 Assume that S_n is true when $n = k$ and deduce that S_n is true when $n = k + 1$.

Step 3 Conclude that S_n is true for all n by the Principle of Mathematical Induction.

EXAMPLE 3 If $f_0(x) = x/(x + 1)$ and $f_{n+1} = f_0 \circ f_n$ for $n = 0, 1, 2, \ldots$, find a formula for $f_n(x)$.

PS Analogy: Try a similar, simpler problem

SOLUTION We start by finding formulas for $f_n(x)$ for the special cases $n = 1, 2$, and 3.

$$f_1(x) = (f_0 \circ f_0)(x) = f_0(f_0(x)) = f_0\left(\frac{x}{x + 1}\right)$$

$$= \frac{\dfrac{x}{x + 1}}{\dfrac{x}{x + 1} + 1} = \frac{\dfrac{x}{x + 1}}{\dfrac{2x + 1}{x + 1}} = \frac{x}{2x + 1}$$

$$f_2(x) = (f_0 \circ f_1)(x) = f_0(f_1(x)) = f_0\left(\frac{x}{2x + 1}\right)$$

$$= \frac{\dfrac{x}{2x + 1}}{\dfrac{x}{2x + 1} + 1} = \frac{\dfrac{x}{2x + 1}}{\dfrac{3x + 1}{2x + 1}} = \frac{x}{3x + 1}$$

$$f_3(x) = (f_0 \circ f_2)(x) = f_0(f_2(x)) = f_0\left(\frac{x}{3x + 1}\right)$$

PS Look for a pattern

$$= \frac{\dfrac{x}{3x + 1}}{\dfrac{x}{3x + 1} + 1} = \frac{\dfrac{x}{3x + 1}}{\dfrac{4x + 1}{3x + 1}} = \frac{x}{4x + 1}$$

We notice a pattern: The coefficient of x in the denominator of $f_n(x)$ is $n + 1$ in the three cases we have computed. So we make the guess that, in general,

$$\boxed{4} \qquad f_n(x) = \frac{x}{(n + 1)x + 1}$$

To prove this, we use the Principle of Mathematical Induction. We have already verified that (4) is true for $n = 1$. Assume that it is true for $n = k$, that is,

$$f_k(x) = \frac{x}{(k + 1)x + 1}$$

Then $\qquad f_{k+1}(x) = (f_0 \circ f_k)(x) = f_0(f_k(x)) = f_0\left(\frac{x}{(k + 1)x + 1}\right)$

$$= \frac{\dfrac{x}{(k + 1)x + 1}}{\dfrac{x}{(k + 1)x + 1} + 1} = \frac{\dfrac{x}{(k + 1)x + 1}}{\dfrac{(k + 2)x + 1}{(k + 1)x + 1}} = \frac{x}{(k + 2)x + 1}$$

This expression shows that (4) is true for $n = k + 1$. Therefore, by mathematical induction, it is true for all positive integers n. ∎

Problems

1. One of the legs of a right triangle has length 4 cm. Express the length of the altitude perpendicular to the hypotenuse as a function of the length of the hypotenuse.

2. The altitude perpendicular to the hypotenuse of a right triangle is 12 cm. Express the length of the hypotenuse as a function of the perimeter.

3. Solve the equation $|2x - 1| - |x + 5| = 3$.

4. Solve the inequality $|x - 1| - |x - 3| \geqslant 5$.

5. Sketch the graph of the function $f(x) = |x^2 - 4|x| + 3|$.

6. Sketch the graph of the function $g(x) = |x^2 - 1| - |x^2 - 4|$.

7. Draw the graph of the equation $x + |x| = y + |y|$.

8. Sketch the region in the plane consisting of all points (x, y) such that
$$|x - y| + |x| - |y| \leqslant 2$$

9. The notation $\max\{a, b, \ldots\}$ means the largest of the numbers $a, b, \ldots$. Sketch the graph of each function.
   (a) $f(x) = \max\{x, 1/x\}$
   (b) $f(x) = \max\{\sin x, \cos x\}$
   (c) $f(x) = \max\{x^2, 2 + x, 2 - x\}$

10. Sketch the region in the plane defined by each of the following equations or inequalities.
    (a) $\max\{x, 2y\} = 1$
    (b) $-1 \leqslant \max\{x, 2y\} \leqslant 1$
    (c) $\max\{x, y^2\} = 1$

11. Evaluate $(\log_2 3)(\log_3 4)(\log_4 5) \cdots (\log_{31} 32)$.

12. (a) Show that the function $f(x) = \ln(x + \sqrt{x^2 + 1})$ is an odd function.
    (b) Find the inverse function of f.

13. Solve the inequality $\ln(x^2 - 2x - 2) \leqslant 0$.

14. Use indirect reasoning to prove that $\log_2 5$ is an irrational number.

15. A driver sets out on a journey. For the first half of the distance she drives at the leisurely pace of 30 mi/h; she drives the second half at 60 mi/h. What is her average speed on this trip?

16. Is it true that $f \circ (g + h) = f \circ g + f \circ h$?

17. Prove that if n is a positive integer, then $7^n - 1$ is divisible by 6.

18. Prove that $1 + 3 + 5 + \cdots + (2n - 1) = n^2$.

19. If $f_0(x) = x^2$ and $f_{n+1}(x) = f_0(f_n(x))$ for $n = 0, 1, 2, \ldots$, find a formula for $f_n(x)$.

20. (a) If $f_0(x) = \dfrac{1}{2 - x}$ and $f_{n+1} = f_0 \circ f_n$ for $n = 0, 1, 2, \ldots$, find an expression for $f_n(x)$ and use mathematical induction to prove it.
    (b) Graph f_0, f_1, f_2, f_3 on the same screen and describe the effects of repeated composition.

2 Limits and Derivatives

The maximum sustainable swimming speed S of salmon depends on the water temperature T. Exercise 58 in Section 2.7 asks you to analyze how S varies as T changes by estimating the derivative of S with respect to T.

IN *A PREVIEW OF CALCULUS* (page 1) we saw how the idea of a limit underlies the various branches of calculus. It is therefore appropriate to begin our study of calculus by investigating limits and their properties. The special type of limit that is used to find tangents and velocities gives rise to the central idea in differential calculus, the derivative.

2.1   The Tangent and Velocity Problems

In this section we see how limits arise when we attempt to find the tangent to a curve or the velocity of an object.

■ The Tangent Problem

The word *tangent* is derived from the Latin word *tangens*, which means "touching." Thus a tangent to a curve is a line that touches the curve. In other words, a tangent line should have the same direction as the curve at the point of contact. How can this idea be made precise?

For a circle we could simply follow Euclid and say that a tangent is a line that intersects the circle once and only once, as in Figure 1(a). For more complicated curves this definition is inadequate. Figure 1(b) shows two lines *l* and *t* passing through a point *P* on a curve *C*. The line *l* intersects *C* only once, but it certainly does not look like what we think of as a tangent. The line *t*, on the other hand, looks like a tangent but it intersects *C* twice.

To be specific, let's look at the problem of trying to find a tangent line *t* to the parabola $y = x^2$ in the following example.

(a)

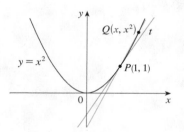

(b)

FIGURE 1

EXAMPLE 1   Find an equation of the tangent line to the parabola $y = x^2$ at the point $P(1, 1)$.

SOLUTION   We will be able to find an equation of the tangent line *t* as soon as we know its slope *m*. The difficulty is that we know only one point, *P*, on *t*, whereas we need two points to compute the slope. But observe that we can compute an approximation to *m* by choosing a nearby point $Q(x, x^2)$ on the parabola (as in Figure 2) and computing the slope m_{PQ} of the secant line *PQ*. [A **secant line**, from the Latin word *secans*, meaning cutting, is a line that cuts (intersects) a curve more than once.]

We choose $x \neq 1$ so that $Q \neq P$. Then

$$m_{PQ} = \frac{x^2 - 1}{x - 1}$$

FIGURE 2

For instance, for the point $Q(1.5, 2.25)$ we have

$$m_{PQ} = \frac{2.25 - 1}{1.5 - 1} = \frac{1.25}{0.5} = 2.5$$

The tables in the margin show the values of m_{PQ} for several values of *x* close to 1. The closer *Q* is to *P*, the closer *x* is to 1 and, it appears from the tables, the closer m_{PQ} is to 2. This suggests that the slope of the tangent line *t* should be $m = 2$.

We say that the slope of the tangent line is the *limit* of the slopes of the secant lines, and we express this symbolically by writing

$$\lim_{Q \to P} m_{PQ} = m \qquad \text{and} \qquad \lim_{x \to 1} \frac{x^2 - 1}{x - 1} = 2$$

Assuming that the slope of the tangent line is indeed 2, we use the point-slope form of the equation of a line $[y - y_1 = m(x - x_1)$, see Appendix B] to write the equation of the tangent line through $(1, 1)$ as

$$y - 1 = 2(x - 1) \qquad \text{or} \qquad y = 2x - 1$$

x	m_{PQ}
2	3
1.5	2.5
1.1	2.1
1.01	2.01
1.001	2.001

x	m_{PQ}
0	1
0.5	1.5
0.9	1.9
0.99	1.99
0.999	1.999

Figure 3 illustrates the limiting process that occurs in this example. As Q approaches P along the parabola, the corresponding secant lines rotate about P and approach the tangent line t.

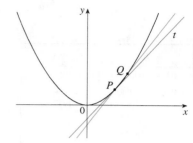

Q approaches P from the right

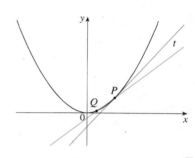

Q approaches P from the left

FIGURE 3

TEC In Visual 2.1 you can see how the process in Figure 3 works for additional functions.

t	Q
0.00	100.00
0.02	81.87
0.04	67.03
0.06	54.88
0.08	44.93
0.10	36.76

Many functions that occur in science are not described by explicit equations; they are defined by experimental data. The next example shows how to estimate the slope of the tangent line to the graph of such a function.

EXAMPLE 2 The flash unit on a camera operates by storing charge on a capacitor and releasing it suddenly when the flash is set off. The data in the table describe the charge Q remaining on the capacitor (measured in microcoulombs) at time t (measured in seconds after the flash goes off). Use the data to draw the graph of this function and estimate the slope of the tangent line at the point where $t = 0.04$. [*Note:* The slope of the tangent line represents the electric current flowing from the capacitor to the flash bulb (measured in microamperes).]

SOLUTION In Figure 4 we plot the given data and use them to sketch a curve that approximates the graph of the function.

FIGURE 4

Given the points $P(0.04, 67.03)$ and $R(0.00, 100.00)$ on the graph, we find that the slope of the secant line PR is

$$m_{PR} = \frac{100.00 - 67.03}{0.00 - 0.04} = -824.25$$

R	m_{PR}
$(0.00, 100.00)$	-824.25
$(0.02, 81.87)$	-742.00
$(0.06, 54.88)$	-607.50
$(0.08, 44.93)$	-552.50
$(0.10, 36.76)$	-504.50

The table at the left shows the results of similar calculations for the slopes of other secant lines. From this table we would expect the slope of the tangent line at $t = 0.04$ to lie somewhere between -742 and -607.5. In fact, the average of the slopes of the two closest secant lines is

$$\tfrac{1}{2}(-742 - 607.5) = -674.75$$

So, by this method, we estimate the slope of the tangent line to be about -675.

Another method is to draw an approximation to the tangent line at P and measure the sides of the triangle ABC, as in Figure 5.

FIGURE 5

This gives an estimate of the slope of the tangent line as

$$-\frac{|AB|}{|BC|} \approx -\frac{80.4 - 53.6}{0.06 - 0.02} = -670$$

The physical meaning of the answer in Example 2 is that the electric current flowing from the capacitor to the flash bulb after 0.04 seconds is about -670 microamperes.

The Velocity Problem

If you watch the speedometer of a car as you travel in city traffic, you see that the speed doesn't stay the same for very long; that is, the velocity of the car is not constant. We assume from watching the speedometer that the car has a definite velocity at each moment, but how is the "instantaneous" velocity defined? Let's investigate the example of a falling ball.

EXAMPLE 3 Suppose that a ball is dropped from the upper observation deck of the CN Tower in Toronto, 450 m above the ground. Find the velocity of the ball after 5 seconds.

SOLUTION Through experiments carried out four centuries ago, Galileo discovered that the distance fallen by any freely falling body is proportional to the square of the time it has been falling. (This model for free fall neglects air resistance.) If the distance fallen

The CN Tower in Toronto was the tallest freestanding building in the world for 32 years.

after t seconds is denoted by $s(t)$ and measured in meters, then Galileo's law is expressed by the equation

$$s(t) = 4.9t^2$$

The difficulty in finding the velocity after 5 seconds is that we are dealing with a single instant of time ($t = 5$), so no time interval is involved. However, we can approximate the desired quantity by computing the average velocity over the brief time interval of a tenth of a second from $t = 5$ to $t = 5.1$:

$$\text{average velocity} = \frac{\text{change in position}}{\text{time elapsed}}$$

$$= \frac{s(5.1) - s(5)}{0.1}$$

$$= \frac{4.9(5.1)^2 - 4.9(5)^2}{0.1} = 49.49 \text{ m/s}$$

The following table shows the results of similar calculations of the average velocity over successively smaller time periods.

Time interval	Average velocity (m/s)
$5 \leqslant t \leqslant 6$	53.9
$5 \leqslant t \leqslant 5.1$	49.49
$5 \leqslant t \leqslant 5.05$	49.245
$5 \leqslant t \leqslant 5.01$	49.049
$5 \leqslant t \leqslant 5.001$	49.0049

It appears that as we shorten the time period, the average velocity is becoming closer to 49 m/s. The **instantaneous velocity** when $t = 5$ is defined to be the limiting value of these average velocities over shorter and shorter time periods that start at $t = 5$. Thus it appears that the (instantaneous) velocity after 5 seconds is

$$v = 49 \text{ m/s}$$

You may have the feeling that the calculations used in solving this problem are very similar to those used earlier in this section to find tangents. In fact, there is a close connection between the tangent problem and the problem of finding velocities. If we draw the graph of the distance function of the ball (as in Figure 6) and we consider the points $P(a, 4.9a^2)$ and $Q(a + h, 4.9(a + h)^2)$ on the graph, then the slope of the secant line PQ is

$$m_{PQ} = \frac{4.9(a + h)^2 - 4.9a^2}{(a + h) - a}$$

which is the same as the average velocity over the time interval $[a, a + h]$. Therefore the velocity at time $t = a$ (the limit of these average velocities as h approaches 0) must be equal to the slope of the tangent line at P (the limit of the slopes of the secant lines).

Examples 1 and 3 show that in order to solve tangent and velocity problems we must be able to find limits. After studying methods for computing limits in the next five sections, we will return to the problems of finding tangents and velocities in Section 2.7.

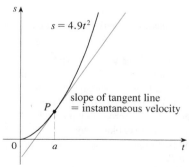

FIGURE 6

2.1 EXERCISES

1. A tank holds 1000 gallons of water, which drains from the bottom of the tank in half an hour. The values in the table show the volume V of water remaining in the tank (in gallons) after t minutes.

t (min)	5	10	15	20	25	30
V (gal)	694	444	250	111	28	0

(a) If P is the point $(15, 250)$ on the graph of V, find the slopes of the secant lines PQ when Q is the point on the graph with $t = 5, 10, 20, 25,$ and 30.
(b) Estimate the slope of the tangent line at P by averaging the slopes of two secant lines.
(c) Use a graph of the function to estimate the slope of the tangent line at P. (This slope represents the rate at which the water is flowing from the tank after 15 minutes.)

2. A cardiac monitor is used to measure the heart rate of a patient after surgery. It compiles the number of heartbeats after t minutes. When the data in the table are graphed, the slope of the tangent line represents the heart rate in beats per minute.

t (min)	36	38	40	42	44
Heartbeats	2530	2661	2806	2948	3080

The monitor estimates this value by calculating the slope of a secant line. Use the data to estimate the patient's heart rate after 42 minutes using the secant line between the points with the given values of t.
(a) $t = 36$ and $t = 42$
(b) $t = 38$ and $t = 42$
(c) $t = 40$ and $t = 42$
(d) $t = 42$ and $t = 44$
What are your conclusions?

3. The point $P(2, -1)$ lies on the curve $y = 1/(1 - x)$.
(a) If Q is the point $(x, 1/(1 - x))$, use your calculator to find the slope of the secant line PQ (correct to six decimal places) for the following values of x:
(i) 1.5   (ii) 1.9   (iii) 1.99   (iv) 1.999
(v) 2.5   (vi) 2.1   (vii) 2.01   (viii) 2.001
(b) Using the results of part (a), guess the value of the slope of the tangent line to the curve at $P(2, -1)$.
(c) Using the slope from part (b), find an equation of the tangent line to the curve at $P(2, -1)$.

4. The point $P(0.5, 0)$ lies on the curve $y = \cos \pi x$.
(a) If Q is the point $(x, \cos \pi x)$, use your calculator to find the slope of the secant line PQ (correct to six decimal places) for the following values of x:
(i) 0   (ii) 0.4   (iii) 0.49
(iv) 0.499   (v) 1   (vi) 0.6
(vii) 0.51   (viii) 0.501

(b) Using the results of part (a), guess the value of the slope of the tangent line to the curve at $P(0.5, 0)$.
(c) Using the slope from part (b), find an equation of the tangent line to the curve at $P(0.5, 0)$.
(d) Sketch the curve, two of the secant lines, and the tangent line.

5. If a ball is thrown into the air with a velocity of 40 ft/s, its height in feet t seconds later is given by $y = 40t - 16t^2$.
(a) Find the average velocity for the time period beginning when $t = 2$ and lasting
(i) 0.5 seconds   (ii) 0.1 seconds
(iii) 0.05 seconds   (iv) 0.01 seconds
(b) Estimate the instantaneous velocity when $t = 2$.

6. If a rock is thrown upward on the planet Mars with a velocity of 10 m/s, its height in meters t seconds later is given by $y = 10t - 1.86t^2$.
(a) Find the average velocity over the given time intervals:
(i) [1, 2]   (ii) [1, 1.5]
(iii) [1, 1.1]   (iv) [1, 1.01]
(v) [1, 1.001]
(b) Estimate the instantaneous velocity when $t = 1$.

7. The table shows the position of a motorcyclist after accelerating from rest.

t (seconds)	0	1	2	3	4	5	6
s (feet)	0	4.9	20.6	46.5	79.2	124.8	176.7

(a) Find the average velocity for each time period:
(i) [2, 4]   (ii) [3, 4]   (iii) [4, 5]   (iv) [4, 6]
(b) Use the graph of s as a function of t to estimate the instantaneous velocity when $t = 3$.

8. The displacement (in centimeters) of a particle moving back and forth along a straight line is given by the equation of motion $s = 2 \sin \pi t + 3 \cos \pi t$, where t is measured in seconds.
(a) Find the average velocity during each time period:
(i) [1, 2]   (ii) [1, 1.1]
(iii) [1, 1.01]   (iv) [1, 1.001]
(b) Estimate the instantaneous velocity of the particle when $t = 1$.

9. The point $P(1, 0)$ lies on the curve $y = \sin(10\pi/x)$.
(a) If Q is the point $(x, \sin(10\pi/x))$, find the slope of the secant line PQ (correct to four decimal places) for $x = 2, 1.5, 1.4, 1.3, 1.2, 1.1, 0.5, 0.6, 0.7, 0.8,$ and 0.9. Do the slopes appear to be approaching a limit?
(b) Use a graph of the curve to explain why the slopes of the secant lines in part (a) are not close to the slope of the tangent line at P.
(c) By choosing appropriate secant lines, estimate the slope of the tangent line at P.

2.2 The Limit of a Function

Having seen in the preceding section how limits arise when we want to find the tangent to a curve or the velocity of an object, we now turn our attention to limits in general and numerical and graphical methods for computing them.

Let's investigate the behavior of the function f defined by $f(x) = x^2 - x + 2$ for values of x near 2. The following table gives values of $f(x)$ for values of x close to 2 but not equal to 2.

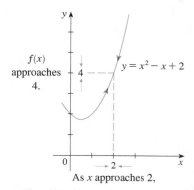

$f(x)$ approaches 4.

As x approaches 2,

FIGURE 1

x	$f(x)$	x	$f(x)$
1.0	2.000000	3.0	8.000000
1.5	2.750000	2.5	5.750000
1.8	3.440000	2.2	4.640000
1.9	3.710000	2.1	4.310000
1.95	3.852500	2.05	4.152500
1.99	3.970100	2.01	4.030100
1.995	3.985025	2.005	4.015025
1.999	3.997001	2.001	4.003001

From the table and the graph of f (a parabola) shown in Figure 1 we see that the closer x is to 2 (on either side of 2), the closer $f(x)$ is to 4. In fact, it appears that we can make the values of $f(x)$ as close as we like to 4 by taking x sufficiently close to 2. We express this by saying "the limit of the function $f(x) = x^2 - x + 2$ as x approaches 2 is equal to 4." The notation for this is

$$\lim_{x \to 2} (x^2 - x + 2) = 4$$

In general, we use the following notation.

---

1 Intuitive Definition of a Limit Suppose $f(x)$ is defined when x is near the number a. (This means that f is defined on some open interval that contains a, except possibly at a itself.) Then we write

$$\lim_{x \to a} f(x) = L$$

and say   "the limit of $f(x)$, as x approaches a, equals L"

if we can make the values of $f(x)$ arbitrarily close to L (as close to L as we like) by restricting x to be sufficiently close to a (on either side of a) but not equal to a.

Roughly speaking, this says that the values of $f(x)$ approach L as x approaches a. In other words, the values of $f(x)$ tend to get closer and closer to the number L as x gets closer and closer to the number a (from either side of a) but $x \neq a$. (A more precise definition will be given in Section 2.4.)

An alternative notation for

$$\lim_{x \to a} f(x) = L$$

is $$f(x) \to L \quad\quad \text{as} \quad\quad x \to a$$

which is usually read "$f(x)$ approaches L as x approaches a."

Notice the phrase "but $x \neq a$" in the definition of limit. This means that in finding the limit of $f(x)$ as x approaches a, we never consider $x = a$. In fact, $f(x)$ need not even be defined when $x = a$. The only thing that matters is how f is defined *near a*.

Figure 2 shows the graphs of three functions. Note that in part (c), $f(a)$ is not defined and in part (b), $f(a) \neq L$. But in each case, regardless of what happens at a, it is true that $\lim_{x \to a} f(x) = L$.

(a)

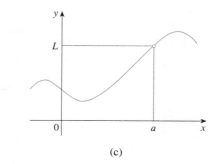

(b)

(c)

FIGURE 2 $\lim\limits_{x \to a} f(x) = L$ in all three cases

EXAMPLE 1 Guess the value of $\lim\limits_{x \to 1} \dfrac{x - 1}{x^2 - 1}$.

SOLUTION Notice that the function $f(x) = (x - 1)/(x^2 - 1)$ is not defined when $x = 1$, but that doesn't matter because the definition of $\lim_{x \to a} f(x)$ says that we consider values of x that are close to a but not equal to a.

The tables at the left give values of $f(x)$ (correct to six decimal places) for values of x that approach 1 (but are not equal to 1). On the basis of the values in the tables, we make the guess that

$$\lim_{x \to 1} \frac{x - 1}{x^2 - 1} = 0.5$$

$x < 1$	$f(x)$
0.5	0.666667
0.9	0.526316
0.99	0.502513
0.999	0.500250
0.9999	0.500025

$x > 1$	$f(x)$
1.5	0.400000
1.1	0.476190
1.01	0.497512
1.001	0.499750
1.0001	0.499975

1	0.5

Example 1 is illustrated by the graph of f in Figure 3. Now let's change f slightly by giving it the value 2 when $x = 1$ and calling the resulting function g:

$$g(x) = \begin{cases} \dfrac{x - 1}{x^2 - 1} & \text{if } x \neq 1 \\ 2 & \text{if } x = 1 \end{cases}$$

This new function g still has the same limit as x approaches 1. (See Figure 4.)

FIGURE 3

FIGURE 4

EXAMPLE 2 Estimate the value of $\displaystyle\lim_{t \to 0} \frac{\sqrt{t^2 + 9} - 3}{t^2}$.

SOLUTION The table lists values of the function for several values of t near 0.

t	$\dfrac{\sqrt{t^2 + 9} - 3}{t^2}$
± 1.0	$0.162277\ldots$
± 0.5	$0.165525\ldots$
± 0.1	$0.166620\ldots$
± 0.05	$0.166655\ldots$
± 0.01	$0.166666\ldots$

As t approaches 0, the values of the function seem to approach $0.1666666\ldots$ and so we guess that

$$\lim_{t \to 0} \frac{\sqrt{t^2 + 9} - 3}{t^2} = \frac{1}{6}$$

t	$\dfrac{\sqrt{t^2 + 9} - 3}{t^2}$
± 0.001	0.166667
± 0.0001	0.166670
± 0.00001	0.167000
± 0.000001	0.000000

In Example 2 what would have happened if we had taken even smaller values of t? The table in the margin shows the results from one calculator; you can see that something strange seems to be happening.

If you try these calculations on your own calculator you might get different values, but eventually you will get the value 0 if you make t sufficiently small. Does this mean that the answer is really 0 instead of $\frac{1}{6}$? No, the value of the limit is $\frac{1}{6}$, as we will show in the next section. The problem is that the calculator gave false values because $\sqrt{t^2 + 9}$ is very close to 3 when t is small. (In fact, when t is sufficiently small, a calculator's value for $\sqrt{t^2 + 9}$ is $3.000\ldots$ to as many digits as the calculator is capable of carrying.)

Something similar happens when we try to graph the function

$$f(t) = \frac{\sqrt{t^2 + 9} - 3}{t^2}$$

www.stewartcalculus.com
For a further explanation of why calculators sometimes give false values, click on *Lies My Calculator and Computer Told Me*. In particular, see the section called *The Perils of Subtraction*.

of Example 2 on a graphing calculator or computer. Parts (a) and (b) of Figure 5 show quite accurate graphs of f, and when we use the trace mode (if available) we can estimate easily that the limit is about $\frac{1}{6}$. But if we zoom in too much, as in parts (c) and (d), then we get inaccurate graphs, again because of rounding errors from the subtraction.

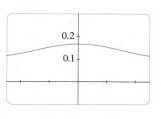

(a) $-5 \leqslant t \leqslant 5$

(b) $-0.1 \leqslant t \leqslant 0.1$

(c) $-10^{-6} \leqslant t \leqslant 10^{-6}$

(d) $-10^{-7} \leqslant t \leqslant 10^{-7}$

FIGURE 5

EXAMPLE 3 Guess the value of $\lim\limits_{x \to 0} \dfrac{\sin x}{x}$.

SOLUTION The function $f(x) = (\sin x)/x$ is not defined when $x = 0$. Using a calculator (and remembering that, if $x \in \mathbb{R}$, $\sin x$ means the sine of the angle whose *radian* measure is x), we construct a table of values correct to eight decimal places. From the table at the left and the graph in Figure 6 we guess that

$$\lim_{x \to 0} \frac{\sin x}{x} = 1$$

This guess is in fact correct, as will be proved in Chapter 3 using a geometric argument.

x	$\dfrac{\sin x}{x}$
± 1.0	0.84147098
± 0.5	0.95885108
± 0.4	0.97354586
± 0.3	0.98506736
± 0.2	0.99334665
± 0.1	0.99833417
± 0.05	0.99958339
± 0.01	0.99998333
± 0.005	0.99999583
± 0.001	0.99999983

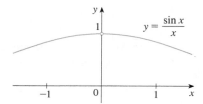

FIGURE 6

Computer Algebra Systems
Computer algebra systems (CAS) have commands that compute limits. In order to avoid the types of pitfalls demonstrated in Examples 2, 4, and 5, they don't find limits by numerical experimentation. Instead, they use more sophisticated techniques such as computing infinite series. If you have access to a CAS, use the limit command to compute the limits in the examples of this section and to check your answers in the exercises of this chapter.

EXAMPLE 4 Investigate $\lim\limits_{x \to 0} \sin \dfrac{\pi}{x}$.

SOLUTION Again the function $f(x) = \sin(\pi/x)$ is undefined at 0. Evaluating the function for some small values of x, we get

$$f(1) = \sin \pi = 0 \qquad\qquad f(\tfrac{1}{2}) = \sin 2\pi = 0$$

$$f(\tfrac{1}{3}) = \sin 3\pi = 0 \qquad\qquad f(\tfrac{1}{4}) = \sin 4\pi = 0$$

$$f(0.1) = \sin 10\pi = 0 \qquad f(0.01) = \sin 100\pi = 0$$

Similarly, $f(0.001) = f(0.0001) = 0$. On the basis of this information we might be tempted to guess that

$$\lim_{x \to 0} \sin \frac{\pi}{x} = 0$$

but this time our guess is wrong. Note that although $f(1/n) = \sin n\pi = 0$ for any integer n, it is also true that $f(x) = 1$ for infinitely many values of x (such as $2/5$ or $2/101$) that approach 0. You can see this from the graph of f shown in Figure 7.

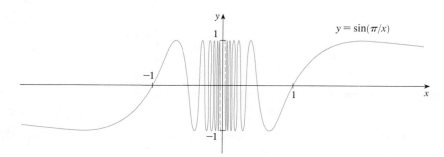

FIGURE 7

The dashed lines near the y-axis indicate that the values of $\sin(\pi/x)$ oscillate between 1 and -1 infinitely often as x approaches 0. (See Exercise 51.)

Since the values of $f(x)$ do not approach a fixed number as x approaches 0,

$$\lim_{x \to 0} \sin \frac{\pi}{x} \quad \text{does not exist}$$

x	$x^3 + \dfrac{\cos 5x}{10{,}000}$
1	1.000028
0.5	0.124920
0.1	0.001088
0.05	0.000222
0.01	0.000101

x	$x^3 + \dfrac{\cos 5x}{10{,}000}$
0.005	0.00010009
0.001	0.00010000

EXAMPLE 5 Find $\displaystyle \lim_{x \to 0} \left(x^3 + \frac{\cos 5x}{10{,}000} \right)$.

SOLUTION As before, we construct a table of values. From the first table in the margin it appears that

$$\lim_{x \to 0} \left(x^3 + \frac{\cos 5x}{10{,}000} \right) = 0$$

But if we persevere with smaller values of x, the second table suggests that

$$\lim_{x \to 0} \left(x^3 + \frac{\cos 5x}{10{,}000} \right) = 0.000100 = \frac{1}{10{,}000}$$

Later we will see that $\lim_{x \to 0} \cos 5x = 1$; then it follows that the limit is 0.0001.

Examples 4 and 5 illustrate some of the pitfalls in guessing the value of a limit. It is easy to guess the wrong value if we use inappropriate values of x, but it is difficult to know when to stop calculating values. And, as the discussion after Example 2 shows, sometimes calculators and computers give the wrong values. In the next section, however, we will develop foolproof methods for calculating limits.

One-Sided Limits

EXAMPLE 6 The Heaviside function H is defined by

$$H(t) = \begin{cases} 0 & \text{if } t < 0 \\ 1 & \text{if } t \geq 0 \end{cases}$$

[This function is named after the electrical engineer Oliver Heaviside (1850–1925) and can be used to describe an electric current that is switched on at time $t = 0$.] Its graph is shown in Figure 8.

As t approaches 0 from the left, $H(t)$ approaches 0. As t approaches 0 from the right, $H(t)$ approaches 1. There is no single number that $H(t)$ approaches as t approaches 0. Therefore $\lim_{t \to 0} H(t)$ does not exist.

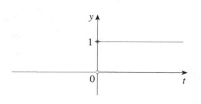

FIGURE 8
The Heaviside function

We noticed in Example 6 that $H(t)$ approaches 0 as t approaches 0 from the left and $H(t)$ approaches 1 as t approaches 0 from the right. We indicate this situation symbolically by writing

$$\lim_{t \to 0^-} H(t) = 0 \qquad \text{and} \qquad \lim_{t \to 0^+} H(t) = 1$$

The notation $t \to 0^-$ indicates that we consider only values of t that are less than 0. Likewise, $t \to 0^+$ indicates that we consider only values of t that are greater than 0.

> **2** **Definition of One-Sided Limits** We write
>
> $$\lim_{x \to a^-} f(x) = L$$
>
> and say the **left-hand limit of $f(x)$ as x approaches a** [or the **limit of $f(x)$ as x approaches a from the left**] is equal to L if we can make the values of $f(x)$ arbitrarily close to L by taking x to be sufficiently close to a with x *less than a.*

Notice that Definition 2 differs from Definition 1 only in that we require x to be less than a. Similarly, if we require that x be greater than a, we get "the **right-hand limit of** $f(x)$ **as** x **approaches** a is equal to L" and we write

$$\lim_{x \to a^+} f(x) = L$$

Thus the notation $x \to a^+$ means that we consider only x *greater than a.* These definitions are illustrated in Figure 9.

FIGURE 9

(a) $\lim\limits_{x \to a^-} f(x) = L$ (b) $\lim\limits_{x \to a^+} f(x) = L$

By comparing Definition 1 with the definitions of one-sided limits, we see that the following is true.

> **3** $\lim\limits_{x \to a} f(x) = L$ if and only if $\lim\limits_{x \to a^-} f(x) = L$ and $\lim\limits_{x \to a^+} f(x) = L$

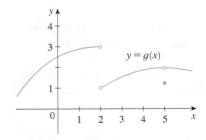

FIGURE 10

EXAMPLE 7 The graph of a function g is shown in Figure 10. Use it to state the values (if they exist) of the following:

(a) $\lim\limits_{x \to 2^-} g(x)$ (b) $\lim\limits_{x \to 2^+} g(x)$ (c) $\lim\limits_{x \to 2} g(x)$

(d) $\lim\limits_{x \to 5^-} g(x)$ (e) $\lim\limits_{x \to 5^+} g(x)$ (f) $\lim\limits_{x \to 5} g(x)$

SOLUTION From the graph we see that the values of $g(x)$ approach 3 as x approaches 2 from the left, but they approach 1 as x approaches 2 from the right. Therefore

(a) $\lim\limits_{x \to 2^-} g(x) = 3$ and   (b) $\lim\limits_{x \to 2^+} g(x) = 1$

(c) Since the left and right limits are different, we conclude from (3) that $\lim_{x \to 2} g(x)$ does not exist.

The graph also shows that

(d) $\lim\limits_{x \to 5^-} g(x) = 2$ and   (e) $\lim\limits_{x \to 5^+} g(x) = 2$

(f) This time the left and right limits are the same and so, by (3), we have

$$\lim_{x \to 5} g(x) = 2$$

Despite this fact, notice that $g(5) \neq 2$.

Infinite Limits

EXAMPLE 8 Find $\lim_{x \to 0} \dfrac{1}{x^2}$ if it exists.

SOLUTION As x becomes close to 0, x^2 also becomes close to 0, and $1/x^2$ becomes very large. (See the table in the margin.) In fact, it appears from the graph of the function $f(x) = 1/x^2$ shown in Figure 11 that the values of $f(x)$ can be made arbitrarily large by taking x close enough to 0. Thus the values of $f(x)$ do not approach a number, so $\lim_{x \to 0} (1/x^2)$ does not exist.

To indicate the kind of behavior exhibited in Example 8, we use the notation

$$\lim_{x \to 0} \frac{1}{x^2} = \infty$$

This does not mean that we are regarding ∞ as a number. Nor does it mean that the limit exists. It simply expresses the particular way in which the limit does not exist: $1/x^2$ can be made as large as we like by taking x close enough to 0.

In general, we write symbolically

$$\lim_{x \to a} f(x) = \infty$$

to indicate that the values of $f(x)$ tend to become larger and larger (or "increase without bound") as x becomes closer and closer to a.

x	$\dfrac{1}{x^2}$
± 1	1
± 0.5	4
± 0.2	25
± 0.1	100
± 0.05	400
± 0.01	10,000
± 0.001	1,000,000

FIGURE 11

> **4** **Intuitive Definition of an Infinite Limit** Let f be a function defined on both sides of a, except possibly at a itself. Then
>
> $$\lim_{x \to a} f(x) = \infty$$
>
> means that the values of $f(x)$ can be made arbitrarily large (as large as we please) by taking x sufficiently close to a, but not equal to a.

Another notation for $\lim_{x \to a} f(x) = \infty$ is

$$f(x) \to \infty \qquad \text{as} \qquad x \to a$$

Again, the symbol ∞ is not a number, but the expression $\lim_{x \to a} f(x) = \infty$ is often read as

"the limit of $f(x)$, as x approaches a, is infinity"

or

"$f(x)$ becomes infinite as x approaches a"

or

"$f(x)$ increases without bound as x approaches a"

This definition is illustrated graphically in Figure 12.

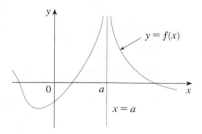

FIGURE 12
$\lim_{x \to a} f(x) = \infty$

When we say a number is "large negative," we mean that it is negative but its magnitude (absolute value) is large.

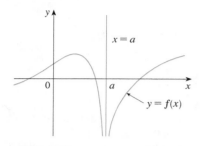

FIGURE 13
$\lim\limits_{x \to a} f(x) = -\infty$

A similar sort of limit, for functions that become large negative as x gets close to a, is defined in Definition 5 and is illustrated in Figure 13.

> **5   Definition** Let f be a function defined on both sides of a, except possibly at a itself. Then
>
> $$\lim_{x \to a} f(x) = -\infty$$
>
> means that the values of $f(x)$ can be made arbitrarily large negative by taking x sufficiently close to a, but not equal to a.

The symbol $\lim_{x \to a} f(x) = -\infty$ can be read as "the limit of $f(x)$, as x approaches a, is negative infinity" or "$f(x)$ decreases without bound as x approaches a." As an example we have

$$\lim_{x \to 0} \left(-\frac{1}{x^2} \right) = -\infty$$

Similar definitions can be given for the one-sided infinite limits

$$\lim_{x \to a^-} f(x) = \infty \qquad\qquad \lim_{x \to a^+} f(x) = \infty$$

$$\lim_{x \to a^-} f(x) = -\infty \qquad\qquad \lim_{x \to a^+} f(x) = -\infty$$

remembering that $x \to a^-$ means that we consider only values of x that are less than a, and similarly $x \to a^+$ means that we consider only $x > a$. Illustrations of these four cases are given in Figure 14.

(a) $\lim\limits_{x \to a^-} f(x) = \infty$

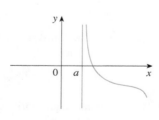

(b) $\lim\limits_{x \to a^+} f(x) = \infty$

(c) $\lim\limits_{x \to a^-} f(x) = -\infty$

(d) $\lim\limits_{x \to a^+} f(x) = -\infty$

FIGURE 14

> **6   Definition** The vertical line $x = a$ is called a **vertical asymptote** of the curve $y = f(x)$ if at least one of the following statements is true:
>
> $$\lim_{x \to a} f(x) = \infty \qquad \lim_{x \to a^-} f(x) = \infty \qquad \lim_{x \to a^+} f(x) = \infty$$
>
> $$\lim_{x \to a} f(x) = -\infty \qquad \lim_{x \to a^-} f(x) = -\infty \qquad \lim_{x \to a^+} f(x) = -\infty$$

For instance, the y-axis is a vertical asymptote of the curve $y = 1/x^2$ because $\lim_{x \to 0} (1/x^2) = \infty$. In Figure 14 the line $x = a$ is a vertical asymptote in each of the four cases shown. In general, knowledge of vertical asymptotes is very useful in sketching graphs.

EXAMPLE 9 Find $\displaystyle\lim_{x \to 3^+} \frac{2x}{x - 3}$ and $\displaystyle\lim_{x \to 3^-} \frac{2x}{x - 3}$.

SOLUTION If x is close to 3 but larger than 3, then the denominator $x - 3$ is a small positive number and $2x$ is close to 6. So the quotient $2x/(x - 3)$ is a large *positive* number. [For instance, if $x = 3.01$ then $2x/(x - 3) = 6.02/0.01 = 602$.] Thus, intuitively, we see that

$$\lim_{x \to 3^+} \frac{2x}{x - 3} = \infty$$

Likewise, if x is close to 3 but smaller than 3, then $x - 3$ is a small negative number but $2x$ is still a positive number (close to 6). So $2x/(x - 3)$ is a numerically large *negative* number. Thus

$$\lim_{x \to 3^-} \frac{2x}{x - 3} = -\infty$$

The graph of the curve $y = 2x/(x - 3)$ is given in Figure 15. The line $x = 3$ is a vertical asymptote. ∎

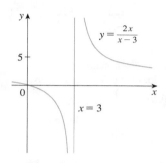

FIGURE 15

EXAMPLE 10 Find the vertical asymptotes of $f(x) = \tan x$.

SOLUTION Because

$$\tan x = \frac{\sin x}{\cos x}$$

there are potential vertical asymptotes where $\cos x = 0$. In fact, since $\cos x \to 0^+$ as $x \to (\pi/2)^-$ and $\cos x \to 0^-$ as $x \to (\pi/2)^+$, whereas $\sin x$ is positive (near 1) when x is near $\pi/2$, we have

$$\lim_{x \to (\pi/2)^-} \tan x = \infty \quad \text{and} \quad \lim_{x \to (\pi/2)^+} \tan x = -\infty$$

This shows that the line $x = \pi/2$ is a vertical asymptote. Similar reasoning shows that the lines $x = \pi/2 + n\pi$, where n is an integer, are all vertical asymptotes of $f(x) = \tan x$. The graph in Figure 16 confirms this. ∎

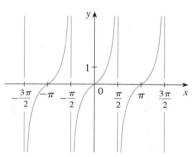

FIGURE 16
$y = \tan x$

Another example of a function whose graph has a vertical asymptote is the natural logarithmic function $y = \ln x$. From Figure 17 we see that

$$\lim_{x \to 0^+} \ln x = -\infty$$

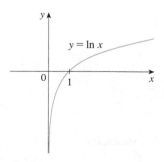

FIGURE 17
The y-axis is a vertical asymptote of the natural logarithmic function.

and so the line $x = 0$ (the y-axis) is a vertical asymptote. In fact, the same is true for $y = \log_b x$ provided that $b > 1$. (See Figures 1.5.11 and 1.5.12.)

2.2   EXERCISES

1. Explain in your own words what is meant by the equation

$$\lim_{x \to 2} f(x) = 5$$

Is it possible for this statement to be true and yet $f(2) = 3$? Explain.

2. Explain what it means to say that

$$\lim_{x \to 1^-} f(x) = 3 \quad \text{and} \quad \lim_{x \to 1^+} f(x) = 7$$

In this situation is it possible that $\lim_{x \to 1} f(x)$ exists? Explain.

3. Explain the meaning of each of the following.

(a) $\lim_{x \to -3} f(x) = \infty$ (b) $\lim_{x \to 4^+} f(x) = -\infty$

4. Use the given graph of f to state the value of each quantity, if it exists. If it does not exist, explain why.

(a) $\lim_{x \to 2^-} f(x)$ (b) $\lim_{x \to 2^+} f(x)$ (c) $\lim_{x \to 2} f(x)$

(d) $f(2)$ (e) $\lim_{x \to 4} f(x)$ (f) $f(4)$

5. For the function f whose graph is given, state the value of each quantity, if it exists. If it does not exist, explain why.

(a) $\lim_{x \to 1} f(x)$ (b) $\lim_{x \to 3^-} f(x)$ (c) $\lim_{x \to 3^+} f(x)$

(d) $\lim_{x \to 3} f(x)$ (e) $f(3)$

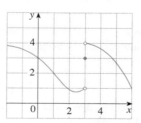

6. For the function h whose graph is given, state the value of each quantity, if it exists. If it does not exist, explain why.

(a) $\lim_{x \to -3^-} h(x)$ (b) $\lim_{x \to -3^+} h(x)$ (c) $\lim_{x \to -3} h(x)$

(d) $h(-3)$ (e) $\lim_{x \to 0^-} h(x)$ (f) $\lim_{x \to 0^+} h(x)$

(g) $\lim_{x \to 0} h(x)$ (h) $h(0)$ (i) $\lim_{x \to 2} h(x)$

(j) $h(2)$ (k) $\lim_{x \to 5^+} h(x)$ (l) $\lim_{x \to 5^-} h(x)$

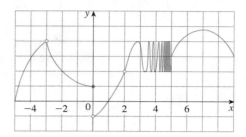

7. For the function g whose graph is given, state the value of each quantity, if it exists. If it does not exist, explain why.

(a) $\lim_{t \to 0^-} g(t)$ (b) $\lim_{t \to 0^+} g(t)$ (c) $\lim_{t \to 0} g(t)$

(d) $\lim_{t \to 2^-} g(t)$ (e) $\lim_{t \to 2^+} g(t)$ (f) $\lim_{t \to 2} g(t)$

(g) $g(2)$ (h) $\lim_{t \to 4} g(t)$

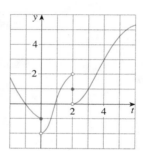

8. For the function A whose graph is shown, state the following.

(a) $\lim_{x \to -3} A(x)$ (b) $\lim_{x \to 2^-} A(x)$

(c) $\lim_{x \to 2^+} A(x)$ (d) $\lim_{x \to -1} A(x)$

(e) The equations of the vertical asymptotes

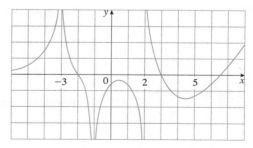

9. For the function f whose graph is shown, state the following.

(a) $\lim_{x \to -7} f(x)$ (b) $\lim_{x \to -3} f(x)$ (c) $\lim_{x \to 0} f(x)$

(d) $\lim_{x \to 6^-} f(x)$ (e) $\lim_{x \to 6^+} f(x)$

(f) The equations of the vertical asymptotes.

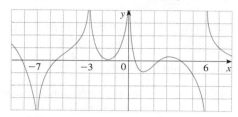

10. A patient receives a 150-mg injection of a drug every 4 hours. The graph shows the amount $f(t)$ of the drug in the bloodstream after t hours. Find

$$\lim_{t \to 12^-} f(t) \qquad \text{and} \qquad \lim_{t \to 12^+} f(t)$$

and explain the significance of these one-sided limits.

11–12 Sketch the graph of the function and use it to determine the values of a for which $\lim_{x \to a} f(x)$ exists.

11. $f(x) = \begin{cases} 1 + x & \text{if } x < -1 \\ x^2 & \text{if } -1 \le x < 1 \\ 2 - x & \text{if } x \ge 1 \end{cases}$

12. $f(x) = \begin{cases} 1 + \sin x & \text{if } x < 0 \\ \cos x & \text{if } 0 \le x \le \pi \\ \sin x & \text{if } x > \pi \end{cases}$

13–14 Use the graph of the function f to state the value of each limit, if it exists. If it does not exist, explain why.

(a) $\lim_{x \to 0^-} f(x)$ (b) $\lim_{x \to 0^+} f(x)$ (c) $\lim_{x \to 0} f(x)$

13. $f(x) = \dfrac{1}{1 + e^{1/x}}$ **14.** $f(x) = \dfrac{x^2 + x}{\sqrt{x^3 + x^2}}$

15–18 Sketch the graph of an example of a function f that satisfies all of the given conditions.

15. $\lim\limits_{x \to 0^-} f(x) = -1, \quad \lim\limits_{x \to 0^+} f(x) = 2, \quad f(0) = 1$

16. $\lim\limits_{x \to 0} f(x) = 1, \quad \lim\limits_{x \to 3^-} f(x) = -2, \quad \lim\limits_{x \to 3^+} f(x) = 2,$
$f(0) = -1, \quad f(3) = 1$

17. $\lim\limits_{x \to 3^+} f(x) = 4, \quad \lim\limits_{x \to 3^-} f(x) = 2, \quad \lim\limits_{x \to -2} f(x) = 2,$
$f(3) = 3, \quad f(-2) = 1$

18. $\lim\limits_{x \to 0^-} f(x) = 2, \quad \lim\limits_{x \to 0^+} f(x) = 0, \quad \lim\limits_{x \to 4^-} f(x) = 3,$
$\lim\limits_{x \to 4^+} f(x) = 0, \quad f(0) = 2, \quad f(4) = 1$

19–22 Guess the value of the limit (if it exists) by evaluating the function at the given numbers (correct to six decimal places).

19. $\lim\limits_{x \to 3} \dfrac{x^2 - 3x}{x^2 - 9}$,

$x = 3.1, 3.05, 3.01, 3.001, 3.0001,$

$2.9, 2.95, 2.99, 2.999, 2.9999$

20. $\lim\limits_{x \to -3} \dfrac{x^2 - 3x}{x^2 - 9}$,

$x = -2.5, -2.9, -2.95, -2.99, -2.999, -2.9999,$
$-3.5, -3.1, -3.05, -3.01, -3.001, -3.0001$

21. $\lim\limits_{t \to 0} \dfrac{e^{5t} - 1}{t}$, $t = \pm 0.5, \pm 0.1, \pm 0.01, \pm 0.001, \pm 0.0001$

22. $\lim\limits_{h \to 0} \dfrac{(2 + h)^5 - 32}{h}$,

$h = \pm 0.5, \pm 0.1, \pm 0.01, \pm 0.001, \pm 0.0001$

23–28 Use a table of values to estimate the value of the limit. If you have a graphing device, use it to confirm your result graphically.

23. $\lim\limits_{x \to 4} \dfrac{\ln x - \ln 4}{x - 4}$ **24.** $\lim\limits_{p \to -1} \dfrac{1 + p^9}{1 + p^{15}}$

25. $\lim\limits_{\theta \to 0} \dfrac{\sin 3\theta}{\tan 2\theta}$ **26.** $\lim\limits_{t \to 0} \dfrac{5^t - 1}{t}$

27. $\lim\limits_{x \to 0^+} x^x$ **28.** $\lim\limits_{x \to 0^+} x^2 \ln x$

29. (a) By graphing the function $f(x) = (\cos 2x - \cos x)/x^2$ and zooming in toward the point where the graph crosses the y-axis, estimate the value of $\lim_{x \to 0} f(x)$.
(b) Check your answer in part (a) by evaluating $f(x)$ for values of x that approach 0.

30. (a) Estimate the value of

$$\lim_{x \to 0} \frac{\sin x}{\sin \pi x}$$

by graphing the function $f(x) = (\sin x)/(\sin \pi x)$. State your answer correct to two decimal places.
(b) Check your answer in part (a) by evaluating $f(x)$ for values of x that approach 0.

31–43 Determine the infinite limit.

31. $\lim\limits_{x \to 5^+} \dfrac{x + 1}{x - 5}$

32. $\lim\limits_{x \to 5^-} \dfrac{x + 1}{x - 5}$

33. $\lim\limits_{x \to 1} \dfrac{2 - x}{(x - 1)^2}$

34. $\lim\limits_{x \to 3^-} \dfrac{\sqrt{x}}{(x - 3)^5}$

35. $\lim\limits_{x \to 3^+} \ln(x^2 - 9)$

36. $\lim\limits_{x \to 0^+} \ln(\sin x)$

37. $\lim\limits_{x \to (\pi/2)^+} \dfrac{1}{x} \sec x$

38. $\lim\limits_{x \to \pi^-} \cot x$

39. $\lim\limits_{x \to 2\pi^-} x \csc x$

40. $\lim\limits_{x \to 2^-} \dfrac{x^2 - 2x}{x^2 - 4x + 4}$

41. $\lim\limits_{x \to 2^+} \dfrac{x^2 - 2x - 8}{x^2 - 5x + 6}$

42. $\lim\limits_{x \to 0^+} \left(\dfrac{1}{x} - \ln x \right)$

43. $\lim\limits_{x \to 0} (\ln x^2 - x^{-2})$

---

44. (a) Find the vertical asymptotes of the function

$$y = \frac{x^2 + 1}{3x - 2x^2}$$

(b) Confirm your answer to part (a) by graphing the function.

45. Determine $\lim\limits_{x \to 1^-} \dfrac{1}{x^3 - 1}$ and $\lim\limits_{x \to 1^+} \dfrac{1}{x^3 - 1}$

(a) by evaluating $f(x) = 1/(x^3 - 1)$ for values of x that approach 1 from the left and from the right,

(b) by reasoning as in Example 9, and

(c) from a graph of f.

46. (a) By graphing the function $f(x) = (\tan 4x)/x$ and zooming in toward the point where the graph crosses the y-axis, estimate the value of $\lim_{x \to 0} f(x)$.

(b) Check your answer in part (a) by evaluating $f(x)$ for values of x that approach 0.

47. (a) Estimate the value of the limit $\lim_{x \to 0} (1 + x)^{1/x}$ to five decimal places. Does this number look familiar?

(b) Illustrate part (a) by graphing the function $y = (1 + x)^{1/x}$.

48. (a) Graph the function $f(x) = e^x + \ln|x - 4|$ for $0 \le x \le 5$. Do you think the graph is an accurate representation of f?

(b) How would you get a graph that represents f better?

49. (a) Evaluate the function $f(x) = x^2 - (2^x/1000)$ for $x = 1, 0.8, 0.6, 0.4, 0.2, 0.1,$ and 0.05, and guess the value of

$$\lim_{x \to 0} \left(x^2 - \frac{2^x}{1000} \right)$$

(b) Evaluate $f(x)$ for $x = 0.04, 0.02, 0.01, 0.005, 0.003,$ and 0.001. Guess again.

50. (a) Evaluate $h(x) = (\tan x - x)/x^3$ for $x = 1, 0.5, 0.1, 0.05, 0.01,$ and 0.005.

(b) Guess the value of $\lim\limits_{x \to 0} \dfrac{\tan x - x}{x^3}$.

(c) Evaluate $h(x)$ for successively smaller values of x until you finally reach a value of 0 for $h(x)$. Are you still confident that your guess in part (b) is correct? Explain why you eventually obtained 0 values. (In Section 4.4 a method for evaluating this limit will be explained.)

(d) Graph the function h in the viewing rectangle $[-1, 1]$ by $[0, 1]$. Then zoom in toward the point where the graph crosses the y-axis to estimate the limit of $h(x)$ as x approaches 0. Continue to zoom in until you observe distortions in the graph of h. Compare with the results of part (c).

51. Graph the function $f(x) = \sin(\pi/x)$ of Example 4 in the viewing rectangle $[-1, 1]$ by $[-1, 1]$. Then zoom in toward the origin several times. Comment on the behavior of this function.

52. Consider the function $f(x) = \tan \dfrac{1}{x}$.

(a) Show that $f(x) = 0$ for $x = \dfrac{1}{\pi}, \dfrac{1}{2\pi}, \dfrac{1}{3\pi}, \ldots$

(b) Show that $f(x) = 1$ for $x = \dfrac{4}{\pi}, \dfrac{4}{5\pi}, \dfrac{4}{9\pi}, \ldots$

(c) What can you conclude about $\lim\limits_{x \to 0^+} \tan \dfrac{1}{x}$?

53. Use a graph to estimate the equations of all the vertical asymptotes of the curve

$$y = \tan(2 \sin x) \qquad -\pi \le x \le \pi$$

Then find the exact equations of these asymptotes.

54. In the theory of relativity, the mass of a particle with velocity v is

$$m = \frac{m_0}{\sqrt{1 - v^2/c^2}}$$

where m_0 is the mass of the particle at rest and c is the speed of light. What happens as $v \to c^-$?

55. (a) Use numerical and graphical evidence to guess the value of the limit

$$\lim_{x \to 1} \frac{x^3 - 1}{\sqrt{x} - 1}$$

(b) How close to 1 does x have to be to ensure that the function in part (a) is within a distance 0.5 of its limit?

2.3   Calculating Limits Using the Limit Laws

In Section 2.2 we used calculators and graphs to guess the values of limits, but we saw that such methods don't always lead to the correct answer. In this section we use the following properties of limits, called the *Limit Laws*, to calculate limits.

Limit Laws Suppose that c is a constant and the limits

$$\lim_{x \to a} f(x) \qquad \text{and} \qquad \lim_{x \to a} g(x)$$

exist. Then

1. $\lim_{x \to a} [f(x) + g(x)] = \lim_{x \to a} f(x) + \lim_{x \to a} g(x)$

2. $\lim_{x \to a} [f(x) - g(x)] = \lim_{x \to a} f(x) - \lim_{x \to a} g(x)$

3. $\lim_{x \to a} [cf(x)] = c \lim_{x \to a} f(x)$

4. $\lim_{x \to a} [f(x) g(x)] = \lim_{x \to a} f(x) \cdot \lim_{x \to a} g(x)$

5. $\lim_{x \to a} \dfrac{f(x)}{g(x)} = \dfrac{\lim\limits_{x \to a} f(x)}{\lim\limits_{x \to a} g(x)} \qquad$ if $\lim\limits_{x \to a} g(x) \neq 0$

These five laws can be stated verbally as follows:

Sum Law

1. The limit of a sum is the sum of the limits.

Difference Law

2. The limit of a difference is the difference of the limits.

Constant Multiple Law

3. The limit of a constant times a function is the constant times the limit of the function.

Product Law

4. The limit of a product is the product of the limits.

Quotient Law

5. The limit of a quotient is the quotient of the limits (provided that the limit of the denominator is not 0).

It is easy to believe that these properties are true. For instance, if $f(x)$ is close to L and $g(x)$ is close to M, it is reasonable to conclude that $f(x) + g(x)$ is close to $L + M$. This gives us an intuitive basis for believing that Law 1 is true. In Section 2.4 we give a precise definition of a limit and use it to prove this law. The proofs of the remaining laws are given in Appendix F.

EXAMPLE 1 Use the Limit Laws and the graphs of f and g in Figure 1 to evaluate the following limits, if they exist.

(a) $\lim\limits_{x \to -2} [f(x) + 5g(x)]$ (b) $\lim\limits_{x \to 1} [f(x)g(x)]$ (c) $\lim\limits_{x \to 2} \dfrac{f(x)}{g(x)}$

SOLUTION

(a) From the graphs of f and g we see that

$$\lim_{x \to -2} f(x) = 1 \qquad \text{and} \qquad \lim_{x \to -2} g(x) = -1$$

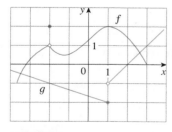

FIGURE 1

Therefore we have

$$\lim_{x \to -2} [f(x) + 5g(x)] = \lim_{x \to -2} f(x) + \lim_{x \to -2} [5g(x)] \qquad \text{(by Limit Law 1)}$$

$$= \lim_{x \to -2} f(x) + 5 \lim_{x \to -2} g(x) \qquad \text{(by Limit Law 3)}$$

$$= 1 + 5(-1) = -4$$

(b) We see that $\lim_{x \to 1} f(x) = 2$. But $\lim_{x \to 1} g(x)$ does not exist because the left and right limits are different:

$$\lim_{x \to 1^-} g(x) = -2 \qquad\qquad \lim_{x \to 1^+} g(x) = -1$$

So we can't use Law 4 for the desired limit. But we *can* use Law 4 for the one-sided limits:

$$\lim_{x \to 1^-} [f(x)g(x)] = \lim_{x \to 1^-} f(x) \cdot \lim_{x \to 1^-} g(x) = 2 \cdot (-2) = -4$$

$$\lim_{x \to 1^+} [f(x)g(x)] = \lim_{x \to 1^+} f(x) \cdot \lim_{x \to 1^+} g(x) = 2 \cdot (-1) = -2$$

The left and right limits aren't equal, so $\lim_{x \to 1} [f(x)g(x)]$ does not exist.

(c) The graphs show that

$$\lim_{x \to 2} f(x) \approx 1.4 \qquad \text{and} \qquad \lim_{x \to 2} g(x) = 0$$

Because the limit of the denominator is 0, we can't use Law 5. The given limit does not exist because the denominator approaches 0 while the numerator approaches a nonzero number. ∎

If we use the Product Law repeatedly with $g(x) = f(x)$, we obtain the following law.

Power Law

> **6.** $\lim_{x \to a} [f(x)]^n = \left[\lim_{x \to a} f(x) \right]^n$ where n is a positive integer

In applying these six limit laws, we need to use two special limits:

> **7.** $\lim_{x \to a} c = c$ **8.** $\lim_{x \to a} x = a$

These limits are obvious from an intuitive point of view (state them in words or draw graphs of $y = c$ and $y = x$), but proofs based on the precise definition are requested in the exercises for Section 2.4.

If we now put $f(x) = x$ in Law 6 and use Law 8, we get another useful special limit.

> **9.** $\lim_{x \to a} x^n = a^n$ where n is a positive integer

A similar limit holds for roots as follows. (For square roots the proof is outlined in Exercise 2.4.37.)

> **10.** $\lim_{x \to a} \sqrt[n]{x} = \sqrt[n]{a}$ where n is a positive integer
>
> (If n is even, we assume that $a > 0$.)

More generally, we have the following law, which is proved in Section 2.5 as a consequence of Law 10.

Root Law

> **11.** $\displaystyle\lim_{x \to a} \sqrt[n]{f(x)} = \sqrt[n]{\lim_{x \to a} f(x)}$ where n is a positive integer
>
> $\left[\text{If } n \text{ is even, we assume that } \displaystyle\lim_{x \to a} f(x) > 0. \right]$

Newton and Limits

Isaac Newton was born on Christmas Day in 1642, the year of Galileo's death. When he entered Cambridge University in 1661 Newton didn't know much mathematics, but he learned quickly by reading Euclid and Descartes and by attending the lectures of Isaac Barrow. Cambridge was closed because of the plague in 1665 and 1666, and Newton returned home to reflect on what he had learned. Those two years were amazingly productive for at that time he made four of his major discoveries: (1) his representation of functions as sums of infinite series, including the binomial theorem; (2) his work on differential and integral calculus; (3) his laws of motion and law of universal gravitation; and (4) his prism experiments on the nature of light and color. Because of a fear of controversy and criticism, he was reluctant to publish his discoveries and it wasn't until 1687, at the urging of the astronomer Halley, that Newton published *Principia Mathematica*. In this work, the greatest scientific treatise ever written, Newton set forth his version of calculus and used it to investigate mechanics, fluid dynamics, and wave motion, and to explain the motion of planets and comets.

The beginnings of calculus are found in the calculations of areas and volumes by ancient Greek scholars such as Eudoxus and Archimedes. Although aspects of the idea of a limit are implicit in their "method of exhaustion," Eudoxus and Archimedes never explicitly formulated the concept of a limit. Likewise, mathematicians such as Cavalieri, Fermat, and Barrow, the immediate precursors of Newton in the development of calculus, did not actually use limits. It was Isaac Newton who was the first to talk explicitly about limits. He explained that the main idea behind limits is that quantities "approach nearer than by any given difference." Newton stated that the limit was the basic concept in calculus, but it was left to later mathematicians like Cauchy to clarify his ideas about limits.

EXAMPLE 2 Evaluate the following limits and justify each step.

(a) $\displaystyle\lim_{x \to 5} (2x^2 - 3x + 4)$ (b) $\displaystyle\lim_{x \to -2} \frac{x^3 + 2x^2 - 1}{5 - 3x}$

SOLUTION

(a) $\displaystyle\lim_{x \to 5} (2x^2 - 3x + 4) = \lim_{x \to 5} (2x^2) - \lim_{x \to 5} (3x) + \lim_{x \to 5} 4$ (by Laws 2 and 1)

$\displaystyle\qquad\qquad = 2 \lim_{x \to 5} x^2 - 3 \lim_{x \to 5} x + \lim_{x \to 5} 4$ (by 3)

$\displaystyle\qquad\qquad = 2(5^2) - 3(5) + 4$ (by 9, 8, and 7)

$\qquad\qquad = 39$

(b) We start by using Law 5, but its use is fully justified only at the final stage when we see that the limits of the numerator and denominator exist and the limit of the denominator is not 0.

$\displaystyle\lim_{x \to -2} \frac{x^3 + 2x^2 - 1}{5 - 3x} = \frac{\displaystyle\lim_{x \to -2} (x^3 + 2x^2 - 1)}{\displaystyle\lim_{x \to -2} (5 - 3x)}$ (by Law 5)

$\displaystyle\qquad = \frac{\displaystyle\lim_{x \to -2} x^3 + 2 \lim_{x \to -2} x^2 - \lim_{x \to -2} 1}{\displaystyle\lim_{x \to -2} 5 - 3 \lim_{x \to -2} x}$ (by 1, 2, and 3)

$\displaystyle\qquad = \frac{(-2)^3 + 2(-2)^2 - 1}{5 - 3(-2)}$ (by 9, 8, and 7)

$\displaystyle\qquad = -\frac{1}{11}$ ■

NOTE If we let $f(x) = 2x^2 - 3x + 4$, then $f(5) = 39$. In other words, we would have gotten the correct answer in Example 2(a) by substituting 5 for x. Similarly, direct substitution provides the correct answer in part (b). The functions in Example 2 are a polynomial and a rational function, respectively, and similar use of the Limit Laws proves that direct substitution always works for such functions (see Exercises 57 and 58). We state this fact as follows.

> **Direct Substitution Property** If f is a polynomial or a rational function and a is in the domain of f, then
>
> $$\lim_{x \to a} f(x) = f(a)$$

Functions with the Direct Substitution Property are called *continuous at a* and will be studied in Section 2.5. However, not all limits can be evaluated by direct substitution, as the following examples show.

EXAMPLE 3 Find $\lim\limits_{x \to 1} \dfrac{x^2 - 1}{x - 1}$.

SOLUTION Let $f(x) = (x^2 - 1)/(x - 1)$. We can't find the limit by substituting $x = 1$ because $f(1)$ isn't defined. Nor can we apply the Quotient Law, because the limit of the denominator is 0. Instead, we need to do some preliminary algebra. We factor the numerator as a difference of squares:

$$\frac{x^2 - 1}{x - 1} = \frac{(x - 1)(x + 1)}{x - 1}$$

Notice that in Example 3 we do not have an infinite limit even though the denominator approaches 0 as $x \to 1$. When both numerator and denominator approach 0, the limit may be infinite or it may be some finite value.

The numerator and denominator have a common factor of $x - 1$. When we take the limit as x approaches 1, we have $x \neq 1$ and so $x - 1 \neq 0$. Therefore we can cancel the common factor and then compute the limit by direct substitution as follows:

$$\lim_{x \to 1} \frac{x^2 - 1}{x - 1} = \lim_{x \to 1} \frac{(x - 1)(x + 1)}{x - 1}$$
$$= \lim_{x \to 1} (x + 1)$$
$$= 1 + 1 = 2$$

The limit in this example arose in Example 2.1.1 when we were trying to find the tangent to the parabola $y = x^2$ at the point $(1, 1)$.   ∎

NOTE In Example 3 we were able to compute the limit by replacing the given function $f(x) = (x^2 - 1)/(x - 1)$ by a simpler function, $g(x) = x + 1$, with the same limit. This is valid because $f(x) = g(x)$ except when $x = 1$, and in computing a limit as x approaches 1 we don't consider what happens when x is actually *equal* to 1. In general, we have the following useful fact.

> If $f(x) = g(x)$ when $x \neq a$, then $\lim\limits_{x \to a} f(x) = \lim\limits_{x \to a} g(x)$, provided the limits exist.

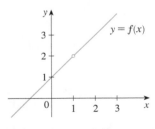

EXAMPLE 4 Find $\lim\limits_{x \to 1} g(x)$ where

$$g(x) = \begin{cases} x + 1 & \text{if } x \neq 1 \\ \pi & \text{if } x = 1 \end{cases}$$

SOLUTION Here g is defined at $x = 1$ and $g(1) = \pi$, but the value of a limit as x *approaches* 1 does not depend on the value of the function *at* 1. Since $g(x) = x + 1$ for $x \neq 1$, we have

$$\lim_{x \to 1} g(x) = \lim_{x \to 1} (x + 1) = 2$$ ∎

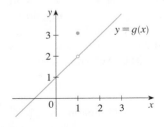

FIGURE 2
The graphs of the functions f (from Example 3) and g (from Example 4)

Note that the values of the functions in Examples 3 and 4 are identical except when $x = 1$ (see Figure 2) and so they have the same limit as x approaches 1.

EXAMPLE 5 Evaluate $\lim\limits_{h \to 0} \dfrac{(3 + h)^2 - 9}{h}$.

SOLUTION If we define

$$F(h) = \frac{(3 + h)^2 - 9}{h}$$

then, as in Example 3, we can't compute $\lim_{h \to 0} F(h)$ by letting $h = 0$ since $F(0)$ is undefined. But if we simplify $F(h)$ algebraically, we find that

$$F(h) = \frac{(9 + 6h + h^2) - 9}{h} = \frac{6h + h^2}{h} = \frac{h(6 + h)}{h} = 6 + h$$

(Recall that we consider only $h \neq 0$ when letting h approach 0.) Thus

$$\lim_{h \to 0} \frac{(3 + h)^2 - 9}{h} = \lim_{h \to 0} (6 + h) = 6$$

■

EXAMPLE 6 Find $\lim\limits_{t \to 0} \dfrac{\sqrt{t^2 + 9} - 3}{t^2}$.

SOLUTION We can't apply the Quotient Law immediately, since the limit of the denominator is 0. Here the preliminary algebra consists of rationalizing the numerator:

$$\lim_{t \to 0} \frac{\sqrt{t^2 + 9} - 3}{t^2} = \lim_{t \to 0} \frac{\sqrt{t^2 + 9} - 3}{t^2} \cdot \frac{\sqrt{t^2 + 9} + 3}{\sqrt{t^2 + 9} + 3}$$

$$= \lim_{t \to 0} \frac{(t^2 + 9) - 9}{t^2 \left(\sqrt{t^2 + 9} + 3 \right)}$$

$$= \lim_{t \to 0} \frac{t^2}{t^2 \left(\sqrt{t^2 + 9} + 3 \right)}$$

$$= \lim_{t \to 0} \frac{1}{\sqrt{t^2 + 9} + 3}$$

$$= \frac{1}{\sqrt{\lim\limits_{t \to 0} (t^2 + 9)} + 3}$$

Here we use several properties of limits (5, 1, 10, 7, 9).

$$= \frac{1}{3 + 3} = \frac{1}{6}$$

This calculation confirms the guess that we made in Example 2.2.2.

■

Some limits are best calculated by first finding the left- and right-hand limits. The following theorem is a reminder of what we discovered in Section 2.2. It says that a two-sided limit exists if and only if both of the one-sided limits exist and are equal.

$$\boxed{\text{1 Theorem} \quad \lim_{x \to a} f(x) = L \qquad \text{if and only if} \qquad \lim_{x \to a^-} f(x) = L = \lim_{x \to a^+} f(x)}$$

When computing one-sided limits, we use the fact that the Limit Laws also hold for one-sided limits.

EXAMPLE 7 Show that $\lim\limits_{x \to 0} |x| = 0$.

SOLUTION Recall that

$$|x| = \begin{cases} x & \text{if } x \geq 0 \\ -x & \text{if } x < 0 \end{cases}$$

Since $|x| = x$ for $x > 0$, we have

$$\lim_{x \to 0^+} |x| = \lim_{x \to 0^+} x = 0$$

For $x < 0$ we have $|x| = -x$ and so

$$\lim_{x \to 0^-} |x| = \lim_{x \to 0^-} (-x) = 0$$

Therefore, by Theorem 1,

$$\lim_{x \to 0} |x| = 0$$

The result of Example 7 looks plausible from Figure 3.

$y = |x|$

FIGURE 3

EXAMPLE 8 Prove that $\lim\limits_{x \to 0} \dfrac{|x|}{x}$ does not exist.

SOLUTION Using the facts that $|x| = x$ when $x > 0$ and $|x| = -x$ when $x < 0$, we have

$$\lim_{x \to 0^+} \frac{|x|}{x} = \lim_{x \to 0^+} \frac{x}{x} = \lim_{x \to 0^+} 1 = 1$$

$$\lim_{x \to 0^-} \frac{|x|}{x} = \lim_{x \to 0^-} \frac{-x}{x} = \lim_{x \to 0^-} (-1) = -1$$

Since the right- and left-hand limits are different, it follows from Theorem 1 that $\lim_{x \to 0} |x|/x$ does not exist. The graph of the function $f(x) = |x|/x$ is shown in Figure 4 and supports the one-sided limits that we found.

$y = \dfrac{|x|}{x}$

FIGURE 4

EXAMPLE 9 If

$$f(x) = \begin{cases} \sqrt{x - 4} & \text{if } x > 4 \\ 8 - 2x & \text{if } x < 4 \end{cases}$$

determine whether $\lim_{x \to 4} f(x)$ exists.

SOLUTION Since $f(x) = \sqrt{x - 4}$ for $x > 4$, we have

$$\lim_{x \to 4^+} f(x) = \lim_{x \to 4^+} \sqrt{x - 4} = \sqrt{4 - 4} = 0$$

Since $f(x) = 8 - 2x$ for $x < 4$, we have

$$\lim_{x \to 4^-} f(x) = \lim_{x \to 4^-} (8 - 2x) = 8 - 2 \cdot 4 = 0$$

The right- and left-hand limits are equal. Thus the limit exists and

$$\lim_{x \to 4} f(x) = 0$$

It is shown in Example 2.4.3 that $\lim_{x \to 0^+} \sqrt{x} = 0$.

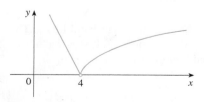

FIGURE 5

The graph of f is shown in Figure 5.

Other notations for $[\![x]\!]$ are $[x]$ and $\lfloor x \rfloor$. The greatest integer function is sometimes called the *floor function*.

FIGURE 6
Greatest integer function

EXAMPLE 10 The **greatest integer function** is defined by $[\![x]\!]$ = the largest integer that is less than or equal to x. (For instance, $[\![4]\!] = 4$, $[\![4.8]\!] = 4$, $[\![\pi]\!] = 3$, $[\![\sqrt{2}]\!] = 1$, $[\![-\frac{1}{2}]\!] = -1$.) Show that $\lim_{x\to 3} [\![x]\!]$ does not exist.

SOLUTION The graph of the greatest integer function is shown in Figure 6. Since $[\![x]\!] = 3$ for $3 \leq x < 4$, we have

$$\lim_{x\to 3^+} [\![x]\!] = \lim_{x\to 3^+} 3 = 3$$

Since $[\![x]\!] = 2$ for $2 \leq x < 3$, we have

$$\lim_{x\to 3^-} [\![x]\!] = \lim_{x\to 3^-} 2 = 2$$

Because these one-sided limits are not equal, $\lim_{x\to 3} [\![x]\!]$ does not exist by Theorem 1. ∎

The next two theorems give two additional properties of limits. Their proofs can be found in Appendix F.

> **2 Theorem** If $f(x) \leq g(x)$ when x is near a (except possibly at a) and the limits of f and g both exist as x approaches a, then
>
> $$\lim_{x\to a} f(x) \leq \lim_{x\to a} g(x)$$

> **3 The Squeeze Theorem** If $f(x) \leq g(x) \leq h(x)$ when x is near a (except possibly at a) and
>
> $$\lim_{x\to a} f(x) = \lim_{x\to a} h(x) = L$$
>
> then
> $$\lim_{x\to a} g(x) = L$$

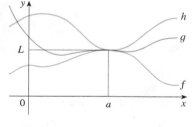

FIGURE 7

The Squeeze Theorem, which is sometimes called the Sandwich Theorem or the Pinching Theorem, is illustrated by Figure 7. It says that if $g(x)$ is squeezed between $f(x)$ and $h(x)$ near a, and if f and h have the same limit L at a, then g is forced to have the same limit L at a.

EXAMPLE 11 Show that $\lim_{x\to 0} x^2 \sin \dfrac{1}{x} = 0$.

SOLUTION First note that we **cannot** use

$$\lim_{x\to 0} x^2 \sin \frac{1}{x} = \lim_{x\to 0} x^2 \cdot \lim_{x\to 0} \sin \frac{1}{x}$$

because $\lim_{x\to 0} \sin(1/x)$ does not exist (see Example 2.2.4).

Instead we apply the Squeeze Theorem, and so we need to find a function f smaller than $g(x) = x^2 \sin(1/x)$ and a function h bigger than g such that both $f(x)$ and $h(x)$ approach 0. To do this we use our knowledge of the sine function. Because the sine of any number lies between -1 and 1, we can write.

$$\boxed{4} \qquad\qquad -1 \le \sin\frac{1}{x} \le 1$$

Any inequality remains true when multiplied by a positive number. We know that $x^2 \ge 0$ for all x and so, multiplying each side of the inequalities in (4) by x^2, we get

$$-x^2 \le x^2 \sin\frac{1}{x} \le x^2$$

as illustrated by Figure 8. We know that

$$\lim_{x\to 0} x^2 = 0 \qquad \text{and} \qquad \lim_{x\to 0} (-x^2) = 0$$

Taking $f(x) = -x^2$, $g(x) = x^2 \sin(1/x)$, and $h(x) = x^2$ in the Squeeze Theorem, we obtain

$$\lim_{x\to 0} x^2 \sin\frac{1}{x} = 0$$ ∎

FIGURE 8
$y = x^2 \sin(1/x)$

2.3   EXERCISES

1. Given that

$$\lim_{x\to 2} f(x) = 4 \qquad \lim_{x\to 2} g(x) = -2 \qquad \lim_{x\to 2} h(x) = 0$$

find the limits that exist. If the limit does not exist, explain why.

(a) $\lim\limits_{x\to 2} [f(x) + 5g(x)]$

(b) $\lim\limits_{x\to 2} [g(x)]^3$

(c) $\lim\limits_{x\to 2} \sqrt{f(x)}$

(d) $\lim\limits_{x\to 2} \dfrac{3f(x)}{g(x)}$

(e) $\lim\limits_{x\to 2} \dfrac{g(x)}{h(x)}$

(f) $\lim\limits_{x\to 2} \dfrac{g(x)h(x)}{f(x)}$

2. The graphs of f and g are given. Use them to evaluate each limit, if it exists. If the limit does not exist, explain why.

(a) $\lim\limits_{x\to 2} [f(x) + g(x)]$

(b) $\lim\limits_{x\to 0} [f(x) - g(x)]$

(c) $\lim\limits_{x\to -1} [f(x)g(x)]$

(d) $\lim\limits_{x\to 3} \dfrac{f(x)}{g(x)}$

(e) $\lim\limits_{x\to 2} [x^2 f(x)]$

(f) $f(-1) + \lim\limits_{x\to -1} g(x)$

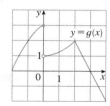

3–9 Evaluate the limit and justify each step by indicating the appropriate Limit Law(s).

3. $\lim\limits_{x\to 3} (5x^3 - 3x^2 + x - 6)$

4. $\lim\limits_{x\to -1} (x^4 - 3x)(x^2 + 5x + 3)$

5. $\lim\limits_{t\to -2} \dfrac{t^4 - 2}{2t^2 - 3t + 2}$

6. $\lim\limits_{u\to -2} \sqrt{u^4 + 3u + 6}$

7. $\lim\limits_{x\to 8} (1 + \sqrt[3]{x})(2 - 6x^2 + x^3)$

8. $\lim\limits_{t\to 2} \left(\dfrac{t^2 - 2}{t^3 - 3t + 5}\right)^2$

9. $\lim\limits_{x\to 2} \sqrt{\dfrac{2x^2 + 1}{3x - 2}}$

---

10. (a) What is wrong with the following equation?

$$\frac{x^2 + x - 6}{x - 2} = x + 3$$

(b) In view of part (a), explain why the equation

$$\lim_{x\to 2} \frac{x^2 + x - 6}{x - 2} = \lim_{x\to 2} (x + 3)$$

is correct.

11–32 Evaluate the limit, if it exists.

11. $\lim\limits_{x\to 5} \dfrac{x^2 - 6x + 5}{x - 5}$

12. $\lim\limits_{x\to -3} \dfrac{x^2 + 3x}{x^2 - x - 12}$

13. $\lim\limits_{x\to 5} \dfrac{x^2 - 5x + 6}{x - 5}$

14. $\lim\limits_{x\to 4} \dfrac{x^2 + 3x}{x^2 - x - 12}$

15. $\lim\limits_{t\to -3} \dfrac{t^2 - 9}{2t^2 + 7t + 3}$

16. $\lim\limits_{x\to -1} \dfrac{2x^2 + 3x + 1}{x^2 - 2x - 3}$

17. $\lim\limits_{h\to 0} \dfrac{(-5 + h)^2 - 25}{h}$

18. $\lim\limits_{h\to 0} \dfrac{(2 + h)^3 - 8}{h}$

19. $\lim\limits_{x \to -2} \dfrac{x+2}{x^3+8}$

20. $\lim\limits_{t \to 1} \dfrac{t^4-1}{t^3-1}$

21. $\lim\limits_{h \to 0} \dfrac{\sqrt{9+h}-3}{h}$

22. $\lim\limits_{u \to 2} \dfrac{\sqrt{4u+1}-3}{u-2}$

23. $\lim\limits_{x \to 3} \dfrac{\frac{1}{x}-\frac{1}{3}}{x-3}$

24. $\lim\limits_{h \to 0} \dfrac{(3+h)^{-1}-3^{-1}}{h}$

25. $\lim\limits_{t \to 0} \dfrac{\sqrt{1+t}-\sqrt{1-t}}{t}$

26. $\lim\limits_{t \to 0} \left(\dfrac{1}{t} - \dfrac{1}{t^2+t} \right)$

27. $\lim\limits_{x \to 16} \dfrac{4-\sqrt{x}}{16x-x^2}$

28. $\lim\limits_{x \to 2} \dfrac{x^2-4x+4}{x^4-3x^2-4}$

29. $\lim\limits_{t \to 0} \left(\dfrac{1}{t\sqrt{1+t}} - \dfrac{1}{t} \right)$

30. $\lim\limits_{x \to -4} \dfrac{\sqrt{x^2+9}-5}{x+4}$

31. $\lim\limits_{h \to 0} \dfrac{(x+h)^3-x^3}{h}$

32. $\lim\limits_{h \to 0} \dfrac{\frac{1}{(x+h)^2} - \frac{1}{x^2}}{h}$

33. (a) Estimate the value of
$$\lim_{x \to 0} \frac{x}{\sqrt{1+3x}-1}$$
by graphing the function $f(x) = x/(\sqrt{1+3x}-1)$.
(b) Make a table of values of $f(x)$ for x close to 0 and guess the value of the limit.
(c) Use the Limit Laws to prove that your guess is correct.

34. (a) Use a graph of
$$f(x) = \frac{\sqrt{3+x}-\sqrt{3}}{x}$$
to estimate the value of $\lim_{x \to 0} f(x)$ to two decimal places.
(b) Use a table of values of $f(x)$ to estimate the limit to four decimal places.
(c) Use the Limit Laws to find the exact value of the limit.

35. Use the Squeeze Theorem to show that $\lim_{x \to 0} (x^2 \cos 20\pi x) = 0$. Illustrate by graphing the functions $f(x) = -x^2$, $g(x) = x^2 \cos 20\pi x$, and $h(x) = x^2$ on the same screen.

36. Use the Squeeze Theorem to show that
$$\lim_{x \to 0} \sqrt{x^3+x^2} \, \sin \frac{\pi}{x} = 0$$
Illustrate by graphing the functions f, g, and h (in the notation of the Squeeze Theorem) on the same screen.

37. If $4x - 9 \leqslant f(x) \leqslant x^2 - 4x + 7$ for $x \geqslant 0$, find $\lim\limits_{x \to 4} f(x)$.

38. If $2x \leqslant g(x) \leqslant x^4 - x^2 + 2$ for all x, evaluate $\lim\limits_{x \to 1} g(x)$.

39. Prove that $\lim\limits_{x \to 0} x^4 \cos \dfrac{2}{x} = 0$.

40. Prove that $\lim\limits_{x \to 0^+} \sqrt{x} \, e^{\sin(\pi/x)} = 0$.

41–46 Find the limit, if it exists. If the limit does not exist, explain why.

41. $\lim\limits_{x \to 3} \left(2x + |x-3| \right)$

42. $\lim\limits_{x \to -6} \dfrac{2x+12}{|x+6|}$

43. $\lim\limits_{x \to 0.5^-} \dfrac{2x-1}{|2x^3-x^2|}$

44. $\lim\limits_{x \to -2} \dfrac{2-|x|}{2+x}$

45. $\lim\limits_{x \to 0^-} \left(\dfrac{1}{x} - \dfrac{1}{|x|} \right)$

46. $\lim\limits_{x \to 0^+} \left(\dfrac{1}{x} - \dfrac{1}{|x|} \right)$

47. The *signum* (or *sign*) *function*, denoted by sgn, is defined by
$$\operatorname{sgn} x = \begin{cases} -1 & \text{if } x < 0 \\ 0 & \text{if } x = 0 \\ 1 & \text{if } x > 0 \end{cases}$$
(a) Sketch the graph of this function.
(b) Find each of the following limits or explain why it does not exist.
   (i) $\lim\limits_{x \to 0^+} \operatorname{sgn} x$ (ii) $\lim\limits_{x \to 0^-} \operatorname{sgn} x$
   (iii) $\lim\limits_{x \to 0} \operatorname{sgn} x$ (iv) $\lim\limits_{x \to 0} |\operatorname{sgn} x|$

48. Let $g(x) = \operatorname{sgn}(\sin x)$.
(a) Find each of the following limits or explain why it does not exist.
   (i) $\lim\limits_{x \to 0^+} g(x)$ (ii) $\lim\limits_{x \to 0^-} g(x)$ (iii) $\lim\limits_{x \to 0} g(x)$
   (iv) $\lim\limits_{x \to \pi^+} g(x)$ (v) $\lim\limits_{x \to \pi^-} g(x)$ (vi) $\lim\limits_{x \to \pi} g(x)$
(b) For which values of a does $\lim_{x \to a} g(x)$ not exist?
(c) Sketch a graph of g.

49. Let $g(x) = \dfrac{x^2+x-6}{|x-2|}$.
(a) Find
   (i) $\lim\limits_{x \to 2^+} g(x)$ (ii) $\lim\limits_{x \to 2^-} g(x)$
(b) Does $\lim_{x \to 2} g(x)$ exist?
(c) Sketch the graph of g.

50. Let
$$f(x) = \begin{cases} x^2+1 & \text{if } x < 1 \\ (x-2)^2 & \text{if } x \geqslant 1 \end{cases}$$
(a) Find $\lim_{x \to 1^-} f(x)$ and $\lim_{x \to 1^+} f(x)$.
(b) Does $\lim_{x \to 1} f(x)$ exist?
(c) Sketch the graph of f.

51. Let
$$B(t) = \begin{cases} 4 - \frac{1}{2}t & \text{if } t < 2 \\ \sqrt{t+c} & \text{if } t \geqslant 2 \end{cases}$$
Find the value of c so that $\lim\limits_{t \to 2} B(t)$ exists.

52. Let

$$g(x) = \begin{cases} x & \text{if } x < 1 \\ 3 & \text{if } x = 1 \\ 2 - x^2 & \text{if } 1 < x \leq 2 \\ x - 3 & \text{if } x > 2 \end{cases}$$

(a) Evaluate each of the following, if it exists.

(i) $\lim_{x \to 1^-} g(x)$ (ii) $\lim_{x \to 1} g(x)$ (iii) $g(1)$

(iv) $\lim_{x \to 2^-} g(x)$ (v) $\lim_{x \to 2^+} g(x)$ (vi) $\lim_{x \to 2} g(x)$

(b) Sketch the graph of g.

53. (a) If the symbol $[\![\;]\!]$ denotes the greatest integer function defined in Example 10, evaluate

(i) $\lim_{x \to -2^+} [\![x]\!]$ (ii) $\lim_{x \to -2} [\![x]\!]$ (iii) $\lim_{x \to -2.4} [\![x]\!]$

(b) If n is an integer, evaluate

(i) $\lim_{x \to n^-} [\![x]\!]$ (ii) $\lim_{x \to n^+} [\![x]\!]$

(c) For what values of a does $\lim_{x \to a} [\![x]\!]$ exist?

54. Let $f(x) = [\![\cos x]\!]$, $-\pi \leq x \leq \pi$.

(a) Sketch the graph of f.

(b) Evaluate each limit, if it exists.

(i) $\lim_{x \to 0} f(x)$ (ii) $\lim_{x \to (\pi/2)^-} f(x)$

(iii) $\lim_{x \to (\pi/2)^+} f(x)$ (iv) $\lim_{x \to \pi/2} f(x)$

(c) For what values of a does $\lim_{x \to a} f(x)$ exist?

55. If $f(x) = [\![x]\!] + [\![-x]\!]$, show that $\lim_{x \to 2} f(x)$ exists but is not equal to $f(2)$.

56. In the theory of relativity, the Lorentz contraction formula

$$L = L_0 \sqrt{1 - v^2/c^2}$$

expresses the length L of an object as a function of its velocity v with respect to an observer, where L_0 is the length of the object at rest and c is the speed of light. Find $\lim_{v \to c^-} L$ and interpret the result. Why is a left-hand limit necessary?

57. If p is a polynomial, show that $\lim_{x \to a} p(x) = p(a)$.

58. If r is a rational function, use Exercise 57 to show that $\lim_{x \to a} r(x) = r(a)$ for every number a in the domain of r.

59. If $\lim_{x \to 1} \dfrac{f(x) - 8}{x - 1} = 10$, find $\lim_{x \to 1} f(x)$.

60. If $\lim_{x \to 0} \dfrac{f(x)}{x^2} = 5$, find the following limits.

(a) $\lim_{x \to 0} f(x)$ (b) $\lim_{x \to 0} \dfrac{f(x)}{x}$

61. If

$$f(x) = \begin{cases} x^2 & \text{if } x \text{ is rational} \\ 0 & \text{if } x \text{ is irrational} \end{cases}$$

prove that $\lim_{x \to 0} f(x) = 0$.

62. Show by means of an example that $\lim_{x \to a} [f(x) + g(x)]$ may exist even though neither $\lim_{x \to a} f(x)$ nor $\lim_{x \to a} g(x)$ exists.

63. Show by means of an example that $\lim_{x \to a} [f(x) g(x)]$ may exist even though neither $\lim_{x \to a} f(x)$ nor $\lim_{x \to a} g(x)$ exists.

64. Evaluate $\lim_{x \to 2} \dfrac{\sqrt{6 - x} - 2}{\sqrt{3 - x} - 1}$.

65. Is there a number a such that

$$\lim_{x \to -2} \dfrac{3x^2 + ax + a + 3}{x^2 + x - 2}$$

exists? If so, find the value of a and the value of the limit.

66. The figure shows a fixed circle C_1 with equation $(x - 1)^2 + y^2 = 1$ and a shrinking circle C_2 with radius r and center the origin. P is the point $(0, r)$, Q is the upper point of intersection of the two circles, and R is the point of intersection of the line PQ and the x-axis. What happens to R as C_2 shrinks, that is, as $r \to 0^+$?

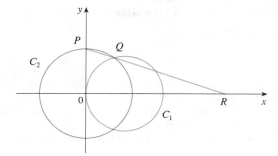

2.4 The Precise Definition of a Limit

The intuitive definition of a limit given in Section 2.2 is inadequate for some purposes because such phrases as "x is close to 2" and "$f(x)$ gets closer and closer to L" are vague. In order to be able to prove conclusively that

$$\lim_{x \to 0} \left(x^3 + \frac{\cos 5x}{10,000} \right) = 0.0001 \qquad \text{or} \qquad \lim_{x \to 0} \frac{\sin x}{x} = 1$$

we must make the definition of a limit precise.

To motivate the precise definition of a limit, let's consider the function

$$f(x) = \begin{cases} 2x - 1 & \text{if } x \neq 3 \\ 6 & \text{if } x = 3 \end{cases}$$

Intuitively, it is clear that when x is close to 3 but $x \neq 3$, then $f(x)$ is close to 5, and so $\lim_{x \to 3} f(x) = 5$.

To obtain more detailed information about how $f(x)$ varies when x is close to 3, we ask the following question:

> How close to 3 does x have to be so that $f(x)$ differs from 5 by less than 0.1?

The distance from x to 3 is $|x - 3|$ and the distance from $f(x)$ to 5 is $|f(x) - 5|$, so our problem is to find a number δ such that

It is traditional to use the Greek letter δ (delta) in this situation.

$$|f(x) - 5| < 0.1 \quad \text{if} \quad |x - 3| < \delta \quad \text{but } x \neq 3$$

If $|x - 3| > 0$, then $x \neq 3$, so an equivalent formulation of our problem is to find a number δ such that

$$|f(x) - 5| < 0.1 \quad \text{if} \quad 0 < |x - 3| < \delta$$

Notice that if $0 < |x - 3| < (0.1)/2 = 0.05$, then

$$|f(x) - 5| = |(2x - 1) - 5| = |2x - 6| = 2|x - 3| < 2(0.05) = 0.1$$

that is, $$|f(x) - 5| < 0.1 \quad \text{if} \quad 0 < |x - 3| < 0.05$$

Thus an answer to the problem is given by $\delta = 0.05$; that is, if x is within a distance of 0.05 from 3, then $f(x)$ will be within a distance of 0.1 from 5.

If we change the number 0.1 in our problem to the smaller number 0.01, then by using the same method we find that $f(x)$ will differ from 5 by less than 0.01 provided that x differs from 3 by less than $(0.01)/2 = 0.005$:

$$|f(x) - 5| < 0.01 \quad \text{if} \quad 0 < |x - 3| < 0.005$$

Similarly,

$$|f(x) - 5| < 0.001 \quad \text{if} \quad 0 < |x - 3| < 0.0005$$

The numbers 0.1, 0.01, and 0.001 that we have considered are *error tolerances* that we might allow. For 5 to be the precise limit of $f(x)$ as x approaches 3, we must not only be able to bring the difference between $f(x)$ and 5 below each of these three numbers; we must be able to bring it below *any* positive number. And, by the same reasoning, we can! If we write ε (the Greek letter epsilon) for an arbitrary positive number, then we find as before that

$$\boxed{1} \qquad |f(x) - 5| < \varepsilon \quad \text{if} \quad 0 < |x - 3| < \delta = \frac{\varepsilon}{2}$$

This is a precise way of saying that $f(x)$ is close to 5 when x is close to 3 because (1) says that we can make the values of $f(x)$ within an arbitrary distance ε from 5 by restricting the values of x to be within a distance $\varepsilon/2$ from 3 (but $x \neq 3$).

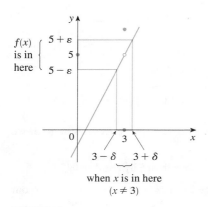

$f(x)$
is in
here $\begin{cases} 5+\varepsilon \\ 5 \\ 5-\varepsilon \end{cases}$

when x is in here
$(x \neq 3)$

FIGURE 1

Note that (1) can be rewritten as follows:

$$\text{if} \quad 3 - \delta < x < 3 + \delta \quad (x \neq 3) \qquad \text{then} \qquad 5 - \varepsilon < f(x) < 5 + \varepsilon$$

and this is illustrated in Figure 1. By taking the values of x ($\neq 3$) to lie in the interval $(3 - \delta, 3 + \delta)$ we can make the values of $f(x)$ lie in the interval $(5 - \varepsilon, 5 + \varepsilon)$.

Using (1) as a model, we give a precise definition of a limit.

> **2** **Precise Definition of a Limit** Let f be a function defined on some open interval that contains the number a, except possibly at a itself. Then we say that the **limit of $f(x)$ as x approaches a is L**, and we write
>
> $$\lim_{x \to a} f(x) = L$$
>
> if for every number $\varepsilon > 0$ there is a number $\delta > 0$ such that
>
> $$\text{if} \quad 0 < |x - a| < \delta \qquad \text{then} \qquad |f(x) - L| < \varepsilon$$

Since $|x - a|$ is the distance from x to a and $|f(x) - L|$ is the distance from $f(x)$ to L, and since ε can be arbitrarily small, the definition of a limit can be expressed in words as follows:

$\lim_{x \to a} f(x) = L$ means that the distance between $f(x)$ and L can be made arbitrarily small by requiring that the distance from x to a be sufficiently small (but not 0).

Alternatively,

$\lim_{x \to a} f(x) = L$ means that the values of $f(x)$ can be made as close as we please to L by requiring x to be close enough to a (but not equal to a).

We can also reformulate Definition 2 in terms of intervals by observing that the inequality $|x - a| < \delta$ is equivalent to $-\delta < x - a < \delta$, which in turn can be written as $a - \delta < x < a + \delta$. Also $0 < |x - a|$ is true if and only if $x - a \neq 0$, that is, $x \neq a$. Similarly, the inequality $|f(x) - L| < \varepsilon$ is equivalent to the pair of inequalities $L - \varepsilon < f(x) < L + \varepsilon$. Therefore, in terms of intervals, Definition 2 can be stated as follows:

$\lim_{x \to a} f(x) = L$ means that for every $\varepsilon > 0$ (no matter how small ε is) we can find $\delta > 0$ such that if x lies in the open interval $(a - \delta, a + \delta)$ and $x \neq a$, then $f(x)$ lies in the open interval $(L - \varepsilon, L + \varepsilon)$.

We interpret this statement geometrically by representing a function by an arrow diagram as in Figure 2, where f maps a subset of $\mathbb{R}$ onto another subset of $\mathbb{R}$.

FIGURE 2

The definition of limit says that if any small interval $(L - \varepsilon, L + \varepsilon)$ is given around L, then we can find an interval $(a - \delta, a + \delta)$ around a such that f maps all the points in $(a - \delta, a + \delta)$ (except possibly a) into the interval $(L - \varepsilon, L + \varepsilon)$. (See Figure 3.)

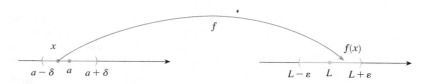

FIGURE 3

Another geometric interpretation of limits can be given in terms of the graph of a function. If $\varepsilon > 0$ is given, then we draw the horizontal lines $y = L + \varepsilon$ and $y = L - \varepsilon$ and the graph of f. (See Figure 4.) If $\lim_{x \to a} f(x) = L$, then we can find a number $\delta > 0$ such that if we restrict x to lie in the interval $(a - \delta, a + \delta)$ and take $x \neq a$, then the curve $y = f(x)$ lies between the lines $y = L - \varepsilon$ and $y = L + \varepsilon$. (See Figure 5.) You can see that if such a δ has been found, then any smaller δ will also work.

It is important to realize that the process illustrated in Figures 4 and 5 must work for *every* positive number ε, no matter how small it is chosen. Figure 6 shows that if a smaller ε is chosen, then a smaller δ may be required.

FIGURE 4

FIGURE 5

FIGURE 6

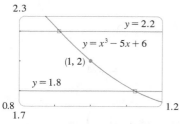

FIGURE 7

EXAMPLE 1 Since $f(x) = x^3 - 5x + 6$ is a polynomial, we know from the Direct Substitution Property that $\lim_{x \to 1} f(x) = f(1) = 1^3 - 5(1) + 6 = 2$. Use a graph to find a number δ such that if x is within δ of 1, then y is within 0.2 of 2, that is,

$$\text{if} \quad |x - 1| < \delta \quad \text{then} \quad |(x^3 - 5x + 6) - 2| < 0.2$$

In other words, find a number δ that corresponds to $\varepsilon = 0.2$ in the definition of a limit for the function $f(x) = x^3 - 5x + 6$ with $a = 1$ and $L = 2$.

SOLUTION A graph of f is shown in Figure 7; we are interested in the region near the point $(1, 2)$. Notice that we can rewrite the inequality

$$|(x^3 - 5x + 6) - 2| < 0.2$$

as

$$-0.2 < (x^3 - 5x + 6) - 2 < 0.2$$

or equivalently

$$1.8 < x^3 - 5x + 6 < 2.2$$

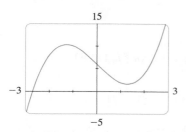

FIGURE 8

So we need to determine the values of x for which the curve $y = x^3 - 5x + 6$ lies between the horizontal lines $y = 1.8$ and $y = 2.2$. Therefore we graph the curves $y = x^3 - 5x + 6$, $y = 1.8$, and $y = 2.2$ near the point $(1, 2)$ in Figure 8. Then we use the cursor to estimate that the x-coordinate of the point of intersection of the line $y = 2.2$ and the curve $y = x^3 - 5x + 6$ is about 0.911. Similarly, $y = x^3 - 5x + 6$ intersects the line $y = 1.8$ when $x \approx 1.124$. So, rounding toward 1 to be safe, we can say that

$$\text{if} \qquad 0.92 < x < 1.12 \qquad \text{then} \qquad 1.8 < x^3 - 5x + 6 < 2.2$$

This interval $(0.92, 1.12)$ is not symmetric about $x = 1$. The distance from $x = 1$ to the left endpoint is $1 - 0.92 = 0.08$ and the distance to the right endpoint is 0.12. We can choose δ to be the smaller of these numbers, that is, $\delta = 0.08$. Then we can rewrite our inequalities in terms of distances as follows:

$$\text{if} \qquad |x - 1| < 0.08 \qquad \text{then} \qquad |(x^3 - 5x + 6) - 2| < 0.2$$

This just says that by keeping x within 0.08 of 1, we are able to keep $f(x)$ within 0.2 of 2.

Although we chose $\delta = 0.08$, any smaller positive value of δ would also have worked. ∎

The graphical procedure in Example 1 gives an illustration of the definition for $\varepsilon = 0.2$, but it does not *prove* that the limit is equal to 2. A proof has to provide a δ for *every* ε.

In proving limit statements it may be helpful to think of the definition of limit as a challenge. First it challenges you with a number ε. Then you must be able to produce a suitable δ. You have to be able to do this for *every* $\varepsilon > 0$, not just a particular ε.

TEC In Module 2.4/2.6 you can explore the precise definition of a limit both graphically and numerically.

Imagine a contest between two people, A and B, and imagine yourself to be B. Person A stipulates that the fixed number L should be approximated by the values of $f(x)$ to within a degree of accuracy ε (say, 0.01). Person B then responds by finding a number δ such that if $0 < |x - a| < \delta$, then $|f(x) - L| < \varepsilon$. Then A may become more exacting and challenge B with a smaller value of ε (say, 0.0001). Again B has to respond by finding a corresponding δ. Usually the smaller the value of ε, the smaller the corresponding value of δ must be. If B always wins, no matter how small A makes ε, then $\lim_{x \to a} f(x) = L$.

EXAMPLE 2 Prove that $\lim_{x \to 3} (4x - 5) = 7$.

SOLUTION

1. *Preliminary analysis of the problem (guessing a value for δ).* Let ε be a given positive number. We want to find a number δ such that

$$\text{if} \qquad 0 < |x - 3| < \delta \qquad \text{then} \qquad |(4x - 5) - 7| < \varepsilon$$

But $|(4x - 5) - 7| = |4x - 12| = |4(x - 3)| = 4|x - 3|$. Therefore we want δ such that

$$\text{if} \qquad 0 < |x - 3| < \delta \qquad \text{then} \qquad 4|x - 3| < \varepsilon$$

that is, $\text{if} \qquad 0 < |x - 3| < \delta \qquad \text{then} \qquad |x - 3| < \dfrac{\varepsilon}{4}$

This suggests that we should choose $\delta = \varepsilon/4$.

2. *Proof (showing that this δ works).* Given $\varepsilon > 0$, choose $\delta = \varepsilon/4$. If $0 < |x - 3| < \delta$, then

$$|(4x - 5) - 7| = |4x - 12| = 4|x - 3| < 4\delta = 4\left(\frac{\varepsilon}{4}\right) = \varepsilon$$

Thus

$$\text{if} \quad 0 < |x - 3| < \delta \quad \text{then} \quad |(4x - 5) - 7| < \varepsilon$$

Therefore, by the definition of a limit,

$$\lim_{x \to 3} (4x - 5) = 7$$

This example is illustrated by Figure 9.

FIGURE 9

Cauchy and Limits

After the invention of calculus in the 17th century, there followed a period of free development of the subject in the 18th century. Mathematicians like the Bernoulli brothers and Euler were eager to exploit the power of calculus and boldly explored the consequences of this new and wonderful mathematical theory without worrying too much about whether their proofs were completely correct.

The 19th century, by contrast, was the Age of Rigor in mathematics. There was a movement to go back to the foundations of the subject—to provide careful definitions and rigorous proofs. At the forefront of this movement was the French mathematician Augustin-Louis Cauchy (1789–1857), who started out as a military engineer before becoming a mathematics professor in Paris. Cauchy took Newton's idea of a limit, which was kept alive in the 18th century by the French mathematician Jean d'Alembert, and made it more precise. His definition of a limit reads as follows: "When the successive values attributed to a variable approach indefinitely a fixed value so as to end by differing from it by as little as one wishes, this last is called the *limit* of all the others." But when Cauchy used this definition in examples and proofs, he often employed delta-epsilon inequalities similar to the ones in this section. A typical Cauchy proof starts with: "Designate by δ and ε two very small numbers;..." He used ε because of the correspondence between epsilon and the French word *erreur* and δ because delta corresponds to *différence*. Later, the German mathematician Karl Weierstrass (1815–1897) stated the definition of a limit exactly as in our Definition 2.

Note that in the solution of Example 2 there were two stages—guessing and proving. We made a preliminary analysis that enabled us to guess a value for δ. But then in the second stage we had to go back and prove in a careful, logical fashion that we had made a correct guess. This procedure is typical of much of mathematics. Sometimes it is necessary to first make an intelligent guess about the answer to a problem and then prove that the guess is correct.

The intuitive definitions of one-sided limits that were given in Section 2.2 can be precisely reformulated as follows.

---

3 **Definition of Left-Hand Limit**

$$\lim_{x \to a^-} f(x) = L$$

if for every number $\varepsilon > 0$ there is a number $\delta > 0$ such that

$$\text{if} \quad a - \delta < x < a \quad \text{then} \quad |f(x) - L| < \varepsilon$$

---

4 **Definition of Right-Hand Limit**

$$\lim_{x \to a^+} f(x) = L$$

if for every number $\varepsilon > 0$ there is a number $\delta > 0$ such that

$$\text{if} \quad a < x < a + \delta \quad \text{then} \quad |f(x) - L| < \varepsilon$$

Notice that Definition 3 is the same as Definition 2 except that x is restricted to lie in the *left* half $(a - \delta, a)$ of the interval $(a - \delta, a + \delta)$. In Definition 4, x is restricted to lie in the *right* half $(a, a + \delta)$ of the interval $(a - \delta, a + \delta)$.

EXAMPLE 3 Use Definition 4 to prove that $\lim\limits_{x \to 0^+} \sqrt{x} = 0$.

SOLUTION

  1. *Guessing a value for* δ. Let ε be a given positive number. Here $a = 0$ and $L = 0$, so we want to find a number δ such that

$$\text{if} \quad 0 < x < \delta \quad \text{then} \quad |\sqrt{x} - 0| < \varepsilon$$

that is, $\quad$ if $\quad 0 < x < \delta \quad$ then $\quad \sqrt{x} < \varepsilon$

or, squaring both sides of the inequality $\sqrt{x} < \varepsilon$, we get

$$\text{if} \quad 0 < x < \delta \quad \text{then} \quad x < \varepsilon^2$$

This suggests that we should choose $\delta = \varepsilon^2$.

  2. *Showing that this* δ *works.* Given $\varepsilon > 0$, let $\delta = \varepsilon^2$. If $0 < x < \delta$, then

$$\sqrt{x} < \sqrt{\delta} = \sqrt{\varepsilon^2} = \varepsilon$$

so $\quad$ $|\sqrt{x} - 0| < \varepsilon$

According to Definition 4, this shows that $\lim_{x \to 0^+} \sqrt{x} = 0$. ∎

EXAMPLE 4 Prove that $\lim\limits_{x \to 3} x^2 = 9$.

SOLUTION

  1. *Guessing a value for* δ. Let $\varepsilon > 0$ be given. We have to find a number $\delta > 0$ such that

$$\text{if} \quad 0 < |x - 3| < \delta \quad \text{then} \quad |x^2 - 9| < \varepsilon$$

To connect $|x^2 - 9|$ with $|x - 3|$ we write $|x^2 - 9| = |(x + 3)(x - 3)|$. Then we want

$$\text{if} \quad 0 < |x - 3| < \delta \quad \text{then} \quad |x + 3||x - 3| < \varepsilon$$

Notice that if we can find a positive constant C such that $|x + 3| < C$, then

$$|x + 3||x - 3| < C|x - 3|$$

and we can make $C|x - 3| < \varepsilon$ by taking $|x - 3| < \varepsilon/C$, so we could choose $\delta = \varepsilon/C$.

  We can find such a number C if we restrict x to lie in some interval centered at 3. In fact, since we are interested only in values of x that are close to 3, it is reasonable to assume that x is within a distance 1 from 3, that is, $|x - 3| < 1$. Then $2 < x < 4$, so $5 < x + 3 < 7$. Thus we have $|x + 3| < 7$, and so $C = 7$ is a suitable choice for the constant.

  But now there are two restrictions on $|x - 3|$, namely

$$|x - 3| < 1 \quad \text{and} \quad |x - 3| < \frac{\varepsilon}{C} = \frac{\varepsilon}{7}$$

To make sure that both of these inequalities are satisfied, we take δ to be the smaller of the two numbers 1 and $\varepsilon/7$. The notation for this is $\delta = \min\{1, \varepsilon/7\}$.

2. *Showing that this δ works.* Given $\varepsilon > 0$, let $\delta = \min\{1, \varepsilon/7\}$. If $0 < |x - 3| < \delta$, then $|x - 3| < 1 \Rightarrow 2 < x < 4 \Rightarrow |x + 3| < 7$ (as in part l). We also have $|x - 3| < \varepsilon/7$, so

$$|x^2 - 9| = |x + 3||x - 3| < 7 \cdot \frac{\varepsilon}{7} = \varepsilon$$

This shows that $\lim_{x \to 3} x^2 = 9$. ∎

As Example 4 shows, it is not always easy to prove that limit statements are true using the ε, δ definition. In fact, if we had been given a more complicated function such as $f(x) = (6x^2 - 8x + 9)/(2x^2 - 1)$, a proof would require a great deal of ingenuity. Fortunately this is unnecessary because the Limit Laws stated in Section 2.3 can be proved using Definition 2, and then the limits of complicated functions can be found rigorously from the Limit Laws without resorting to the definition directly.

For instance, we prove the Sum Law: If $\lim_{x \to a} f(x) = L$ and $\lim_{x \to a} g(x) = M$ both exist, then

$$\lim_{x \to a} [f(x) + g(x)] = L + M$$

The remaining laws are proved in the exercises and in Appendix F.

PROOF OF THE SUM LAW Let $\varepsilon > 0$ be given. We must find $\delta > 0$ such that

$$\text{if} \quad 0 < |x - a| < \delta \quad \text{then} \quad |f(x) + g(x) - (L + M)| < \varepsilon$$

Triangle Inequality:

$$|a + b| \leq |a| + |b|$$

(See Appendix A).

Using the Triangle Inequality we can write

$$\boxed{5} \qquad |f(x) + g(x) - (L + M)| = |(f(x) - L) + (g(x) - M)|$$

$$\leq |f(x) - L| + |g(x) - M|$$

We make $|f(x) + g(x) - (L + M)|$ less than ε by making each of the terms $|f(x) - L|$ and $|g(x) - M|$ less than $\varepsilon/2$.

Since $\varepsilon/2 > 0$ and $\lim_{x \to a} f(x) = L$, there exists a number $\delta_1 > 0$ such that

$$\text{if} \quad 0 < |x - a| < \delta_1 \quad \text{then} \quad |f(x) - L| < \frac{\varepsilon}{2}$$

Similarly, since $\lim_{x \to a} g(x) = M$, there exists a number $\delta_2 > 0$ such that

$$\text{if} \quad 0 < |x - a| < \delta_2 \quad \text{then} \quad |g(x) - M| < \frac{\varepsilon}{2}$$

Let $\delta = \min\{\delta_1, \delta_2\}$, the smaller of the numbers δ_1 and δ_2. Notice that

$$\text{if} \quad 0 < |x - a| < \delta \quad \text{then} \quad 0 < |x - a| < \delta_1 \quad \text{and} \quad 0 < |x - a| < \delta_2$$

and so

$$|f(x) - L| < \frac{\varepsilon}{2} \quad \text{and} \quad |g(x) - M| < \frac{\varepsilon}{2}$$

Therefore, by (5),

$$|f(x) + g(x) - (L + M)| \leq |f(x) - L| + |g(x) - M|$$

$$< \frac{\varepsilon}{2} + \frac{\varepsilon}{2} = \varepsilon$$

To summarize,

$$\text{if} \qquad 0 < |x - a| < \delta \qquad \text{then} \qquad |f(x) + g(x) - (L + M)| < \varepsilon$$

Thus, by the definition of a limit,

$$\lim_{x \to a} [f(x) + g(x)] = L + M \qquad \blacksquare$$

■ Infinite Limits

Infinite limits can also be defined in a precise way. The following is a precise version of Definition 2.2.4.

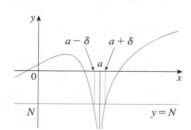

FIGURE 10

> **6** **Precise Definition of an Infinite Limit** Let f be a function defined on some open interval that contains the number a, except possibly at a itself. Then
>
> $$\lim_{x \to a} f(x) = \infty$$
>
> means that for every positive number M there is a positive number δ such that
>
> $$\text{if} \qquad 0 < |x - a| < \delta \qquad \text{then} \qquad f(x) > M$$

This says that the values of $f(x)$ can be made arbitrarily large (larger than any given number M) by requiring x to be close enough to a (within a distance δ, where δ depends on M, but with $x \neq a$). A geometric illustration is shown in Figure 10.

Given any horizontal line $y = M$, we can find a number $\delta > 0$ such that if we restrict x to lie in the interval $(a - \delta, a + \delta)$ but $x \neq a$, then the curve $y = f(x)$ lies above the line $y = M$. You can see that if a larger M is chosen, then a smaller δ may be required.

EXAMPLE 5 Use Definition 6 to prove that $\lim\limits_{x \to 0} \dfrac{1}{x^2} = \infty$.

SOLUTION Let M be a given positive number. We want to find a number δ such that

$$\text{if} \qquad 0 < |x| < \delta \qquad \text{then} \qquad 1/x^2 > M$$

But

$$\frac{1}{x^2} > M \quad \Longleftrightarrow \quad x^2 < \frac{1}{M} \quad \Longleftrightarrow \quad \sqrt{x^2} < \sqrt{\frac{1}{M}} \quad \Longleftrightarrow \quad |x| < \frac{1}{\sqrt{M}}$$

So if we choose $\delta = 1/\sqrt{M}$ and $0 < |x| < \delta = 1/\sqrt{M}$, then $1/x^2 > M$. This shows that $1/x^2 \to \infty$ as $x \to 0$. $\qquad \blacksquare$

Similarly, the following is a precise version of Definition 2.2.5. It is illustrated by Figure 11.

FIGURE 11

> **7** **Definition** Let f be a function defined on some open interval that contains the number a, except possibly at a itself. Then
>
> $$\lim_{x \to a} f(x) = -\infty$$
>
> means that for every negative number N there is a positive number δ such that
>
> $$\text{if} \qquad 0 < |x - a| < \delta \qquad \text{then} \qquad f(x) < N$$

2.4 EXERCISES

1. Use the given graph of f to find a number δ such that

if $\quad |x - 1| < \delta \quad$ then $\quad |f(x) - 1| < 0.2$

2. Use the given graph of f to find a number δ such that

if $\quad 0 < |x - 3| < \delta \quad$ then $\quad |f(x) - 2| < 0.5$

3. Use the given graph of $f(x) = \sqrt{x}$ to find a number δ such that

if $\quad |x - 4| < \delta \quad$ then $\quad |\sqrt{x} - 2| < 0.4$

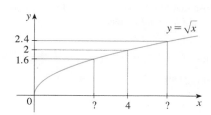

4. Use the given graph of $f(x) = x^2$ to find a number δ such that

if $\quad |x - 1| < \delta \quad$ then $\quad |x^2 - 1| < \frac{1}{2}$

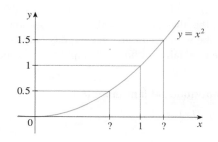

5. Use a graph to find a number δ such that

if $\quad \left| x - \dfrac{\pi}{4} \right| < \delta \quad$ then $\quad |\tan x - 1| < 0.2$

6. Use a graph to find a number δ such that

if $\quad |x - 1| < \delta \quad$ then $\quad \left| \dfrac{2x}{x^2 + 4} - 0.4 \right| < 0.1$

7. For the limit

$$\lim_{x \to 2} (x^3 - 3x + 4) = 6$$

illustrate Definition 2 by finding values of δ that correspond to $\varepsilon = 0.2$ and $\varepsilon = 0.1$.

8. For the limit

$$\lim_{x \to 0} \frac{e^{2x} - 1}{x} = 2$$

illustrate Definition 2 by finding values of δ that correspond to $\varepsilon = 0.5$ and $\varepsilon = 0.1$.

9. (a) Use a graph to find a number δ such that

if $\quad 2 < x < 2 + \delta \quad$ then $\quad \dfrac{1}{\ln(x - 1)} > 100$

(b) What limit does part (a) suggest is true?

10. Given that $\lim_{x \to \pi} \csc^2 x = \infty$, illustrate Definition 6 by finding values of δ that correspond to (a) $M = 500$ and (b) $M = 1000$.

11. A machinist is required to manufacture a circular metal disk with area 1000 cm².
(a) What radius produces such a disk?
(b) If the machinist is allowed an error tolerance of ± 5 cm² in the area of the disk, how close to the ideal radius in part (a) must the machinist control the radius?
(c) In terms of the ε, δ definition of $\lim_{x \to a} f(x) = L$, what is x? What is $f(x)$? What is a? What is L? What value of ε is given? What is the corresponding value of δ?

12. A crystal growth furnace is used in research to determine how best to manufacture crystals used in electronic components for the space shuttle. For proper growth of the crystal, the temperature must be controlled accurately by adjusting the input power. Suppose the relationship is given by

$$T(w) = 0.1w^2 + 2.155w + 20$$

where T is the temperature in degrees Celsius and w is the power input in watts.
(a) How much power is needed to maintain the temperature at 200°C?

(b) If the temperature is allowed to vary from 200°C by up to ±1°C, what range of wattage is allowed for the input power?

(c) In terms of the ε, δ definition of $\lim_{x \to a} f(x) = L$, what is x? What is $f(x)$? What is a? What is L? What value of ε is given? What is the corresponding value of δ?

13. (a) Find a number δ such that if $|x - 2| < \delta$, then $|4x - 8| < \varepsilon$, where $\varepsilon = 0.1$.
(b) Repeat part (a) with $\varepsilon = 0.01$.

14. Given that $\lim_{x \to 2}(5x - 7) = 3$, illustrate Definition 2 by finding values of δ that correspond to $\varepsilon = 0.1$, $\varepsilon = 0.05$, and $\varepsilon = 0.01$.

15–18 Prove the statement using the ε, δ definition of a limit and illustrate with a diagram like Figure 9.

15. $\lim_{x \to 3} \left(1 + \frac{1}{3}x\right) = 2$

16. $\lim_{x \to 4} (2x - 5) = 3$

17. $\lim_{x \to -3} (1 - 4x) = 13$

18. $\lim_{x \to -2} (3x + 5) = -1$

19–32 Prove the statement using the ε, δ definition of a limit.

19. $\lim_{x \to 1} \frac{2 + 4x}{3} = 2$

20. $\lim_{x \to 10} \left(3 - \frac{4}{5}x\right) = -5$

21. $\lim_{x \to 4} \frac{x^2 - 2x - 8}{x - 4} = 6$

22. $\lim_{x \to -1.5} \frac{9 - 4x^2}{3 + 2x} = 6$

23. $\lim_{x \to a} x = a$

24. $\lim_{x \to a} c = c$

25. $\lim_{x \to 0} x^2 = 0$

26. $\lim_{x \to 0} x^3 = 0$

27. $\lim_{x \to 0} |x| = 0$

28. $\lim_{x \to -6^+} \sqrt[8]{6 + x} = 0$

29. $\lim_{x \to 2} (x^2 - 4x + 5) = 1$

30. $\lim_{x \to 2} (x^2 + 2x - 7) = 1$

31. $\lim_{x \to -2} (x^2 - 1) = 3$

32. $\lim_{x \to 2} x^3 = 8$

33. Verify that another possible choice of δ for showing that $\lim_{x \to 3} x^2 = 9$ in Example 4 is $\delta = \min\{2, \varepsilon/8\}$.

34. Verify, by a geometric argument, that the largest possible choice of δ for showing that $\lim_{x \to 3} x^2 = 9$ is $\delta = \sqrt{9 + \varepsilon} - 3$.

35. (a) For the limit $\lim_{x \to 1} (x^3 + x + 1) = 3$, use a graph to find a value of δ that corresponds to $\varepsilon = 0.4$.
(b) By using a computer algebra system to solve the cubic equation $x^3 + x + 1 = 3 + \varepsilon$, find the largest possible value of δ that works for any given $\varepsilon > 0$.
(c) Put $\varepsilon = 0.4$ in your answer to part (b) and compare with your answer to part (a).

36. Prove that $\lim_{x \to 2} \frac{1}{x} = \frac{1}{2}$.

37. Prove that $\lim_{x \to a} \sqrt{x} = \sqrt{a}$ if $a > 0$.

$$\left[\textit{Hint: Use } |\sqrt{x} - \sqrt{a}| = \frac{|x - a|}{\sqrt{x} + \sqrt{a}}. \right]$$

38. If H is the Heaviside function defined in Example 2.2.6, prove, using Definition 2, that $\lim_{t \to 0} H(t)$ does not exist. [*Hint:* Use an indirect proof as follows. Suppose that the limit is L. Take $\varepsilon = \frac{1}{2}$ in the definition of a limit and try to arrive at a contradiction.]

39. If the function f is defined by

$$f(x) = \begin{cases} 0 & \text{if } x \text{ is rational} \\ 1 & \text{if } x \text{ is irrational} \end{cases}$$

prove that $\lim_{x \to 0} f(x)$ does not exist.

40. By comparing Definitions 2, 3, and 4, prove Theorem 2.3.1.

41. How close to -3 do we have to take x so that

$$\frac{1}{(x + 3)^4} > 10,000$$

42. Prove, using Definition 6, that $\lim_{x \to -3} \frac{1}{(x + 3)^4} = \infty$.

43. Prove that $\lim_{x \to 0^+} \ln x = -\infty$.

44. Suppose that $\lim_{x \to a} f(x) = \infty$ and $\lim_{x \to a} g(x) = c$, where c is a real number. Prove each statement.
(a) $\lim_{x \to a} [f(x) + g(x)] = \infty$
(b) $\lim_{x \to a} [f(x) g(x)] = \infty$ if $c > 0$
(c) $\lim_{x \to a} [f(x) g(x)] = -\infty$ if $c < 0$

2.5 Continuity

We noticed in Section 2.3 that the limit of a function as x approaches a can often be found simply by calculating the value of the function at a. Functions with this property are called *continuous at a*. We will see that the mathematical definition of continuity corresponds closely with the meaning of the word *continuity* in everyday language. (A continuous process is one that takes place gradually, without interruption or abrupt change.)

1 **Definition** A function f is **continuous at a number a** if

$$\lim_{x \to a} f(x) = f(a)$$

As illustrated in Figure 1, if f is continuous, then the points $(x, f(x))$ on the graph of f approach the point $(a, f(a))$ on the graph. So there is no gap in the curve.

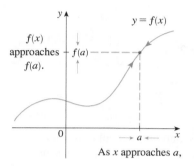

$f(x)$ approaches $f(a)$.

As x approaches a,

FIGURE 1

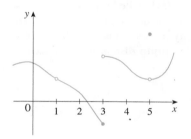

FIGURE 2

Notice that Definition 1 implicitly requires three things if f is continuous at a:

1. $f(a)$ is defined (that is, a is in the domain of f)

2. $\lim\limits_{x \to a} f(x)$ exists

3. $\lim\limits_{x \to a} f(x) = f(a)$

The definition says that f is continuous at a if $f(x)$ approaches $f(a)$ as x approaches a. Thus a continuous function f has the property that a small change in x produces only a small change in $f(x)$. In fact, the change in $f(x)$ can be kept as small as we please by keeping the change in x sufficiently small.

If f is defined near a (in other words, f is defined on an open interval containing a, except perhaps at a), we say that f is **discontinuous at a** (or f has a **discontinuity** at a) if f is not continuous at a.

Physical phenomena are usually continuous. For instance, the displacement or velocity of a vehicle varies continuously with time, as does a person's height. But discontinuities do occur in such situations as electric currents. [See Example 2.2.6, where the Heaviside function is discontinuous at 0 because $\lim_{t \to 0} H(t)$ does not exist.]

Geometrically, you can think of a function that is continuous at every number in an interval as a function whose graph has no break in it: the graph can be drawn without removing your pen from the paper.

EXAMPLE 1 Figure 2 shows the graph of a function f. At which numbers is f discontinuous? Why?

SOLUTION It looks as if there is a discontinuity when $a = 1$ because the graph has a break there. The official reason that f is discontinuous at 1 is that $f(1)$ is not defined.

The graph also has a break when $a = 3$, but the reason for the discontinuity is different. Here, $f(3)$ is defined, but $\lim_{x \to 3} f(x)$ does not exist (because the left and right limits are different). So f is discontinuous at 3.

What about $a = 5$? Here, $f(5)$ is defined and $\lim_{x \to 5} f(x)$ exists (because the left and right limits are the same). But

$$\lim_{x \to 5} f(x) \neq f(5)$$

So f is discontinuous at 5. ∎

Now let's see how to detect discontinuities when a function is defined by a formula.

EXAMPLE 2 Where are each of the following functions discontinuous?

(a) $f(x) = \dfrac{x^2 - x - 2}{x - 2}$

(b) $f(x) = \begin{cases} \dfrac{1}{x^2} & \text{if } x \neq 0 \\ 1 & \text{if } x = 0 \end{cases}$

(c) $f(x) = \begin{cases} \dfrac{x^2 - x - 2}{x - 2} & \text{if } x \neq 2 \\ 1 & \text{if } x = 2 \end{cases}$

(d) $f(x) = [\![x]\!]$

SOLUTION
(a) Notice that $f(2)$ is not defined, so f is discontinuous at 2. Later we'll see why f is continuous at all other numbers.

(b) Here $f(0) = 1$ is defined but

$$\lim_{x \to 0} f(x) = \lim_{x \to 0} \frac{1}{x^2}$$

does not exist. (See Example 2.2.8.) So f is discontinuous at 0.

(c) Here $f(2) = 1$ is defined and

$$\lim_{x \to 2} f(x) = \lim_{x \to 2} \frac{x^2 - x - 2}{x - 2} = \lim_{x \to 2} \frac{(x - 2)(x + 1)}{x - 2} = \lim_{x \to 2} (x + 1) = 3$$

exists. But

$$\lim_{x \to 2} f(x) \ne f(2)$$

so f is not continuous at 2.

(d) The greatest integer function $f(x) = [\![x]\!]$ has discontinuities at all of the integers because $\lim_{x \to n} [\![x]\!]$ does not exist if n is an integer. (See Example 2.3.10 and Exercise 2.3.53.)   ∎

   Figure 3 shows the graphs of the functions in Example 2. In each case the graph can't be drawn without lifting the pen from the paper because a hole or break or jump occurs in the graph. The kind of discontinuity illustrated in parts (a) and (c) is called **removable** because we could remove the discontinuity by redefining f at just the single number 2. [The function $g(x) = x + 1$ is continuous.] The discontinuity in part (b) is called an **infinite discontinuity**. The discontinuities in part (d) are called **jump discontinuities** because the function "jumps" from one value to another.

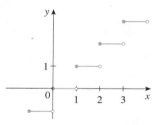

(a) $f(x) = \dfrac{x^2 - x - 2}{x - 2}$

(b) $f(x) = \begin{cases} \dfrac{1}{x^2} & \text{if } x \ne 0 \\ 1 & \text{if } x = 0 \end{cases}$

(c) $f(x) = \begin{cases} \dfrac{x^2 - x - 2}{x - 2} & \text{if } x \ne 2 \\ 1 & \text{if } x = 2 \end{cases}$

(d) $f(x) = [\![x]\!]$

FIGURE 3
Graphs of the functions in Example 2

---

2   **Definition**   A function f is **continuous from the right at a number** a if

$$\lim_{x \to a^+} f(x) = f(a)$$

and f is **continuous from the left at** a if

$$\lim_{x \to a^-} f(x) = f(a)$$

EXAMPLE 3   At each integer n, the function $f(x) = [\![x]\!]$ [see Figure 3(d)] is continuous from the right but discontinuous from the left because

$$\lim_{x \to n^+} f(x) = \lim_{x \to n^+} [\![x]\!] = n = f(n)$$

but
$$\lim_{x \to n^-} f(x) = \lim_{x \to n^-} [\![x]\!] = n - 1 \neq f(n)$$
■

---

3 **Definition** A function f is **continuous on an interval** if it is continuous at every number in the interval. (If f is defined only on one side of an endpoint of the interval, we understand *continuous* at the endpoint to mean *continuous from the right* or *continuous from the left*.)

EXAMPLE 4 Show that the function $f(x) = 1 - \sqrt{1 - x^2}$ is continuous on the interval $[-1, 1]$.

SOLUTION If $-1 < a < 1$, then using the Limit Laws, we have

$$\lim_{x \to a} f(x) = \lim_{x \to a} \left(1 - \sqrt{1 - x^2}\right)$$

$$= 1 - \lim_{x \to a} \sqrt{1 - x^2} \qquad \text{(by Laws 2 and 7)}$$

$$= 1 - \sqrt{\lim_{x \to a} (1 - x^2)} \qquad \text{(by 11)}$$

$$= 1 - \sqrt{1 - a^2} \qquad \text{(by 2, 7, and 9)}$$

$$= f(a)$$

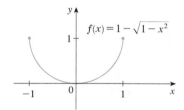

$f(x) = 1 - \sqrt{1 - x^2}$

FIGURE 4

Thus, by Definition 1, f is continuous at a if $-1 < a < 1$. Similar calculations show that

$$\lim_{x \to -1^+} f(x) = 1 = f(-1) \qquad \text{and} \qquad \lim_{x \to 1^-} f(x) = 1 = f(1)$$

so f is continuous from the right at -1 and continuous from the left at 1. Therefore, according to Definition 3, f is continuous on $[-1, 1]$.

The graph of f is sketched in Figure 4. It is the lower half of the circle

$$x^2 + (y - 1)^2 = 1$$
■

Instead of always using Definitions 1, 2, and 3 to verify the continuity of a function as we did in Example 4, it is often convenient to use the next theorem, which shows how to build up complicated continuous functions from simple ones.

---

4 **Theorem** If f and g are continuous at a and c is a constant, then the following functions are also continuous at a:

1. $f + g$ 2. $f - g$ 3. cf

4. fg 5. $\dfrac{f}{g}$ if $g(a) \neq 0$

PROOF Each of the five parts of this theorem follows from the corresponding Limit Law in Section 2.3. For instance, we give the proof of part 1. Since f and g are continuous at a, we have

$$\lim_{x \to a} f(x) = f(a) \qquad \text{and} \qquad \lim_{x \to a} g(x) = g(a)$$

Therefore

$$\lim_{x \to a} (f + g)(x) = \lim_{x \to a} [f(x) + g(x)]$$

$$= \lim_{x \to a} f(x) + \lim_{x \to a} g(x) \qquad \text{(by Law 1)}$$

$$= f(a) + g(a)$$

$$= (f + g)(a)$$

This shows that $f + g$ is continuous at a.   ■

It follows from Theorem 4 and Definition 3 that if f and g are continuous on an interval, then so are the functions $f + g$, $f - g$, cf, fg, and (if g is never 0) f/g. The following theorem was stated in Section 2.3 as the Direct Substitution Property.

---

5 **Theorem**

(a) Any polynomial is continuous everywhere; that is, it is continuous on $\mathbb{R} = (-\infty, \infty)$.

(b) Any rational function is continuous wherever it is defined; that is, it is continuous on its domain.

PROOF

(a) A polynomial is a function of the form

$$P(x) = c_n x^n + c_{n-1} x^{n-1} + \cdots + c_1 x + c_0$$

where $c_0, c_1, \ldots, c_n$ are constants. We know that

$$\lim_{x \to a} c_0 = c_0 \qquad \text{(by Law 7)}$$

and

$$\lim_{x \to a} x^m = a^m \qquad m = 1, 2, \ldots, n \qquad \text{(by 9)}$$

This equation is precisely the statement that the function $f(x) = x^m$ is a continuous function. Thus, by part 3 of Theorem 4, the function $g(x) = cx^m$ is continuous. Since P is a sum of functions of this form and a constant function, it follows from part 1 of Theorem 4 that P is continuous.

(b) A rational function is a function of the form

$$f(x) = \frac{P(x)}{Q(x)}$$

where P and Q are polynomials. The domain of f is $D = \{x \in \mathbb{R} \mid Q(x) \neq 0\}$. We know from part (a) that P and Q are continuous everywhere. Thus, by part 5 of Theorem 4, f is continuous at every number in D.   ■

As an illustration of Theorem 5, observe that the volume of a sphere varies continuously with its radius because the formula $V(r) = \frac{4}{3}\pi r^3$ shows that V is a polynomial function of r. Likewise, if a ball is thrown vertically into the air with a velocity of 50 ft/s, then the height of the ball in feet t seconds later is given by the formula $h = 50t - 16t^2$. Again this is a polynomial function, so the height is a continuous function of the elapsed time, as we might expect.

Knowledge of which functions are continuous enables us to evaluate some limits very quickly, as the following example shows. Compare it with Example 2.3.2(b).

EXAMPLE 5 Find $\displaystyle\lim_{x \to -2} \frac{x^3 + 2x^2 - 1}{5 - 3x}$.

SOLUTION The function

$$f(x) = \frac{x^3 + 2x^2 - 1}{5 - 3x}$$

is rational, so by Theorem 5 it is continuous on its domain, which is $\left\{ x \mid x \neq \frac{5}{3} \right\}$. Therefore

$$\lim_{x \to -2} \frac{x^3 + 2x^2 - 1}{5 - 3x} = \lim_{x \to -2} f(x) = f(-2)$$

$$= \frac{(-2)^3 + 2(-2)^2 - 1}{5 - 3(-2)} = -\frac{1}{11} \qquad \blacksquare$$

It turns out that most of the familiar functions are continuous at every number in their domains. For instance, Limit Law 10 (page 96) is exactly the statement that root functions are continuous.

From the appearance of the graphs of the sine and cosine functions (Figure 1.2.18), we would certainly guess that they are continuous. We know from the definitions of $\sin \theta$ and $\cos \theta$ that the coordinates of the point P in Figure 5 are $(\cos \theta, \sin \theta)$. As $\theta \to 0$, we see that P approaches the point $(1, 0)$ and so $\cos \theta \to 1$ and $\sin \theta \to 0$. Thus

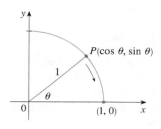

FIGURE 5

$$\boxed{6} \qquad \lim_{\theta \to 0} \cos \theta = 1 \qquad \lim_{\theta \to 0} \sin \theta = 0$$

Another way to establish the limits in (6) is to use the Squeeze Theorem with the inequality $\sin \theta < \theta$ (for $\theta > 0$), which is proved in Section 3.3.

Since $\cos 0 = 1$ and $\sin 0 = 0$, the equations in (6) assert that the cosine and sine functions are continuous at 0. The addition formulas for cosine and sine can then be used to deduce that these functions are continuous everywhere (see Exercises 64 and 65).

It follows from part 5 of Theorem 4 that

$$\tan x = \frac{\sin x}{\cos x}$$

is continuous except where $\cos x = 0$. This happens when x is an odd integer multiple of $\pi/2$, so $y = \tan x$ has infinite discontinuities when $x = \pm\pi/2, \pm3\pi/2, \pm5\pi/2$, and so on (see Figure 6).

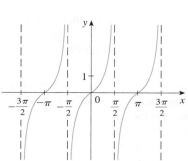

FIGURE 6
$y = \tan x$

The inverse function of any continuous one-to-one function is also continuous. (This fact is proved in Appendix F, but our geometric intuition makes it seem plausible: The graph of f^{-1} is obtained by reflecting the graph of f about the line $y = x$. So if the graph of f has no break in it, neither does the graph of f^{-1}.) Thus the inverse trigonometric functions are continuous.

In Section 1.4 we defined the exponential function $y = b^x$ so as to fill in the holes in the graph of $y = b^x$ where x is rational. In other words, the very definition of $y = b^x$ makes it a continuous function on $\mathbb{R}$. Therefore its inverse function $y = \log_b x$ is continuous on $(0, \infty)$.

The inverse trigonometric functions are reviewed in Section 1.5.

> **7 Theorem** The following types of functions are continuous at every number in their domains:
>
> - polynomials
> - rational functions
> - root functions
> - trigonometric functions
> - inverse trigonometric functions
> - exponential functions
> - logarithmic functions

EXAMPLE 6 Where is the function $f(x) = \dfrac{\ln x + \tan^{-1}x}{x^2 - 1}$ continuous?

SOLUTION We know from Theorem 7 that the function $y = \ln x$ is continuous for $x > 0$ and $y = \tan^{-1}x$ is continuous on $\mathbb{R}$. Thus, by part 1 of Theorem 4, $y = \ln x + \tan^{-1}x$ is continuous on $(0, \infty)$. The denominator, $y = x^2 - 1$, is a polynomial, so it is continuous everywhere. Therefore, by part 5 of Theorem 4, f is continuous at all positive numbers x except where $x^2 - 1 = 0 \iff x = \pm 1$. So f is continuous on the intervals $(0, 1)$ and $(1, \infty)$. ∎

EXAMPLE 7 Evaluate $\displaystyle\lim_{x \to \pi} \dfrac{\sin x}{2 + \cos x}$.

SOLUTION Theorem 7 tells us that $y = \sin x$ is continuous. The function in the denominator, $y = 2 + \cos x$, is the sum of two continuous functions and is therefore continuous. Notice that this function is never 0 because $\cos x \geqslant -1$ for all x and so $2 + \cos x > 0$ everywhere. Thus the ratio

$$f(x) = \frac{\sin x}{2 + \cos x}$$

is continuous everywhere. Hence, by the definition of a continuous function,

$$\lim_{x \to \pi} \frac{\sin x}{2 + \cos x} = \lim_{x \to \pi} f(x) = f(\pi) = \frac{\sin \pi}{2 + \cos \pi} = \frac{0}{2 - 1} = 0 \qquad ∎$$

Another way of combining continuous functions f and g to get a new continuous function is to form the composite function $f \circ g$. This fact is a consequence of the following theorem.

This theorem says that a limit symbol can be moved through a function symbol if the function is continuous and the limit exists. In other words, the order of these two symbols can be reversed.

> **8 Theorem** If f is continuous at b and $\displaystyle\lim_{x \to a} g(x) = b$, then $\displaystyle\lim_{x \to a} f(g(x)) = f(b)$. In other words,
>
> $$\lim_{x \to a} f(g(x)) = f\left(\lim_{x \to a} g(x) \right)$$

Intuitively, Theorem 8 is reasonable because if x is close to a, then $g(x)$ is close to b, and since f is continuous at b, if $g(x)$ is close to b, then $f(g(x))$ is close to $f(b)$. A proof of Theorem 8 is given in Appendix F.

EXAMPLE 8 Evaluate $\displaystyle\lim_{x \to 1} \arcsin\left(\dfrac{1 - \sqrt{x}}{1 - x}\right)$.

SOLUTION Because arcsin is a continuous function, we can apply Theorem 8:

$$\lim_{x \to 1} \arcsin\left(\frac{1 - \sqrt{x}}{1 - x}\right) = \arcsin\left(\lim_{x \to 1}\frac{1 - \sqrt{x}}{1 - x}\right)$$

$$= \arcsin\left(\lim_{x \to 1}\frac{1 - \sqrt{x}}{(1 - \sqrt{x})(1 + \sqrt{x})}\right)$$

$$= \arcsin\left(\lim_{x \to 1}\frac{1}{1 + \sqrt{x}}\right)$$

$$= \arcsin\frac{1}{2} = \frac{\pi}{6}$$

Let's now apply Theorem 8 in the special case where $f(x) = \sqrt[n]{x}$, with n being a positive integer. Then

$$f(g(x)) = \sqrt[n]{g(x)}$$

and

$$f\left(\lim_{x \to a} g(x)\right) = \sqrt[n]{\lim_{x \to a} g(x)}$$

If we put these expressions into Theorem 8, we get

$$\lim_{x \to a} \sqrt[n]{g(x)} = \sqrt[n]{\lim_{x \to a} g(x)}$$

and so Limit Law 11 has now been proved. (We assume that the roots exist.)

> **9 Theorem** If g is continuous at a and f is continuous at $g(a)$, then the composite function $f \circ g$ given by $(f \circ g)(x) = f(g(x))$ is continuous at a.

This theorem is often expressed informally by saying "a continuous function of a continuous function is a continuous function."

PROOF Since g is continuous at a, we have

$$\lim_{x \to a} g(x) = g(a)$$

Since f is continuous at $b = g(a)$, we can apply Theorem 8 to obtain

$$\lim_{x \to a} f(g(x)) = f(g(a))$$

which is precisely the statement that the function $h(x) = f(g(x))$ is continuous at a; that is, $f \circ g$ is continuous at a. ∎

EXAMPLE 9 Where are the following functions continuous?

(a) $h(x) = \sin(x^2)$ (b) $F(x) = \ln(1 + \cos x)$

SOLUTION

(a) We have $h(x) = f(g(x))$, where

$$g(x) = x^2 \quad \text{and} \quad f(x) = \sin x$$

Now g is continuous on $\mathbb{R}$ since it is a polynomial, and f is also continuous everywhere. Thus $h = f \circ g$ is continuous on $\mathbb{R}$ by Theorem 9.

(b) We know from Theorem 7 that $f(x) = \ln x$ is continuous and $g(x) = 1 + \cos x$ is continuous (because both $y = 1$ and $y = \cos x$ are continuous). Therefore, by Theorem 9, $F(x) = f(g(x))$ is continuous wherever it is defined. Now $\ln(1 + \cos x)$ is defined when $1 + \cos x > 0$. So it is undefined when $\cos x = -1$, and this happens when $x = \pm\pi, \pm3\pi, \ldots$. Thus F has discontinuities when x is an odd multiple of π and is continuous on the intervals between these values (see Figure 7). ∎

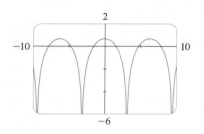

FIGURE 7
$y = \ln(1 + \cos x)$

An important property of continuous functions is expressed by the following theorem, whose proof is found in more advanced books on calculus.

---

10 **The Intermediate Value Theorem** Suppose that f is continuous on the closed interval $[a, b]$ and let N be any number between $f(a)$ and $f(b)$, where $f(a) \neq f(b)$. Then there exists a number c in (a, b) such that $f(c) = N$.

The Intermediate Value Theorem states that a continuous function takes on every intermediate value between the function values $f(a)$ and $f(b)$. It is illustrated by Figure 8. Note that the value N can be taken on once [as in part (a)] or more than once [as in part (b)].

(a)

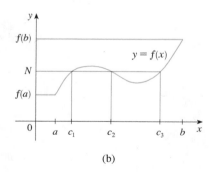

(b)

FIGURE 8

If we think of a continuous function as a function whose graph has no hole or break, then it is easy to believe that the Intermediate Value Theorem is true. In geometric terms it says that if any horizontal line $y = N$ is given between $y = f(a)$ and $y = f(b)$ as in Figure 9, then the graph of f can't jump over the line. It must intersect $y = N$ somewhere.

It is important that the function f in Theorem 10 be continuous. The Intermediate Value Theorem is not true in general for discontinuous functions (see Exercise 50).

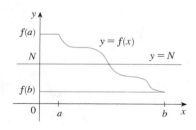

FIGURE 9

One use of the Intermediate Value Theorem is in locating roots of equations as in the following example.

EXAMPLE 10 Show that there is a root of the equation

$$4x^3 - 6x^2 + 3x - 2 = 0$$

between 1 and 2.

SOLUTION Let $f(x) = 4x^3 - 6x^2 + 3x - 2$. We are looking for a solution of the given equation, that is, a number c between 1 and 2 such that $f(c) = 0$. Therefore we take $a = 1$, $b = 2$, and $N = 0$ in Theorem 10. We have

$$f(1) = 4 - 6 + 3 - 2 = -1 < 0$$

and

$$f(2) = 32 - 24 + 6 - 2 = 12 > 0$$

Thus $f(1) < 0 < f(2)$; that is, $N = 0$ is a number between $f(1)$ and $f(2)$. Now f is continuous since it is a polynomial, so the Intermediate Value Theorem says there is a number c between 1 and 2 such that $f(c) = 0$. In other words, the equation $4x^3 - 6x^2 + 3x - 2 = 0$ has at least one root c in the interval $(1, 2)$.

In fact, we can locate a root more precisely by using the Intermediate Value Theorem again. Since

$$f(1.2) = -0.128 < 0 \quad \text{and} \quad f(1.3) = 0.548 > 0$$

a root must lie between 1.2 and 1.3. A calculator gives, by trial and error,

$$f(1.22) = -0.007008 < 0 \quad \text{and} \quad f(1.23) = 0.056068 > 0$$

so a root lies in the interval $(1.22, 1.23)$. ∎

We can use a graphing calculator or computer to illustrate the use of the Intermediate Value Theorem in Example 10. Figure 10 shows the graph of f in the viewing rectangle $[-1, 3]$ by $[-3, 3]$ and you can see that the graph crosses the x-axis between 1 and 2. Figure 11 shows the result of zooming in to the viewing rectangle $[1.2, 1.3]$ by $[-0.2, 0.2]$.

FIGURE 10

FIGURE 11

In fact, the Intermediate Value Theorem plays a role in the very way these graphing devices work. A computer calculates a finite number of points on the graph and turns on the pixels that contain these calculated points. It assumes that the function is continuous and takes on all the intermediate values between two consecutive points. The computer therefore "connects the dots" by turning on the intermediate pixels.

2.5   EXERCISES

1. Write an equation that expresses the fact that a function f is continuous at the number 4.

2. If f is continuous on $(-\infty, \infty)$, what can you say about its graph?

3. (a) From the graph of f, state the numbers at which f is discontinuous and explain why.
   (b) For each of the numbers stated in part (a), determine whether f is continuous from the right, or from the left, or neither.

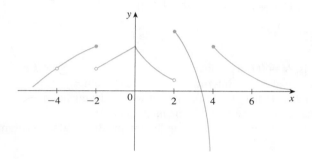

4. From the graph of g, state the intervals on which g is continuous.

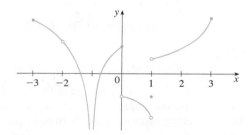

5–8 Sketch the graph of a function f that is continuous except for the stated discontinuity.

5. Discontinuous, but continuous from the right, at 2

6. Discontinuities at -1 and 4, but continuous from the left at -1 and from the right at 4

7. Removable discontinuity at 3, jump discontinuity at 5

8. Neither left nor right continuous at -2, continuous only from the left at 2

9. The toll T charged for driving on a certain stretch of a toll road is $5 except during rush hours (between 7 AM and 10 AM and between 4 PM and 7 PM) when the toll is $7.
   (a) Sketch a graph of T as a function of the time t, measured in hours past midnight.
   (b) Discuss the discontinuities of this function and their significance to someone who uses the road.

10. Explain why each function is continuous or discontinuous.
   (a) The temperature at a specific location as a function of time
   (b) The temperature at a specific time as a function of the distance due west from New York City
   (c) The altitude above sea level as a function of the distance due west from New York City
   (d) The cost of a taxi ride as a function of the distance traveled
   (e) The current in the circuit for the lights in a room as a function of time

11–14 Use the definition of continuity and the properties of limits to show that the function is continuous at the given number a.

11. $f(x) = (x + 2x^3)^4$, $a = -1$

12. $g(t) = \dfrac{t^2 + 5t}{2t + 1}$, $a = 2$

13. $p(v) = 2\sqrt{3v^2 + 1}$, $a = 1$

14. $f(x) = 3x^4 - 5x + \sqrt[3]{x^2 + 4}$, $a = 2$

15–16 Use the definition of continuity and the properties of limits to show that the function is continuous on the given interval.

15. $f(x) = x + \sqrt{x - 4}$, $[4, \infty)$

16. $g(x) = \dfrac{x - 1}{3x + 6}$, $(-\infty, -2)$

17–22 Explain why the function is discontinuous at the given number a. Sketch the graph of the function.

17. $f(x) = \dfrac{1}{x + 2}$ $\qquad\qquad\qquad\qquad$ $a = -2$

18. $f(x) = \begin{cases} \dfrac{1}{x + 2} & \text{if } x \neq -2 \\ 1 & \text{if } x = -2 \end{cases}$ $\qquad$ $a = -2$

19. $f(x) = \begin{cases} x + 3 & \text{if } x \leq -1 \\ 2^x & \text{if } x > -1 \end{cases}$ $\qquad$ $a = -1$

20. $f(x) = \begin{cases} \dfrac{x^2 - x}{x^2 - 1} & \text{if } x \neq 1 \\ 1 & \text{if } x = 1 \end{cases}$ $\qquad$ $a = 1$

21. $f(x) = \begin{cases} \cos x & \text{if } x < 0 \\ 0 & \text{if } x = 0 \\ 1 - x^2 & \text{if } x > 0 \end{cases}$ $\qquad$ $a = 0$

22. $f(x) = \begin{cases} \dfrac{2x^2 - 5x - 3}{x - 3} & \text{if } x \neq 3 \\ 6 & \text{if } x = 3 \end{cases}$ $\qquad$ $a = 3$

23–24 How would you "remove the discontinuity" of f? In other words, how would you define $f(2)$ in order to make f continuous at 2?

23. $f(x) = \dfrac{x^2 - x - 2}{x - 2}$

24. $f(x) = \dfrac{x^3 - 8}{x^2 - 4}$

25–32 Explain, using Theorems 4, 5, 7, and 9, why the function is continuous at every number in its domain. State the domain.

25. $F(x) = \dfrac{2x^2 - x - 1}{x^2 + 1}$

26. $G(x) = \dfrac{x^2 + 1}{2x^2 - x - 1}$

27. $Q(x) = \dfrac{\sqrt[3]{x - 2}}{x^3 - 2}$

28. $R(t) = \dfrac{e^{\sin t}}{2 + \cos \pi t}$

29. $A(t) = \arcsin(1 + 2t)$

30. $B(x) = \dfrac{\tan x}{\sqrt{4 - x^2}}$

31. $M(x) = \sqrt{1 + \dfrac{1}{x}}$

32. $N(r) = \tan^{-1}(1 + e^{-r^2})$

33–34 Locate the discontinuities of the function and illustrate by graphing.

33. $y = \dfrac{1}{1 + e^{1/x}}$

34. $y = \ln(\tan^2 x)$

35–38 Use continuity to evaluate the limit.

35. $\lim\limits_{x \to 2} x\sqrt{20 - x^2}$

36. $\lim\limits_{x \to \pi} \sin(x + \sin x)$

37. $\lim\limits_{x \to 1} \ln\left(\dfrac{5 - x^2}{1 + x}\right)$

38. $\lim\limits_{x \to 4} 3^{\sqrt{x^2 - 2x - 4}}$

39–40 Show that f is continuous on $(-\infty, \infty)$.

39. $f(x) = \begin{cases} 1 - x^2 & \text{if } x \leq 1 \\ \ln x & \text{if } x > 1 \end{cases}$

40. $f(x) = \begin{cases} \sin x & \text{if } x < \pi/4 \\ \cos x & \text{if } x \geq \pi/4 \end{cases}$

41–43 Find the numbers at which f is discontinuous. At which of these numbers is f continuous from the right, from the left, or neither? Sketch the graph of f.

41. $f(x) = \begin{cases} x^2 & \text{if } x < -1 \\ x & \text{if } -1 \leq x < 1 \\ 1/x & \text{if } x \geq 1 \end{cases}$

42. $f(x) = \begin{cases} 2^x & \text{if } x \leq 1 \\ 3 - x & \text{if } 1 < x \leq 4 \\ \sqrt{x} & \text{if } x > 4 \end{cases}$

43. $f(x) = \begin{cases} x + 2 & \text{if } x < 0 \\ e^x & \text{if } 0 \leq x \leq 1 \\ 2 - x & \text{if } x > 1 \end{cases}$

44. The gravitational force exerted by the planet Earth on a unit mass at a distance r from the center of the planet is

$$F(r) = \begin{cases} \dfrac{GMr}{R^3} & \text{if } r < R \\[2mm] \dfrac{GM}{r^2} & \text{if } r \geq R \end{cases}$$

where M is the mass of Earth, R is its radius, and G is the gravitational constant. Is F a continuous function of r?

45. For what value of the constant c is the function f continuous on $(-\infty, \infty)$?

$$f(x) = \begin{cases} cx^2 + 2x & \text{if } x < 2 \\ x^3 - cx & \text{if } x \geq 2 \end{cases}$$

46. Find the values of a and b that make f continuous everywhere.

$$f(x) = \begin{cases} \dfrac{x^2 - 4}{x - 2} & \text{if } x < 2 \\[2mm] ax^2 - bx + 3 & \text{if } 2 \leq x < 3 \\[1mm] 2x - a + b & \text{if } x \geq 3 \end{cases}$$

47. Suppose f and g are continuous functions such that $g(2) = 6$ and $\lim\limits_{x \to 2} [3f(x) + f(x)g(x)] = 36$. Find $f(2)$.

48. Let $f(x) = 1/x$ and $g(x) = 1/x^2$.
(a) Find $(f \circ g)(x)$.
(b) Is $f \circ g$ continuous everywhere? Explain.

49. Which of the following functions f has a removable discontinuity at a? If the discontinuity is removable, find a function g that agrees with f for $x \neq a$ and is continuous at a.

(a) $f(x) = \dfrac{x^4 - 1}{x - 1}$, $\quad a = 1$

(b) $f(x) = \dfrac{x^3 - x^2 - 2x}{x - 2}$, $\quad a = 2$

(c) $f(x) = [\![\sin x]\!]$, $\quad a = \pi$

50. Suppose that a function f is continuous on $[0, 1]$ except at 0.25 and that $f(0) = 1$ and $f(1) = 3$. Let $N = 2$. Sketch two possible graphs of f, one showing that f might not satisfy the conclusion of the Intermediate Value Theorem and one showing that f might still satisfy the conclusion of the Intermediate Value Theorem (even though it doesn't satisfy the hypothesis).

51. If $f(x) = x^2 + 10 \sin x$, show that there is a number c such that $f(c) = 1000$.

52. Suppose f is continuous on $[1, 5]$ and the only solutions of the equation $f(x) = 6$ are $x = 1$ and $x = 4$. If $f(2) = 8$, explain why $f(3) > 6$.

53–56 Use the Intermediate Value Theorem to show that there is a root of the given equation in the specified interval.

53. $x^4 + x - 3 = 0$, $(1, 2)$

54. $\ln x = x - \sqrt{x}$, $(2, 3)$

55. $e^x = 3 - 2x$, $(0, 1)$

56. $\sin x = x^2 - x$, $(1, 2)$

57–58 (a) Prove that the equation has at least one real root.
(b) Use your calculator to find an interval of length 0.01 that contains a root.

57. $\cos x = x^3$

58. $\ln x = 3 - 2x$

59–60 (a) Prove that the equation has at least one real root.
(b) Use your graphing device to find the root correct to three decimal places.

59. $100e^{-x/100} = 0.01x^2$

60. $\arctan x = 1 - x$

61–62 Prove, without graphing, that the graph of the function has at least two x-intercepts in the specified interval.

61. $y = \sin x^3$, $(1, 2)$

62. $y = x^2 - 3 + 1/x$, $(0, 2)$

63. Prove that f is continuous at a if and only if

$$\lim_{h \to 0} f(a + h) = f(a)$$

64. To prove that sine is continuous, we need to show that $\lim_{x \to a} \sin x = \sin a$ for every real number a. By Exercise 63 an equivalent statement is that

$$\lim_{h \to 0} \sin(a + h) = \sin a$$

Use (6) to show that this is true.

65. Prove that cosine is a continuous function.

66. (a) Prove Theorem 4, part 3.
(b) Prove Theorem 4, part 5.

67. For what values of x is f continuous?

$$f(x) = \begin{cases} 0 & \text{if } x \text{ is rational} \\ 1 & \text{if } x \text{ is irrational} \end{cases}$$

68. For what values of x is g continuous?

$$g(x) = \begin{cases} 0 & \text{if } x \text{ is rational} \\ x & \text{if } x \text{ is irrational} \end{cases}$$

69. Is there a number that is exactly 1 more than its cube?

70. If a and b are positive numbers, prove that the equation

$$\frac{a}{x^3 + 2x^2 - 1} + \frac{b}{x^3 + x - 2} = 0$$

has at least one solution in the interval $(-1, 1)$.

71. Show that the function

$$f(x) = \begin{cases} x^4 \sin(1/x) & \text{if } x \neq 0 \\ 0 & \text{if } x = 0 \end{cases}$$

is continuous on $(-\infty, \infty)$.

72. (a) Show that the absolute value function $F(x) = |x|$ is continuous everywhere.
(b) Prove that if f is a continuous function on an interval, then so is $|f|$.
(c) Is the converse of the statement in part (b) also true? In other words, if $|f|$ is continuous, does it follow that f is continuous? If so, prove it. If not, find a counterexample.

73. A Tibetan monk leaves the monastery at 7:00 AM and takes his usual path to the top of the mountain, arriving at 7:00 PM. The following morning, he starts at 7:00 AM at the top and takes the same path back, arriving at the monastery at 7:00 PM. Use the Intermediate Value Theorem to show that there is a point on the path that the monk will cross at exactly the same time of day on both days.

2.6   Limits at Infinity; Horizontal Asymptotes

In Sections 2.2 and 2.4 we investigated infinite limits and vertical asymptotes. There we let x approach a number and the result was that the values of y became arbitrarily large (positive or negative). In this section we let x become arbitrarily large (positive or negative) and see what happens to y.

Let's begin by investigating the behavior of the function f defined by

$$f(x) = \frac{x^2 - 1}{x^2 + 1}$$

x	$f(x)$
0	-1
± 1	0
± 2	0.600000
± 3	0.800000
± 4	0.882353
± 5	0.923077
± 10	0.980198
± 50	0.999200
± 100	0.999800
± 1000	0.999998

as x becomes large. The table at the left gives values of this function correct to six decimal places, and the graph of f has been drawn by a computer in Figure 1.

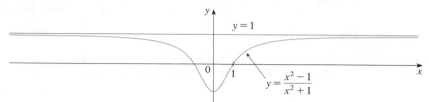

FIGURE 1

As x grows larger and larger you can see that the values of $f(x)$ get closer and closer to 1. (The graph of f approaches the horizontal line $y = 1$ as we look to the right.) In fact, it seems that we can make the values of $f(x)$ as close as we like to 1 by taking x sufficiently large. This situation is expressed symbolically by writing

$$\lim_{x \to \infty} \frac{x^2 - 1}{x^2 + 1} = 1$$

In general, we use the notation

$$\lim_{x \to \infty} f(x) = L$$

to indicate that the values of $f(x)$ approach L as x becomes larger and larger.

> **1 Intuitive Definition of a Limit at Infinity** Let f be a function defined on some interval (a, ∞). Then
>
> $$\lim_{x \to \infty} f(x) = L$$
>
> means that the values of $f(x)$ can be made arbitrarily close to L by requiring x to be sufficiently large.

Another notation for $\lim_{x \to \infty} f(x) = L$ is

$$f(x) \to L \quad \text{as} \quad x \to \infty$$

The symbol ∞ does not represent a number. Nonetheless, the expression $\lim_{x \to \infty} f(x) = L$ is often read as

"the limit of $f(x)$, as x approaches infinity, is L"

or "the limit of $f(x)$, as x becomes infinite, is L"

or "the limit of $f(x)$, as x increases without bound, is L"

The meaning of such phrases is given by Definition 1. A more precise definition, similar to the ε, δ definition of Section 2.4, is given at the end of this section.

Geometric illustrations of Definition 1 are shown in Figure 2. Notice that there are many ways for the graph of f to approach the line $y = L$ (which is called a *horizontal asymptote*) as we look to the far right of each graph.

Referring back to Figure 1, we see that for numerically large negative values of x, the values of $f(x)$ are close to 1. By letting x decrease through negative values without bound, we can make $f(x)$ as close to 1 as we like. This is expressed by writing

$$\lim_{x \to -\infty} \frac{x^2 - 1}{x^2 + 1} = 1$$

FIGURE 2
Examples illustrating $\lim_{x \to \infty} f(x) = L$

The general definition is as follows.

> **2 Definition** Let f be a function defined on some interval $(-\infty, a)$. Then
>
> $$\lim_{x \to -\infty} f(x) = L$$
>
> means that the values of $f(x)$ can be made arbitrarily close to L by requiring x to be sufficiently large negative.

Again, the symbol $-\infty$ does not represent a number, but the expression $\lim_{x \to -\infty} f(x) = L$ is often read as

"the limit of $f(x)$, as x approaches negative infinity, is L"

Definition 2 is illustrated in Figure 3. Notice that the graph approaches the line $y = L$ as we look to the far left of each graph.

FIGURE 3
Examples illustrating $\lim\limits_{x \to -\infty} f(x) = L$

> **3 Definition** The line $y = L$ is called a **horizontal asymptote** of the curve $y = f(x)$ if either
>
> $$\lim_{x \to \infty} f(x) = L \qquad \text{or} \qquad \lim_{x \to -\infty} f(x) = L$$

For instance, the curve illustrated in Figure 1 has the line $y = 1$ as a horizontal asymptote because

$$\lim_{x \to \infty} \frac{x^2 - 1}{x^2 + 1} = 1$$

An example of a curve with two horizontal asymptotes is $y = \tan^{-1}x$. (See Figure 4.) In fact,

> **4**
> $$\lim_{x \to -\infty} \tan^{-1}x = -\frac{\pi}{2} \qquad \lim_{x \to \infty} \tan^{-1}x = \frac{\pi}{2}$$

so both of the lines $y = -\pi/2$ and $y = \pi/2$ are horizontal asymptotes. (This follows from the fact that the lines $x = \pm\pi/2$ are vertical asymptotes of the graph of the tangent function.)

FIGURE 4
$y = \tan^{-1}x$

EXAMPLE 1 Find the infinite limits, limits at infinity, and asymptotes for the function f whose graph is shown in Figure 5.

SOLUTION We see that the values of $f(x)$ become large as $x \to -1$ from both sides, so

$$\lim_{x \to -1} f(x) = \infty$$

Notice that $f(x)$ becomes large negative as x approaches 2 from the left, but large positive as x approaches 2 from the right. So

$$\lim_{x \to 2^-} f(x) = -\infty \qquad \text{and} \qquad \lim_{x \to 2^+} f(x) = \infty$$

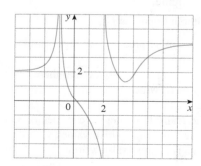

FIGURE 5

Thus both of the lines $x = -1$ and $x = 2$ are vertical asymptotes.

As x becomes large, it appears that $f(x)$ approaches 4. But as x decreases through negative values, $f(x)$ approaches 2. So

$$\lim_{x \to \infty} f(x) = 4 \qquad \text{and} \qquad \lim_{x \to -\infty} f(x) = 2$$

This means that both $y = 4$ and $y = 2$ are horizontal asymptotes. ■

EXAMPLE 2 Find $\displaystyle\lim_{x \to \infty} \frac{1}{x}$ and $\displaystyle\lim_{x \to -\infty} \frac{1}{x}$.

SOLUTION Observe that when x is large, $1/x$ is small. For instance,

$$\frac{1}{100} = 0.01 \qquad \frac{1}{10,000} = 0.0001 \qquad \frac{1}{1,000,000} = 0.000001$$

In fact, by taking x large enough, we can make $1/x$ as close to 0 as we please. Therefore, according to Definition 1, we have

$$\lim_{x \to \infty} \frac{1}{x} = 0$$

Similar reasoning shows that when x is large negative, $1/x$ is small negative, so we also have

$$\lim_{x \to -\infty} \frac{1}{x} = 0$$

It follows that the line $y = 0$ (the x-axis) is a horizontal asymptote of the curve $y = 1/x$. (This is an equilateral hyperbola; see Figure 6.) ■

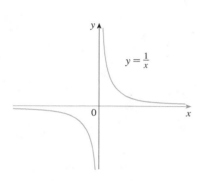

FIGURE 6

$\displaystyle\lim_{x \to \infty} \frac{1}{x} = 0, \ \lim_{x \to -\infty} \frac{1}{x} = 0$

Most of the Limit Laws that were given in Section 2.3 also hold for limits at infinity. It can be proved that *the Limit Laws listed in Section 2.3 (with the exception of Laws 9 and 10) are also valid if "$x \to a$" is replaced by "$x \to \infty$" or "$x \to -\infty$."* In particular, if we combine Laws 6 and 11 with the results of Example 2, we obtain the following important rule for calculating limits.

---

5 **Theorem** If $r > 0$ is a rational number, then

$$\lim_{x \to \infty} \frac{1}{x^r} = 0$$

If $r > 0$ is a rational number such that x^r is defined for all x, then

$$\lim_{x \to -\infty} \frac{1}{x^r} = 0$$

EXAMPLE 3 Evaluate

$$\lim_{x \to \infty} \frac{3x^2 - x - 2}{5x^2 + 4x + 1}$$

and indicate which properties of limits are used at each stage.

SOLUTION As x becomes large, both numerator and denominator become large, so it isn't obvious what happens to their ratio. We need to do some preliminary algebra.

To evaluate the limit at infinity of any rational function, we first divide both the numerator and denominator by the highest power of x that occurs in the denominator. (We may assume that $x \neq 0$, since we are interested only in large values of x.) In this case the highest power of x in the denominator is x^2, so we have

$$\lim_{x \to \infty} \frac{3x^2 - x - 2}{5x^2 + 4x + 1} = \lim_{x \to \infty} \frac{\dfrac{3x^2 - x - 2}{x^2}}{\dfrac{5x^2 + 4x + 1}{x^2}} = \lim_{x \to \infty} \frac{3 - \dfrac{1}{x} - \dfrac{2}{x^2}}{5 + \dfrac{4}{x} + \dfrac{1}{x^2}}$$

$$= \frac{\displaystyle\lim_{x \to \infty}\left(3 - \frac{1}{x} - \frac{2}{x^2}\right)}{\displaystyle\lim_{x \to \infty}\left(5 + \frac{4}{x} + \frac{1}{x^2}\right)} \qquad \text{(by Limit Law 5)}$$

$$= \frac{\displaystyle\lim_{x \to \infty} 3 - \lim_{x \to \infty} \frac{1}{x} - 2 \lim_{x \to \infty} \frac{1}{x^2}}{\displaystyle\lim_{x \to \infty} 5 + 4 \lim_{x \to \infty} \frac{1}{x} + \lim_{x \to \infty} \frac{1}{x^2}} \qquad \text{(by 1, 2, and 3)}$$

$$= \frac{3 - 0 - 0}{5 + 0 + 0} \qquad \text{(by 7 and Theorem 5)}$$

$$= \frac{3}{5}$$

A similar calculation shows that the limit as $x \to -\infty$ is also $\frac{3}{5}$. Figure 7 illustrates the results of these calculations by showing how the graph of the given rational function approaches the horizontal asymptote $y = \frac{3}{5} = 0.6$. ∎

FIGURE 7
$$y = \frac{3x^2 - x - 2}{5x^2 + 4x + 1}$$

EXAMPLE 4   Find the horizontal and vertical asymptotes of the graph of the function

$$f(x) = \frac{\sqrt{2x^2 + 1}}{3x - 5}$$

SOLUTION   Dividing both numerator and denominator by x and using the properties of limits, we have

$$\lim_{x \to \infty} \frac{\sqrt{2x^2 + 1}}{3x - 5} = \lim_{x \to \infty} \frac{\dfrac{\sqrt{2x^2 + 1}}{x}}{\dfrac{3x - 5}{x}} = \lim_{x \to \infty} \frac{\sqrt{\dfrac{2x^2 + 1}{x^2}}}{\dfrac{3x - 5}{x}} \qquad \text{(since } \sqrt{x^2} = x \text{ for } x > 0\text{)}$$

$$= \frac{\displaystyle\lim_{x \to \infty} \sqrt{2 + \frac{1}{x^2}}}{\displaystyle\lim_{x \to \infty}\left(3 - \frac{5}{x}\right)} = \frac{\sqrt{\displaystyle\lim_{x \to \infty} 2 + \lim_{x \to \infty} \frac{1}{x^2}}}{\displaystyle\lim_{x \to \infty} 3 - 5 \lim_{x \to \infty} \frac{1}{x}} = \frac{\sqrt{2 + 0}}{3 - 5 \cdot 0} = \frac{\sqrt{2}}{3}$$

Therefore the line $y = \sqrt{2}/3$ is a horizontal asymptote of the graph of f.

In computing the limit as $x \to -\infty$, we must remember that for $x < 0$, we have $\sqrt{x^2} = |x| = -x$. So when we divide the numerator by x, for $x < 0$ we get

$$\frac{\sqrt{2x^2 + 1}}{x} = \frac{\sqrt{2x^2 + 1}}{-\sqrt{x^2}} = -\sqrt{\frac{2x^2 + 1}{x^2}} = -\sqrt{2 + \frac{1}{x^2}}$$

Therefore

$$\lim_{x \to -\infty} \frac{\sqrt{2x^2 + 1}}{3x - 5} = \lim_{x \to -\infty} \frac{-\sqrt{2 + \dfrac{1}{x^2}}}{3 - \dfrac{5}{x}} = \frac{-\sqrt{2 + \lim\limits_{x \to -\infty} \dfrac{1}{x^2}}}{3 - 5 \lim\limits_{x \to -\infty} \dfrac{1}{x}} = -\frac{\sqrt{2}}{3}$$

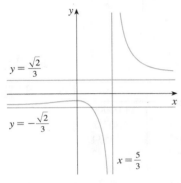

$$y = \frac{\sqrt{2}}{3}$$

$$y = -\frac{\sqrt{2}}{3}$$

$$x = \frac{5}{3}$$

FIGURE 8

$$y = \frac{\sqrt{2x^2 + 1}}{3x - 5}$$

Thus the line $y = -\sqrt{2}/3$ is also a horizontal asymptote.

A vertical asymptote is likely to occur when the denominator, $3x - 5$, is 0, that is, when $x = \frac{5}{3}$. If x is close to $\frac{5}{3}$ and $x > \frac{5}{3}$, then the denominator is close to 0 and $3x - 5$ is positive. The numerator $\sqrt{2x^2 + 1}$ is always positive, so $f(x)$ is positive. Therefore

$$\lim_{x \to (5/3)^+} \frac{\sqrt{2x^2 + 1}}{3x - 5} = \infty$$

(Notice that the numerator does *not* approach 0 as $x \to 5/3$).

If x is close to $\frac{5}{3}$ but $x < \frac{5}{3}$, then $3x - 5 < 0$ and so $f(x)$ is large negative. Thus

$$\lim_{x \to (5/3)^-} \frac{\sqrt{2x^2 + 1}}{3x - 5} = -\infty$$

The vertical asymptote is $x = \frac{5}{3}$. All three asymptotes are shown in Figure 8. ■

EXAMPLE 5 Compute $\displaystyle\lim_{x \to \infty} \left(\sqrt{x^2 + 1} - x\right)$.

SOLUTION Because both $\sqrt{x^2 + 1}$ and x are large when x is large, it's difficult to see what happens to their difference, so we use algebra to rewrite the function. We first multiply numerator and denominator by the conjugate radical:

We can think of the given function as having a denominator of 1.

$$\lim_{x \to \infty} \left(\sqrt{x^2 + 1} - x\right) = \lim_{x \to \infty} \left(\sqrt{x^2 + 1} - x\right) \cdot \frac{\sqrt{x^2 + 1} + x}{\sqrt{x^2 + 1} + x}$$

$$= \lim_{x \to \infty} \frac{(x^2 + 1) - x^2}{\sqrt{x^2 + 1} + x} = \lim_{x \to \infty} \frac{1}{\sqrt{x^2 + 1} + x}$$

Notice that the denominator of this last expression $\left(\sqrt{x^2 + 1} + x\right)$ becomes large as $x \to \infty$ (it's bigger than x). So

$$\lim_{x \to \infty} \left(\sqrt{x^2 + 1} - x\right) = \lim_{x \to \infty} \frac{1}{\sqrt{x^2 + 1} + x} = 0$$

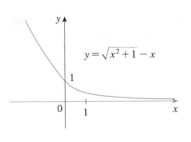

$$y = \sqrt{x^2 + 1} - x$$

FIGURE 9

Figure 9 illustrates this result. ■

EXAMPLE 6 Evaluate $\lim\limits_{x\to 2^+} \arctan\left(\dfrac{1}{x-2}\right)$.

SOLUTION If we let $t = 1/(x-2)$, we know that $t \to \infty$ as $x \to 2^+$. Therefore, by the second equation in (4), we have

$$\lim_{x\to 2^+} \arctan\left(\frac{1}{x-2}\right) = \lim_{t\to\infty} \arctan t = \frac{\pi}{2}$$

The graph of the natural exponential function $y = e^x$ has the line $y = 0$ (the x-axis) as a horizontal asymptote. (The same is true of any exponential function with base $b > 1$.) In fact, from the graph in Figure 10 and the corresponding table of values, we see that

$$\boxed{6}$$

$$\lim_{x\to -\infty} e^x = 0$$

Notice that the values of e^x approach 0 very rapidly.

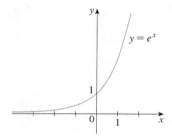

x	e^x
0	1.00000
-1	0.36788
-2	0.13534
-3	0.04979
-5	0.00674
-8	0.00034
-10	0.00005

FIGURE 10

EXAMPLE 7 Evaluate $\lim\limits_{x\to 0^-} e^{1/x}$.

PS The problem-solving strategy for Examples 6 and 7 is *introducing something extra* (see page 71). Here, the something extra, the auxiliary aid, is the new variable t.

SOLUTION If we let $t = 1/x$, we know that $t \to -\infty$ as $x \to 0^-$. Therefore, by (6),

$$\lim_{x\to 0^-} e^{1/x} = \lim_{t\to -\infty} e^t = 0$$

(See Exercise 81.)

EXAMPLE 8 Evaluate $\lim\limits_{x\to\infty} \sin x$.

SOLUTION As x increases, the values of $\sin x$ oscillate between 1 and -1 infinitely often and so they don't approach any definite number. Thus $\lim_{x\to\infty} \sin x$ does not exist.

■ Infinite Limits at Infinity

The notation

$$\lim_{x\to\infty} f(x) = \infty$$

is used to indicate that the values of $f(x)$ become large as x becomes large. Similar meanings are attached to the following symbols:

$$\lim_{x \to -\infty} f(x) = \infty \qquad \lim_{x \to \infty} f(x) = -\infty \qquad \lim_{x \to -\infty} f(x) = -\infty$$

EXAMPLE 9 Find $\lim_{x \to \infty} x^3$ and $\lim_{x \to -\infty} x^3$.

SOLUTION When x becomes large, x^3 also becomes large. For instance,

$$10^3 = 1000 \qquad 100^3 = 1{,}000{,}000 \qquad 1000^3 = 1{,}000{,}000{,}000$$

In fact, we can make x^3 as big as we like by requiring x to be large enough. Therefore we can write

$$\lim_{x \to \infty} x^3 = \infty$$

Similarly, when x is large negative, so is x^3. Thus

$$\lim_{x \to -\infty} x^3 = -\infty$$

These limit statements can also be seen from the graph of $y = x^3$ in Figure 11.   ■

Looking at Figure 10 we see that

$$\lim_{x \to \infty} e^x = \infty$$

but, as Figure 12 demonstrates, $y = e^x$ becomes large as $x \to \infty$ at a much faster rate than $y = x^3$.

EXAMPLE 10 Find $\lim_{x \to \infty} (x^2 - x)$.

⊘ **SOLUTION** It would be **wrong** to write

$$\lim_{x \to \infty} (x^2 - x) = \lim_{x \to \infty} x^2 - \lim_{x \to \infty} x = \infty - \infty$$

The Limit Laws can't be applied to infinite limits because ∞ is not a number ($\infty - \infty$ can't be defined). However, we *can* write

$$\lim_{x \to \infty} (x^2 - x) = \lim_{x \to \infty} x(x - 1) = \infty$$

because both x and $x - 1$ become arbitrarily large and so their product does too.   ■

EXAMPLE 11 Find $\lim_{x \to \infty} \dfrac{x^2 + x}{3 - x}$.

SOLUTION As in Example 3, we divide the numerator and denominator by the highest power of x in the denominator, which is just x:

$$\lim_{x \to \infty} \frac{x^2 + x}{3 - x} = \lim_{x \to \infty} \frac{x + 1}{\dfrac{3}{x} - 1} = -\infty$$

because $x + 1 \to \infty$ and $3/x - 1 \to 0 - 1 = -1$ as $x \to \infty$.   ■

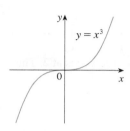

FIGURE 11

$\lim_{x \to \infty} x^3 = \infty, \ \lim_{x \to -\infty} x^3 = -\infty$

FIGURE 12

e^x is much larger than x^3 when x is large.

The next example shows that by using infinite limits at infinity, together with intercepts, we can get a rough idea of the graph of a polynomial without having to plot a large number of points.

EXAMPLE 12 Sketch the graph of $y = (x - 2)^4(x + 1)^3(x - 1)$ by finding its intercepts and its limits as $x \to \infty$ and as $x \to -\infty$.

SOLUTION The y-intercept is $f(0) = (-2)^4(1)^3(-1) = -16$ and the x-intercepts are found by setting $y = 0$: $x = 2, -1, 1$. Notice that since $(x - 2)^4$ is never negative, the function doesn't change sign at 2; thus the graph doesn't cross the x-axis at 2. The graph crosses the axis at -1 and 1.

When x is large positive, all three factors are large, so

$$\lim_{x \to \infty} (x - 2)^4(x + 1)^3(x - 1) = \infty$$

When x is large negative, the first factor is large positive and the second and third factors are both large negative, so

$$\lim_{x \to -\infty} (x - 2)^4(x + 1)^3(x - 1) = \infty$$

Combining this information, we give a rough sketch of the graph in Figure 13. ∎

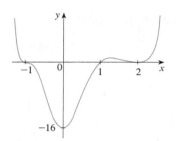

FIGURE 13
$y = (x - 2)^4(x + 1)^3(x - 1)$

▪ Precise Definitions

Definition 1 can be stated precisely as follows.

---

7 Precise Definition of a Limit at Infinity Let f be a function defined on some interval (a, ∞). Then

$$\lim_{x \to \infty} f(x) = L$$

means that for every $\varepsilon > 0$ there is a corresponding number N such that

$$\text{if} \quad x > N \quad \text{then} \quad |f(x) - L| < \varepsilon$$

In words, this says that the values of $f(x)$ can be made arbitrarily close to L (within a distance ε, where ε is any positive number) by requiring x to be sufficiently large (larger than N, where N depends on ε). Graphically it says that by keeping x large enough (larger than some number N) we can make the graph of f lie between the given horizontal lines $y = L - \varepsilon$ and $y = L + \varepsilon$ as in Figure 14. This must be true no matter how small we choose ε.

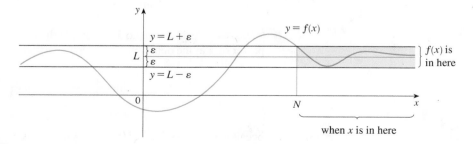

FIGURE 14
$\lim_{x \to \infty} f(x) = L$

Figure 15 shows that if a smaller value of ε is chosen, then a larger value of N may be required.

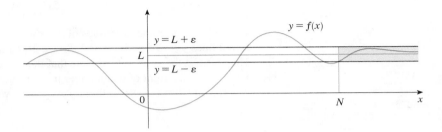

FIGURE 15
$\lim\limits_{x \to \infty} f(x) = L$

Similarly, a precise version of Definition 2 is given by Definition 8, which is illustrated in Figure 16.

---

8 Definition Let f be a function defined on some interval $(-\infty, a)$. Then

$$\lim\limits_{x \to -\infty} f(x) = L$$

means that for every $\varepsilon > 0$ there is a corresponding number N such that

$$\text{if} \quad x < N \quad \text{then} \quad |f(x) - L| < \varepsilon$$

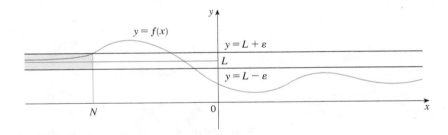

FIGURE 16
$\lim\limits_{x \to -\infty} f(x) = L$

In Example 3 we calculated that

$$\lim\limits_{x \to \infty} \frac{3x^2 - x - 2}{5x^2 + 4x + 1} = \frac{3}{5}$$

In the next example we use a graphing device to relate this statement to Definition 7 with $L = \frac{3}{5} = 0.6$ and $\varepsilon = 0.1$.

TEC In Module 2.4/2.6 you can explore the precise definition of a limit both graphically and numerically.

EXAMPLE 13 Use a graph to find a number N such that

$$\text{if} \quad x > N \quad \text{then} \quad \left| \frac{3x^2 - x - 2}{5x^2 + 4x + 1} - 0.6 \right| < 0.1$$

SOLUTION We rewrite the given inequality as

$$0.5 < \frac{3x^2 - x - 2}{5x^2 + 4x + 1} < 0.7$$

We need to determine the values of x for which the given curve lies between the horizontal lines $y = 0.5$ and $y = 0.7$. So we graph the curve and these lines in Figure 17. Then we use the cursor to estimate that the curve crosses the line $y = 0.5$ when $x \approx 6.7$. To the right of this number it seems that the curve stays between the lines $y = 0.5$ and $y = 0.7$. Rounding up to be safe, we can say that

$$\text{if} \quad x > 7 \quad \text{then} \quad \left| \frac{3x^2 - x - 2}{5x^2 + 4x + 1} - 0.6 \right| < 0.1$$

In other words, for $\varepsilon = 0.1$ we can choose $N = 7$ (or any larger number) in Definition 7.

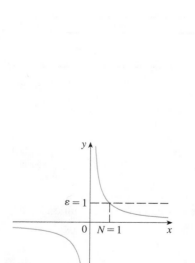

FIGURE 17

EXAMPLE 14 Use Definition 7 to prove that $\lim\limits_{x \to \infty} \dfrac{1}{x} = 0$.

SOLUTION Given $\varepsilon > 0$, we want to find N such that

$$\text{if} \quad x > N \quad \text{then} \quad \left| \frac{1}{x} - 0 \right| < \varepsilon$$

In computing the limit we may assume that $x > 0$. Then $1/x < \varepsilon \iff x > 1/\varepsilon$. Let's choose $N = 1/\varepsilon$. So

$$\text{if} \quad x > N = \frac{1}{\varepsilon} \quad \text{then} \quad \left| \frac{1}{x} - 0 \right| = \frac{1}{x} < \varepsilon$$

Therefore, by Definition 7,

$$\lim_{x \to \infty} \frac{1}{x} = 0$$

Figure 18 illustrates the proof by showing some values of ε and the corresponding values of N.

FIGURE 18

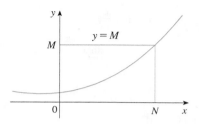

FIGURE 19
$\lim\limits_{x \to \infty} f(x) = \infty$

Finally we note that an infinite limit at infinity can be defined as follows. The geometric illustration is given in Figure 19.

> **9** **Definition of an Infinite Limit at Infinity** Let f be a function defined on some interval (a, ∞). Then
> $$\lim_{x \to \infty} f(x) = \infty$$
> means that for every positive number M there is a corresponding positive number N such that
> $$\text{if} \quad x > N \quad \text{then} \quad f(x) > M$$

Similar definitions apply when the symbol ∞ is replaced by $-\infty$. (See Exercise 80.)

2.6 EXERCISES

1. Explain in your own words the meaning of each of the following.
   (a) $\lim\limits_{x \to \infty} f(x) = 5$ (b) $\lim\limits_{x \to -\infty} f(x) = 3$

2. (a) Can the graph of $y = f(x)$ intersect a vertical asymptote? Can it intersect a horizontal asymptote? Illustrate by sketching graphs.
   (b) How many horizontal asymptotes can the graph of $y = f(x)$ have? Sketch graphs to illustrate the possibilities.

3. For the function f whose graph is given, state the following.
   (a) $\lim\limits_{x \to \infty} f(x)$ (b) $\lim\limits_{x \to -\infty} f(x)$
   (c) $\lim\limits_{x \to 1} f(x)$ (d) $\lim\limits_{x \to 3} f(x)$
   (e) The equations of the asymptotes

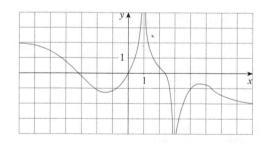

4. For the function g whose graph is given, state the following.
   (a) $\lim\limits_{x \to \infty} g(x)$ (b) $\lim\limits_{x \to -\infty} g(x)$
   (c) $\lim\limits_{x \to 0} g(x)$ (d) $\lim\limits_{x \to 2^-} g(x)$
   (e) $\lim\limits_{x \to 2^+} g(x)$ (f) The equations of the asymptotes

5–10 Sketch the graph of an example of a function f that satisfies all of the given conditions.

5. $\lim\limits_{x \to 0} f(x) = -\infty$, $\lim\limits_{x \to -\infty} f(x) = 5$, $\lim\limits_{x \to \infty} f(x) = -5$

6. $\lim\limits_{x \to 2} f(x) = \infty$, $\lim\limits_{x \to -2^+} f(x) = \infty$, $\lim\limits_{x \to -2^-} f(x) = -\infty$,
   $\lim\limits_{x \to -\infty} f(x) = 0$, $\lim\limits_{x \to \infty} f(x) = 0$, $f(0) = 0$

7. $\lim\limits_{x \to 2} f(x) = -\infty$, $\lim\limits_{x \to \infty} f(x) = \infty$, $\lim\limits_{x \to -\infty} f(x) = 0$,
   $\lim\limits_{x \to 0^+} f(x) = \infty$, $\lim\limits_{x \to 0^-} f(x) = -\infty$

8. $\lim\limits_{x \to \infty} f(x) = 3$, $\lim\limits_{x \to 2^-} f(x) = \infty$, $\lim\limits_{x \to 2^+} f(x) = -\infty$, f is odd

9. $f(0) = 3$, $\lim\limits_{x \to 0^-} f(x) = 4$, $\lim\limits_{x \to 0^+} f(x) = 2$,
   $\lim\limits_{x \to -\infty} f(x) = -\infty$, $\lim\limits_{x \to 4^-} f(x) = -\infty$, $\lim\limits_{x \to 4^+} f(x) = \infty$,
   $\lim\limits_{x \to \infty} f(x) = 3$

10. $\lim\limits_{x \to 3} f(x) = -\infty$, $\lim\limits_{x \to \infty} f(x) = 2$, $f(0) = 0$, f is even

11. Guess the value of the limit

$$\lim_{x \to \infty} \frac{x^2}{2^x}$$

by evaluating the function $f(x) = x^2/2^x$ for $x = 0, 1, 2, 3,$ 4, 5, 6, 7, 8, 9, 10, 20, 50, and 100. Then use a graph of f to support your guess.

12. (a) Use a graph of

$$f(x) = \left(1 - \frac{2}{x}\right)^x$$

to estimate the value of $\lim_{x \to \infty} f(x)$ correct to two decimal places.
(b) Use a table of values of $f(x)$ to estimate the limit to four decimal places.

13–14 Evaluate the limit and justify each step by indicating the appropriate properties of limits.

13. $\lim_{x \to \infty} \dfrac{2x^2 - 7}{5x^2 + x - 3}$

14. $\lim_{x \to \infty} \sqrt{\dfrac{9x^3 + 8x - 4}{3 - 5x + x^3}}$

15–42 Find the limit or show that it does not exist.

15. $\lim_{x \to \infty} \dfrac{3x - 2}{2x + 1}$

16. $\lim_{x \to \infty} \dfrac{1 - x^2}{x^3 - x + 1}$

17. $\lim_{x \to -\infty} \dfrac{x - 2}{x^2 + 1}$

18. $\lim_{x \to -\infty} \dfrac{4x^3 + 6x^2 - 2}{2x^3 - 4x + 5}$

19. $\lim_{t \to \infty} \dfrac{\sqrt{t} + t^2}{2t - t^2}$

20. $\lim_{t \to \infty} \dfrac{t - t\sqrt{t}}{2t^{3/2} + 3t - 5}$

21. $\lim_{x \to \infty} \dfrac{(2x^2 + 1)^2}{(x - 1)^2(x^2 + x)}$

22. $\lim_{x \to \infty} \dfrac{x^2}{\sqrt{x^4 + 1}}$

23. $\lim_{x \to \infty} \dfrac{\sqrt{1 + 4x^6}}{2 - x^3}$

24. $\lim_{x \to -\infty} \dfrac{\sqrt{1 + 4x^6}}{2 - x^3}$

25. $\lim_{x \to \infty} \dfrac{\sqrt{x} + 3x^2}{4x - 1}$

26. $\lim_{x \to \infty} \dfrac{x + 3x^2}{4x - 1}$

27. $\lim_{x \to \infty} \left(\sqrt{9x^2 + x} - 3x\right)$

28. $\lim_{x \to -\infty} \left(\sqrt{4x^2 + 3x} + 2x\right)$

29. $\lim_{x \to \infty} \left(\sqrt{x^2 + ax} - \sqrt{x^2 + bx}\right)$

30. $\lim_{x \to \infty} \sqrt{x^2 + 1}$

31. $\lim_{x \to \infty} \dfrac{x^4 - 3x^2 + x}{x^3 - x + 2}$

32. $\lim_{x \to \infty} (e^{-x} + 2\cos 3x)$

33. $\lim_{x \to -\infty} (x^2 + 2x^7)$

34. $\lim_{x \to -\infty} \dfrac{1 + x^6}{x^4 + 1}$

35. $\lim_{x \to \infty} \arctan(e^x)$

36. $\lim_{x \to \infty} \dfrac{e^{3x} - e^{-3x}}{e^{3x} + e^{-3x}}$

37. $\lim_{x \to \infty} \dfrac{1 - e^x}{1 + 2e^x}$

38. $\lim_{x \to \infty} \dfrac{\sin^2 x}{x^2 + 1}$

39. $\lim_{x \to \infty} (e^{-2x} \cos x)$

40. $\lim_{x \to 0^+} \tan^{-1}(\ln x)$

41. $\lim_{x \to \infty} [\ln(1 + x^2) - \ln(1 + x)]$

42. $\lim_{x \to \infty} [\ln(2 + x) - \ln(1 + x)]$

43. (a) For $f(x) = \dfrac{x}{\ln x}$ find each of the following limits.

   (i) $\lim_{x \to 0^+} f(x)$ (ii) $\lim_{x \to 1^-} f(x)$ (iii) $\lim_{x \to 1^+} f(x)$

(b) Use a table of values to estimate $\lim_{x \to \infty} f(x)$.
(c) Use the information from parts (a) and (b) to make a rough sketch of the graph of f.

44. For $f(x) = \dfrac{2}{x} - \dfrac{1}{\ln x}$ find each of the following limits.

(a) $\lim_{x \to \infty} f(x)$ (b) $\lim_{x \to 0^+} f(x)$

(c) $\lim_{x \to 1^-} f(x)$ (d) $\lim_{x \to 1^+} f(x)$

(e) Use the information from parts (a)–(d) to make a rough sketch of the graph of f.

45. (a) Estimate the value of

$$\lim_{x \to -\infty} \left(\sqrt{x^2 + x + 1} + x\right)$$

by graphing the function $f(x) = \sqrt{x^2 + x + 1} + x$.
(b) Use a table of values of $f(x)$ to guess the value of the limit.
(c) Prove that your guess is correct.

46. (a) Use a graph of

$$f(x) = \sqrt{3x^2 + 8x + 6} - \sqrt{3x^2 + 3x + 1}$$

to estimate the value of $\lim_{x \to \infty} f(x)$ to one decimal place.
(b) Use a table of values of $f(x)$ to estimate the limit to four decimal places.
(c) Find the exact value of the limit.

47–52 Find the horizontal and vertical asymptotes of each curve. If you have a graphing device, check your work by graphing the curve and estimating the asymptotes.

47. $y = \dfrac{5 + 4x}{x + 3}$

48. $y = \dfrac{2x^2 + 1}{3x^2 + 2x - 1}$

49. $y = \dfrac{2x^2 + x - 1}{x^2 + x - 2}$

50. $y = \dfrac{1 + x^4}{x^2 - x^4}$

51. $y = \dfrac{x^3 - x}{x^2 - 6x + 5}$ **52.** $y = \dfrac{2e^x}{e^x - 5}$

---

53. Estimate the horizontal asymptote of the function

$$f(x) = \frac{3x^3 + 500x^2}{x^3 + 500x^2 + 100x + 2000}$$

by graphing f for $-10 \le x \le 10$. Then calculate the equation of the asymptote by evaluating the limit. How do you explain the discrepancy?

54. (a) Graph the function

$$f(x) = \frac{\sqrt{2x^2 + 1}}{3x - 5}$$

How many horizontal and vertical asymptotes do you observe? Use the graph to estimate the values of the limits

$$\lim_{x \to \infty} \frac{\sqrt{2x^2 + 1}}{3x - 5} \quad \text{and} \quad \lim_{x \to -\infty} \frac{\sqrt{2x^2 + 1}}{3x - 5}$$

(b) By calculating values of $f(x)$, give numerical estimates of the limits in part (a).

(c) Calculate the exact values of the limits in part (a). Did you get the same value or different values for these two limits? [In view of your answer to part (a), you might have to check your calculation for the second limit.]

55. Let P and Q be polynomials. Find

$$\lim_{x \to \infty} \frac{P(x)}{Q(x)}$$

if the degree of P is (a) less than the degree of Q and (b) greater than the degree of Q.

56. Make a rough sketch of the curve $y = x^n$ (n an integer) for the following five cases:

(i) $n = 0$ (ii) $n > 0$, n odd
(iii) $n > 0$, n even   (iv) $n < 0$, n odd
(v) $n < 0$, n even

Then use these sketches to find the following limits.

(a) $\lim\limits_{x \to 0^+} x^n$ (b) $\lim\limits_{x \to 0^-} x^n$

(c) $\lim\limits_{x \to \infty} x^n$ (d) $\lim\limits_{x \to -\infty} x^n$

57. Find a formula for a function f that satisfies the following conditions:

$$\lim_{x \to \pm\infty} f(x) = 0, \quad \lim_{x \to 0} f(x) = -\infty, \quad f(2) = 0,$$

$$\lim_{x \to 3^-} f(x) = \infty, \quad \lim_{x \to 3^+} f(x) = -\infty$$

58. Find a formula for a function that has vertical asymptotes $x = 1$ and $x = 3$ and horizontal asymptote $y = 1$.

59. A function f is a ratio of quadratic functions and has a vertical asymptote $x = 4$ and just one x-intercept, $x = 1$.

It is known that f has a removable discontinuity at $x = -1$ and $\lim_{x \to -1} f(x) = 2$. Evaluate

(a) $f(0)$ (b) $\lim\limits_{x \to \infty} f(x)$

60–64 Find the limits as $x \to \infty$ and as $x \to -\infty$. Use this information, together with intercepts, to give a rough sketch of the graph as in Example 12.

60. $y = 2x^3 - x^4$ **61.** $y = x^4 - x^6$

62. $y = x^3(x + 2)^2(x - 1)$

63. $y = (3 - x)(1 + x)^2(1 - x)^4$

64. $y = x^2(x^2 - 1)^2(x + 2)$

---

65. (a) Use the Squeeze Theorem to evaluate $\lim\limits_{x \to \infty} \dfrac{\sin x}{x}$.

(b) Graph $f(x) = (\sin x)/x$. How many times does the graph cross the asymptote?

66. By the *end behavior* of a function we mean the behavior of its values as $x \to \infty$ and as $x \to -\infty$.

(a) Describe and compare the end behavior of the functions

$$P(x) = 3x^5 - 5x^3 + 2x \qquad Q(x) = 3x^5$$

by graphing both functions in the viewing rectangles $[-2, 2]$ by $[-2, 2]$ and $[-10, 10]$ by $[-10{,}000, 10{,}000]$.

(b) Two functions are said to have the *same end behavior* if their ratio approaches 1 as $x \to \infty$. Show that P and Q have the same end behavior.

67. Find $\lim_{x \to \infty} f(x)$ if, for all $x > 1$,

$$\frac{10e^x - 21}{2e^x} < f(x) < \frac{5\sqrt{x}}{\sqrt{x - 1}}$$

68. (a) A tank contains 5000 L of pure water. Brine that contains 30 g of salt per liter of water is pumped into the tank at a rate of 25 L/min. Show that the concentration of salt after t minutes (in grams per liter) is

$$C(t) = \frac{30t}{200 + t}$$

(b) What happens to the concentration as $t \to \infty$?

69. In Chapter 9 we will be able to show, under certain assumptions, that the velocity $v(t)$ of a falling raindrop at time t is

$$v(t) = v^*(1 - e^{-gt/v^*})$$

where g is the acceleration due to gravity and v^* is the *terminal velocity* of the raindrop.

(a) Find $\lim_{t \to \infty} v(t)$.

(b) Graph $v(t)$ if $v^* = 1$ m/s and $g = 9.8$ m/s^2. How long does it take for the velocity of the raindrop to reach 99% of its terminal velocity?

70. (a) By graphing $y = e^{-x/10}$ and $y = 0.1$ on a common screen, discover how large you need to make x so that $e^{-x/10} < 0.1$.

(b) Can you solve part (a) without using a graphing device?

71. Use a graph to find a number N such that

$$\text{if} \quad x > N \quad \text{then} \quad \left| \frac{3x^2 + 1}{2x^2 + x + 1} - 1.5 \right| < 0.05$$

72. For the limit

$$\lim_{x \to \infty} \frac{1 - 3x}{\sqrt{x^2 + 1}} = -3$$

illustrate Definition 7 by finding values of N that correspond to $\varepsilon = 0.1$ and $\varepsilon = 0.05$.

73. For the limit

$$\lim_{x \to -\infty} \frac{1 - 3x}{\sqrt{x^2 + 1}} = 3$$

illustrate Definition 8 by finding values of N that correspond to $\varepsilon = 0.1$ and $\varepsilon = 0.05$.

74. For the limit

$$\lim_{x \to \infty} \sqrt{x} \ln x = \infty$$

illustrate Definition 9 by finding a value of N that corresponds to $M = 100$.

75. (a) How large do we have to take x so that $1/x^2 < 0.0001$?

(b) Taking $r = 2$ in Theorem 5, we have the statement

$$\lim_{x \to \infty} \frac{1}{x^2} = 0$$

Prove this directly using Definition 7.

76. (a) How large do we have to take x so that $1/\sqrt{x} < 0.0001$?

(b) Taking $r = \frac{1}{2}$ in Theorem 5, we have the statement

$$\lim_{x \to \infty} \frac{1}{\sqrt{x}} = 0$$

Prove this directly using Definition 7.

77. Use Definition 8 to prove that $\lim\limits_{x \to -\infty} \dfrac{1}{x} = 0$.

78. Prove, using Definition 9, that $\lim\limits_{x \to \infty} x^3 = \infty$.

79. Use Definition 9 to prove that $\lim\limits_{x \to \infty} e^x = \infty$.

80. Formulate a precise definition of

$$\lim_{x \to -\infty} f(x) = -\infty$$

Then use your definition to prove that

$$\lim_{x \to -\infty} (1 + x^3) = -\infty$$

81. (a) Prove that

$$\lim_{x \to \infty} f(x) = \lim_{t \to 0^+} f(1/t)$$

and $\quad \lim\limits_{x \to -\infty} f(x) = \lim\limits_{t \to 0^-} f(1/t)$

if these limits exist.

(b) Use part (a) and Exercise 65 to find

$$\lim_{x \to 0^+} x \sin \frac{1}{x}$$

2.7 Derivatives and Rates of Change

The problem of finding the tangent line to a curve and the problem of finding the velocity of an object both involve finding the same type of limit, as we saw in Section 2.1. This special type of limit is called a *derivative* and we will see that it can be interpreted as a rate of change in any of the natural or social sciences or engineering.

■ Tangents

If a curve C has equation $y = f(x)$ and we want to find the tangent line to C at the point $P(a, f(a))$, then we consider a nearby point $Q(x, f(x))$, where $x \neq a$, and compute the slope of the secant line PQ:

$$m_{PQ} = \frac{f(x) - f(a)}{x - a}$$

Then we let Q approach P along the curve C by letting x approach a. If m_{PQ} approaches

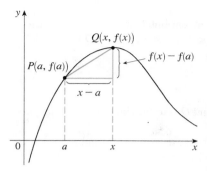

a number m, then we define the *tangent* t to be the line through P with slope m. (This amounts to saying that the tangent line is the limiting position of the secant line PQ as Q approaches P. See Figure 1.)

> **1 Definition** The **tangent line** to the curve $y = f(x)$ at the point $P(a, f(a))$ is the line through P with slope
>
> $$m = \lim_{x \to a} \frac{f(x) - f(a)}{x - a}$$
>
> provided that this limit exists.

In our first example we confirm the guess we made in Example 2.1.1.

EXAMPLE 1 Find an equation of the tangent line to the parabola $y = x^2$ at the point $P(1, 1)$.

SOLUTION Here we have $a = 1$ and $f(x) = x^2$, so the slope is

$$m = \lim_{x \to 1} \frac{f(x) - f(1)}{x - 1} = \lim_{x \to 1} \frac{x^2 - 1}{x - 1}$$

$$= \lim_{x \to 1} \frac{(x - 1)(x + 1)}{x - 1}$$

$$= \lim_{x \to 1} (x + 1) = 1 + 1 = 2$$

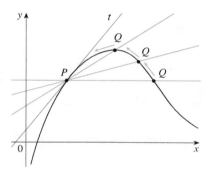

FIGURE 1

Point-slope form for a line through the point (x_1, y_1) with slope m:

$$y - y_1 = m(x - x_1)$$

Using the point-slope form of the equation of a line, we find that an equation of the tangent line at $(1, 1)$ is

$$y - 1 = 2(x - 1) \qquad \text{or} \qquad y = 2x - 1 \qquad ■$$

We sometimes refer to the slope of the tangent line to a curve at a point as the **slope of the curve** at the point. The idea is that if we zoom in far enough toward the point, the curve looks almost like a straight line. Figure 2 illustrates this procedure for the curve $y = x^2$ in Example 1. The more we zoom in, the more the parabola looks like a line. In other words, the curve becomes almost indistinguishable from its tangent line.

TEC Visual 2.7 shows an animation of Figure 2.

FIGURE 2 Zooming in toward the point $(1, 1)$ on the parabola $y = x^2$

FIGURE 3

There is another expression for the slope of a tangent line that is sometimes easier to use. If $h = x - a$, then $x = a + h$ and so the slope of the secant line PQ is

$$m_{PQ} = \frac{f(a + h) - f(a)}{h}$$

(See Figure 3 where the case $h > 0$ is illustrated and Q is to the right of P. If it happened that $h < 0$, however, Q would be to the left of P.)

Notice that as x approaches a, h approaches 0 (because $h = x - a$) and so the expression for the slope of the tangent line in Definition 1 becomes

$$\boxed{2} \qquad m = \lim_{h \to 0} \frac{f(a + h) - f(a)}{h}$$

EXAMPLE 2 Find an equation of the tangent line to the hyperbola $y = 3/x$ at the point $(3, 1)$.

SOLUTION Let $f(x) = 3/x$. Then, by Equation 2, the slope of the tangent at $(3, 1)$ is

$$m = \lim_{h \to 0} \frac{f(3 + h) - f(3)}{h}$$

$$= \lim_{h \to 0} \frac{\dfrac{3}{3 + h} - 1}{h} = \lim_{h \to 0} \frac{\dfrac{3 - (3 + h)}{3 + h}}{h}$$

$$= \lim_{h \to 0} \frac{-h}{h(3 + h)} = \lim_{h \to 0} -\frac{1}{3 + h} = -\frac{1}{3}$$

Therefore an equation of the tangent at the point $(3, 1)$ is

$$y - 1 = -\tfrac{1}{3}(x - 3)$$

which simplifies to

$$x + 3y - 6 = 0$$

The hyperbola and its tangent are shown in Figure 4.

FIGURE 4

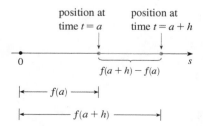

FIGURE 5

Velocities

In Section 2.1 we investigated the motion of a ball dropped from the CN Tower and defined its velocity to be the limiting value of average velocities over shorter and shorter time periods.

In general, suppose an object moves along a straight line according to an equation of motion $s = f(t)$, where s is the displacement (directed distance) of the object from the origin at time t. The function f that describes the motion is called the **position function** of the object. In the time interval from $t = a$ to $t = a + h$ the change in position is $f(a + h) - f(a)$. (See Figure 5.)

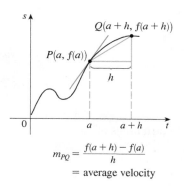

$$m_{PQ} = \frac{f(a+h) - f(a)}{h}$$
$$= \text{average velocity}$$

FIGURE 6

The average velocity over this time interval is

$$\text{average velocity} = \frac{\text{displacement}}{\text{time}} = \frac{f(a+h) - f(a)}{h}$$

which is the same as the slope of the secant line PQ in Figure 6.

Now suppose we compute the average velocities over shorter and shorter time intervals $[a, a + h]$. In other words, we let h approach 0. As in the example of the falling ball, we define the **velocity** (or **instantaneous velocity**) $v(a)$ at time $t = a$ to be the limit of these average velocities:

$$\boxed{3} \qquad v(a) = \lim_{h \to 0} \frac{f(a+h) - f(a)}{h}$$

This means that the velocity at time $t = a$ is equal to the slope of the tangent line at P (compare Equations 2 and 3).

Now that we know how to compute limits, let's reconsider the problem of the falling ball.

EXAMPLE 3 Suppose that a ball is dropped from the upper observation deck of the CN Tower, 450 m above the ground.
(a) What is the velocity of the ball after 5 seconds?
(b) How fast is the ball traveling when it hits the ground?

Recall from Section 2.1: The distance (in meters) fallen after t seconds is $4.9t^2$.

SOLUTION We will need to find the velocity both when $t = 5$ and when the ball hits the ground, so it's efficient to start by finding the velocity at a general time t. Using the equation of motion $s = f(t) = 4.9t^2$, we have

$$v(t) = \lim_{h \to 0} \frac{f(t+h) - f(t)}{h} = \lim_{h \to 0} \frac{4.9(t+h)^2 - 4.9t^2}{h}$$

$$= \lim_{h \to 0} \frac{4.9(t^2 + 2th + h^2 - t^2)}{h} = \lim_{h \to 0} \frac{4.9(2th + h^2)}{h}$$

$$= \lim_{h \to 0} \frac{4.9h(2t + h)}{h} = \lim_{h \to 0} 4.9(2t + h) = 9.8t$$

(a) The velocity after 5 seconds is $v(5) = (9.8)(5) = 49$ m/s.

(b) Since the observation deck is 450 m above the ground, the ball will hit the ground at the time t when $s(t) = 450$, that is,

$$4.9t^2 = 450$$

This gives

$$t^2 = \frac{450}{4.9} \qquad \text{and} \qquad t = \sqrt{\frac{450}{4.9}} \approx 9.6 \text{ s}$$

The velocity of the ball as it hits the ground is therefore

$$v\left(\sqrt{\frac{450}{4.9}}\right) = 9.8\sqrt{\frac{450}{4.9}} \approx 94 \text{ m/s}$$

∎

Derivatives

We have seen that the same type of limit arises in finding the slope of a tangent line (Equation 2) or the velocity of an object (Equation 3). In fact, limits of the form

$$\lim_{h \to 0} \frac{f(a + h) - f(a)}{h}$$

arise whenever we calculate a rate of change in any of the sciences or engineering, such as a rate of reaction in chemistry or a marginal cost in economics. Since this type of limit occurs so widely, it is given a special name and notation.

$f'(a)$ is read "f prime of a."

> **4** **Definition** The **derivative of a function f at a number a**, denoted by $f'(a)$, is
>
> $$f'(a) = \lim_{h \to 0} \frac{f(a + h) - f(a)}{h}$$
>
> if this limit exists.

If we write $x = a + h$, then we have $h = x - a$ and h approaches 0 if and only if x approaches a. Therefore an equivalent way of stating the definition of the derivative, as we saw in finding tangent lines, is

5

$$f'(a) = \lim_{x \to a} \frac{f(x) - f(a)}{x - a}$$

EXAMPLE 4

Find the derivative of the function $f(x) = x^2 - 8x + 9$ at the number a.

SOLUTION From Definition 4 we have

Definitions 4 and 5 are equivalent, so we can use either one to compute the derivative. In practice, Definition 4 often leads to simpler computations.

$$f'(a) = \lim_{h \to 0} \frac{f(a + h) - f(a)}{h}$$

$$= \lim_{h \to 0} \frac{[(a + h)^2 - 8(a + h) + 9] - [a^2 - 8a + 9]}{h}$$

$$= \lim_{h \to 0} \frac{a^2 + 2ah + h^2 - 8a - 8h + 9 - a^2 + 8a - 9}{h}$$

$$= \lim_{h \to 0} \frac{2ah + h^2 - 8h}{h} = \lim_{h \to 0} (2a + h - 8)$$

$$= 2a - 8$$

∎

We defined the tangent line to the curve $y = f(x)$ at the point $P(a, f(a))$ to be the line that passes through P and has slope m given by Equation 1 or 2. Since, by Definition 4, this is the same as the derivative $f'(a)$, we can now say the following.

> The tangent line to $y = f(x)$ at $(a, f(a))$ is the line through $(a, f(a))$ whose slope is equal to $f'(a)$, the derivative of f at a.

If we use the point-slope form of the equation of a line, we can write an equation of the tangent line to the curve $y = f(x)$ at the point $(a, f(a))$:

$$y - f(a) = f'(a)(x - a)$$

EXAMPLE 5 Find an equation of the tangent line to the parabola $y = x^2 - 8x + 9$ at the point $(3, -6)$.

SOLUTION From Example 4 we know that the derivative of $f(x) = x^2 - 8x + 9$ at the number a is $f'(a) = 2a - 8$. Therefore the slope of the tangent line at $(3, -6)$ is $f'(3) = 2(3) - 8 = -2$. Thus an equation of the tangent line, shown in Figure 7, is

$$y - (-6) = (-2)(x - 3) \qquad \text{or} \qquad y = -2x \qquad \blacksquare$$

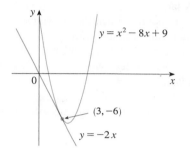

FIGURE 7

Rates of Change

Suppose y is a quantity that depends on another quantity x. Thus y is a function of x and we write $y = f(x)$. If x changes from x_1 to x_2, then the change in x (also called the **increment** of x) is

$$\Delta x = x_2 - x_1$$

and the corresponding change in y is

$$\Delta y = f(x_2) - f(x_1)$$

The difference quotient

$$\frac{\Delta y}{\Delta x} = \frac{f(x_2) - f(x_1)}{x_2 - x_1}$$

is called the **average rate of change of y with respect to x** over the interval $[x_1, x_2]$ and can be interpreted as the slope of the secant line PQ in Figure 8.

By analogy with velocity, we consider the average rate of change over smaller and smaller intervals by letting x_2 approach x_1 and therefore letting Δx approach 0. The limit of these average rates of change is called the **(instantaneous) rate of change of y with respect to x** at $x = x_1$, which (as in the case of velocity) is interpreted as the slope of the tangent to the curve $y = f(x)$ at $P(x_1, f(x_1))$:

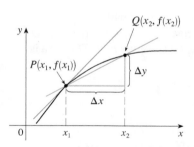

average rate of change = m_{PQ}

instantaneous rate of change = slope of tangent at P

FIGURE 8

> **6**   instantaneous rate of change $= \lim\limits_{\Delta x \to 0} \dfrac{\Delta y}{\Delta x} = \lim\limits_{x_2 \to x_1} \dfrac{f(x_2) - f(x_1)}{x_2 - x_1}$

We recognize this limit as being the derivative $f'(x_1)$.

We know that one interpretation of the derivative $f'(a)$ is as the slope of the tangent line to the curve $y = f(x)$ when $x = a$. We now have a second interpretation:

> The derivative $f'(a)$ is the instantaneous rate of change of $y = f(x)$ with respect to x when $x = a$.

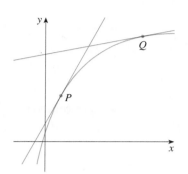

FIGURE 9
The y-values are changing rapidly at P and slowly at Q.

The connection with the first interpretation is that if we sketch the curve $y = f(x)$, then the instantaneous rate of change is the slope of the tangent to this curve at the point where $x = a$. This means that when the derivative is large (and therefore the curve is steep, as at the point P in Figure 9), the y-values change rapidly. When the derivative is small, the curve is relatively flat (as at point Q) and the y-values change slowly.

In particular, if $s = f(t)$ is the position function of a particle that moves along a straight line, then $f'(a)$ is the rate of change of the displacement s with respect to the time t. In other words, $f'(a)$ *is the velocity of the particle at time $t = a$.* The **speed** of the particle is the absolute value of the velocity, that is, $|f'(a)|$.

In the next example we discuss the meaning of the derivative of a function that is defined verbally.

EXAMPLE 6 A manufacturer produces bolts of a fabric with a fixed width. The cost of producing x yards of this fabric is $C = f(x)$ dollars.
(a) What is the meaning of the derivative $f'(x)$? What are its units?
(b) In practical terms, what does it mean to say that $f'(1000) = 9$?
(c) Which do you think is greater, $f'(50)$ or $f'(500)$? What about $f'(5000)$?

SOLUTION
(a) The derivative $f'(x)$ is the instantaneous rate of change of C with respect to x; that is, $f'(x)$ means the rate of change of the production cost with respect to the number of yards produced. (Economists call this rate of change the *marginal cost.* This idea is discussed in more detail in Sections 3.7 and 4.7.)

Because

$$f'(x) = \lim_{\Delta x \to 0} \frac{\Delta C}{\Delta x}$$

the units for $f'(x)$ are the same as the units for the difference quotient $\Delta C/\Delta x$. Since ΔC is measured in dollars and Δx in yards, it follows that the units for $f'(x)$ are dollars per yard.

(b) The statement that $f'(1000) = 9$ means that, after 1000 yards of fabric have been manufactured, the rate at which the production cost is increasing is \$9/yard. (When $x = 1000$, C is increasing 9 times as fast as x.)

Since $\Delta x = 1$ is small compared with $x = 1000$, we could use the approximation

Here we are assuming that the cost function is well behaved; in other words, $C(x)$ doesn't oscillate rapidly near $x = 1000$.

$$f'(1000) \approx \frac{\Delta C}{\Delta x} = \frac{\Delta C}{1} = \Delta C$$

and say that the cost of manufacturing the 1000th yard (or the 1001st) is about \$9.

(c) The rate at which the production cost is increasing (per yard) is probably lower when $x = 500$ than when $x = 50$ (the cost of making the 500th yard is less than the cost of the 50th yard) because of economies of scale. (The manufacturer makes more efficient use of the fixed costs of production.) So

$$f'(50) > f'(500)$$

But, as production expands, the resulting large-scale operation might become ineffi-
cient and there might be overtime costs. Thus it is possible that the rate of increase of
costs will eventually start to rise. So it may happen that

$$f'(5000) > f'(500)$$ ∎

In the following example we estimate the rate of change of the national debt with
respect to time. Here the function is defined not by a formula but by a table of values.

t	$D(t)$
1985	1945.9
1990	3364.8
1995	4988.7
2000	5662.2
2005	8170.4
2010	14,025.2

Source: US Dept. of the Treasury

EXAMPLE 7 Let $D(t)$ be the US national debt at time t. The table in the margin gives
approximate values of this function by providing end of year estimates, in billions of
dollars, from 1985 to 2010. Interpret and estimate the value of $D'(2000)$.

SOLUTION The derivative $D'(2000)$ means the rate of change of D with respect to t
when $t = 2000$, that is, the rate of increase of the national debt in 2000.
    According to Equation 5,

$$D'(2000) = \lim_{t \to 2000} \frac{D(t) - D(2000)}{t - 2000}$$

So we compute and tabulate values of the difference quotient (the average rates of
change) as follows.

t	Time interval	Average rate of change $= \dfrac{D(t) - D(2000)}{t - 2000}$
1985	[1985, 2000]	247.75
1990	[1990, 2000]	229.74
1995	[1995, 2000]	134.70
2005	[2000, 2005]	501.64
2010	[2000, 2010]	836.30

A Note On Units

The units for the average rate of change
$\Delta D/\Delta t$ are the units for ΔD divided by
the units for Δt, namely billions of dollars per year. The instantaneous rate of
change is the limit of the average rates
of change, so it is measured in the same
units: billions of dollars per year.

From this table we see that $D'(2000)$ lies somewhere between 134.70 and 501.64 billion
dollars per year. [Here we are making the reasonable assumption that the debt didn't
fluctuate wildly between 1995 and 2005.] We estimate that the rate of increase of the
national debt of the United States in 2000 was the average of these two numbers, namely

$$D'(2000) \approx 318 \text{ billion dollars per year}$$

Another method would be to plot the debt function and estimate the slope of the
tangent line when $t = 2000$.   ∎

In Examples 3, 6, and 7 we saw three specific examples of rates of change: the velocity of an object is the rate of change of displacement with respect to time; marginal cost
is the rate of change of production cost with respect to the number of items produced; the
rate of change of the debt with respect to time is of interest in economics. Here is a small
sample of other rates of change: In physics, the rate of change of work with respect to
time is called *power*. Chemists who study a chemical reaction are interested in the rate of
change in the concentration of a reactant with respect to time (called the *rate of reaction*).

A biologist is interested in the rate of change of the population of a colony of bacteria with respect to time. In fact, the computation of rates of change is important in all of the natural sciences, in engineering, and even in the social sciences. Further examples will be given in Section 3.7.

All these rates of change are derivatives and can therefore be interpreted as slopes of tangents. This gives added significance to the solution of the tangent problem. Whenever we solve a problem involving tangent lines, we are not just solving a problem in geometry. We are also implicitly solving a great variety of problems involving rates of change in science and engineering.

2.7 EXERCISES

1. A curve has equation $y = f(x)$.
  (a) Write an expression for the slope of the secant line through the points $P(3, f(3))$ and $Q(x, f(x))$.
  (b) Write an expression for the slope of the tangent line at P.

2. Graph the curve $y = e^x$ in the viewing rectangles $[-1, 1]$ by $[0, 2]$, $[-0.5, 0.5]$ by $[0.5, 1.5]$, and $[-0.1, 0.1]$ by $[0.9, 1.1]$. What do you notice about the curve as you zoom in toward the point $(0, 1)$?

3. (a) Find the slope of the tangent line to the parabola $y = 4x - x^2$ at the point $(1, 3)$
      (i) using Definition 1   (ii) using Equation 2
  (b) Find an equation of the tangent line in part (a).
  (c) Graph the parabola and the tangent line. As a check on your work, zoom in toward the point $(1, 3)$ until the parabola and the tangent line are indistinguishable.

4. (a) Find the slope of the tangent line to the curve $y = x - x^3$ at the point $(1, 0)$
      (i) using Definition 1   (ii) using Equation 2
  (b) Find an equation of the tangent line in part (a).
  (c) Graph the curve and the tangent line in successively smaller viewing rectangles centered at $(1, 0)$ until the curve and the line appear to coincide.

5–8 Find an equation of the tangent line to the curve at the given point.

5. $y = 4x - 3x^2$, $(2, -4)$ **6.** $y = x^3 - 3x + 1$, $(2, 3)$

7. $y = \sqrt{x}$, $(1, 1)$ **8.** $y = \dfrac{2x + 1}{x + 2}$, $(1, 1)$

9. (a) Find the slope of the tangent to the curve $y = 3 + 4x^2 - 2x^3$ at the point where $x = a$.
  (b) Find equations of the tangent lines at the points $(1, 5)$ and $(2, 3)$.
  (c) Graph the curve and both tangents on a common screen.

10. (a) Find the slope of the tangent to the curve $y = 1/\sqrt{x}$ at the point where $x = a$.

  (b) Find equations of the tangent lines at the points $(1, 1)$ and $\left(4, \frac{1}{2}\right)$.
  (c) Graph the curve and both tangents on a common screen.

11. (a) A particle starts by moving to the right along a horizontal line; the graph of its position function is shown in the figure. When is the particle moving to the right? Moving to the left? Standing still?
  (b) Draw a graph of the velocity function.

12. Shown are graphs of the position functions of two runners, A and B, who run a 100-meter race and finish in a tie.

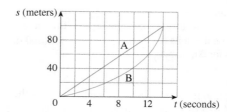

  (a) Describe and compare how the runners run the race.
  (b) At what time is the distance between the runners the greatest?
  (c) At what time do they have the same velocity?

13. If a ball is thrown into the air with a velocity of 40 ft/s, its height (in feet) after t seconds is given by $y = 40t - 16t^2$. Find the velocity when $t = 2$.

14. If a rock is thrown upward on the planet Mars with a velocity of 10 m/s, its height (in meters) after t seconds is given by $H = 10t - 1.86t^2$.
   (a) Find the velocity of the rock after one second.
   (b) Find the velocity of the rock when $t = a$.
   (c) When will the rock hit the surface?
   (d) With what velocity will the rock hit the surface?

15. The displacement (in meters) of a particle moving in a straight line is given by the equation of motion $s = 1/t^2$, where t is measured in seconds. Find the velocity of the particle at times $t = a$, $t = 1$, $t = 2$, and $t = 3$.

16. The displacement (in feet) of a particle moving in a straight line is given by $s = \frac{1}{2}t^2 - 6t + 23$, where t is measured in seconds.
   (a) Find the average velocity over each time interval:
      (i) $[4, 8]$ (ii) $[6, 8]$
      (iii) $[8, 10]$ (iv) $[8, 12]$
   (b) Find the instantaneous velocity when $t = 8$.
   (c) Draw the graph of s as a function of t and draw the secant lines whose slopes are the average velocities in part (a). Then draw the tangent line whose slope is the instantaneous velocity in part (b).

17. For the function g whose graph is given, arrange the following numbers in increasing order and explain your reasoning:

$$0 \qquad g'(-2) \qquad g'(0) \qquad g'(2) \qquad g'(4)$$

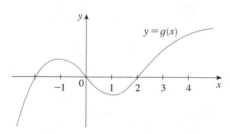

18. The graph of a function f is shown.
   (a) Find the average rate of change of f on the interval $[20, 60]$.
   (b) Identify an interval on which the average rate of change of f is 0.
   (c) Which interval gives a larger average rate of change, $[40, 60]$ or $[40, 70]$?
   (d) Compute $\dfrac{f(40) - f(10)}{40 - 10}$; what does this value represent geometrically?

19. For the function f graphed in Exercise 18:
   (a) Estimate the value of $f'(50)$.
   (b) Is $f'(10) > f'(30)$?
   (c) Is $f'(60) > \dfrac{f(80) - f(40)}{80 - 40}$? Explain.

20. Find an equation of the tangent line to the graph of $y = g(x)$ at $x = 5$ if $g(5) = -3$ and $g'(5) = 4$.

21. If an equation of the tangent line to the curve $y = f(x)$ at the point where $a = 2$ is $y = 4x - 5$, find $f(2)$ and $f'(2)$.

22. If the tangent line to $y = f(x)$ at $(4, 3)$ passes through the point $(0, 2)$, find $f(4)$ and $f'(4)$.

23. Sketch the graph of a function f for which $f(0) = 0$, $f'(0) = 3$, $f'(1) = 0$, and $f'(2) = -1$.

24. Sketch the graph of a function g for which $g(0) = g(2) = g(4) = 0$, $g'(1) = g'(3) = 0$, $g'(0) = g'(4) = 1$, $g'(2) = -1$, $\lim_{x \to \infty} g(x) = \infty$, and $\lim_{x \to -\infty} g(x) = -\infty$.

25. Sketch the graph of a function g that is continuous on its domain $(-5, 5)$ and where $g(0) = 1$, $g'(0) = 1$, $g'(-2) = 0$, $\lim_{x \to -5^+} g(x) = \infty$, and $\lim_{x \to 5^-} g(x) = 3$.

26. Sketch the graph of a function f where the domain is $(-2, 2)$, $f'(0) = -2$, $\lim_{x \to 2^-} f(x) = \infty$, f is continuous at all numbers in its domain except ± 1, and f is odd.

27. If $f(x) = 3x^2 - x^3$, find $f'(1)$ and use it to find an equation of the tangent line to the curve $y = 3x^2 - x^3$ at the point $(1, 2)$.

28. If $g(x) = x^4 - 2$, find $g'(1)$ and use it to find an equation of the tangent line to the curve $y = x^4 - 2$ at the point $(1, -1)$.

29. (a) If $F(x) = 5x/(1 + x^2)$, find $F'(2)$ and use it to find an equation of the tangent line to the curve $y = 5x/(1 + x^2)$ at the point $(2, 2)$.
   (b) Illustrate part (a) by graphing the curve and the tangent line on the same screen.

30. (a) If $G(x) = 4x^2 - x^3$, find $G'(a)$ and use it to find equations of the tangent lines to the curve $y = 4x^2 - x^3$ at the points $(2, 8)$ and $(3, 9)$.
   (b) Illustrate part (a) by graphing the curve and the tangent lines on the same screen.

31–36 Find $f'(a)$.

31. $f(x) = 3x^2 - 4x + 1$ **32.** $f(t) = 2t^3 + t$

33. $f(t) = \dfrac{2t + 1}{t + 3}$ **34.** $f(x) = x^{-2}$

35. $f(x) = \sqrt{1 - 2x}$ **36.** $f(x) = \dfrac{4}{\sqrt{1 - x}}$

37–42 Each limit represents the derivative of some function f at some number a. State such an f and a in each case.

37. $\lim\limits_{h \to 0} \dfrac{\sqrt{9 + h} - 3}{h}$ **38.** $\lim\limits_{h \to 0} \dfrac{e^{-2+h} - e^{-2}}{h}$

39. $\lim\limits_{x \to 2} \dfrac{x^6 - 64}{x - 2}$

40. $\lim\limits_{x \to 1/4} \dfrac{\frac{1}{x} - 4}{x - \frac{1}{4}}$

41. $\lim\limits_{h \to 0} \dfrac{\cos(\pi + h) + 1}{h}$

42. $\lim\limits_{\theta \to \pi/6} \dfrac{\sin\theta - \frac{1}{2}}{\theta - \pi/6}$

43–44 A particle moves along a straight line with equation of motion $s = f(t)$, where s is measured in meters and t in seconds. Find the velocity and the speed when $t = 4$.

43. $f(t) = 80t - 6t^2$

44. $f(t) = 10 + \dfrac{45}{t + 1}$

---

45. A warm can of soda is placed in a cold refrigerator. Sketch the graph of the temperature of the soda as a function of time. Is the initial rate of change of temperature greater or less than the rate of change after an hour?

46. A roast turkey is taken from an oven when its temperature has reached 185°F and is placed on a table in a room where the temperature is 75°F. The graph shows how the temperature of the turkey decreases and eventually approaches room temperature. By measuring the slope of the tangent, estimate the rate of change of the temperature after an hour.

47. Researchers measured the average blood alcohol concentration $C(t)$ of eight men starting one hour after consumption of 30 mL of ethanol (corresponding to two alcoholic drinks).

t (hours)	1.0	1.5	2.0	2.5	3.0
$C(t)$ (mg/mL)	0.33	0.24	0.18	0.12	0.07

(a) Find the average rate of change of C with respect to t over each time interval:
   (i) $[1.0, 2.0]$ (ii) $[1.5, 2.0]$
   (iii) $[2.0, 2.5]$ (iv) $[2.0, 3.0]$
   In each case, include the units.
(b) Estimate the instantaneous rate of change at $t = 2$ and interpret your result. What are the units?

Source: Adapted from P. Wilkinson et al., "Pharmacokinetics of Ethanol after Oral Administration in the Fasting State," *Journal of Pharmacokinetics and Biopharmaceutics* 5 (1977): 207–24.

48. The number N of locations of a popular coffeehouse chain is given in the table. (The numbers of locations as of October 1 are given.)

Year	2004	2006	2008	2010	2012
N	8569	12,440	16,680	16,858	18,066

(a) Find the average rate of growth
   (i) from 2006 to 2008
   (ii) from 2008 to 2010
   In each case, include the units. What can you conclude?
(b) Estimate the instantaneous rate of growth in 2010 by taking the average of two average rates of change. What are its units?
(c) Estimate the instantaneous rate of growth in 2010 by measuring the slope of a tangent.

49. The table shows world average daily oil consumption from 1985 to 2010 measured in thousands of barrels per day.
(a) Compute and interpret the average rate of change from 1990 to 2005. What are the units?
(b) Estimate the instantaneous rate of change in 2000 by taking the average of two average rates of change. What are its units?

Years since 1985	Thousands of barrels of oil per day
0	60,083
5	66,533
10	70,099
15	76,784
20	84,077
25	87,302

Source: US Energy Information Administration

50. The table shows values of the viral load $V(t)$ in HIV patient 303, measured in RNA copies/mL, t days after ABT-538 treatment was begun.

t	4	8	11	15	22
$V(t)$	53	18	9.4	5.2	3.6

(a) Find the average rate of change of V with respect to t over each time interval:
   (i) $[4, 11]$ (ii) $[8, 11]$
   (iii) $[11, 15]$ (iv) $[11, 22]$
   What are the units?
(b) Estimate and interpret the value of the derivative $V'(11)$.

Source: Adapted from D. Ho et al., "Rapid Turnover of Plasma Virions and CD4 Lymphocytes in HIV-1 Infection," *Nature* 373 (1995): 123–26.

51. The cost (in dollars) of producing x units of a certain commodity is $C(x) = 5000 + 10x + 0.05x^2$.

    (a) Find the average rate of change of C with respect to x when the production level is changed

        (i) from $x = 100$ to $x = 105$

        (ii) from $x = 100$ to $x = 101$

    (b) Find the instantaneous rate of change of C with respect to x when $x = 100$. (This is called the *marginal cost*. Its significance will be explained in Section 3.7.)

52. If a cylindrical tank holds 100,000 gallons of water, which can be drained from the bottom of the tank in an hour, then Torricelli's Law gives the volume V of water remaining in the tank after t minutes as

$$V(t) = 100{,}000\left(1 - \tfrac{1}{60}t\right)^2 \qquad 0 \le t \le 60$$

Find the rate at which the water is flowing out of the tank (the instantaneous rate of change of V with respect to t) as a function of t. What are its units? For times $t = 0$, 10, 20, 30, 40, 50, and 60 min, find the flow rate and the amount of water remaining in the tank. Summarize your findings in a sentence or two. At what time is the flow rate the greatest? The least?

53. The cost of producing x ounces of gold from a new gold mine is $C = f(x)$ dollars.

    (a) What is the meaning of the derivative $f'(x)$? What are its units?

    (b) What does the statement $f'(800) = 17$ mean?

    (c) Do you think the values of $f'(x)$ will increase or decrease in the short term? What about the long term? Explain.

54. The number of bacteria after t hours in a controlled laboratory experiment is $n = f(t)$.

    (a) What is the meaning of the derivative $f'(5)$? What are its units?

    (b) Suppose there is an unlimited amount of space and nutrients for the bacteria. Which do you think is larger, $f'(5)$ or $f'(10)$? If the supply of nutrients is limited, would that affect your conclusion? Explain.

55. Let $H(t)$ be the daily cost (in dollars) to heat an office building when the outside temperature is t degrees Fahrenheit.

    (a) What is the meaning of $H'(58)$? What are its units?

    (b) Would you expect $H'(58)$ to be positive or negative? Explain.

56. The quantity (in pounds) of a gourmet ground coffee that is sold by a coffee company at a price of p dollars per pound is $Q = f(p)$.

    (a) What is the meaning of the derivative $f'(8)$? What are its units?

    (b) Is $f'(8)$ positive or negative? Explain.

57. The quantity of oxygen that can dissolve in water depends on the temperature of the water. (So thermal pollution influences the oxygen content of water.) The graph shows how oxygen solubility S varies as a function of the water temperature T.

    (a) What is the meaning of the derivative $S'(T)$? What are its units?

    (b) Estimate the value of $S'(16)$ and interpret it.

Source: C. Kupchella et al., *Environmental Science: Living Within the System of Nature*, 2d ed. (Boston: Allyn and Bacon, 1989).

58. The graph shows the influence of the temperature T on the maximum sustainable swimming speed S of Coho salmon.

    (a) What is the meaning of the derivative $S'(T)$? What are its units?

    (b) Estimate the values of $S'(15)$ and $S'(25)$ and interpret them.

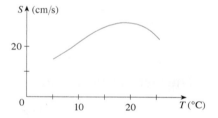

59–60 Determine whether $f'(0)$ exists.

59. $f(x) = \begin{cases} x \sin \dfrac{1}{x} & \text{if } x \ne 0 \\ 0 & \text{if } x = 0 \end{cases}$

60. $f(x) = \begin{cases} x^2 \sin \dfrac{1}{x} & \text{if } x \ne 0 \\ 0 & \text{if } x = 0 \end{cases}$

61. (a) Graph the function $f(x) = \sin x - \frac{1}{1000}\sin(1000x)$ in the viewing rectangle $[-2\pi, 2\pi]$ by $[-4, 4]$. What slope does the graph appear to have at the origin?

    (b) Zoom in to the viewing window $[-0.4, 0.4]$ by $[-0.25, 0.25]$ and estimate the value of $f'(0)$. Does this agree with your answer from part (a)?

    (c) Now zoom in to the viewing window $[-0.008, 0.008]$ by $[-0.005, 0.005]$. Do you wish to revise your estimate for $f'(0)$?

WRITING PROJECT   **EARLY METHODS FOR FINDING TANGENTS**

The first person to formulate explicitly the ideas of limits and derivatives was Sir Isaac Newton in the 1660s. But Newton acknowledged that "If I have seen further than other men, it is because I have stood on the shoulders of giants." Two of those giants were Pierre Fermat (1601–1665) and Newton's mentor at Cambridge, Isaac Barrow (1630–1677). Newton was familiar with the methods that these men used to find tangent lines, and their methods played a role in Newton's eventual formulation of calculus.

The following references contain explanations of these methods. Read one or more of the references and write a report comparing the methods of either Fermat or Barrow to modern methods. In particular, use the method of Section 2.7 to find an equation of the tangent line to the curve $y = x^3 + 2x$ at the point $(1, 3)$ and show how either Fermat or Barrow would have solved the same problem. Although you used derivatives and they did not, point out similarities between the methods.

1. Carl Boyer and Uta Merzbach, *A History of Mathematics* (New York: Wiley, 1989), pp. 389, 432.

2. C. H. Edwards, *The Historical Development of the Calculus* (New York: Springer-Verlag, 1979), pp. 124, 132.

3. Howard Eves, *An Introduction to the History of Mathematics*, 6th ed. (New York: Saunders, 1990), pp. 391, 395.

4. Morris Kline, *Mathematical Thought from Ancient to Modern Times* (New York: Oxford University Press, 1972), pp. 344, 346.

2.8   The Derivative as a Function

In the preceding section we considered the derivative of a function f at a fixed number a:

$$\boxed{1}\qquad f'(a) = \lim_{h \to 0} \frac{f(a + h) - f(a)}{h}$$

Here we change our point of view and let the number a vary. If we replace a in Equation 1 by a variable x, we obtain

$$\boxed{2}\qquad f'(x) = \lim_{h \to 0} \frac{f(x + h) - f(x)}{h}$$

Given any number x for which this limit exists, we assign to x the number $f'(x)$. So we can regard f' as a new function, called the **derivative of f** and defined by Equation 2. We know that the value of f' at x, $f'(x)$, can be interpreted geometrically as the slope of the tangent line to the graph of f at the point $(x, f(x))$.

The function f' is called the derivative of f because it has been "derived" from f by the limiting operation in Equation 2. The domain of f' is the set $\{x \mid f'(x) \text{ exists}\}$ and may be smaller than the domain of f.

EXAMPLE 1 The graph of a function f is given in Figure 1. Use it to sketch the graph of the derivative f'.

SOLUTION We can estimate the value of the derivative at any value of x by drawing the tangent at the point $(x, f(x))$ and estimating its slope. For instance, for $x = 5$ we draw the tangent at P in Figure 2(a) and estimate its slope to be about $\frac{3}{2}$, so $f'(5) \approx 1.5$. This allows us to plot the point $P'(5, 1.5)$ on the graph of f' directly beneath P. (The slope of the graph of f becomes the y-value on the graph of f'.) Repeating this procedure at several points, we get the graph shown in Figure 2(b). Notice that the tangents at A, B, and C are horizontal, so the derivative is 0 there and the graph of f' crosses the x-axis (where $y = 0$) at the points A', B', and C', directly beneath A, B, and C. Between A and B the tangents have positive slope, so $f'(x)$ is positive there. (The graph is above the x-axis.) But between B and C the tangents have negative slope, so $f'(x)$ is negative there.

FIGURE 1

TEC Visual 2.8 shows an animation of Figure 2 for several functions.

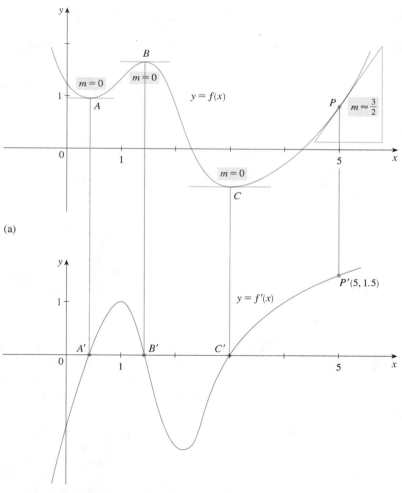

(a)

(b)

FIGURE 2

EXAMPLE 2
(a) If $f(x) = x^3 - x$, find a formula for $f'(x)$.
(b) Illustrate this formula by comparing the graphs of f and f'.

SOLUTION
(a) When using Equation 2 to compute a derivative, we must remember that the variable is h and that x is temporarily regarded as a constant during the calculation of the limit.

$$f'(x) = \lim_{h \to 0} \frac{f(x + h) - f(x)}{h} = \lim_{h \to 0} \frac{[(x + h)^3 - (x + h)] - [x^3 - x]}{h}$$

$$= \lim_{h \to 0} \frac{x^3 + 3x^2h + 3xh^2 + h^3 - x - h - x^3 + x}{h}$$

$$= \lim_{h \to 0} \frac{3x^2h + 3xh^2 + h^3 - h}{h}$$

$$= \lim_{h \to 0} (3x^2 + 3xh + h^2 - 1) = 3x^2 - 1$$

(b) We use a graphing device to graph f and f' in Figure 3. Notice that $f'(x) = 0$ when f has horizontal tangents and $f'(x)$ is positive when the tangents have positive slope. So these graphs serve as a check on our work in part (a).

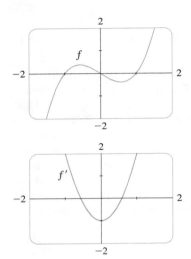

FIGURE 3

EXAMPLE 3 If $f(x) = \sqrt{x}$, find the derivative of f. State the domain of f'.

SOLUTION

$$f'(x) = \lim_{h \to 0} \frac{f(x + h) - f(x)}{h}$$

$$= \lim_{h \to 0} \frac{\sqrt{x + h} - \sqrt{x}}{h}$$

$$= \lim_{h \to 0} \left(\frac{\sqrt{x + h} - \sqrt{x}}{h} \cdot \frac{\sqrt{x + h} + \sqrt{x}}{\sqrt{x + h} + \sqrt{x}} \right) \quad \text{(Rationalize the numerator.)}$$

$$= \lim_{h \to 0} \frac{(x + h) - x}{h(\sqrt{x + h} + \sqrt{x})} = \lim_{h \to 0} \frac{h}{h(\sqrt{x + h} + \sqrt{x})}$$

$$= \lim_{h \to 0} \frac{1}{\sqrt{x + h} + \sqrt{x}} = \frac{1}{\sqrt{x} + \sqrt{x}} = \frac{1}{2\sqrt{x}}$$

We see that $f'(x)$ exists if $x > 0$, so the domain of f' is $(0, \infty)$. This is slightly smaller than the domain of f, which is $[0, \infty)$.

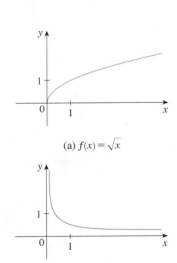

(a) $f(x) = \sqrt{x}$

(b) $f'(x) = \dfrac{1}{2\sqrt{x}}$

FIGURE 4

Let's check to see that the result of Example 3 is reasonable by looking at the graphs of f and f' in Figure 4. When x is close to 0, $\sqrt{x}$ is also close to 0, so $f'(x) = 1/(2\sqrt{x})$ is very large and this corresponds to the steep tangent lines near $(0, 0)$ in Figure 4(a) and the large values of $f'(x)$ just to the right of 0 in Figure 4(b). When x is large, $f'(x)$ is very small and this corresponds to the flatter tangent lines at the far right of the graph of f and the horizontal asymptote of the graph of f'.

EXAMPLE 4 Find f' if $f(x) = \dfrac{1 - x}{2 + x}$.

SOLUTION

$$f'(x) = \lim_{h \to 0} \frac{f(x + h) - f(x)}{h}$$

$$= \lim_{h \to 0} \frac{\dfrac{1 - (x + h)}{2 + (x + h)} - \dfrac{1 - x}{2 + x}}{h}$$

$$= \lim_{h \to 0} \frac{(1 - x - h)(2 + x) - (1 - x)(2 + x + h)}{h(2 + x + h)(2 + x)}$$

$$= \lim_{h \to 0} \frac{(2 - x - 2h - x^2 - xh) - (2 - x + h - x^2 - xh)}{h(2 + x + h)(2 + x)}$$

$$= \lim_{h \to 0} \frac{-3h}{h(2 + x + h)(2 + x)} = \lim_{h \to 0} \frac{-3}{(2 + x + h)(2 + x)} = -\frac{3}{(2 + x)^2} \quad \blacksquare$$

$$\frac{\dfrac{a}{b} - \dfrac{c}{d}}{e} = \frac{ad - bc}{bd} \cdot \frac{1}{e}$$

Leibniz

Gottfried Wilhelm Leibniz was born in Leipzig in 1646 and studied law, theology, philosophy, and mathematics at the university there, graduating with a bachelor's degree at age 17. After earning his doctorate in law at age 20, Leibniz entered the diplomatic service and spent most of his life traveling to the capitals of Europe on political missions. In particular, he worked to avert a French military threat against Germany and attempted to reconcile the Catholic and Protestant churches.

His serious study of mathematics did not begin until 1672 while he was on a diplomatic mission in Paris. There he built a calculating machine and met scientists, like Huygens, who directed his attention to the latest developments in mathematics and science. Leibniz sought to develop a symbolic logic and system of notation that would simplify logical reasoning. In particular, the version of calculus that he published in 1684 established the notation and the rules for finding derivatives that we use today.

Unfortunately, a dreadful priority dispute arose in the 1690s between the followers of Newton and those of Leibniz as to who had invented calculus first. Leibniz was even accused of plagiarism by members of the Royal Society in England. The truth is that each man invented calculus independently. Newton arrived at his version of calculus first but, because of his fear of controversy, did not publish it immediately. So Leibniz's 1684 account of calculus was the first to be published.

■ **Other Notations**

If we use the traditional notation $y = f(x)$ to indicate that the independent variable is x and the dependent variable is y, then some common alternative notations for the derivative are as follows:

$$f'(x) = y' = \frac{dy}{dx} = \frac{df}{dx} = \frac{d}{dx} f(x) = Df(x) = D_x f(x)$$

The symbols D and d/dx are called **differentiation operators** because they indicate the operation of **differentiation**, which is the process of calculating a derivative.

The symbol dy/dx, which was introduced by Leibniz, should not be regarded as a ratio (for the time being); it is simply a synonym for $f'(x)$. Nonetheless, it is a very useful and suggestive notation, especially when used in conjunction with increment notation. Referring to Equation 2.7.6, we can rewrite the definition of derivative in Leibniz notation in the form

$$\frac{dy}{dx} = \lim_{\Delta x \to 0} \frac{\Delta y}{\Delta x}$$

If we want to indicate the value of a derivative dy/dx in Leibniz notation at a specific number a, we use the notation

$$\frac{dy}{dx}\bigg|_{x=a} \qquad \text{or} \qquad \frac{dy}{dx}\bigg]_{x=a}$$

which is a synonym for $f'(a)$. The vertical bar means "evaluate at."

> **3   Definition** A function f is **differentiable at a** if $f'(a)$ exists. It is **differentiable on an open interval** (a, b) [or (a, ∞) or $(-\infty, a)$ or $(-\infty, \infty)$] if it is differentiable at every number in the interval.

EXAMPLE 5 Where is the function $f(x) = |x|$ differentiable?

SOLUTION If $x > 0$, then $|x| = x$ and we can choose h small enough that $x + h > 0$ and hence $|x + h| = x + h$. Therefore, for $x > 0$, we have

$$f'(x) = \lim_{h \to 0} \frac{|x + h| - |x|}{h} = \lim_{h \to 0} \frac{(x + h) - x}{h}$$

$$= \lim_{h \to 0} \frac{h}{h} = \lim_{h \to 0} 1 = 1$$

and so f is differentiable for any $x > 0$.

Similarly, for $x < 0$ we have $|x| = -x$ and h can be chosen small enough that $x + h < 0$ and so $|x + h| = -(x + h)$. Therefore, for $x < 0$,

$$f'(x) = \lim_{h \to 0} \frac{|x + h| - |x|}{h} = \lim_{h \to 0} \frac{-(x + h) - (-x)}{h}$$

$$= \lim_{h \to 0} \frac{-h}{h} = \lim_{h \to 0} (-1) = -1$$

and so f is differentiable for any $x < 0$.

For $x = 0$ we have to investigate

$$f'(0) = \lim_{h \to 0} \frac{f(0 + h) - f(0)}{h}$$

$$= \lim_{h \to 0} \frac{|0 + h| - |0|}{h} = \lim_{h \to 0} \frac{|h|}{h} \qquad \text{(if it exists)}$$

Let's compute the left and right limits separately:

$$\lim_{h \to 0^+} \frac{|h|}{h} = \lim_{h \to 0^+} \frac{h}{h} = \lim_{h \to 0^+} 1 = 1$$

and $$\lim_{h \to 0^-} \frac{|h|}{h} = \lim_{h \to 0^-} \frac{-h}{h} = \lim_{h \to 0^-} (-1) = -1$$

Since these limits are different, $f'(0)$ does not exist. Thus f is differentiable at all x except 0.

A formula for f' is given by

$$f'(x) = \begin{cases} 1 & \text{if } x > 0 \\ -1 & \text{if } x < 0 \end{cases}$$

and its graph is shown in Figure 5(b). The fact that $f'(0)$ does not exist is reflected geometrically in the fact that the curve $y = |x|$ does not have a tangent line at $(0, 0)$. [See Figure 5(a).]

(a) $y = f(x) = |x|$

(b) $y = f'(x)$

FIGURE 5

Both continuity and differentiability are desirable properties for a function to have. The following theorem shows how these properties are related.

4 Theorem If f is differentiable at a, then f is continuous at a.

PROOF To prove that f is continuous at a, we have to show that $\lim_{x \to a} f(x) = f(a)$. We do this by showing that the difference $f(x) - f(a)$ approaches 0.

The given information is that f is differentiable at a, that is,

$$f'(a) = \lim_{x \to a} \frac{f(x) - f(a)}{x - a}$$

PS An important aspect of problem solving is trying to find a connection between the given and the unknown. See Step 2 (Think of a Plan) in *Principles of Problem Solving* on page 71.

exists (see Equation 2.7.5). To connect the given and the unknown, we divide and multiply $f(x) - f(a)$ by $x - a$ (which we can do when $x \neq a$):

$$f(x) - f(a) = \frac{f(x) - f(a)}{x - a} (x - a)$$

Thus, using the Product Law and (2.7.5), we can write

$$\lim_{x \to a} [f(x) - f(a)] = \lim_{x \to a} \frac{f(x) - f(a)}{x - a} (x - a)$$

$$= \lim_{x \to a} \frac{f(x) - f(a)}{x - a} \cdot \lim_{x \to a} (x - a)$$

$$= f'(a) \cdot 0 = 0$$

To use what we have just proved, we start with $f(x)$ and add and subtract $f(a)$:

$$\lim_{x \to a} f(x) = \lim_{x \to a} [f(a) + (f(x) - f(a))]$$

$$= \lim_{x \to a} f(a) + \lim_{x \to a} [f(x) - f(a)]$$

$$= f(a) + 0 = f(a)$$

Therefore f is continuous at a. ∎

NOTE The converse of Theorem 4 is false; that is, there are functions that are continuous but not differentiable. For instance, the function $f(x) = |x|$ is continuous at 0 because

$$\lim_{x \to 0} f(x) = \lim_{x \to 0} |x| = 0 = f(0)$$

(See Example 2.3.7.) But in Example 5 we showed that f is not differentiable at 0.

■ How Can a Function Fail To Be Differentiable?

We saw that the function $y = |x|$ in Example 5 is not differentiable at 0 and Figure 5(a) shows that its graph changes direction abruptly when $x = 0$. In general, if the graph of a function f has a "corner" or "kink" in it, then the graph of f has no tangent at this point and f is not differentiable there. [In trying to compute $f'(a)$, we find that the left and right limits are different.]

Theorem 4 gives another way for a function not to have a derivative. It says that if f is not continuous at a, then f is not differentiable at a. So at any discontinuity (for instance, a jump discontinuity) f fails to be differentiable.

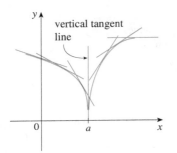

FIGURE 6

A third possibility is that the curve has a **vertical tangent line** when $x = a$; that is, f is continuous at a and

$$\lim_{x \to a} |f'(x)| = \infty$$

This means that the tangent lines become steeper and steeper as $x \to a$. Figure 6 shows one way that this can happen; Figure 7(c) shows another. Figure 7 illustrates the three possibilities that we have discussed.

(a) A corner   (b) A discontinuity   (c) A vertical tangent

FIGURE 7
Three ways for f not to be
differentiable at a

A graphing calculator or computer provides another way of looking at differentiability. If f is differentiable at a, then when we zoom in toward the point $(a, f(a))$ the graph straightens out and appears more and more like a line. (See Figure 8. We saw a specific example of this in Figure 2.7.2.) But no matter how much we zoom in toward a point like the ones in Figures 6 and 7(a), we can't eliminate the sharp point or corner (see Figure 9).

FIGURE 8
f is differentiable at a.

FIGURE 9
f is not differentiable at a.

■ **Higher Derivatives**

If f is a differentiable function, then its derivative f' is also a function, so f' may have a derivative of its own, denoted by $(f')' = f''$. This new function f'' is called the **second derivative** of f because it is the derivative of the derivative of f. Using Leibniz notation, we write the second derivative of $y = f(x)$ as

$$\underbrace{\frac{d}{dx}}_{\substack{\text{derivative} \\ \text{of}}} \underbrace{\left(\frac{dy}{dx} \right)}_{\substack{\text{first} \\ \text{derivative}}} = \underbrace{\frac{d^2y}{dx^2}}_{\substack{\text{second} \\ \text{derivative}}}$$

EXAMPLE 6 If $f(x) = x^3 - x$, find and interpret $f''(x)$.

SOLUTION In Example 2 we found that the first derivative is $f'(x) = 3x^2 - 1$. So the second derivative is

$$f''(x) = (f')'(x) = \lim_{h \to 0} \frac{f'(x+h) - f'(x)}{h}$$

$$= \lim_{h \to 0} \frac{[3(x+h)^2 - 1] - [3x^2 - 1]}{h}$$

$$= \lim_{h \to 0} \frac{3x^2 + 6xh + 3h^2 - 1 - 3x^2 + 1}{h}$$

$$= \lim_{h \to 0} (6x + 3h) = 6x$$

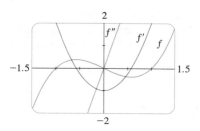

FIGURE 10

TEC In Module 2.8 you can see how changing the coefficients of a polynomial f affects the appearance of the graphs of f, f', and f''.

The graphs of f, f', and f'' are shown in Figure 10.

We can interpret $f''(x)$ as the slope of the curve $y = f'(x)$ at the point $(x, f'(x))$. In other words, it is the rate of change of the slope of the original curve $y = f(x)$.

Notice from Figure 10 that $f''(x)$ is negative when $y = f'(x)$ has negative slope and positive when $y = f'(x)$ has positive slope. So the graphs serve as a check on our calculations. ∎

In general, we can interpret a second derivative as a rate of change of a rate of change. The most familiar example of this is *acceleration*, which we define as follows.

If $s = s(t)$ is the position function of an object that moves in a straight line, we know that its first derivative represents the velocity $v(t)$ of the object as a function of time:

$$v(t) = s'(t) = \frac{ds}{dt}$$

The instantaneous rate of change of velocity with respect to time is called the **acceleration** $a(t)$ of the object. Thus the acceleration function is the derivative of the velocity function and is therefore the second derivative of the position function:

$$a(t) = v'(t) = s''(t)$$

or, in Leibniz notation,

$$a = \frac{dv}{dt} = \frac{d^2s}{dt^2}$$

Acceleration is the change in velocity you feel when speeding up or slowing down in a car.

The **third derivative** f''' is the derivative of the second derivative: $f''' = (f'')'$. So $f'''(x)$ can be interpreted as the slope of the curve $y = f''(x)$ or as the rate of change of $f''(x)$. If $y = f(x)$, then alternative notations for the third derivative are

$$y''' = f'''(x) = \frac{d}{dx}\left(\frac{d^2y}{dx^2}\right) = \frac{d^3y}{dx^3}$$

We can also interpret the third derivative physically in the case where the function is the position function $s = s(t)$ of an object that moves along a straight line. Because $s''' = (s'')' = a'$, the third derivative of the position function is the derivative of the acceleration function and is called the **jerk**:

$$j = \frac{da}{dt} = \frac{d^3s}{dt^3}$$

Thus the jerk j is the rate of change of acceleration. It is aptly named because a large jerk means a sudden change in acceleration, which causes an abrupt movement in a vehicle.

The differentiation process can be continued. The fourth derivative f'''' is usually denoted by $f^{(4)}$. In general, the nth derivative of f is denoted by $f^{(n)}$ and is obtained from f by differentiating n times. If $y = f(x)$, we write

$$y^{(n)} = f^{(n)}(x) = \frac{d^n y}{dx^n}$$

EXAMPLE 7 If $f(x) = x^3 - x$, find $f'''(x)$ and $f^{(4)}(x)$.

SOLUTION In Example 6 we found that $f''(x) = 6x$. The graph of the second derivative has equation $y = 6x$ and so it is a straight line with slope 6. Since the derivative $f'''(x)$ is the slope of $f''(x)$, we have

$$f'''(x) = 6$$

for all values of x. So f''' is a constant function and its graph is a horizontal line. Therefore, for all values of x,

$$f^{(4)}(x) = 0$$ ■

We have seen that one application of second and third derivatives occurs in analyzing the motion of objects using acceleration and jerk. We will investigate another application of second derivatives in Section 4.3, where we show how knowledge of f'' gives us information about the shape of the graph of f. In Chapter 11 we will see how second and higher derivatives enable us to represent functions as sums of infinite series.

2.8   EXERCISES

1–2 Use the given graph to estimate the value of each derivative. Then sketch the graph of f'.

1. (a) $f'(-3)$ (b) $f'(-2)$ (c) $f'(-1)$ (d) $f'(0)$
    (e) $f'(1)$ (f) $f'(2)$ (g) $f'(3)$

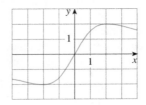

2. (a) $f'(0)$ (b) $f'(1)$ (c) $f'(2)$ (d) $f'(3)$
    (e) $f'(4)$ (f) $f'(5)$ (g) $f'(6)$ (h) $f'(7)$

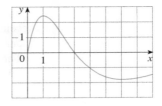

3. Match the graph of each function in (a)–(d) with the graph of its derivative in I–IV. Give reasons for your choices.

(a)

(b)

(c)

(d)

I

II

III

IV

4–11 Trace or copy the graph of the given function f. (Assume that the axes have equal scales.) Then use the method of Example 1 to sketch the graph of f' below it.

4.

5.

6.

7.

8.

9.

10.

11.
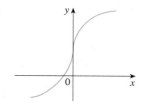

12. Shown is the graph of the population function $P(t)$ for yeast cells in a laboratory culture. Use the method of Example 1 to graph the derivative $P'(t)$. What does the graph of P' tell us about the yeast population?

13. A rechargeable battery is plugged into a charger. The graph shows $C(t)$, the percentage of full capacity that the battery reaches as a function of time t elapsed (in hours).
(a) What is the meaning of the derivative $C'(t)$?
(b) Sketch the graph of $C'(t)$. What does the graph tell you?

14. The graph (from the US Department of Energy) shows how driving speed affects gas mileage. Fuel economy F is measured in miles per gallon and speed v is measured in miles per hour.
   (a) What is the meaning of the derivative $F'(v)$?
   (b) Sketch the graph of $F'(v)$.
   (c) At what speed should you drive if you want to save on gas?

15. The graph shows how the average age of first marriage of Japanese men varied in the last half of the 20th century. Sketch the graph of the derivative function $M'(t)$. During which years was the derivative negative?

16–18 Make a careful sketch of the graph of f and below it sketch the graph of f' in the same manner as in Exercises 4–11. Can you guess a formula for $f'(x)$ from its graph?

16. $f(x) = \sin x$ **17.** $f(x) = e^x$ **18.** $f(x) = \ln x$

19. Let $f(x) = x^2$.
   (a) Estimate the values of $f'(0)$, $f'(\frac{1}{2})$, $f'(1)$, and $f'(2)$ by using a graphing device to zoom in on the graph of f.
   (b) Use symmetry to deduce the values of $f'(-\frac{1}{2})$, $f'(-1)$, and $f'(-2)$.
   (c) Use the results from parts (a) and (b) to guess a formula for $f'(x)$.
   (d) Use the definition of derivative to prove that your guess in part (c) is correct.

20. Let $f(x) = x^3$.
   (a) Estimate the values of $f'(0)$, $f'(\frac{1}{2})$, $f'(1)$, $f'(2)$, and $f'(3)$ by using a graphing device to zoom in on the graph of f.

(b) Use symmetry to deduce the values of $f'(-\frac{1}{2})$, $f'(-1)$, $f'(-2)$, and $f'(-3)$.
   (c) Use the values from parts (a) and (b) to graph f'.
   (d) Guess a formula for $f'(x)$.
   (e) Use the definition of derivative to prove that your guess in part (d) is correct.

21–31 Find the derivative of the function using the definition of derivative. State the domain of the function and the domain of its derivative.

21. $f(x) = 3x - 8$ **22.** $f(x) = mx + b$

23. $f(t) = 2.5t^2 + 6t$ **24.** $f(x) = 4 + 8x - 5x^2$

25. $f(x) = x^2 - 2x^3$ **26.** $g(t) = \dfrac{1}{\sqrt{t}}$

27. $g(x) = \sqrt{9 - x}$ **28.** $f(x) = \dfrac{x^2 - 1}{2x - 3}$

29. $G(t) = \dfrac{1 - 2t}{3 + t}$ **30.** $f(x) = x^{3/2}$

31. $f(x) = x^4$

---

32. (a) Sketch the graph of $f(x) = \sqrt{6 - x}$ by starting with the graph of $y = \sqrt{x}$ and using the transformations of Section 1.3.
   (b) Use the graph from part (a) to sketch the graph of f'.
   (c) Use the definition of a derivative to find $f'(x)$. What are the domains of f and f'?
   (d) Use a graphing device to graph f' and compare with your sketch in part (b).

33. (a) If $f(x) = x^4 + 2x$, find $f'(x)$.
   (b) Check to see that your answer to part (a) is reasonable by comparing the graphs of f and f'.

34. (a) If $f(x) = x + 1/x$, find $f'(x)$.
   (b) Check to see that your answer to part (a) is reasonable by comparing the graphs of f and f'.

35. The unemployment rate $U(t)$ varies with time. The table gives the percentage of unemployed in the US labor force from 2003 to 2012.
   (a) What is the meaning of $U'(t)$? What are its units?
   (b) Construct a table of estimated values for $U'(t)$.

t	$U(t)$	t	$U(t)$
2003	6.0	2008	5.8
2004	5.5	2009	9.3
2005	5.1	2010	9.6
2006	4.6	2011	8.9
2007	4.6	2012	8.1

Source: US Bureau of Labor Statistics

36. The table gives the number $N(t)$, measured in thousands, of minimally invasive cosmetic surgery procedures performed in the United States for various years t.

t	$N(t)$ (thousands)
2000	5,500
2002	4,897
2004	7,470
2006	9,138
2008	10,897
2010	11,561
2012	13,035

Source: American Society of Plastic Surgeons

(a) What is the meaning of $N'(t)$? What are its units?
(b) Construct a table of estimated values for $N'(t)$.
(c) Graph N and N'.
(d) How would it be possible to get more accurate values for $N'(t)$?

37. The table gives the height as time passes of a typical pine tree grown for lumber at a managed site.

Tree age (years)	14	21	28	35	42	49
Height (feet)	41	54	64	72	78	83

Source: Arkansas Forestry Commission

If $H(t)$ is the height of the tree after t years, construct a table of estimated values for H' and sketch its graph.

38. Water temperature affects the growth rate of brook trout. The table shows the amount of weight gained by brook trout after 24 days in various water temperatures.

Temperature (°C)	15.5	17.7	20.0	22.4	24.4
Weight gained (g)	37.2	31.0	19.8	9.7	−9.8

If $W(x)$ is the weight gain at temperature x, construct a table of estimated values for W' and sketch its graph. What are the units for $W'(x)$?

Source: Adapted from J. Chadwick Jr., "Temperature Effects on Growth and Stress Physiology of Brook Trout: Implications for Climate Change Impacts on an Iconic Cold-Water Fish." *Masters Theses.* Paper 897. 2012. scholarworks.umass.edu/theses/897.

39. Let P represent the percentage of a city's electrical power that is produced by solar panels t years after January 1, 2000.
(a) What does dP/dt represent in this context?
(b) Interpret the statement

$$\left. \frac{dP}{dt} \right|_{t=2} = 3.5$$

40. Suppose N is the number of people in the United States who travel by car to another state for a vacation this year when the average price of gasoline is p dollars per gallon. Do you expect dN/dp to be positive or negative? Explain.

41–44 The graph of f is given. State, with reasons, the numbers at which f is *not* differentiable.

41.   **42.**

43.   **44.**

45. Graph the function $f(x) = x + \sqrt{|x|}$. Zoom in repeatedly, first toward the point $(-1, 0)$ and then toward the origin. What is different about the behavior of f in the vicinity of these two points? What do you conclude about the differentiability of f?

46. Zoom in toward the points $(1, 0)$, $(0, 1)$, and $(-1, 0)$ on the graph of the function $g(x) = (x^2 - 1)^{2/3}$. What do you notice? Account for what you see in terms of the differentiability of g.

47–48 The graphs of a function f and its derivative f' are shown. Which is bigger, $f'(-1)$ or $f''(1)$?

47.

48.

49. The figure shows the graphs of f, f', and f''. Identify each curve, and explain your choices.

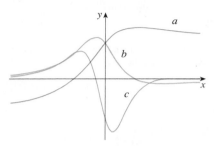

50. The figure shows graphs of f, f', f'', and f'''. Identify each curve, and explain your choices.

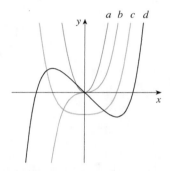

51. The figure shows the graphs of three functions. One is the position function of a car, one is the velocity of the car, and one is its acceleration. Identify each curve, and explain your choices.

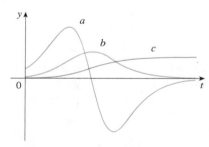

52. The figure shows the graphs of four functions. One is the position function of a car, one is the velocity of the car, one is its acceleration, and one is its jerk. Identify each curve, and explain your choices.

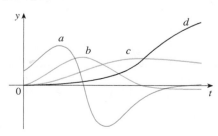

53–54 Use the definition of a derivative to find $f'(x)$ and $f''(x)$. Then graph f, f', and f'' on a common screen and check to see if your answers are reasonable.

53. $f(x) = 3x^2 + 2x + 1$ **54.** $f(x) = x^3 - 3x$

55. If $f(x) = 2x^2 - x^3$, find $f'(x)$, $f''(x)$, $f'''(x)$, and $f^{(4)}(x)$. Graph f, f', f'', and f''' on a common screen. Are the graphs consistent with the geometric interpretations of these derivatives?

56. (a) The graph of a position function of a car is shown, where s is measured in feet and t in seconds. Use it to graph the velocity and acceleration of the car. What is the acceleration at $t = 10$ seconds?

(b) Use the acceleration curve from part (a) to estimate the jerk at $t = 10$ seconds. What are the units for jerk?

57. Let $f(x) = \sqrt[3]{x}$.
(a) If $a \neq 0$, use Equation 2.7.5 to find $f'(a)$.
(b) Show that $f'(0)$ does not exist.
(c) Show that $y = \sqrt[3]{x}$ has a vertical tangent line at $(0, 0)$. (Recall the shape of the graph of f. See Figure 1.2.13.)

58. (a) If $g(x) = x^{2/3}$, show that $g'(0)$ does not exist.
(b) If $a \neq 0$, find $g'(a)$.
(c) Show that $y = x^{2/3}$ has a vertical tangent line at $(0, 0)$.
(d) Illustrate part (c) by graphing $y = x^{2/3}$.

59. Show that the function $f(x) = |x - 6|$ is not differentiable at 6. Find a formula for f' and sketch its graph.

60. Where is the greatest integer function $f(x) = [\![x]\!]$ not differentiable? Find a formula for f' and sketch its graph.

61. (a) Sketch the graph of the function $f(x) = x|x|$.
(b) For what values of x is f differentiable?
(c) Find a formula for f'.

62. (a) Sketch the graph of the function $g(x) = x + |x|$.
(b) For what values of x is g differentiable?
(c) Find a formula for g'.

63. Recall that a function f is called *even* if $f(-x) = f(x)$ for all x in its domain and *odd* if $f(-x) = -f(x)$ for all such x. Prove each of the following.
(a) The derivative of an even function is an odd function.
(b) The derivative of an odd function is an even function.

64. The **left-hand** and **right-hand derivatives** of f at a are defined by

$$f'_-(a) = \lim_{h \to 0^-} \frac{f(a + h) - f(a)}{h}$$

and

$$f'_+(a) = \lim_{h \to 0^+} \frac{f(a + h) - f(a)}{h}$$

if these limits exist. Then $f'(a)$ exists if and only if these one-sided derivatives exist and are equal.

(a) Find $f'_-(4)$ and $f'_+(4)$ for the function

$$f(x) = \begin{cases} 0 & \text{if } x \leq 0 \\ 5 - x & \text{if } 0 < x < 4 \\ \dfrac{1}{5 - x} & \text{if } x \geq 4 \end{cases}$$

(b) Sketch the graph of f.
(c) Where is f discontinuous?
(d) Where is f not differentiable?

65. Nick starts jogging and runs faster and faster for 3 mintues, then he walks for 5 minutes. He stops at an intersection for 2 minutes, runs fairly quickly for 5 minutes, then walks for 4 minutes.

(a) Sketch a possible graph of the distance s Nick has covered after t minutes.
(b) Sketch a graph of ds/dt.

66. When you turn on a hot-water faucet, the temperature T of the water depends on how long the water has been running.

(a) Sketch a possible graph of T as a function of the time t that has elapsed since the faucet was turned on.
(b) Describe how the rate of change of T with respect to t varies as t increases.
(c) Sketch a graph of the derivative of T.

67. Let ℓ be the tangent line to the parabola $y = x^2$ at the point $(1, 1)$. The *angle of inclination* of ℓ is the angle ϕ that ℓ makes with the positive direction of the x-axis. Calculate ϕ correct to the nearest degree.

2 REVIEW

CONCEPT CHECK

Answers to the Concept Check can be found on the back endpapers.

1. Explain what each of the following means and illustrate with a sketch.

(a) $\lim_{x \to a} f(x) = L$ (b) $\lim_{x \to a^+} f(x) = L$ (c) $\lim_{x \to a^-} f(x) = L$
(d) $\lim_{x \to a} f(x) = \infty$ (e) $\lim_{x \to \infty} f(x) = L$

2. Describe several ways in which a limit can fail to exist. Illustrate with sketches.

3. State the following Limit Laws.

(a) Sum Law
(b) Difference Law
(c) Constant Multiple Law
(d) Product Law
(e) Quotient Law
(f) Power Law
(g) Root Law

4. What does the Squeeze Theorem say?

5. (a) What does it mean to say that the line $x = a$ is a vertical asymptote of the curve $y = f(x)$? Draw curves to illustrate the various possibilities.
(b) What does it mean to say that the line $y = L$ is a horizontal asymptote of the curve $y = f(x)$? Draw curves to illustrate the various possibilities.

6. Which of the following curves have vertical asymptotes? Which have horizontal asymptotes?

(a) $y = x^4$ (b) $y = \sin x$ (c) $y = \tan x$
(d) $y = \tan^{-1}x$ (e) $y = e^x$ (f) $y = \ln x$
(g) $y = 1/x$ (h) $y = \sqrt{x}$

7. (a) What does it mean for f to be continuous at a?
(b) What does it mean for f to be continuous on the interval $(-\infty, \infty)$? What can you say about the graph of such a function?

8. (a) Give examples of functions that are continuous on $[-1, 1]$.
(b) Give an example of a function that is not continuous on $[0, 1]$.

9. What does the Intermediate Value Theorem say?

10. Write an expression for the slope of the tangent line to the curve $y = f(x)$ at the point $(a, f(a))$.

11. Suppose an object moves along a straight line with position $f(t)$ at time t. Write an expression for the instantaneous velocity of the object at time $t = a$. How can you interpret this velocity in terms of the graph of f?

12. If $y = f(x)$ and x changes from x_1 to x_2, write expressions for the following.

(a) The average rate of change of y with respect to x over the interval $[x_1, x_2]$.
(b) The instantaneous rate of change of y with respect to x at $x = x_1$.

13. Define the derivative $f'(a)$. Discuss two ways of interpreting this number.

14. Define the second derivative of f. If $f(t)$ is the position function of a particle, how can you interpret the second derivative?

15. (a) What does it mean for f to be differentiable at a?

(b) What is the relation between the differentiability and continuity of a function?

(c) Sketch the graph of a function that is continuous but not differentiable at $a = 2$.

16. Describe several ways in which a function can fail to be differentiable. Illustrate with sketches.

TRUE-FALSE QUIZ

Determine whether the statement is true or false. If it is true, explain why. If it is false, explain why or give an example that disproves the statement.

1. $\lim\limits_{x \to 4} \left(\dfrac{2x}{x-4} - \dfrac{8}{x-4} \right) = \lim\limits_{x \to 4} \dfrac{2x}{x-4} - \lim\limits_{x \to 4} \dfrac{8}{x-4}$

2. $\lim\limits_{x \to 1} \dfrac{x^2 + 6x - 7}{x^2 + 5x - 6} = \dfrac{\lim\limits_{x \to 1}(x^2 + 6x - 7)}{\lim\limits_{x \to 1}(x^2 + 5x - 6)}$

3. $\lim\limits_{x \to 1} \dfrac{x - 3}{x^2 + 2x - 4} = \dfrac{\lim\limits_{x \to 1}(x - 3)}{\lim\limits_{x \to 1}(x^2 + 2x - 4)}$

4. $\dfrac{x^2 - 9}{x - 3} = x + 3$

5. $\lim\limits_{x \to 3} \dfrac{x^2 - 9}{x - 3} = \lim\limits_{x \to 3}(x + 3)$

6. If $\lim_{x \to 5} f(x) = 2$ and $\lim_{x \to 5} g(x) = 0$, then $\lim_{x \to 5}[f(x)/g(x)]$ does not exist.

7. If $\lim_{x \to 5} f(x) = 0$ and $\lim_{x \to 5} g(x) = 0$, then $\lim_{x \to 5}[f(x)/g(x)]$ does not exist.

8. If neither $\lim_{x \to a} f(x)$ nor $\lim_{x \to a} g(x)$ exists, then $\lim_{x \to a}[f(x) + g(x)]$ does not exist.

9. If $\lim_{x \to a} f(x)$ exists but $\lim_{x \to a} g(x)$ does not exist, then $\lim_{x \to a}[f(x) + g(x)]$ does not exist.

10. If $\lim_{x \to 6}[f(x)\,g(x)]$ exists, then the limit must be $f(6)\,g(6)$.

11. If p is a polynomial, then $\lim_{x \to b} p(x) = p(b)$.

12. If $\lim_{x \to 0} f(x) = \infty$ and $\lim_{x \to 0} g(x) = \infty$, then $\lim_{x \to 0}[f(x) - g(x)] = 0$.

13. A function can have two different horizontal asymptotes.

14. If f has domain $[0, \infty)$ and has no horizontal asymptote, then $\lim_{x \to \infty} f(x) = \infty$ or $\lim_{x \to \infty} f(x) = -\infty$.

15. If the line $x = 1$ is a vertical asymptote of $y = f(x)$, then f is not defined at 1.

16. If $f(1) > 0$ and $f(3) < 0$, then there exists a number c between 1 and 3 such that $f(c) = 0$.

17. If f is continuous at 5 and $f(5) = 2$ and $f(4) = 3$, then $\lim_{x \to 2} f(4x^2 - 11) = 2$.

18. If f is continuous on $[-1, 1]$ and $f(-1) = 4$ and $f(1) = 3$, then there exists a number r such that $|r| < 1$ and $f(r) = \pi$.

19. Let f be a function such that $\lim_{x \to 0} f(x) = 6$. Then there exists a positive number δ such that if $0 < |x| < \delta$, then $|f(x) - 6| < 1$.

20. If $f(x) > 1$ for all x and $\lim_{x \to 0} f(x)$ exists, then $\lim_{x \to 0} f(x) > 1$.

21. If f is continuous at a, then f is differentiable at a.

22. If $f'(r)$ exists, then $\lim_{x \to r} f(x) = f(r)$.

23. $\dfrac{d^2y}{dx^2} = \left(\dfrac{dy}{dx} \right)^2$

24. The equation $x^{10} - 10x^2 + 5 = 0$ has a root in the interval $(0, 2)$.

25. If f is continuous at a, so is $|f|$.

26. If $|f|$ is continuous at a, so is f.

EXERCISES

1. The graph of f is given.

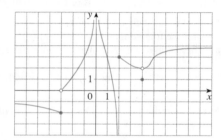

(a) Find each limit, or explain why it does not exist.

(i) $\lim\limits_{x \to 2^+} f(x)$ (ii) $\lim\limits_{x \to -3^+} f(x)$ (iii) $\lim\limits_{x \to -3} f(x)$

(iv) $\lim\limits_{x \to 4} f(x)$ (v) $\lim\limits_{x \to 0} f(x)$ (vi) $\lim\limits_{x \to 2^-} f(x)$

(vii) $\lim\limits_{x \to \infty} f(x)$ (viii) $\lim\limits_{x \to -\infty} f(x)$

(b) State the equations of the horizontal asymptotes.

(c) State the equations of the vertical asymptotes.

(d) At what numbers is f discontinuous? Explain.

2. Sketch the graph of a function f that satisfies all of the following conditions:

$\lim\limits_{x \to -\infty} f(x) = -2$, $\lim\limits_{x \to \infty} f(x) = 0$, $\lim\limits_{x \to -3} f(x) = \infty$,

$\lim\limits_{x \to 3^-} f(x) = -\infty$, $\lim\limits_{x \to 3^+} f(x) = 2$,

f is continuous from the right at 3

3–20 Find the limit.

3. $\displaystyle\lim_{x\to1} e^{x^3-x}$

4. $\displaystyle\lim_{x\to3} \frac{x^2-9}{x^2+2x-3}$

5. $\displaystyle\lim_{x\to-3} \frac{x^2-9}{x^2+2x-3}$

6. $\displaystyle\lim_{x\to1^+} \frac{x^2-9}{x^2+2x-3}$

7. $\displaystyle\lim_{h\to0} \frac{(h-1)^3+1}{h}$

8. $\displaystyle\lim_{t\to2} \frac{t^2-4}{t^3-8}$

9. $\displaystyle\lim_{r\to9} \frac{\sqrt{r}}{(r-9)^4}$

10. $\displaystyle\lim_{v\to4^+} \frac{4-v}{|4-v|}$

11. $\displaystyle\lim_{u\to1} \frac{u^4-1}{u^3+5u^2-6u}$

12. $\displaystyle\lim_{x\to3} \frac{\sqrt{x+6}-x}{x^3-3x^2}$

13. $\displaystyle\lim_{x\to\infty} \frac{\sqrt{x^2-9}}{2x-6}$

14. $\displaystyle\lim_{x\to-\infty} \frac{\sqrt{x^2-9}}{2x-6}$

15. $\displaystyle\lim_{x\to\pi^-} \ln(\sin x)$

16. $\displaystyle\lim_{x\to-\infty} \frac{1-2x^2-x^4}{5+x-3x^4}$

17. $\displaystyle\lim_{x\to\infty} \left(\sqrt{x^2+4x+1}-x\right)$ **18.** $\displaystyle\lim_{x\to\infty} e^{x-x^2}$

19. $\displaystyle\lim_{x\to0^+} \tan^{-1}(1/x)$

20. $\displaystyle\lim_{x\to1} \left(\frac{1}{x-1}+\frac{1}{x^2-3x+2}\right)$

21–22 Use graphs to discover the asymptotes of the curve. Then prove what you have discovered.

21. $y = \dfrac{\cos^2 x}{x^2}$

22. $y = \sqrt{x^2+x+1}-\sqrt{x^2-x}$

23. If $2x-1 \le f(x) \le x^2$ for $0 < x < 3$, find $\lim_{x\to1} f(x)$.

24. Prove that $\lim_{x\to0} x^2 \cos(1/x^2) = 0$.

25–28 Prove the statement using the precise definition of a limit.

25. $\displaystyle\lim_{x\to2} (14-5x) = 4$

26. $\displaystyle\lim_{x\to0} \sqrt[3]{x} = 0$

27. $\displaystyle\lim_{x\to2} (x^2-3x) = -2$

28. $\displaystyle\lim_{x\to4^+} \frac{2}{\sqrt{x-4}} = \infty$

29. Let
$$f(x) = \begin{cases} \sqrt{-x} & \text{if } x < 0 \\ 3-x & \text{if } 0 \le x < 3 \\ (x-3)^2 & \text{if } x > 3 \end{cases}$$

(a) Evaluate each limit, if it exists.

  (i) $\displaystyle\lim_{x\to0^+} f(x)$ (ii) $\displaystyle\lim_{x\to0^-} f(x)$ (iii) $\displaystyle\lim_{x\to0} f(x)$

  (iv) $\displaystyle\lim_{x\to3^-} f(x)$ (v) $\displaystyle\lim_{x\to3^+} f(x)$ (vi) $\displaystyle\lim_{x\to3} f(x)$

(b) Where is f discontinuous?

(c) Sketch the graph of f.

30. Let
$$g(x) = \begin{cases} 2x-x^2 & \text{if } 0 \le x \le 2 \\ 2-x & \text{if } 2 < x \le 3 \\ x-4 & \text{if } 3 < x < 4 \\ \pi & \text{if } x \ge 4 \end{cases}$$

(a) For each of the numbers 2, 3, and 4, discover whether g is continuous from the left, continuous from the right, or continuous at the number.

(b) Sketch the graph of g.

31–32 Show that the function is continuous on its domain. State the domain.

31. $h(x) = xe^{\sin x}$

32. $g(x) = \dfrac{\sqrt{x^2-9}}{x^2-2}$

33–34 Use the Intermediate Value Theorem to show that there is a root of the equation in the given interval.

33. $x^5 - x^3 + 3x - 5 = 0$, $(1, 2)$

34. $\cos\sqrt{x} = e^x - 2$, $(0, 1)$

35. (a) Find the slope of the tangent line to the curve $y = 9 - 2x^2$ at the point $(2, 1)$.

(b) Find an equation of this tangent line.

36. Find equations of the tangent lines to the curve
$$y = \frac{2}{1-3x}$$
at the points with x-coordinates 0 and -1.

37. The displacement (in meters) of an object moving in a straight line is given by $s = 1 + 2t + \frac{1}{4}t^2$, where t is measured in seconds.

(a) Find the average velocity over each time period.

  (i) $[1, 3]$ (ii) $[1, 2]$ (iii) $[1, 1.5]$ (iv) $[1, 1.1]$

(b) Find the instantaneous velocity when $t = 1$.

38. According to Boyle's Law, if the temperature of a confined gas is held fixed, then the product of the pressure P and the volume V is a constant. Suppose that, for a certain gas, $PV = 800$, where P is measured in pounds per square inch and V is measured in cubic inches.

(a) Find the average rate of change of P as V increases from 200 in³ to 250 in³.

(b) Express V as a function of P and show that the instantaneous rate of change of V with respect to P is inversely proportional to the square of P.

39. (a) Use the definition of a derivative to find $f'(2)$, where $f(x) = x^3 - 2x$.

(b) Find an equation of the tangent line to the curve $y = x^3 - 2x$ at the point $(2, 4)$.

(c) Illustrate part (b) by graphing the curve and the tangent line on the same screen.

40. Find a function f and a number a such that
$$\lim_{h\to0} \frac{(2+h)^6-64}{h} = f'(a)$$

41. The total cost of repaying a student loan at an interest rate of $r\%$ per year is $C = f(r)$.

(a) What is the meaning of the derivative $f'(r)$? What are its units?

(b) What does the statement $f'(10) = 1200$ mean?
(c) Is $f'(r)$ always positive or does it change sign?

42–44 Trace or copy the graph of the function. Then sketch a graph of its derivative directly beneath.

42.

43.

44.

45. (a) If $f(x) = \sqrt{3 - 5x}$, use the definition of a derivative to find $f'(x)$.
(b) Find the domains of f and f'.
(c) Graph f and f' on a common screen. Compare the graphs to see whether your answer to part (a) is reasonable.

46. (a) Find the asymptotes of the graph of $f(x) = \dfrac{4 - x}{3 + x}$ and use them to sketch the graph.
(b) Use your graph from part (a) to sketch the graph of f'.
(c) Use the definition of a derivative to find $f'(x)$.
(d) Use a graphing device to graph f' and compare with your sketch in part (b).

47. The graph of f is shown. State, with reasons, the numbers at which f is not differentiable.

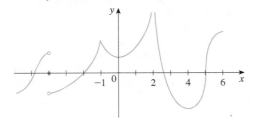

48. The figure shows the graphs of f, f', and f''. Identify each curve, and explain your choices.

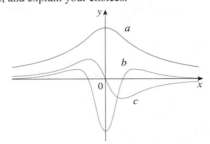

49. Sketch the graph of a function f that satisfies all of the following conditions: The domain of f is all real numbers except 0, $\lim_{x \to 0^-} f(x) = 1$, $\lim_{x \to 0^+} f(x) = 0$, $f'(x) > 0$ for all x in the domain of f, $\lim_{x \to -\infty} f'(x) = 0$, $\lim_{x \to \infty} f'(x) = 1$.

50. Let $P(t)$ be the percentage of Americans under the age of 18 at time t. The table gives values of this function in census years from 1950 to 2010.

t	$P(t)$	t	$P(t)$
1950	31.1	1990	25.7
1960	35.7	2000	25.7
1970	34.0	2010	24.0
1980	28.0		

(a) What is the meaning of $P'(t)$? What are its units?
(b) Construct a table of estimated values for $P'(t)$.
(c) Graph P and P'.
(d) How would it be possible to get more accurate values for $P'(t)$?

51. Let $B(t)$ be the number of US $20 bills in circulation at time t. The table gives values of this function from 1990 to 2010, as of December 31, in billions. Interpret and estimate the value of $B'(2000)$.

t	1990	1995	2000	2005	2010
$B(t)$	3.45	4.21	4.93	5.77	6.53

52. The *total fertility rate* at time t, denoted by $F(t)$, is an estimate of the average number of children born to each woman (assuming that current birth rates remain constant). The graph of the total fertility rate in the United States shows the fluctuations from 1940 to 2010.
(a) Estimate the values of $F'(1950)$, $F'(1965)$, and $F'(1987)$.
(b) What are the meanings of these derivatives?
(c) Can you suggest reasons for the values of these derivatives?

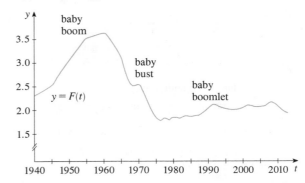

53. Suppose that $|f(x)| \leq g(x)$ for all x, where $\lim_{x \to a} g(x) = 0$. Find $\lim_{x \to a} f(x)$.

54. Let $f(x) = [\![x]\!] + [\![-x]\!]$.
(a) For what values of a does $\lim_{x \to a} f(x)$ exist?
(b) At what numbers is f discontinuous?

Problems Plus

In our discussion of the principles of problem solving we considered the problem-solving strategy of *introducing something extra* (see page 71). In the following example we show how this principle is sometimes useful when we evaluate limits. The idea is to change the variable—to introduce a new variable that is related to the original variable—in such a way as to make the problem simpler. Later, in Section 5.5, we will make more extensive use of this general idea.

EXAMPLE 1 Evaluate $\lim\limits_{x \to 0} \dfrac{\sqrt[3]{1 + cx} - 1}{x}$, where c is a constant.

SOLUTION As it stands, this limit looks challenging. In Section 2.3 we evaluated several limits in which both numerator and denominator approached 0. There our strategy was to perform some sort of algebraic manipulation that led to a simplifying cancellation, but here it's not clear what kind of algebra is necessary.

So we introduce a new variable t by the equation

$$t = \sqrt[3]{1 + cx}$$

We also need to express x in terms of t, so we solve this equation:

$$t^3 = 1 + cx \qquad x = \frac{t^3 - 1}{c} \quad \text{(if } c \neq 0\text{)}$$

Notice that $x \to 0$ is equivalent to $t \to 1$. This allows us to convert the given limit into one involving the variable t:

$$\lim_{x \to 0} \frac{\sqrt[3]{1 + cx} - 1}{x} = \lim_{t \to 1} \frac{t - 1}{(t^3 - 1)/c}$$

$$= \lim_{t \to 1} \frac{c(t - 1)}{t^3 - 1}$$

The change of variable allowed us to replace a relatively complicated limit by a simpler one of a type that we have seen before. Factoring the denominator as a difference of cubes, we get

$$\lim_{t \to 1} \frac{c(t - 1)}{t^3 - 1} = \lim_{t \to 1} \frac{c(t - 1)}{(t - 1)(t^2 + t + 1)}$$

$$= \lim_{t \to 1} \frac{c}{t^2 + t + 1} = \frac{c}{3}$$

In making the change of variable we had to rule out the case $c = 0$. But if $c = 0$, the function is 0 for all nonzero x and so its limit is 0. Therefore, in all cases, the limit is $c/3$. ∎

The following problems are meant to test and challenge your problem-solving skills. Some of them require a considerable amount of time to think through, so don't be discouraged if you can't solve them right away. If you get stuck, you might find it helpful to refer to the discussion of the principles of problem solving on page 71.

Problems

1. Evaluate $\lim\limits_{x \to 1} \dfrac{\sqrt[3]{x} - 1}{\sqrt{x} - 1}$.

2. Find numbers a and b such that $\lim\limits_{x \to 0} \dfrac{\sqrt{ax + b} - 2}{x} = 1$.

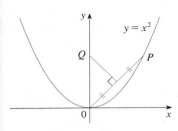

FIGURE FOR PROBLEM 4

3. Evaluate $\lim\limits_{x \to 0} \dfrac{|2x - 1| - |2x + 1|}{x}$.

4. The figure shows a point P on the parabola $y = x^2$ and the point Q where the perpendicular bisector of OP intersects the y-axis. As P approaches the origin along the parabola, what happens to Q? Does it have a limiting position? If so, find it.

5. Evaluate the following limits, if they exist, where $[\![x]\!]$ denotes the greatest integer function.

   (a) $\lim\limits_{x \to 0} \dfrac{[\![x]\!]}{x}$
                                      (b) $\lim\limits_{x \to 0} x \, [\![1/x]\!]$

6. Sketch the region in the plane defined by each of the following equations.

   (a) $[\![x]\!]^2 + [\![y]\!]^2 = 1$
                           (b) $[\![x]\!]^2 - [\![y]\!]^2 = 3$
   (c) $[\![x + y]\!]^2 = 1$
                           (d) $[\![x]\!] + [\![y]\!] = 1$

7. Find all values of a such that f is continuous on $\mathbb{R}$:

$$f(x) = \begin{cases} x + 1 & \text{if } x \leqslant a \\ x^2 & \text{if } x > a \end{cases}$$

8. A **fixed point** of a function f is a number c in its domain such that $f(c) = c$. (The function doesn't move c; it stays fixed.)

   (a) Sketch the graph of a continuous function with domain $[0, 1]$ whose range also lies in $[0, 1]$. Locate a fixed point of f.

   (b) Try to draw the graph of a continuous function with domain $[0, 1]$ and range in $[0, 1]$ that does *not* have a fixed point. What is the obstacle?

   (c) Use the Intermediate Value Theorem to prove that any continuous function with domain $[0, 1]$ and range in $[0, 1]$ must have a fixed point.

9. If $\lim_{x \to a} [f(x) + g(x)] = 2$ and $\lim_{x \to a} [f(x) - g(x)] = 1$, find $\lim_{x \to a} [f(x) g(x)]$.

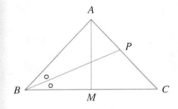

FIGURE FOR PROBLEM 10

10. (a) The figure shows an isosceles triangle ABC with $\angle B = \angle C$. The bisector of angle B intersects the side AC at the point P. Suppose that the base BC remains fixed but the altitude $|AM|$ of the triangle approaches 0, so A approaches the midpoint M of BC. What happens to P during this process? Does it have a limiting position? If so, find it.

   (b) Try to sketch the path traced out by P during this process. Then find an equation of this curve and use this equation to sketch the curve.

11. (a) If we start from $0°$ latitude and proceed in a westerly direction, we can let $T(x)$ denote the temperature at the point x at any given time. Assuming that T is a continuous function of x, show that at any fixed time there are at least two diametrically opposite points on the equator that have exactly the same temperature.

   (b) Does the result in part (a) hold for points lying on any circle on the earth's surface?

   (c) Does the result in part (a) hold for barometric pressure and for altitude above sea level?

12. If f is a differentiable function and $g(x) = xf(x)$, use the definition of a derivative to show that $g'(x) = xf'(x) + f(x)$.

13. Suppose f is a function that satisfies the equation

$$f(x + y) = f(x) + f(y) + x^2 y + xy^2$$

for all real numbers x and y. Suppose also that

$$\lim_{x \to 0} \frac{f(x)}{x} = 1$$

   (a) Find $f(0)$.
       (b) Find $f'(0)$.
       (c) Find $f'(x)$.

14. Suppose f is a function with the property that $|f(x)| \leqslant x^2$ for all x. Show that $f(0) = 0$. Then show that $f'(0) = 0$.

3 Differentiation Rules

© Mechanik / Shutterstock.com

In the project on page 208 you will calculate the distance from an airport runway at which a pilot should start descent for a smooth landing.

WE HAVE SEEN HOW TO INTERPRET derivatives as slopes and rates of change. We have seen how to estimate derivatives of functions given by tables of values. We have learned how to graph derivatives of functions that are defined graphically. We have used the definition of a derivative to calculate the derivatives of functions defined by formulas. But it would be tedious if we always had to use the definition, so in this chapter we develop rules for finding derivatives without having to use the definition directly. These differentiation rules enable us to calculate with relative ease the derivatives of polynomials, rational functions, algebraic functions, exponential and logarithmic functions, and trigonometric and inverse trigonometric functions. We then use these rules to solve problems involving rates of change and the approximation of functions.

3.1   Derivatives of Polynomials and Exponential Functions

In this section we learn how to differentiate constant functions, power functions, polynomials, and exponential functions.

Let's start with the simplest of all functions, the constant function $f(x) = c$. The graph of this function is the horizontal line $y = c$, which has slope 0, so we must have $f'(x) = 0$. (See Figure 1.) A formal proof, from the definition of a derivative, is also easy:

$$f'(x) = \lim_{h \to 0} \frac{f(x + h) - f(x)}{h} = \lim_{h \to 0} \frac{c - c}{h} = \lim_{h \to 0} 0 = 0$$

In Leibniz notation, we write this rule as follows.

FIGURE 1
The graph of $f(x) = c$ is the line $y = c$, so $f'(x) = 0$.

Derivative of a Constant Function

$$\frac{d}{dx}(c) = 0$$

▪ Power Functions

We next look at the functions $f(x) = x^n$, where n is a positive integer. If $n = 1$, the graph of $f(x) = x$ is the line $y = x$, which has slope 1. (See Figure 2.) So

$$\boxed{1} \qquad \frac{d}{dx}(x) = 1$$

(You can also verify Equation 1 from the definition of a derivative.) We have already investigated the cases $n = 2$ and $n = 3$. In fact, in Section 2.8 (Exercises 19 and 20) we found that

$$\boxed{2} \qquad \frac{d}{dx}(x^2) = 2x \qquad \frac{d}{dx}(x^3) = 3x^2$$

For $n = 4$ we find the derivative of $f(x) = x^4$ as follows:

$$f'(x) = \lim_{h \to 0} \frac{f(x + h) - f(x)}{h} = \lim_{h \to 0} \frac{(x + h)^4 - x^4}{h}$$

$$= \lim_{h \to 0} \frac{x^4 + 4x^3h + 6x^2h^2 + 4xh^3 + h^4 - x^4}{h}$$

$$= \lim_{h \to 0} \frac{4x^3h + 6x^2h^2 + 4xh^3 + h^4}{h}$$

$$= \lim_{h \to 0} (4x^3 + 6x^2h + 4xh^2 + h^3) = 4x^3$$

Thus

$$\boxed{3} \qquad \frac{d}{dx}(x^4) = 4x^3$$

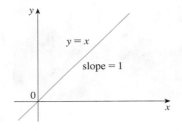

FIGURE 2
The graph of $f(x) = x$ is the line $y = x$, so $f'(x) = 1$.

Comparing the equations in (1), (2), and (3), we see a pattern emerging. It seems to be a reasonable guess that, when n is a positive integer, $(d/dx)(x^n) = nx^{n-1}$. This turns out to be true.

The Power Rule If n is a positive integer, then

$$\frac{d}{dx}(x^n) = nx^{n-1}$$

FIRST PROOF The formula

$$x^n - a^n = (x - a)(x^{n-1} + x^{n-2}a + \cdots + xa^{n-2} + a^{n-1})$$

can be verified simply by multiplying out the right-hand side (or by summing the second factor as a geometric series). If $f(x) = x^n$, we can use Equation 2.7.5 for $f'(a)$ and the equation above to write

$$f'(a) = \lim_{x \to a} \frac{f(x) - f(a)}{x - a} = \lim_{x \to a} \frac{x^n - a^n}{x - a}$$

$$= \lim_{x \to a} (x^{n-1} + x^{n-2}a + \cdots + xa^{n-2} + a^{n-1})$$

$$= a^{n-1} + a^{n-2}a + \cdots + aa^{n-2} + a^{n-1}$$

$$= na^{n-1}$$

SECOND PROOF

$$f'(x) = \lim_{h \to 0} \frac{f(x + h) - f(x)}{h} = \lim_{h \to 0} \frac{(x + h)^n - x^n}{h}$$

The Binomial Theorem is given on Reference Page 1.

In finding the derivative of x^4 we had to expand $(x + h)^4$. Here we need to expand $(x + h)^n$ and we use the Binomial Theorem to do so:

$$f'(x) = \lim_{h \to 0} \frac{\left[x^n + nx^{n-1}h + \dfrac{n(n - 1)}{2}x^{n-2}h^2 + \cdots + nxh^{n-1} + h^n \right] - x^n}{h}$$

$$= \lim_{h \to 0} \frac{nx^{n-1}h + \dfrac{n(n - 1)}{2}x^{n-2}h^2 + \cdots + nxh^{n-1} + h^n}{h}$$

$$= \lim_{h \to 0} \left[nx^{n-1} + \dfrac{n(n - 1)}{2}x^{n-2}h + \cdots + nxh^{n-2} + h^{n-1} \right]$$

$$= nx^{n-1}$$

because every term except the first has h as a factor and therefore approaches 0. ∎

We illustrate the Power Rule using various notations in Example 1.

EXAMPLE 1

(a) If $f(x) = x^6$, then $f'(x) = 6x^5$.   (b) If $y = x^{1000}$, then $y' = 1000x^{999}$.

(c) If $y = t^4$, then $\dfrac{dy}{dt} = 4t^3$.   (d) $\dfrac{d}{dr}(r^3) = 3r^2$ ∎

What about power functions with negative integer exponents? In Exercise 65 we ask you to verify from the definition of a derivative that

$$\frac{d}{dx}\left(\frac{1}{x}\right) = -\frac{1}{x^2}$$

We can rewrite this equation as

$$\frac{d}{dx}(x^{-1}) = (-1)x^{-2}$$

and so the Power Rule is true when $n = -1$. In fact, we will show in the next section [Exercise 3.2.64(c)] that it holds for all negative integers.

What if the exponent is a fraction? In Example 2.8.3 we found that

$$\frac{d}{dx}\sqrt{x} = \frac{1}{2\sqrt{x}}$$

which can be written as

$$\frac{d}{dx}(x^{1/2}) = \tfrac{1}{2}x^{-1/2}$$

This shows that the Power Rule is true even when $n = \tfrac{1}{2}$. In fact, we will show in Section 3.6 that it is true for all real numbers n.

The Power Rule (General Version)   If n is any real number, then

$$\frac{d}{dx}(x^n) = nx^{n-1}$$

EXAMPLE 2   Differentiate:

(a) $f(x) = \dfrac{1}{x^2}$ (b) $y = \sqrt[3]{x^2}$

SOLUTION   In each case we rewrite the function as a power of x.

(a) Since $f(x) = x^{-2}$, we use the Power Rule with $n = -2$:

$$f'(x) = \frac{d}{dx}(x^{-2}) = -2x^{-2-1} = -2x^{-3} = -\frac{2}{x^3}$$

(b) $$\frac{dy}{dx} = \frac{d}{dx}\left(\sqrt[3]{x^2}\right) = \frac{d}{dx}(x^{2/3}) = \tfrac{2}{3}x^{(2/3)-1} = \tfrac{2}{3}x^{-1/3}$$ ∎

Observe from Figure 3 that the function y in Example 2(b) is increasing when y' is positive and is decreasing when y' is negative. In Chapter 4 we will prove that, in general, *a function increases when its derivative is positive and decreases when its derivative is negative.*

Figure 3 shows the function y in Example 2(b) and its derivative y'. Notice that y is not differentiable at 0 (y' is not defined there).

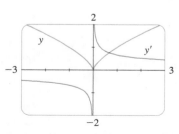

FIGURE 3
$y = \sqrt[3]{x^2}$

The Power Rule enables us to find tangent lines without having to resort to the definition of a derivative. It also enables us to find *normal lines*. The **normal line** to a curve C at a point P is the line through P that is perpendicular to the tangent line at P. (In the study of optics, one needs to consider the angle between a light ray and the normal line to a lens.)

EXAMPLE 3 Find equations of the tangent line and normal line to the curve $y = x\sqrt{x}$ at the point $(1, 1)$. Illustrate by graphing the curve and these lines.

SOLUTION The derivative of $f(x) = x\sqrt{x} = xx^{1/2} = x^{3/2}$ is

$$f'(x) = \tfrac{3}{2}x^{(3/2)-1} = \tfrac{3}{2}x^{1/2} = \tfrac{3}{2}\sqrt{x}$$

So the slope of the tangent line at $(1, 1)$ is $f'(1) = \tfrac{3}{2}$. Therefore an equation of the tangent line is

$$y - 1 = \tfrac{3}{2}(x - 1) \qquad \text{or} \qquad y = \tfrac{3}{2}x - \tfrac{1}{2}$$

The normal line is perpendicular to the tangent line, so its slope is the negative reciprocal of $\tfrac{3}{2}$, that is, $-\tfrac{2}{3}$. Thus an equation of the normal line is

$$y - 1 = -\tfrac{2}{3}(x - 1) \qquad \text{or} \qquad y = -\tfrac{2}{3}x + \tfrac{5}{3}$$

We graph the curve and its tangent line and normal line in Figure 4. ∎

FIGURE 4
$y = x\sqrt{x}$

New Derivatives from Old

When new functions are formed from old functions by addition, subtraction, or multiplication by a constant, their derivatives can be calculated in terms of derivatives of the old functions. In particular, the following formula says that *the derivative of a constant times a function is the constant times the derivative of the function.*

> **The Constant Multiple Rule** If c is a constant and f is a differentiable function, then
> $$\frac{d}{dx}[cf(x)] = c\frac{d}{dx}f(x)$$

PROOF Let $g(x) = cf(x)$. Then

$$g'(x) = \lim_{h \to 0} \frac{g(x + h) - g(x)}{h} = \lim_{h \to 0} \frac{cf(x + h) - cf(x)}{h}$$

$$= \lim_{h \to 0} c\left[\frac{f(x + h) - f(x)}{h}\right]$$

$$= c \lim_{h \to 0} \frac{f(x + h) - f(x)}{h} \qquad \text{(by Limit Law 3)}$$

$$= cf'(x)$$

∎

Geometric Interpretation of the Constant Multiple Rule

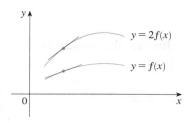

$y = 2f(x)$

$y = f(x)$

Multiplying by $c = 2$ stretches the graph vertically by a factor of 2. All the rises have been doubled but the runs stay the same. So the slopes are doubled too.

EXAMPLE 4

(a) $\dfrac{d}{dx}(3x^4) = 3\dfrac{d}{dx}(x^4) = 3(4x^3) = 12x^3$

(b) $\dfrac{d}{dx}(-x) = \dfrac{d}{dx}[(-1)x] = (-1)\dfrac{d}{dx}(x) = -1(1) = -1$ ∎

The next rule tells us that *the derivative of a sum of functions is the sum of the derivatives.*

Using prime notation, we can write the Sum Rule as

$$(f + g)' = f' + g'$$

> **The Sum Rule**   If f and g are both differentiable, then
>
> $$\frac{d}{dx}[f(x) + g(x)] = \frac{d}{dx}f(x) + \frac{d}{dx}g(x)$$

PROOF   Let $F(x) = f(x) + g(x)$. Then

$$F'(x) = \lim_{h \to 0} \frac{F(x + h) - F(x)}{h}$$

$$= \lim_{h \to 0} \frac{[f(x + h) + g(x + h)] - [f(x) + g(x)]}{h}$$

$$= \lim_{h \to 0} \left[\frac{f(x + h) - f(x)}{h} + \frac{g(x + h) - g(x)}{h} \right]$$

$$= \lim_{h \to 0} \frac{f(x + h) - f(x)}{h} + \lim_{h \to 0} \frac{g(x + h) - g(x)}{h} \qquad \text{(by Limit Law 1)}$$

$$= f'(x) + g'(x)$$ ∎

The Sum Rule can be extended to the sum of any number of functions. For instance, using this theorem twice, we get

$$(f + g + h)' = [(f + g) + h]' = (f + g)' + h' = f' + g' + h'$$

By writing $f - g$ as $f + (-1)g$ and applying the Sum Rule and the Constant Multiple Rule, we get the following formula.

> **The Difference Rule**   If f and g are both differentiable, then
>
> $$\frac{d}{dx}[f(x) - g(x)] = \frac{d}{dx}f(x) - \frac{d}{dx}g(x)$$

The Constant Multiple Rule, the Sum Rule, and the Difference Rule can be combined with the Power Rule to differentiate any polynomial, as the following examples demonstrate.

EXAMPLE 5

$$\frac{d}{dx}(x^8 + 12x^5 - 4x^4 + 10x^3 - 6x + 5)$$

$$= \frac{d}{dx}(x^8) + 12\frac{d}{dx}(x^5) - 4\frac{d}{dx}(x^4) + 10\frac{d}{dx}(x^3) - 6\frac{d}{dx}(x) + \frac{d}{dx}(5)$$

$$= 8x^7 + 12(5x^4) - 4(4x^3) + 10(3x^2) - 6(1) + 0$$

$$= 8x^7 + 60x^4 - 16x^3 + 30x^2 - 6$$

EXAMPLE 6 Find the points on the curve $y = x^4 - 6x^2 + 4$ where the tangent line is horizontal.

SOLUTION Horizontal tangents occur where the derivative is zero. We have

$$\frac{dy}{dx} = \frac{d}{dx}(x^4) - 6\frac{d}{dx}(x^2) + \frac{d}{dx}(4)$$

$$= 4x^3 - 12x + 0 = 4x(x^2 - 3)$$

Thus $dy/dx = 0$ if $x = 0$ or $x^2 - 3 = 0$, that is, $x = \pm\sqrt{3}$. So the given curve has horizontal tangents when $x = 0$, $\sqrt{3}$, and $-\sqrt{3}$. The corresponding points are $(0, 4)$, $\left(\sqrt{3}, -5\right)$, and $\left(-\sqrt{3}, -5\right)$. (See Figure 5.)

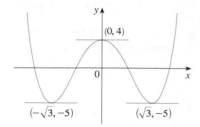

FIGURE 5
The curve $y = x^4 - 6x^2 + 4$ and its horizontal tangents

EXAMPLE 7 The equation of motion of a particle is $s = 2t^3 - 5t^2 + 3t + 4$, where s is measured in centimeters and t in seconds. Find the acceleration as a function of time. What is the acceleration after 2 seconds?

SOLUTION The velocity and acceleration are

$$v(t) = \frac{ds}{dt} = 6t^2 - 10t + 3$$

$$a(t) = \frac{dv}{dt} = 12t - 10$$

The acceleration after 2 s is $a(2) = 14 \text{ cm/s}^2$.

■ **Exponential Functions**

Let's try to compute the derivative of the exponential function $f(x) = b^x$ using the definition of a derivative:

$$f'(x) = \lim_{h \to 0} \frac{f(x + h) - f(x)}{h} = \lim_{h \to 0} \frac{b^{x+h} - b^x}{h}$$

$$= \lim_{h \to 0} \frac{b^x b^h - b^x}{h} = \lim_{h \to 0} \frac{b^x(b^h - 1)}{h}$$

The factor b^x doesn't depend on h, so we can take it in front of the limit:

$$f'(x) = b^x \lim_{h \to 0} \frac{b^h - 1}{h}$$

Notice that the limit is the value of the derivative of f at 0, that is,

$$\lim_{h \to 0} \frac{b^h - 1}{h} = f'(0)$$

Therefore we have shown that if the exponential function $f(x) = b^x$ is differentiable at 0, then it is differentiable everywhere and

[4]

$$f'(x) = f'(0)b^x$$

This equation says that *the rate of change of any exponential function is proportional to the function itself.* (The slope is proportional to the height.)

Numerical evidence for the existence of $f'(0)$ is given in the table at the left for the cases $b = 2$ and $b = 3$. (Values are stated correct to four decimal places.) It appears that the limits exist and

h	$\dfrac{2^h - 1}{h}$	$\dfrac{3^h - 1}{h}$
0.1	0.7177	1.1612
0.01	0.6956	1.1047
0.001	0.6934	1.0992
0.0001	0.6932	1.0987

$$\text{for } b = 2, \quad f'(0) = \lim_{h \to 0} \frac{2^h - 1}{h} \approx 0.69$$

$$\text{for } b = 3, \quad f'(0) = \lim_{h \to 0} \frac{3^h - 1}{h} \approx 1.10$$

In fact, it can be proved that these limits exist and, correct to six decimal places, the values are

$$\frac{d}{dx}(2^x)\bigg|_{x=0} \approx 0.693147 \qquad \frac{d}{dx}(3^x)\bigg|_{x=0} \approx 1.098612$$

Thus, from Equation 4, we have

[5]

$$\frac{d}{dx}(2^x) \approx (0.69)2^x \qquad \frac{d}{dx}(3^x) \approx (1.10)3^x$$

Of all possible choices for the base b in Equation 4, the simplest differentiation formula occurs when $f'(0) = 1$. In view of the estimates of $f'(0)$ for $b = 2$ and $b = 3$, it seems reasonable that there is a number b between 2 and 3 for which $f'(0) = 1$. It is traditional to denote this value by the letter e. (In fact, that is how we introduced e in Section 1.4.) Thus we have the following definition.

In Exercise 1 we will see that e lies between 2.7 and 2.8. Later we will be able to show that, correct to five decimal places,

$$e \approx 2.71828$$

> **Definition of the Number e**
>
> e is the number such that $\displaystyle\lim_{h \to 0} \frac{e^h - 1}{h} = 1$

Geometrically, this means that of all the possible exponential functions $y = b^x$, the function $f(x) = e^x$ is the one whose tangent line at $(0, 1)$ has a slope $f'(0)$ that is exactly 1. (See Figures 6 and 7.)

FIGURE 6

FIGURE 7

If we put $b = e$ and, therefore, $f'(0) = 1$ in Equation 4, it becomes the following important differentiation formula.

> **Derivative of the Natural Exponential Function**
>
> $$\frac{d}{dx}(e^x) = e^x$$

Thus the exponential function $f(x) = e^x$ has the property that it is its own derivative. The geometrical significance of this fact is that the slope of a tangent line to the curve $y = e^x$ is equal to the y-coordinate of the point (see Figure 7).

EXAMPLE 8 If $f(x) = e^x - x$, find f' and f''. Compare the graphs of f and f'.

SOLUTION Using the Difference Rule, we have

$$f'(x) = \frac{d}{dx}(e^x - x) = \frac{d}{dx}(e^x) - \frac{d}{dx}(x) = e^x - 1$$

In Section 2.8 we defined the second derivative as the derivative of f', so

$$f''(x) = \frac{d}{dx}(e^x - 1) = \frac{d}{dx}(e^x) - \frac{d}{dx}(1) = e^x$$

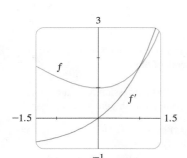

FIGURE 8

The function f and its derivative f' are graphed in Figure 8. Notice that f has a horizontal tangent when $x = 0$; this corresponds to the fact that $f'(0) = 0$. Notice also that, for $x > 0$, $f'(x)$ is positive and f is increasing. When $x < 0$, $f'(x)$ is negative and f is decreasing. ∎

EXAMPLE 9 At what point on the curve $y = e^x$ is the tangent line parallel to the line $y = 2x$?

SOLUTION Since $y = e^x$, we have $y' = e^x$. Let the x-coordinate of the point in question be a. Then the slope of the tangent line at that point is e^a. This tangent line will be parallel to the line $y = 2x$ if it has the same slope, that is, 2. Equating slopes, we get

$$e^a = 2 \qquad a = \ln 2$$

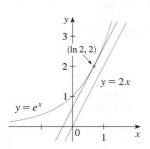

FIGURE 9

Therefore the required point is $(a, e^a) = (\ln 2, 2)$. (See Figure 9.) ∎

3.1 EXERCISES

1. (a) How is the number e defined?

(b) Use a calculator to estimate the values of the limits

$$\lim_{h \to 0} \frac{2.7^h - 1}{h} \quad \text{and} \quad \lim_{h \to 0} \frac{2.8^h - 1}{h}$$

correct to two decimal places. What can you conclude about the value of e?

2. (a) Sketch, by hand, the graph of the function $f(x) = e^x$, paying particular attention to how the graph crosses the y-axis. What fact allows you to do this?

(b) What types of functions are $f(x) = e^x$ and $g(x) = x^e$? Compare the differentiation formulas for f and g.

(c) Which of the two functions in part (b) grows more rapidly when x is large?

3–32 Differentiate the function.

3. $f(x) = 2^{40}$

4. $f(x) = e^5$

5. $f(x) = 5.2x + 2.3$

6. $g(x) = \frac{7}{4}x^2 - 3x + 12$

7. $f(t) = 2t^3 - 3t^2 - 4t$

8. $f(t) = 1.4t^5 - 2.5t^2 + 6.7$

9. $g(x) = x^2(1 - 2x)$

10. $H(u) = (3u - 1)(u + 2)$

11. $g(t) = 2t^{-3/4}$

12. $B(y) = cy^{-6}$

13. $F(r) = \dfrac{5}{r^3}$

14. $y = x^{5/3} - x^{2/3}$

15. $R(a) = (3a + 1)^2$

16. $h(t) = \sqrt[4]{t} - 4e^t$

17. $S(p) = \sqrt{p} - p$

18. $y = \sqrt[3]{x}\,(2 + x)$

19. $y = 3e^x + \dfrac{4}{\sqrt[3]{x}}$

20. $S(R) = 4\pi R^2$

21. $h(u) = Au^3 + Bu^2 + Cu$

22. $y = \dfrac{\sqrt{x} + x}{x^2}$

23. $y = \dfrac{x^2 + 4x + 3}{\sqrt{x}}$

24. $G(t) = \sqrt{5t} + \dfrac{\sqrt{7}}{t}$

25. $j(x) = x^{2.4} + e^{2.4}$

26. $k(r) = e^r + r^e$

27. $G(q) = (1 + q^{-1})^2$

28. $F(z) = \dfrac{A + Bz + Cz^2}{z^2}$

29. $f(v) = \dfrac{\sqrt[3]{v} - 2ve^v}{v}$

30. $D(t) = \dfrac{1 + 16t^2}{(4t)^3}$

31. $z = \dfrac{A}{y^{10}} + Be^y$

32. $y = e^{x+1} + 1$

33–36 Find an equation of the tangent line to the curve at the given point.

33. $y = 2x^3 - x^2 + 2$, $\quad (1, 3)$

34. $y = 2e^x + x$, $\quad (0, 2)$

35. $y = x + \dfrac{2}{x}$, $\quad (2, 3)$

36. $y = \sqrt[4]{x} - x$, $\quad (1, 0)$

37–38 Find equations of the tangent line and normal line to the curve at the given point.

37. $y = x^4 + 2e^x$, $\quad (0, 2)$

38. $y^2 = x^3$, $\quad (1, 1)$

39–40 Find an equation of the tangent line to the curve at the given point. Illustrate by graphing the curve and the tangent line on the same screen.

39. $y = 3x^2 - x^3$, $\quad (1, 2)$

40. $y = x - \sqrt{x}$, $\quad (1, 0)$

41–42 Find $f'(x)$. Compare the graphs of f and f' and use them to explain why your answer is reasonable.

41. $f(x) = x^4 - 2x^3 + x^2$

42. $f(x) = x^5 - 2x^3 + x - 1$

43. (a) Graph the function

$$f(x) = x^4 - 3x^3 - 6x^2 + 7x + 30$$

in the viewing rectangle $[-3, 5]$ by $[-10, 50]$.

(b) Using the graph in part (a) to estimate slopes, make a rough sketch, by hand, of the graph of f'. (See Example 2.8.1.)

(c) Calculate $f'(x)$ and use this expression, with a graphing device, to graph f'. Compare with your sketch in part (b).

44. (a) Graph the function $g(x) = e^x - 3x^2$ in the viewing rectangle $[-1, 4]$ by $[-8, 8]$.

(b) Using the graph in part (a) to estimate slopes, make a rough sketch, by hand, of the graph of g'. (See Example 2.8.1.)

(c) Calculate $g'(x)$ and use this expression, with a graphing device, to graph g'. Compare with your sketch in part (b).

45–46 Find the first and second derivatives of the function.

45. $f(x) = 0.001x^5 - 0.02x^3$

46. $G(r) = \sqrt{r} + \sqrt[3]{r}$

47–48 Find the first and second derivatives of the function. Check to see that your answers are reasonable by comparing the graphs of f, f', and f''.

47. $f(x) = 2x - 5x^{3/4}$

48. $f(x) = e^x - x^3$

49. The equation of motion of a particle is $s = t^3 - 3t$, where s is in meters and t is in seconds. Find
(a) the velocity and acceleration as functions of t,
(b) the acceleration after 2 s, and
(c) the acceleration when the velocity is 0.

50. The equation of motion of a particle is
$s = t^4 - 2t^3 + t^2 - t$, where s is in meters and t is in seconds.
(a) Find the velocity and acceleration as functions of t.
(b) Find the acceleration after 1 s.
(c) Graph the position, velocity, and acceleration functions on the same screen.

51. Biologists have proposed a cubic polynomial to model the length L of Alaskan rockfish at age A:

$$L = 0.0155A^3 - 0.372A^2 + 3.95A + 1.21$$

where L is measured in inches and A in years. Calculate

$$\left.\frac{dL}{dA}\right|_{A=12}$$

and interpret your answer.

52. The number of tree species S in a given area A in the Pasoh Forest Reserve in Malaysia has been modeled by the power function

$$S(A) = 0.882A^{0.842}$$

where A is measured in square meters. Find $S'(100)$ and interpret your answer.

Source: Adapted from K. Kochummen et al., "Floristic Composition of Pasoh Forest Reserve, A Lowland Rain Forest in Peninsular Malaysia," *Journal of Tropical Forest Science* 3 (1991):1–13.

53. Boyle's Law states that when a sample of gas is compressed at a constant temperature, the pressure P of the gas is inversely proportional to the volume V of the gas.
(a) Suppose that the pressure of a sample of air that occupies 0.106 m^3 at 25°C is 50 kPa. Write V as a function of P.
(b) Calculate dV/dP when $P = 50$ kPa. What is the meaning of the derivative? What are its units?

54. Car tires need to be inflated properly because overinflation or underinflation can cause premature tread wear. The data in the table show tire life L (in thousands of miles) for a certain type of tire at various pressures P (in lb/in^2).

P	26	28	31	35	38	42	45
L	50	66	78	81	74	70	59

(a) Use a calculator to model tire life with a quadratic function of the pressure.
(b) Use the model to estimate dL/dP when $P = 30$ and when $P = 40$. What is the meaning of the derivative? What are the units? What is the significance of the signs of the derivatives?

55. Find the points on the curve $y = 2x^3 + 3x^2 - 12x + 1$ where the tangent is horizontal.

56. For what value of x does the graph of $f(x) = e^x - 2x$ have a horizontal tangent?

57. Show that the curve $y = 2e^x + 3x + 5x^3$ has no tangent line with slope 2.

58. Find an equation of the tangent line to the curve $y = x^4 + 1$ that is parallel to the line $32x - y = 15$.

59. Find equations of both lines that are tangent to the curve $y = x^3 - 3x^2 + 3x - 3$ and are parallel to the line $3x - y = 15$.

60. At what point on the curve $y = 1 + 2e^x - 3x$ is the tangent line parallel to the line $3x - y = 5$? Illustrate by graphing the curve and both lines.

61. Find an equation of the normal line to the curve $y = \sqrt{x}$ that is parallel to the line $2x + y = 1$.

62. Where does the normal line to the parabola $y = x^2 - 1$ at the point $(-1, 0)$ intersect the parabola a second time? Illustrate with a sketch.

63. Draw a diagram to show that there are two tangent lines to the parabola $y = x^2$ that pass through the point $(0, -4)$. Find the coordinates of the points where these tangent lines intersect the parabola.

64. (a) Find equations of both lines through the point $(2, -3)$ that are tangent to the parabola $y = x^2 + x$.
(b) Show that there is no line through the point $(2, 7)$ that is tangent to the parabola. Then draw a diagram to see why.

65. Use the definition of a derivative to show that if $f(x) = 1/x$, then $f'(x) = -1/x^2$. (This proves the Power Rule for the case $n = -1$.)

66. Find the nth derivative of each function by calculating the first few derivatives and observing the pattern that occurs.
(a) $f(x) = x^n$ (b) $f(x) = 1/x$

67. Find a second-degree polynomial P such that $P(2) = 5$, $P'(2) = 3$, and $P''(2) = 2$.

68. The equation $y'' + y' - 2y = x^2$ is called a **differential equation** because it involves an unknown function y and its derivatives y' and y''. Find constants A, B, and C such that the function $y = Ax^2 + Bx + C$ satisfies this equation. (Differential equations will be studied in detail in Chapter 9.)

69. Find a cubic function $y = ax^3 + bx^2 + cx + d$ whose graph has horizontal tangents at the points $(-2, 6)$ and $(2, 0)$.

70. Find a parabola with equation $y = ax^2 + bx + c$ that has slope 4 at $x = 1$, slope -8 at $x = -1$, and passes through the point $(2, 15)$.

71. Let

$$f(x) = \begin{cases} x^2 + 1 & \text{if } x < 1 \\ x + 1 & \text{if } x \geq 1 \end{cases}$$

Is f differentiable at 1? Sketch the graphs of f and f'.

72. At what numbers is the following function g differentiable?

$$g(x) = \begin{cases} 2x & \text{if } x \leq 0 \\ 2x - x^2 & \text{if } 0 < x < 2 \\ 2 - x & \text{if } x \geq 2 \end{cases}$$

Give a formula for g' and sketch the graphs of g and g'.

73. (a) For what values of x is the function $f(x) = |x^2 - 9|$ differentiable? Find a formula for f'.
(b) Sketch the graphs of f and f'.

74. Where is the function $h(x) = |x - 1| + |x + 2|$ differentiable? Give a formula for h' and sketch the graphs of h and h'.

75. Find the parabola with equation $y = ax^2 + bx$ whose tangent line at $(1, 1)$ has equation $y = 3x - 2$.

76. Suppose the curve $y = x^4 + ax^3 + bx^2 + cx + d$ has a tangent line when $x = 0$ with equation $y = 2x + 1$ and a tangent line when $x = 1$ with equation $y = 2 - 3x$. Find the values of a, b, c, and d.

77. For what values of a and b is the line $2x + y = b$ tangent to the parabola $y = ax^2$ when $x = 2$?

78. Find the value of c such that the line $y = \frac{3}{2}x + 6$ is tangent to the curve $y = c\sqrt{x}$.

79. What is the value of c such that the line $y = 2x + 3$ is tangent to the parabola $y = cx^2$?

80. The graph of any quadratic function $f(x) = ax^2 + bx + c$ is a parabola. Prove that the average of the slopes of the tangent lines to the parabola at the endpoints of any interval $[p, q]$ equals the slope of the tangent line at the midpoint of the interval.

81. Let

$$f(x) = \begin{cases} x^2 & \text{if } x \leq 2 \\ mx + b & \text{if } x > 2 \end{cases}$$

Find the values of m and b that make f differentiable everywhere.

82. A tangent line is drawn to the hyperbola $xy = c$ at a point P.
(a) Show that the midpoint of the line segment cut from this tangent line by the coordinate axes is P.
(b) Show that the triangle formed by the tangent line and the coordinate axes always has the same area, no matter where P is located on the hyperbola.

83. Evaluate $\displaystyle\lim_{x \to 1} \frac{x^{1000} - 1}{x - 1}$.

84. Draw a diagram showing two perpendicular lines that intersect on the y-axis and are both tangent to the parabola $y = x^2$. Where do these lines intersect?

85. If $c > \frac{1}{2}$, how many lines through the point $(0, c)$ are normal lines to the parabola $y = x^2$? What if $c \leq \frac{1}{2}$?

86. Sketch the parabolas $y = x^2$ and $y = x^2 - 2x + 2$. Do you think there is a line that is tangent to both curves? If so, find its equation. If not, why not?

APPLIED PROJECT   **BUILDING A BETTER ROLLER COASTER**

Suppose you are asked to design the first ascent and drop for a new roller coaster. By studying photographs of your favorite coasters, you decide to make the slope of the ascent 0.8 and the slope of the drop -1.6. You decide to connect these two straight stretches $y = L_1(x)$ and $y = L_2(x)$ with part of a parabola $y = f(x) = ax^2 + bx + c$, where x and $f(x)$ are measured in feet. For the track to be smooth there can't be abrupt changes in direction, so you want the linear segments L_1 and L_2 to be tangent to the parabola at the transition points P and Q. (See the figure.) To simplify the equations, you decide to place the origin at P.

1. (a) Suppose the horizontal distance between P and Q is 100 ft. Write equations in a, b, and c that will ensure that the track is smooth at the transition points.
(b) Solve the equations in part (a) for a, b, and c to find a formula for $f(x)$.
(c) Plot L_1, f, and L_2 to verify graphically that the transitions are smooth.
(d) Find the difference in elevation between P and Q.

2. The solution in Problem 1 might *look* smooth, but it might not *feel* smooth because the piecewise defined function [consisting of $L_1(x)$ for $x < 0$, $f(x)$ for $0 \leq x \leq 100$, and

$L_2(x)$ for $x > 100$] doesn't have a continuous second derivative. So you decide to improve the design by using a quadratic function $q(x) = ax^2 + bx + c$ only on the interval $10 \le x \le 90$ and connecting it to the linear functions by means of two cubic functions:

$$g(x) = kx^3 + lx^2 + mx + n \qquad 0 \le x < 10$$

$$h(x) = px^3 + qx^2 + rx + s \qquad 90 < x \le 100$$

(a) Write a system of equations in 11 unknowns that ensure that the functions and their first two derivatives agree at the transition points.

CAS (b) Solve the equations in part (a) with a computer algebra system to find formulas for $q(x)$, $g(x)$, and $h(x)$.

(c) Plot L_1, g, q, h, and L_2, and compare with the plot in Problem 1(c).

3.2 The Product and Quotient Rules

The formulas of this section enable us to differentiate new functions formed from old functions by multiplication or division.

■ The Product Rule

By analogy with the Sum and Difference Rules, one might be tempted to guess, as Leibniz did three centuries ago, that the derivative of a product is the product of the derivatives. We can see, however, that this guess is wrong by looking at a particular example. Let $f(x) = x$ and $g(x) = x^2$. Then the Power Rule gives $f'(x) = 1$ and $g'(x) = 2x$. But $(fg)(x) = x^3$, so $(fg)'(x) = 3x^2$. Thus $(fg)' \ne f'g'$. The correct formula was discovered by Leibniz (soon after his false start) and is called the Product Rule.

Before stating the Product Rule, let's see how we might discover it. We start by assuming that $u = f(x)$ and $v = g(x)$ are both positive differentiable functions. Then we can interpret the product uv as an area of a rectangle (see Figure 1). If x changes by an amount Δx, then the corresponding changes in u and v are

$$\Delta u = f(x + \Delta x) - f(x) \qquad \Delta v = g(x + \Delta x) - g(x)$$

and the new value of the product, $(u + \Delta u)(v + \Delta v)$, can be interpreted as the area of the large rectangle in Figure 1 (provided that Δu and Δv happen to be positive).

The change in the area of the rectangle is

1 $$\Delta(uv) = (u + \Delta u)(v + \Delta v) - uv = u\,\Delta v + v\,\Delta u + \Delta u\,\Delta v$$

$$= \text{the sum of the three shaded areas}$$

If we divide by Δx, we get

$$\frac{\Delta(uv)}{\Delta x} = u\,\frac{\Delta v}{\Delta x} + v\,\frac{\Delta u}{\Delta x} + \Delta u\,\frac{\Delta v}{\Delta x}$$

FIGURE 1
The geometry of the Product Rule

Recall that in Leibniz notation the definition of a derivative can be written as

$$\frac{dy}{dx} = \lim_{\Delta x \to 0} \frac{\Delta y}{\Delta x}$$

If we now let $\Delta x \to 0$, we get the derivative of uv:

$$\frac{d}{dx}(uv) = \lim_{\Delta x \to 0} \frac{\Delta(uv)}{\Delta x} = \lim_{\Delta x \to 0} \left(u \frac{\Delta v}{\Delta x} + v \frac{\Delta u}{\Delta x} + \Delta u \frac{\Delta v}{\Delta x} \right)$$

$$= u \lim_{\Delta x \to 0} \frac{\Delta v}{\Delta x} + v \lim_{\Delta x \to 0} \frac{\Delta u}{\Delta x} + \left(\lim_{\Delta x \to 0} \Delta u \right)\left(\lim_{\Delta x \to 0} \frac{\Delta v}{\Delta x} \right)$$

$$= u \frac{dv}{dx} + v \frac{du}{dx} + 0 \cdot \frac{dv}{dx}$$

$$\boxed{2} \qquad \frac{d}{dx}(uv) = u \frac{dv}{dx} + v \frac{du}{dx}$$

(Notice that $\Delta u \to 0$ as $\Delta x \to 0$ since f is differentiable and therefore continuous.)

Although we started by assuming (for the geometric interpretation) that all the quantities are positive, we notice that Equation 1 is always true. (The algebra is valid whether u, v, Δu, and Δv are positive or negative.) So we have proved Equation 2, known as the Product Rule, for all differentiable functions u and v.

In prime notation:

$$(fg)' = fg' + gf'$$

> **The Product Rule** If f and g are both differentiable, then
>
> $$\frac{d}{dx}[f(x)g(x)] = f(x)\frac{d}{dx}[g(x)] + g(x)\frac{d}{dx}[f(x)]$$

In words, the Product Rule says that *the derivative of a product of two functions is the first function times the derivative of the second function plus the second function times the derivative of the first function.*

EXAMPLE 1
(a) If $f(x) = xe^x$, find $f'(x)$.
(b) Find the nth derivative, $f^{(n)}(x)$.

SOLUTION
(a) By the Product Rule, we have

$$f'(x) = \frac{d}{dx}(xe^x)$$

$$= x \frac{d}{dx}(e^x) + e^x \frac{d}{dx}(x)$$

$$= xe^x + e^x \cdot 1 = (x + 1)e^x$$

(b) Using the Product Rule a second time, we get

$$f''(x) = \frac{d}{dx}[(x + 1)e^x]$$

$$= (x + 1)\frac{d}{dx}(e^x) + e^x \frac{d}{dx}(x + 1)$$

$$= (x + 1)e^x + e^x \cdot 1 = (x + 2)e^x$$

Figure 2 shows the graphs of the function f of Example 1 and its derivative f'. Notice that $f'(x)$ is positive when f is increasing and negative when f is decreasing.

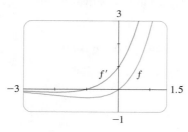

FIGURE 2

Further applications of the Product Rule give

$$f'''(x) = (x + 3)e^x \qquad f^{(4)}(x) = (x + 4)e^x$$

In fact, each successive differentiation adds another term e^x, so

$$f^{(n)}(x) = (x + n)e^x$$

In Example 2, a and b are constants. It is customary in mathematics to use letters near the beginning of the alphabet to represent constants and letters near the end of the alphabet to represent variables.

EXAMPLE 2 Differentiate the function $f(t) = \sqrt{t}\,(a + bt)$.

SOLUTION 1 Using the Product Rule, we have

$$f'(t) = \sqrt{t}\,\frac{d}{dt}(a + bt) + (a + bt)\frac{d}{dt}\left(\sqrt{t}\,\right)$$

$$= \sqrt{t} \cdot b + (a + bt) \cdot \tfrac{1}{2}t^{-1/2}$$

$$= b\sqrt{t} + \frac{a + bt}{2\sqrt{t}} = \frac{a + 3bt}{2\sqrt{t}}$$

SOLUTION 2 If we first use the laws of exponents to rewrite $f(t)$, then we can proceed directly without using the Product Rule.

$$f(t) = a\sqrt{t} + bt\sqrt{t} = at^{1/2} + bt^{3/2}$$

$$f'(t) = \tfrac{1}{2}at^{-1/2} + \tfrac{3}{2}bt^{1/2}$$

which is equivalent to the answer given in Solution 1.

Example 2 shows that it is sometimes easier to simplify a product of functions before differentiating than to use the Product Rule. In Example 1, however, the Product Rule is the only possible method.

EXAMPLE 3 If $f(x) = \sqrt{x}\,g(x)$, where $g(4) = 2$ and $g'(4) = 3$, find $f'(4)$.

SOLUTION Applying the Product Rule, we get

$$f'(x) = \frac{d}{dx}\left[\sqrt{x}\,g(x)\right] = \sqrt{x}\,\frac{d}{dx}[g(x)] + g(x)\frac{d}{dx}\left[\sqrt{x}\,\right]$$

$$= \sqrt{x}\,g'(x) + g(x) \cdot \tfrac{1}{2}x^{-1/2}$$

$$= \sqrt{x}\,g'(x) + \frac{g(x)}{2\sqrt{x}}$$

So

$$f'(4) = \sqrt{4}\,g'(4) + \frac{g(4)}{2\sqrt{4}} = 2 \cdot 3 + \frac{2}{2 \cdot 2} = 6.5$$

The Quotient Rule

We find a rule for differentiating the quotient of two differentiable functions $u = f(x)$ and $v = g(x)$ in much the same way that we found the Product Rule. If x, u, and v change by amounts Δx, Δu, and Δv, then the corresponding change in the quotient u/v is

$$\Delta\!\left(\frac{u}{v}\right) = \frac{u + \Delta u}{v + \Delta v} - \frac{u}{v} = \frac{(u + \Delta u)v - u(v + \Delta v)}{v(v + \Delta v)}$$

$$= \frac{v\,\Delta u - u\,\Delta v}{v(v + \Delta v)}$$

so

$$\frac{d}{dx}\!\left(\frac{u}{v}\right) = \lim_{\Delta x \to 0}\frac{\Delta(u/v)}{\Delta x} = \lim_{\Delta x \to 0}\frac{v\,\dfrac{\Delta u}{\Delta x} - u\,\dfrac{\Delta v}{\Delta x}}{v(v + \Delta v)}$$

As $\Delta x \to 0$, $\Delta v \to 0$ also, because $v = g(x)$ is differentiable and therefore continuous. Thus, using the Limit Laws, we get

$$\frac{d}{dx}\!\left(\frac{u}{v}\right) = \frac{v \lim\limits_{\Delta x \to 0}\dfrac{\Delta u}{\Delta x} - u \lim\limits_{\Delta x \to 0}\dfrac{\Delta v}{\Delta x}}{v \lim\limits_{\Delta x \to 0}(v + \Delta v)} = \frac{v\,\dfrac{du}{dx} - u\,\dfrac{dv}{dx}}{v^2}$$

In prime notation:

$$\left(\frac{f}{g}\right)' = \frac{gf' - fg'}{g^2}$$

The Quotient Rule If f and g are differentiable, then

$$\frac{d}{dx}\!\left[\frac{f(x)}{g(x)}\right] = \frac{g(x)\dfrac{d}{dx}[f(x)] - f(x)\dfrac{d}{dx}[g(x)]}{[g(x)]^2}$$

In words, the Quotient Rule says that the *derivative of a quotient is the denominator times the derivative of the numerator minus the numerator times the derivative of the denominator, all divided by the square of the denominator.*

The Quotient Rule and the other differentiation formulas enable us to compute the derivative of any rational function, as the next example illustrates.

We can use a graphing device to check that the answer to Example 4 is plausible. Figure 3 shows the graphs of the function of Example 4 and its derivative. Notice that when y grows rapidly (near -2), y' is large. And when y grows slowly, y' is near 0.

FIGURE 3

EXAMPLE 4 Let $y = \dfrac{x^2 + x - 2}{x^3 + 6}$. Then

$$y' = \frac{(x^3 + 6)\dfrac{d}{dx}(x^2 + x - 2) - (x^2 + x - 2)\dfrac{d}{dx}(x^3 + 6)}{(x^3 + 6)^2}$$

$$= \frac{(x^3 + 6)(2x + 1) - (x^2 + x - 2)(3x^2)}{(x^3 + 6)^2}$$

$$= \frac{(2x^4 + x^3 + 12x + 6) - (3x^4 + 3x^3 - 6x^2)}{(x^3 + 6)^2}$$

$$= \frac{-x^4 - 2x^3 + 6x^2 + 12x + 6}{(x^3 + 6)^2} \qquad \blacksquare$$

EXAMPLE 5 Find an equation of the tangent line to the curve $y = e^x/(1 + x^2)$ at the point $\left(1, \frac{1}{2}e\right)$.

SOLUTION According to the Quotient Rule, we have

$$\frac{dy}{dx} = \frac{(1 + x^2)\dfrac{d}{dx}(e^x) - e^x \dfrac{d}{dx}(1 + x^2)}{(1 + x^2)^2}$$

$$= \frac{(1 + x^2)e^x - e^x(2x)}{(1 + x^2)^2} = \frac{e^x(1 - 2x + x^2)}{(1 + x^2)^2}$$

$$= \frac{e^x(1 - x)^2}{(1 + x^2)^2}$$

So the slope of the tangent line at $\left(1, \frac{1}{2}e\right)$ is

$$\left.\frac{dy}{dx}\right|_{x=1} = 0$$

This means that the tangent line at $\left(1, \frac{1}{2}e\right)$ is horizontal and its equation is $y = \frac{1}{2}e$. $\left[\text{See Figure 4. Notice that the function is increasing and crosses its tangent line at }\left(1, \frac{1}{2}e\right).\right]$

FIGURE 4

NOTE Don't use the Quotient Rule *every* time you see a quotient. Sometimes it's easier to rewrite a quotient first to put it in a form that is simpler for the purpose of differentiation. For instance, although it is possible to differentiate the function

$$F(x) = \frac{3x^2 + 2\sqrt{x}}{x}$$

using the Quotient Rule, it is much easier to perform the division first and write the function as

$$F(x) = 3x + 2x^{-1/2}$$

before differentiating.

We summarize the differentiation formulas we have learned so far as follows.

Table of Differentiation Formulas

$$\frac{d}{dx}(c) = 0 \qquad \frac{d}{dx}(x^n) = nx^{n-1} \qquad \frac{d}{dx}(e^x) = e^x$$

$$(cf)' = cf' \qquad (f + g)' = f' + g' \qquad (f - g)' = f' - g'$$

$$(fg)' = fg' + gf' \qquad \left(\frac{f}{g}\right)' = \frac{gf' - fg'}{g^2}$$

3.2 EXERCISES

1. Find the derivative of $f(x) = (1 + 2x^2)(x - x^2)$ in two ways: by using the Product Rule and by performing the multiplication first. Do your answers agree?

2. Find the derivative of the function

$$F(x) = \frac{x^4 - 5x^3 + \sqrt{x}}{x^2}$$

in two ways: by using the Quotient Rule and by simplifying first. Show that your answers are equivalent. Which method do you prefer?

3–26 Differentiate.

3. $f(x) = (3x^2 - 5x)e^x$

4. $g(x) = \left(x + 2\sqrt{x}\right)e^x$

5. $y = \dfrac{x}{e^x}$

6. $y = \dfrac{e^x}{1 - e^x}$

7. $g(x) = \dfrac{1 + 2x}{3 - 4x}$

8. $G(x) = \dfrac{x^2 - 2}{2x + 1}$

9. $H(u) = \left(u - \sqrt{u}\right)\left(u + \sqrt{u}\right)$

10. $J(v) = (v^3 - 2v)(v^{-4} + v^{-2})$

11. $F(y) = \left(\dfrac{1}{y^2} - \dfrac{3}{y^4}\right)(y + 5y^3)$

12. $f(z) = (1 - e^z)(z + e^z)$

13. $y = \dfrac{x^2 + 1}{x^3 - 1}$

14. $y = \dfrac{\sqrt{x}}{2 + x}$

15. $y = \dfrac{t^3 + 3t}{t^2 - 4t + 3}$

16. $y = \dfrac{1}{t^3 + 2t^2 - 1}$

17. $y = e^p(p + p\sqrt{p})$

18. $h(r) = \dfrac{ae^r}{b + e^r}$

19. $y = \dfrac{s - \sqrt{s}}{s^2}$

20. $y = (z^2 + e^z)\sqrt{z}$

21. $f(t) = \dfrac{\sqrt[3]{t}}{t - 3}$

22. $V(t) = \dfrac{4 + t}{te^t}$

23. $f(x) = \dfrac{x^2 e^x}{x^2 + e^x}$

24. $F(t) = \dfrac{At}{Bt^2 + Ct^3}$

25. $f(x) = \dfrac{x}{x + \dfrac{c}{x}}$

26. $f(x) = \dfrac{ax + b}{cx + d}$

27–30 Find $f'(x)$ and $f''(x)$.

27. $f(x) = (x^3 + 1)e^x$

28. $f(x) = \sqrt{x}\,e^x$

29. $f(x) = \dfrac{x^2}{1 + e^x}$

30. $f(x) = \dfrac{x}{x^2 - 1}$

31–32 Find an equation of the tangent line to the given curve at the specified point.

31. $y = \dfrac{x^2 - 1}{x^2 + x + 1}$, $(1, 0)$

32. $y = \dfrac{1 + x}{1 + e^x}$, $\left(0, \frac{1}{2}\right)$

33–34 Find equations of the tangent line and normal line to the given curve at the specified point.

33. $y = 2xe^x$, $(0, 0)$

34. $y = \dfrac{2x}{x^2 + 1}$, $(1, 1)$

35. (a) The curve $y = 1/(1 + x^2)$ is called a **witch of Maria Agnesi**. Find an equation of the tangent line to this curve at the point $\left(-1, \frac{1}{2}\right)$.
   (b) Illustrate part (a) by graphing the curve and the tangent line on the same screen.

36. (a) The curve $y = x/(1 + x^2)$ is called a **serpentine**. Find an equation of the tangent line to this curve at the point $(3, 0.3)$.
   (b) Illustrate part (a) by graphing the curve and the tangent line on the same screen.

37. (a) If $f(x) = (x^3 - x)e^x$, find $f'(x)$.
   (b) Check to see that your answer to part (a) is reasonable by comparing the graphs of f and f'.

38. (a) If $f(x) = e^x/(2x^2 + x + 1)$, find $f'(x)$.
   (b) Check to see that your answer to part (a) is reasonable by comparing the graphs of f and f'.

39. (a) If $f(x) = (x^2 - 1)/(x^2 + 1)$, find $f'(x)$ and $f''(x)$.
   (b) Check to see that your answers to part (a) are reasonable by comparing the graphs of f, f', and f''.

40. (a) If $f(x) = (x^2 - 1)e^x$, find $f'(x)$ and $f''(x)$.
   (b) Check to see that your answers to part (a) are reasonable by comparing the graphs of f, f', and f''.

41. If $f(x) = x^2/(1 + x)$, find $f''(1)$.

42. If $g(x) = x/e^x$, find $g^{(n)}(x)$.

43. Suppose that $f(5) = 1$, $f'(5) = 6$, $g(5) = -3$, and $g'(5) = 2$. Find the following values.
   (a) $(fg)'(5)$ (b) $(f/g)'(5)$ (c) $(g/f)'(5)$

44. Suppose that $f(4) = 2$, $g(4) = 5$, $f'(4) = 6$, and $g'(4) = -3$. Find $h'(4)$.
   (a) $h(x) = 3f(x) + 8g(x)$ (b) $h(x) = f(x)g(x)$
   (c) $h(x) = \dfrac{f(x)}{g(x)}$ (d) $h(x) = \dfrac{g(x)}{f(x) + g(x)}$

45. If $f(x) = e^x g(x)$, where $g(0) = 2$ and $g'(0) = 5$, find $f'(0)$.

46. If $h(2) = 4$ and $h'(2) = -3$, find

$$\frac{d}{dx}\left(\frac{h(x)}{x}\right)\bigg|_{x=2}$$

47. If $g(x) = xf(x)$, where $f(3) = 4$ and $f'(3) = -2$, find an equation of the tangent line to the graph of g at the point where $x = 3$.

48. If $f(2) = 10$ and $f'(x) = x^2 f(x)$ for all x, find $f''(2)$.

49. If f and g are the functions whose graphs are shown, let $u(x) = f(x)g(x)$ and $v(x) = f(x)/g(x)$.
(a) Find $u'(1)$.   (b) Find $v'(5)$.

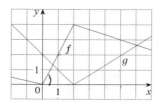

50. Let $P(x) = F(x)G(x)$ and $Q(x) = F(x)/G(x)$, where F and G are the functions whose graphs are shown.
(a) Find $P'(2)$.   (b) Find $Q'(7)$.

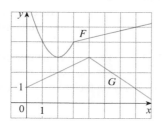

51. If g is a differentiable function, find an expression for the derivative of each of the following functions.
(a) $y = xg(x)$ (b) $y = \dfrac{x}{g(x)}$
(c) $y = \dfrac{g(x)}{x}$

52. If f is a differentiable function, find an expression for the derivative of each of the following functions.
(a) $y = x^2 f(x)$ (b) $y = \dfrac{f(x)}{x^2}$
(c) $y = \dfrac{x^2}{f(x)}$ (d) $y = \dfrac{1 + xf(x)}{\sqrt{x}}$

53. How many tangent lines to the curve $y = x/(x + 1)$ pass through the point $(1, 2)$? At which points do these tangent lines touch the curve?

54. Find equations of the tangent lines to the curve

$$y = \frac{x - 1}{x + 1}$$

that are parallel to the line $x - 2y = 2$.

55. Find $R'(0)$, where

$$R(x) = \frac{x - 3x^3 + 5x^5}{1 + 3x^3 + 6x^6 + 9x^9}$$

Hint: Instead of finding $R'(x)$ first, let $f(x)$ be the numerator and $g(x)$ the denominator of $R(x)$ and compute $R'(0)$ from $f(0)$, $f'(0)$, $g(0)$, and $g'(0)$.

56. Use the method of Exercise 55 to compute $Q'(0)$, where

$$Q(x) = \frac{1 + x + x^2 + xe^x}{1 - x + x^2 - xe^x}$$

57. In this exercise we estimate the rate at which the total personal income is rising in the Richmond-Petersburg, Virginia, metropolitan area. In 1999, the population of this area was 961,400, and the population was increasing at roughly 9200 people per year. The average annual income was $30,593 per capita, and this average was increasing at about $1400 per year (a little above the national average of about $1225 yearly). Use the Product Rule and these figures to estimate the rate at which total personal income was rising in the Richmond-Petersburg area in 1999. Explain the meaning of each term in the Product Rule.

58. A manufacturer produces bolts of a fabric with a fixed width. The quantity q of this fabric (measured in yards) that is sold is a function of the selling price p (in dollars per yard), so we can write $q = f(p)$. Then the total revenue earned with selling price p is $R(p) = pf(p)$.
(a) What does it mean to say that $f(20) = 10,000$ and $f'(20) = -350$?
(b) Assuming the values in part (a), find $R'(20)$ and interpret your answer.

59. The Michaelis-Menten equation for the enzyme chymotrypsin is

$$v = \frac{0.14[S]}{0.015 + [S]}$$

where v is the rate of an enzymatic reaction and $[S]$ is the concentration of a substrate S. Calculate $dv/d[S]$ and interpret it.

60. The *biomass* $B(t)$ of a fish population is the total mass of the members of the population at time t. It is the product of the number of individuals $N(t)$ in the population and the average mass $M(t)$ of a fish at time t. In the case of guppies, breeding occurs continually. Suppose that at time $t = 4$ weeks the population is 820 guppies and is growing at a rate of 50 guppies per week, while the average mass is 1.2 g and is increasing at a rate of 0.14 g/week. At what rate is the biomass increasing when $t = 4$?

61. (a) Use the Product Rule twice to prove that if f, g, and h are differentiable, then $(fgh)' = f'gh + fg'h + fgh'$.

(b) Taking $f = g = h$ in part (a), show that

$$\frac{d}{dx}[f(x)]^3 = 3[f(x)]^2 f'(x)$$

(c) Use part (b) to differentiate $y = e^{3x}$.

62. (a) If $F(x) = f(x)\,g(x)$, where f and g have derivatives of all orders, show that $F'' = f''g + 2f'g' + fg''$.

(b) Find similar formulas for F''' and $F^{(4)}$.

(c) Guess a formula for $F^{(n)}$.

63. Find expressions for the first five derivatives of $f(x) = x^2 e^x$. Do you see a pattern in these expressions? Guess a formula for $f^{(n)}(x)$ and prove it using mathematical induction.

64. (a) If g is differentiable, the **Reciprocal Rule** says that

$$\frac{d}{dx}\left[\frac{1}{g(x)}\right] = -\frac{g'(x)}{[g(x)]^2}$$

Use the Quotient Rule to prove the Reciprocal Rule.

(b) Use the Reciprocal Rule to differentiate the function in Exercise 16.

(c) Use the Reciprocal Rule to verify that the Power Rule is valid for negative integers, that is,

$$\frac{d}{dx}(x^{-n}) = -nx^{-n-1}$$

for all positive integers n.

3.3   Derivatives of Trigonometric Functions

A review of the trigonometric functions is given in Appendix D.

Before starting this section, you might need to review the trigonometric functions. In particular, it is important to remember that when we talk about the function f defined for all real numbers x by

$$f(x) = \sin x$$

it is understood that $\sin x$ means the sine of the angle whose *radian* measure is x. A similar convention holds for the other trigonometric functions cos, tan, csc, sec, and cot. Recall from Section 2.5 that all of the trigonometric functions are continuous at every number in their domains.

If we sketch the graph of the function $f(x) = \sin x$ and use the interpretation of $f'(x)$ as the slope of the tangent to the sine curve in order to sketch the graph of f' (see Exercise 2.8.16), then it looks as if the graph of f' may be the same as the cosine curve (see Figure 1).

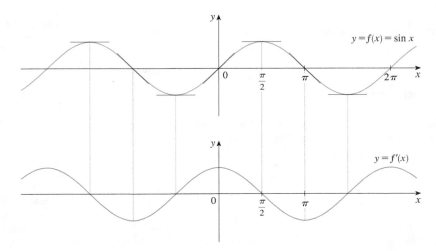

TEC Visual 3.3 shows an animation of Figure 1.

FIGURE 1

Let's try to confirm our guess that if $f(x) = \sin x$, then $f'(x) = \cos x$. From the definition of a derivative, we have

$$f'(x) = \lim_{h \to 0} \frac{f(x + h) - f(x)}{h} = \lim_{h \to 0} \frac{\sin(x + h) - \sin x}{h}$$

We have used the addition formula for sine. See Appendix D.

$$= \lim_{h \to 0} \frac{\sin x \cos h + \cos x \sin h - \sin x}{h}$$

$$= \lim_{h \to 0} \left[\frac{\sin x \cos h - \sin x}{h} + \frac{\cos x \sin h}{h} \right]$$

$$= \lim_{h \to 0} \left[\sin x \left(\frac{\cos h - 1}{h} \right) + \cos x \left(\frac{\sin h}{h} \right) \right]$$

$$\boxed{1} \qquad = \lim_{h \to 0} \sin x \cdot \lim_{h \to 0} \frac{\cos h - 1}{h} + \lim_{h \to 0} \cos x \cdot \lim_{h \to 0} \frac{\sin h}{h}$$

Two of these four limits are easy to evaluate. Since we regard x as a constant when computing a limit as $h \to 0$, we have

$$\lim_{h \to 0} \sin x = \sin x \qquad \text{and} \qquad \lim_{h \to 0} \cos x = \cos x$$

The limit of $(\sin h)/h$ is not so obvious. In Example 2.2.3 we made the guess, on the basis of numerical and graphical evidence, that

$$\boxed{2} \qquad \boxed{\lim_{\theta \to 0} \frac{\sin \theta}{\theta} = 1}$$

(a)

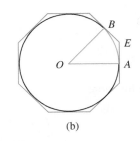

(b)

FIGURE 2

We now use a geometric argument to prove Equation 2. Assume first that θ lies between 0 and $\pi/2$. Figure 2(a) shows a sector of a circle with center O, central angle θ, and radius 1. BC is drawn perpendicular to OA. By the definition of radian measure, we have arc $AB = \theta$. Also $|BC| = |OB| \sin \theta = \sin \theta$. From the diagram we see that

$$|BC| < |AB| < \text{arc } AB$$

Therefore $\qquad\qquad \sin \theta < \theta \qquad$ so $\qquad \dfrac{\sin \theta}{\theta} < 1$

Let the tangent lines at A and B intersect at E. You can see from Figure 2(b) that the circumference of a circle is smaller than the length of a circumscribed polygon, and so arc $AB < |AE| + |EB|$. Thus

$$\theta = \text{arc } AB < |AE| + |EB|$$

$$< |AE| + |ED|$$

$$= |AD| = |OA| \tan \theta$$

$$= \tan \theta$$

(In Appendix F the inequality $\theta \leqslant \tan \theta$ is proved directly from the definition of the length of an arc without resorting to geometric intuition as we did here.) Therefore we have

$$\theta < \frac{\sin \theta}{\cos \theta}$$

so

$$\cos \theta < \frac{\sin \theta}{\theta} < 1$$

We know that $\lim_{\theta \to 0} 1 = 1$ and $\lim_{\theta \to 0} \cos \theta = 1$, so by the Squeeze Theorem, we have

$$\lim_{\theta \to 0^+} \frac{\sin \theta}{\theta} = 1$$

But the function $(\sin \theta)/\theta$ is an even function, so its right and left limits must be equal. Hence, we have

$$\lim_{\theta \to 0} \frac{\sin \theta}{\theta} = 1$$

so we have proved Equation 2.

We can deduce the value of the remaining limit in (1) as follows:

We multiply numerator and denominator by $\cos \theta + 1$ in order to put the function in a form in which we can use the limits we know.

$$\lim_{\theta \to 0} \frac{\cos \theta - 1}{\theta} = \lim_{\theta \to 0} \left(\frac{\cos \theta - 1}{\theta} \cdot \frac{\cos \theta + 1}{\cos \theta + 1} \right) = \lim_{\theta \to 0} \frac{\cos^2 \theta - 1}{\theta \, (\cos \theta + 1)}$$

$$= \lim_{\theta \to 0} \frac{-\sin^2 \theta}{\theta \, (\cos \theta + 1)} = -\lim_{\theta \to 0} \left(\frac{\sin \theta}{\theta} \cdot \frac{\sin \theta}{\cos \theta + 1} \right)$$

$$= -\lim_{\theta \to 0} \frac{\sin \theta}{\theta} \cdot \lim_{\theta \to 0} \frac{\sin \theta}{\cos \theta + 1}$$

$$= -1 \cdot \left(\frac{0}{1 + 1} \right) = 0 \qquad \text{(by Equation 2)}$$

3

$$\lim_{\theta \to 0} \frac{\cos \theta - 1}{\theta} = 0$$

If we now put the limits (2) and (3) in (1), we get

$$f'(x) = \lim_{h \to 0} \sin x \cdot \lim_{h \to 0} \frac{\cos h - 1}{h} + \lim_{h \to 0} \cos x \cdot \lim_{h \to 0} \frac{\sin h}{h}$$

$$= (\sin x) \cdot 0 + (\cos x) \cdot 1 = \cos x$$

So we have proved the formula for the derivative of the sine function:

4

$$\frac{d}{dx} (\sin x) = \cos x$$

EXAMPLE 1 Differentiate $y = x^2 \sin x$.

SOLUTION Using the Product Rule and Formula 4, we have

$$\frac{dy}{dx} = x^2 \frac{d}{dx}(\sin x) + \sin x \frac{d}{dx}(x^2)$$

$$= x^2 \cos x + 2x \sin x \qquad \blacksquare$$

Figure 3 shows the graphs of the function of Example 1 and its derivative. Notice that $y' = 0$ whenever y has a horizontal tangent.

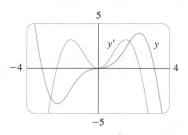

FIGURE 3

Using the same methods as in the proof of Formula 4, one can prove (see Exercise 20) that

$$\boxed{5} \qquad \boxed{\frac{d}{dx}(\cos x) = -\sin x}$$

The tangent function can also be differentiated by using the definition of a derivative, but it is easier to use the Quotient Rule together with Formulas 4 and 5:

$$\frac{d}{dx}(\tan x) = \frac{d}{dx}\left(\frac{\sin x}{\cos x}\right)$$

$$= \frac{\cos x \dfrac{d}{dx}(\sin x) - \sin x \dfrac{d}{dx}(\cos x)}{\cos^2 x}$$

$$= \frac{\cos x \cdot \cos x - \sin x(-\sin x)}{\cos^2 x}$$

$$= \frac{\cos^2 x + \sin^2 x}{\cos^2 x}$$

$$= \frac{1}{\cos^2 x} = \sec^2 x$$

$$\boxed{6} \qquad \boxed{\frac{d}{dx}(\tan x) = \sec^2 x}$$

The derivatives of the remaining trigonometric functions, csc, sec, and cot, can also be found easily using the Quotient Rule (see Exercises 17–19). We collect all the differentiation formulas for trigonometric functions in the following table. Remember that they are valid only when x is measured in radians.

When you memorize this table, it is helpful to notice that the minus signs go with the derivatives of the "cofunctions," that is, cosine, cosecant, and cotangent.

Derivatives of Trigonometric Functions	
$\dfrac{d}{dx}(\sin x) = \cos x$	$\dfrac{d}{dx}(\csc x) = -\csc x \cot x$
$\dfrac{d}{dx}(\cos x) = -\sin x$	$\dfrac{d}{dx}(\sec x) = \sec x \tan x$
$\dfrac{d}{dx}(\tan x) = \sec^2 x$	$\dfrac{d}{dx}(\cot x) = -\csc^2 x$

EXAMPLE 2 Differentiate $f(x) = \dfrac{\sec x}{1 + \tan x}$. For what values of x does the graph of f have a horizontal tangent?

SOLUTION The Quotient Rule gives

$$f'(x) = \frac{(1 + \tan x)\dfrac{d}{dx}(\sec x) - \sec x \dfrac{d}{dx}(1 + \tan x)}{(1 + \tan x)^2}$$

$$= \frac{(1 + \tan x)\sec x \tan x - \sec x \cdot \sec^2 x}{(1 + \tan x)^2}$$

$$= \frac{\sec x (\tan x + \tan^2 x - \sec^2 x)}{(1 + \tan x)^2}$$

$$= \frac{\sec x (\tan x - 1)}{(1 + \tan x)^2}$$

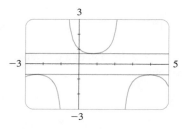

FIGURE 4
The horizontal tangents in Example 2

In simplifying the answer we have used the identity $\tan^2 x + 1 = \sec^2 x$.

Since $\sec x$ is never 0, we see that $f'(x) = 0$ when $\tan x = 1$, and this occurs when $x = n\pi + \pi/4$, where n is an integer (see Figure 4). ∎

Trigonometric functions are often used in modeling real-world phenomena. In particular, vibrations, waves, elastic motions, and other quantities that vary in a periodic manner can be described using trigonometric functions. In the following example we discuss an instance of simple harmonic motion.

EXAMPLE 3 An object at the end of a vertical spring is stretched 4 cm beyond its rest position and released at time $t = 0$. (See Figure 5 and note that the downward direction is positive.) Its position at time t is

$$s = f(t) = 4\cos t$$

FIGURE 5

Find the velocity and acceleration at time t and use them to analyze the motion of the object.

SOLUTION The velocity and acceleration are

$$v = \frac{ds}{dt} = \frac{d}{dt}(4\cos t) = 4\frac{d}{dt}(\cos t) = -4\sin t$$

$$a = \frac{dv}{dt} = \frac{d}{dt}(-4\sin t) = -4\frac{d}{dt}(\sin t) = -4\cos t$$

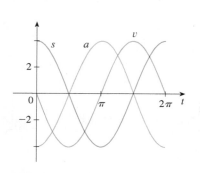

FIGURE 6

The object oscillates from the lowest point ($s = 4$ cm) to the highest point ($s = -4$ cm). The period of the oscillation is 2π, the period of $\cos t$.

The speed is $|v| = 4|\sin t|$, which is greatest when $|\sin t| = 1$, that is, when $\cos t = 0$. So the object moves fastest as it passes through its equilibrium position ($s = 0$). Its speed is 0 when $\sin t = 0$, that is, at the high and low points.

The acceleration $a = -4\cos t = 0$ when $s = 0$. It has greatest magnitude at the high and low points. See the graphs in Figure 6. ∎

EXAMPLE 4 Find the 27th derivative of $\cos x$.

SOLUTION The first few derivatives of $f(x) = \cos x$ are as follows:

$$f'(x) = -\sin x$$

$$f''(x) = -\cos x$$

$$f'''(x) = \sin x$$

$$f^{(4)}(x) = \cos x$$

$$f^{(5)}(x) = -\sin x$$

PS Look for a pattern.

We see that the successive derivatives occur in a cycle of length 4 and, in particular, $f^{(n)}(x) = \cos x$ whenever n is a multiple of 4. Therefore

$$f^{(24)}(x) = \cos x$$

and, differentiating three more times, we have

$$f^{(27)}(x) = \sin x$$ ∎

Our main use for the limit in Equation 2 has been to prove the differentiation formula for the sine function. But this limit is also useful in finding certain other trigonometric limits, as the following two examples show.

EXAMPLE 5 Find $\displaystyle\lim_{x \to 0} \frac{\sin 7x}{4x}$.

SOLUTION In order to apply Equation 2, we first rewrite the function by multiplying and dividing by 7:

$$\frac{\sin 7x}{4x} = \frac{7}{4}\left(\frac{\sin 7x}{7x}\right)$$

Note that $\sin 7x \neq 7\sin x$.

If we let $\theta = 7x$, then $\theta \to 0$ as $x \to 0$, so by Equation 2 we have

$$\lim_{x \to 0} \frac{\sin 7x}{4x} = \frac{7}{4}\lim_{x \to 0}\left(\frac{\sin 7x}{7x}\right)$$

$$= \frac{7}{4}\lim_{\theta \to 0}\frac{\sin \theta}{\theta} = \frac{7}{4} \cdot 1 = \frac{7}{4}$$ ∎

EXAMPLE 6 Calculate $\displaystyle\lim_{x \to 0} x \cot x$.

SOLUTION Here we divide numerator and denominator by x:

$$\lim_{x \to 0} x \cot x = \lim_{x \to 0}\frac{x \cos x}{\sin x}$$

$$= \lim_{x \to 0}\frac{\cos x}{\dfrac{\sin x}{x}} = \frac{\displaystyle\lim_{x \to 0}\cos x}{\displaystyle\lim_{x \to 0}\dfrac{\sin x}{x}}$$

$$= \frac{\cos 0}{1} \qquad \text{(by the continuity of cosine and Equation 2)}$$

$$= 1$$ ∎

3.3 EXERCISES

1–16 Differentiate.

1. $f(x) = x^2 \sin x$

2. $f(x) = x \cos x + 2 \tan x$

3. $f(x) = e^x \cos x$

4. $y = 2 \sec x - \csc x$

5. $y = \sec \theta \, \tan \theta$

6. $g(\theta) = e^\theta (\tan \theta - \theta)$

7. $y = c \cos t + t^2 \sin t$

8. $f(t) = \dfrac{\cot t}{e^t}$

9. $y = \dfrac{x}{2 - \tan x}$

10. $y = \sin \theta \cos \theta$

11. $f(\theta) = \dfrac{\sin \theta}{1 + \cos \theta}$

12. $y = \dfrac{\cos x}{1 - \sin x}$

13. $y = \dfrac{t \sin t}{1 + t}$

14. $y = \dfrac{\sin t}{1 + \tan t}$

15. $f(\theta) = \theta \cos \theta \sin \theta$

16. $f(t) = t e^t \cot t$

17. Prove that $\dfrac{d}{dx}(\csc x) = -\csc x \cot x$.

18. Prove that $\dfrac{d}{dx}(\sec x) = \sec x \tan x$.

19. Prove that $\dfrac{d}{dx}(\cot x) = -\csc^2 x$.

20. Prove, using the definition of derivative, that if $f(x) = \cos x$, then $f'(x) = -\sin x$.

21–24 Find an equation of the tangent line to the curve at the given point.

21. $y = \sin x + \cos x$, $(0, 1)$ **22.** $y = e^x \cos x$, $(0, 1)$

23. $y = \cos x - \sin x$, $(\pi, -1)$ **24.** $y = x + \tan x$, (π, π)

25. (a) Find an equation of the tangent line to the curve $y = 2x \sin x$ at the point $(\pi/2, \pi)$.
    (b) Illustrate part (a) by graphing the curve and the tangent line on the same screen.

26. (a) Find an equation of the tangent line to the curve $y = 3x + 6 \cos x$ at the point $(\pi/3, \pi + 3)$.
    (b) Illustrate part (a) by graphing the curve and the tangent line on the same screen.

27. (a) If $f(x) = \sec x - x$, find $f'(x)$.
    (b) Check to see that your answer to part (a) is reasonable by graphing both f and f' for $|x| < \pi/2$.

28. (a) If $f(x) = e^x \cos x$, find $f'(x)$ and $f''(x)$.
    (b) Check to see that your answers to part (a) are reasonable by graphing f, f', and f''.

29. If $H(\theta) = \theta \sin \theta$, find $H'(\theta)$ and $H''(\theta)$.

30. If $f(t) = \sec t$, find $f''(\pi/4)$.

31. (a) Use the Quotient Rule to differentiate the function
$$f(x) = \frac{\tan x - 1}{\sec x}$$
    (b) Simplify the expression for $f(x)$ by writing it in terms of $\sin x$ and $\cos x$, and then find $f'(x)$.
    (c) Show that your answers to parts (a) and (b) are equivalent.

32. Suppose $f(\pi/3) = 4$ and $f'(\pi/3) = -2$, and let $g(x) = f(x) \sin x$ and $h(x) = (\cos x)/f(x)$. Find
    (a) $g'(\pi/3)$ (b) $h'(\pi/3)$

33–34 For what values of x does the graph of f have a horizontal tangent?

33. $f(x) = x + 2 \sin x$ **34.** $f(x) = e^x \cos x$

35. A mass on a spring vibrates horizontally on a smooth level surface (see the figure). Its equation of motion is $x(t) = 8 \sin t$, where t is in seconds and x in centimeters.
    (a) Find the velocity and acceleration at time t.
    (b) Find the position, velocity, and acceleration of the mass at time $t = 2\pi/3$. In what direction is it moving at that time?

36. An elastic band is hung on a hook and a mass is hung on the lower end of the band. When the mass is pulled downward and then released, it vibrates vertically. The equation of motion is $s = 2 \cos t + 3 \sin t$, $t \geqslant 0$, where s is measured in centimeters and t in seconds. (Take the positive direction to be downward.)
    (a) Find the velocity and acceleration at time t.
    (b) Graph the velocity and acceleration functions.
    (c) When does the mass pass through the equilibrium position for the first time?
    (d) How far from its equilibrium position does the mass travel?
    (e) When is the speed the greatest?

37. A ladder 10 ft long rests against a vertical wall. Let θ be the angle between the top of the ladder and the wall and let x be the distance from the bottom of the ladder to the wall. If the bottom of the ladder slides away from the wall, how fast does x change with respect to θ when $\theta = \pi/3$?

38. An object with weight W is dragged along a horizontal plane by a force acting along a rope attached to the object.

If the rope makes an angle θ with the plane, then the magnitude of the force is

$$F = \frac{\mu W}{\mu \sin\theta + \cos\theta}$$

where μ is a constant called the *coefficient of friction*.
(a) Find the rate of change of F with respect to θ.
(b) When is this rate of change equal to 0?
(c) If $W = 50$ lb and $\mu = 0.6$, draw the graph of F as a function of θ and use it to locate the value of θ for which $dF/d\theta = 0$. Is the value consistent with your answer to part (b)?

39–50 Find the limit.

39. $\displaystyle\lim_{x\to 0} \frac{\sin 5x}{3x}$

40. $\displaystyle\lim_{x\to 0} \frac{\sin x}{\sin \pi x}$

41. $\displaystyle\lim_{t\to 0} \frac{\tan 6t}{\sin 2t}$

42. $\displaystyle\lim_{\theta\to 0} \frac{\cos\theta - 1}{\sin\theta}$

43. $\displaystyle\lim_{x\to 0} \frac{\sin 3x}{5x^3 - 4x}$

44. $\displaystyle\lim_{x\to 0} \frac{\sin 3x \sin 5x}{x^2}$

45. $\displaystyle\lim_{\theta\to 0} \frac{\sin\theta}{\theta + \tan\theta}$

46. $\displaystyle\lim_{x\to 0} \csc x \sin(\sin x)$

47. $\displaystyle\lim_{\theta\to 0} \frac{\cos\theta - 1}{2\theta^2}$

48. $\displaystyle\lim_{x\to 0} \frac{\sin(x^2)}{x}$

49. $\displaystyle\lim_{x\to \pi/4} \frac{1 - \tan x}{\sin x - \cos x}$

50. $\displaystyle\lim_{x\to 1} \frac{\sin(x - 1)}{x^2 + x - 2}$

51–52 Find the given derivative by finding the first few derivatives and observing the pattern that occurs.

51. $\dfrac{d^{99}}{dx^{99}} (\sin x)$

52. $\dfrac{d^{35}}{dx^{35}} (x \sin x)$

53. Find constants A and B such that the function $y = A \sin x + B \cos x$ satisfies the differential equation $y'' + y' - 2y = \sin x$.

54. (a) Evaluate $\displaystyle\lim_{x\to\infty} x \sin\frac{1}{x}$.

(b) Evaluate $\displaystyle\lim_{x\to 0} x \sin\frac{1}{x}$.

(c) Illustrate parts (a) and (b) by graphing $y = x \sin(1/x)$.

55. Differentiate each trigonometric identity to obtain a new (or familiar) identity.

(a) $\tan x = \dfrac{\sin x}{\cos x}$

(b) $\sec x = \dfrac{1}{\cos x}$

(c) $\sin x + \cos x = \dfrac{1 + \cot x}{\csc x}$

56. A semicircle with diameter PQ sits on an isosceles triangle PQR to form a region shaped like a two-dimensional ice-cream cone, as shown in the figure. If $A(\theta)$ is the area of the semicircle and $B(\theta)$ is the area of the triangle, find

$$\lim_{\theta\to 0^+} \frac{A(\theta)}{B(\theta)}$$

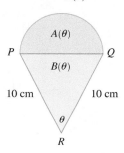

57. The figure shows a circular arc of length s and a chord of length d, both subtended by a central angle θ. Find

$$\lim_{\theta\to 0^+} \frac{s}{d}$$

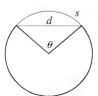

58. Let $f(x) = \dfrac{x}{\sqrt{1 - \cos 2x}}$.

(a) Graph f. What type of discontinuity does it appear to have at 0?

(b) Calculate the left and right limits of f at 0. Do these values confirm your answer to part (a)?

3.4 The Chain Rule

Suppose you are asked to differentiate the function

$$F(x) = \sqrt{x^2 + 1}$$

The differentiation formulas you learned in the previous sections of this chapter do not enable you to calculate $F'(x)$.

See Section 1.3 for a review of composite functions.

Observe that F is a composite function. In fact, if we let $y = f(u) = \sqrt{u}$ and let $u = g(x) = x^2 + 1$, then we can write $y = F(x) = f(g(x))$, that is, $F = f \circ g$. We know how to differentiate both f and g, so it would be useful to have a rule that tells us how to find the derivative of $F = f \circ g$ in terms of the derivatives of f and g.

It turns out that the derivative of the composite function $f \circ g$ is the product of the derivatives of f and g. This fact is one of the most important of the differentiation rules and is called the *Chain Rule*. It seems plausible if we interpret derivatives as rates of change. Regard du/dx as the rate of change of u with respect to x, dy/du as the rate of change of y with respect to u, and dy/dx as the rate of change of y with respect to x. If u changes twice as fast as x and y changes three times as fast as u, then it seems reasonable that y changes six times as fast as x, and so we expect that

$$\frac{dy}{dx} = \frac{dy}{du}\frac{du}{dx}$$

> **The Chain Rule** If g is differentiable at x and f is differentiable at $g(x)$, then the composite function $F = f \circ g$ defined by $F(x) = f(g(x))$ is differentiable at x and F' is given by the product
>
> $$F'(x) = f'(g(x)) \cdot g'(x)$$
>
> In Leibniz notation, if $y = f(u)$ and $u = g(x)$ are both differentiable functions, then
>
> $$\frac{dy}{dx} = \frac{dy}{du}\frac{du}{dx}$$

James Gregory

The first person to formulate the Chain Rule was the Scottish mathematician James Gregory (1638–1675), who also designed the first practical reflecting telescope. Gregory discovered the basic ideas of calculus at about the same time as Newton. He became the first Professor of Mathematics at the University of St. Andrews and later held the same position at the University of Edinburgh. But one year after accepting that position he died at the age of 36.

COMMENTS ON THE PROOF OF THE CHAIN RULE Let Δu be the change in u corresponding to a change of Δx in x, that is,

$$\Delta u = g(x + \Delta x) - g(x)$$

Then the corresponding change in y is

$$\Delta y = f(u + \Delta u) - f(u)$$

It is tempting to write

$$\frac{dy}{dx} = \lim_{\Delta x \to 0} \frac{\Delta y}{\Delta x}$$

$$\boxed{1} \qquad = \lim_{\Delta x \to 0} \frac{\Delta y}{\Delta u} \cdot \frac{\Delta u}{\Delta x}$$

$$= \lim_{\Delta x \to 0} \frac{\Delta y}{\Delta u} \cdot \lim_{\Delta x \to 0} \frac{\Delta u}{\Delta x}$$

$$= \lim_{\Delta u \to 0} \frac{\Delta y}{\Delta u} \cdot \lim_{\Delta x \to 0} \frac{\Delta u}{\Delta x} \qquad \text{(Note that } \Delta u \to 0 \text{ as } \Delta x \to 0 \text{ since } g \text{ is continuous.)}$$

$$= \frac{dy}{du}\frac{du}{dx}$$

The only flaw in this reasoning is that in (1) it might happen that $\Delta u = 0$ (even when $\Delta x \neq 0$) and, of course, we can't divide by 0. Nonetheless, this reasoning does at least *suggest* that the Chain Rule is true. A full proof of the Chain Rule is given at the end of this section. ∎

The Chain Rule can be written either in the prime notation

$$\boxed{2} \qquad\qquad (f \circ g)'(x) = f'(g(x)) \cdot g'(x)$$

or, if $y = f(u)$ and $u = g(x)$, in Leibniz notation:

$$\boxed{3} \qquad\qquad \frac{dy}{dx} = \frac{dy}{du}\,\frac{du}{dx}$$

Equation 3 is easy to remember because if dy/du and du/dx were quotients, then we could cancel du. Remember, however, that du has not been defined and du/dx should not be thought of as an actual quotient.

EXAMPLE 1 Find $F'(x)$ if $F(x) = \sqrt{x^2 + 1}$.

SOLUTION 1 (using Equation 2): At the beginning of this section we expressed F as $F(x) = (f \circ g)(x) = f(g(x))$ where $f(u) = \sqrt{u}$ and $g(x) = x^2 + 1$. Since

$$f'(u) = \tfrac{1}{2}u^{-1/2} = \frac{1}{2\sqrt{u}} \qquad \text{and} \qquad g'(x) = 2x$$

we have
$$F'(x) = f'(g(x)) \cdot g'(x)$$

$$= \frac{1}{2\sqrt{x^2 + 1}} \cdot 2x = \frac{x}{\sqrt{x^2 + 1}}$$

SOLUTION 2 (using Equation 3): If we let $u = x^2 + 1$ and $y = \sqrt{u}$, then

$$F'(x) = \frac{dy}{du}\,\frac{du}{dx} = \frac{1}{2\sqrt{u}}\,(2x) = \frac{1}{2\sqrt{x^2 + 1}}(2x) = \frac{x}{\sqrt{x^2 + 1}}$$ ∎

When using Formula 3 we should bear in mind that dy/dx refers to the derivative of y when y is considered as a function of x (called the *derivative of y with respect to x*), whereas dy/du refers to the derivative of y when considered as a function of u (the derivative of y with respect to u). For instance, in Example 1, y can be considered as a function of x $\left(y = \sqrt{x^2 + 1}\right)$ and also as a function of u $\left(y = \sqrt{u}\,\right)$. Note that

$$\frac{dy}{dx} = F'(x) = \frac{x}{\sqrt{x^2 + 1}} \qquad \text{whereas} \qquad \frac{dy}{du} = f'(u) = \frac{1}{2\sqrt{u}}$$

NOTE In using the Chain Rule we work from the outside to the inside. Formula 2 says that *we differentiate the outer function f [at the inner function $g(x)$] and then we multiply by the derivative of the inner function.*

$$\frac{d}{dx} \quad \underbrace{f}_{\substack{\text{outer} \\ \text{function}}} \quad \underbrace{(g(x))}_{\substack{\text{evaluated} \\ \text{at inner} \\ \text{function}}} \quad = \quad \underbrace{f'}_{\substack{\text{derivative} \\ \text{of outer} \\ \text{function}}} \quad \underbrace{(g(x))}_{\substack{\text{evaluated} \\ \text{at inner} \\ \text{function}}} \quad \cdot \quad \underbrace{g'(x)}_{\substack{\text{derivative} \\ \text{of inner} \\ \text{function}}}$$

EXAMPLE 2 Differentiate (a) $y = \sin(x^2)$ and (b) $y = \sin^2 x$.

SOLUTION

(a) If $y = \sin(x^2)$, then the outer function is the sine function and the inner function is the squaring function, so the Chain Rule gives

$$\frac{dy}{dx} = \underbrace{\frac{d}{dx}}_{} \quad \underbrace{\sin}_{\substack{\text{outer} \\ \text{function}}} \quad \underbrace{(x^2)}_{\substack{\text{evaluated} \\ \text{at inner} \\ \text{function}}} = \underbrace{\cos}_{\substack{\text{derivative} \\ \text{of outer} \\ \text{function}}} \quad \underbrace{(x^2)}_{\substack{\text{evaluated} \\ \text{at inner} \\ \text{function}}} \cdot \underbrace{2x}_{\substack{\text{derivative} \\ \text{of inner} \\ \text{function}}}$$

$$= 2x \cos(x^2)$$

(b) Note that $\sin^2 x = (\sin x)^2$. Here the outer function is the squaring function and the inner function is the sine function. So

$$\frac{dy}{dx} = \frac{d}{dx} \underbrace{(\sin x)^2}_{\substack{\text{inner} \\ \text{function}}} = \underbrace{2}_{\substack{\text{derivative} \\ \text{of outer} \\ \text{function}}} \cdot \underbrace{(\sin x)}_{\substack{\text{evaluated} \\ \text{at inner} \\ \text{function}}} \cdot \underbrace{\cos x}_{\substack{\text{derivative} \\ \text{of inner} \\ \text{function}}}$$

See Reference Page 2 or Appendix D.

The answer can be left as $2 \sin x \cos x$ or written as $\sin 2x$ (by a trigonometric identity known as the double-angle formula). ∎

In Example 2(a) we combined the Chain Rule with the rule for differentiating the sine function. In general, if $y = \sin u$, where u is a differentiable function of x, then, by the Chain Rule,

$$\frac{dy}{dx} = \frac{dy}{du} \frac{du}{dx} = \cos u \frac{du}{dx}$$

Thus

$$\frac{d}{dx}(\sin u) = \cos u \frac{du}{dx}$$

In a similar fashion, all of the formulas for differentiating trigonometric functions can be combined with the Chain Rule.

Let's make explicit the special case of the Chain Rule where the outer function f is a power function. If $y = [g(x)]^n$, then we can write $y = f(u) = u^n$ where $u = g(x)$. By using the Chain Rule and then the Power Rule, we get

$$\frac{dy}{dx} = \frac{dy}{du} \frac{du}{dx} = nu^{n-1} \frac{du}{dx} = n[g(x)]^{n-1} g'(x)$$

---

4 **The Power Rule Combined with the Chain Rule** If n is any real number and $u = g(x)$ is differentiable, then

$$\frac{d}{dx}(u^n) = nu^{n-1} \frac{du}{dx}$$

Alternatively,

$$\frac{d}{dx}[g(x)]^n = n[g(x)]^{n-1} \cdot g'(x)$$

Notice that the derivative in Example 1 could be calculated by taking $n = \frac{1}{2}$ in Rule 4.

EXAMPLE 3 Differentiate $y = (x^3 - 1)^{100}$.

SOLUTION Taking $u = g(x) = x^3 - 1$ and $n = 100$ in (4), we have

$$\frac{dy}{dx} = \frac{d}{dx}(x^3 - 1)^{100} = 100(x^3 - 1)^{99}\frac{d}{dx}(x^3 - 1)$$

$$= 100(x^3 - 1)^{99} \cdot 3x^2 = 300x^2(x^3 - 1)^{99}$$

EXAMPLE 4 Find $f'(x)$ if $f(x) = \dfrac{1}{\sqrt[3]{x^2 + x + 1}}$.

SOLUTION First rewrite f: $f(x) = (x^2 + x + 1)^{-1/3}$

Thus $$f'(x) = -\tfrac{1}{3}(x^2 + x + 1)^{-4/3}\frac{d}{dx}(x^2 + x + 1)$$

$$= -\tfrac{1}{3}(x^2 + x + 1)^{-4/3}(2x + 1)$$

EXAMPLE 5 Find the derivative of the function

$$g(t) = \left(\frac{t - 2}{2t + 1}\right)^9$$

SOLUTION Combining the Power Rule, Chain Rule, and Quotient Rule, we get

$$g'(t) = 9\left(\frac{t - 2}{2t + 1}\right)^8 \frac{d}{dt}\left(\frac{t - 2}{2t + 1}\right)$$

$$= 9\left(\frac{t - 2}{2t + 1}\right)^8 \frac{(2t + 1) \cdot 1 - 2(t - 2)}{(2t + 1)^2} = \frac{45(t - 2)^8}{(2t + 1)^{10}}$$

EXAMPLE 6 Differentiate $y = (2x + 1)^5(x^3 - x + 1)^4$.

SOLUTION In this example we must use the Product Rule before using the Chain Rule:

$$\frac{dy}{dx} = (2x + 1)^5\frac{d}{dx}(x^3 - x + 1)^4 + (x^3 - x + 1)^4\frac{d}{dx}(2x + 1)^5$$

$$= (2x + 1)^5 \cdot 4(x^3 - x + 1)^3\frac{d}{dx}(x^3 - x + 1)$$

$$+ (x^3 - x + 1)^4 \cdot 5(2x + 1)^4\frac{d}{dx}(2x + 1)$$

$$= 4(2x + 1)^5(x^3 - x + 1)^3(3x^2 - 1) + 5(x^3 - x + 1)^4(2x + 1)^4 \cdot 2$$

Noticing that each term has the common factor $2(2x + 1)^4(x^3 - x + 1)^3$, we could factor it out and write the answer as

$$\frac{dy}{dx} = 2(2x + 1)^4(x^3 - x + 1)^3(17x^3 + 6x^2 - 9x + 3)$$

The graphs of the functions y and y' in Example 6 are shown in Figure 1. Notice that y' is large when y increases rapidly and $y' = 0$ when y has a horizontal tangent. So our answer appears to be reasonable.

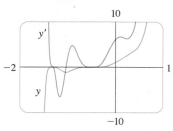

FIGURE 1

EXAMPLE 7 Differentiate $y = e^{\sin x}$.

SOLUTION Here the inner function is $g(x) = \sin x$ and the outer function is the exponential function $f(x) = e^x$. So, by the Chain Rule,

$$\frac{dy}{dx} = \frac{d}{dx}(e^{\sin x}) = e^{\sin x}\frac{d}{dx}(\sin x) = e^{\sin x}\cos x \qquad \blacksquare$$

More generally, the Chain Rule gives

$$\frac{d}{dx}(e^u) = e^u \frac{du}{dx}$$

We can use the Chain Rule to differentiate an exponential function with any base $b > 0$. Recall from Section 1.5 that $b = e^{\ln b}$. So

$$b^x = (e^{\ln b})^x = e^{(\ln b)x}$$

and the Chain Rule gives

$$\frac{d}{dx}(b^x) = \frac{d}{dx}(e^{(\ln b)x}) = e^{(\ln b)x}\frac{d}{dx}(\ln b)x$$

$$= e^{(\ln b)x} \cdot \ln b = b^x \ln b$$

because $\ln b$ is a constant. So we have the formula

Don't confuse Formula 5 (where x is the *exponent*) with the Power Rule (where x is the *base*):

$$\frac{d}{dx}(x^n) = nx^{n-1}$$

$\boxed{5}$
$$\frac{d}{dx}(b^x) = b^x \ln b$$

In particular, if $b = 2$, we get

$\boxed{6}$
$$\frac{d}{dx}(2^x) = 2^x \ln 2$$

In Section 3.1 we gave the estimate

$$\frac{d}{dx}(2^x) \approx (0.69)2^x$$

This is consistent with the exact formula (6) because $\ln 2 \approx 0.693147$.

The reason for the name "Chain Rule" becomes clear when we make a longer chain by adding another link. Suppose that $y = f(u)$, $u = g(x)$, and $x = h(t)$, where f, g, and h are differentiable functions. Then, to compute the derivative of y with respect to t, we use the Chain Rule twice:

$$\frac{dy}{dt} = \frac{dy}{dx}\frac{dx}{dt} = \frac{dy}{du}\frac{du}{dx}\frac{dx}{dt}$$

EXAMPLE 8 If $f(x) = \sin(\cos(\tan x))$, then

$$f'(x) = \cos(\cos(\tan x))\frac{d}{dx}\cos(\tan x)$$

$$= \cos(\cos(\tan x))[-\sin(\tan x)]\frac{d}{dx}(\tan x)$$

$$= -\cos(\cos(\tan x))\sin(\tan x)\sec^2 x$$

Notice that we used the Chain Rule twice. $\qquad \blacksquare$

EXAMPLE 9 Differentiate $y = e^{\sec 3\theta}$.

SOLUTION The outer function is the exponential function, the middle function is the secant function, and the inner function is the tripling function. So we have

$$\frac{dy}{d\theta} = e^{\sec 3\theta} \frac{d}{d\theta} (\sec 3\theta)$$

$$= e^{\sec 3\theta} \sec 3\theta \tan 3\theta \frac{d}{d\theta} (3\theta)$$

$$= 3e^{\sec 3\theta} \sec 3\theta \tan 3\theta$$

■ How to Prove the Chain Rule

Recall that if $y = f(x)$ and x changes from a to $a + \Delta x$, we define the increment of y as

$$\Delta y = f(a + \Delta x) - f(a)$$

According to the definition of a derivative, we have

$$\lim_{\Delta x \to 0} \frac{\Delta y}{\Delta x} = f'(a)$$

So if we denote by ε the difference between the difference quotient and the derivative, we obtain

$$\lim_{\Delta x \to 0} \varepsilon = \lim_{\Delta x \to 0} \left(\frac{\Delta y}{\Delta x} - f'(a) \right) = f'(a) - f'(a) = 0$$

But

$$\varepsilon = \frac{\Delta y}{\Delta x} - f'(a) \qquad \Rightarrow \qquad \Delta y = f'(a) \, \Delta x + \varepsilon \, \Delta x$$

If we define ε to be 0 when $\Delta x = 0$, then ε becomes a continuous function of Δx. Thus, for a differentiable function f, we can write

$$\boxed{7} \qquad \Delta y = f'(a) \, \Delta x + \varepsilon \, \Delta x \qquad \text{where} \qquad \varepsilon \to 0 \text{ as } \Delta x \to 0$$

and ε is a continuous function of Δx. This property of differentiable functions is what enables us to prove the Chain Rule.

PROOF OF THE CHAIN RULE Suppose $u = g(x)$ is differentiable at a and $y = f(u)$ is differentiable at $b = g(a)$. If Δx is an increment in x and Δu and Δy are the corresponding increments in u and y, then we can use Equation 7 to write

$$\boxed{8} \qquad \Delta u = g'(a) \, \Delta x + \varepsilon_1 \, \Delta x = [g'(a) + \varepsilon_1] \, \Delta x$$

where $\varepsilon_1 \to 0$ as $\Delta x \to 0$. Similarly

$$\boxed{9} \qquad \Delta y = f'(b) \, \Delta u + \varepsilon_2 \, \Delta u = [f'(b) + \varepsilon_2] \, \Delta u$$

where $\varepsilon_2 \to 0$ as $\Delta u \to 0$. If we now substitute the expression for Δu from Equation 8 into Equation 9, we get

$$\Delta y = [f'(b) + \varepsilon_2][g'(a) + \varepsilon_1] \, \Delta x$$

so
$$\frac{\Delta y}{\Delta x} = [f'(b) + \varepsilon_2][g'(a) + \varepsilon_1]$$

As $\Delta x \to 0$, Equation 8 shows that $\Delta u \to 0$. So both $\varepsilon_1 \to 0$ and $\varepsilon_2 \to 0$ as $\Delta x \to 0$. Therefore

$$\frac{dy}{dx} = \lim_{\Delta x \to 0} \frac{\Delta y}{\Delta x} = \lim_{\Delta x \to 0} [f'(b) + \varepsilon_2][g'(a) + \varepsilon_1]$$

$$= f'(b)\, g'(a) = f'(g(a))\, g'(a)$$

This proves the Chain Rule. ∎

3.4 EXERCISES

1–6 Write the composite function in the form $f(g(x))$. [Identify the inner function $u = g(x)$ and the outer function $y = f(u)$.] Then find the derivative dy/dx.

1. $y = \sqrt[3]{1 + 4x}$

2. $y = (2x^3 + 5)^4$

3. $y = \tan \pi x$

4. $y = \sin(\cot x)$

5. $y = e^{\sqrt{x}}$

6. $y = \sqrt{2 - e^x}$

7–46 Find the derivative of the function.

7. $F(x) = (5x^6 + 2x^3)^4$

8. $F(x) = (1 + x + x^2)^{99}$

9. $f(x) = \sqrt{5x + 1}$

10. $f(x) = \dfrac{1}{\sqrt[3]{x^2 - 1}}$

11. $f(\theta) = \cos(\theta^2)$

12. $g(\theta) = \cos^2\theta$

13. $y = x^2 e^{-3x}$

14. $f(t) = t \sin \pi t$

15. $f(t) = e^{at} \sin bt$

16. $g(x) = e^{x^2 - x}$

17. $f(x) = (2x - 3)^4 (x^2 + x + 1)^5$

18. $g(x) = (x^2 + 1)^3 (x^2 + 2)^6$

19. $h(t) = (t + 1)^{2/3} (2t^2 - 1)^3$

20. $F(t) = (3t - 1)^4 (2t + 1)^{-3}$

21. $y = \sqrt{\dfrac{x}{x + 1}}$

22. $y = \left(x + \dfrac{1}{x}\right)^5$

23. $y = e^{\tan\theta}$

24. $f(t) = 2^{t^3}$

25. $g(u) = \left(\dfrac{u^3 - 1}{u^3 + 1}\right)^8$

26. $s(t) = \sqrt{\dfrac{1 + \sin t}{1 + \cos t}}$

27. $r(t) = 10^{2\sqrt{t}}$

28. $f(z) = e^{z/(z-1)}$

29. $H(r) = \dfrac{(r^2 - 1)^3}{(2r + 1)^5}$

30. $J(\theta) = \tan^2(n\theta)$

31. $F(t) = e^{t\sin 2t}$

32. $F(t) = \dfrac{t^2}{\sqrt{t^3 + 1}}$

33. $G(x) = 4^{C/x}$

34. $U(y) = \left(\dfrac{y^4 + 1}{y^2 + 1}\right)^5$

35. $y = \cos\left(\dfrac{1 - e^{2x}}{1 + e^{2x}}\right)$

36. $y = x^2 e^{-1/x}$

37. $y = \cot^2(\sin\theta)$

38. $y = \sqrt{1 + xe^{-2x}}$

39. $f(t) = \tan(\sec(\cos t))$

40. $y = e^{\sin 2x} + \sin(e^{2x})$

41. $f(t) = \sin^2(e^{\sin^2 t})$

42. $y = \sqrt{x + \sqrt{x + \sqrt{x}}}$

43. $g(x) = (2ra^{rx} + n)^p$

44. $y = 2^{3^{4^x}}$

45. $y = \cos\sqrt{\sin(\tan \pi x)}$

46. $y = [x + (x + \sin^2 x)^3]^4$

47–50 Find y' and y''.

47. $y = \cos(\sin 3\theta)$

48. $y = \dfrac{1}{(1 + \tan x)^2}$

49. $y = \sqrt{1 - \sec t}$

50. $y = e^{e^x}$

51–54 Find an equation of the tangent line to the curve at the given point.

51. $y = 2^x$, $\ (0, 1)$

52. $y = \sqrt{1 + x^3}$, $\ (2, 3)$

53. $y = \sin(\sin x)$, $\ (\pi, 0)$

54. $y = xe^{-x^2}$, $\ (0, 0)$

55. (a) Find an equation of the tangent line to the curve $y = 2/(1 + e^{-x})$ at the point $(0, 1)$.
(b) Illustrate part (a) by graphing the curve and the tangent line on the same screen.

56. (a) The curve $y = |x|/\sqrt{2 - x^2}$ is called a *bullet-nose curve*. Find an equation of the tangent line to this curve at the point $(1, 1)$.

(b) Illustrate part (a) by graphing the curve and the tangent line on the same screen.

57. (a) If $f(x) = x\sqrt{2 - x^2}$, find $f'(x)$.

(b) Check to see that your answer to part (a) is reasonable by comparing the graphs of f and f'.

58. The function $f(x) = \sin(x + \sin 2x)$, $0 \leqslant x \leqslant \pi$, arises in applications to frequency modulation (FM) synthesis.

(a) Use a graph of f produced by a calculator to make a rough sketch of the graph of f'.

(b) Calculate $f'(x)$ and use this expression, with a calculator, to graph f'. Compare with your sketch in part (a).

59. Find all points on the graph of the function $f(x) = 2\sin x + \sin^2 x$ at which the tangent line is horizontal.

60. At what point on the curve $y = \sqrt{1 + 2x}$ is the tangent line perpendicular to the line $6x + 2y = 1$?

61. If $F(x) = f(g(x))$, where $f(-2) = 8$, $f'(-2) = 4$, $f'(5) = 3$, $g(5) = -2$, and $g'(5) = 6$, find $F'(5)$.

62. If $h(x) = \sqrt{4 + 3f(x)}$, where $f(1) = 7$ and $f'(1) = 4$, find $h'(1)$.

63. A table of values for f, g, f', and g' is given.

x	$f(x)$	$g(x)$	$f'(x)$	$g'(x)$
1	3	2	4	6
2	1	8	5	7
3	7	2	7	9

(a) If $h(x) = f(g(x))$, find $h'(1)$.

(b) If $H(x) = g(f(x))$, find $H'(1)$.

64. Let f and g be the functions in Exercise 63.

(a) If $F(x) = f(f(x))$, find $F'(2)$.

(b) If $G(x) = g(g(x))$, find $G'(3)$.

65. If f and g are the functions whose graphs are shown, let $u(x) = f(g(x))$, $v(x) = g(f(x))$, and $w(x) = g(g(x))$. Find each derivative, if it exists. If it does not exist, explain why.

(a) $u'(1)$ (b) $v'(1)$ (c) $w'(1)$

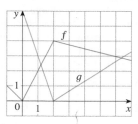

66. If f is the function whose graph is shown, let $h(x) = f(f(x))$ and $g(x) = f(x^2)$. Use the graph of f to estimate the value of each derivative.

(a) $h'(2)$ (b) $g'(2)$

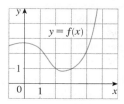

67. If $g(x) = \sqrt{f(x)}$, where the graph of f is shown, evaluate $g'(3)$.

68. Suppose f is differentiable on $\mathbb{R}$ and α is a real number. Let $F(x) = f(x^\alpha)$ and $G(x) = [f(x)]^\alpha$. Find expressions for (a) $F'(x)$ and (b) $G'(x)$.

69. Suppose f is differentiable on $\mathbb{R}$. Let $F(x) = f(e^x)$ and $G(x) = e^{f(x)}$. Find expressions for (a) $F'(x)$ and (b) $G'(x)$.

70. Let $g(x) = e^{cx} + f(x)$ and $h(x) = e^{kx}f(x)$, where $f(0) = 3$, $f'(0) = 5$, and $f''(0) = -2$.

(a) Find $g'(0)$ and $g''(0)$ in terms of c.

(b) In terms of k, find an equation of the tangent line to the graph of h at the point where $x = 0$.

71. Let $r(x) = f(g(h(x)))$, where $h(1) = 2$, $g(2) = 3$, $h'(1) = 4$, $g'(2) = 5$, and $f'(3) = 6$. Find $r'(1)$.

72. If g is a twice differentiable function and $f(x) = xg(x^2)$, find f'' in terms of g, g', and g''.

73. If $F(x) = f(3f(4f(x)))$, where $f(0) = 0$ and $f'(0) = 2$, find $F'(0)$.

74. If $F(x) = f(xf(xf(x)))$, where $f(1) = 2$, $f(2) = 3$, $f'(1) = 4$, $f'(2) = 5$, and $f'(3) = 6$, find $F'(1)$.

75. Show that the function $y = e^{2x}(A\cos 3x + B\sin 3x)$ satisfies the differential equation $y'' - 4y' + 13y = 0$.

76. For what values of r does the function $y = e^{rx}$ satisfy the differential equation $y'' - 4y' + y = 0$?

77. Find the 50th derivative of $y = \cos 2x$.

78. Find the 1000th derivative of $f(x) = xe^{-x}$.

79. The displacement of a particle on a vibrating string is given by the equation $s(t) = 10 + \frac{1}{4}\sin(10\pi t)$ where s is measured in centimeters and t in seconds. Find the velocity of the particle after t seconds.

80. If the equation of motion of a particle is given by $s = A\cos(\omega t + \delta)$, the particle is said to undergo *simple harmonic motion*.
  (a) Find the velocity of the particle at time t.
  (b) When is the velocity 0?

81. A Cepheid variable star is a star whose brightness alternately increases and decreases. The most easily visible such star is Delta Cephei, for which the interval between times of maximum brightness is 5.4 days. The average brightness of this star is 4.0 and its brightness changes by ± 0.35. In view of these data, the brightness of Delta Cephei at time t, where t is measured in days, has been modeled by the function

$$B(t) = 4.0 + 0.35\sin\left(\frac{2\pi t}{5.4}\right)$$

  (a) Find the rate of change of the brightness after t days.
  (b) Find, correct to two decimal places, the rate of increase after one day.

82. In Example 1.3.4 we arrived at a model for the length of daylight (in hours) in Philadelphia on the tth day of the year:

$$L(t) = 12 + 2.8\sin\left[\frac{2\pi}{365}(t - 80)\right]$$

Use this model to compare how the number of hours of daylight is increasing in Philadelphia on March 21 and May 21.

83. The motion of a spring that is subject to a frictional force or a damping force (such as a shock absorber in a car) is often modeled by the product of an exponential function and a sine or cosine function. Suppose the equation of motion of a point on such a spring is

$$s(t) = 2e^{-1.5t}\sin 2\pi t$$

where s is measured in centimeters and t in seconds. Find the velocity after t seconds and graph both the position and velocity functions for $0 \le t \le 2$.

84. Under certain circumstances a rumor spreads according to the equation

$$p(t) = \frac{1}{1 + ae^{-kt}}$$

where $p(t)$ is the proportion of the population that has heard the rumor at time t and a and k are positive constants. [In Section 9.4 we will see that this is a reasonable equation for $p(t)$.]
  (a) Find $\lim_{t \to \infty} p(t)$.
  (b) Find the rate of spread of the rumor.

  (c) Graph p for the case $a = 10$, $k = 0.5$ with t measured in hours. Use the graph to estimate how long it will take for 80% of the population to hear the rumor.

85. The average blood alcohol concentration (BAC) of eight male subjects was measured after consumption of 15 mL of ethanol (corresponding to one alcoholic drink). The resulting data were modeled by the concentration function

$$C(t) = 0.0225te^{-0.0467t}$$

where t is measured in minutes after consumption and C is measured in mg/mL.
  (a) How rapidly was the BAC increasing after 10 minutes?
  (b) How rapidly was it decreasing half an hour later?

Source: Adapted from P. Wilkinson et al., "Pharmacokinetics of Ethanol after Oral Administration in the Fasting State," *Journal of Pharmacokinetics and Biopharmaceutics* 5 (1977): 207–24.

86. In Section 1.4 we modeled the world population from 1900 to 2010 with the exponential function

$$P(t) = (1436.53) \cdot (1.01395)^t$$

where $t = 0$ corresponds to the year 1900 and $P(t)$ is measured in millions. According to this model, what was the rate of increase of world population in 1920? In 1950? In 2000?

87. A particle moves along a straight line with displacement $s(t)$, velocity $v(t)$, and acceleration $a(t)$. Show that

$$a(t) = v(t)\frac{dv}{ds}$$

Explain the difference between the meanings of the derivatives dv/dt and dv/ds.

88. Air is being pumped into a spherical weather balloon. At any time t, the volume of the balloon is $V(t)$ and its radius is $r(t)$.
  (a) What do the derivatives dV/dr and dV/dt represent?
  (b) Express dV/dt in terms of dr/dt.

89. The flash unit on a camera operates by storing charge on a capacitor and releasing it suddenly when the flash is set off. The following data describe the charge Q remaining on the capacitor (measured in microcoulombs, μC) at time t (measured in seconds).

t	0.00	0.02	0.04	0.06	0.08	0.10
Q	100.00	81.87	67.03	54.88	44.93	36.76

  (a) Use a graphing calculator or computer to find an exponential model for the charge.
  (b) The derivative $Q'(t)$ represents the electric current (measured in microamperes, μA) flowing from the capacitor to the flash bulb. Use part (a) to estimate the current when $t = 0.04$ s. Compare with the result of Example 2.1.2.

90. The table gives the US population from 1790 to 1860.

Year	Population	Year	Population
1790	3,929,000	1830	12,861,000
1800	5,308,000	1840	17,063,000
1810	7,240,000	1850	23,192,000
1820	9,639,000	1860	31,443,000

(a) Use a graphing calculator or computer to fit an exponential function to the data. Graph the data points and the exponential model. How good is the fit?

(b) Estimate the rates of population growth in 1800 and 1850 by averaging slopes of secant lines.

(c) Use the exponential model in part (a) to estimate the rates of growth in 1800 and 1850. Compare these estimates with the ones in part (b).

(d) Use the exponential model to predict the population in 1870. Compare with the actual population of 38,558,000. Can you explain the discrepancy?

CAS 91. Computer algebra systems have commands that differentiate functions, but the form of the answer may not be convenient and so further commands may be necessary to simplify the answer.

(a) Use a CAS to find the derivative in Example 5 and compare with the answer in that example. Then use the simplify command and compare again.

(b) Use a CAS to find the derivative in Example 6. What happens if you use the simplify command? What happens if you use the factor command? Which form of the answer would be best for locating horizontal tangents?

CAS 92. (a) Use a CAS to differentiate the function

$$f(x) = \sqrt{\frac{x^4 - x + 1}{x^4 + x + 1}}$$

and to simplify the result.

(b) Where does the graph of f have horizontal tangents?

(c) Graph f and f' on the same screen. Are the graphs consistent with your answer to part (b)?

93. Use the Chain Rule to prove the following.

(a) The derivative of an even function is an odd function.

(b) The derivative of an odd function is an even function.

94. Use the Chain Rule and the Product Rule to give an alternative proof of the Quotient Rule.
[*Hint:* Write $f(x)/g(x) = f(x)[g(x)]^{-1}$.]

95. (a) If n is a positive integer, prove that

$$\frac{d}{dx}(\sin^n x \cos nx) = n \sin^{n-1} x \cos(n + 1)x$$

(b) Find a formula for the derivative of $y = \cos^n x \cos nx$ that is similar to the one in part (a).

96. Suppose $y = f(x)$ is a curve that always lies above the x-axis and never has a horizontal tangent, where f is differentiable everywhere. For what value of y is the rate of change of y^5 with respect to x eighty times the rate of change of y with respect to x?

97. Use the Chain Rule to show that if θ is measured in degrees, then

$$\frac{d}{d\theta}(\sin \theta) = \frac{\pi}{180} \cos \theta$$

(This gives one reason for the convention that radian measure is always used when dealing with trigonometric functions in calculus: the differentiation formulas would not be as simple if we used degree measure.)

98. (a) Write $|x| = \sqrt{x^2}$ and use the Chain Rule to show that

$$\frac{d}{dx}|x| = \frac{x}{|x|}$$

(b) If $f(x) = |\sin x|$, find $f'(x)$ and sketch the graphs of f and f'. Where is f not differentiable?

(c) If $g(x) = \sin|x|$, find $g'(x)$ and sketch the graphs of g and g'. Where is g not differentiable?

99. If $y = f(u)$ and $u = g(x)$, where f and g are twice differentiable functions, show that

$$\frac{d^2y}{dx^2} = \frac{d^2y}{du^2}\left(\frac{du}{dx}\right)^2 + \frac{dy}{du}\frac{d^2u}{dx^2}$$

100. If $y = f(u)$ and $u = g(x)$, where f and g possess third derivatives, find a formula for d^3y/dx^3 similar to the one given in Exercise 99.

APPLIED PROJECT   WHERE SHOULD A PILOT START DESCENT?

An approach path for an aircraft landing is shown in the figure and satisfies the following conditions:

(i) The cruising altitude is h when descent starts at a horizontal distance ℓ from touchdown at the origin.

(ii) The pilot must maintain a constant horizontal speed v throughout descent.

(iii) The absolute value of the vertical acceleration should not exceed a constant k (which is much less than the acceleration due to gravity).

1. Find a cubic polynomial $P(x) = ax^3 + bx^2 + cx + d$ that satisfies condition (i) by imposing suitable conditions on $P(x)$ and $P'(x)$ at the start of descent and at touchdown.

2. Use conditions (ii) and (iii) to show that

$$\frac{6hv^2}{\ell^2} \leqslant k$$

3. Suppose that an airline decides not to allow vertical acceleration of a plane to exceed $k = 860$ mi/h². If the cruising altitude of a plane is 35,000 ft and the speed is 300 mi/h, how far away from the airport should the pilot start descent?

4. Graph the approach path if the conditions stated in Problem 3 are satisfied.

3.5   Implicit Differentiation

The functions that we have met so far can be described by expressing one variable explicitly in terms of another variable—for example,

$$y = \sqrt{x^3 + 1} \qquad \text{or} \qquad y = x \sin x$$

or, in general, $y = f(x)$. Some functions, however, are defined implicitly by a relation between x and y such as

$$\boxed{1} \qquad\qquad x^2 + y^2 = 25$$

or

$$\boxed{2} \qquad\qquad x^3 + y^3 = 6xy$$

In some cases it is possible to solve such an equation for y as an explicit function (or several functions) of x. For instance, if we solve Equation 1 for y, we get $y = \pm\sqrt{25 - x^2}$, so two of the functions determined by the implicit Equation l are $f(x) = \sqrt{25 - x^2}$ and $g(x) = -\sqrt{25 - x^2}$. The graphs of f and g are the upper and lower semicircles of the circle $x^2 + y^2 = 25$. (See Figure 1.)

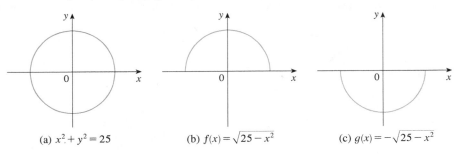

FIGURE 1   (a) $x^2 + y^2 = 25$ (b) $f(x) = \sqrt{25 - x^2}$ (c) $g(x) = -\sqrt{25 - x^2}$

It's not easy to solve Equation 2 for y explicitly as a function of x by hand. (A computer algebra system has no trouble, but the expressions it obtains are very complicated.) Nonetheless, (2) is the equation of a curve called the **folium of Descartes** shown in Figure 2 and it implicitly defines y as several functions of x. The graphs of three such functions are shown in Figure 3. When we say that f is a function defined implicitly by Equation 2, we mean that the equation

$$x^3 + [f(x)]^3 = 6xf(x)$$

is true for all values of x in the domain of f.

FIGURE 2 The folium of Descartes

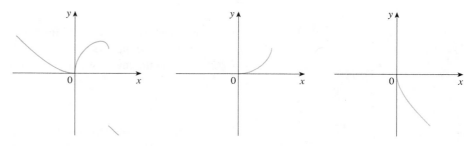

FIGURE 3 Graphs of three functions defined by the folium of Descartes

Fortunately, we don't need to solve an equation for y in terms of x in order to find the derivative of y. Instead we can use the method of **implicit differentiation**. This consists of differentiating both sides of the equation with respect to x and then solving the resulting equation for y'. In the examples and exercises of this section it is always assumed that the given equation determines y implicitly as a differentiable function of x so that the method of implicit differentiation can be applied.

EXAMPLE 1

(a) If $x^2 + y^2 = 25$, find $\dfrac{dy}{dx}$.

(b) Find an equation of the tangent to the circle $x^2 + y^2 = 25$ at the point $(3, 4)$.

SOLUTION 1

(a) Differentiate both sides of the equation $x^2 + y^2 = 25$:

$$\frac{d}{dx}(x^2 + y^2) = \frac{d}{dx}(25)$$

$$\frac{d}{dx}(x^2) + \frac{d}{dx}(y^2) = 0$$

Remembering that y is a function of x and using the Chain Rule, we have

$$\frac{d}{dx}(y^2) = \frac{d}{dy}(y^2)\frac{dy}{dx} = 2y\frac{dy}{dx}$$

Thus

$$2x + 2y\frac{dy}{dx} = 0$$

Now we solve this equation for dy/dx:

$$\frac{dy}{dx} = -\frac{x}{y}$$

(b) At the point $(3, 4)$ we have $x = 3$ and $y = 4$, so

$$\frac{dy}{dx} = -\frac{3}{4}$$

An equation of the tangent to the circle at $(3, 4)$ is therefore

$$y - 4 = -\tfrac{3}{4}(x - 3) \qquad \text{or} \qquad 3x + 4y = 25$$

SOLUTION 2

(b) Solving the equation $x^2 + y^2 = 25$ for y, we get $y = \pm\sqrt{25 - x^2}$. The point $(3, 4)$ lies on the upper semicircle $y = \sqrt{25 - x^2}$ and so we consider the function $f(x) = \sqrt{25 - x^2}$. Differentiating f using the Chain Rule, we have

$$f'(x) = \tfrac{1}{2}(25 - x^2)^{-1/2}\frac{d}{dx}(25 - x^2)$$

$$= \tfrac{1}{2}(25 - x^2)^{-1/2}(-2x) = -\frac{x}{\sqrt{25 - x^2}}$$

Example 1 illustrates that even when it is possible to solve an equation explicitly for y in terms of x, it may be easier to use implicit differentiation.

So

$$f'(3) = -\frac{3}{\sqrt{25 - 3^2}} = -\frac{3}{4}$$

and, as in Solution 1, an equation of the tangent is $3x + 4y = 25$.   ∎

NOTE 1 The expression $dy/dx = -x/y$ in Solution 1 gives the derivative in terms of both x and y. It is correct no matter which function y is determined by the given equation. For instance, for $y = f(x) = \sqrt{25 - x^2}$ we have

$$\frac{dy}{dx} = -\frac{x}{y} = -\frac{x}{\sqrt{25 - x^2}}$$

whereas for $y = g(x) = -\sqrt{25 - x^2}$ we have

$$\frac{dy}{dx} = -\frac{x}{y} = -\frac{x}{-\sqrt{25 - x^2}} = \frac{x}{\sqrt{25 - x^2}}$$

EXAMPLE 2

(a) Find y' if $x^3 + y^3 = 6xy$.

(b) Find the tangent to the folium of Descartes $x^3 + y^3 = 6xy$ at the point $(3, 3)$.

(c) At what point in the first quadrant is the tangent line horizontal?

SOLUTION

(a) Differentiating both sides of $x^3 + y^3 = 6xy$ with respect to x, regarding y as a function of x, and using the Chain Rule on the term y^3 and the Product Rule on the term $6xy$, we get

$$3x^2 + 3y^2y' = 6xy' + 6y$$

or

$$x^2 + y^2y' = 2xy' + 2y$$

We now solve for y':

$$y^2 y' - 2xy' = 2y - x^2$$

$$(y^2 - 2x)y' = 2y - x^2$$

$$y' = \frac{2y - x^2}{y^2 - 2x}$$

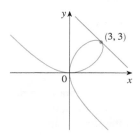

FIGURE 4

(b) When $x = y = 3$,

$$y' = \frac{2 \cdot 3 - 3^2}{3^2 - 2 \cdot 3} = -1$$

and a glance at Figure 4 confirms that this is a reasonable value for the slope at (3, 3). So an equation of the tangent to the folium at (3, 3) is

$$y - 3 = -1(x - 3) \qquad \text{or} \qquad x + y = 6$$

(c) The tangent line is horizontal if $y' = 0$. Using the expression for y' from part (a), we see that $y' = 0$ when $2y - x^2 = 0$ (provided that $y^2 - 2x \neq 0$). Substituting $y = \frac{1}{2}x^2$ in the equation of the curve, we get

$$x^3 + \left(\tfrac{1}{2}x^2\right)^3 = 6x\left(\tfrac{1}{2}x^2\right)$$

which simplifies to $x^6 = 16x^3$. Since $x \neq 0$ in the first quadrant, we have $x^3 = 16$. If $x = 16^{1/3} = 2^{4/3}$, then $y = \frac{1}{2}(2^{8/3}) = 2^{5/3}$. Thus the tangent is horizontal at $(2^{4/3}, 2^{5/3})$, which is approximately (2.5198, 3.1748). Looking at Figure 5, we see that our answer is reasonable. ■

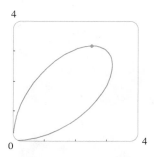

FIGURE 5

NOTE 2 There is a formula for the three roots of a cubic equation that is like the quadratic formula but much more complicated. If we use this formula (or a computer algebra system) to solve the equation $x^3 + y^3 = 6xy$ for y in terms of x, we get three functions determined by the equation:

$$y = f(x) = \sqrt[3]{-\tfrac{1}{2}x^3 + \sqrt{\tfrac{1}{4}x^6 - 8x^3}} + \sqrt[3]{-\tfrac{1}{2}x^3 - \sqrt{\tfrac{1}{4}x^6 - 8x^3}}$$

and

$$y = \tfrac{1}{2}\left[-f(x) \pm \sqrt{-3}\left(\sqrt[3]{-\tfrac{1}{2}x^3 + \sqrt{\tfrac{1}{4}x^6 - 8x^3}} - \sqrt[3]{-\tfrac{1}{2}x^3 - \sqrt{\tfrac{1}{4}x^6 - 8x^3}}\right)\right]$$

(These are the three functions whose graphs are shown in Figure 3.) You can see that the method of implicit differentiation saves an enormous amount of work in cases such as this. Moreover, implicit differentiation works just as easily for equations such as

$$y^5 + 3x^2 y^2 + 5x^4 = 12$$

for which it is *impossible* to find a similar expression for y in terms of x.

EXAMPLE 3 Find y' if $\sin(x + y) = y^2 \cos x$.

SOLUTION Differentiating implicitly with respect to x and remembering that y is a function of x, we get

$$\cos(x + y) \cdot (1 + y') = y^2(-\sin x) + (\cos x)(2yy')$$

Abel and Galois

The Norwegian mathematician Niels Abel proved in 1824 that no general formula can be given for the roots of a fifth-degree equation in terms of radicals. Later the French mathematician Evariste Galois proved that it is impossible to find a general formula for the roots of an nth-degree equation (in terms of algebraic operations on the coefficients) if n is any integer larger than 4.

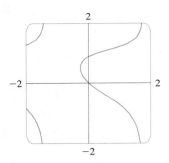

FIGURE 6

(Note that we have used the Chain Rule on the left side and the Product Rule and Chain Rule on the right side.) If we collect the terms that involve y', we get

$$\cos(x + y) + y^2 \sin x = (2y \cos x)y' - \cos(x + y) \cdot y'$$

So

$$y' = \frac{y^2 \sin x + \cos(x + y)}{2y \cos x - \cos(x + y)}$$

Figure 6, drawn with the implicit-plotting command of a computer algebra system, shows part of the curve $\sin(x + y) = y^2 \cos x$. As a check on our calculation, notice that $y' = -1$ when $x = y = 0$ and it appears from the graph that the slope is approximately -1 at the origin. ∎

Figures 7, 8, and 9 show three more curves produced by a computer algebra system with an implicit-plotting command. In Exercises 41–42 you will have an opportunity to create and examine unusual curves of this nature.

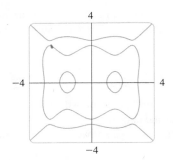

FIGURE 7
$(x^2 - 1)(x^2 - 4)(x^2 - 9)$
$= y^2(y^2 - 4)(y^2 - 9)$

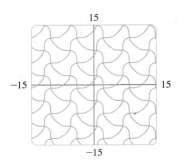

FIGURE 8
$\cos(x - \sin y) = \sin(y - \sin x)$

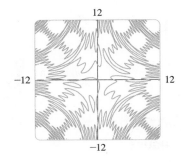

FIGURE 9
$\sin(xy) = \sin x + \sin y$

The following example shows how to find the second derivative of a function that is defined implicitly.

EXAMPLE 4 Find y'' if $x^4 + y^4 = 16$.

SOLUTION Differentiating the equation implicitly with respect to x, we get

$$4x^3 + 4y^3y' = 0$$

Solving for y' gives

$$\boxed{3} \qquad y' = -\frac{x^3}{y^3}$$

To find y'' we differentiate this expression for y' using the Quotient Rule and remembering that y is a function of x:

$$y'' = \frac{d}{dx}\left(-\frac{x^3}{y^3}\right) = -\frac{y^3\,(d/dx)(x^3) - x^3\,(d/dx)(y^3)}{(y^3)^2}$$

$$= -\frac{y^3 \cdot 3x^2 - x^3(3y^2y')}{y^6}$$

If we now substitute Equation 3 into this expression, we get

$$y'' = -\frac{3x^2y^3 - 3x^3y^2\left(-\dfrac{x^3}{y^3}\right)}{y^6}$$

$$= -\frac{3(x^2y^4 + x^6)}{y^7} = -\frac{3x^2(y^4 + x^4)}{y^7}$$

But the values of x and y must satisfy the original equation $x^4 + y^4 = 16$. So the answer simplifies to

$$y'' = -\frac{3x^2(16)}{y^7} = -48\frac{x^2}{y^7}$$

Figure 10 shows the graph of the curve $x^4 + y^4 = 16$ of Example 4. Notice that it's a stretched and flattened version of the circle $x^2 + y^2 = 4$. For this reason it's sometimes called a *fat circle*. It starts out very steep on the left but quickly becomes very flat. This can be seen from the expression

$$y' = -\frac{x^3}{y^3} = -\left(\frac{x}{y}\right)^3$$

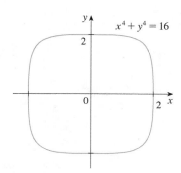

FIGURE 10

■ Derivatives of Inverse Trigonometric Functions

The inverse trigonometric functions were reviewed in Section 1.5. We discussed their continuity in Section 2.5 and their asymptotes in Section 2.6. Here we use implicit differentiation to find the derivatives of the inverse trigonometric functions, assuming that these functions are differentiable. [In fact, if f is any one-to-one differentiable function, it can be proved that its inverse function f^{-1} is also differentiable, except where its tangents are vertical. This is plausible because the graph of a differentiable function has no corner or kink and so if we reflect it about $y = x$, the graph of its inverse function also has no corner or kink.]

Recall the definition of the arcsine function:

$$y = \sin^{-1}x \qquad \text{means} \qquad \sin y = x \quad \text{and} \quad -\frac{\pi}{2} \leq y \leq \frac{\pi}{2}$$

Differentiating $\sin y = x$ implicitly with respect to x, we obtain

$$\cos y \frac{dy}{dx} = 1 \qquad \text{or} \qquad \frac{dy}{dx} = \frac{1}{\cos y}$$

Now $\cos y \geq 0$, since $-\pi/2 \leq y \leq \pi/2$, so

$$\cos y = \sqrt{1 - \sin^2 y} = \sqrt{1 - x^2}$$

The same method can be used to find a formula for the derivative of *any* inverse function. See Exercise 77.

Therefore

$$\frac{dy}{dx} = \frac{1}{\cos y} = \frac{1}{\sqrt{1 - x^2}}$$

$$\frac{d}{dx}(\sin^{-1}x) = \frac{1}{\sqrt{1 - x^2}}$$

Figure 11 shows the graph of $f(x) = \tan^{-1}x$ and its derivative $f'(x) = 1/(1 + x^2)$. Notice that f is increasing and $f'(x)$ is always positive. The fact that $\tan^{-1}x \to \pm\pi/2$ as $x \to \pm\infty$ is reflected in the fact that $f'(x) \to 0$ as $x \to \pm\infty$.

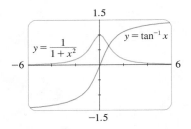

FIGURE 11

The formula for the derivative of the arctangent function is derived in a similar way. If $y = \tan^{-1}x$, then $\tan y = x$. Differentiating this latter equation implicitly with respect to x, we have

$$\sec^2 y \, \frac{dy}{dx} = 1$$

$$\frac{dy}{dx} = \frac{1}{\sec^2 y} = \frac{1}{1 + \tan^2 y} = \frac{1}{1 + x^2}$$

$$\frac{d}{dx}(\tan^{-1}x) = \frac{1}{1 + x^2}$$

EXAMPLE 5 Differentiate (a) $y = \dfrac{1}{\sin^{-1}x}$ and (b) $f(x) = x \arctan \sqrt{x}$.

SOLUTION

(a)
$$\frac{dy}{dx} = \frac{d}{dx}(\sin^{-1}x)^{-1} = -(\sin^{-1}x)^{-2}\frac{d}{dx}(\sin^{-1}x)$$

$$= -\frac{1}{(\sin^{-1}x)^2\sqrt{1 - x^2}}$$

Recall that arctan x is an alternative notation for $\tan^{-1}x$.

(b)
$$f'(x) = x\,\frac{1}{1 + (\sqrt{x})^2}\left(\tfrac{1}{2}x^{-1/2}\right) + \arctan\sqrt{x}$$

$$= \frac{\sqrt{x}}{2(1 + x)} + \arctan\sqrt{x}$$

The inverse trigonometric functions that occur most frequently are the ones that we have just discussed. The derivatives of the remaining four are given in the following table. The proofs of the formulas are left as exercises.

The formulas for the derivatives of $\csc^{-1}x$ and $\sec^{-1}x$ depend on the definitions that are used for these functions. See Exercise 64.

Derivatives of Inverse Trigonometric Functions

$$\frac{d}{dx}(\sin^{-1}x) = \frac{1}{\sqrt{1 - x^2}} \qquad \frac{d}{dx}(\csc^{-1}x) = -\frac{1}{x\sqrt{x^2 - 1}}$$

$$\frac{d}{dx}(\cos^{-1}x) = -\frac{1}{\sqrt{1 - x^2}} \qquad \frac{d}{dx}(\sec^{-1}x) = \frac{1}{x\sqrt{x^2 - 1}}$$

$$\frac{d}{dx}(\tan^{-1}x) = \frac{1}{1 + x^2} \qquad \frac{d}{dx}(\cot^{-1}x) = -\frac{1}{1 + x^2}$$

3.5 EXERCISES

1–4
(a) Find y' by implicit differentiation.
(b) Solve the equation explicitly for y and differentiate to get y' in terms of x.
(c) Check that your solutions to parts (a) and (b) are consistent by substituting the expression for y into your solution for part (a).

1. $9x^2 - y^2 = 1$ **2.** $2x^2 + x + xy = 1$

3. $\sqrt{x} + \sqrt{y} = 1$ **4.** $\dfrac{2}{x} - \dfrac{1}{y} = 4$

5–20 Find dy/dx by implicit differentiation.

5. $x^2 - 4xy + y^2 = 4$ **6.** $2x^2 + xy - y^2 = 2$

7. $x^4 + x^2y^2 + y^3 = 5$ **8.** $x^3 - xy^2 + y^3 = 1$

9. $\dfrac{x^2}{x + y} = y^2 + 1$ **10.** $xe^y = x - y$

11. $y \cos x = x^2 + y^2$ **12.** $\cos(xy) = 1 + \sin y$

13. $\sqrt{x + y} = x^4 + y^4$ **14.** $e^y \sin x = x + xy$

15. $e^{x/y} = x - y$ **16.** $xy = \sqrt{x^2 + y^2}$

17. $\tan^{-1}(x^2y) = x + xy^2$ **18.** $x \sin y + y \sin x = 1$

19. $\sin(xy) = \cos(x + y)$ **20.** $\tan(x - y) = \dfrac{y}{1 + x^2}$

---

21. If $f(x) + x^2[f(x)]^3 = 10$ and $f(1) = 2$, find $f'(1)$.

22. If $g(x) + x \sin g(x) = x^2$, find $g'(0)$.

23–24 Regard y as the independent variable and x as the dependent variable and use implicit differentiation to find dx/dy.

23. $x^4y^2 - x^3y + 2xy^3 = 0$ **24.** $y \sec x = x \tan y$

25–32 Use implicit differentiation to find an equation of the tangent line to the curve at the given point.

25. $y \sin 2x = x \cos 2y$, $(\pi/2, \pi/4)$

26. $\sin(x + y) = 2x - 2y$, (π, π)

27. $x^2 - xy - y^2 = 1$, $(2, 1)$ (hyperbola)

28. $x^2 + 2xy + 4y^2 = 12$, $(2, 1)$ (ellipse)

29. $x^2 + y^2 = (2x^2 + 2y^2 - x)^2$, $\left(0, \frac{1}{2}\right)$,   (cardioid)

30. $x^{2/3} + y^{2/3} = 4$, $\left(-3\sqrt{3}, 1\right)$,   (astroid)

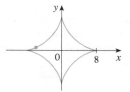

31. $2(x^2 + y^2)^2 = 25(x^2 - y^2)$, $(3, 1)$,   (lemniscate)

32. $y^2(y^2 - 4) = x^2(x^2 - 5)$, $(0, -2)$,   (devil's curve)

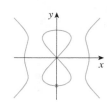

---

33. (a) The curve with equation $y^2 = 5x^4 - x^2$ is called a **kampyle of Eudoxus**. Find an equation of the tangent line to this curve at the point $(1, 2)$.
(b) Illustrate part (a) by graphing the curve and the tangent line on a common screen. (If your graphing device will graph implicitly defined curves, then use that capability. If not, you can still graph this curve by graphing its upper and lower halves separately.)

34. (a) The curve with equation $y^2 = x^3 + 3x^2$ is called the **Tschirnhausen cubic**. Find an equation of the tangent line to this curve at the point $(1, -2)$.
(b) At what points does this curve have horizontal tangents?
(c) Illustrate parts (a) and (b) by graphing the curve and the tangent lines on a common screen.

35–38 Find y'' by implicit differentiation.

35. $x^2 + 4y^2 = 4$

36. $x^2 + xy + y^2 = 3$

37. $\sin y + \cos x = 1$

38. $x^3 - y^3 = 7$

---

39. If $xy + e^y = e$, find the value of y'' at the point where $x = 0$.

40. If $x^2 + xy + y^3 = 1$, find the value of y''' at the point where $x = 1$.

CAS 41. Fanciful shapes can be created by using the implicit plotting capabilities of computer algebra systems.
   (a) Graph the curve with equation
   $$y(y^2 - 1)(y - 2) = x(x - 1)(x - 2)$$
   At how many points does this curve have horizontal tangents? Estimate the x-coordinates of these points.
   (b) Find equations of the tangent lines at the points $(0, 1)$ and $(0, 2)$.
   (c) Find the exact x-coordinates of the points in part (a).
   (d) Create even more fanciful curves by modifying the equation in part (a).

CAS 42. (a) The curve with equation
   $$2y^3 + y^2 - y^5 = x^4 - 2x^3 + x^2$$
   has been likened to a bouncing wagon. Use a computer algebra system to graph this curve and discover why.
   (b) At how many points does this curve have horizontal tangent lines? Find the x-coordinates of these points.

43. Find the points on the lemniscate in Exercise 31 where the tangent is horizontal.

44. Show by implicit differentiation that the tangent to the ellipse
$$\frac{x^2}{a^2} + \frac{y^2}{b^2} = 1$$
at the point (x_0, y_0) is
$$\frac{x_0 x}{a^2} + \frac{y_0 y}{b^2} = 1$$

45. Find an equation of the tangent line to the hyperbola
$$\frac{x^2}{a^2} - \frac{y^2}{b^2} = 1$$
at the point (x_0, y_0).

46. Show that the sum of the x- and y-intercepts of any tangent line to the curve $\sqrt{x} + \sqrt{y} = \sqrt{c}$ is equal to c.

47. Show, using implicit differentiation, that any tangent line at a point P to a circle with center O is perpendicular to the radius OP.

48. The Power Rule can be proved using implicit differentiation for the case where n is a rational number, $n = p/q$, and $y = f(x) = x^n$ is assumed beforehand to be a differentiable

function. If $y = x^{p/q}$, then $y^q = x^p$. Use implicit differentiation to show that
$$y' = \frac{p}{q} x^{(p/q)-1}$$

49–60 Find the derivative of the function. Simplify where possible.

49. $y = (\tan^{-1} x)^2$

50. $y = \tan^{-1}(x^2)$

51. $y = \sin^{-1}(2x + 1)$

52. $g(x) = \arccos \sqrt{x}$

53. $F(x) = x \sec^{-1}(x^3)$

54. $y = \tan^{-1}\left(x - \sqrt{1 + x^2}\right)$

55. $h(t) = \cot^{-1}(t) + \cot^{-1}(1/t)$

56. $R(t) = \arcsin(1/t)$

57. $y = x \sin^{-1} x + \sqrt{1 - x^2}$

58. $y = \cos^{-1}(\sin^{-1} t)$

59. $y = \arccos\left(\dfrac{b + a \cos x}{a + b \cos x}\right), \quad 0 \le x \le \pi, \ a > b > 0$

60. $y = \arctan \sqrt{\dfrac{1 - x}{1 + x}}$

61–62 Find $f'(x)$. Check that your answer is reasonable by comparing the graphs of f and f'.

61. $f(x) = \sqrt{1 - x^2} \arcsin x$

62. $f(x) = \arctan(x^2 - x)$

63. Prove the formula for $(d/dx)(\cos^{-1} x)$ by the same method as for $(d/dx)(\sin^{-1} x)$.

64. (a) One way of defining $\sec^{-1} x$ is to say that
   $y = \sec^{-1} x \iff \sec y = x$ and $0 \le y < \pi/2$ or
   $\pi \le y < 3\pi/2$. Show that, with this definition,
   $$\frac{d}{dx}(\sec^{-1} x) = \frac{1}{x\sqrt{x^2 - 1}}$$
   (b) Another way of defining $\sec^{-1} x$ that is sometimes used is to say that $y = \sec^{-1} x \iff \sec y = x$ and $0 \le y \le \pi$, $y \ne \pi/2$. Show that, with this definition,
   $$\frac{d}{dx}(\sec^{-1} x) = \frac{1}{|x|\sqrt{x^2 - 1}}$$

65–68 Two curves are **orthogonal** if their tangent lines are perpendicular at each point of intersection. Show that the given families of curves are **orthogonal trajectories** of each other; that is, every curve in one family is orthogonal to every curve in the other family. Sketch both families of curves on the same axes.

65. $x^2 + y^2 = r^2, \quad ax + by = 0$

66. $x^2 + y^2 = ax, \quad x^2 + y^2 = by$

67. $y = cx^2, \quad x^2 + 2y^2 = k$

68. $y = ax^3, \quad x^2 + 3y^2 = b$

69. Show that the ellipse $x^2/a^2 + y^2/b^2 = 1$ and the hyperbola $x^2/A^2 - y^2/B^2 = 1$ are orthogonal trajectories if $A^2 < a^2$ and $a^2 - b^2 = A^2 + B^2$ (so the ellipse and hyperbola have the same foci).

70. Find the value of the number a such that the families of curves $y = (x + c)^{-1}$ and $y = a(x + k)^{1/3}$ are orthogonal trajectories.

71. (a) The *van der Waals equation* for n moles of a gas is

$$\left(P + \frac{n^2 a}{V^2}\right)(V - nb) = nRT$$

where P is the pressure, V is the volume, and T is the temperature of the gas. The constant R is the universal gas constant and a and b are positive constants that are characteristic of a particular gas. If T remains constant, use implicit differentiation to find dV/dP.

(b) Find the rate of change of volume with respect to pressure of 1 mole of carbon dioxide at a volume of $V = 10$ L and a pressure of $P = 2.5$ atm. Use $a = 3.592$ L^2-atm/mole^2 and $b = 0.04267$ L/mole.

72. (a) Use implicit differentiation to find y' if

$$x^2 + xy + y^2 + 1 = 0$$

(b) Plot the curve in part (a). What do you see? Prove that what you see is correct.

(c) In view of part (b), what can you say about the expression for y' that you found in part (a)?

73. The equation $x^2 - xy + y^2 = 3$ represents a "rotated ellipse," that is, an ellipse whose axes are not parallel to the coordinate axes. Find the points at which this ellipse crosses the x-axis and show that the tangent lines at these points are parallel.

74. (a) Where does the normal line to the ellipse $x^2 - xy + y^2 = 3$ at the point $(-1, 1)$ intersect the ellipse a second time?

(b) Illustrate part (a) by graphing the ellipse and the normal line.

75. Find all points on the curve $x^2 y^2 + xy = 2$ where the slope of the tangent line is -1.

76. Find equations of both the tangent lines to the ellipse $x^2 + 4y^2 = 36$ that pass through the point $(12, 3)$.

77. (a) Suppose f is a one-to-one differentiable function and its inverse function f^{-1} is also differentiable. Use implicit differentiation to show that

$$(f^{-1})'(x) = \frac{1}{f'(f^{-1}(x))}$$

provided that the denominator is not 0.

(b) If $f(4) = 5$ and $f'(4) = \frac{2}{3}$, find $(f^{-1})'(5)$.

78. (a) Show that $f(x) = x + e^x$ is one-to-one.

(b) What is the value of $f^{-1}(1)$?

(c) Use the formula from Exercise 77(a) to find $(f^{-1})'(1)$.

79. The **Bessel function** of order 0, $y = J(x)$, satisfies the differential equation $xy'' + y' + xy = 0$ for all values of x and its value at 0 is $J(0) = 1$.

(a) Find $J'(0)$.

(b) Use implicit differentiation to find $J''(0)$.

80. The figure shows a lamp located three units to the right of the y-axis and a shadow created by the elliptical region $x^2 + 4y^2 \leq 5$. If the point $(-5, 0)$ is on the edge of the shadow, how far above the x-axis is the lamp located?

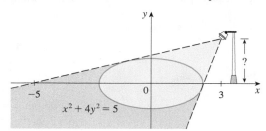

LABORATORY PROJECT [CAS] FAMILIES OF IMPLICIT CURVES

In this project you will explore the changing shapes of implicitly defined curves as you vary the constants in a family, and determine which features are common to all members of the family.

1. Consider the family of curves

$$y^2 - 2x^2(x + 8) = c[(y + 1)^2(y + 9) - x^2]$$

(a) By graphing the curves with $c = 0$ and $c = 2$, determine how many points of intersection there are. (You might have to zoom in to find all of them.)

(b) Now add the curves with $c = 5$ and $c = 10$ to your graphs in part (a). What do you notice? What about other values of c?

2. (a) Graph several members of the family of curves

$$x^2 + y^2 + cx^2y^2 = 1$$

Describe how the graph changes as you change the value of c.

(b) What happens to the curve when $c = -1$? Describe what appears on the screen. Can you prove it algebraically?

(c) Find y' by implicit differentiation. For the case $c = -1$, is your expression for y' consistent with what you discovered in part (b)?

3.6   Derivatives of Logarithmic Functions

In this section we use implicit differentiation to find the derivatives of the logarithmic functions $y = \log_b x$ and, in particular, the natural logarithmic function $y = \ln x$. [It can be proved that logarithmic functions are differentiable; this is certainly plausible from their graphs (see Figure 1.5.12).]

$\boxed{1}$

$$\frac{d}{dx}(\log_b x) = \frac{1}{x \ln b}$$

PROOF Let $y = \log_b x$. Then

$$b^y = x$$

Differentiating this equation implicitly with respect to x, using Formula 3.4.5, we get

Formula 3.4.5 says that

$$\frac{d}{dx}(b^x) = b^x \ln b$$

$$b^y(\ln b)\frac{dy}{dx} = 1$$

and so

$$\frac{dy}{dx} = \frac{1}{b^y \ln b} = \frac{1}{x \ln b}$$ ∎

If we put $b = e$ in Formula 1, then the factor $\ln b$ on the right side becomes $\ln e = 1$ and we get the formula for the derivative of the natural logarithmic function $\log_e x = \ln x$:

$\boxed{2}$

$$\frac{d}{dx}(\ln x) = \frac{1}{x}$$

By comparing Formulas 1 and 2, we see one of the main reasons that natural logarithms (logarithms with base e) are used in calculus: The differentiation formula is simplest when $b = e$ because $\ln e = 1$.

EXAMPLE 1 Differentiate $y = \ln(x^3 + 1)$.

SOLUTION To use the Chain Rule, we let $u = x^3 + 1$. Then $y = \ln u$, so

$$\frac{dy}{dx} = \frac{dy}{du}\frac{du}{dx} = \frac{1}{u}\frac{du}{dx}$$

$$= \frac{1}{x^3 + 1}(3x^2) = \frac{3x^2}{x^3 + 1}$$ ∎

In general, if we combine Formula 2 with the Chain Rule as in Example 1, we get

$$\boxed{3} \qquad \boxed{\frac{d}{dx}(\ln u) = \frac{1}{u}\frac{du}{dx}} \qquad \text{or} \qquad \boxed{\frac{d}{dx}[\ln g(x)] = \frac{g'(x)}{g(x)}}$$

EXAMPLE 2 Find $\dfrac{d}{dx}\ln(\sin x)$.

SOLUTION Using (3), we have

$$\frac{d}{dx}\ln(\sin x) = \frac{1}{\sin x}\frac{d}{dx}(\sin x) = \frac{1}{\sin x}\cos x = \cot x$$ ∎

EXAMPLE 3 Differentiate $f(x) = \sqrt{\ln x}$.

SOLUTION This time the logarithm is the inner function, so the Chain Rule gives

$$f'(x) = \tfrac{1}{2}(\ln x)^{-1/2}\frac{d}{dx}(\ln x) = \frac{1}{2\sqrt{\ln x}}\cdot\frac{1}{x} = \frac{1}{2x\sqrt{\ln x}}$$ ∎

EXAMPLE 4 Differentiate $f(x) = \log_{10}(2 + \sin x)$.

SOLUTION Using Formula 1 with $b = 10$, we have

$$f'(x) = \frac{d}{dx}\log_{10}(2 + \sin x)$$

$$= \frac{1}{(2 + \sin x)\ln 10}\frac{d}{dx}(2 + \sin x)$$

$$= \frac{\cos x}{(2 + \sin x)\ln 10}$$ ∎

Figure 1 shows the graph of the function f of Example 5 together with the graph of its derivative. It gives a visual check on our calculation. Notice that $f'(x)$ is large negative when f is rapidly decreasing.

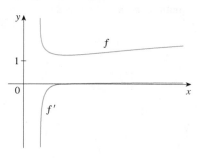

FIGURE 1

EXAMPLE 5 Find $\dfrac{d}{dx}\ln\dfrac{x + 1}{\sqrt{x - 2}}$.

SOLUTION 1

$$\frac{d}{dx}\ln\frac{x + 1}{\sqrt{x - 2}} = \frac{1}{\dfrac{x + 1}{\sqrt{x - 2}}}\frac{d}{dx}\frac{x + 1}{\sqrt{x - 2}}$$

$$= \frac{\sqrt{x - 2}}{x + 1}\cdot\frac{\sqrt{x - 2}\cdot 1 - (x + 1)(\tfrac{1}{2})(x - 2)^{-1/2}}{x - 2}$$

$$= \frac{x - 2 - \tfrac{1}{2}(x + 1)}{(x + 1)(x - 2)}$$

$$= \frac{x - 5}{2(x + 1)(x - 2)}$$

SOLUTION 2 If we first simplify the given function using the laws of logarithms, then the differentiation becomes easier:

$$\frac{d}{dx} \ln \frac{x+1}{\sqrt{x-2}} = \frac{d}{dx}\left[\ln(x+1) - \tfrac{1}{2}\ln(x-2)\right]$$

$$= \frac{1}{x+1} - \frac{1}{2}\left(\frac{1}{x-2}\right)$$

(This answer can be left as written, but if we used a common denominator we would see that it gives the same answer as in Solution 1.)   ■

Figure 2 shows the graph of the function $f(x) = \ln|x|$ in Example 6 and its derivative $f'(x) = 1/x$. Notice that when x is small, the graph of $y = \ln|x|$ is steep and so $f'(x)$ is large (positive or negative).

EXAMPLE 6 Find $f'(x)$ if $f(x) = \ln|x|$.

SOLUTION Since

$$f(x) = \begin{cases} \ln x & \text{if } x > 0 \\ \ln(-x) & \text{if } x < 0 \end{cases}$$

it follows that

$$f'(x) = \begin{cases} \dfrac{1}{x} & \text{if } x > 0 \\[2ex] \dfrac{1}{-x}(-1) = \dfrac{1}{x} & \text{if } x < 0 \end{cases}$$

Thus $f'(x) = 1/x$ for all $x \neq 0$.   ■

The result of Example 6 is worth remembering:

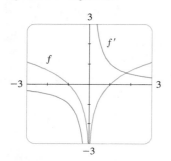

FIGURE 2

4

$$\frac{d}{dx}\ln|x| = \frac{1}{x}$$

■ Logarithmic Differentiation

The calculation of derivatives of complicated functions involving products, quotients, or powers can often be simplified by taking logarithms. The method used in the following example is called **logarithmic differentiation**.

EXAMPLE 7 Differentiate $y = \dfrac{x^{3/4}\sqrt{x^2+1}}{(3x+2)^5}$.

SOLUTION We take logarithms of both sides of the equation and use the Laws of Logarithms to simplify:

$$\ln y = \tfrac{3}{4}\ln x + \tfrac{1}{2}\ln(x^2+1) - 5\ln(3x+2)$$

Differentiating implicitly with respect to x gives

$$\frac{1}{y}\frac{dy}{dx} = \frac{3}{4}\cdot\frac{1}{x} + \frac{1}{2}\cdot\frac{2x}{x^2+1} - 5\cdot\frac{3}{3x+2}$$

Solving for dy/dx, we get

$$\frac{dy}{dx} = y\left(\frac{3}{4x} + \frac{x}{x^2+1} - \frac{15}{3x+2}\right)$$

If we hadn't used logarithmic differentiation in Example 7, we would have had to use both the Quotient Rule and the Product Rule. The resulting calculation would have been horrendous.

Because we have an explicit expression for y, we can substitute and write

$$\frac{dy}{dx} = \frac{x^{3/4}\sqrt{x^2 + 1}}{(3x + 2)^5}\left(\frac{3}{4x} + \frac{x}{x^2 + 1} - \frac{15}{3x + 2}\right)$$

∎

Steps in Logarithmic Differentiation

1. Take natural logarithms of both sides of an equation $y = f(x)$ and use the Laws of Logarithms to simplify.
2. Differentiate implicitly with respect to x.
3. Solve the resulting equation for y'.

If $f(x) < 0$ for some values of x, then $\ln f(x)$ is not defined, but we can write $|y| = |f(x)|$ and use Equation 4. We illustrate this procedure by proving the general version of the Power Rule, as promised in Section 3.1.

The Power Rule If n is any real number and $f(x) = x^n$, then

$$f'(x) = nx^{n-1}$$

If $x = 0$, we can show that $f'(0) = 0$ for $n > 1$ directly from the definition of a derivative.

PROOF Let $y = x^n$ and use logarithmic differentiation:

$$\ln|y| = \ln|x|^n = n\ln|x| \qquad x \neq 0$$

Therefore

$$\frac{y'}{y} = \frac{n}{x}$$

Hence

$$y' = n\frac{y}{x} = n\frac{x^n}{x} = nx^{n-1}$$

∎

 You should distinguish carefully between the Power Rule $[(x^n)' = nx^{n-1}]$, where the base is variable and the exponent is constant, and the rule for differentiating exponential functions $[(b^x)' = b^x \ln b]$, where the base is constant and the exponent is variable. In general there are four cases for exponents and bases:

Constant base, constant exponent

1. $\dfrac{d}{dx}(b^n) = 0$ (b and n are constants)

Variable base, constant exponent

2. $\dfrac{d}{dx}[f(x)]^n = n[f(x)]^{n-1}f'(x)$

Constant base, variable exponent

3. $\dfrac{d}{dx}[b^{g(x)}] = b^{g(x)}(\ln b)g'(x)$

Variable base, variable exponent

4. To find $(d/dx)[f(x)]^{g(x)}$, logaritmic differentiation can be used, as in the next example.

EXAMPLE 8 Differentiate $y = x^{\sqrt{x}}$.

SOLUTION 1 Since both the base and the exponent are variable, we use logarithmic differentiation:

$$\ln y = \ln x^{\sqrt{x}} = \sqrt{x} \ln x$$

$$\frac{y'}{y} = \sqrt{x} \cdot \frac{1}{x} + (\ln x) \frac{1}{2\sqrt{x}}$$

$$y' = y\left(\frac{1}{\sqrt{x}} + \frac{\ln x}{2\sqrt{x}}\right) = x^{\sqrt{x}}\left(\frac{2 + \ln x}{2\sqrt{x}}\right)$$

SOLUTION 2 Another method is to write $x^{\sqrt{x}} = (e^{\ln x})^{\sqrt{x}}$:

$$\frac{d}{dx}\left(x^{\sqrt{x}}\right) = \frac{d}{dx}\left(e^{\sqrt{x}\ln x}\right) = e^{\sqrt{x}\ln x} \frac{d}{dx}\left(\sqrt{x} \ln x\right)$$

$$= x^{\sqrt{x}}\left(\frac{2 + \ln x}{2\sqrt{x}}\right) \qquad \text{(as in Solution 1)} \qquad \blacksquare$$

Figure 3 illustrates Example 8 by showing the graphs of $f(x) = x^{\sqrt{x}}$ and its derivative.

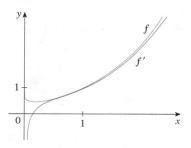

FIGURE 3

▨ The Number e as a Limit

We have shown that if $f(x) = \ln x$, then $f'(x) = 1/x$. Thus $f'(1) = 1$. We now use this fact to express the number e as a limit.

From the definition of a derivative as a limit, we have

$$f'(1) = \lim_{h \to 0} \frac{f(1 + h) - f(1)}{h} = \lim_{x \to 0} \frac{f(1 + x) - f(1)}{x}$$

$$= \lim_{x \to 0} \frac{\ln(1 + x) - \ln 1}{x} = \lim_{x \to 0} \frac{1}{x} \ln(1 + x)$$

$$= \lim_{x \to 0} \ln(1 + x)^{1/x}$$

Because $f'(1) = 1$, we have

$$\lim_{x \to 0} \ln(1 + x)^{1/x} = 1$$

Then, by Theorem 2.5.8 and the continuity of the exponential function, we have

$$e = e^1 = e^{\lim_{x \to 0} \ln(1+x)^{1/x}} = \lim_{x \to 0} e^{\ln(1+x)^{1/x}} = \lim_{x \to 0} (1 + x)^{1/x}$$

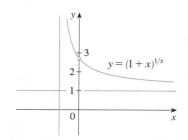

$$y = (1 + x)^{1/x}$$

FIGURE 4

$$\boxed{5} \qquad \boxed{e = \lim_{x \to 0} (1 + x)^{1/x}}$$

Formula 5 is illustrated by the graph of the function $y = (1 + x)^{1/x}$ in Figure 4 and a table of values for small values of x. This illustrates the fact that, correct to seven decimal places,

$$e \approx 2.7182818$$

If we put $n = 1/x$ in Formula 5, then $n \to \infty$ as $x \to 0^+$ and so an alternative expression for e is

$$\boxed{6} \qquad \boxed{e = \lim_{n \to \infty} \left(1 + \frac{1}{n}\right)^n}$$

x	$(1 + x)^{1/x}$
0.1	2.59374246
0.01	2.70481383
0.001	2.71692393
0.0001	2.71814593
0.00001	2.71826824
0.000001	2.71828047
0.0000001	2.71828169
0.00000001	2.71828181

3.6   EXERCISES

1. Explain why the natural logarithmic function $y = \ln x$ is used much more frequently in calculus than the other logarithmic functions $y = \log_b x$.

2–22 Differentiate the function.

2. $f(x) = x \ln x - x$

3. $f(x) = \sin(\ln x)$

4. $f(x) = \ln(\sin^2 x)$

5. $f(x) = \ln \dfrac{1}{x}$

6. $y = \dfrac{1}{\ln x}$

7. $f(x) = \log_{10}(1 + \cos x)$

8. $f(x) = \log_{10}\sqrt{x}$

9. $g(x) = \ln(xe^{-2x})$

10. $g(t) = \sqrt{1 + \ln t}$

11. $F(t) = (\ln t)^2 \sin t$

12. $h(x) = \ln\left(x + \sqrt{x^2 - 1}\right)$

13. $G(y) = \ln \dfrac{(2y + 1)^5}{\sqrt{y^2 + 1}}$

14. $P(v) = \dfrac{\ln v}{1 - v}$

15. $F(s) = \ln \ln s$

16. $y = \ln|1 + t - t^3|$

17. $T(z) = 2^z \log_2 z$

18. $y = \ln(\csc x - \cot x)$

19. $y = \ln(e^{-x} + xe^{-x})$

20. $H(z) = \ln \sqrt{\dfrac{a^2 - z^2}{a^2 + z^2}}$

21. $y = \tan[\ln(ax + b)]$

22. $y = \log_2(x \log_5 x)$

23–26 Find y' and y''.

23. $y = \sqrt{x}\, \ln x$

24. $y = \dfrac{\ln x}{1 + \ln x}$

25. $y = \ln|\sec x|$

26. $y = \ln(1 + \ln x)$

27–30 Differentiate f and find the domain of f.

27. $f(x) = \dfrac{x}{1 - \ln(x - 1)}$

28. $f(x) = \sqrt{2 + \ln x}$

29. $f(x) = \ln(x^2 - 2x)$

30. $f(x) = \ln \ln \ln x$

31. If $f(x) = \ln(x + \ln x)$, find $f'(1)$.

32. If $f(x) = \cos(\ln x^2)$, find $f'(1)$.

33–34 Find an equation of the tangent line to the curve at the given point.

33. $y = \ln(x^2 - 3x + 1)$, $(3, 0)$

34. $y = x^2 \ln x$, $(1, 0)$

35. If $f(x) = \sin x + \ln x$, find $f'(x)$. Check that your answer is reasonable by comparing the graphs of f and f'.

36. Find equations of the tangent lines to the curve $y = (\ln x)/x$ at the points $(1, 0)$ and $(e, 1/e)$. Illustrate by graphing the curve and its tangent lines.

37. Let $f(x) = cx + \ln(\cos x)$. For what value of c is $f'(\pi/4) = 6$?

38. Let $f(x) = \log_b(3x^2 - 2)$. For what value of b is $f'(1) = 3$?

39–50 Use logarithmic differentiation to find the derivative of the function.

39. $y = (x^2 + 2)^2(x^4 + 4)^4$

40. $y = \dfrac{e^{-x} \cos^2 x}{x^2 + x + 1}$

41. $y = \sqrt{\dfrac{x - 1}{x^4 + 1}}$

42. $y = \sqrt{x}\, e^{x^2 - x}(x + 1)^{2/3}$

43. $y = x^x$

44. $y = x^{\cos x}$

45. $y = x^{\sin x}$

46. $y = \sqrt{x}^{\,x}$

47. $y = (\cos x)^x$

48. $y = (\sin x)^{\ln x}$

49. $y = (\tan x)^{1/x}$

50. $y = (\ln x)^{\cos x}$

51. Find y' if $y = \ln(x^2 + y^2)$.

52. Find y' if $x^y = y^x$.

53. Find a formula for $f^{(n)}(x)$ if $f(x) = \ln(x - 1)$.

54. Find $\dfrac{d^9}{dx^9}(x^8 \ln x)$.

55. Use the definition of derivative to prove that

$$\lim_{x \to 0} \frac{\ln(1 + x)}{x} = 1$$

56. Show that $\displaystyle\lim_{n \to \infty}\left(1 + \frac{x}{n}\right)^n = e^x$ for any $x > 0$.

3.7   Rates of Change in the Natural and Social Sciences

We know that if $y = f(x)$, then the derivative dy/dx can be interpreted as the rate of change of y with respect to x. In this section we examine some of the applications of this idea to physics, chemistry, biology, economics, and other sciences.

Let's recall from Section 2.7 the basic idea behind rates of change. If x changes from x_1 to x_2, then the change in x is

$$\Delta x = x_2 - x_1$$

and the corresponding change in y is

$$\Delta y = f(x_2) - f(x_1)$$

The difference quotient

$$\frac{\Delta y}{\Delta x} = \frac{f(x_2) - f(x_1)}{x_2 - x_1}$$

is **the average rate of change of y with respect to x** over the interval $[x_1, x_2]$ and can be interpreted as the slope of the secant line PQ in Figure 1. Its limit as $\Delta x \to 0$ is the derivative $f'(x_1)$, which can therefore be interpreted as the **instantaneous rate of change of y with respect to x** or the slope of the tangent line at $P(x_1, f(x_1))$. Using Leibniz notation, we write the process in the form

$$\frac{dy}{dx} = \lim_{\Delta x \to 0} \frac{\Delta y}{\Delta x}$$

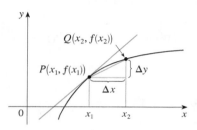

m_{PQ} = average rate of change
$m = f'(x_1)$ = instantaneous rate of change

FIGURE 1

Whenever the function $y = f(x)$ has a specific interpretation in one of the sciences, its derivative will have a specific interpretation as a rate of change. (As we discussed in Section 2.7, the units for dy/dx are the units for y divided by the units for x.) We now look at some of these interpretations in the natural and social sciences.

■ Physics

If $s = f(t)$ is the position function of a particle that is moving in a straight line, then $\Delta s/\Delta t$ represents the average velocity over a time period Δt, and $v = ds/dt$ represents the instantaneous **velocity** (the rate of change of displacement with respect to time). The instantaneous rate of change of velocity with respect to time is **acceleration**: $a(t) = v'(t) = s''(t)$. This was discussed in Sections 2.7 and 2.8, but now that we know the differentiation formulas, we are able to solve problems involving the motion of objects more easily.

EXAMPLE 1 The position of a particle is given by the equation

$$s = f(t) = t^3 - 6t^2 + 9t$$

where t is measured in seconds and s in meters.
(a) Find the velocity at time t.
(b) What is the velocity after 2 s? After 4 s?
(c) When is the particle at rest?
(d) When is the particle moving forward (that is, in the positive direction)?
(e) Draw a diagram to represent the motion of the particle.
(f) Find the total distance traveled by the particle during the first five seconds.
(g) Find the acceleration at time t and after 4 s.

(h) Graph the position, velocity, and acceleration functions for $0 \leqslant t \leqslant 5$.

(i) When is the particle speeding up? When is it slowing down?

SOLUTION

(a) The velocity function is the derivative of the position function.

$$s = f(t) = t^3 - 6t^2 + 9t$$

$$v(t) = \frac{ds}{dt} = 3t^2 - 12t + 9$$

(b) The velocity after 2 s means the instantaneous velocity when $t = 2$, that is,

$$v(2) = \frac{ds}{dt}\bigg|_{t=2} = 3(2)^2 - 12(2) + 9 = -3 \text{ m/s}$$

The velocity after 4 s is

$$v(4) = 3(4)^2 - 12(4) + 9 = 9 \text{ m/s}$$

(c) The particle is at rest when $v(t) = 0$, that is,

$$3t^2 - 12t + 9 = 3(t^2 - 4t + 3) = 3(t - 1)(t - 3) = 0$$

and this is true when $t = 1$ or $t = 3$. Thus the particle is at rest after 1 s and after 3 s.

(d) The particle moves in the positive direction when $v(t) > 0$, that is,

$$3t^2 - 12t + 9 = 3(t - 1)(t - 3) > 0$$

This inequality is true when both factors are positive $(t > 3)$ or when both factors are negative $(t < 1)$. Thus the particle moves in the positive direction in the time intervals $t < 1$ and $t > 3$. It moves backward (in the negative direction) when $1 < t < 3$.

(e) Using the information from part (d) we make a schematic sketch in Figure 2 of the motion of the particle back and forth along a line (the s-axis).

$t = 3$
$s = 0$

$t = 0$ $t = 1$ s
$s = 0$ $s = 4$

FIGURE 2

(f) Because of what we learned in parts (d) and (e), we need to calculate the distances traveled during the time intervals [0, 1], [1, 3], and [3, 5] separately.

The distance traveled in the first second is

$$|f(1) - f(0)| = |4 - 0| = 4 \text{ m}$$

From $t = 1$ to $t = 3$ the distance traveled is

$$|f(3) - f(1)| = |0 - 4| = 4 \text{ m}$$

From $t = 3$ to $t = 5$ the distance traveled is

$$|f(5) - f(3)| = |20 - 0| = 20 \text{ m}$$

The total distance is $4 + 4 + 20 = 28$ m.

(g) The acceleration is the derivative of the velocity function:

$$a(t) = \frac{d^2s}{dt^2} = \frac{dv}{dt} = 6t - 12$$

$$a(4) = 6(4) - 12 = 12 \text{ m/s}^2$$

(h) Figure 3 shows the graphs of s, v, and a.

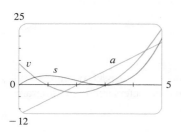

25

v a

s

0   5

-12

FIGURE 3

(i) The particle speeds up when the velocity is positive and increasing (v and a are both positive) and also when the velocity is negative and decreasing (v and a are both negative). In other words, the particle speeds up when the velocity and acceleration have the same sign. (The particle is pushed in the same direction it is moving.) From Figure 3 we see that this happens when $1 < t < 2$ and when $t > 3$. The particle slows down when v and a have opposite signs, that is, when $0 \leqslant t < 1$ and when $2 < t < 3$. Figure 4 summarizes the motion of the particle.

TEC In Module 3.7 you can see an animation of Figure 4 with an expression for s that you can choose yourself.

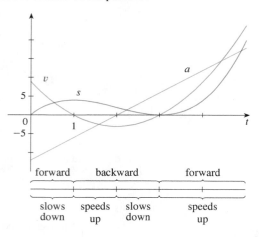

FIGURE 4

EXAMPLE 2 If a rod or piece of wire is homogeneous, then its linear density is uniform and is defined as the mass per unit length ($\rho = m/l$) and measured in kilograms per meter. Suppose, however, that the rod is not homogeneous but that its mass measured from its left end to a point x is $m = f(x)$, as shown in Figure 5.

FIGURE 5   This part of the rod has mass $f(x)$.

The mass of the part of the rod that lies between $x = x_1$ and $x = x_2$ is given by $\Delta m = f(x_2) - f(x_1)$, so the average density of that part of the rod is

$$\text{average density} = \frac{\Delta m}{\Delta x} = \frac{f(x_2) - f(x_1)}{x_2 - x_1}$$

If we now let $\Delta x \rightarrow 0$ (that is, $x_2 \rightarrow x_1$), we are computing the average density over smaller and smaller intervals. The **linear density** ρ at x_1 is the limit of these average densities as $\Delta x \rightarrow 0$; that is, the linear density is the rate of change of mass with respect to length. Symbolically,

$$\rho = \lim_{\Delta x \rightarrow 0} \frac{\Delta m}{\Delta x} = \frac{dm}{dx}$$

Thus the linear density of the rod is the derivative of mass with respect to length.

For instance, if $m = f(x) = \sqrt{x}$, where x is measured in meters and m in kilograms, then the average density of the part of the rod given by $1 \leqslant x \leqslant 1.2$ is

$$\frac{\Delta m}{\Delta x} = \frac{f(1.2) - f(1)}{1.2 - 1} = \frac{\sqrt{1.2} - 1}{0.2} \approx 0.48 \text{ kg/m}$$

while the density right at $x = 1$ is

$$\rho = \left.\frac{dm}{dx}\right|_{x=1} = \left.\frac{1}{2\sqrt{x}}\right|_{x=1} = 0.50 \text{ kg/m}$$ ∎

FIGURE 6

EXAMPLE 3 A current exists whenever electric charges move. Figure 6 shows part of a wire and electrons moving through a plane surface, shaded red. If ΔQ is the net charge that passes through this surface during a time period Δt, then the average current during this time interval is defined as

$$\text{average current} = \frac{\Delta Q}{\Delta t} = \frac{Q_2 - Q_1}{t_2 - t_1}$$

If we take the limit of this average current over smaller and smaller time intervals, we get what is called the **current** I at a given time t_1:

$$I = \lim_{\Delta t \to 0} \frac{\Delta Q}{\Delta t} = \frac{dQ}{dt}$$

Thus the current is the rate at which charge flows through a surface. It is measured in units of charge per unit time (often coulombs per second, called amperes).   ∎

Velocity, density, and current are not the only rates of change that are important in physics. Others include power (the rate at which work is done), the rate of heat flow, temperature gradient (the rate of change of temperature with respect to position), and the rate of decay of a radioactive substance in nuclear physics.

■ Chemistry

EXAMPLE 4 A chemical reaction results in the formation of one or more substances (called *products*) from one or more starting materials (called *reactants*). For instance, the "equation"

$$2H_2 + O_2 \to 2H_2O$$

indicates that two molecules of hydrogen and one molecule of oxygen form two molecules of water. Let's consider the reaction

$$A + B \to C$$

where A and B are the reactants and C is the product. The **concentration** of a reactant A is the number of moles ($1 \text{ mole} = 6.022 \times 10^{23}$ molecules) per liter and is denoted by [A]. The concentration varies during a reaction, so [A], [B], and [C] are all functions of time (t). The average rate of reaction of the product C over a time interval $t_1 \leqslant t \leqslant t_2$ is

$$\frac{\Delta[C]}{\Delta t} = \frac{[C](t_2) - [C](t_1)}{t_2 - t_1}$$

But chemists are more interested in the **instantaneous rate of reaction**, which is obtained by taking the limit of the average rate of reaction as the time interval Δt approaches 0:

$$\text{rate of reaction} = \lim_{\Delta t \to 0} \frac{\Delta[C]}{\Delta t} = \frac{d[C]}{dt}$$

Since the concentration of the product increases as the reaction proceeds, the derivative $d[C]/dt$ will be positive, and so the rate of reaction of C is positive. The concentrations of the reactants, however, decrease during the reaction, so, to make the rates of reaction of A and B positive numbers, we put minus signs in front of the derivatives $d[A]/dt$ and $d[B]/dt$. Since [A] and [B] each decrease at the same rate that [C] increases, we have

$$\text{rate of reaction} = \frac{d[C]}{dt} = -\frac{d[A]}{dt} = -\frac{d[B]}{dt}$$

More generally, it turns out that for a reaction of the form

$$a\text{A} + b\text{B} \rightarrow c\text{C} + d\text{D}$$

we have

$$-\frac{1}{a}\frac{d[A]}{dt} = -\frac{1}{b}\frac{d[B]}{dt} = \frac{1}{c}\frac{d[C]}{dt} = \frac{1}{d}\frac{d[D]}{dt}$$

The rate of reaction can be determined from data and graphical methods. In some cases there are explicit formulas for the concentrations as functions of time, which enable us to compute the rate of reaction (see Exercise 24). ∎

EXAMPLE 5 One of the quantities of interest in thermodynamics is compressibility. If a given substance is kept at a constant temperature, then its volume V depends on its pressure P. We can consider the rate of change of volume with respect to pressure—namely, the derivative dV/dP. As P increases, V decreases, so $dV/dP < 0$. The **compressibility** is defined by introducing a minus sign and dividing this derivative by the volume V:

$$\text{isothermal compressibility} = \beta = -\frac{1}{V}\frac{dV}{dP}$$

Thus β measures how fast, per unit volume, the volume of a substance decreases as the pressure on it increases at constant temperature.

For instance, the volume V (in cubic meters) of a sample of air at 25°C was found to be related to the pressure P (in kilopascals) by the equation

$$V = \frac{5.3}{P}$$

The rate of change of V with respect to P when $P = 50$ kPa is

$$\frac{dV}{dP}\bigg|_{P=50} = -\frac{5.3}{P^2}\bigg|_{P=50}$$

$$= -\frac{5.3}{2500} = -0.00212 \text{ m}^3/\text{kPa}$$

The compressibility at that pressure is

$$\beta = -\frac{1}{V}\frac{dV}{dP}\bigg|_{P=50} = \frac{0.00212}{\dfrac{5.3}{50}} = 0.02 \text{ (m}^3/\text{kPa)}/\text{m}^3 \qquad ∎$$

■ Biology

EXAMPLE 6 Let $n = f(t)$ be the number of individuals in an animal or plant population at time t. The change in the population size between the times $t = t_1$ and $t = t_2$ is $\Delta n = f(t_2) - f(t_1)$, and so the average rate of growth during the time period $t_1 \leqslant t \leqslant t_2$ is

$$\text{average rate of growth} = \frac{\Delta n}{\Delta t} = \frac{f(t_2) - f(t_1)}{t_2 - t_1}$$

The **instantaneous rate of growth** is obtained from this average rate of growth by letting the time period Δt approach 0:

$$\text{growth rate} = \lim_{\Delta t \to 0} \frac{\Delta n}{\Delta t} = \frac{dn}{dt}$$

Strictly speaking, this is not quite accurate because the actual graph of a population function $n = f(t)$ would be a step function that is discontinuous whenever a birth or death occurs and therefore not differentiable. However, for a large animal or plant population, we can replace the graph by a smooth approximating curve as in Figure 7.

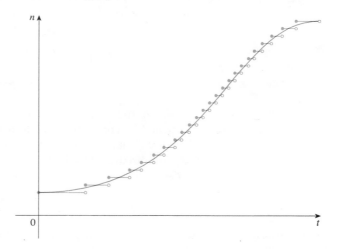

FIGURE 7
A smooth curve approximating
a growth function

E. coli bacteria are about 2 micrometers (μm) long and 0.75 μm wide. The image was produced with a scanning electron microscope.

To be more specific, consider a population of bacteria in a homogeneous nutrient medium. Suppose that by sampling the population at certain intervals it is determined that the population doubles every hour. If the initial population is n_0 and the time t is measured in hours, then

$$f(1) = 2f(0) = 2n_0$$

$$f(2) = 2f(1) = 2^2 n_0$$

$$f(3) = 2f(2) = 2^3 n_0$$

and, in general,

$$f(t) = 2^t n_0$$

The population function is $n = n_0 2^t$.

In Section 3.4 we showed that

$$\frac{d}{dx}(b^x) = b^x \ln b$$

So the rate of growth of the bacteria population at time t is

$$\frac{dn}{dt} = \frac{d}{dt}(n_0 2^t) = n_0 2^t \ln 2$$

For example, suppose that we start with an initial population of $n_0 = 100$ bacteria. Then the rate of growth after 4 hours is

$$\left.\frac{dn}{dt}\right|_{t=4} = 100 \cdot 2^4 \ln 2 = 1600 \ln 2 \approx 1109$$

This means that, after 4 hours, the bacteria population is growing at a rate of about 1109 bacteria per hour.

EXAMPLE 7 When we consider the flow of blood through a blood vessel, such as a vein or artery, we can model the shape of the blood vessel by a cylindrical tube with radius R and length l as illustrated in Figure 8.

FIGURE 8
Blood flow in an artery

Because of friction at the walls of the tube, the velocity v of the blood is greatest along the central axis of the tube and decreases as the distance r from the axis increases until v becomes 0 at the wall. The relationship between v and r is given by the **law of laminar flow** discovered by the French physician Jean-Louis-Marie Poiseuille in 1840. This law states that

$$\boxed{1} \qquad\qquad v = \frac{P}{4\eta l}(R^2 - r^2)$$

For more detailed information, see W. Nichols and M. O'Rourke (eds.), *McDonald's Blood Flow in Arteries: Theoretical, Experimental, and Clinical Principles*, 5th ed. (New York, 2005).

where η is the viscosity of the blood and P is the pressure difference between the ends of the tube. If P and l are constant, then v is a function of r with domain $[0, R]$.

The average rate of change of the velocity as we move from $r = r_1$ outward to $r = r_2$ is given by

$$\frac{\Delta v}{\Delta r} = \frac{v(r_2) - v(r_1)}{r_2 - r_1}$$

and if we let $\Delta r \to 0$, we obtain the **velocity gradient**, that is, the instantaneous rate of change of velocity with respect to r:

$$\text{velocity gradient} = \lim_{\Delta r \to 0} \frac{\Delta v}{\Delta r} = \frac{dv}{dr}$$

Using Equation 1, we obtain

$$\frac{dv}{dr} = \frac{P}{4\eta l}(0 - 2r) = -\frac{Pr}{2\eta l}$$

For one of the smaller human arteries we can take $\eta = 0.027$, $R = 0.008$ cm, $l = 2$ cm, and $P = 4000$ dynes/cm^2, which gives

$$v = \frac{4000}{4(0.027)2}(0.000064 - r^2)$$

$$\approx 1.85 \times 10^4(6.4 \times 10^{-5} - r^2)$$

At $r = 0.002$ cm the blood is flowing at a speed of

$$v(0.002) \approx 1.85 \times 10^4(64 \times 10^{-6} - 4 \times 10^{-6})$$

$$= 1.11 \text{ cm/s}$$

and the velocity gradient at that point is

$$\left.\frac{dv}{dr}\right|_{r=0.002} = -\frac{4000(0.002)}{2(0.027)2} \approx -74 \text{ (cm/s)/cm}$$

To get a feeling for what this statement means, let's change our units from centimeters to micrometers (1 cm $= 10{,}000$ μm). Then the radius of the artery is 80 μm. The velocity at the central axis is 11,850 μm/s, which decreases to 11,110 μm/s at a distance of $r = 20$ μm. The fact that $dv/dr = -74$ (μm/s)/μm means that, when $r = 20$ μm, the velocity is decreasing at a rate of about 74 μm/s for each micrometer that we proceed away from the center.   ■

■ Economics

EXAMPLE 8 Suppose $C(x)$ is the total cost that a company incurs in producing x units of a certain commodity. The function C is called a **cost function**. If the number of items produced is increased from x_1 to x_2, then the additional cost is $\Delta C = C(x_2) - C(x_1)$, and the average rate of change of the cost is

$$\frac{\Delta C}{\Delta x} = \frac{C(x_2) - C(x_1)}{x_2 - x_1} = \frac{C(x_1 + \Delta x) - C(x_1)}{\Delta x}$$

The limit of this quantity as $\Delta x \to 0$, that is, the instantaneous rate of change of cost with respect to the number of items produced, is called the **marginal cost** by economists:

$$\text{marginal cost} = \lim_{\Delta x \to 0} \frac{\Delta C}{\Delta x} = \frac{dC}{dx}$$

[Since x often takes on only integer values, it may not make literal sense to let Δx approach 0, but we can always replace $C(x)$ by a smooth approximating function as in Example 6.]

Taking $\Delta x = 1$ and n large (so that Δx is small compared to n), we have

$$C'(n) \approx C(n + 1) - C(n)$$

Thus the marginal cost of producing n units is approximately equal to the cost of producing one more unit [the $(n + 1)$st unit].

It is often appropriate to represent a total cost function by a polynomial

$$C(x) = a + bx + cx^2 + dx^3$$

where a represents the overhead cost (rent, heat, maintenance) and the other terms represent the cost of raw materials, labor, and so on. (The cost of raw materials may be proportional to x, but labor costs might depend partly on higher powers of x because of overtime costs and inefficiencies involved in large-scale operations.)

For instance, suppose a company has estimated that the cost (in dollars) of producing x items is

$$C(x) = 10{,}000 + 5x + 0.01x^2$$

Then the marginal cost function is

$$C'(x) = 5 + 0.02x$$

The marginal cost at the production level of 500 items is

$$C'(500) = 5 + 0.02(500) = \$15/\text{item}$$

This gives the rate at which costs are increasing with respect to the production level when $x = 500$ and predicts the cost of the 501st item.

The actual cost of producing the 501st item is

$$C(501) - C(500) = [10{,}000 + 5(501) + 0.01(501)^2]$$
$$- [10{,}000 + 5(500) + 0.01(500)^2]$$
$$= \$15.01$$

Notice that $C'(500) \approx C(501) - C(500)$. ∎

Economists also study marginal demand, marginal revenue, and marginal profit, which are the derivatives of the demand, revenue, and profit functions. These will be considered in Chapter 4 after we have developed techniques for finding the maximum and minimum values of functions.

Other Sciences

Rates of change occur in all the sciences. A geologist is interested in knowing the rate at which an intruded body of molten rock cools by conduction of heat into surrounding rocks. An engineer wants to know the rate at which water flows into or out of a reservoir. An urban geographer is interested in the rate of change of the population density in a city as the distance from the city center increases. A meteorologist is concerned with the rate of change of atmospheric pressure with respect to height (see Exercise 3.8.19).

In psychology, those interested in learning theory study the so-called learning curve, which graphs the performance $P(t)$ of someone learning a skill as a function of the training time t. Of particular interest is the rate at which performance improves as time passes, that is, dP/dt.

In sociology, differential calculus is used in analyzing the spread of rumors (or innovations or fads or fashions). If $p(t)$ denotes the proportion of a population that knows a rumor by time t, then the derivative dp/dt represents the rate of spread of the rumor (see Exercise 3.4.84).

A Single Idea, Many Interpretations

Velocity, density, current, power, and temperature gradient in physics; rate of reaction and compressibility in chemistry; rate of growth and blood velocity gradient in biology;

marginal cost and marginal profit in economics; rate of heat flow in geology; rate of improvement of performance in psychology; rate of spread of a rumor in sociology—these are all special cases of a single mathematical concept, the derivative.

This is an illustration of the fact that part of the power of mathematics lies in its abstractness. A single abstract mathematical concept (such as the derivative) can have different interpretations in each of the sciences. When we develop the properties of the mathematical concept once and for all, we can then turn around and apply these results to all of the sciences. This is much more efficient than developing properties of special concepts in each separate science. The French mathematician Joseph Fourier (1768–1830) put it succinctly: "Mathematics compares the most diverse phenomena and discovers the secret analogies that unite them."

3.7   EXERCISES

1–4 A particle moves according to a law of motion $s = f(t)$, $t \geqslant 0$, where t is measured in seconds and s in feet.
(a) Find the velocity at time t.
(b) What is the velocity after 1 second?
(c) When is the particle at rest?
(d) When is the particle moving in the positive direction?
(e) Find the total distance traveled during the first 6 seconds.
(f) Draw a diagram like Figure 2 to illustrate the motion of the particle.
(g) Find the acceleration at time t and after 1 second.
(h) Graph the position, velocity, and acceleration functions for $0 \leqslant t \leqslant 6$.
(i) When is the particle speeding up? When is it slowing down?

1. $f(t) = t^3 - 8t^2 + 24t$ **2.** $f(t) = \dfrac{9t}{t^2 + 9}$

3. $f(t) = \sin(\pi t/2)$ **4.** $f(t) = t^2 e^{-t}$

5. Graphs of the *velocity* functions of two particles are shown, where t is measured in seconds. When is each particle speeding up? When is it slowing down? Explain.

(a)   (b)

6. Graphs of the *position* functions of two particles are shown, where t is measured in seconds. When is each particle speeding up? When is it slowing down? Explain.

(a)   (b)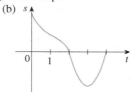

7. The height (in meters) of a projectile shot vertically upward from a point 2 m above ground level with an initial velocity of 24.5 m/s is $h = 2 + 24.5t - 4.9t^2$ after t seconds.
(a) Find the velocity after 2 s and after 4 s.
(b) When does the projectile reach its maximum height?
(c) What is the maximum height?
(d) When does it hit the ground?
(e) With what velocity does it hit the ground?

8. If a ball is thrown vertically upward with a velocity of 80 ft/s, then its height after t seconds is $s = 80t - 16t^2$.
(a) What is the maximum height reached by the ball?
(b) What is the velocity of the ball when it is 96 ft above the ground on its way up? On its way down?

9. If a rock is thrown vertically upward from the surface of Mars with velocity 15 m/s, its height after t seconds is $h = 15t - 1.86t^2$.
(a) What is the velocity of the rock after 2 s?
(b) What is the velocity of the rock when its height is 25 m on its way up? On its way down?

10. A particle moves with position function

$$s = t^4 - 4t^3 - 20t^2 + 20t \qquad t \geqslant 0$$

(a) At what time does the particle have a velocity of 20 m/s?
(b) At what time is the acceleration 0? What is the significance of this value of t?

11. (a) A company makes computer chips from square wafers of silicon. It wants to keep the side length of a wafer very close to 15 mm and it wants to know how the area $A(x)$ of a wafer changes when the side length x changes. Find $A'(15)$ and explain its meaning in this situation.
(b) Show that the rate of change of the area of a square with respect to its side length is half its perimeter. Try to explain geometrically why this is true by drawing a square whose side length x is increased by an amount Δx. How can you approximate the resulting change in area ΔA if Δx is small?

12. (a) Sodium chlorate crystals are easy to grow in the shape of cubes by allowing a solution of water and sodium chlorate to evaporate slowly. If V is the volume of such a cube with side length x, calculate dV/dx when $x = 3$ mm and explain its meaning.

(b) Show that the rate of change of the volume of a cube with respect to its edge length is equal to half the surface area of the cube. Explain geometrically why this result is true by arguing by analogy with Exercise 11(b).

13. (a) Find the average rate of change of the area of a circle with respect to its radius r as r changes from
(i) 2 to 3   (ii) 2 to 2.5   (iii) 2 to 2.1

(b) Find the instantaneous rate of change when $r = 2$.

(c) Show that the rate of change of the area of a circle with respect to its radius (at any r) is equal to the circumference of the circle. Try to explain geometrically why this is true by drawing a circle whose radius is increased by an amount Δr. How can you approximate the resulting change in area ΔA if Δr is small?

14. A stone is dropped into a lake, creating a circular ripple that travels outward at a speed of 60 cm/s. Find the rate at which the area within the circle is increasing after (a) 1 s, (b) 3 s, and (c) 5 s. What can you conclude?

15. A spherical balloon is being inflated. Find the rate of increase of the surface area ($S = 4\pi r^2$) with respect to the radius r when r is (a) 1 ft, (b) 2 ft, and (c) 3 ft. What conclusion can you make?

16. (a) The volume of a growing spherical cell is $V = \frac{4}{3}\pi r^3$, where the radius r is measured in micrometers (1 μm $= 10^{-6}$ m). Find the average rate of change of V with respect to r when r changes from
(i) 5 to 8 μm   (ii) 5 to 6 μm   (iii) 5 to 5.1 μm

(b) Find the instantaneous rate of change of V with respect to r when $r = 5$ μm.

(c) Show that the rate of change of the volume of a sphere with respect to its radius is equal to its surface area. Explain geometrically why this result is true. Argue by analogy with Exercise 13(c).

17. The mass of the part of a metal rod that lies between its left end and a point x meters to the right is $3x^2$ kg. Find the linear density (see Example 2) when x is (a) 1 m, (b) 2 m, and (c) 3 m. Where is the density the highest? The lowest?

18. If a tank holds 5000 gallons of water, which drains from the bottom of the tank in 40 minutes, then Torricelli's Law gives the volume V of water remaining in the tank after t minutes as

$$V = 5000\left(1 - \tfrac{1}{40}t\right)^2 \qquad 0 \leqslant t \leqslant 40$$

Find the rate at which water is draining from the tank after (a) 5 min, (b) 10 min, (c) 20 min, and (d) 40 min. At what time is the water flowing out the fastest? The slowest? Summarize your findings.

19. The quantity of charge Q in coulombs (C) that has passed through a point in a wire up to time t (measured in seconds) is given by $Q(t) = t^3 - 2t^2 + 6t + 2$. Find the current when (a) $t = 0.5$ s and (b) $t = 1$ s. [See Example 3. The unit of current is an ampere (1 A $= 1$ C/s).] At what time is the current lowest?

20. Newton's Law of Gravitation says that the magnitude F of the force exerted by a body of mass m on a body of mass M is

$$F = \frac{GmM}{r^2}$$

where G is the gravitational constant and r is the distance between the bodies.

(a) Find dF/dr and explain its meaning. What does the minus sign indicate?

(b) Suppose it is known that the earth attracts an object with a force that decreases at the rate of 2 N/km when $r = 20{,}000$ km. How fast does this force change when $r = 10{,}000$ km?

21. The force F acting on a body with mass m and velocity v is the rate of change of momentum: $F = (d/dt)(mv)$. If m is constant, this becomes $F = ma$, where $a = dv/dt$ is the acceleration. But in the theory of relativity the mass of a particle varies with v as follows: $m = m_0/\sqrt{1 - v^2/c^2}$, where m_0 is the mass of the particle at rest and c is the speed of light. Show that

$$F = \frac{m_0 a}{(1 - v^2/c^2)^{3/2}}$$

22. Some of the highest tides in the world occur in the Bay of Fundy on the Atlantic Coast of Canada. At Hopewell Cape the water depth at low tide is about 2.0 m and at high tide it is about 12.0 m. The natural period of oscillation is a little more than 12 hours and on June 30, 2009, high tide occurred at 6:45 AM. This helps explain the following model for the water depth D (in meters) as a function of the time t (in hours after midnight) on that day:

$$D(t) = 7 + 5\cos[0.503(t - 6.75)]$$

How fast was the tide rising (or falling) at the following times?
(a) 3:00 AM   (b) 6:00 AM
(c) 9:00 AM   (d) Noon

23. Boyle's Law states that when a sample of gas is compressed at a constant temperature, the product of the pressure and the volume remains constant: $PV = C$.

(a) Find the rate of change of volume with respect to pressure.

(b) A sample of gas is in a container at low pressure and is steadily compressed at constant temperature for 10 minutes. Is the volume decreasing more rapidly at the beginning or the end of the 10 minutes? Explain.

(c) Prove that the isothermal compressibility (see Example 5) is given by $\beta = 1/P$.

24. If, in Example 4, one molecule of the product C is formed from one molecule of the reactant A and one molecule of the reactant B, and the initial concentrations of A and B have a common value $[A] = [B] = a$ moles/L, then

$$[C] = a^2kt/(akt + 1)$$

where k is a constant.
(a) Find the rate of reaction at time t.
(b) Show that if $x = [C]$, then

$$\frac{dx}{dt} = k(a - x)^2$$

(c) What happens to the concentration as $t \to \infty$?
(d) What happens to the rate of reaction as $t \to \infty$?
(e) What do the results of parts (c) and (d) mean in practical terms?

25. In Example 6 we considered a bacteria population that doubles every hour. Suppose that another population of bacteria triples every hour and starts with 400 bacteria. Find an expression for the number n of bacteria after t hours and use it to estimate the rate of growth of the bacteria population after 2.5 hours.

26. The number of yeast cells in a laboratory culture increases rapidly initially but levels off eventually. The population is modeled by the function

$$n = f(t) = \frac{a}{1 + be^{-0.7t}}$$

where t is measured in hours. At time $t = 0$ the population is 20 cells and is increasing at a rate of 12 cells/hour. Find the values of a and b. According to this model, what happens to the yeast population in the long run?

27. The table gives the population of the world $P(t)$, in millions, where t is measured in years and $t = 0$ corresponds to the year 1900.

t	Population (millions)	t	Population (millions)
0	1650	60	3040
10	1750	70	3710
20	1860	80	4450
30	2070	90	5280
40	2300	100	6080
50	2560	110	6870

(a) Estimate the rate of population growth in 1920 and in 1980 by averaging the slopes of two secant lines.
(b) Use a graphing device to find a cubic function (a third-degree polynomial) that models the data.
(c) Use your model in part (b) to find a model for the rate of population growth.
(d) Use part (c) to estimate the rates of growth in 1920 and 1980. Compare with your estimates in part (a).

(e) In Section 1.1 we modeled $P(t)$ with the exponential function

$$f(t) = (1.43653 \times 10^9) \cdot (1.01395)^t$$

Use this model to find a model for the rate of population growth.
(f) Use your model in part (e) to estimate the rate of growth in 1920 and 1980. Compare with your estimates in parts (a) and (d).
(g) Estimate the rate of growth in 1985.

28. The table shows how the average age of first marriage of Japanese women has varied since 1950.

t	$A(t)$	t	$A(t)$
1950	23.0	1985	25.5
1955	23.8	1990	25.9
1960	24.4	1995	26.3
1965	24.5	2000	27.0
1970	24.2	2005	28.0
1975	24.7	2010	28.8
1980	25.2		

(a) Use a graphing calculator or computer to model these data with a fourth-degree polynomial.
(b) Use part (a) to find a model for $A'(t)$.
(c) Estimate the rate of change of marriage age for women in 1990.
(d) Graph the data points and the models for A and A'.

29. Refer to the law of laminar flow given in Example 7. Consider a blood vessel with radius 0.01 cm, length 3 cm, pressure difference 3000 dynes/cm^2, and viscosity $\eta = 0.027$.
(a) Find the velocity of the blood along the center-line $r = 0$, at radius $r = 0.005$ cm, and at the wall $r = R = 0.01$ cm.
(b) Find the velocity gradient at $r = 0$, $r = 0.005$, and $r = 0.01$.
(c) Where is the velocity the greatest? Where is the velocity changing most?

30. The frequency of vibrations of a vibrating violin string is given by

$$f = \frac{1}{2L}\sqrt{\frac{T}{\rho}}$$

where L is the length of the string, T is its tension, and ρ is its linear density. [See Chapter 11 in D. E. Hall, *Musical Acoustics*, 3rd ed. (Pacific Grove, CA: Brooks/Cole, 2002).]
(a) Find the rate of change of the frequency with respect to
   (i) the length (when T and ρ are constant),
   (ii) the tension (when L and ρ are constant), and
   (iii) the linear density (when L and T are constant).
(b) The pitch of a note (how high or low the note sounds) is determined by the frequency f. (The higher the frequency, the higher the pitch.) Use the signs of the

derivatives in part (a) to determine what happens to the pitch of a note
   (i) when the effective length of a string is decreased by placing a finger on the string so a shorter portion of the string vibrates,
   (ii) when the tension is increased by turning a tuning peg,
   (iii) when the linear density is increased by switching to another string.

31. Suppose that the cost (in dollars) for a company to produce x pairs of a new line of jeans is

$$C(x) = 2000 + 3x + 0.01x^2 + 0.0002x^3$$

   (a) Find the marginal cost function.
   (b) Find $C'(100)$ and explain its meaning. What does it predict?
   (c) Compare $C'(100)$ with the cost of manufacturing the 101st pair of jeans.

32. The cost function for a certain commodity is

$$C(q) = 84 + 0.16q - 0.0006q^2 + 0.000003q^3$$

   (a) Find and interpret $C'(100)$.
   (b) Compare $C'(100)$ with the cost of producing the 101st item.

33. If $p(x)$ is the total value of the production when there are x workers in a plant, then the *average productivity* of the workforce at the plant is

$$A(x) = \frac{p(x)}{x}$$

   (a) Find $A'(x)$. Why does the company want to hire more workers if $A'(x) > 0$?
   (b) Show that $A'(x) > 0$ if $p'(x)$ is greater than the average productivity.

34. If R denotes the reaction of the body to some stimulus of strength x, the *sensitivity* S is defined to be the rate of change of the reaction with respect to x. A particular example is that when the brightness x of a light source is increased, the eye reacts by decreasing the area R of the pupil. The experimental formula

$$R = \frac{40 + 24x^{0.4}}{1 + 4x^{0.4}}$$

has been used to model the dependence of R on x when R is measured in square millimeters and x is measured in appropriate units of brightness.
   (a) Find the sensitivity.
   (b) Illustrate part (a) by graphing both R and S as functions of x. Comment on the values of R and S at low levels of brightness. Is this what you would expect?

35. Patients undergo dialysis treatment to remove urea from their blood when their kidneys are not functioning properly. Blood is diverted from the patient through a machine that filters out urea. Under certain conditions, the duration of dialysis required, given that the initial urea concentration is $c > 1$, is

given by the equation

$$t = \ln\left(\frac{3c + \sqrt{9c^2 - 8c}}{2} \right)$$

Calculate the derivative of t with respect to c and interpret it.

36. Invasive species often display a wave of advance as they colonize new areas. Mathematical models based on random dispersal and reproduction have demonstrated that the speed with which such waves move is given by the function $f(r) = 2\sqrt{Dr}$, where r is the reproductive rate of individuals and D is a parameter quantifying dispersal. Calculate the derivative of the wave speed with respect to the reproductive rate r and explain its meaning.

37. The gas law for an ideal gas at absolute temperature T (in kelvins), pressure P (in atmospheres), and volume V (in liters) is $PV = nRT$, where n is the number of moles of the gas and $R = 0.0821$ is the gas constant. Suppose that, at a certain instant, $P = 8.0$ atm and is increasing at a rate of 0.10 atm/min and $V = 10$ L and is decreasing at a rate of 0.15 L/min. Find the rate of change of T with respect to time at that instant if $n = 10$ mol.

38. In a fish farm, a population of fish is introduced into a pond and harvested regularly. A model for the rate of change of the fish population is given by the equation

$$\frac{dP}{dt} = r_0\left(1 - \frac{P(t)}{P_c} \right)P(t) - \beta P(t)$$

where r_0 is the birth rate of the fish, P_c is the maximum population that the pond can sustain (called the *carrying capacity*), and β is the percentage of the population that is harvested.
   (a) What value of dP/dt corresponds to a stable population?
   (b) If the pond can sustain 10,000 fish, the birth rate is 5%, and the harvesting rate is 4%, find the stable population level.
   (c) What happens if β is raised to 5%?

39. In the study of ecosystems, *predator-prey models* are often used to study the interaction between species. Consider populations of tundra wolves, given by $W(t)$, and caribou, given by $C(t)$, in northern Canada. The interaction has been modeled by the equations

$$\frac{dC}{dt} = aC - bCW \qquad \frac{dW}{dt} = -cW + dCW$$

   (a) What values of dC/dt and dW/dt correspond to stable populations?
   (b) How would the statement "The caribou go extinct" be represented mathematically?
   (c) Suppose that $a = 0.05$, $b = 0.001$, $c = 0.05$, and $d = 0.0001$. Find all population pairs (C, W) that lead to stable populations. According to this model, is it possible for the two species to live in balance or will one or both species become extinct?

3.8   Exponential Growth and Decay

In many natural phenomena, quantities grow or decay at a rate proportional to their size. For instance, if $y = f(t)$ is the number of individuals in a population of animals or bacteria at time t, then it seems reasonable to expect that the rate of growth $f'(t)$ is proportional to the population $f(t)$; that is, $f'(t) = kf(t)$ for some constant k. Indeed, under ideal conditions (unlimited environment, adequate nutrition, immunity to disease) the mathematical model given by the equation $f'(t) = kf(t)$ predicts what actually happens fairly accurately. Another example occurs in nuclear physics where the mass of a radioactive substance decays at a rate proportional to the mass. In chemistry, the rate of a unimolecular first-order reaction is proportional to the concentration of the substance. In finance, the value of a savings account with continuously compounded interest increases at a rate proportional to that value.

In general, if $y(t)$ is the value of a quantity y at time t and if the rate of change of y with respect to t is proportional to its size $y(t)$ at any time, then

$$\boxed{1}\qquad \frac{dy}{dt} = ky$$

where k is a constant. Equation 1 is sometimes called the **law of natural growth** (if $k > 0$) or the **law of natural decay** (if $k < 0$). It is called a **differential equation** because it involves an unknown function y and its derivative dy/dt.

It's not hard to think of a solution of Equation 1. This equation asks us to find a function whose derivative is a constant multiple of itself. We have met such functions in this chapter. Any exponential function of the form $y(t) = Ce^{kt}$, where C is a constant, satisfies

$$y'(t) = C(ke^{kt}) = k(Ce^{kt}) = ky(t)$$

We will see in Section 9.4 that *any* function that satisfies $dy/dt = ky$ must be of the form $y = Ce^{kt}$. To see the significance of the constant C, we observe that

$$y(0) = Ce^{k \cdot 0} = C$$

Therefore C is the initial value of the function.

> **2**   **Theorem**   The only solutions of the differential equation $dy/dt = ky$ are the exponential functions
>
> $$y(t) = y(0)e^{kt}$$

■ Population Growth

What is the significance of the proportionality constant k? In the context of population growth, where $P(t)$ is the size of a population at time t, we can write

$$\boxed{3}\qquad \frac{dP}{dt} = kP \qquad \text{or} \qquad \frac{1}{P}\frac{dP}{dt} = k$$

The quantity

$$\frac{1}{P}\frac{dP}{dt}$$

is the growth rate divided by the population size; it is called the **relative growth rate**.

According to (3), instead of saying "the growth rate is proportional to population size" we could say "the relative growth rate is constant." Then (2) says that a population with constant relative growth rate must grow exponentially. Notice that the relative growth rate k appears as the coefficient of t in the exponential function Ce^{kt}. For instance, if

$$\frac{dP}{dt} = 0.02P$$

and t is measured in years, then the relative growth rate is $k = 0.02$ and the population grows at a relative rate of 2% per year. If the population at time 0 is P_0, then the expression for the population is

$$P(t) = P_0 e^{0.02t}$$

EXAMPLE 1 Use the fact that the world population was 2560 million in 1950 and 3040 million in 1960 to model the population of the world in the second half of the 20th century. (Assume that the growth rate is proportional to the population size.) What is the relative growth rate? Use the model to estimate the world population in 1993 and to predict the population in the year 2020.

SOLUTION We measure the time t in years and let $t = 0$ in the year 1950. We measure the population $P(t)$ in millions of people. Then $P(0) = 2560$ and $P(10) = 3040$. Since we are assuming that $dP/dt = kP$, Theorem 2 gives

$$P(t) = P(0)e^{kt} = 2560e^{kt}$$

$$P(10) = 2560e^{10k} = 3040$$

$$k = \frac{1}{10}\ln\frac{3040}{2560} \approx 0.017185$$

The relative growth rate is about 1.7% per year and the model is

$$P(t) = 2560e^{0.017185t}$$

We estimate that the world population in 1993 was

$$P(43) = 2560e^{0.017185(43)} \approx 5360 \text{ million}$$

The model predicts that the population in 2020 will be

$$P(70) = 2560e^{0.017185(70)} \approx 8524 \text{ million}$$

The graph in Figure 1 shows that the model is fairly accurate to the end of the 20th century (the dots represent the actual population), so the estimate for 1993 is quite reliable. But the prediction for 2020 is riskier.

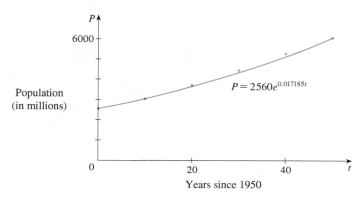

Population
(in millions)

$P = 2560e^{0.017185t}$

Years since 1950

FIGURE 1
A model for world population growth
in the second half of the 20th century

Radioactive Decay

Radioactive substances decay by spontaneously emitting radiation. If $m(t)$ is the mass remaining from an initial mass m_0 of the substance after time t, then the relative decay rate

$$-\frac{1}{m}\frac{dm}{dt}$$

has been found experimentally to be constant. (Since dm/dt is negative, the relative decay rate is positive.) It follows that

$$\frac{dm}{dt} = km$$

where k is a negative constant. In other words, radioactive substances decay at a rate proportional to the remaining mass. This means that we can use (2) to show that the mass decays exponentially:

$$m(t) = m_0 e^{kt}$$

Physicists express the rate of decay in terms of **half-life**, the time required for half of any given quantity to decay.

EXAMPLE 2 The half-life of radium-226 is 1590 years.
(a) A sample of radium-226 has a mass of 100 mg. Find a formula for the mass of the sample that remains after t years.
(b) Find the mass after 1000 years correct to the nearest milligram.
(c) When will the mass be reduced to 30 mg?

SOLUTION
(a) Let $m(t)$ be the mass of radium-226 (in milligrams) that remains after t years. Then $dm/dt = km$ and $m(0) = 100$, so (2) gives

$$m(t) = m(0)e^{kt} = 100e^{kt}$$

In order to determine the value of k, we use the fact that $m(1590) = \frac{1}{2}(100)$. Thus

$$100e^{1590k} = 50 \qquad \text{so} \qquad e^{1590k} = \frac{1}{2}$$

and

$$1590k = \ln \frac{1}{2} = -\ln 2$$

$$k = -\frac{\ln 2}{1590}$$

Therefore

$$m(t) = 100e^{-(\ln 2)t/1590}$$

We could use the fact that $e^{\ln 2} = 2$ to write the expression for $m(t)$ in the alternative form

$$m(t) = 100 \times 2^{-t/1590}$$

(b) The mass after 1000 years is

$$m(1000) = 100e^{-(\ln 2)1000/1590} \approx 65 \text{ mg}$$

(c) We want to find the value of t such that $m(t) = 30$, that is,

$$100e^{-(\ln 2)t/1590} = 30 \qquad \text{or} \qquad e^{-(\ln 2)t/1590} = 0.3$$

FIGURE 2

We solve this equation for t by taking the natural logarithm of both sides:

$$-\frac{\ln 2}{1590} t = \ln 0.3$$

Thus

$$t = -1590 \, \frac{\ln 0.3}{\ln 2} \approx 2762 \text{ years} \quad \blacksquare$$

As a check on our work in Example 2, we use a graphing device to draw the graph of $m(t)$ in Figure 2 together with the horizontal line $m = 30$. These curves intersect when $t \approx 2800$, and this agrees with the answer to part (c).

■ Newton's Law of Cooling

Newton's Law of Cooling states that the rate of cooling of an object is proportional to the temperature difference between the object and its surroundings, provided that this difference is not too large. (This law also applies to warming.) If we let $T(t)$ be the temperature of the object at time t and T_s be the temperature of the surroundings, then we can formulate Newton's Law of Cooling as a differential equation:

$$\frac{dT}{dt} = k(T - T_s)$$

where k is a constant. This equation is not quite the same as Equation 1, so we make the change of variable $y(t) = T(t) - T_s$. Because T_s is constant, we have $y'(t) = T'(t)$ and so the equation becomes

$$\frac{dy}{dt} = ky$$

We can then use (2) to find an expression for y, from which we can find T.

EXAMPLE 3 A bottle of soda pop at room temperature (72°F) is placed in a refrigerator where the temperature is 44°F. After half an hour the soda pop has cooled to 61°F.
(a) What is the temperature of the soda pop after another half hour?
(b) How long does it take for the soda pop to cool to 50°F?

SOLUTION
(a) Let $T(t)$ be the temperature of the soda after t minutes. The surrounding temperature is $T_s = 44°F$, so Newton's Law of Cooling states that

$$\frac{dT}{dt} = k(T - 44)$$

If we let $y = T - 44$, then $y(0) = T(0) - 44 = 72 - 44 = 28$, so y satisfies

$$\frac{dy}{dt} = ky \qquad y(0) = 28$$

and by (2) we have

$$y(t) = y(0)e^{kt} = 28e^{kt}$$

We are given that $T(30) = 61$, so $y(30) = 61 - 44 = 17$ and

$$28e^{30k} = 17 \qquad e^{30k} = \tfrac{17}{28}$$

Taking logarithms, we have

$$k = \frac{\ln\left(\frac{17}{28}\right)}{30} \approx -0.01663$$

Thus

$$y(t) = 28e^{-0.01663t}$$

$$T(t) = 44 + 28e^{-0.01663t}$$

$$T(60) = 44 + 28e^{-0.01663(60)} \approx 54.3$$

So after another half hour the pop has cooled to about 54°F.

(b) We have $T(t) = 50$ when

$$44 + 28e^{-0.01663t} = 50$$

$$e^{-0.01663t} = \tfrac{6}{28}$$

$$t = \frac{\ln\left(\frac{6}{28}\right)}{-0.01663} \approx 92.6$$

The pop cools to 50°F after about 1 hour 33 minutes.

Notice that in Example 3, we have

$$\lim_{t\to\infty} T(t) = \lim_{t\to\infty}\left(44 + 28e^{-0.01663t}\right) = 44 + 28 \cdot 0 = 44$$

FIGURE 3

which is to be expected. The graph of the temperature function is shown in Figure 3.

Continuously Compounded Interest

EXAMPLE 4 If $1000 is invested at 6% interest, compounded annually, then after 1 year the investment is worth $1000(1.06) = $1060, after 2 years it's worth $[1000(1.06)]1.06 = $1123.60, and after t years it's worth $1000(1.06)^t$. In general, if an amount A_0 is invested at an interest rate r ($r = 0.06$ in this example), then after t years it's worth $A_0(1 + r)^t$. Usually, however, interest is compounded more frequently, say, n times a year. Then in each compounding period the interest rate is r/n and there are nt compounding periods in t years, so the value of the investment is

$$A_0\left(1 + \frac{r}{n}\right)^{nt}$$

For instance, after 3 years at 6% interest a $1000 investment will be worth

$$\$1000(1.06)^3 = \$1191.02 \quad \text{with annual compounding}$$

$$\$1000(1.03)^6 = \$1194.05 \quad \text{with semiannual compounding}$$

$$\$1000(1.015)^{12} = \$1195.62 \quad \text{with quarterly compounding}$$

$$\$1000(1.005)^{36} = \$1196.68 \quad \text{with monthly compounding}$$

$$\$1000\left(1 + \frac{0.06}{365}\right)^{365 \cdot 3} = \$1197.20 \quad \text{with daily compounding}$$

You can see that the interest paid increases as the number of compounding periods (n) increases. If we let $n \to \infty$, then we will be compounding the interest **continuously** and the value of the investment will be

$$A(t) = \lim_{n \to \infty} A_0 \left(1 + \frac{r}{n}\right)^{nt}$$

$$= \lim_{n \to \infty} A_0 \left[\left(1 + \frac{r}{n}\right)^{n/r}\right]^{rt}$$

$$= A_0 \left[\lim_{n \to \infty} \left(1 + \frac{r}{n}\right)^{n/r}\right]^{rt}$$

$$= A_0 \left[\lim_{m \to \infty} \left(1 + \frac{1}{m}\right)^{m}\right]^{rt} \qquad \text{(where } m = n/r)$$

But the limit in this expression is equal to the number e (see Equation 3.6.6). So with continuous compounding of interest at interest rate r, the amount after t years is

$$A(t) = A_0 e^{rt}$$

If we differentiate this equation, we get

$$\frac{dA}{dt} = rA_0 e^{rt} = rA(t)$$

which says that, with continuous compounding of interest, the rate of increase of an investment is proportional to its size.

Returning to the example of $1000 invested for 3 years at 6% interest, we see that with continuous compounding of interest the value of the investment will be

$$A(3) = \$1000 e^{(0.06)3} = \$1197.22$$

Notice how close this is to the amount we calculated for daily compounding, $1197.20. But the amount is easier to compute if we use continuous compounding. ■

3.8   EXERCISES

1. A population of protozoa develops with a constant relative growth rate of 0.7944 per member per day. On day zero the population consists of two members. Find the population size after six days.

2. A common inhabitant of human intestines is the bacterium *Escherichia coli*, named after the German pediatrician Theodor Escherich, who identified it in 1885. A cell of this bacterium in a nutrient-broth medium divides into two cells every 20 minutes. The initial population of a culture is 50 cells.
(a) Find the relative growth rate.
(b) Find an expression for the number of cells after t hours.

(c) Find the number of cells after 6 hours.
(d) Find the rate of growth after 6 hours.
(e) When will the population reach a million cells?

3. A bacteria culture initially contains 100 cells and grows at a rate proportional to its size. After an hour the population has increased to 420.
(a) Find an expression for the number of bacteria after t hours.
(b) Find the number of bacteria after 3 hours.
(c) Find the rate of growth after 3 hours.
(d) When will the population reach 10,000?

4. A bacteria culture grows with constant relative growth rate. The bacteria count was 400 after 2 hours and 25,600 after 6 hours.
  (a) What is the relative growth rate? Express your answer as a percentage.
  (b) What was the intitial size of the culture?
  (c) Find an expression for the number of bacteria after t hours.
  (d) Find the number of cells after 4.5 hours.
  (e) Find the rate of growth after 4.5 hours.
  (f) When will the population reach 50,000?

5. The table gives estimates of the world population, in millions, from 1750 to 2000.

Year	Population	Year	Population
1750	790	1900	1650
1800	980	1950	2560
1850	1260	2000	6080

  (a) Use the exponential model and the population figures for 1750 and 1800 to predict the world population in 1900 and 1950. Compare with the actual figures.
  (b) Use the exponential model and the population figures for 1850 and 1900 to predict the world population in 1950. Compare with the actual population.
  (c) Use the exponential model and the population figures for 1900 and 1950 to predict the world population in 2000. Compare with the actual population and try to explain the discrepancy.

6. The table gives the population of Indonesia, in millions, for the second half of the 20th century.

Year	Population
1950	83
1960	100
1970	122
1980	150
1990	182
2000	214

  (a) Assuming the population grows at a rate proportional to its size, use the census figures for 1950 and 1960 to predict the population in 1980. Compare with the actual figure.
  (b) Use the census figures for 1960 and 1980 to predict the population in 2000. Compare with the actual population.
  (c) Use the census figures for 1980 and 2000 to predict the population in 2010 and compare with the actual population of 243 million.
  (d) Use the model in part (c) to predict the population in 2020. Do you think the prediction will be too high or too low? Why?

7. Experiments show that if the chemical reaction

$$N_2O_5 \rightarrow 2NO_2 + \tfrac{1}{2}O_2$$

takes place at 45°C, the rate of reaction of dinitrogen pentoxide is proportional to its concentration as follows:

$$-\frac{d[N_2O_5]}{dt} = 0.0005[N_2O_5]$$

(See Example 3.7.4.)
  (a) Find an expression for the concentration $[N_2O_5]$ after t seconds if the initial concentration is C.
  (b) How long will the reaction take to reduce the concentration of N_2O_5 to 90% of its original value?

8. Strontium-90 has a half-life of 28 days.
  (a) A sample has a mass of 50 mg initially. Find a formula for the mass remaining after t days.
  (b) Find the mass remaining after 40 days.
  (c) How long does it take the sample to decay to a mass of 2 mg?
  (d) Sketch the graph of the mass function.

9. The half-life of cesium-137 is 30 years. Suppose we have a 100-mg sample.
  (a) Find the mass that remains after t years.
  (b) How much of the sample remains after 100 years?
  (c) After how long will only 1 mg remain?

10. A sample of tritium-3 decayed to 94.5% of its original amount after a year.
  (a) What is the half-life of tritium-3?
  (b) How long would it take the sample to decay to 20% of its original amount?

11. Scientists can determine the age of ancient objects by the method of *radiocarbon dating*. The bombardment of the upper atmosphere by cosmic rays converts nitrogen to a radioactive isotope of carbon, ^{14}C, with a half-life of about 5730 years. Vegetation absorbs carbon dioxide through the atmosphere and animal life assimilates ^{14}C through food chains. When a plant or animal dies, it stops replacing its carbon and the amount of ^{14}C begins to decrease through radioactive decay. Therefore the level of radioactivity must also decay exponentially.
     A discovery revealed a parchment fragment that had about 74% as much ^{14}C radioactivity as does plant material on the earth today. Estimate the age of the parchment.

12. Dinosaur fossils are too old to be reliably dated using carbon-14. (See Exercise 11.) Suppose we had a 68-million-year-old dinosaur fossil. What fraction of the living dinosaur's ^{14}C would be remaining today? Suppose the minimum detectable amount is 0.1%. What is the maximum age of a fossil that we could date using ^{14}C?

13. Dinosaur fossils are often dated by using an element other than carbon, such as potassium-40, that has a longer half-life (in this case, approximately 1.25 billion years). Suppose the minimum detectable amount is 0.1% and a dinosaur is dated with ^{40}K to be 68 million years old. Is such a dating possible? In other words, what is the maximum age of a fossil that we could date using ^{40}K?

14. A curve passes through the point (0, 5) and has the property that the slope of the curve at every point P is twice the y-coordinate of P. What is the equation of the curve?

15. A roast turkey is taken from an oven when its temperature has reached 185°F and is placed on a table in a room where the temperature is 75°F.
   (a) If the temperature of the turkey is 150°F after half an hour, what is the temperature after 45 minutes?
   (b) When will the turkey have cooled to 100°F?

16. In a murder investigation, the temperature of the corpse was 32.5°C at 1:30 PM and 30.3°C an hour later. Normal body temperature is 37.0°C and the temperature of the surroundings was 20.0°C. When did the murder take place?

17. When a cold drink is taken from a refrigerator, its temperature is 5°C. After 25 minutes in a 20°C room its temperature has increased to 10°C.
   (a) What is the temperature of the drink after 50 minutes?
   (b) When will its temperature be 15°C?

18. A freshly brewed cup of coffee has temperature 95°C in a 20°C room. When its temperature is 70°C, it is cooling at a rate of 1°C per minute. When does this occur?

19. The rate of change of atmospheric pressure P with respect to altitude h is proportional to P, provided that the temperature is constant. At 15°C the pressure is 101.3 kPa at sea level and 87.14 kPa at $h = 1000$ m.
   (a) What is the pressure at an altitude of 3000 m?
   (b) What is the pressure at the top of Mount McKinley, at an altitude of 6187 m?

20. (a) If $1000 is borrowed at 8% interest, find the amounts due at the end of 3 years if the interest is compounded (i) annually, (ii) quarterly, (iii) monthly, (iv) weekly, (v) daily, (vi) hourly, and (vii) continuously.
   (b) Suppose $1000 is borrowed and the interest is compounded continuously. If $A(t)$ is the amount due after t years, where $0 \le t \le 3$, graph $A(t)$ for each of the interest rates 6%, 8%, and 10% on a common screen.

21. (a) If $3000 is invested at 5% interest, find the value of the investment at the end of 5 years if the interest is compounded (i) annually, (ii) semiannually, (iii) monthly, (iv) weekly, (v) daily, and (vi) continuously.
   (b) If $A(t)$ is the amount of the investment at time t for the case of continuous compounding, write a differential equation and an initial condition satisfied by $A(t)$.

22. (a) How long will it take an investment to double in value if the interest rate is 6% compounded continuously?
   (b) What is the equivalent annual interest rate?

APPLIED PROJECT **CONTROLLING RED BLOOD CELL LOSS DURING SURGERY**

A typical volume of blood in the human body is about 5 L. A certain percentage of that volume (called the *hematocrit*) consists of red blood cells (RBCs); typically the hematocrit is about 45% in males. Suppose that a surgery takes four hours and a male patient bleeds 2.5 L of blood. During surgery the patient's blood volume is maintained at 5 L by injection of saline solution, which mixes quickly with the blood but dilutes it so that the hematocrit decreases as time passes.

1. Assuming that the rate of RBC loss is proportional to the volume of RBCs, determine the patient's volume of RBCs by the end of the operation.

2. A procedure called *acute normovolemic hemodilution* (ANH) has been developed to minimize RBC loss during surgery. In this procedure blood is extracted from the patient before the operation and replaced with saline solution. This dilutes the patient's blood, resulting in fewer RBCs being lost during the bleeding. The extracted blood is then returned to the patient after surgery. Only a certain amount of blood can be extracted, however, because the RBC concentration can never be allowed to drop below 25% during surgery. What is the maximum amount of blood that can be extracted in the ANH procedure for the surgery described in this project?

3. What is the RBC loss without the ANH procedure? What is the loss if the procedure is carried out with the volume calculated in Problem 2?

3.9   Related Rates

If we are pumping air into a balloon, both the volume and the radius of the balloon are increasing and their rates of increase are related to each other. But it is much easier to measure directly the rate of increase of the volume than the rate of increase of the radius.

In a related rates problem the idea is to compute the rate of change of one quantity in terms of the rate of change of another quantity (which may be more easily measured). The procedure is to find an equation that relates the two quantities and then use the Chain Rule to differentiate both sides with respect to time.

EXAMPLE 1 Air is being pumped into a spherical balloon so that its volume increases at a rate of 100 cm³/s. How fast is the radius of the balloon increasing when the diameter is 50 cm?

SOLUTION We start by identifying two things:

the *given information:*

the rate of increase of the volume of air is 100 cm³/s

and the *unknown:*

the rate of increase of the radius when the diameter is 50 cm

PS According to the Principles of Problem Solving discussed on page 71, the first step is to understand the problem. This includes reading the problem carefully, identifying the given and the unknown, and introducing suitable notation.

In order to express these quantities mathematically, we introduce some suggestive *notation:*

Let V be the volume of the balloon and let r be its radius.

The key thing to remember is that rates of change are derivatives. In this problem, the volume and the radius are both functions of the time t. The rate of increase of the volume with respect to time is the derivative dV/dt, and the rate of increase of the radius is dr/dt. We can therefore restate the given and the unknown as follows:

$$\text{Given:} \qquad \frac{dV}{dt} = 100 \text{ cm}^3/\text{s}$$

$$\text{Unknown:} \qquad \frac{dr}{dt} \quad \text{when } r = 25 \text{ cm}$$

PS The second stage of problem solving is to think of a plan for connecting the given and the unknown.

In order to connect dV/dt and dr/dt, we first relate V and r by the formula for the volume of a sphere:

$$V = \tfrac{4}{3}\pi r^3$$

In order to use the given information, we differentiate each side of this equation with respect to t. To differentiate the right side, we need to use the Chain Rule:

$$\frac{dV}{dt} = \frac{dV}{dr}\frac{dr}{dt} = 4\pi r^2 \frac{dr}{dt}$$

Now we solve for the unknown quantity:

Notice that, although dV/dt is constant, dr/dt is *not* constant.

$$\frac{dr}{dt} = \frac{1}{4\pi r^2}\frac{dV}{dt}$$

If we put $r = 25$ and $dV/dt = 100$ in this equation, we obtain

$$\frac{dr}{dt} = \frac{1}{4\pi(25)^2} 100 = \frac{1}{25\pi}$$

The radius of the balloon is increasing at the rate of $1/(25\pi) \approx 0.0127$ cm/s. ∎

EXAMPLE 2 A ladder 10 ft long rests against a vertical wall. If the bottom of the ladder slides away from the wall at a rate of 1 ft/s, how fast is the top of the ladder sliding down the wall when the bottom of the ladder is 6 ft from the wall?

SOLUTION We first draw a diagram and label it as in Figure 1. Let x feet be the distance from the bottom of the ladder to the wall and y feet the distance from the top of the ladder to the ground. Note that x and y are both functions of t (time, measured in seconds).

We are given that $dx/dt = 1$ ft/s and we are asked to find dy/dt when $x = 6$ ft (see Figure 2). In this problem, the relationship between x and y is given by the Pythagorean Theorem:

$$x^2 + y^2 = 100$$

Differentiating each side with respect to t using the Chain Rule, we have

$$2x\frac{dx}{dt} + 2y\frac{dy}{dt} = 0$$

and solving this equation for the desired rate, we obtain

$$\frac{dy}{dt} = -\frac{x}{y}\frac{dx}{dt}$$

When $x = 6$, the Pythagorean Theorem gives $y = 8$ and so, substituting these values and $dx/dt = 1$, we have

$$\frac{dy}{dt} = -\frac{6}{8}(1) = -\frac{3}{4} \text{ ft/s}$$

The fact that dy/dt is negative means that the distance from the top of the ladder to the ground is *decreasing* at a rate of $\frac{3}{4}$ ft/s. In other words, the top of the ladder is sliding down the wall at a rate of $\frac{3}{4}$ ft/s. ∎

EXAMPLE 3 A water tank has the shape of an inverted circular cone with base radius 2 m and height 4 m. If water is being pumped into the tank at a rate of 2 m³/min, find the rate at which the water level is rising when the water is 3 m deep.

SOLUTION We first sketch the cone and label it as in Figure 3. Let V, r, and h be the volume of the water, the radius of the surface, and the height of the water at time t, where t is measured in minutes.

We are given that $dV/dt = 2$ m³/min and we are asked to find dh/dt when h is 3 m. The quantities V and h are related by the equation

$$V = \tfrac{1}{3}\pi r^2 h$$

but it is very useful to express V as a function of h alone. In order to eliminate r, we use

FIGURE 1

FIGURE 2

FIGURE 3

the similar triangles in Figure 3 to write

$$\frac{r}{h} = \frac{2}{4} \qquad r = \frac{h}{2}$$

and the expression for V becomes

$$V = \frac{1}{3}\pi\left(\frac{h}{2}\right)^2 h = \frac{\pi}{12}h^3$$

Now we can differentiate each side with respect to t:

$$\frac{dV}{dt} = \frac{\pi}{4}h^2\frac{dh}{dt}$$

so

$$\frac{dh}{dt} = \frac{4}{\pi h^2}\frac{dV}{dt}$$

Substituting $h = 3$ m and $dV/dt = 2$ m^3/min, we have

$$\frac{dh}{dt} = \frac{4}{\pi(3)^2} \cdot 2 = \frac{8}{9\pi}$$

The water level is rising at a rate of $8/(9\pi) \approx 0.28$ m/min. ∎

PS Look back: What have we learned from Examples 1–3 that will help us solve future problems?

⊘ **WARNING** A common error is to substitute the given numerical information (for quantities that vary with time) too early. This should be done only *after* the differentiation. (Step 7 follows Step 6.) For instance, in Example 3 we dealt with general values of h until we finally substituted $h = 3$ at the last stage. (If we had put $h = 3$ earlier, we would have gotten $dV/dt = 0$, which is clearly wrong.)

Problem Solving Strategy It is useful to recall some of the problem-solving principles from page 71 and adapt them to related rates in light of our experience in Examples 1–3:

1. Read the problem carefully.
2. Draw a diagram if possible.
3. Introduce notation. Assign symbols to all quantities that are functions of time.
4. Express the given information and the required rate in terms of derivatives.
5. Write an equation that relates the various quantities of the problem. If necessary, use the geometry of the situation to eliminate one of the variables by substitution (as in Example 3).
6. Use the Chain Rule to differentiate both sides of the equation with respect to t.
7. Substitute the given information into the resulting equation and solve for the unknown rate.

The following examples are further illustrations of the strategy.

EXAMPLE 4 Car A is traveling west at 50 mi/h and car B is traveling north at 60 mi/h. Both are headed for the intersection of the two roads. At what rate are the cars approaching each other when car A is 0.3 mi and car B is 0.4 mi from the intersection?

SOLUTION We draw Figure 4, where C is the intersection of the roads. At a given time t, let x be the distance from car A to C, let y be the distance from car B to C, and let z be the distance between the cars, where x, y, and z are measured in miles.

We are given that $dx/dt = -50$ mi/h and $dy/dt = -60$ mi/h. (The derivatives are negative because x and y are decreasing.) We are asked to find dz/dt. The equation that

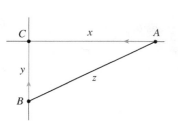

FIGURE 4

relates x, y, and z is given by the Pythagorean Theorem:

$$z^2 = x^2 + y^2$$

Differentiating each side with respect to t, we have

$$2z \frac{dz}{dt} = 2x \frac{dx}{dt} + 2y \frac{dy}{dt}$$

$$\frac{dz}{dt} = \frac{1}{z} \left(x \frac{dx}{dt} + y \frac{dy}{dt} \right)$$

When $x = 0.3$ mi and $y = 0.4$ mi, the Pythagorean Theorem gives $z = 0.5$ mi, so

$$\frac{dz}{dt} = \frac{1}{0.5} [0.3(-50) + 0.4(-60)]$$

$$= -78 \text{ mi/h}$$

The cars are approaching each other at a rate of 78 mi/h.

EXAMPLE 5 A man walks along a straight path at a speed of 4 ft/s. A searchlight is located on the ground 20 ft from the path and is kept focused on the man. At what rate is the searchlight rotating when the man is 15 ft from the point on the path closest to the searchlight?

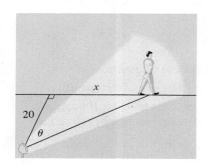

SOLUTION We draw Figure 5 and let x be the distance from the man to the point on the path closest to the searchlight. We let θ be the angle between the beam of the search-light and the perpendicular to the path.

We are given that $dx/dt = 4$ ft/s and are asked to find $d\theta/dt$ when $x = 15$. The equation that relates x and θ can be written from Figure 5:

$$\frac{x}{20} = \tan \theta \qquad x = 20 \tan \theta$$

Differentiating each side with respect to t, we get

$$\frac{dx}{dt} = 20 \sec^2\theta \, \frac{d\theta}{dt}$$

so

$$\frac{d\theta}{dt} = \frac{1}{20} \cos^2\theta \, \frac{dx}{dt}$$

$$= \frac{1}{20} \cos^2\theta \, (4) = \frac{1}{5} \cos^2\theta$$

When $x = 15$, the length of the beam is 25, so $\cos \theta = \frac{4}{5}$ and

$$\frac{d\theta}{dt} = \frac{1}{5} \left(\frac{4}{5} \right)^2 = \frac{16}{125} = 0.128$$

The searchlight is rotating at a rate of 0.128 rad/s.

FIGURE 5

3.9   EXERCISES

1. If V is the volume of a cube with edge length x and the cube expands as time passes, find dV/dt in terms of dx/dt.

2. (a) If A is the area of a circle with radius r and the circle expands as time passes, find dA/dt in terms of dr/dt.
(b) Suppose oil spills from a ruptured tanker and spreads in a circular pattern. If the radius of the oil spill increases at a constant rate of 1 m/s, how fast is the area of the spill increasing when the radius is 30 m?

3. Each side of a square is increasing at a rate of 6 cm/s. At what rate is the area of the square increasing when the area of the square is 16 cm²?

4. The length of a rectangle is increasing at a rate of 8 cm/s and its width is increasing at a rate of 3 cm/s. When the length is 20 cm and the width is 10 cm, how fast is the area of the rectangle increasing?

5. A cylindrical tank with radius 5 m is being filled with water at a rate of 3 m³/min. How fast is the height of the water increasing?

6. The radius of a sphere is increasing at a rate of 4 mm/s. How fast is the volume increasing when the diameter is 80 mm?

7. The radius of a spherical ball is increasing at a rate of 2 cm/min. At what rate is the surface area of the ball increasing when the radius is 8 cm?

8. The area of a triangle with sides of lengths a and b and contained angle θ is
$$A = \tfrac{1}{2}ab \sin \theta$$
(a) If $a = 2$ cm, $b = 3$ cm, and θ increases at a rate of 0.2 rad/min, how fast is the area increasing when $\theta = \pi/3$?
(b) If $a = 2$ cm, b increases at a rate of 1.5 cm/min, and θ increases at a rate of 0.2 rad/min, how fast is the area increasing when $b = 3$ cm and $\theta = \pi/3$?
(c) If a increases at a rate of 2.5 cm/min, b increases at a rate of 1.5 cm/min, and θ increases at a rate of 0.2 rad/min, how fast is the area increasing when $a = 2$ cm, $b = 3$ cm, and $\theta = \pi/3$?

9. Suppose $y = \sqrt{2x + 1}$, where x and y are functions of t.
(a) If $dx/dt = 3$, find dy/dt when $x = 4$.
(b) If $dy/dt = 5$, find dx/dt when $x = 12$.

10. Suppose $4x^2 + 9y^2 = 36$, where x and y are functions of t.
(a) If $dy/dt = \tfrac{1}{3}$, find dx/dt when $x = 2$ and $y = \tfrac{2}{3}\sqrt{5}$.
(b) If $dx/dt = 3$, find dy/dt when $x = -2$ and $y = \tfrac{2}{3}\sqrt{5}$.

11. If $x^2 + y^2 + z^2 = 9$, $dx/dt = 5$, and $dy/dt = 4$, find dz/dt when $(x, y, z) = (2, 2, 1)$.

12. A particle is moving along a hyperbola $xy = 8$. As it reaches the point $(4, 2)$, the y-coordinate is decreasing at a rate of

3 cm/s. How fast is the x-coordinate of the point changing at that instant?

13–16
(a) What quantities are given in the problem?
(b) What is the unknown?
(c) Draw a picture of the situation for any time t.
(d) Write an equation that relates the quantities.
(e) Finish solving the problem.

13. A plane flying horizontally at an altitude of 1 mi and a speed of 500 mi/h passes directly over a radar station. Find the rate at which the distance from the plane to the station is increasing when it is 2 mi away from the station.

14. If a snowball melts so that its surface area decreases at a rate of 1 cm²/min, find the rate at which the diameter decreases when the diameter is 10 cm.

15. A street light is mounted at the top of a 15-ft-tall pole. A man 6 ft tall walks away from the pole with a speed of 5 ft/s along a straight path. How fast is the tip of his shadow moving when he is 40 ft from the pole?

16. At noon, ship A is 150 km west of ship B. Ship A is sailing east at 35 km/h and ship B is sailing north at 25 km/h. How fast is the distance between the ships changing at 4:00 PM?

17. Two cars start moving from the same point. One travels south at 60 mi/h and the other travels west at 25 mi/h. At what rate is the distance between the cars increasing two hours later?

18. A spotlight on the ground shines on a wall 12 m away. If a man 2 m tall walks from the spotlight toward the building at a speed of 1.6 m/s, how fast is the length of his shadow on the building decreasing when he is 4 m from the building?

19. A man starts walking north at 4 ft/s from a point P. Five minutes later a woman starts walking south at 5 ft/s from a point 500 ft due east of P. At what rate are the people moving apart 15 min after the woman starts walking?

20. A baseball diamond is a square with side 90 ft. A batter hits the ball and runs toward first base with a speed of 24 ft/s.
(a) At what rate is his distance from second base decreasing when he is halfway to first base?
(b) At what rate is his distance from third base increasing at the same moment?

90 ft

21. The altitude of a triangle is increasing at a rate of 1 cm/min while the area of the triangle is increasing at a rate of 2 cm²/min. At what rate is the base of the triangle changing when the altitude is 10 cm and the area is 100 cm²?

22. A boat is pulled into a dock by a rope attached to the bow of the boat and passing through a pulley on the dock that is 1 m higher than the bow of the boat. If the rope is pulled in at a rate of 1 m/s, how fast is the boat approaching the dock when it is 8 m from the dock?

23. At noon, ship A is 100 km west of ship B. Ship A is sailing south at 35 km/h and ship B is sailing north at 25 km/h. How fast is the distance between the ships changing at 4:00 PM?

24. A particle moves along the curve $y = 2 \sin(\pi x/2)$. As the particle passes through the point $\left(\frac{1}{3}, 1\right)$, its x-coordinate increases at a rate of $\sqrt{10}$ cm/s. How fast is the distance from the particle to the origin changing at this instant?

25. Water is leaking out of an inverted conical tank at a rate of 10,000 cm³/min at the same time that water is being pumped into the tank at a constant rate. The tank has height 6 m and the diameter at the top is 4 m. If the water level is rising at a rate of 20 cm/min when the height of the water is 2 m, find the rate at which water is being pumped into the tank.

26. A trough is 10 ft long and its ends have the shape of isosceles triangles that are 3 ft across at the top and have a height of 1 ft. If the trough is being filled with water at a rate of 12 ft³/min, how fast is the water level rising when the water is 6 inches deep?

27. A water trough is 10 m long and a cross-section has the shape of an isosceles trapezoid that is 30 cm wide at the bottom, 80 cm wide at the top, and has height 50 cm. If the trough is being filled with water at the rate of 0.2 m³/min, how fast is the water level rising when the water is 30 cm deep?

28. A swimming pool is 20 ft wide, 40 ft long, 3 ft deep at the shallow end, and 9 ft deep at its deepest point. A cross-section is shown in the figure. If the pool is being filled at a rate of 0.8 ft³/min, how fast is the water level rising when the depth at the deepest point is 5 ft?

29. Gravel is being dumped from a conveyor belt at a rate of 30 ft³/min, and its coarseness is such that it forms a pile in the shape of a cone whose base diameter and height are always equal. How fast is the height of the pile increasing when the pile is 10 ft high?

30. A kite 100 ft above the ground moves horizontally at a speed of 8 ft/s. At what rate is the angle between the string and the horizontal decreasing when 200 ft of string has been let out?

31. The sides of an equilateral triangle are increasing at a rate of 10 cm/min. At what rate is the area of the triangle increasing when the sides are 30 cm long?

32. How fast is the angle between the ladder and the ground changing in Example 2 when the bottom of the ladder is 6 ft from the wall?

33. The top of a ladder slides down a vertical wall at a rate of 0.15 m/s. At the moment when the bottom of the ladder is 3 m from the wall, it slides away from the wall at a rate of 0.2 m/s. How long is the ladder?

34. According to the model we used to solve Example 2, what happens as the top of the ladder approaches the ground? Is the model appropriate for small values of y?

35. If the minute hand of a clock has length r (in centimeters), find the rate at which it sweeps out area as a function of r.

36. A faucet is filling a hemispherical basin of diameter 60 cm with water at a rate of 2 L/min. Find the rate at which the water is rising in the basin when it is half full. [Use the following facts: 1 L is 1000 cm³. The volume of the portion of a sphere with radius r from the bottom to a height h is $V = \pi\left(rh^2 - \frac{1}{3}h^3\right)$, as we will show in Chapter 6.]

37. Boyle's Law states that when a sample of gas is compressed at a constant temperature, the pressure P and volume V satisfy the equation $PV = C$, where C is a constant. Suppose that at a certain instant the volume is 600 cm³, the pressure is 150 kPa, and the pressure is increasing at a rate of 20 kPa/min. At what rate is the volume decreasing at this instant?

38. When air expands adiabatically (without gaining or losing heat), its pressure P and volume V are related by the equation $PV^{1.4} = C$, where C is a constant. Suppose that at a certain instant the volume is 400 cm³ and the pressure is 80 kPa and is decreasing at a rate of 10 kPa/min. At what rate is the volume increasing at this instant?

39. If two resistors with resistances R_1 and R_2 are connected in parallel, as in the figure, then the total resistance R, measured in ohms (Ω), is given by

$$\frac{1}{R} = \frac{1}{R_1} + \frac{1}{R_2}$$

If R_1 and R_2 are increasing at rates of 0.3 Ω/s and 0.2 Ω/s, respectively, how fast is R changing when $R_1 = 80\ \Omega$ and $R_2 = 100\ \Omega$?

40. Brain weight B as a function of body weight W in fish has been modeled by the power function $B = 0.007W^{2/3}$, where B and W are measured in grams. A model for body weight as a function of body length L (measured in centimeters) is $W = 0.12L^{2.53}$. If, over 10 million years, the average length of a certain species of fish evolved from 15 cm to 20 cm at a constant rate, how fast was this species' brain growing when the average length was 18 cm?

41. Two sides of a triangle have lengths 12 m and 15 m. The angle between them is increasing at a rate of 2°/min. How fast is the length of the third side increasing when the angle between the sides of fixed length is 60°?

42. Two carts, A and B, are connected by a rope 39 ft long that passes over a pulley P (see the figure). The point Q is on the floor 12 ft directly beneath P and between the carts. Cart A is being pulled away from Q at a speed of 2 ft/s. How fast is cart B moving toward Q at the instant when cart A is 5 ft from Q?

43. A television camera is positioned 4000 ft from the base of a rocket launching pad. The angle of elevation of the camera has to change at the correct rate in order to keep the rocket in sight. Also, the mechanism for focusing the camera has to take into account the increasing distance from the camera to the rising rocket. Let's assume the rocket rises vertically and its speed is 600 ft/s when it has risen 3000 ft.
  (a) How fast is the distance from the television camera to the rocket changing at that moment?
  (b) If the television camera is always kept aimed at the rocket, how fast is the camera's angle of elevation changing at that same moment?

44. A lighthouse is located on a small island 3 km away from the nearest point P on a straight shoreline and its light makes four revolutions per minute. How fast is the beam of light moving along the shoreline when it is 1 km from P?

45. A plane flies horizontally at an altitude of 5 km and passes directly over a tracking telescope on the ground. When the angle of elevation is $\pi/3$, this angle is decreasing at a rate of $\pi/6$ rad/min. How fast is the plane traveling at that time?

46. A Ferris wheel with a radius of 10 m is rotating at a rate of one revolution every 2 minutes. How fast is a rider rising when his seat is 16 m above ground level?

47. A plane flying with a constant speed of 300 km/h passes over a ground radar station at an altitude of 1 km and climbs at an angle of 30°. At what rate is the distance from the plane to the radar station increasing a minute later?

48. Two people start from the same point. One walks east at 3 mi/h and the other walks northeast at 2 mi/h. How fast is the distance between the people changing after 15 minutes?

49. A runner sprints around a circular track of radius 100 m at a constant speed of 7 m/s. The runner's friend is standing at a distance 200 m from the center of the track. How fast is the distance between the friends changing when the distance between them is 200 m?

50. The minute hand on a watch is 8 mm long and the hour hand is 4 mm long. How fast is the distance between the tips of the hands changing at one o'clock?

3.10 Linear Approximations and Differentials

We have seen that a curve lies very close to its tangent line near the point of tangency. In fact, by zooming in toward a point on the graph of a differentiable function, we noticed that the graph looks more and more like its tangent line. (See Figure 2.7.2.) This observation is the basis for a method of finding approximate values of functions.

The idea is that it might be easy to calculate a value $f(a)$ of a function, but difficult (or even impossible) to compute nearby values of f. So we settle for the easily computed

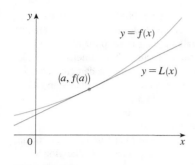

FIGURE 1

values of the linear function L whose graph is the tangent line of f at $(a, f(a))$. (See Figure 1.)

In other words, we use the tangent line at $(a, f(a))$ as an approximation to the curve $y = f(x)$ when x is near a. An equation of this tangent line is

$$y = f(a) + f'(a)(x - a)$$

and the approximation

$$\boxed{1} \qquad f(x) \approx f(a) + f'(a)(x - a)$$

is called the **linear approximation** or **tangent line approximation** of f at a. The linear function whose graph is this tangent line, that is,

$$\boxed{2} \qquad L(x) = f(a) + f'(a)(x - a)$$

is called the **linearization** of f at a.

EXAMPLE 1 Find the linearization of the function $f(x) = \sqrt{x + 3}$ at $a = 1$ and use it to approximate the numbers $\sqrt{3.98}$ and $\sqrt{4.05}$. Are these approximations overestimates or underestimates?

SOLUTION The derivative of $f(x) = (x + 3)^{1/2}$ is

$$f'(x) = \tfrac{1}{2}(x + 3)^{-1/2} = \frac{1}{2\sqrt{x + 3}}$$

and so we have $f(1) = 2$ and $f'(1) = \tfrac{1}{4}$. Putting these values into Equation 2, we see that the linearization is

$$L(x) = f(1) + f'(1)(x - 1) = 2 + \tfrac{1}{4}(x - 1) = \frac{7}{4} + \frac{x}{4}$$

The corresponding linear approximation (1) is

$$\sqrt{x + 3} \approx \frac{7}{4} + \frac{x}{4} \qquad \text{(when } x \text{ is near 1)}$$

In particular, we have

$$\sqrt{3.98} \approx \tfrac{7}{4} + \tfrac{0.98}{4} = 1.995 \qquad \text{and} \qquad \sqrt{4.05} \approx \tfrac{7}{4} + \tfrac{1.05}{4} = 2.0125$$

FIGURE 2

The linear approximation is illustrated in Figure 2. We see that, indeed, the tangent line approximation is a good approximation to the given function when x is near 1. We also see that our approximations are overestimates because the tangent line lies above the curve.

Of course, a calculator could give us approximations for $\sqrt{3.98}$ and $\sqrt{4.05}$, but the linear approximation gives an approximation *over an entire interval*.

In the following table we compare the estimates from the linear approximation in Example 1 with the true values. Notice from this table, and also from Figure 2, that the tangent line approximation gives good estimates when x is close to 1 but the accuracy of the approximation deteriorates when x is farther away from 1.

	x	From $L(x)$	Actual value
$\sqrt{3.9}$	0.9	1.975	1.97484176...
$\sqrt{3.98}$	0.98	1.995	1.99499373...
$\sqrt{4}$	1	2	2.00000000...
$\sqrt{4.05}$	1.05	2.0125	2.01246117...
$\sqrt{4.1}$	1.1	2.025	2.02484567...
$\sqrt{5}$	2	2.25	2.23606797...
$\sqrt{6}$	3	2.5	2.44948974...

How good is the approximation that we obtained in Example 1? The next example shows that by using a graphing calculator or computer we can determine an interval throughout which a linear approximation provides a specified accuracy.

EXAMPLE 2 For what values of x is the linear approximation

$$\sqrt{x+3} \approx \frac{7}{4} + \frac{x}{4}$$

accurate to within 0.5? What about accuracy to within 0.1?

SOLUTION Accuracy to within 0.5 means that the functions should differ by less than 0.5:

$$\left| \sqrt{x+3} - \left(\frac{7}{4} + \frac{x}{4} \right) \right| < 0.5$$

Equivalently, we could write

$$\sqrt{x+3} - 0.5 < \frac{7}{4} + \frac{x}{4} < \sqrt{x+3} + 0.5$$

This says that the linear approximation should lie between the curves obtained by shifting the curve $y = \sqrt{x+3}$ upward and downward by an amount 0.5. Figure 3 shows the tangent line $y = (7+x)/4$ intersecting the upper curve $y = \sqrt{x+3} + 0.5$ at P and Q. Zooming in and using the cursor, we estimate that the x-coordinate of P is about -2.66 and the x-coordinate of Q is about 8.66. Thus we see from the graph that the approximation

$$\sqrt{x+3} \approx \frac{7}{4} + \frac{x}{4}$$

is accurate to within 0.5 when $-2.6 < x < 8.6$. (We have rounded to be safe.)

Similarly, from Figure 4 we see that the approximation is accurate to within 0.1 when $-1.1 < x < 3.9$.

FIGURE 3

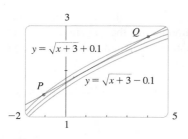

FIGURE 4

■ Applications to Physics

Linear approximations are often used in physics. In analyzing the consequences of an equation, a physicist sometimes needs to simplify a function by replacing it with its linear approximation. For instance, in deriving a formula for the period of a pendulum, physics textbooks obtain the expression $a_T = -g \sin \theta$ for tangential acceleration and then replace $\sin \theta$ by θ with the remark that $\sin \theta$ is very close to θ if θ is not too large. [See, for example, *Physics: Calculus*, 2d ed., by Eugene Hecht (Pacific Grove, CA: Brooks/ Cole, 2000), p. 431.] You can verify that the linearization of the function $f(x) = \sin x$ at $a = 0$ is $L(x) = x$ and so the linear approximation at 0 is

$$\sin x \approx x$$

(see Exercise 42). So, in effect, the derivation of the formula for the period of a pendulum uses the tangent line approximation for the sine function.

Another example occurs in the theory of optics, where light rays that arrive at shallow angles relative to the optical axis are called *paraxial rays*. In paraxial (or Gaussian) optics, both $\sin \theta$ and $\cos \theta$ are replaced by their linearizations. In other words, the linear approximations

$$\sin \theta \approx \theta \qquad \text{and} \qquad \cos \theta \approx 1$$

are used because θ is close to 0. The results of calculations made with these approximations became the basic theoretical tool used to design lenses. [See *Optics,* 4th ed., by Eugene Hecht (San Francisco, 2002), p. 154.]

In Section 11.11 we will present several other applications of the idea of linear approximations to physics and engineering.

Differentials

The ideas behind linear approximations are sometimes formulated in the terminology and notation of *differentials*. If $y = f(x)$, where f is a differentiable function, then the **differential** dx is an independent variable; that is, dx can be given the value of any real number. The **differential** dy is then defined in terms of dx by the equation

If $dx \neq 0$, we can divide both sides of Equation 3 by dx to obtain

$$\frac{dy}{dx} = f'(x)$$

We have seen similar equations before, but now the left side can genuinely be interpreted as a ratio of differentials.

$$\boxed{3} \qquad dy = f'(x)\,dx$$

So dy is a dependent variable; it depends on the values of x and dx. If dx is given a specific value and x is taken to be some specific number in the domain of f, then the numerical value of dy is determined.

The geometric meaning of differentials is shown in Figure 5. Let $P(x, f(x))$ and $Q(x + \Delta x, f(x + \Delta x))$ be points on the graph of f and let $dx = \Delta x$. The corresponding change in y is

$$\Delta y = f(x + \Delta x) - f(x)$$

The slope of the tangent line PR is the derivative $f'(x)$. Thus the directed distance from S to R is $f'(x)\,dx = dy$. Therefore dy represents the amount that the tangent line rises or falls (the change in the linearization), whereas Δy represents the amount that the curve $y = f(x)$ rises or falls when x changes by an amount dx.

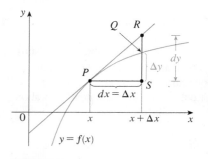

FIGURE 5

EXAMPLE 3 Compare the values of Δy and dy if $y = f(x) = x^3 + x^2 - 2x + 1$ and x changes (a) from 2 to 2.05 and (b) from 2 to 2.01.

SOLUTION
(a) We have

$$f(2) = 2^3 + 2^2 - 2(2) + 1 = 9$$

$$f(2.05) = (2.05)^3 + (2.05)^2 - 2(2.05) + 1 = 9.717625$$

$$\Delta y = f(2.05) - f(2) = 0.717625$$

Figure 6 shows the function in Example 3 and a comparison of dy and Δy when $a = 2$. The viewing rectangle is $[1.8, 2.5]$ by $[6, 18]$.

FIGURE 6

In general,
$$dy = f'(x)\,dx = (3x^2 + 2x - 2)\,dx$$

When $x = 2$ and $dx = \Delta x = 0.05$, this becomes
$$dy = [3(2)^2 + 2(2) - 2]0.05 = 0.7$$

(b)
$$f(2.01) = (2.01)^3 + (2.01)^2 - 2(2.01) + 1 = 9.140701$$

$$\Delta y = f(2.01) - f(2) = 0.140701$$

When $dx = \Delta x = 0.01$,
$$dy = [3(2)^2 + 2(2) - 2]0.01 = 0.14 \qquad \blacksquare$$

Notice that the approximation $\Delta y \approx dy$ becomes better as Δx becomes smaller in Example 3. Notice also that dy was easier to compute than Δy. For more complicated functions it may be impossible to compute Δy exactly. In such cases the approximation by differentials is especially useful.

In the notation of differentials, the linear approximation (1) can be written as
$$f(a + dx) \approx f(a) + dy$$

For instance, for the function $f(x) = \sqrt{x + 3}$ in Example 1, we have
$$dy = f'(x)\,dx = \frac{dx}{2\sqrt{x + 3}}$$

If $a = 1$ and $dx = \Delta x = 0.05$, then
$$dy = \frac{0.05}{2\sqrt{1 + 3}} = 0.0125$$

and
$$\sqrt{4.05} = f(1.05) \approx f(1) + dy = 2.0125$$

just as we found in Example 1.

Our final example illustrates the use of differentials in estimating the errors that occur because of approximate measurements.

EXAMPLE 4 The radius of a sphere was measured and found to be 21 cm with a possible error in measurement of at most 0.05 cm. What is the maximum error in using this value of the radius to compute the volume of the sphere?

SOLUTION If the radius of the sphere is r, then its volume is $V = \frac{4}{3}\pi r^3$. If the error in the measured value of r is denoted by $dr = \Delta r$, then the corresponding error in the calculated value of V is ΔV, which can be approximated by the differential
$$dV = 4\pi r^2\,dr$$

When $r = 21$ and $dr = 0.05$, this becomes
$$dV = 4\pi(21)^2 0.05 \approx 277$$

The maximum error in the calculated volume is about 277 cm^3. $\qquad \blacksquare$

NOTE Although the possible error in Example 4 may appear to be rather large, a better picture of the error is given by the **relative error**, which is computed by dividing the error by the total volume:

$$\frac{\Delta V}{V} \approx \frac{dV}{V} = \frac{4\pi r^2 \, dr}{\frac{4}{3}\pi r^3} = 3\frac{dr}{r}$$

Thus the relative error in the volume is about three times the relative error in the radius. In Example 4 the relative error in the radius is approximately $dr/r = 0.05/21 \approx 0.0024$ and it produces a relative error of about 0.007 in the volume. The errors could also be expressed as **percentage errors** of 0.24% in the radius and 0.7% in the volume.

3.10 EXERCISES

1–4 Find the linearization $L(x)$ of the function at a.

1. $f(x) = x^3 - x^2 + 3$, $a = -2$

2. $f(x) = \sin x$, $a = \pi/6$

3. $f(x) = \sqrt{x}$, $a = 4$

4. $f(x) = 2^x$, $a = 0$

5. Find the linear approximation of the function $f(x) = \sqrt{1 - x}$ at $a = 0$ and use it to approximate the numbers $\sqrt{0.9}$ and $\sqrt{0.99}$. Illustrate by graphing f and the tangent line.

6. Find the linear approximation of the function $g(x) = \sqrt[3]{1 + x}$ at $a = 0$ and use it to approximate the numbers $\sqrt[3]{0.95}$ and $\sqrt[3]{1.1}$. Illustrate by graphing g and the tangent line.

7–10 Verify the given linear approximation at $a = 0$. Then determine the values of x for which the linear approximation is accurate to within 0.1.

7. $\ln(1 + x) \approx x$

8. $(1 + x)^{-3} \approx 1 - 3x$

9. $\sqrt[4]{1 + 2x} \approx 1 + \frac{1}{2}x$

10. $e^x \cos x \approx 1 + x$

11–14 Find the differential of each function.

11. (a) $y = xe^{-4x}$ (b) $y = \sqrt{1 - t^4}$

12. (a) $y = \dfrac{1 + 2u}{1 + 3u}$ (b) $y = \theta^2 \sin 2\theta$

13. (a) $y = \tan\sqrt{t}$ (b) $y = \dfrac{1 - v^2}{1 + v^2}$

14. (a) $y = \ln(\sin\theta)$ (b) $y = \dfrac{e^x}{1 - e^x}$

15–18 (a) Find the differential dy and (b) evaluate dy for the given values of x and dx.

15. $y = e^{x/10}$, $x = 0$, $dx = 0.1$

16. $y = \cos\pi x$, $x = \frac{1}{3}$, $dx = -0.02$

17. $y = \sqrt{3 + x^2}$, $x = 1$, $dx = -0.1$

18. $y = \dfrac{x + 1}{x - 1}$, $x = 2$, $dx = 0.05$

19–22 Compute Δy and dy for the given values of x and $dx = \Delta x$. Then sketch a diagram like Figure 5 showing the line segments with lengths dx, dy, and Δy.

19. $y = x^2 - 4x$, $x = 3$, $\Delta x = 0.5$

20. $y = x - x^3$, $x = 0$, $\Delta x = -0.3$

21. $y = \sqrt{x - 2}$, $x = 3$, $\Delta x = 0.8$

22. $y = e^x$, $x = 0$, $\Delta x = 0.5$

23–28 Use a linear approximation (or differentials) to estimate the given number.

23. $(1.999)^4$

24. $1/4.002$

25. $\sqrt[3]{1001}$

26. $\sqrt{100.5}$

27. $e^{0.1}$

28. $\cos 29°$

29–31 Explain, in terms of linear approximations or differentials, why the approximation is reasonable.

29. $\sec 0.08 \approx 1$

30. $\sqrt{4.02} \approx 2.005$

31. $\dfrac{1}{9.98} \approx 0.1002$

32. Let $\quad f(x) = (x - 1)^2 \qquad g(x) = e^{-2x}$

and $\qquad h(x) = 1 + \ln(1 - 2x)$

(a) Find the linearizations of f, g, and h at $a = 0$. What do you notice? How do you explain what happened?

(b) Graph f, g, and h and their linear approximations. For which function is the linear approximation best? For which is it worst? Explain.

33. The edge of a cube was found to be 30 cm with a possible error in measurement of 0.1 cm. Use differentials to estimate the maximum possible error, relative error, and percentage error in computing (a) the volume of the cube and (b) the surface area of the cube.

34. The radius of a circular disk is given as 24 cm with a maximum error in measurement of 0.2 cm.

(a) Use differentials to estimate the maximum error in the calculated area of the disk.

(b) What is the relative error? What is the percentage error?

35. The circumference of a sphere was measured to be 84 cm with a possible error of 0.5 cm.

(a) Use differentials to estimate the maximum error in the calculated surface area. What is the relative error?

(b) Use differentials to estimate the maximum error in the calculated volume. What is the relative error?

36. Use differentials to estimate the amount of paint needed to apply a coat of paint 0.05 cm thick to a hemispherical dome with diameter 50 m.

37. (a) Use differentials to find a formula for the approximate volume of a thin cylindrical shell with height h, inner radius r, and thickness Δr.

(b) What is the error involved in using the formula from part (a)?

38. One side of a right triangle is known to be 20 cm long and the opposite angle is measured as 30°, with a possible error of $\pm 1°$.

(a) Use differentials to estimate the error in computing the length of the hypotenuse.

(b) What is the percentage error?

39. If a current I passes through a resistor with resistance R, Ohm's Law states that the voltage drop is $V = RI$. If V is constant and R is measured with a certain error, use differentials to show that the relative error in calculating I is approximately the same (in magnitude) as the relative error in R.

40. When blood flows along a blood vessel, the flux F (the volume of blood per unit time that flows past a given point) is proportional to the fourth power of the radius R of the blood vessel:

$$F = kR^4$$

(This is known as Poiseuille's Law; we will show why it is true in Section 8.4.) A partially clogged artery can be expanded by an operation called angioplasty, in which a balloon-tipped catheter is inflated inside the artery in order to widen it and restore the normal blood flow.

Show that the relative change in F is about four times the relative change in R. How will a 5% increase in the radius affect the flow of blood?

41. Establish the following rules for working with differentials (where c denotes a constant and u and v are functions of x).

(a) $dc = 0$
(b) $d(cu) = c\,du$
(c) $d(u + v) = du + dv$
(d) $d(uv) = u\,dv + v\,du$
(e) $d\left(\dfrac{u}{v}\right) = \dfrac{v\,du - u\,dv}{v^2}$
(f) $d(x^n) = nx^{n-1}\,dx$

42. On page 431 of *Physics: Calculus*, 2d ed., by Eugene Hecht (Pacific Grove, CA: Brooks/Cole, 2000), in the course of deriving the formula $T = 2\pi\sqrt{L/g}$ for the period of a pendulum of length L, the author obtains the equation $a_T = -g\sin\theta$ for the tangential acceleration of the bob of the pendulum. He then says, "for small angles, the value of θ in radians is very nearly the value of $\sin\theta$; they differ by less than 2% out to about 20°."

(a) Verify the linear approximation at 0 for the sine function:

$$\sin x \approx x$$

(b) Use a graphing device to determine the values of x for which $\sin x$ and x differ by less than 2%. Then verify Hecht's statement by converting from radians to degrees.

43. Suppose that the only information we have about a function f is that $f(1) = 5$ and the graph of its *derivative* is as shown.

(a) Use a linear approximation to estimate $f(0.9)$ and $f(1.1)$.

(b) Are your estimates in part (a) too large or too small? Explain.

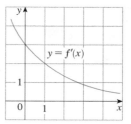

44. Suppose that we don't have a formula for $g(x)$ but we know that $g(2) = -4$ and $g'(x) = \sqrt{x^2 + 5}$ for all x.

(a) Use a linear approximation to estimate $g(1.95)$ and $g(2.05)$.

(b) Are your estimates in part (a) too large or too small? Explain.

LABORATORY PROJECT   ⊞ TAYLOR POLYNOMIALS

The tangent line approximation $L(x)$ is the best first-degree (linear) approximation to $f(x)$ near $x = a$ because $f(x)$ and $L(x)$ have the same rate of change (derivative) at a. For a better approximation than a linear one, let's try a second-degree (quadratic) approximation $P(x)$. In other words, we approximate a curve by a parabola instead of by a straight line. To make sure that the approximation is a good one, we stipulate the following:

  (i) $P(a) = f(a)$ (P and f should have the same value at a.)
  (ii) $P'(a) = f'(a)$ (P and f should have the same rate of change at a.)
  (iii) $P''(a) = f''(a)$ (The slopes of P and f should change at the same rate at a.)

1. Find the quadratic approximation $P(x) = A + Bx + Cx^2$ to the function $f(x) = \cos x$ that satisfies conditions (i), (ii), and (iii) with $a = 0$. Graph P, f, and the linear approximation $L(x) = 1$ on a common screen. Comment on how well the functions P and L approximate f.

2. Determine the values of x for which the quadratic approximation $f(x) \approx P(x)$ in Problem 1 is accurate to within 0.1. [*Hint:* Graph $y = P(x)$, $y = \cos x - 0.1$, and $y = \cos x + 0.1$ on a common screen.]

3. To approximate a function f by a quadratic function P near a number a, it is best to write P in the form

$$P(x) = A + B(x - a) + C(x - a)^2$$

   Show that the quadratic function that satisfies conditions (i), (ii), and (iii) is

$$P(x) = f(a) + f'(a)(x - a) + \tfrac{1}{2}f''(a)(x - a)^2$$

4. Find the quadratic approximation to $f(x) = \sqrt{x + 3}$ near $a = 1$. Graph f, the quadratic approximation, and the linear approximation from Example 3.10.2 on a common screen. What do you conclude?

5. Instead of being satisfied with a linear or quadratic approximation to $f(x)$ near $x = a$, let's try to find better approximations with higher-degree polynomials. We look for an nth-degree polynomial

$$T_n(x) = c_0 + c_1(x - a) + c_2(x - a)^2 + c_3(x - a)^3 + \cdots + c_n(x - a)^n$$

   such that T_n and its first n derivatives have the same values at $x = a$ as f and its first n derivatives. By differentiating repeatedly and setting $x = a$, show that these conditions are satisfied if $c_0 = f(a)$, $c_1 = f'(a)$, $c_2 = \tfrac{1}{2}f''(a)$, and in general

$$c_k = \frac{f^{(k)}(a)}{k!}$$

   where $k! = 1 \cdot 2 \cdot 3 \cdot 4 \cdot \cdots \cdot k$. The resulting polynomial

$$T_n(x) = f(a) + f'(a)(x - a) + \frac{f''(a)}{2!}(x - a)^2 + \cdots + \frac{f^{(n)}(a)}{n!}(x - a)^n$$

   is called the **nth-degree Taylor polynomial of f centered at a**.

6. Find the 8th-degree Taylor polynomial centered at $a = 0$ for the function $f(x) = \cos x$. Graph f together with the Taylor polynomials T_2, T_4, T_6, T_8 in the viewing rectangle $[-5, 5]$ by $[-1.4, 1.4]$ and comment on how well they approximate f.

3.11 Hyperbolic Functions

Certain even and odd combinations of the exponential functions e^x and e^{-x} arise so frequently in mathematics and its applications that they deserve to be given special names. In many ways they are analogous to the trigonometric functions, and they have the same relationship to the hyperbola that the trigonometric functions have to the circle. For this reason they are collectively called **hyperbolic functions** and individually called **hyperbolic sine**, **hyperbolic cosine**, and so on.

Definition of the Hyperbolic Functions

$$\sinh x = \frac{e^x - e^{-x}}{2} \qquad\qquad \operatorname{csch} x = \frac{1}{\sinh x}$$

$$\cosh x = \frac{e^x + e^{-x}}{2} \qquad\qquad \operatorname{sech} x = \frac{1}{\cosh x}$$

$$\tanh x = \frac{\sinh x}{\cosh x} \qquad\qquad \coth x = \frac{\cosh x}{\sinh x}$$

The graphs of hyperbolic sine and cosine can be sketched using graphical addition as in Figures 1 and 2.

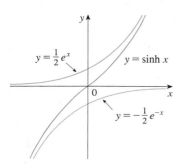

FIGURE 1
$y = \sinh x = \frac{1}{2}e^x - \frac{1}{2}e^{-x}$

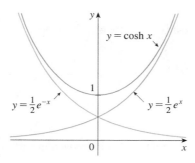

FIGURE 2
$y = \cosh x = \frac{1}{2}e^x + \frac{1}{2}e^{-x}$

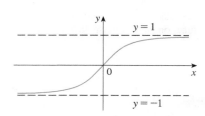

FIGURE 3
$y = \tanh x$

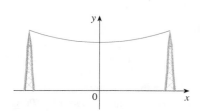

FIGURE 4
A catenary $y = c + a\cosh(x/a)$

Note that sinh has domain $\mathbb{R}$ and range $\mathbb{R}$, while cosh has domain $\mathbb{R}$ and range $[1, \infty)$. The graph of tanh is shown in Figure 3. It has the horizontal asymptotes $y = \pm 1$. (See Exercise 23.)

Some of the mathematical uses of hyperbolic functions will be seen in Chapter 7. Applications to science and engineering occur whenever an entity such as light, velocity, electricity, or radioactivity is gradually absorbed or extinguished, for the decay can be represented by hyperbolic functions. The most famous application is the use of hyperbolic cosine to describe the shape of a hanging wire. It can be proved that if a heavy flexible cable (such as a telephone or power line) is suspended between two points at the same height, then it takes the shape of a curve with equation $y = c + a\cosh(x/a)$ called a *catenary* (see Figure 4). (The Latin word *catena* means "chain.")

FIGURE 5
Idealized ocean wave

Another application of hyperbolic functions occurs in the description of ocean waves: The velocity of a water wave with length L moving across a body of water with depth d is modeled by the function

$$v = \sqrt{\frac{gL}{2\pi} \tanh\left(\frac{2\pi d}{L}\right)}$$

where g is the acceleration due to gravity. (See Figure 5 and Exercise 49.)

The hyperbolic functions satisfy a number of identities that are similar to well-known trigonometric identities. We list some of them here and leave most of the proofs to the exercises.

The Gateway Arch in St. Louis was designed using a hyperbolic cosine function (see Exercise 48).

Hyperbolic Identities

$$\sinh(-x) = -\sinh x \qquad\qquad \cosh(-x) = \cosh x$$

$$\cosh^2 x - \sinh^2 x = 1 \qquad\qquad 1 - \tanh^2 x = \operatorname{sech}^2 x$$

$$\sinh(x + y) = \sinh x \cosh y + \cosh x \sinh y$$

$$\cosh(x + y) = \cosh x \cosh y + \sinh x \sinh y$$

EXAMPLE 1 Prove (a) $\cosh^2 x - \sinh^2 x = 1$ and (b) $1 - \tanh^2 x = \operatorname{sech}^2 x$.

SOLUTION

(a)
$$\cosh^2 x - \sinh^2 x = \left(\frac{e^x + e^{-x}}{2}\right)^2 - \left(\frac{e^x - e^{-x}}{2}\right)^2$$

$$= \frac{e^{2x} + 2 + e^{-2x}}{4} - \frac{e^{2x} - 2 + e^{-2x}}{4}$$

$$= \frac{4}{4} = 1$$

(b) We start with the identity proved in part (a):

$$\cosh^2 x - \sinh^2 x = 1$$

If we divide both sides by $\cosh^2 x$, we get

$$1 - \frac{\sinh^2 x}{\cosh^2 x} = \frac{1}{\cosh^2 x}$$

or

$$1 - \tanh^2 x = \operatorname{sech}^2 x \qquad\qquad ■$$

The identity proved in Example 1(a) gives a clue to the reason for the name "hyperbolic" functions:

If t is any real number, then the point $P(\cos t, \sin t)$ lies on the unit circle $x^2 + y^2 = 1$ because $\cos^2 t + \sin^2 t = 1$. In fact, t can be interpreted as the radian measure of $\angle POQ$ in Figure 6. For this reason the trigonometric functions are sometimes called *circular* functions.

FIGURE 6

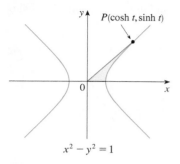

$P(\cosh t, \sinh t)$

$x^2 - y^2 = 1$

FIGURE 7

Likewise, if t is any real number, then the point $P(\cosh t, \sinh t)$ lies on the right branch of the hyperbola $x^2 - y^2 = 1$ because $\cosh^2 t - \sinh^2 t = 1$ and $\cosh t \geqslant 1$. This time, t does not represent the measure of an angle. However, it turns out that t represents twice the area of the shaded hyperbolic sector in Figure 7, just as in the trigonometric case t represents twice the area of the shaded circular sector in Figure 6.

The derivatives of the hyperbolic functions are easily computed. For example,

$$\frac{d}{dx}(\sinh x) = \frac{d}{dx}\left(\frac{e^x - e^{-x}}{2}\right) = \frac{e^x + e^{-x}}{2} = \cosh x$$

We list the differentiation formulas for the hyperbolic functions as Table 1. The remaining proofs are left as exercises. Note the analogy with the differentiation formulas for trigonometric functions, but beware that the signs are different in some cases.

---

1 **Derivatives of Hyperbolic Functions**

$$\frac{d}{dx}(\sinh x) = \cosh x \qquad\qquad \frac{d}{dx}(\operatorname{csch} x) = -\operatorname{csch} x \coth x$$

$$\frac{d}{dx}(\cosh x) = \sinh x \qquad\qquad \frac{d}{dx}(\operatorname{sech} x) = -\operatorname{sech} x \tanh x$$

$$\frac{d}{dx}(\tanh x) = \operatorname{sech}^2 x \qquad\qquad \frac{d}{dx}(\coth x) = -\operatorname{csch}^2 x$$

EXAMPLE 2 Any of these differentiation rules can be combined with the Chain Rule. For instance,

$$\frac{d}{dx}\left(\cosh\sqrt{x}\right) = \sinh\sqrt{x} \cdot \frac{d}{dx}\sqrt{x} = \frac{\sinh\sqrt{x}}{2\sqrt{x}}$$

■

■ Inverse Hyperbolic Functions

You can see from Figures 1 and 3 that sinh and tanh are one-to-one functions and so they have inverse functions denoted by $\sinh^{-1}$ and $\tanh^{-1}$. Figure 2 shows that cosh is not one-to-one, but when restricted to the domain $[0, \infty)$ it becomes one-to-one. The inverse hyperbolic cosine function is defined as the inverse of this restricted function.

2

$$y = \sinh^{-1}x \iff \sinh y = x$$

$$y = \cosh^{-1}x \iff \cosh y = x \quad \text{and} \quad y \geqslant 0$$

$$y = \tanh^{-1}x \iff \tanh y = x$$

The remaining inverse hyperbolic functions are defined similarly (see Exercise 28).

We can sketch the graphs of $\sinh^{-1}$, $\cosh^{-1}$, and $\tanh^{-1}$ in Figures 8, 9, and 10 by using Figures 1, 2, and 3.

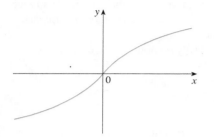

FIGURE 8 $y = \sinh^{-1}x$
domain $= \mathbb{R}$ range $= \mathbb{R}$

FIGURE 9 $y = \cosh^{-1}x$
domain $= [1, \infty)$ range $= [0, \infty)$

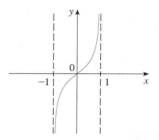

FIGURE 10 $y = \tanh^{-1}x$
domain $= (-1, 1)$ range $= \mathbb{R}$

Since the hyperbolic functions are defined in terms of exponential functions, it's not surprising to learn that the inverse hyperbolic functions can be expressed in terms of logarithms. In particular, we have:

Formula 3 is proved in Example 3. The proofs of Formulas 4 and 5 are requested in Exercises 26 and 27.

$$\boxed{3} \qquad \sinh^{-1}x = \ln\left(x + \sqrt{x^2 + 1}\right) \qquad x \in \mathbb{R}$$

$$\boxed{4} \qquad \cosh^{-1}x = \ln\left(x + \sqrt{x^2 - 1}\right) \qquad x \geq 1$$

$$\boxed{5} \qquad \tanh^{-1}x = \tfrac{1}{2}\ln\left(\frac{1 + x}{1 - x}\right) \qquad -1 < x < 1$$

EXAMPLE 3 Show that $\sinh^{-1}x = \ln\left(x + \sqrt{x^2 + 1}\right)$.

SOLUTION Let $y = \sinh^{-1}x$. Then

$$x = \sinh y = \frac{e^y - e^{-y}}{2}$$

so

$$e^y - 2x - e^{-y} = 0$$

or, multiplying by e^y,

$$e^{2y} - 2xe^y - 1 = 0$$

This is really a quadratic equation in e^y:

$$(e^y)^2 - 2x(e^y) - 1 = 0$$

Solving by the quadratic formula, we get

$$e^y = \frac{2x \pm \sqrt{4x^2 + 4}}{2} = x \pm \sqrt{x^2 + 1}$$

Note that $e^y > 0$, but $x - \sqrt{x^2 + 1} < 0$ (because $x < \sqrt{x^2 + 1}$). Thus the minus sign is inadmissible and we have

$$e^y = x + \sqrt{x^2 + 1}$$

Therefore $\qquad\qquad\qquad\qquad y = \ln(e^y) = \ln\!\left(x + \sqrt{x^2 + 1}\right)$

This shows that $\qquad\qquad\qquad \sinh^{-1}x = \ln\!\left(x + \sqrt{x^2 + 1}\right)$

(See Exercise 25 for another method.)

---

6 Derivatives of Inverse Hyperbolic Functions

Notice that the formulas for the derivatives of $\tanh^{-1}x$ and $\coth^{-1}x$ appear to be identical. But the domains of these functions have no numbers in common: $\tanh^{-1}x$ is defined for $|x| < 1$, whereas $\coth^{-1}x$ is defined for $|x| > 1$.

$$\frac{d}{dx}(\sinh^{-1}x) = \frac{1}{\sqrt{1 + x^2}} \qquad\qquad \frac{d}{dx}(\operatorname{csch}^{-1}x) = -\frac{1}{|x|\sqrt{x^2 + 1}}$$

$$\frac{d}{dx}(\cosh^{-1}x) = \frac{1}{\sqrt{x^2 - 1}} \qquad\qquad \frac{d}{dx}(\operatorname{sech}^{-1}x) = -\frac{1}{x\sqrt{1 - x^2}}$$

$$\frac{d}{dx}(\tanh^{-1}x) = \frac{1}{1 - x^2} \qquad\qquad \frac{d}{dx}(\coth^{-1}x) = \frac{1}{1 - x^2}$$

The inverse hyperbolic functions are all differentiable because the hyperbolic functions are differentiable. The formulas in Table 6 can be proved either by the method for inverse functions or by differentiating Formulas 3, 4, and 5.

EXAMPLE 4 Prove that $\dfrac{d}{dx}(\sinh^{-1}x) = \dfrac{1}{\sqrt{1 + x^2}}$.

SOLUTION 1 Let $y = \sinh^{-1}x$. Then $\sinh y = x$. If we differentiate this equation implicitly with respect to x, we get

$$\cosh y \, \frac{dy}{dx} = 1$$

Since $\cosh^2 y - \sinh^2 y = 1$ and $\cosh y \geqslant 0$, we have $\cosh y = \sqrt{1 + \sinh^2 y}$, so

$$\frac{dy}{dx} = \frac{1}{\cosh y} = \frac{1}{\sqrt{1 + \sinh^2 y}} = \frac{1}{\sqrt{1 + x^2}}$$

SOLUTION 2 From Equation 3 (proved in Example 3), we have

$$\frac{d}{dx}(\sinh^{-1}x) = \frac{d}{dx}\ln\!\left(x + \sqrt{x^2 + 1}\right)$$

$$= \frac{1}{x + \sqrt{x^2 + 1}} \frac{d}{dx}\left(x + \sqrt{x^2 + 1}\right)$$

$$= \frac{1}{x + \sqrt{x^2 + 1}}\left(1 + \frac{x}{\sqrt{x^2 + 1}}\right)$$

$$= \frac{\sqrt{x^2 + 1} + x}{\left(x + \sqrt{x^2 + 1}\right)\sqrt{x^2 + 1}}$$

$$= \frac{1}{\sqrt{x^2 + 1}}$$

EXAMPLE 5 Find $\dfrac{d}{dx}[\tanh^{-1}(\sin x)]$.

SOLUTION Using Table 6 and the Chain Rule, we have

$$\frac{d}{dx}[\tanh^{-1}(\sin x)] = \frac{1}{1 - (\sin x)^2}\frac{d}{dx}(\sin x)$$

$$= \frac{1}{1 - \sin^2 x}\cos x = \frac{\cos x}{\cos^2 x} = \sec x \qquad \blacksquare$$

3.11 EXERCISES

1–6 Find the numerical value of each expression.

1. (a) $\sinh 0$ (b) $\cosh 0$

2. (a) $\tanh 0$ (b) $\tanh 1$

3. (a) $\cosh(\ln 5)$ (b) $\cosh 5$

4. (a) $\sinh 4$ (b) $\sinh(\ln 4)$

5. (a) $\operatorname{sech} 0$ (b) $\cosh^{-1} 1$

6. (a) $\sinh 1$ (b) $\sinh^{-1} 1$

7–19 Prove the identity.

7. $\sinh(-x) = -\sinh x$
(This shows that sinh is an odd function.)

8. $\cosh(-x) = \cosh x$
(This shows that cosh is an even function.)

9. $\cosh x + \sinh x = e^x$

10. $\cosh x - \sinh x = e^{-x}$

11. $\sinh(x + y) = \sinh x \cosh y + \cosh x \sinh y$

12. $\cosh(x + y) = \cosh x \cosh y + \sinh x \sinh y$

13. $\coth^2 x - 1 = \operatorname{csch}^2 x$

14. $\tanh(x + y) = \dfrac{\tanh x + \tanh y}{1 + \tanh x \tanh y}$

15. $\sinh 2x = 2 \sinh x \cosh x$

16. $\cosh 2x = \cosh^2 x + \sinh^2 x$

17. $\tanh(\ln x) = \dfrac{x^2 - 1}{x^2 + 1}$

18. $\dfrac{1 + \tanh x}{1 - \tanh x} = e^{2x}$

19. $(\cosh x + \sinh x)^n = \cosh nx + \sinh nx$
(n any real number)

20. If $\tanh x = \frac{12}{13}$, find the values of the other hyperbolic functions at x.

21. If $\cosh x = \frac{5}{3}$ and $x > 0$, find the values of the other hyperbolic functions at x.

22. (a) Use the graphs of sinh, cosh, and tanh in Figures 1–3 to draw the graphs of csch, sech, and coth.
(b) Check the graphs that you sketched in part (a) by using a graphing device to produce them.

23. Use the definitions of the hyperbolic functions to find each of the following limits.
(a) $\lim\limits_{x \to \infty} \tanh x$ (b) $\lim\limits_{x \to -\infty} \tanh x$
(c) $\lim\limits_{x \to \infty} \sinh x$ (d) $\lim\limits_{x \to -\infty} \sinh x$
(e) $\lim\limits_{x \to \infty} \operatorname{sech} x$ (f) $\lim\limits_{x \to \infty} \coth x$
(g) $\lim\limits_{x \to 0^+} \coth x$ (h) $\lim\limits_{x \to 0^-} \coth x$
(i) $\lim\limits_{x \to -\infty} \operatorname{csch} x$ (j) $\lim\limits_{x \to \infty} \dfrac{\sinh x}{e^x}$

24. Prove the formulas given in Table 1 for the derivatives of the functions (a) cosh, (b) tanh, (c) csch, (d) sech, and (e) coth.

25. Give an alternative solution to Example 3 by letting $y = \sinh^{-1} x$ and then using Exercise 9 and Example 1(a) with x replaced by y.

26. Prove Equation 4.

27. Prove Equation 5 using (a) the method of Example 3 and (b) Exercise 18 with x replaced by y.

28. For each of the following functions (i) give a definition like those in (2), (ii) sketch the graph, and (iii) find a formula similar to Equation 3.
(a) csch^{-1} (b) sech^{-1} (c) $\coth^{-1}$

29. Prove the formulas given in Table 6 for the derivatives of the following functions.
(a) $\cosh^{-1}$ (b) $\tanh^{-1}$ (c) csch^{-1}
(d) sech^{-1} (e) $\coth^{-1}$

30–45 Find the derivative. Simplify where possible.

30. $f(x) = e^x \cosh x$

31. $f(x) = \tanh \sqrt{x}$ **32.** $g(x) = \sinh^2 x$

33. $h(x) = \sinh(x^2)$

34. $F(t) = \ln(\sinh t)$

35. $G(t) = \sinh(\ln t)$

36. $y = \operatorname{sech} x\,(1 + \ln \operatorname{sech} x)$

37. $y = e^{\cosh 3x}$

38. $f(t) = \dfrac{1 + \sinh t}{1 - \sinh t}$

39. $g(t) = t \coth \sqrt{t^2 + 1}$

40. $y = \sinh^{-1}(\tan x)$

41. $y = \cosh^{-1}\sqrt{x}$

42. $y = x \tanh^{-1}x + \ln \sqrt{1 - x^2}$

43. $y = x \sinh^{-1}(x/3) - \sqrt{9 + x^2}$

44. $y = \operatorname{sech}^{-1}(e^{-x})$

45. $y = \coth^{-1}(\sec x)$

---

46. Show that $\dfrac{d}{dx}\sqrt[4]{\dfrac{1 + \tanh x}{1 - \tanh x}} = \tfrac{1}{2}e^{x/2}$.

47. Show that $\dfrac{d}{dx}\arctan(\tanh x) = \operatorname{sech} 2x$.

48. The Gateway Arch in St. Louis was designed by Eero Saarinen and was constructed using the equation
$$y = 211.49 - 20.96 \cosh 0.03291765x$$
for the central curve of the arch, where x and y are measured in meters and $|x| \le 91.20$.
   (a) Graph the central curve.
   (b) What is the height of the arch at its center?
   (c) At what points is the height 100 m?
   (d) What is the slope of the arch at the points in part (c)?

49. If a water wave with length L moves with velocity v in a body of water with depth d, then
$$v = \sqrt{\dfrac{gL}{2\pi}\tanh\!\left(\dfrac{2\pi d}{L}\right)}$$
where g is the acceleration due to gravity. (See Figure 5.) Explain why the approximation
$$v \approx \sqrt{\dfrac{gL}{2\pi}}$$
is appropriate in deep water.

50. A flexible cable always hangs in the shape of a catenary $y = c + a \cosh(x/a)$, where c and a are constants and $a > 0$ (see Figure 4 and Exercise 52). Graph several members of the family of functions $y = a \cosh(x/a)$. How does the graph change as a varies?

51. A telephone line hangs between two poles 14 m apart in the shape of the catenary $y = 20 \cosh(x/20) - 15$, where x and y are measured in meters.
   (a) Find the slope of this curve where it meets the right pole.
   (b) Find the angle θ between the line and the pole.

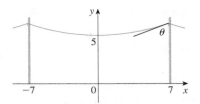

52. Using principles from physics it can be shown that when a cable is hung between two poles, it takes the shape of a curve $y = f(x)$ that satisfies the differential equation
$$\dfrac{d^2 y}{dx^2} = \dfrac{\rho g}{T}\sqrt{1 + \left(\dfrac{dy}{dx}\right)^2}$$
where ρ is the linear density of the cable, g is the acceleration due to gravity, T is the tension in the cable at its lowest point, and the coordinate system is chosen appropriately. Verify that the function
$$y = f(x) = \dfrac{T}{\rho g}\cosh\!\left(\dfrac{\rho g x}{T}\right)$$
is a solution of this differential equation.

53. A cable with linear density $\rho = 2$ kg/m is strung from the tops of two poles that are 200 m apart.
   (a) Use Exercise 52 to find the tension T so that the cable is 60 m above the ground at its lowest point. How tall are the poles?
   (b) If the tension is doubled, what is the new low point of the cable? How tall are the poles now?

54. A model for the velocity of a falling object after time t is
$$v(t) = \sqrt{\dfrac{mg}{k}}\tanh\!\left(t\sqrt{\dfrac{gk}{m}}\right)$$
where m is the mass of the object, $g = 9.8$ m/s^2 is the acceleration due to gravity, k is a constant, t is measured in seconds, and v in m/s.
   (a) Calculate the terminal velocity of the object, that is, $\lim_{t\to\infty} v(t)$.
   (b) If a person falls from a building, the value of the constant k depends on his or her position. For a "belly-to-earth" position, $k = 0.515$ kg/s, but for a "feet-first" position, $k = 0.067$ kg/s. If a 60-kg person falls in belly-to-earth position, what is the terminal velocity? What about feet-first?

Source: L. Long et al., "How Terminal Is Terminal Velocity?" *American Mathematical Monthly* 113 (2006): 752–55.

55. (a) Show that any function of the form
$$y = A \sinh mx + B \cosh mx$$
satisfies the differential equation $y'' = m^2 y$.
   (b) Find $y = y(x)$ such that $y'' = 9y$, $y(0) = -4$, and $y'(0) = 6$.

56. If $x = \ln(\sec\theta + \tan\theta)$, show that $\sec\theta = \cosh x$.

57. At what point of the curve $y = \cosh x$ does the tangent have slope 1?

58. Investigate the family of functions

$$f_n(x) = \tanh(n \sin x)$$

where n is a positive integer. Describe what happens to the graph of f_n when n becomes large.

59. Show that if $a \neq 0$ and $b \neq 0$, then there exist numbers α and β such that $ae^x + be^{-x}$ equals either

$$\alpha \sinh(x + \beta) \qquad \text{or} \qquad \alpha \cosh(x + \beta)$$

In other words, almost every function of the form $f(x) = ae^x + be^{-x}$ is a shifted and stretched hyperbolic sine or cosine function.

3 REVIEW

CONCEPT CHECK

Answers to the Concept Check can be found on the back endpapers.

1. State each differentiation rule both in symbols and in words.
   (a) The Power Rule
   (b) The Constant Multiple Rule
   (c) The Sum Rule
   (d) The Difference Rule
   (e) The Product Rule
   (f) The Quotient Rule
   (g) The Chain Rule

2. State the derivative of each function.
   (a) $y = x^n$
   (b) $y = e^x$
   (c) $y = b^x$
   (d) $y = \ln x$
   (e) $y = \log_b x$
   (f) $y = \sin x$
   (g) $y = \cos x$
   (h) $y = \tan x$
   (i) $y = \csc x$
   (j) $y = \sec x$
   (k) $y = \cot x$
   (l) $y = \sin^{-1}x$
   (m) $y = \cos^{-1}x$
   (n) $y = \tan^{-1}x$
   (o) $y = \sinh x$
   (p) $y = \cosh x$
   (q) $y = \tanh x$
   (r) $y = \sinh^{-1}x$
   (s) $y = \cosh^{-1}x$
   (t) $y = \tanh^{-1}x$

3. (a) How is the number e defined?
   (b) Express e as a limit.
   (c) Why is the natural exponential function $y = e^x$ used more often in calculus than the other exponential functions $y = b^x$?

   (d) Why is the natural logarithmic function $y = \ln x$ used more often in calculus than the other logarithmic functions $y = \log_b x$?

4. (a) Explain how implicit differentiation works.
   (b) Explain how logarithmic differentiation works.

5. Give several examples of how the derivative can be interpreted as a rate of change in physics, chemistry, biology, economics, or other sciences.

6. (a) Write a differential equation that expresses the law of natural growth.
   (b) Under what circumstances is this an appropriate model for population growth?
   (c) What are the solutions of this equation?

7. (a) Write an expression for the linearization of f at a.
   (b) If $y = f(x)$, write an expression for the differential dy.
   (c) If $dx = \Delta x$, draw a picture showing the geometric meanings of Δy and dy.

TRUE-FALSE QUIZ

Determine whether the statement is true or false. If it is true, explain why. If it is false, explain why or give an example that disproves the statement.

1. If f and g are differentiable, then

$$\frac{d}{dx}[f(x) + g(x)] = f'(x) + g'(x)$$

2. If f and g are differentiable, then

$$\frac{d}{dx}[f(x)g(x)] = f'(x)g'(x)$$

3. If f and g are differentiable, then

$$\frac{d}{dx}[f(g(x))] = f'(g(x))g'(x)$$

4. If f is differentiable, then $\dfrac{d}{dx}\sqrt{f(x)} = \dfrac{f'(x)}{2\sqrt{f(x)}}$.

5. If f is differentiable, then $\dfrac{d}{dx}f(\sqrt{x}) = \dfrac{f'(x)}{2\sqrt{x}}$.

6. If $y = e^2$, then $y' = 2e$.

7. $\dfrac{d}{dx}(10^x) = x10^{x-1}$

8. $\dfrac{d}{dx}(\ln 10) = \dfrac{1}{10}$

9. $\dfrac{d}{dx}(\tan^2 x) = \dfrac{d}{dx}(\sec^2 x)$

10. $\dfrac{d}{dx}|x^2 + x| = |2x + 1|$

11. The derivative of a polynomial is a polynomial.

12. If $f(x) = (x^6 - x^4)^5$, then $f^{(31)}(x) = 0$.

13. The derivative of a rational function is a rational function.

14. An equation of the tangent line to the parabola $y = x^2$ at $(-2, 4)$ is $y - 4 = 2x(x + 2)$.

15. If $g(x) = x^5$, then $\displaystyle\lim_{x \to 2} \frac{g(x) - g(2)}{x - 2} = 80$

EXERCISES

1–50 Calculate y'.

1. $y = (x^2 + x^3)^4$

2. $y = \dfrac{1}{\sqrt{x}} - \dfrac{1}{\sqrt[5]{x^3}}$

3. $y = \dfrac{x^2 - x + 2}{\sqrt{x}}$

4. $y = \dfrac{\tan x}{1 + \cos x}$

5. $y = x^2 \sin \pi x$

6. $y = x \cos^{-1} x$

7. $y = \dfrac{t^4 - 1}{t^4 + 1}$

8. $xe^y = y \sin x$

9. $y = \ln(x \ln x)$

10. $y = e^{mx} \cos nx$

11. $y = \sqrt{x} \cos \sqrt{x}$

12. $y = (\arcsin 2x)^2$

13. $y = \dfrac{e^{1/x}}{x^2}$

14. $y = \ln \sec x$

15. $y + x \cos y = x^2 y$

16. $y = \left(\dfrac{u - 1}{u^2 + u + 1}\right)^4$

17. $y = \sqrt{\arctan x}$

18. $y = \cot(\csc x)$

19. $y = \tan\left(\dfrac{t}{1 + t^2}\right)$

20. $y = e^{x \sec x}$

21. $y = 3^{x \ln x}$

22. $y = \sec(1 + x^2)$

23. $y = (1 - x^{-1})^{-1}$

24. $y = 1/\sqrt[3]{x + \sqrt{x}}$

25. $\sin(xy) = x^2 - y$

26. $y = \sqrt{\sin \sqrt{x}}$

27. $y = \log_5(1 + 2x)$

28. $y = (\cos x)^x$

29. $y = \ln \sin x - \tfrac{1}{2} \sin^2 x$

30. $y = \dfrac{(x^2 + 1)^4}{(2x + 1)^3(3x - 1)^5}$

31. $y = x \tan^{-1}(4x)$

32. $y = e^{\cos x} + \cos(e^x)$

33. $y = \ln|\sec 5x + \tan 5x|$

34. $y = 10^{\tan \pi \theta}$

35. $y = \cot(3x^2 + 5)$

36. $y = \sqrt{t \ln(t^4)}$

37. $y = \sin\left(\tan \sqrt{1 + x^3}\right)$

38. $y = \arctan\left(\arcsin \sqrt{x}\right)$

39. $y = \tan^2(\sin \theta)$

40. $xe^y = y - 1$

41. $y = \dfrac{\sqrt{x + 1}\,(2 - x)^5}{(x + 3)^7}$

42. $y = \dfrac{(x + \lambda)^4}{x^4 + \lambda^4}$

43. $y = x \sinh(x^2)$

44. $y = \dfrac{\sin mx}{x}$

45. $y = \ln(\cosh 3x)$

46. $y = \ln\left|\dfrac{x^2 - 4}{2x + 5}\right|$

47. $y = \cosh^{-1}(\sinh x)$

48. $y = x \tanh^{-1}\sqrt{x}$

49. $y = \cos\left(e^{\sqrt{\tan 3x}}\right)$

50. $y = \sin^2\left(\cos \sqrt{\sin \pi x}\right)$

51. If $f(t) = \sqrt{4t + 1}$, find $f''(2)$.

52. If $g(\theta) = \theta \sin \theta$, find $g''(\pi/6)$.

53. Find y'' if $x^6 + y^6 = 1$.

54. Find $f^{(n)}(x)$ if $f(x) = 1/(2 - x)$.

55. Use mathematical induction (page 72) to show that if $f(x) = xe^x$, then $f^{(n)}(x) = (x + n)e^x$.

56. Evaluate $\displaystyle\lim_{t \to 0} \frac{t^3}{\tan^3(2t)}$.

57–59 Find an equation of the tangent to the curve at the given point.

57. $y = 4 \sin^2 x$, $(\pi/6, 1)$

58. $y = \dfrac{x^2 - 1}{x^2 + 1}$, $(0, -1)$

59. $y = \sqrt{1 + 4 \sin x}$, $(0, 1)$

60–61 Find equations of the tangent line and normal line to the curve at the given point.

60. $x^2 + 4xy + y^2 = 13$, $(2, 1)$

61. $y = (2 + x)e^{-x}$, $(0, 2)$

62. If $f(x) = xe^{\sin x}$, find $f'(x)$. Graph f and f' on the same screen and comment.

63. (a) If $f(x) = x\sqrt{5 - x}$, find $f'(x)$.
(b) Find equations of the tangent lines to the curve $y = x\sqrt{5 - x}$ at the points $(1, 2)$ and $(4, 4)$.
(c) Illustrate part (b) by graphing the curve and tangent lines on the same screen.
(d) Check to see that your answer to part (a) is reasonable by comparing the graphs of f and f'.

64. (a) If $f(x) = 4x - \tan x$, $-\pi/2 < x < \pi/2$, find f' and f''.
(b) Check to see that your answers to part (a) are reasonable by comparing the graphs of f, f', and f''.

65. At what points on the curve $y = \sin x + \cos x$, $0 \le x \le 2\pi$, is the tangent line horizontal?

66. Find the points on the ellipse $x^2 + 2y^2 = 1$ where the tangent line has slope 1.

67. If $f(x) = (x - a)(x - b)(x - c)$, show that

$$\frac{f'(x)}{f(x)} = \frac{1}{x - a} + \frac{1}{x - b} + \frac{1}{x - c}$$

68. (a) By differentiating the double-angle formula

$$\cos 2x = \cos^2 x - \sin^2 x$$

obtain the double-angle formula for the sine function.

(b) By differentiating the addition formula

$$\sin(x + a) = \sin x \cos a + \cos x \sin a$$

obtain the addition formula for the cosine function.

69. Suppose that

$$f(1) = 2 \quad f'(1) = 3 \quad f(2) = 1 \quad f'(2) = 2$$
$$g(1) = 3 \quad g'(1) = 1 \quad g(2) = 1 \quad g'(2) = 4$$

(a) If $S(x) = f(x) + g(x)$, find $S'(1)$.
(b) If $P(x) = f(x)g(x)$, find $P'(2)$.
(c) If $Q(x) = f(x)/g(x)$, find $Q'(1)$.
(d) If $C(x) = f(g(x))$, find $C'(2)$.

70. If f and g are the functions whose graphs are shown, let $P(x) = f(x)g(x)$, $Q(x) = f(x)/g(x)$, and $C(x) = f(g(x))$. Find (a) $P'(2)$, (b) $Q'(2)$, and (c) $C'(2)$.

71–78 Find f' in terms of g'.

71. $f(x) = x^2 g(x)$

72. $f(x) = g(x^2)$

73. $f(x) = [g(x)]^2$

74. $f(x) = g(g(x))$

75. $f(x) = g(e^x)$

76. $f(x) = e^{g(x)}$

77. $f(x) = \ln |g(x)|$

78. $f(x) = g(\ln x)$

79–81 Find h' in terms of f' and g'.

79. $h(x) = \dfrac{f(x)g(x)}{f(x) + g(x)}$

80. $h(x) = \sqrt{\dfrac{f(x)}{g(x)}}$

81. $h(x) = f(g(\sin 4x))$

82. (a) Graph the function $f(x) = x - 2 \sin x$ in the viewing rectangle $[0, 8]$ by $[-2, 8]$.
(b) On which interval is the average rate of change larger: $[1, 2]$ or $[2, 3]$?
(c) At which value of x is the instantaneous rate of change larger: $x = 2$ or $x = 5$?
(d) Check your visual estimates in part (c) by computing $f'(x)$ and comparing the numerical values of $f'(2)$ and $f'(5)$.

83. At what point on the curve $y = [\ln(x + 4)]^2$ is the tangent horizontal?

84. (a) Find an equation of the tangent to the curve $y = e^x$ that is parallel to the line $x - 4y = 1$.
(b) Find an equation of the tangent to the curve $y = e^x$ that passes through the origin.

85. Find a parabola $y = ax^2 + bx + c$ that passes through the point $(1, 4)$ and whose tangent lines at $x = -1$ and $x = 5$ have slopes 6 and -2, respectively.

86. The function $C(t) = K(e^{-at} - e^{-bt})$, where a, b, and K are positive constants and $b > a$, is used to model the concentration at time t of a drug injected into the bloodstream.
(a) Show that $\lim_{t \to \infty} C(t) = 0$.
(b) Find $C'(t)$, the rate of change of drug concentration in the blood.
(c) When is this rate equal to 0?

87. An equation of motion of the form $s = Ae^{-ct}\cos(\omega t + \delta)$ represents damped oscillation of an object. Find the velocity and acceleration of the object.

88. A particle moves along a horizontal line so that its coordinate at time t is $x = \sqrt{b^2 + c^2 t^2}$, $t \ge 0$, where b and c are positive constants.
(a) Find the velocity and acceleration functions.
(b) Show that the particle always moves in the positive direction.

89. A particle moves on a vertical line so that its coordinate at time t is $y = t^3 - 12t + 3$, $t \ge 0$.
(a) Find the velocity and acceleration functions.
(b) When is the particle moving upward and when is it moving downward?
(c) Find the distance that the particle travels in the time interval $0 \le t \le 3$.
(d) Graph the position, velocity, and acceleration functions for $0 \le t \le 3$.
(e) When is the particle speeding up? When is it slowing down?

90. The volume of a right circular cone is $V = \frac{1}{3}\pi r^2 h$, where r is the radius of the base and h is the height.
(a) Find the rate of change of the volume with respect to the height if the radius is constant.

(b) Find the rate of change of the volume with respect to the radius if the height is constant.

91. The mass of part of a wire is $x\left(1 + \sqrt{x}\right)$ kilograms, where x is measured in meters from one end of the wire. Find the linear density of the wire when $x = 4$ m.

92. The cost, in dollars, of producing x units of a certain commodity is

$$C(x) = 920 + 2x - 0.02x^2 + 0.00007x^3$$

(a) Find the marginal cost function.
(b) Find $C'(100)$ and explain its meaning.
(c) Compare $C'(100)$ with the cost of producing the 101st item.

93. A bacteria culture contains 200 cells initially and grows at a rate proportional to its size. After half an hour the population has increased to 360 cells.
(a) Find the number of bacteria after t hours.
(b) Find the number of bacteria after 4 hours.
(c) Find the rate of growth after 4 hours.
(d) When will the population reach 10,000?

94. Cobalt-60 has a half-life of 5.24 years.
(a) Find the mass that remains from a 100-mg sample after 20 years.
(b) How long would it take for the mass to decay to 1 mg?

95. Let $C(t)$ be the concentration of a drug in the bloodstream. As the body eliminates the drug, $C(t)$ decreases at a rate that is proportional to the amount of the drug that is present at the time. Thus $C'(t) = -kC(t)$, where k is a positive number called the *elimination constant* of the drug.
(a) If C_0 is the concentration at time $t = 0$, find the concentration at time t.
(b) If the body eliminates half the drug in 30 hours, how long does it take to eliminate 90% of the drug?

96. A cup of hot chocolate has temperature 80°C in a room kept at 20°C. After half an hour the hot chocolate cools to 60°C.
(a) What is the temperature of the chocolate after another half hour?
(b) When will the chocolate have cooled to 40°C?

97. The volume of a cube is increasing at a rate of 10 cm³/min. How fast is the surface area increasing when the length of an edge is 30 cm?

98. A paper cup has the shape of a cone with height 10 cm and radius 3 cm (at the top). If water is poured into the cup at a rate of 2 cm³/s, how fast is the water level rising when the water is 5 cm deep?

99. A balloon is rising at a constant speed of 5 ft/s. A boy is cycling along a straight road at a speed of 15 ft/s. When he passes under the balloon, it is 45 ft above him. How fast is the distance between the boy and the balloon increasing 3 s later?

100. A waterskier skis over the ramp shown in the figure at a speed of 30 ft/s. How fast is she rising as she leaves the ramp?

101. The angle of elevation of the sun is decreasing at a rate of 0.25 rad/h. How fast is the shadow cast by a 400-ft-tall building increasing when the angle of elevation of the sun is $\pi/6$?

102. (a) Find the linear approximation to $f(x) = \sqrt{25 - x^2}$ near 3.
(b) Illustrate part (a) by graphing f and the linear approximation.
(c) For what values of x is the linear approximation accurate to within 0.1?

103. (a) Find the linearization of $f(x) = \sqrt[3]{1 + 3x}$ at $a = 0$. State the corresponding linear approximation and use it to give an approximate value for $\sqrt[3]{1.03}$.
(b) Determine the values of x for which the linear approximation given in part (a) is accurate to within 0.1.

104. Evaluate dy if $y = x^3 - 2x^2 + 1$, $x = 2$, and $dx = 0.2$.

105. A window has the shape of a square surmounted by a semicircle. The base of the window is measured as having width 60 cm with a possible error in measurement of 0.1 cm. Use differentials to estimate the maximum error possible in computing the area of the window.

106–108 Express the limit as a derivative and evaluate.

106. $\displaystyle\lim_{x \to 1} \frac{x^{17} - 1}{x - 1}$

107. $\displaystyle\lim_{h \to 0} \frac{\sqrt[4]{16 + h} - 2}{h}$

108. $\displaystyle\lim_{\theta \to \pi/3} \frac{\cos\theta - 0.5}{\theta - \pi/3}$

109. Evaluate $\displaystyle\lim_{x \to 0} \frac{\sqrt{1 + \tan x} - \sqrt{1 + \sin x}}{x^3}$.

110. Suppose f is a differentiable function such that $f(g(x)) = x$ and $f'(x) = 1 + [f(x)]^2$. Show that $g'(x) = 1/(1 + x^2)$.

111. Find $f'(x)$ if it is known that

$$\frac{d}{dx}[f(2x)] = x^2$$

112. Show that the length of the portion of any tangent line to the astroid $x^{2/3} + y^{2/3} = a^{2/3}$ cut off by the coordinate axes is constant.

Problems Plus

Before you look at the examples, cover up the solutions and try them yourself first.

EXAMPLE 1   How many lines are tangent to both of the parabolas $y = -1 - x^2$ and $y = 1 + x^2$? Find the coordinates of the points at which these tangents touch the parabolas.

SOLUTION To gain insight into this problem, it is essential to draw a diagram. So we sketch the parabolas $y = 1 + x^2$ (which is the standard parabola $y = x^2$ shifted 1 unit upward) and $y = -1 - x^2$ (which is obtained by reflecting the first parabola about the x-axis). If we try to draw a line tangent to both parabolas, we soon discover that there are only two possibilities, as illustrated in Figure 1.

Let P be a point at which one of these tangents touches the upper parabola and let a be its x-coordinate. (The choice of notation for the unknown is important. Of course we could have used b or c or x_0 or x_1 instead of a. However, it's not advisable to use x in place of a because that x could be confused with the variable x in the equation of the parabola.) Then, since P lies on the parabola $y = 1 + x^2$, its y-coordinate must be $1 + a^2$. Because of the symmetry shown in Figure 1, the coordinates of the point Q where the tangent touches the lower parabola must be $(-a, -(1 + a^2))$.

To use the given information that the line is a tangent, we equate the slope of the line PQ to the slope of the tangent line at P. We have

$$m_{PQ} = \frac{1 + a^2 - (-1 - a^2)}{a - (-a)} = \frac{1 + a^2}{a}$$

If $f(x) = 1 + x^2$, then the slope of the tangent line at P is $f'(a) = 2a$. Thus the condition that we need to use is that

$$\frac{1 + a^2}{a} = 2a$$

Solving this equation, we get $1 + a^2 = 2a^2$, so $a^2 = 1$ and $a = \pm 1$. Therefore the points are $(1, 2)$ and $(-1, -2)$. By symmetry, the two remaining points are $(-1, 2)$ and $(1, -2)$. ∎

EXAMPLE 2   For what values of c does the equation $\ln x = cx^2$ have exactly one solution?

SOLUTION One of the most important principles of problem solving is to draw a diagram, even if the problem as stated doesn't explicitly mention a geometric situation. Our present problem can be reformulated geometrically as follows: For what values of c does the curve $y = \ln x$ intersect the curve $y = cx^2$ in exactly one point?

Let's start by graphing $y = \ln x$ and $y = cx^2$ for various values of c. We know that, for $c \neq 0$, $y = cx^2$ is a parabola that opens upward if $c > 0$ and downward if $c < 0$. Figure 2 shows the parabolas $y = cx^2$ for several positive values of c. Most of them don't intersect $y = \ln x$ at all and one intersects twice. We have the feeling that there must be a value of c (somewhere between 0.1 and 0.3) for which the curves intersect exactly once, as in Figure 3.

To find that particular value of c, we let a be the x-coordinate of the single point of intersection. In other words, $\ln a = ca^2$, so a is the unique solution of the given equation. We see from Figure 3 that the curves just touch, so they have a common tangent line when $x = a$. That means the curves $y = \ln x$ and $y = cx^2$ have the same slope when $x = a$. Therefore

$$\frac{1}{a} = 2ca$$

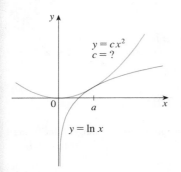

FIGURE 1

FIGURE 2

FIGURE 3

Solving the equations $\ln a = ca^2$ and $1/a = 2ca$, we get

$$\ln a = ca^2 = c \cdot \frac{1}{2c} = \frac{1}{2}$$

Thus $a = e^{1/2}$ and

$$c = \frac{\ln a}{a^2} = \frac{\ln e^{1/2}}{e} = \frac{1}{2e}$$

For negative values of c we have the situation illustrated in Figure 4: All parabolas $y = cx^2$ with negative values of c intersect $y = \ln x$ exactly once. And let's not forget about $c = 0$: The curve $y = 0x^2 = 0$ is just the x-axis, which intersects $y = \ln x$ exactly once.

To summarize, the required values of c are $c = 1/(2e)$ and $c \le 0$. ∎

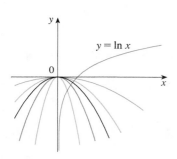

FIGURE 4

Problems

1. Find points P and Q on the parabola $y = 1 - x^2$ so that the triangle ABC formed by the x-axis and the tangent lines at P and Q is an equilateral triangle. (See the figure.)

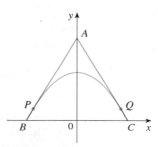

2. Find the point where the curves $y = x^3 - 3x + 4$ and $y = 3(x^2 - x)$ are tangent to each other, that is, have a common tangent line. Illustrate by sketching both curves and the common tangent.

3. Show that the tangent lines to the parabola $y = ax^2 + bx + c$ at any two points with x-coordinates p and q must intersect at a point whose x-coordinate is halfway between p and q.

4. Show that

$$\frac{d}{dx}\left(\frac{\sin^2 x}{1 + \cot x} + \frac{\cos^2 x}{1 + \tan x} \right) = -\cos 2x$$

5. If $f(x) = \lim\limits_{t \to x} \dfrac{\sec t - \sec x}{t - x}$, find the value of $f'(\pi/4)$.

6. Find the values of the constants a and b such that

$$\lim_{x \to 0} \frac{\sqrt[3]{ax + b} - 2}{x} = \frac{5}{12}$$

7. Show that $\sin^{-1}(\tanh x) = \tan^{-1}(\sinh x)$.

8. A car is traveling at night along a highway shaped like a parabola with its vertex at the origin (see the figure). The car starts at a point 100 m west and 100 m north of the origin and travels in an easterly direction. There is a statue located 100 m east and 50 m north of the origin. At what point on the highway will the car's headlights illuminate the statue?

FIGURE FOR PROBLEM 8

9. Prove that $\dfrac{d^n}{dx^n}(\sin^4 x + \cos^4 x) = 4^{n-1}\cos(4x + n\pi/2)$.

10. If f is differentiable at a, where $a > 0$, evaluate the following limit in terms of $f'(a)$:

$$\lim_{x \to a} \frac{f(x) - f(a)}{\sqrt{x} - \sqrt{a}}$$

11. The figure shows a circle with radius 1 inscribed in the parabola $y = x^2$. Find the center of the circle.

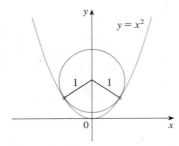

12. Find all values of c such that the parabolas $y = 4x^2$ and $x = c + 2y^2$ intersect each other at right angles.

13. How many lines are tangent to both of the circles $x^2 + y^2 = 4$ and $x^2 + (y - 3)^2 = 1$? At what points do these tangent lines touch the circles?

14. If $f(x) = \dfrac{x^{46} + x^{45} + 2}{1 + x}$, calculate $f^{(46)}(3)$. Express your answer using factorial notation: $n! = 1 \cdot 2 \cdot 3 \cdot \cdots \cdot (n - 1) \cdot n$.

15. The figure shows a rotating wheel with radius 40 cm and a connecting rod AP with length 1.2 m. The pin P slides back and forth along the x-axis as the wheel rotates counterclockwise at a rate of 360 revolutions per minute.
   (a) Find the angular velocity of the connecting rod, $d\alpha/dt$, in radians per second, when $\theta = \pi/3$.
   (b) Express the distance $x = |OP|$ in terms of θ.
   (c) Find an expression for the velocity of the pin P in terms of θ.

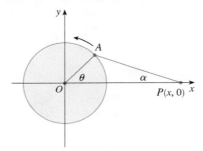

16. Tangent lines T_1 and T_2 are drawn at two points P_1 and P_2 on the parabola $y = x^2$ and they intersect at a point P. Another tangent line T is drawn at a point between P_1 and P_2; it intersects T_1 at Q_1 and T_2 at Q_2. Show that

$$\frac{|PQ_1|}{|PP_1|} + \frac{|PQ_2|}{|PP_2|} = 1$$

17. Show that

$$\frac{d^n}{dx^n} (e^{ax} \sin bx) = r^n e^{ax} \sin(bx + n\theta)$$

where a and b are positive numbers, $r^2 = a^2 + b^2$, and $\theta = \tan^{-1}(b/a)$.

18. Evaluate $\displaystyle\lim_{x \to \pi} \frac{e^{\sin x} - 1}{x - \pi}$.

19. Let T and N be the tangent and normal lines to the ellipse $x^2/9 + y^2/4 = 1$ at any point P on the ellipse in the first quadrant. Let x_T and y_T be the x- and y-intercepts of T and x_N and y_N be the intercepts of N. As P moves along the ellipse in the first quadrant (but not on the axes), what values can x_T, y_T, x_N, and y_N take on? First try to guess the answers just by looking at the figure. Then use calculus to solve the problem and see how good your intuition is.

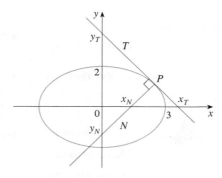

20. Evaluate $\displaystyle\lim_{x \to 0} \frac{\sin(3 + x)^2 - \sin 9}{x}$.

21. (a) Use the identity for $\tan(x - y)$ (see Equation 14b in Appendix D) to show that if two lines L_1 and L_2 intersect at an angle α, then

$$\tan \alpha = \frac{m_2 - m_1}{1 + m_1 m_2}$$

where m_1 and m_2 are the slopes of L_1 and L_2, respectively.

(b) The **angle between the curves** C_1 and C_2 at a point of intersection P is defined to be the angle between the tangent lines to C_1 and C_2 at P (if these tangent lines exist). Use part (a) to find, correct to the nearest degree, the angle between each pair of curves at each point of intersection.
   (i) $y = x^2$ and $y = (x - 2)^2$
   (ii) $x^2 - y^2 = 3$ and $x^2 - 4x + y^2 + 3 = 0$

22. Let $P(x_1, y_1)$ be a point on the parabola $y^2 = 4px$ with focus $F(p, 0)$. Let α be the angle between the parabola and the line segment FP, and let β be the angle between the horizontal line $y = y_1$ and the parabola as in the figure. Prove that $\alpha = \beta$. (Thus, by a principle of geometrical optics, light from a source placed at F will be reflected along a line parallel to the x-axis. This explains why *paraboloids,* the surfaces obtained by rotating parabolas about their axes, are used as the shape of some automobile headlights and mirrors for telescopes.)

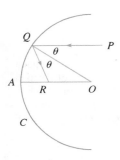

Q θ P

θ

A R O

C

FIGURE FOR PROBLEM 23

23. Suppose that we replace the parabolic mirror of Problem 22 by a spherical mirror. Although the mirror has no focus, we can show the existence of an *approximate* focus. In the figure, C is a semicircle with center O. A ray of light coming in toward the mirror parallel to the axis along the line PQ will be reflected to the point R on the axis so that $\angle PQO = \angle OQR$ (the angle of incidence is equal to the angle of reflection). What happens to the point R as P is taken closer and closer to the axis?

24. If f and g are differentiable functions with $f(0) = g(0) = 0$ and $g'(0) \neq 0$, show that

$$\lim_{x \to 0} \frac{f(x)}{g(x)} = \frac{f'(0)}{g'(0)}$$

25. Evaluate $\displaystyle\lim_{x \to 0} \frac{\sin(a + 2x) - 2\sin(a + x) + \sin a}{x^2}$.

CAS 26. (a) The cubic function $f(x) = x(x - 2)(x - 6)$ has three distinct zeros: 0, 2, and 6. Graph f and its tangent lines at the *average* of each pair of zeros. What do you notice?
    (b) Suppose the cubic function $f(x) = (x - a)(x - b)(x - c)$ has three distinct zeros: a, b, and c. Prove, with the help of a computer algebra system, that a tangent line drawn at the average of the zeros a and b intersects the graph of f at the third zero.

27. For what value of k does the equation $e^{2x} = k\sqrt{x}$ have exactly one solution?

28. For which positive numbers a is it true that $a^x \geq 1 + x$ for all x?

29. If

$$y = \frac{x}{\sqrt{a^2 - 1}} - \frac{2}{\sqrt{a^2 - 1}} \arctan \frac{\sin x}{a + \sqrt{a^2 - 1} + \cos x}$$

   show that $y' = \dfrac{1}{a + \cos x}$.

30. Given an ellipse $x^2/a^2 + y^2/b^2 = 1$, where $a \neq b$, find the equation of the set of all points from which there are two tangents to the curve whose slopes are (a) reciprocals and (b) negative reciprocals.

31. Find the two points on the curve $y = x^4 - 2x^2 - x$ that have a common tangent line.

32. Suppose that three points on the parabola $y = x^2$ have the property that their normal lines intersect at a common point. Show that the sum of their x-coordinates is 0.

33. A *lattice point* in the plane is a point with integer coordinates. Suppose that circles with radius r are drawn using all lattice points as centers. Find the smallest value of r such that any line with slope $\frac{2}{5}$ intersects some of these circles.

34. A cone of radius r centimeters and height h centimeters is lowered point first at a rate of 1 cm/s into a tall cylinder of radius R centimeters that is partially filled with water. How fast is the water level rising at the instant the cone is completely submerged?

35. A container in the shape of an inverted cone has height 16 cm and radius 5 cm at the top. It is partially filled with a liquid that oozes through the sides at a rate proportional to the area of the container that is in contact with the liquid. (The surface area of a cone is $\pi r l$, where r is the radius and l is the slant height.) If we pour the liquid into the container at a rate of 2 cm³/min, then the height of the liquid decreases at a rate of 0.3 cm/min when the height is 10 cm. If our goal is to keep the liquid at a constant height of 10 cm, at what rate should we pour the liquid into the container?

4 Applications of Differentiation

© Tatiana Makotra / Shutterstock.com

When we view the world around us, the light entering the eye near the center of the pupil is perceived brighter than light entering closer to the edges of the pupil. This phenomenon, known as the *Stiles–Crawford effect*, is explored as the pupil changes in radius in Exercise 80 on page 313.

WE HAVE ALREADY INVESTIGATED SOME of the applications of derivatives, but now that we know the differentiation rules we are in a better position to pursue the applications of differentiation in greater depth. Here we learn how derivatives affect the shape of a graph of a function and, in particular, how they help us locate maximum and minimum values of functions. Many practical problems require us to minimize a cost or maximize an area or somehow find the best possible outcome of a situation. In particular, we will be able to investigate the optimal shape of a can and to explain the location of rainbows in the sky.

4.1   Maximum and Minimum Values

Some of the most important applications of differential calculus are *optimization problems*, in which we are required to find the optimal (best) way of doing something. Here are examples of such problems that we will solve in this chapter:

- What is the shape of a can that minimizes manufacturing costs?
- What is the maximum acceleration of a space shuttle? (This is an important question to the astronauts who have to withstand the effects of acceleration.)
- What is the radius of a contracted windpipe that expels air most rapidly during a cough?
- At what angle should blood vessels branch so as to minimize the energy expended by the heart in pumping blood?

These problems can be reduced to finding the maximum or minimum values of a function. Let's first explain exactly what we mean by maximum and minimum values.

We see that the highest point on the graph of the function f shown in Figure 1 is the point $(3, 5)$. In other words, the largest value of f is $f(3) = 5$. Likewise, the smallest value is $f(6) = 2$. We say that $f(3) = 5$ is the *absolute maximum* of f and $f(6) = 2$ is the *absolute minimum*. In general, we use the following definition.

FIGURE 1

1   Definition   Let c be a number in the domain D of a function f. Then $f(c)$ is the
- **absolute maximum** value of f on D if $f(c) \geqslant f(x)$ for all x in D.
- **absolute minimum** value of f on D if $f(c) \leqslant f(x)$ for all x in D.

An absolute maximum or minimum is sometimes called a **global** maximum or minimum. The maximum and minimum values of f are called **extreme values** of f.

Figure 2 shows the graph of a function f with absolute maximum at d and absolute minimum at a. Note that $(d, f(d))$ is the highest point on the graph and $(a, f(a))$ is the lowest point. In Figure 2, if we consider only values of x near b [for instance, if we restrict our attention to the interval (a, c)], then $f(b)$ is the largest of those values of $f(x)$ and is called a *local maximum value* of f. Likewise, $f(c)$ is called a *local minimum value* of f because $f(c) \leqslant f(x)$ for x near c [in the interval (b, d), for instance]. The function f also has a local minimum at e. In general, we have the following definition.

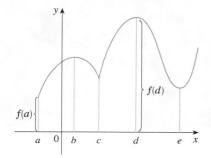

FIGURE 2
Abs min $f(a)$, abs max $f(d)$,
loc min $f(c), f(e)$, loc max $f(b), f(d)$

2   Definition   The number $f(c)$ is a
- **local maximum** value of f if $f(c) \geqslant f(x)$ when x is near c.
- **local minimum** value of f if $f(c) \leqslant f(x)$ when x is near c.

In Definition 2 (and elsewhere), if we say that something is true **near** c, we mean that it is true on some open interval containing c. For instance, in Figure 3 we see that $f(4) = 5$ is a local minimum because it's the smallest value of f on the interval I. It's not the absolute minimum because $f(x)$ takes smaller values when x is near 12 (in the interval K, for instance). In fact $f(12) = 3$ is both a local minimum and the absolute minimum. Similarly, $f(8) = 7$ is a local maximum, but not the absolute maximum because f takes larger values near 1.

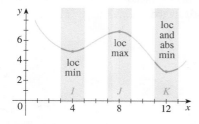

FIGURE 3

EXAMPLE 1 The function $f(x) = \cos x$ takes on its (local and absolute) maximum value of 1 infinitely many times, since $\cos 2n\pi = 1$ for any integer n and $-1 \leq \cos x \leq 1$ for all x. (See Figure 4.) Likewise, $\cos(2n + 1)\pi = -1$ is its minimum value, where n is any integer.

FIGURE 4

$y = \cos x$

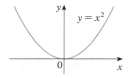

FIGURE 5

Mimimum value 0, no maximum

EXAMPLE 2 If $f(x) = x^2$, then $f(x) \geq f(0)$ because $x^2 \geq 0$ for all x. Therefore $f(0) = 0$ is the absolute (and local) minimum value of f. This corresponds to the fact that the origin is the lowest point on the parabola $y = x^2$. (See Figure 5.) However, there is no highest point on the parabola and so this function has no maximum value.

EXAMPLE 3 From the graph of the function $f(x) = x^3$, shown in Figure 6, we see that this function has neither an absolute maximum value nor an absolute minimum value. In fact, it has no local extreme values either.

EXAMPLE 4 The graph of the function

$$f(x) = 3x^4 - 16x^3 + 18x^2 \qquad -1 \leq x \leq 4$$

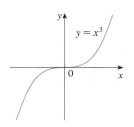

FIGURE 6

No mimimum, no maximum

is shown in Figure 7. You can see that $f(1) = 5$ is a local maximum, whereas the absolute maximum is $f(-1) = 37$. (This absolute maximum is not a local maximum because it occurs at an endpoint.) Also, $f(0) = 0$ is a local minimum and $f(3) = -27$ is both a local and an absolute minimum. Note that f has neither a local nor an absolute maximum at $x = 4$.

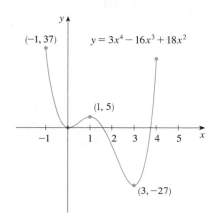

FIGURE 7

We have seen that some functions have extreme values, whereas others do not. The following theorem gives conditions under which a function is guaranteed to possess extreme values.

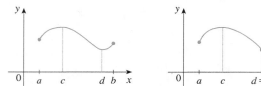

3 **The Extreme Value Theorem** If f is continuous on a closed interval $[a, b]$, then f attains an absolute maximum value $f(c)$ and an absolute minimum value $f(d)$ at some numbers c and d in $[a, b]$.

The Extreme Value Theorem is illustrated in Figure 8. Note that an extreme value can be taken on more than once. Although the Extreme Value Theorem is intuitively very plausible, it is difficult to prove and so we omit the proof.

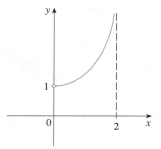

FIGURE 8 Functions continuous on a closed interval always attain extreme values.

Figures 9 and 10 show that a function need not possess extreme values if either hypothesis (continuity or closed interval) is omitted from the Extreme Value Theorem.

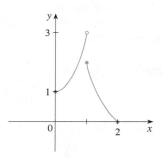

FIGURE 9
This function has minimum value
$f(2) = 0$, but no maximum value.

FIGURE 10
This continuous function g has
no maximum or minimum.

The function f whose graph is shown in Figure 9 is defined on the closed interval $[0, 2]$ but has no maximum value. (Notice that the range of f is $[0, 3)$. The function takes on values arbitrarily close to 3, but never actually attains the value 3.) This does not contradict the Extreme Value Theorem because f is not continuous. [Nonetheless, a discontinuous function *could* have maximum and minimum values. See Exercise 13(b).]

The function g shown in Figure 10 is continuous on the open interval $(0, 2)$ but has neither a maximum nor a minimum value. [The range of g is $(1, \infty)$. The function takes on arbitrarily large values.] This does not contradict the Extreme Value Theorem because the interval $(0, 2)$ is not closed.

The Extreme Value Theorem says that a continuous function on a closed interval has a maximum value and a minimum value, but it does not tell us how to find these extreme values. Notice in Figure 8 that the absolute maximum and minimum values that are *between* a and b occur at local maximum or minimum values, so we start by looking for local extreme values.

Figure 11 shows the graph of a function f with a local maximum at c and a local minimum at d. It appears that at the maximum and minimum points the tangent lines are horizontal and therefore each has slope 0. We know that the derivative is the slope of the

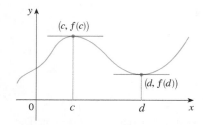

FIGURE 11

tangent line, so it appears that $f'(c) = 0$ and $f'(d) = 0$. The following theorem says that this is always true for differentiable functions.

Fermat

Fermat's Theorem is named after Pierre Fermat (1601–1665), a French lawyer who took up mathematics as a hobby. Despite his amateur status, Fermat was one of the two inventors of analytic geometry (Descartes was the other). His methods for finding tangents to curves and maximum and minimum values (before the invention of limits and derivatives) made him a forerunner of Newton in the creation of differential calculus.

4 **Fermat's Theorem** If f has a local maximum or minimum at c, and if $f'(c)$ exists, then $f'(c) = 0$.

PROOF Suppose, for the sake of definiteness, that f has a local maximum at c. Then, according to Definition 2, $f(c) \geq f(x)$ if x is sufficiently close to c. This implies that if h is sufficiently close to 0, with h being positive or negative, then

$$f(c) \geq f(c + h)$$

and therefore

5
$$f(c + h) - f(c) \leq 0$$

We can divide both sides of an inequality by a positive number. Thus, if $h > 0$ and h is sufficiently small, we have

$$\frac{f(c + h) - f(c)}{h} \leq 0$$

Taking the right-hand limit of both sides of this inequality (using Theorem 2.3.2), we get

$$\lim_{h \to 0^+} \frac{f(c + h) - f(c)}{h} \leq \lim_{h \to 0^+} 0 = 0$$

But since $f'(c)$ exists, we have

$$f'(c) = \lim_{h \to 0} \frac{f(c + h) - f(c)}{h} = \lim_{h \to 0^+} \frac{f(c + h) - f(c)}{h}$$

and so we have shown that $f'(c) \leq 0$.

If $h < 0$, then the direction of the inequality (5) is reversed when we divide by h:

$$\frac{f(c + h) - f(c)}{h} \geq 0 \qquad h < 0$$

So, taking the left-hand limit, we have

$$f'(c) = \lim_{h \to 0} \frac{f(c + h) - f(c)}{h} = \lim_{h \to 0^-} \frac{f(c + h) - f(c)}{h} \geq 0$$

We have shown that $f'(c) \geq 0$ and also that $f'(c) \leq 0$. Since both of these inequalities must be true, the only possibility is that $f'(c) = 0$.

We have proved Fermat's Theorem for the case of a local maximum. The case of a local minimum can be proved in a similar manner, or we could use Exercise 78 to deduce it from the case we have just proved (see Exercise 79). ∎

The following examples caution us against reading too much into Fermat's Theorem: We can't expect to locate extreme values simply by setting $f'(x) = 0$ and solving for x.

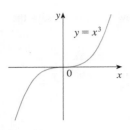

FIGURE 12
If $f(x) = x^3$, then $f'(0) = 0$,
but f has no maximum
or minimum.

FIGURE 13
If $f(x) = |x|$, then $f(0) = 0$ is
a minimum value, but $f'(0)$ does
not exist.

EXAMPLE 5 If $f(x) = x^3$, then $f'(x) = 3x^2$, so $f'(0) = 0$. But f has no maximum or minimum at 0, as you can see from its graph in Figure 12. (Or observe that $x^3 > 0$ for $x > 0$ but $x^3 < 0$ for $x < 0$.) The fact that $f'(0) = 0$ simply means that the curve $y = x^3$ has a horizontal tangent at $(0, 0)$. Instead of having a maximum or minimum at $(0, 0)$, the curve crosses its horizontal tangent there. ∎

EXAMPLE 6 The function $f(x) = |x|$ has its (local and absolute) minimum value at 0, but that value can't be found by setting $f'(x) = 0$ because, as was shown in Example 2.8.5, $f'(0)$ does not exist. (See Figure 13.) ∎

⊘ **WARNING** Examples 5 and 6 show that we must be careful when using Fermat's Theorem. Example 5 demonstrates that even when $f'(c) = 0$ there need not be a maximum or minimum at c. (In other words, the converse of Fermat's Theorem is false in general.) Furthermore, there may be an extreme value even when $f'(c)$ does not exist (as in Example 6).

Fermat's Theorem does suggest that we should at least *start* looking for extreme values of f at the numbers c where $f'(c) = 0$ or where $f'(c)$ does not exist. Such numbers are given a special name.

> **6** **Definition** A **critical number** of a function f is a number c in the domain of f such that either $f'(c) = 0$ or $f'(c)$ does not exist.

EXAMPLE 7 Find the critical numbers of $f(x) = x^{3/5}(4 - x)$.

SOLUTION The Product Rule gives

Figure 14 shows a graph of the function f in Example 7. It supports our answer because there is a horizontal tangent when $x = 1.5$ [where $f'(x) = 0$] and a vertical tangent when $x = 0$ [where $f'(x)$ is undefined].

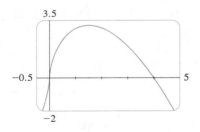

FIGURE 14

$$f'(x) = x^{3/5}(-1) + (4 - x)(\tfrac{3}{5}x^{-2/5}) = -x^{3/5} + \frac{3(4 - x)}{5x^{2/5}}$$

$$= \frac{-5x + 3(4 - x)}{5x^{2/5}} = \frac{12 - 8x}{5x^{2/5}}$$

[The same result could be obtained by first writing $f(x) = 4x^{3/5} - x^{8/5}$.] Therefore $f'(x) = 0$ if $12 - 8x = 0$, that is, $x = \tfrac{3}{2}$, and $f'(x)$ does not exist when $x = 0$. Thus the critical numbers are $\tfrac{3}{2}$ and 0. ∎

In terms of critical numbers, Fermat's Theorem can be rephrased as follows (compare Definition 6 with Theorem 4):

> **7** If f has a local maximum or minimum at c, then c is a critical number of f.

To find an absolute maximum or minimum of a continuous function on a closed interval, we note that either it is local [in which case it occurs at a critical number by (7)] or it occurs at an endpoint of the interval, as we see from the examples in Figure 8. Thus the following three-step procedure always works.

> **The Closed Interval Method** To find the *absolute* maximum and minimum values of a continuous function f on a closed interval $[a, b]$:
>
> **1.** Find the values of f at the critical numbers of f in (a, b).
>
> **2.** Find the values of f at the endpoints of the interval.
>
> **3.** The largest of the values from Steps 1 and 2 is the absolute maximum value; the smallest of these values is the absolute minimum value.

EXAMPLE 8 Find the absolute maximum and minimum values of the function

$$f(x) = x^3 - 3x^2 + 1 \qquad -\tfrac{1}{2} \leqslant x \leqslant 4$$

SOLUTION Since f is continuous on $\left[-\tfrac{1}{2}, 4\right]$, we can use the Closed Interval Method:

$$f(x) = x^3 - 3x^2 + 1$$

$$f'(x) = 3x^2 - 6x = 3x(x - 2)$$

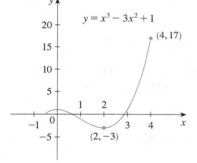

$y = x^3 - 3x^2 + 1$

Since $f'(x)$ exists for all x, the only critical numbers of f occur when $f'(x) = 0$, that is, $x = 0$ or $x = 2$. Notice that each of these critical numbers lies in the interval $\left(-\tfrac{1}{2}, 4\right)$. The values of f at these critical numbers are

$$f(0) = 1 \qquad f(2) = -3$$

The values of f at the endpoints of the interval are

$$f\left(-\tfrac{1}{2}\right) = \tfrac{1}{8} \qquad f(4) = 17$$

Comparing these four numbers, we see that the absolute maximum value is $f(4) = 17$ and the absolute minimum value is $f(2) = -3$.

Note that in this example the absolute maximum occurs at an endpoint, whereas the absolute minimum occurs at a critical number. The graph of f is sketched in Figure 15. ∎

FIGURE 15

If you have a graphing calculator or a computer with graphing software, it is possible to estimate maximum and minimum values very easily. But, as the next example shows, calculus is needed to find the *exact* values.

EXAMPLE 9

(a) Use a graphing device to estimate the absolute minimum and maximum values of the function $f(x) = x - 2 \sin x$, $0 \leqslant x \leqslant 2\pi$.

(b) Use calculus to find the exact minimum and maximum values.

SOLUTION

(a) Figure 16 shows a graph of f in the viewing rectangle $[0, 2\pi]$ by $[-1, 8]$. By moving the cursor close to the maximum point, we see that the y-coordinates don't change very much in the vicinity of the maximum. The absolute maximum value is about 6.97 and it occurs when $x \approx 5.2$. Similarly, by moving the cursor close to the minimum point, we see that the absolute minimum value is about -0.68 and it occurs when $x \approx 1.0$. It is possible to get more accurate estimates by zooming in toward the

FIGURE 16

maximum and minimum points (or using a built-in maximum or minimum feature), but instead let's use calculus.

(b) The function $f(x) = x - 2 \sin x$ is continuous on $[0, 2\pi]$. Since $f'(x) = 1 - 2 \cos x$, we have $f'(x) = 0$ when $\cos x = \frac{1}{2}$ and this occurs when $x = \pi/3$ or $5\pi/3$. The values of f at these critical numbers are

$$f(\pi/3) = \frac{\pi}{3} - 2 \sin \frac{\pi}{3} = \frac{\pi}{3} - \sqrt{3} \approx -0.684853$$

and
$$f(5\pi/3) = \frac{5\pi}{3} - 2 \sin \frac{5\pi}{3} = \frac{5\pi}{3} + \sqrt{3} \approx 6.968039$$

The values of f at the endpoints are

$$f(0) = 0 \qquad \text{and} \qquad f(2\pi) = 2\pi \approx 6.28$$

Comparing these four numbers and using the Closed Interval Method, we see that the absolute minimum value is $f(\pi/3) = \pi/3 - \sqrt{3}$ and the absolute maximum value is $f(5\pi/3) = 5\pi/3 + \sqrt{3}$. The values from part (a) serve as a check on our work. ∎

NASA

EXAMPLE 10 The Hubble Space Telescope was deployed on April 24, 1990, by the space shuttle *Discovery*. A model for the velocity of the shuttle during this mission, from liftoff at $t = 0$ until the solid rocket boosters were jettisoned at $t = 126$ seconds, is given by

$$v(t) = 0.001302t^3 - 0.09029t^2 + 23.61t - 3.083$$

(in feet per second). Using this model, estimate the absolute maximum and minimum values of the *acceleration* of the shuttle between liftoff and the jettisoning of the boosters.

SOLUTION We are asked for the extreme values not of the given velocity function, but rather of the acceleration function. So we first need to differentiate to find the acceleration:

$$a(t) = v'(t) = \frac{d}{dt}(0.001302t^3 - 0.09029t^2 + 23.61t - 3.083)$$

$$= 0.003906t^2 - 0.18058t + 23.61$$

We now apply the Closed Interval Method to the continuous function a on the interval $0 \leqslant t \leqslant 126$. Its derivative is

$$a'(t) = 0.007812t - 0.18058$$

The only critical number occurs when $a'(t) = 0$:

$$t_1 = \frac{0.18058}{0.007812} \approx 23.12$$

Evaluating $a(t)$ at the critical number and at the endpoints, we have

$$a(0) = 23.61 \qquad a(t_1) \approx 21.52 \qquad a(126) \approx 62.87$$

So the maximum acceleration is about 62.87 ft/s^2 and the minimum acceleration is about 21.52 ft/s^2. ∎

4.1 EXERCISES

1. Explain the difference between an absolute minimum and a local minimum.

2. Suppose f is a continuous function defined on a closed interval $[a, b]$.
  (a) What theorem guarantees the existence of an absolute maximum value and an absolute minimum value for f?
  (b) What steps would you take to find those maximum and minimum values?

3–4 For each of the numbers a, b, c, d, r, and s, state whether the function whose graph is shown has an absolute maximum or minimum, a local maximum or minimum, or neither a maximum nor a minimum.

3.

4.

5–6 Use the graph to state the absolute and local maximum and minimum values of the function.

5.

6.

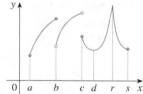

7–10 Sketch the graph of a function f that is continuous on $[1, 5]$ and has the given properties.

7. Absolute maximum at 5, absolute minimum at 2, local maximum at 3, local minima at 2 and 4

8. Absolute maximum at 4, absolute minimum at 5, local maximum at 2, local minimum at 3

9. Absolute minimum at 3, absolute maximum at 4, local maximum at 2

10. Absolute maximum at 2, absolute minimum at 5, 4 is a critical number but there is no local maximum or minimum there.

11. (a) Sketch the graph of a function that has a local maximum at 2 and is differentiable at 2.
  (b) Sketch the graph of a function that has a local maximum at 2 and is continuous but not differentiable at 2.

  (c) Sketch the graph of a function that has a local maximum at 2 and is not continuous at 2.

12. (a) Sketch the graph of a function on $[-1, 2]$ that has an absolute maximum but no local maximum.
  (b) Sketch the graph of a function on $[-1, 2]$ that has a local maximum but no absolute maximum.

13. (a) Sketch the graph of a function on $[-1, 2]$ that has an absolute maximum but no absolute minimum.
  (b) Sketch the graph of a function on $[-1, 2]$ that is discontinuous but has both an absolute maximum and an absolute minimum.

14. (a) Sketch the graph of a function that has two local maxima, one local minimum, and no absolute minimum.
  (b) Sketch the graph of a function that has three local minima, two local maxima, and seven critical numbers.

15–28 Sketch the graph of f by hand and use your sketch to find the absolute and local maximum and minimum values of f. (Use the graphs and transformations of Sections 1.2 and 1.3.)

15. $f(x) = \frac{1}{2}(3x - 1), \quad x \le 3$

16. $f(x) = 2 - \frac{1}{3}x, \quad x \ge -2$

17. $f(x) = 1/x, \quad x \ge 1$

18. $f(x) = 1/x, \quad 1 < x < 3$

19. $f(x) = \sin x, \quad 0 \le x < \pi/2$

20. $f(x) = \sin x, \quad 0 < x \le \pi/2$

21. $f(x) = \sin x, \quad -\pi/2 \le x \le \pi/2$

22. $f(t) = \cos t, \quad -3\pi/2 \le t \le 3\pi/2$

23. $f(x) = \ln x, \quad 0 < x \le 2$

24. $f(x) = |x|$

25. $f(x) = 1 - \sqrt{x}$

26. $f(x) = e^x$

27. $f(x) = \begin{cases} x^2 & \text{if } -1 \le x \le 0 \\ 2 - 3x & \text{if } 0 < x \le 1 \end{cases}$

28. $f(x) = \begin{cases} 2x + 1 & \text{if } 0 \le x < 1 \\ 4 - 2x & \text{if } 1 \le x \le 3 \end{cases}$

29–44 Find the critical numbers of the function.

29. $f(x) = 4 + \frac{1}{3}x - \frac{1}{2}x^2$

30. $f(x) = x^3 + 6x^2 - 15x$

31. $f(x) = 2x^3 - 3x^2 - 36x$

32. $f(x) = 2x^3 + x^2 + 2x$

33. $g(t) = t^4 + t^3 + t^2 + 1$

34. $g(t) = |3t - 4|$

35. $g(y) = \dfrac{y - 1}{y^2 - y + 1}$

36. $h(p) = \dfrac{p - 1}{p^2 + 4}$

37. $h(t) = t^{3/4} - 2t^{1/4}$

38. $g(x) = \sqrt[3]{4 - x^2}$

39. $F(x) = x^{4/5}(x - 4)^2$

40. $g(\theta) = 4\theta - \tan\theta$

41. $f(\theta) = 2\cos\theta + \sin^2\theta$

42. $h(t) = 3t - \arcsin t$

43. $f(x) = x^2 e^{-3x}$

44. $f(x) = x^{-2}\ln x$

45–46 A formula for the *derivative* of a function f is given. How many critical numbers does f have?

45. $f'(x) = 5e^{-0.1|x|}\sin x - 1$ **46.** $f'(x) = \dfrac{100\cos^2 x}{10 + x^2} - 1$

47–62 Find the absolute maximum and absolute minimum values of f on the given interval.

47. $f(x) = 12 + 4x - x^2$, $[0, 5]$

48. $f(x) = 5 + 54x - 2x^3$, $[0, 4]$

49. $f(x) = 2x^3 - 3x^2 - 12x + 1$, $[-2, 3]$

50. $f(x) = x^3 - 6x^2 + 5$, $[-3, 5]$

51. $f(x) = 3x^4 - 4x^3 - 12x^2 + 1$, $[-2, 3]$

52. $f(t) = (t^2 - 4)^3$, $[-2, 3]$

53. $f(x) = x + \dfrac{1}{x}$, $[0.2, 4]$

54. $f(x) = \dfrac{x}{x^2 - x + 1}$, $[0, 3]$

55. $f(t) = t - \sqrt[3]{t}$, $[-1, 4]$

56. $f(t) = \dfrac{\sqrt{t}}{1 + t^2}$, $[0, 2]$

57. $f(t) = 2\cos t + \sin 2t$, $[0, \pi/2]$

58. $f(t) = t + \cot(t/2)$, $[\pi/4, 7\pi/4]$

59. $f(x) = x^{-2}\ln x$, $\left[\frac{1}{2}, 4\right]$

60. $f(x) = xe^{x/2}$, $[-3, 1]$

61. $f(x) = \ln(x^2 + x + 1)$, $[-1, 1]$

62. $f(x) = x - 2\tan^{-1}x$, $[0, 4]$

63. If a and b are positive numbers, find the maximum value of $f(x) = x^a(1 - x)^b$, $0 \le x \le 1$.

64. Use a graph to estimate the critical numbers of $f(x) = |1 + 5x - x^3|$ correct to one decimal place.

65–68

(a) Use a graph to estimate the absolute maximum and minimum values of the function to two decimal places.

(b) Use calculus to find the exact maximum and minimum values.

65. $f(x) = x^5 - x^3 + 2$, $-1 \le x \le 1$

66. $f(x) = e^x + e^{-2x}$, $0 \le x \le 1$

67. $f(x) = x\sqrt{x - x^2}$

68. $f(x) = x - 2\cos x$, $-2 \le x \le 0$

69. After the consumption of an alcoholic beverage, the concentration of alcohol in the bloodstream (blood alcohol concentration, or BAC) surges as the alcohol is absorbed, followed by a gradual decline as the alcohol is metabolized. The function

$$C(t) = 1.35te^{-2.802t}$$

models the average BAC, measured in mg/mL, of a group of eight male subjects t hours after rapid consumption of 15 mL of ethanol (corresponding to one alcoholic drink). What is the maximum average BAC during the first 3 hours? When does it occur?

Source: Adapted from P. Wilkinson et al., "Pharmacokinetics of Ethanol after Oral Administration in the Fasting State," *Journal of Pharmacokinetics and Biopharmaceutics* 5 (1977): 207–24.

70. After an antibiotic tablet is taken, the concentration of the antibiotic in the bloodstream is modeled by the function

$$C(t) = 8(e^{-0.4t} - e^{-0.6t})$$

where the time t is measured in hours and C is measured in µg/mL. What is the maximum concentration of the antibiotic during the first 12 hours?

71. Between 0°C and 30°C, the volume V (in cubic centimeters) of 1 kg of water at a temperature T is given approximately by the formula

$$V = 999.87 - 0.06426T + 0.0085043T^2 - 0.0000679T^3$$

Find the temperature at which water has its maximum density.

72. An object with weight W is dragged along a horizontal plane by a force acting along a rope attached to the object. If the rope makes an angle θ with the plane, then the magnitude of the force is

$$F = \frac{\mu W}{\mu\sin\theta + \cos\theta}$$

where μ is a positive constant called the *coefficient of friction* and where $0 \le \theta \le \pi/2$. Show that F is minimized when $\tan\theta = \mu$.

73. The water level, measured in feet above mean sea level, of Lake Lanier in Georgia, USA, during 2012 can be modeled by the function

$$L(t) = 0.01441t^3 - 0.4177t^2 + 2.703t + 1060.1$$

where t is measured in months since January 1, 2012. Estimate when the water level was highest during 2012.

74. On May 7, 1992, the space shuttle *Endeavour* was launched on mission STS-49, the purpose of which was to install a new perigee kick motor in an Intelsat communications satellite. The table gives the velocity data for the shuttle between liftoff and the jettisoning of the solid rocket boosters.

(a) Use a graphing calculator or computer to find the cubic polynomial that best models the velocity of the shuttle for the time interval $t \in [0, 125]$. Then graph this polynomial.

(b) Find a model for the acceleration of the shuttle and use it to estimate the maximum and minimum values of the acceleration during the first 125 seconds.

Event	Time (s)	Velocity (ft/s)
Launch	0	0
Begin roll maneuver	10	185
End roll maneuver	15	319
Throttle to 89%	20	447
Throttle to 67%	32	742
Throttle to 104%	59	1325
Maximum dynamic pressure	62	1445
Solid rocket booster separation	125	4151

75. When a foreign object lodged in the trachea (windpipe) forces a person to cough, the diaphragm thrusts upward causing an increase in pressure in the lungs. This is accompanied by a contraction of the trachea, making a narrower channel for the expelled air to flow through. For a given amount of air to escape in a fixed time, it must move faster through the narrower channel than the wider one. The greater the velocity of the air-stream, the greater the force on the foreign object. X rays show that the radius of the circular tracheal tube contracts to about two-thirds of its normal radius during a cough. According to a mathematical model of coughing, the velocity v of the air-stream is related to the radius r of the trachea by the equation

$$v(r) = k(r_0 - r)r^2 \qquad \tfrac{1}{2}r_0 \leq r \leq r_0$$

where k is a constant and r_0 is the normal radius of the trachea.

The restriction on r is due to the fact that the tracheal wall stiffens under pressure and a contraction greater than $\tfrac{1}{2}r_0$ is prevented (otherwise the person would suffocate).

(a) Determine the value of r in the interval $\left[\tfrac{1}{2}r_0, r_0\right]$ at which v has an absolute maximum. How does this compare with experimental evidence?

(b) What is the absolute maximum value of v on the interval?

(c) Sketch the graph of v on the interval $[0, r_0]$.

76. Show that 5 is a critical number of the function

$$g(x) = 2 + (x - 5)^3$$

but g does not have a local extreme value at 5.

77. Prove that the function

$$f(x) = x^{101} + x^{51} + x + 1$$

has neither a local maximum nor a local minimum.

78. If f has a local minimum value at c, show that the function $g(x) = -f(x)$ has a local maximum value at c.

79. Prove Fermat's Theorem for the case in which f has a local minimum at c.

80. A cubic function is a polynomial of degree 3; that is, it has the form $f(x) = ax^3 + bx^2 + cx + d$, where $a \neq 0$.

(a) Show that a cubic function can have two, one, or no critical number(s). Give examples and sketches to illustrate the three possibilities.

(b) How many local extreme values can a cubic function have?

APPLIED PROJECT   THE CALCULUS OF RAINBOWS

Rainbows are created when raindrops scatter sunlight. They have fascinated mankind since ancient times and have inspired attempts at scientific explanation since the time of Aristotle. In this project we use the ideas of Descartes and Newton to explain the shape, location, and colors of rainbows.

Formation of the primary rainbow

1. The figure shows a ray of sunlight entering a spherical raindrop at A. Some of the light is reflected, but the line AB shows the path of the part that enters the drop. Notice that the light is refracted toward the normal line AO and in fact Snell's Law says that $\sin \alpha = k \sin \beta$, where α is the angle of incidence, β is the angle of refraction, and $k \approx \tfrac{4}{3}$ is the index of refraction for water. At B some of the light passes through the drop and is refracted into the air, but the line BC shows the part that is reflected. (The angle of inci-dence equals the angle of reflection.) When the ray reaches C, part of it is reflected, but for the time being we are more interested in the part that leaves the raindrop at C. (Notice that it is refracted away from the normal line.) The *angle of deviation* $D(\alpha)$ is the amount of clockwise rotation that the ray has undergone during this three-stage process. Thus

$$D(\alpha) = (\alpha - \beta) + (\pi - 2\beta) + (\alpha - \beta) = \pi + 2\alpha - 4\beta$$

Show that the minimum value of the deviation is $D(\alpha) \approx 138°$ and occurs when $\alpha \approx 59.4°$.

rays from sun

138°

rays from sun

42°

observer

The significance of the minimum deviation is that when $\alpha \approx 59.4°$ we have $D'(\alpha) \approx 0$, so $\Delta D / \Delta \alpha \approx 0$. This means that many rays with $\alpha \approx 59.4°$ become deviated by approximately the same amount. It is the *concentration* of rays coming from near the direction of minimum deviation that creates the brightness of the primary rainbow. The figure at the left shows that the angle of elevation from the observer up to the highest point on the rainbow is $180° - 138° = 42°$. (This angle is called the *rainbow angle*.)

2. Problem 1 explains the location of the primary rainbow, but how do we explain the colors? Sunlight comprises a range of wavelengths, from the red range through orange, yellow, green, blue, indigo, and violet. As Newton discovered in his prism experiments of 1666, the index of refraction is different for each color. (The effect is called *dispersion*.) For red light the refractive index is $k \approx 1.3318$, whereas for violet light it is $k \approx 1.3435$. By repeating the calculation of Problem 1 for these values of k, show that the rainbow angle is about 42.3° for the red bow and 40.6° for the violet bow. So the rainbow really consists of seven individual bows corresponding to the seven colors.

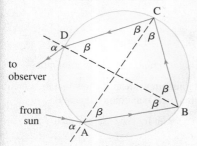

C

β β

D

α β

to
observer

β

from
sun

β

β B

α A

Formation of the secondary rainbow

3. Perhaps you have seen a fainter secondary rainbow above the primary bow. That results from the part of a ray that enters a raindrop and is refracted at A, reflected twice (at B and C), and refracted as it leaves the drop at D (see the figure at the left). This time the deviation angle $D(\alpha)$ is the total amount of counterclockwise rotation that the ray undergoes in this four-stage process. Show that

$$D(\alpha) = 2\alpha - 6\beta + 2\pi$$

and $D(\alpha)$ has a minimum value when

$$\cos \alpha = \sqrt{\frac{k^2 - 1}{8}}$$

Taking $k = \frac{4}{3}$, show that the minimum deviation is about 129° and so the rainbow angle for the secondary rainbow is about 51°, as shown in the figure at the left.

4. Show that the colors in the secondary rainbow appear in the opposite order from those in the primary rainbow.

42° 51°

4.2   The Mean Value Theorem

We will see that many of the results of this chapter depend on one central fact, which is called the Mean Value Theorem. But to arrive at the Mean Value Theorem we first need the following result.

Rolle

Rolle's Theorem was first published in 1691 by the French mathematician Michel Rolle (1652–1719) in a book entitled *Méthode pour resoudre les Egalitez.* He was a vocal critic of the methods of his day and attacked calculus as being a "collection of ingenious fallacies." Later, however, he became convinced of the essential correctness of the methods of calculus.

Rolle's Theorem Let f be a function that satisfies the following three hypotheses:

1. f is continuous on the closed interval $[a, b]$.

2. f is differentiable on the open interval (a, b).

3. $f(a) = f(b)$

Then there is a number c in (a, b) such that $f'(c) = 0$.

Before giving the proof let's take a look at the graphs of some typical functions that satisfy the three hypotheses. Figure 1 shows the graphs of four such functions. In each case it appears that there is at least one point $(c, f(c))$ on the graph where the tangent is horizontal and therefore $f'(c) = 0$. Thus Rolle's Theorem is plausible.

(a)

(b)

(c)

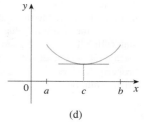

(d)

FIGURE 1

PS Take cases

PROOF There are three cases:

CASE I $f(x) = k$, **a constant**
Then $f'(x) = 0$, so the number c can be taken to be *any* number in (a, b).

CASE II $f(x) > f(a)$ **for some x in (a, b)** [as in Figure 1(b) or (c)]
By the Extreme Value Theorem (which we can apply by hypothesis 1), f has a maximum value somewhere in $[a, b]$. Since $f(a) = f(b)$, it must attain this maximum value at a number c in the open interval (a, b). Then f has a *local* maximum at c and, by hypothesis 2, f is differentiable at c. Therefore $f'(c) = 0$ by Fermat's Theorem.

CASE III $f(x) < f(a)$ **for some x in (a, b)** [as in Figure 1(c) or (d)]
By the Extreme Value Theorem, f has a minimum value in $[a, b]$ and, since $f(a) = f(b)$, it attains this minimum value at a number c in (a, b). Again $f'(c) = 0$ by Fermat's Theorem. ∎

EXAMPLE 1 Let's apply Rolle's Theorem to the position function $s = f(t)$ of a moving object. If the object is in the same place at two different instants $t = a$ and $t = b$, then $f(a) = f(b)$. Rolle's Theorem says that there is some instant of time $t = c$ between a and b when $f'(c) = 0$; that is, the velocity is 0. (In particular, you can see that this is true when a ball is thrown directly upward.) ∎

EXAMPLE 2 Prove that the equation $x^3 + x - 1 = 0$ has exactly one real root.

SOLUTION First we use the Intermediate Value Theorem (2.5.10) to show that a root exists. Let $f(x) = x^3 + x - 1$. Then $f(0) = -1 < 0$ and $f(1) = 1 > 0$. Since f is a

Figure 2 shows a graph of the function $f(x) = x^3 + x - 1$ discussed in Example 2. Rolle's Theorem shows that, no matter how much we enlarge the viewing rectangle, we can never find a second x-intercept.

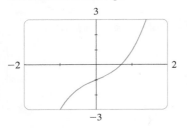

FIGURE 2

The Mean Value Theorem is an example of what is called an existence theorem. Like the Intermediate Value Theorem, the Extreme Value Theorem, and Rolle's Theorem, it guarantees that there *exists* a number with a certain property, but it doesn't tell us how to find the number.

polynomial, it is continuous, so the Intermediate Value Theorem states that there is a number c between 0 and 1 such that $f(c) = 0$. Thus the given equation has a root.

To show that the equation has no other real root, we use Rolle's Theorem and argue by contradiction. Suppose that it had two roots a and b. Then $f(a) = 0 = f(b)$ and, since f is a polynomial, it is differentiable on (a, b) and continuous on $[a, b]$. Thus, by Rolle's Theorem, there is a number c between a and b such that $f'(c) = 0$. But

$$f'(x) = 3x^2 + 1 \geqslant 1 \qquad \text{for all } x$$

(since $x^2 \geqslant 0$) so $f'(x)$ can never be 0. This gives a contradiction. Therefore the equation can't have two real roots. ∎

Our main use of Rolle's Theorem is in proving the following important theorem, which was first stated by another French mathematician, Joseph-Louis Lagrange.

The Mean Value Theorem   Let f be a function that satisfies the following hypotheses:

1. f is continuous on the closed interval $[a, b]$.

2. f is differentiable on the open interval (a, b).

Then there is a number c in (a, b) such that

$$\boxed{1} \qquad f'(c) = \frac{f(b) - f(a)}{b - a}$$

or, equivalently,

$$\boxed{2} \qquad f(b) - f(a) = f'(c)(b - a)$$

Before proving this theorem, we can see that it is reasonable by interpreting it geometrically. Figures 3 and 4 show the points $A(a, f(a))$ and $B(b, f(b))$ on the graphs of two differentiable functions. The slope of the secant line AB is

$$\boxed{3} \qquad m_{AB} = \frac{f(b) - f(a)}{b - a}$$

which is the same expression as on the right side of Equation 1. Since $f'(c)$ is the slope of the tangent line at the point $(c, f(c))$, the Mean Value Theorem, in the form given by Equation 1, says that there is at least one point $P(c, f(c))$ on the graph where the slope of the tangent line is the same as the slope of the secant line AB. In other words, there is a point P where the tangent line is parallel to the secant line AB. (Imagine a line far away that stays parallel to AB while moving toward AB until it touches the graph for the first time.)

FIGURE 3

FIGURE 4

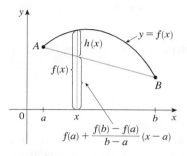

FIGURE 5

Lagrange and the Mean Value Theorem

The Mean Value Theorem was first formulated by Joseph-Louis Lagrange (1736–1813), born in Italy of a French father and an Italian mother. He was a child prodigy and became a professor in Turin at the tender age of 19. Lagrange made great contributions to number theory, theory of functions, theory of equations, and analytical and celestial mechanics. In particular, he applied calculus to the analysis of the stability of the solar system. At the invitation of Frederick the Great, he succeeded Euler at the Berlin Academy and, when Frederick died, Lagrange accepted King Louis XVI's invitation to Paris, where he was given apartments in the Louvre and became a professor at the Ecole Polytechnique. Despite all the trappings of luxury and fame, he was a kind and quiet man, living only for science.

PROOF We apply Rolle's Theorem to a new function h defined as the difference between f and the function whose graph is the secant line AB. Using Equation 3 and the point-slope equation of a line, we see that the equation of the line AB can be written as

$$y - f(a) = \frac{f(b) - f(a)}{b - a}(x - a)$$

or as

$$y = f(a) + \frac{f(b) - f(a)}{b - a}(x - a)$$

So, as shown in Figure 5,

$$\boxed{4} \qquad h(x) = f(x) - f(a) - \frac{f(b) - f(a)}{b - a}(x - a)$$

First we must verify that h satisfies the three hypotheses of Rolle's Theorem.

1. The function h is continuous on $[a, b]$ because it is the sum of f and a first-degree polynomial, both of which are continuous.

2. The function h is differentiable on (a, b) because both f and the first-degree polynomial are differentiable. In fact, we can compute h' directly from Equation 4:

$$h'(x) = f'(x) - \frac{f(b) - f(a)}{b - a}$$

(Note that $f(a)$ and $[f(b) - f(a)]/(b - a)$ are constants.)

3. $$h(a) = f(a) - f(a) - \frac{f(b) - f(a)}{b - a}(a - a) = 0$$

$$h(b) = f(b) - f(a) - \frac{f(b) - f(a)}{b - a}(b - a)$$

$$= f(b) - f(a) - [f(b) - f(a)] = 0$$

Therefore $h(a) = h(b)$.

Since h satisfies the hypotheses of Rolle's Theorem, that theorem says there is a number c in (a, b) such that $h'(c) = 0$. Therefore

$$0 = h'(c) = f'(c) - \frac{f(b) - f(a)}{b - a}$$

and so

$$f'(c) = \frac{f(b) - f(a)}{b - a} \qquad \blacksquare$$

EXAMPLE 3 To illustrate the Mean Value Theorem with a specific function, let's consider $f(x) = x^3 - x$, $a = 0$, $b = 2$. Since f is a polynomial, it is continuous and differentiable for all x, so it is certainly continuous on $[0, 2]$ and differentiable on $(0, 2)$. Therefore, by the Mean Value Theorem, there is a number c in $(0, 2)$ such that

$$f(2) - f(0) = f'(c)(2 - 0)$$

Now $f(2) = 6$, $f(0) = 0$, and $f'(x) = 3x^2 - 1$, so this equation becomes

$$6 = (3c^2 - 1)2 = 6c^2 - 2$$

which gives $c^2 = \frac{4}{3}$, that is, $c = \pm 2/\sqrt{3}$. But c must lie in $(0, 2)$, so $c = 2/\sqrt{3}$.

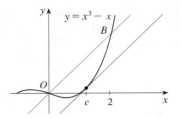

FIGURE 6

Figure 6 illustrates this calculation: The tangent line at this value of c is parallel to the secant line OB. ∎

EXAMPLE 4 If an object moves in a straight line with position function $s = f(t)$, then the average velocity between $t = a$ and $t = b$ is

$$\frac{f(b) - f(a)}{b - a}$$

and the velocity at $t = c$ is $f'(c)$. Thus the Mean Value Theorem (in the form of Equation 1) tells us that at some time $t = c$ between a and b the instantaneous velocity $f'(c)$ is equal to that average velocity. For instance, if a car traveled 180 km in 2 hours, then the speedometer must have read 90 km/h at least once.

In general, the Mean Value Theorem can be interpreted as saying that there is a number at which the instantaneous rate of change is equal to the average rate of change over an interval. ∎

The main significance of the Mean Value Theorem is that it enables us to obtain information about a function from information about its derivative. The next example provides an instance of this principle.

EXAMPLE 5 Suppose that $f(0) = -3$ and $f'(x) \leq 5$ for all values of x. How large can $f(2)$ possibly be?

SOLUTION We are given that f is differentiable (and therefore continuous) everywhere. In particular, we can apply the Mean Value Theorem on the interval $[0, 2]$. There exists a number c such that

$$f(2) - f(0) = f'(c)(2 - 0)$$

so

$$f(2) = f(0) + 2f'(c) = -3 + 2f'(c)$$

We are given that $f'(x) \leq 5$ for all x, so in particular we know that $f'(c) \leq 5$. Multiplying both sides of this inequality by 2, we have $2f'(c) \leq 10$, so

$$f(2) = -3 + 2f'(c) \leq -3 + 10 = 7$$

The largest possible value for $f(2)$ is 7. ∎

The Mean Value Theorem can be used to establish some of the basic facts of differential calculus. One of these basic facts is the following theorem. Others will be found in the following sections.

> **5   Theorem** If $f'(x) = 0$ for all x in an interval (a, b), then f is constant on (a, b).

PROOF Let x_1 and x_2 be any two numbers in (a, b) with $x_1 < x_2$. Since f is differentiable on (a, b), it must be differentiable on (x_1, x_2) and continuous on $[x_1, x_2]$. By applying the Mean Value Theorem to f on the interval $[x_1, x_2]$, we get a number c such that $x_1 < c < x_2$ and

$$\boxed{6} \qquad f(x_2) - f(x_1) = f'(c)(x_2 - x_1)$$

Since $f'(x) = 0$ for all x, we have $f'(c) = 0$, and so Equation 6 becomes

$$f(x_2) - f(x_1) = 0 \qquad \text{or} \qquad f(x_2) = f(x_1)$$

Therefore f has the same value at *any* two numbers x_1 and x_2 in (a, b). This means that f is constant on (a, b). ∎

Corollary 7 says that if two functions have the same derivatives on an interval, then their graphs must be vertical translations of each other there. In other words, the graphs have the same shape, but could be shifted up or down.

> **7 Corollary** If $f'(x) = g'(x)$ for all x in an interval (a, b), then $f - g$ is constant on (a, b); that is, $f(x) = g(x) + c$ where c is a constant.

PROOF Let $F(x) = f(x) - g(x)$. Then

$$F'(x) = f'(x) - g'(x) = 0$$

for all x in (a, b). Thus, by Theorem 5, F is constant; that is, $f - g$ is constant. ∎

NOTE Care must be taken in applying Theorem 5. Let

$$f(x) = \frac{x}{|x|} = \begin{cases} 1 & \text{if } x > 0 \\ -1 & \text{if } x < 0 \end{cases}$$

The domain of f is $D = \{x \mid x \neq 0\}$ and $f'(x) = 0$ for all x in D. But f is obviously not a constant function. This does not contradict Theorem 5 because D is not an interval. Notice that f is constant on the interval $(0, \infty)$ and also on the interval $(-\infty, 0)$.

EXAMPLE 6 Prove the identity $\tan^{-1}x + \cot^{-1}x = \pi/2$.

SOLUTION Although calculus isn't needed to prove this identity, the proof using calculus is quite simple. If $f(x) = \tan^{-1}x + \cot^{-1}x$, then

$$f'(x) = \frac{1}{1 + x^2} - \frac{1}{1 + x^2} = 0$$

for all values of x. Therefore $f(x) = C$, a constant. To determine the value of C, we put $x = 1$ [because we can evaluate $f(1)$ exactly]. Then

$$C = f(1) = \tan^{-1}1 + \cot^{-1}1 = \frac{\pi}{4} + \frac{\pi}{4} = \frac{\pi}{2}$$

Thus $\tan^{-1}x + \cot^{-1}x = \pi/2$. ∎

4.2 EXERCISES

1. The graph of a function f is shown. Verify that f satisfies the hypotheses of Rolle's Theorem on the interval $[0, 8]$. Then estimate the value(s) of c that satisfy the conclusion of Rolle's Theorem on that interval.

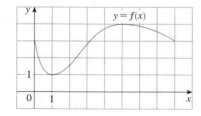

2. Draw the graph of a function defined on $[0, 8]$ such that $f(0) = f(8) = 3$ and the function does not satisfy the conclusion of Rolle's Theorem on $[0, 8]$.

3. The graph of a function g is shown.

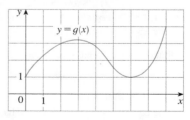

(a) Verify that g satisfies the hypotheses of the Mean Value Theorem on the interval $[0, 8]$.

(b) Estimate the value(s) of c that satisfy the conclusion of the Mean Value Theorem on the interval $[0, 8]$.

(c) Estimate the value(s) of c that satisfy the conclusion of the Mean Value Theorem on the interval $[2, 6]$.

4. Draw the graph of a function that is continuous on $[0, 8]$ where $f(0) = 1$ and $f(8) = 4$ and that does not satisfy the conclusion of the Mean Value Theorem on $[0, 8]$.

5–8 Verify that the function satisfies the three hypotheses of Rolle's Theorem on the given interval. Then find all numbers c that satisfy the conclusion of Rolle's Theorem.

5. $f(x) = 2x^2 - 4x + 5,$ $[-1, 3]$

6. $f(x) = x^3 - 2x^2 - 4x + 2,$ $[-2, 2]$

7. $f(x) = \sin(x/2),$ $[\pi/2, 3\pi/2]$

8. $f(x) = x + 1/x,$ $\left[\frac{1}{2}, 2\right]$

9. Let $f(x) = 1 - x^{2/3}$. Show that $f(-1) = f(1)$ but there is no number c in $(-1, 1)$ such that $f'(c) = 0$. Why does this not contradict Rolle's Theorem?

10. Let $f(x) = \tan x$. Show that $f(0) = f(\pi)$ but there is no number c in $(0, \pi)$ such that $f'(c) = 0$. Why does this not contradict Rolle's Theorem?

11–14 Verify that the function satisfies the hypotheses of the Mean Value Theorem on the given interval. Then find all numbers c that satisfy the conclusion of the Mean Value Theorem.

11. $f(x) = 2x^2 - 3x + 1,$ $[0, 2]$

12. $f(x) = x^3 - 3x + 2,$ $[-2, 2]$

13. $f(x) = \ln x,$ $[1, 4]$ 14. $f(x) = 1/x,$ $[1, 3]$

15–16 Find the number c that satisfies the conclusion of the Mean Value Theorem on the given interval. Graph the function, the secant line through the endpoints, and the tangent line at $(c, f(c))$. Are the secant line and the tangent line parallel?

15. $f(x) = \sqrt{x},$ $[0, 4]$ 16. $f(x) = e^{-x},$ $[0, 2]$

17. Let $f(x) = (x - 3)^{-2}$. Show that there is no value of c in $(1, 4)$ such that $f(4) - f(1) = f'(c)(4 - 1)$. Why does this not contradict the Mean Value Theorem?

18. Let $f(x) = 2 - |2x - 1|$. Show that there is no value of c such that $f(3) - f(0) = f'(c)(3 - 0)$. Why does this not contradict the Mean Value Theorem?

19–20 Show that the equation has exactly one real root.

19. $2x + \cos x = 0$ 20. $x^3 + e^x = 0$

21. Show that the equation $x^3 - 15x + c = 0$ has at most one root in the interval $[-2, 2]$.

22. Show that the equation $x^4 + 4x + c = 0$ has at most two real roots.

23. (a) Show that a polynomial of degree 3 has at most three real roots.

(b) Show that a polynomial of degree n has at most n real roots.

24. (a) Suppose that f is differentiable on $\mathbb{R}$ and has two roots. Show that f' has at least one root.
   (b) Suppose f is twice differentiable on $\mathbb{R}$ and has three roots. Show that f'' has at least one real root.
   (c) Can you generalize parts (a) and (b)?

25. If $f(1) = 10$ and $f'(x) \geqslant 2$ for $1 \leqslant x \leqslant 4$, how small can $f(4)$ possibly be?

26. Suppose that $3 \leqslant f'(x) \leqslant 5$ for all values of x. Show that $18 \leqslant f(8) - f(2) \leqslant 30$.

27. Does there exist a function f such that $f(0) = -1$, $f(2) = 4$, and $f'(x) \leqslant 2$ for all x?

28. Suppose that f and g are continuous on $[a, b]$ and differentiable on (a, b). Suppose also that $f(a) = g(a)$ and $f'(x) < g'(x)$ for $a < x < b$. Prove that $f(b) < g(b)$. [*Hint:* Apply the Mean Value Theorem to the function $h = f - g$.]

29. Show that $\sin x < x$ if $0 < x < 2\pi$.

30. Suppose f is an odd function and is differentiable everywhere. Prove that for every positive number b, there exists a number c in $(-b, b)$ such that $f'(c) = f(b)/b$.

31. Use the Mean Value Theorem to prove the inequality

$$|\sin a - \sin b| \leqslant |a - b| \text{for all } a \text{ and } b$$

32. If $f'(x) = c$ (c a constant) for all x, use Corollary 7 to show that $f(x) = cx + d$ for some constant d.

33. Let $f(x) = 1/x$ and

$$g(x) = \begin{cases} \dfrac{1}{x} & \text{if } x > 0 \\ 1 + \dfrac{1}{x} & \text{if } x < 0 \end{cases}$$

Show that $f'(x) = g'(x)$ for all x in their domains. Can we conclude from Corollary 7 that $f - g$ is constant?

34. Use the method of Example 6 to prove the identity

$$2 \sin^{-1}x = \cos^{-1}(1 - 2x^2)   x \geqslant 0$$

35. Prove the identity $\arcsin \dfrac{x - 1}{x + 1} = 2 \arctan \sqrt{x} - \dfrac{\pi}{2}$.

36. At 2:00 PM a car's speedometer reads 30 mi/h. At 2:10 PM it reads 50 mi/h. Show that at some time between 2:00 and 2:10 the acceleration is exactly 120 mi/h^2.

37. Two runners start a race at the same time and finish in a tie. Prove that at some time during the race they have the same speed. [*Hint:* Consider $f(t) = g(t) - h(t)$, where g and h are the position functions of the two runners.]

38. A number a is called a **fixed point** of a function f if $f(a) = a$. Prove that if $f'(x) \neq 1$ for all real numbers x, then f has at most one fixed point.

4.3   How Derivatives Affect the Shape of a Graph

Many of the applications of calculus depend on our ability to deduce facts about a function f from information concerning its derivatives. Because $f'(x)$ represents the slope of the curve $y = f(x)$ at the point $(x, f(x))$, it tells us the direction in which the curve proceeds at each point. So it is reasonable to expect that information about $f'(x)$ will provide us with information about $f(x)$.

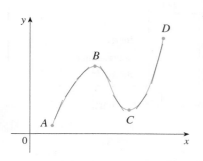

FIGURE 1

■ What Does f' Say About f?

To see how the derivative of f can tell us where a function is increasing or decreasing, look at Figure 1. (Increasing functions and decreasing functions were defined in Section 1.1.) Between A and B and between C and D, the tangent lines have positive slope and so $f'(x) > 0$. Between B and C, the tangent lines have negative slope and so $f'(x) < 0$. Thus it appears that f increases when $f'(x)$ is positive and decreases when $f'(x)$ is negative. To prove that this is always the case, we use the Mean Value Theorem.

Let's abbreviate the name of this test to the I/D Test.

> **Increasing/Decreasing Test**
>
> (a) If $f'(x) > 0$ on an interval, then f is increasing on that interval.
>
> (b) If $f'(x) < 0$ on an interval, then f is decreasing on that interval.

PROOF

(a) Let x_1 and x_2 be any two numbers in the interval with $x_1 < x_2$. According to the definition of an increasing function (page 19), we have to show that $f(x_1) < f(x_2)$.

Because we are given that $f'(x) > 0$, we know that f is differentiable on $[x_1, x_2]$. So, by the Mean Value Theorem, there is a number c between x_1 and x_2 such that

$$\boxed{1} \qquad f(x_2) - f(x_1) = f'(c)(x_2 - x_1)$$

Now $f'(c) > 0$ by assumption and $x_2 - x_1 > 0$ because $x_1 < x_2$. Thus the right side of Equation 1 is positive, and so

$$f(x_2) - f(x_1) > 0 \qquad \text{or} \qquad f(x_1) < f(x_2)$$

This shows that f is increasing.

Part (b) is proved similarly. ■

EXAMPLE 1 Find where the function $f(x) = 3x^4 - 4x^3 - 12x^2 + 5$ is increasing and where it is decreasing.

SOLUTION We start by differentiating f:

$$f'(x) = 12x^3 - 12x^2 - 24x = 12x(x - 2)(x + 1)$$

To use the I/D Test we have to know where $f'(x) > 0$ and where $f'(x) < 0$. To solve these inequalities we first find where $f'(x) = 0$, namely at $x = 0, 2$, and -1. These are the critical numbers of f, and they divide the domain into four intervals (see the number line at the left). Within each interval, $f'(x)$ must be always positive or always negative. (See Examples 3 and 4 in Appendix A.) We can determine which is the case for each interval from the signs of the three factors of $f'(x)$, namely, $12x$, $x - 2$, and $x + 1$, as shown in the following chart. A plus sign indicates that the given expression is positive, and a minus sign indicates that it is negative. The last column of the chart gives the conclusion based on the I/D Test. For instance, $f'(x) < 0$ for $0 < x < 2$, so f is decreasing on $(0, 2)$. (It would also be true to say that f is decreasing on the closed interval $[0, 2]$.)

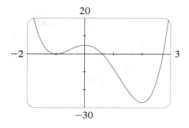

FIGURE 2

Interval	$12x$	$x - 2$	$x + 1$	$f'(x)$	f
$x < -1$	$-$	$-$	$-$	$-$	decreasing on $(-\infty, -1)$
$-1 < x < 0$	$-$	$-$	$+$	$+$	increasing on $(-1, 0)$
$0 < x < 2$	$+$	$-$	$+$	$-$	decreasing on $(0, 2)$
$x > 2$	$+$	$+$	$+$	$+$	increasing on $(2, \infty)$

The graph of f shown in Figure 2 confirms the information in the chart.   ∎

Local Extreme Values

Recall from Section 4.1 that if f has a local maximum or minimum at c, then c must be a critical number of f (by Fermat's Theorem), but not every critical number gives rise to a maximum or a minimum. We therefore need a test that will tell us whether or not f has a local maximum or minimum at a critical number.

You can see from Figure 2 that $f(0) = 5$ is a local maximum value of f because f increases on $(-1, 0)$ and decreases on $(0, 2)$. Or, in terms of derivatives, $f'(x) > 0$ for $-1 < x < 0$ and $f'(x) < 0$ for $0 < x < 2$. In other words, the sign of $f'(x)$ changes from positive to negative at 0. This observation is the basis of the following test.

> **The First Derivative Test** Suppose that c is a critical number of a continuous function f.
>
> (a) If f' changes from positive to negative at c, then f has a local maximum at c.
>
> (b) If f' changes from negative to positive at c, then f has a local minimum at c.
>
> (c) If f' is positive to the left and right of c, or negative to the left and right of c, then f has no local maximum or minimum at c.

The First Derivative Test is a consequence of the I/D Test. In part (a), for instance, since the sign of $f'(x)$ changes from positive to negative at c, f is increasing to the left of c and decreasing to the right of c. It follows that f has a local maximum at c.

It is easy to remember the First Derivative Test by visualizing diagrams such as those in Figure 3.

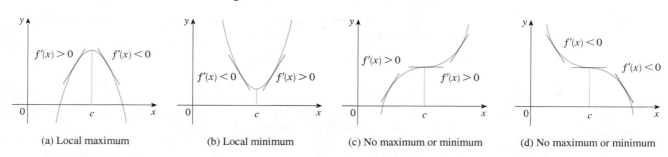

(a) Local maximum   (b) Local minimum   (c) No maximum or minimum   (d) No maximum or minimum

FIGURE 3

EXAMPLE 2 Find the local minimum and maximum values of the function f in Example 1.

SOLUTION From the chart in the solution to Example 1 we see that $f'(x)$ changes from negative to positive at -1, so $f(-1) = 0$ is a local minimum value by the First Derivative Test. Similarly, f' changes from negative to positive at 2, so $f(2) = -27$ is also a local minimum value. As noted previously, $f(0) = 5$ is a local maximum value because $f'(x)$ changes from positive to negative at 0. ∎

EXAMPLE 3 Find the local maximum and minimum values of the function
$$g(x) = x + 2 \sin x \qquad 0 \leqslant x \leqslant 2\pi$$

SOLUTION As in Example 1, we start by finding the critical numbers. The derivative is:
$$g'(x) = 1 + 2 \cos x$$
so $g'(x) = 0$ when $\cos x = -\frac{1}{2}$. The solutions of this equation are $2\pi/3$ and $4\pi/3$. Because g is differentiable everywhere, the only critical numbers are $2\pi/3$ and $4\pi/3$. We split the domain into intervals according to the critical numbers. Within each interval, $g'(x)$ is either always positive or always negative and so we analyze g in the following chart.

The $+$ signs in the chart come from the fact that $g'(x) > 0$ when $\cos x > -\frac{1}{2}$. From the graph of $y = \cos x$, this is true in the indicated intervals.

Interval	$g'(x) = 1 + 2 \cos x$	g
$0 < x < 2\pi/3$	$+$	increasing on $(0, 2\pi/3)$
$2\pi/3 < x < 4\pi/3$	$-$	decreasing on $(2\pi/3, 4\pi/3)$
$4\pi/3 < x < 2\pi$	$+$	increasing on $(4\pi/3, 2\pi)$

Because $g'(x)$ changes from positive to negative at $2\pi/3$, the First Derivative Test tells us that there is a local maximum at $2\pi/3$ and the local maximum value is
$$g(2\pi/3) = \frac{2\pi}{3} + 2 \sin \frac{2\pi}{3} = \frac{2\pi}{3} + 2\left(\frac{\sqrt{3}}{2}\right) = \frac{2\pi}{3} + \sqrt{3} \approx 3.83$$

Likewise, $g'(x)$ changes from negative to positive at $4\pi/3$ and so
$$g(4\pi/3) = \frac{4\pi}{3} + 2 \sin \frac{4\pi}{3} = \frac{4\pi}{3} + 2\left(-\frac{\sqrt{3}}{2}\right) = \frac{4\pi}{3} - \sqrt{3} \approx 2.46$$

is a local minimum value. The graph of g in Figure 4 supports our conclusion. ∎

FIGURE 4
$g(x) = x + 2 \sin x$

◼ What Does f'' Say About f?

Figure 5 shows the graphs of two increasing functions on (a, b). Both graphs join point A to point B but they look different because they bend in different directions. How can we distinguish between these two types of behavior?

FIGURE 5   (a)   (b)

(a) Concave upward

(b) Concave downward

FIGURE 6

In Figure 6 tangents to these curves have been drawn at several points. In (a) the curve lies above the tangents and f is called *concave upward* on (a, b). In (b) the curve lies below the tangents and g is called *concave downward* on (a, b).

> **Definition** If the graph of f lies above all of its tangents on an interval I, then it is called **concave upward** on I. If the graph of f lies below all of its tangents on I, it is called **concave downward** on I.

Figure 7 shows the graph of a function that is concave upward (abbreviated CU) on the intervals (b, c), (d, e), and (e, p) and concave downward (CD) on the intervals (a, b), (c, d), and (p, q).

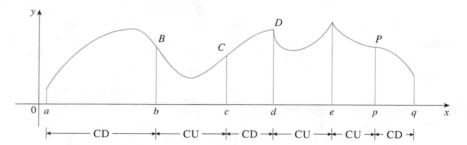

FIGURE 7

Let's see how the second derivative helps determine the intervals of concavity. Looking at Figure 6(a), you can see that, going from left to right, the slope of the tangent increases. This means that the derivative f' is an increasing function and therefore its derivative f'' is positive. Likewise, in Figure 6(b) the slope of the tangent decreases from left to right, so f' decreases and therefore f'' is negative. This reasoning can be reversed and suggests that the following theorem is true. A proof is given in Appendix F with the help of the Mean Value Theorem.

> **Concavity Test**
> (a) If $f''(x) > 0$ for all x in I, then the graph of f is concave upward on I.
> (b) If $f''(x) < 0$ for all x in I, then the graph of f is concave downward on I.

EXAMPLE 4 Figure 8 shows a population graph for Cyprian honeybees raised in an apiary. How does the rate of population increase change over time? When is this rate highest? Over what intervals is P concave upward or concave downward?

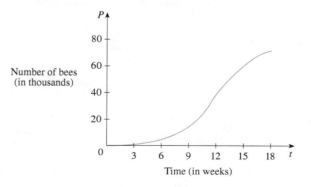

FIGURE 8

SOLUTION By looking at the slope of the curve as t increases, we see that the rate of increase of the population is initially very small, then gets larger until it reaches a maximum at about $t = 12$ weeks, and decreases as the population begins to level off. As the population approaches its maximum value of about 75,000 (called the *carrying capacity*), the rate of increase, $P'(t)$, approaches 0. The curve appears to be concave upward on $(0, 12)$ and concave downward on $(12, 18)$. ■

In Example 4, the population curve changed from concave upward to concave downward at approximately the point $(12, 38{,}000)$. This point is called an *inflection point* of the curve. The significance of this point is that the rate of population increase has its maximum value there. In general, an inflection point is a point where a curve changes its direction of concavity.

> **Definition** A point P on a curve $y = f(x)$ is called an **inflection point** if f is continuous there and the curve changes from concave upward to concave downward or from concave downward to concave upward at P.

For instance, in Figure 7, B, C, D, and P are the points of inflection. Notice that if a curve has a tangent at a point of inflection, then the curve crosses its tangent there.

In view of the Concavity Test, there is a point of inflection at any point where the second derivative changes sign.

EXAMPLE 5 Sketch a possible graph of a function f that satisfies the following conditions:

(i) $f'(x) > 0$ on $(-\infty, 1)$, $f'(x) < 0$ on $(1, \infty)$
(ii) $f''(x) > 0$ on $(-\infty, -2)$ and $(2, \infty)$, $f''(x) < 0$ on $(-2, 2)$
(iii) $\lim\limits_{x \to -\infty} f(x) = -2$, $\lim\limits_{x \to \infty} f(x) = 0$

SOLUTION Condition (i) tells us that f is increasing on $(-\infty, 1)$ and decreasing on $(1, \infty)$. Condition (ii) says that f is concave upward on $(-\infty, -2)$ and $(2, \infty)$, and concave downward on $(-2, 2)$. From condition (iii) we know that the graph of f has two horizontal asymptotes: $y = -2$ (to the left) and $y = 0$ (to the right).

We first draw the horizontal asymptote $y = -2$ as a dashed line (see Figure 9). We then draw the graph of f approaching this asymptote at the far left, increasing to its maximum point at $x = 1$, and decreasing toward the x-axis as at the far right. We also make sure that the graph has inflection points when $x = -2$ and 2. Notice that we made the curve bend upward for $x < -2$ and $x > 2$, and bend downward when x is between -2 and 2. ■

Another application of the second derivative is the following test for identifying local maximum and minimum values. It is a consequence of the Concavity Test, and serves as an alternative to the First Derivative Test.

> **The Second Derivative Test** Suppose f'' is continuous near c.
>
> (a) If $f'(c) = 0$ and $f''(c) > 0$, then f has a local minimum at c.
> (b) If $f'(c) = 0$ and $f''(c) < 0$, then f has a local maximum at c.

For instance, part (a) is true because $f''(x) > 0$ near c and so f is concave upward near c. This means that the graph of f lies *above* its horizontal tangent at c and so f has a local minimum at c. (See Figure 10.)

FIGURE 9

FIGURE 10
$f''(c) > 0$, f is concave upward

EXAMPLE 6 Discuss the curve $y = x^4 - 4x^3$ with respect to concavity, points of inflection, and local maxima and minima. Use this information to sketch the curve.

SOLUTION If $f(x) = x^4 - 4x^3$, then

$$f'(x) = 4x^3 - 12x^2 = 4x^2(x - 3)$$

$$f''(x) = 12x^2 - 24x = 12x(x - 2)$$

To find the critical numbers we set $f'(x) = 0$ and obtain $x = 0$ and $x = 3$. (Note that f' is a polynomial and hence defined everywhere.) To use the Second Derivative Test we evaluate f'' at these critical numbers:

$$f''(0) = 0 \qquad f''(3) = 36 > 0$$

Since $f'(3) = 0$ and $f''(3) > 0$, $f(3) = -27$ is a local minimum. [In fact, the expression for $f'(x)$ shows that f decreases to the left of 3 and increases to the right of 3.] Since $f''(0) = 0$, the Second Derivative Test gives no information about the critical number 0. But since $f'(x) < 0$ for $x < 0$ and also for $0 < x < 3$, the First Derivative Test tells us that f does not have a local maximum or minimum at 0.

Since $f''(x) = 0$ when $x = 0$ or 2, we divide the real line into intervals with these numbers as endpoints and complete the following chart.

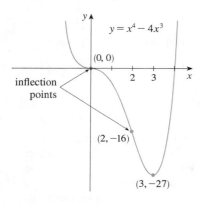

$y = x^4 - 4x^3$

(0, 0)

inflection points

(2, −16)

(3, −27)

FIGURE 11

Interval	$f''(x) = 12x(x - 2)$	Concavity
$(-\infty, 0)$	$+$	upward
$(0, 2)$	$-$	downward
$(2, \infty)$	$+$	upward

The point $(0, 0)$ is an inflection point since the curve changes from concave upward to concave downward there. Also $(2, -16)$ is an inflection point since the curve changes from concave downward to concave upward there.

Using the local minimum, the intervals of concavity, and the inflection points, we sketch the curve in Figure 11. ∎

NOTE The Second Derivative Test is inconclusive when $f''(c) = 0$. In other words, at such a point there might be a maximum, there might be a minimum, or there might be neither (as in Example 6). This test also fails when $f''(c)$ does not exist. In such cases the First Derivative Test must be used. In fact, even when both tests apply, the First Derivative Test is often the easier one to use.

EXAMPLE 7 Sketch the graph of the function $f(x) = x^{2/3}(6 - x)^{1/3}$.

SOLUTION Calculation of the first two derivatives gives

Use the differentiation rules to check these calculations.

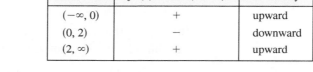

$$f'(x) = \frac{4 - x}{x^{1/3}(6 - x)^{2/3}} \qquad f''(x) = \frac{-8}{x^{4/3}(6 - x)^{5/3}}$$

Since $f'(x) = 0$ when $x = 4$ and $f'(x)$ does not exist when $x = 0$ or $x = 6$, the critical numbers are 0, 4, and 6.

Try reproducing the graph in Figure 12 with a graphing calculator or computer. Some machines produce the complete graph, some produce only the portion to the right of the *y*-axis, and some produce only the portion between $x = 0$ and $x = 6$. For an explanation and cure, see Example 7 in "Graphing Calculators and Computers" at www.stewartcalculus.com. An equivalent expression that gives the correct graph is

$$y = (x^2)^{1/3} \cdot \frac{6 - x}{|6 - x|}|6 - x|^{1/3}$$

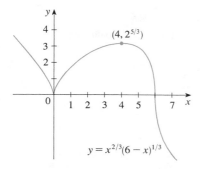

FIGURE 12

TEC In Module 4.3 you can practice using information about f', f'', and asymptotes to determine the shape of the graph of f.

Interval	$4 - x$	$x^{1/3}$	$(6 - x)^{2/3}$	$f'(x)$	f
$x < 0$	+	−	+	−	decreasing on $(-\infty, 0)$
$0 < x < 4$	+	+	+	+	increasing on $(0, 4)$
$4 < x < 6$	−	+	+	−	decreasing on $(4, 6)$
$x > 6$	−	+	+	−	decreasing on $(6, \infty)$

To find the local extreme values we use the First Derivative Test. Since f' changes from negative to positive at 0, $f(0) = 0$ is a local minimum. Since f' changes from positive to negative at 4, $f(4) = 2^{5/3}$ is a local maximum. The sign of f' does not change at 6, so there is no minimum or maximum there. (The Second Derivative Test could be used at 4 but not at 0 or 6 since f'' does not exist at either of these numbers.)

Looking at the expression for $f''(x)$ and noting that $x^{4/3} \geq 0$ for all x, we have $f''(x) < 0$ for $x < 0$ and for $0 < x < 6$ and $f''(x) > 0$ for $x > 6$. So f is concave downward on $(-\infty, 0)$ and $(0, 6)$ and concave upward on $(6, \infty)$, and the only inflection point is $(6, 0)$. The graph is sketched in Figure 12. Note that the curve has vertical tangents at $(0, 0)$ and $(6, 0)$ because $|f'(x)| \to \infty$ as $x \to 0$ and as $x \to 6$.   ■

EXAMPLE 8   Use the first and second derivatives of $f(x) = e^{1/x}$, together with asymptotes, to sketch its graph.

SOLUTION   Notice that the domain of f is $\{x \mid x \neq 0\}$, so we check for vertical asymptotes by computing the left and right limits as $x \to 0$. As $x \to 0^+$, we know that $t = 1/x \to \infty$, so

$$\lim_{x \to 0^+} e^{1/x} = \lim_{t \to \infty} e^t = \infty$$

and this shows that $x = 0$ is a vertical asymptote. As $x \to 0^-$, we have $t = 1/x \to -\infty$, so

$$\lim_{x \to 0^-} e^{1/x} = \lim_{t \to -\infty} e^t = 0$$

As $x \to \pm\infty$, we have $1/x \to 0$ and so

$$\lim_{x \to \pm\infty} e^{1/x} = e^0 = 1$$

This shows that $y = 1$ is a horizontal asymptote (both to the left and right).

Now let's compute the derivative. The Chain Rule gives

$$f'(x) = -\frac{e^{1/x}}{x^2}$$

Since $e^{1/x} > 0$ and $x^2 > 0$ for all $x \neq 0$, we have $f'(x) < 0$ for all $x \neq 0$. Thus f is decreasing on $(-\infty, 0)$ and on $(0, \infty)$. There is no critical number, so the function has no local maximum or minimum. The second derivative is

$$f''(x) = -\frac{x^2 e^{1/x}(-1/x^2) - e^{1/x}(2x)}{x^4} = \frac{e^{1/x}(2x + 1)}{x^4}$$

Since $e^{1/x} > 0$ and $x^4 > 0$, we have $f''(x) > 0$ when $x > -\frac{1}{2}$ $(x \neq 0)$ and $f''(x) < 0$ when $x < -\frac{1}{2}$. So the curve is concave downward on $\left(-\infty, -\frac{1}{2}\right)$ and concave upward on $\left(-\frac{1}{2}, 0\right)$ and on $(0, \infty)$. The inflection point is $\left(-\frac{1}{2}, e^{-2}\right)$.

To sketch the graph of f we first draw the horizontal asymptote $y = 1$ (as a dashed line), together with the parts of the curve near the asymptotes in a preliminary sketch

[Figure 13(a)]. These parts reflect the information concerning limits and the fact that f is decreasing on both $(-\infty, 0)$ and $(0, \infty)$. Notice that we have indicated that $f(x) \to 0$ as $x \to 0^-$ even though $f(0)$ does not exist. In Figure 13(b) we finish the sketch by incorporating the information concerning concavity and the inflection point. In Figure 13(c) we check our work with a graphing device.

(a) Preliminary sketch

(b) Finished sketch

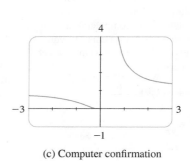

(c) Computer confirmation

FIGURE 13

4.3 EXERCISES

1–2 Use the given graph of f to find the following.
(a) The open intervals on which f is increasing.
(b) The open intervals on which f is decreasing.
(c) The open intervals on which f is concave upward.
(d) The open intervals on which f is concave downward.
(e) The coordinates of the points of inflection.

1.

2.

---

3. Suppose you are given a formula for a function f.
  (a) How do you determine where f is increasing or decreasing?
  (b) How do you determine where the graph of f is concave upward or concave downward?
  (c) How do you locate inflection points?

4. (a) State the First Derivative Test.
  (b) State the Second Derivative Test. Under what circumstances is it inconclusive? What do you do if it fails?

5–6 The graph of the *derivative* f' of a function f is shown.
(a) On what intervals is f increasing or decreasing?

(b) At what values of x does f have a local maximum or minimum?

5.

6.

---

7. In each part state the x-coordinates of the inflection points of f. Give reasons for your answers.
  (a) The curve is the graph of f.
  (b) The curve is the graph of f'.
  (c) The curve is the graph of f''.

8. The graph of the first derivative f' of a function f is shown.
   (a) On what intervals is f increasing? Explain.
   (b) At what values of x does f have a local maximum or minimum? Explain.
   (c) On what intervals is f concave upward or concave downward? Explain.
   (d) What are the x-coordinates of the inflection points of f? Why?

9–18
(a) Find the intervals on which f is increasing or decreasing.
(b) Find the local maximum and minimum values of f.
(c) Find the intervals of concavity and the inflection points.

9. $f(x) = x^3 - 3x^2 - 9x + 4$

10. $f(x) = 2x^3 - 9x^2 + 12x - 3$

11. $f(x) = x^4 - 2x^2 + 3$ **12.** $f(x) = \dfrac{x}{x^2 + 1}$

13. $f(x) = \sin x + \cos x, \quad 0 \le x \le 2\pi$

14. $f(x) = \cos^2 x - 2\sin x, \quad 0 \le x \le 2\pi$

15. $f(x) = e^{2x} + e^{-x}$ **16.** $f(x) = x^2 \ln x$

17. $f(x) = x^2 - x - \ln x$ **18.** $f(x) = x^4 e^{-x}$

19–21 Find the local maximum and minimum values of f using both the First and Second Derivative Tests. Which method do you prefer?

19. $f(x) = 1 + 3x^2 - 2x^3$ **20.** $f(x) = \dfrac{x^2}{x - 1}$

21. $f(x) = \sqrt{x} - \sqrt[4]{x}$

---

22. (a) Find the critical numbers of $f(x) = x^4(x - 1)^3$.
   (b) What does the Second Derivative Test tell you about the behavior of f at these critical numbers?
   (c) What does the First Derivative Test tell you?

23. Suppose f'' is continuous on $(-\infty, \infty)$.
   (a) If $f'(2) = 0$ and $f''(2) = -5$, what can you say about f?
   (b) If $f'(6) = 0$ and $f''(6) = 0$, what can you say about f?

24–31 Sketch the graph of a function that satisfies all of the given conditions.

24. (a) $f'(x) < 0$ and $f''(x) < 0$ for all x
   (b) $f'(x) > 0$ and $f''(x) > 0$ for all x

25. (a) $f'(x) > 0$ and $f''(x) < 0$ for all x
   (b) $f'(x) < 0$ and $f''(x) > 0$ for all x

26. Vertical asymptote $x = 0$, $\quad f'(x) > 0$ if $x < -2$,
   $f'(x) < 0$ if $x > -2 \ (x \ne 0)$,
   $f''(x) < 0$ if $x < 0$, $\quad f''(x) > 0$ if $x > 0$

27. $f'(0) = f'(2) = f'(4) = 0$,
   $f'(x) > 0$ if $x < 0$ or $2 < x < 4$,
   $f'(x) < 0$ if $0 < x < 2$ or $x > 4$,
   $f''(x) > 0$ if $1 < x < 3$, $\quad f''(x) < 0$ if $x < 1$ or $x > 3$

28. $f'(x) > 0$ for all $x \ne 1$, $\quad$ vertical asymptote $x = 1$,
   $f''(x) > 0$ if $x < 1$ or $x > 3$, $\quad f''(x) < 0$ if $1 < x < 3$

29. $f'(5) = 0$, $\quad f'(x) < 0$ when $x < 5$,
   $f'(x) > 0$ when $x > 5$, $\quad f''(2) = 0$, $\quad f''(8) = 0$,
   $f''(x) < 0$ when $x < 2$ or $x > 8$,
   $f''(x) > 0$ for $2 < x < 8$, $\quad \lim\limits_{x \to \infty} f(x) = 3$, $\quad \lim\limits_{x \to -\infty} f(x) = 3$

30. $f'(0) = f'(4) = 0$, $\quad f'(x) = 1$ if $x < -1$,
   $f'(x) > 0$ if $0 < x < 2$,
   $f'(x) < 0$ if $-1 < x < 0$ or $2 < x < 4$ or $x > 4$,
   $\lim\limits_{x \to 2^-} f'(x) = \infty$, $\quad \lim\limits_{x \to 2^+} f'(x) = -\infty$,
   $f''(x) > 0$ if $-1 < x < 2$ or $2 < x < 4$,
   $f''(x) < 0$ if $x > 4$

31. $f'(x) > 0$ if $x \ne 2$, $\quad f''(x) > 0$ if $x < 2$,
   $f''(x) < 0$ if $x > 2$, $\quad f$ has inflection point $(2, 5)$,
   $\lim\limits_{x \to \infty} f(x) = 8$, $\quad \lim\limits_{x \to -\infty} f(x) = 0$

---

32. Suppose $f(3) = 2$, $f'(3) = \frac{1}{2}$, and $f'(x) > 0$ and $f''(x) < 0$ for all x.
   (a) Sketch a possible graph for f.
   (b) How many solutions does the equation $f(x) = 0$ have? Why?
   (c) Is it possible that $f'(2) = \frac{1}{3}$? Why?

33. Suppose f is a continuous function where $f(x) > 0$ for all x, $f(0) = 4$, $\quad f'(x) > 0$ if $x < 0$ or $x > 2$, $\quad f'(x) < 0$ if $0 < x < 2$, $\quad f''(-1) = f''(1) = 0$, $\quad f''(x) > 0$ if $x < -1$ or $x > 1$, $\quad f''(x) < 0$ if $-1 < x < 1$.
   (a) Can f have an absolute maximum? If so, sketch a possible graph of f. If not, explain why.
   (b) Can f have an absolute minimum? If so, sketch a possible graph of f. If not, explain why.
   (c) Sketch a possible graph for f that does *not* achieve an absolute minimum.

34. The graph of a function $y = f(x)$ is shown. At which point(s) are the following true?
   (a) $\dfrac{dy}{dx}$ and $\dfrac{d^2y}{dx^2}$ are both positive.
   (b) $\dfrac{dy}{dx}$ and $\dfrac{d^2y}{dx^2}$ are both negative.
   (c) $\dfrac{dy}{dx}$ is negative but $\dfrac{d^2y}{dx^2}$ is positive.

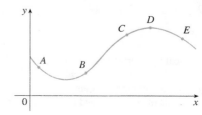

35–36 The graph of the derivative f' of a continuous function f is shown.
(a) On what intervals is f increasing? Decreasing?
(b) At what values of x does f have a local maximum? Local minimum?
(c) On what intervals is f concave upward? Concave downward?
(d) State the x-coordinate(s) of the point(s) of inflection.
(e) Assuming that $f(0) = 0$, sketch a graph of f.

35.

36.

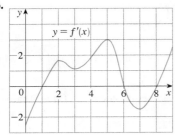

37–48
(a) Find the intervals of increase or decrease.
(b) Find the local maximum and minimum values.
(c) Find the intervals of concavity and the inflection points.
(d) Use the information from parts (a)–(c) to sketch the graph. Check your work with a graphing device if you have one.

37. $f(x) = x^3 - 12x + 2$

38. $f(x) = 36x + 3x^2 - 2x^3$

39. $f(x) = \frac{1}{2}x^4 - 4x^2 + 3$

40. $g(x) = 200 + 8x^3 + x^4$

41. $h(x) = (x + 1)^5 - 5x - 2$

42. $h(x) = 5x^3 - 3x^5$

43. $F(x) = x\sqrt{6 - x}$

44. $G(x) = 5x^{2/3} - 2x^{5/3}$

45. $C(x) = x^{1/3}(x + 4)$

46. $f(x) = \ln(x^2 + 9)$

47. $f(\theta) = 2\cos\theta + \cos^2\theta, \quad 0 \le \theta \le 2\pi$

48. $S(x) = x - \sin x, \quad 0 \le x \le 4\pi$

49–56
(a) Find the vertical and horizontal asymptotes.
(b) Find the intervals of increase or decrease.
(c) Find the local maximum and minimum values.
(d) Find the intervals of concavity and the inflection points.

(e) Use the information from parts (a)–(d) to sketch the graph of f.

49. $f(x) = 1 + \dfrac{1}{x} - \dfrac{1}{x^2}$

50. $f(x) = \dfrac{x^2 - 4}{x^2 + 4}$

51. $f(x) = \sqrt{x^2 + 1} - x$

52. $f(x) = \dfrac{e^x}{1 - e^x}$

53. $f(x) = e^{-x^2}$

54. $f(x) = x - \frac{1}{6}x^2 - \frac{2}{3}\ln x$

55. $f(x) = \ln(1 - \ln x)$

56. $f(x) = e^{\arctan x}$

57. Suppose the derivative of a function f is $f'(x) = (x + 1)^2(x - 3)^5(x - 6)^4$. On what interval is f increasing?

58. Use the methods of this section to sketch the curve $y = x^3 - 3a^2x + 2a^3$, where a is a positive constant. What do the members of this family of curves have in common? How do they differ from each other?

⊞ 59–60
(a) Use a graph of f to estimate the maximum and minimum values. Then find the exact values.
(b) Estimate the value of x at which f increases most rapidly. Then find the exact value.

59. $f(x) = \dfrac{x + 1}{\sqrt{x^2 + 1}}$

60. $f(x) = x^2 e^{-x}$

⊞ 61–62
(a) Use a graph of f to give a rough estimate of the intervals of concavity and the coordinates of the points of inflection.
(b) Use a graph of f'' to give better estimates.

61. $f(x) = \sin 2x + \sin 4x, \quad 0 \le x \le \pi$

62. $f(x) = (x - 1)^2(x + 1)^3$

CAS 63–64 Estimate the intervals of concavity to one decimal place by using a computer algebra system to compute and graph f''.

63. $f(x) = \dfrac{x^4 + x^3 + 1}{\sqrt{x^2 + x + 1}}$

64. $f(x) = \dfrac{x^2 \tan^{-1} x}{1 + x^3}$

65. A graph of a population of yeast cells in a new laboratory culture as a function of time is shown.

(a) Describe how the rate of population increase varies.

(b) When is this rate highest?

(c) On what intervals is the population function concave upward or downward?

(d) Estimate the coordinates of the inflection point.

66. In an episode of *The Simpsons* television show, Homer reads from a newspaper and announces "Here's good news! According to this eye-catching article, SAT scores are declining at a slower rate." Interpret Homer's statement in terms of a function and its first and second derivatives.

67. The president announces that the national deficit is increasing, but at a decreasing rate. Interpret this statement in terms of a function and its first and second derivatives.

68. Let $f(t)$ be the temperature at time t where you live and suppose that at time $t = 3$ you feel uncomfortably hot. How do you feel about the given data in each case?

(a) $f'(3) = 2, \quad f''(3) = 4$

(b) $f'(3) = 2, \quad f''(3) = -4$

(c) $f'(3) = -2, \quad f''(3) = 4$

(d) $f'(3) = -2, \quad f''(3) = -4$

69. Let $K(t)$ be a measure of the knowledge you gain by studying for a test for t hours. Which do you think is larger, $K(8) - K(7)$ or $K(3) - K(2)$? Is the graph of K concave upward or concave downward? Why?

70. Coffee is being poured into the mug shown in the figure at a constant rate (measured in volume per unit time). Sketch a rough graph of the depth of the coffee in the mug as a function of time. Account for the shape of the graph in terms of concavity. What is the significance of the inflection point?

71. A *drug response curve* describes the level of medication in the bloodstream after a drug is administered. A surge function $S(t) = At^p e^{-kt}$ is often used to model the response curve, reflecting an initial surge in the drug level and then a more gradual decline. If, for a particular drug, $A = 0.01$, $p = 4$, $k = 0.07$, and t is measured in minutes, estimate the times corresponding to the inflection points and explain their significance. If you have a graphing device, use it to graph the drug response curve.

72. The family of bell-shaped curves

$$y = \frac{1}{\sigma\sqrt{2\pi}} e^{-(x-\mu)^2/(2\sigma^2)}$$

occurs in probability and statistics, where it is called the *normal density function*. The constant μ is called the *mean* and

the positive constant σ is called the *standard deviation*. For simplicity, let's scale the function so as to remove the factor $1/(\sigma\sqrt{2\pi})$ and let's analyze the special case where $\mu = 0$. So we study the function

$$f(x) = e^{-x^2/(2\sigma^2)}$$

(a) Find the asymptote, maximum value, and inflection points of f.

(b) What role does σ play in the shape of the curve?

(c) Illustrate by graphing four members of this family on the same screen.

73. Find a cubic function $f(x) = ax^3 + bx^2 + cx + d$ that has a local maximum value of 3 at $x = -2$ and a local minimum value of 0 at $x = 1$.

74. For what values of the numbers a and b does the function

$$f(x) = axe^{bx^2}$$

have the maximum value $f(2) = 1$?

75. (a) If the function $f(x) = x^3 + ax^2 + bx$ has the local minimum value $-\frac{2}{9}\sqrt{3}$ at $x = 1/\sqrt{3}$, what are the values of a and b?

(b) Which of the tangent lines to the curve in part (a) has the smallest slope?

76. For what values of a and b is (2, 2.5) an inflection point of the curve $x^2y + ax + by = 0$? What additional inflection points does the curve have?

77. Show that the curve $y = (1 + x)/(1 + x^2)$ has three points of inflection and they all lie on one straight line.

78. Show that the curves $y = e^{-x}$ and $y = -e^{-x}$ touch the curve $y = e^{-x} \sin x$ at its inflection points.

79. Show that the inflection points of the curve $y = x \sin x$ lie on the curve $y^2(x^2 + 4) = 4x^2$.

80–82 Assume that all of the functions are twice differentiable and the second derivatives are never 0.

80. (a) If f and g are concave upward on I, show that $f + g$ is concave upward on I.

(b) If f is positive and concave upward on I, show that the function $g(x) = [f(x)]^2$ is concave upward on I.

81. (a) If f and g are positive, increasing, concave upward functions on I, show that the product function fg is concave upward on I.

(b) Show that part (a) remains true if f and g are both decreasing.

(c) Suppose f is increasing and g is decreasing. Show, by giving three examples, that fg may be concave upward, concave downward, or linear. Why doesn't the argument in parts (a) and (b) work in this case?

82. Suppose f and g are both concave upward on $(-\infty, \infty)$. Under what condition on f will the composite function $h(x) = f(g(x))$ be concave upward?

83. Show that $\tan x > x$ for $0 < x < \pi/2$. [*Hint:* Show that $f(x) = \tan x - x$ is increasing on $(0, \pi/2)$.]

84. (a) Show that $e^x \geqslant 1 + x$ for $x \geqslant 0$.
   (b) Deduce that $e^x \geqslant 1 + x + \frac{1}{2}x^2$ for $x \geqslant 0$.
   (c) Use mathematical induction to prove that for $x \geqslant 0$ and any positive integer n,

$$e^x \geqslant 1 + x + \frac{x^2}{2!} + \cdots + \frac{x^n}{n!}$$

85. Show that a cubic function (a third-degree polynomial) always has exactly one point of inflection. If its graph has three x-intercepts x_1, x_2, and x_3, show that the x-coordinate of the inflection point is $(x_1 + x_2 + x_3)/3$.

86. For what values of c does the polynomial $P(x) = x^4 + cx^3 + x^2$ have two inflection points? One inflection point? None? Illustrate by graphing P for several values of c. How does the graph change as c decreases?

87. Prove that if $(c, f(c))$ is a point of inflection of the graph of f and f'' exists in an open interval that contains c, then $f''(c) = 0$. [*Hint:* Apply the First Derivative Test and Fermat's Theorem to the function $g = f'$.]

88. Show that if $f(x) = x^4$, then $f''(0) = 0$, but $(0, 0)$ is not an inflection point of the graph of f.

89. Show that the function $g(x) = x|x|$ has an inflection point at $(0, 0)$ but $g''(0)$ does not exist.

90. Suppose that f''' is continuous and $f'(c) = f''(c) = 0$, but $f'''(c) > 0$. Does f have a local maximum or minimum at c? Does f have a point of inflection at c?

91. Suppose f is differentiable on an interval I and $f'(x) > 0$ for all numbers x in I except for a single number c. Prove that f is increasing on the entire interval I.

92. For what values of c is the function

$$f(x) = cx + \frac{1}{x^2 + 3}$$

increasing on $(-\infty, \infty)$?

93. The three cases in the First Derivative Test cover the situations one commonly encounters but do not exhaust all possibilities. Consider the functions f, g, and h whose values at 0 are all 0 and, for $x \neq 0$,

$$f(x) = x^4 \sin \frac{1}{x} \qquad g(x) = x^4\left(2 + \sin \frac{1}{x}\right)$$

$$h(x) = x^4\left(-2 + \sin \frac{1}{x}\right)$$

   (a) Show that 0 is a critical number of all three functions but their derivatives change sign infinitely often on both sides of 0.
   (b) Show that f has neither a local maximum nor a local minimum at 0, g has a local minimum, and h has a local maximum.

4.4 Indeterminate Forms and l'Hospital's Rule

Suppose we are trying to analyze the behavior of the function

$$F(x) = \frac{\ln x}{x - 1}$$

Although F is not defined when $x = 1$, we need to know how F behaves *near* 1. In particular, we would like to know the value of the limit

$$\boxed{1} \qquad \lim_{x \to 1} \frac{\ln x}{x - 1}$$

In computing this limit we can't apply Law 5 of limits (the limit of a quotient is the quotient of the limits, see Section 2.3) because the limit of the denominator is 0. In fact, although the limit in (1) exists, its value is not obvious because both numerator and denominator approach 0 and $\frac{0}{0}$ is not defined.

In general, if we have a limit of the form

$$\lim_{x \to a} \frac{f(x)}{g(x)}$$

where both $f(x) \to 0$ and $g(x) \to 0$ as $x \to a$, then this limit may or may not exist and is called an **indeterminate form of type $\frac{0}{0}$**. We met some limits of this type in Chapter 2. For rational functions, we can cancel common factors:

$$\lim_{x \to 1} \frac{x^2 - x}{x^2 - 1} = \lim_{x \to 1} \frac{x(x - 1)}{(x + 1)(x - 1)} = \lim_{x \to 1} \frac{x}{x + 1} = \frac{1}{2}$$

We used a geometric argument to show that

$$\lim_{x \to 0} \frac{\sin x}{x} = 1$$

But these methods do not work for limits such as (1), so in this section we introduce a systematic method, known as *l'Hospital's Rule*, for the evaluation of indeterminate forms.

Another situation in which a limit is not obvious occurs when we look for a horizontal asymptote of F and need to evaluate the limit

$$\boxed{2} \qquad \lim_{x \to \infty} \frac{\ln x}{x - 1}$$

It isn't obvious how to evaluate this limit because both numerator and denominator become large as $x \to \infty$. There is a struggle between numerator and denominator. If the numerator wins, the limit will be ∞ (the numerator was increasing significantly faster than the denominator); if the denominator wins, the answer will be 0. Or there may be some compromise, in which case the answer will be some finite positive number.

In general, if we have a limit of the form

$$\lim_{x \to a} \frac{f(x)}{g(x)}$$

where both $f(x) \to \infty$ (or $-\infty$) and $g(x) \to \infty$ (or $-\infty$), then the limit may or may not exist and is called an **indeterminate form of type** ∞ / ∞. We saw in Section 2.6 that this type of limit can be evaluated for certain functions, including rational functions, by dividing numerator and denominator by the highest power of x that occurs in the denominator. For instance,

$$\lim_{x \to \infty} \frac{x^2 - 1}{2x^2 + 1} = \lim_{x \to \infty} \frac{1 - \dfrac{1}{x^2}}{2 + \dfrac{1}{x^2}} = \frac{1 - 0}{2 + 0} = \frac{1}{2}$$

This method does not work for limits such as (2), but l'Hospital's Rule also applies to this type of indeterminate form.

FIGURE 1

Figure 1 suggests visually why l'Hospital's Rule might be true. The first graph shows two differentiable functions f and g, each of which approaches 0 as $x \to a$. If we were to zoom in toward the point $(a, 0)$, the graphs would start to look almost linear. But if the functions actually *were* linear, as in the second graph, then their ratio would be

$$\frac{m_1(x - a)}{m_2(x - a)} = \frac{m_1}{m_2}$$

which is the ratio of their derivatives. This suggests that

$$\lim_{x \to a} \frac{f(x)}{g(x)} = \lim_{x \to a} \frac{f'(x)}{g'(x)}$$

L'Hospital's Rule Suppose f and g are differentiable and $g'(x) \neq 0$ on an open interval I that contains a (except possibly at a). Suppose that

$$\lim_{x \to a} f(x) = 0 \qquad \text{and} \qquad \lim_{x \to a} g(x) = 0$$

or that

$$\lim_{x \to a} f(x) = \pm\infty \qquad \text{and} \qquad \lim_{x \to a} g(x) = \pm\infty$$

(In other words, we have an indeterminate form of type $\frac{0}{0}$ or ∞/∞.) Then

$$\lim_{x \to a} \frac{f(x)}{g(x)} = \lim_{x \to a} \frac{f'(x)}{g'(x)}$$

if the limit on the right side exists (or is ∞ or $-\infty$).

NOTE 1 L'Hospital's Rule says that the limit of a quotient of functions is equal to the limit of the quotient of their derivatives, provided that the given conditions are satisfied.

L'Hospital

L'Hospital's Rule is named after a French nobleman, the Marquis de l'Hospital (1661–1704), but was discovered by a Swiss mathematician, John Bernoulli (1667–1748). You might sometimes see l'Hospital spelled as l'Hôpital, but he spelled his own name l'Hospital, as was common in the 17th century. See Exercise 83 for the example that the Marquis used to illustrate his rule. See the project on page 314 for further historical details.

It is especially important to verify the conditions regarding the limits of f and g before using l'Hospital's Rule.

NOTE 2 L'Hospital's Rule is also valid for one-sided limits and for limits at infinity or negative infinity; that is, "$x \to a$" can be replaced by any of the symbols $x \to a^+$, $x \to a^-$, $x \to \infty$, or $x \to -\infty$.

NOTE 3 For the special case in which $f(a) = g(a) = 0$, f' and g' are continuous, and $g'(a) \neq 0$, it is easy to see why l'Hospital's Rule is true. In fact, using the alternative form of the definition of a derivative, we have

$$\lim_{x \to a} \frac{f'(x)}{g'(x)} = \frac{f'(a)}{g'(a)} = \frac{\displaystyle\lim_{x \to a} \frac{f(x) - f(a)}{x - a}}{\displaystyle\lim_{x \to a} \frac{g(x) - g(a)}{x - a}} = \lim_{x \to a} \frac{\dfrac{f(x) - f(a)}{x - a}}{\dfrac{g(x) - g(a)}{x - a}}$$

$$= \lim_{x \to a} \frac{f(x) - f(a)}{g(x) - g(a)} = \lim_{x \to a} \frac{f(x)}{g(x)} \qquad [\text{since } f(a) = g(a) = 0]$$

It is more difficult to prove the general version of l'Hospital's Rule. See Appendix F.

EXAMPLE 1 Find $\displaystyle\lim_{x \to 1} \frac{\ln x}{x - 1}$.

SOLUTION Since

$$\lim_{x \to 1} \ln x = \ln 1 = 0 \qquad \text{and} \qquad \lim_{x \to 1} (x - 1) = 0$$

the limit is an indeterminate form of type $\frac{0}{0}$, so we can apply l'Hospital's Rule:

$$\lim_{x \to 1} \frac{\ln x}{x - 1} = \lim_{x \to 1} \frac{\dfrac{d}{dx}(\ln x)}{\dfrac{d}{dx}(x - 1)} = \lim_{x \to 1} \frac{1/x}{1}$$

$$= \lim_{x \to 1} \frac{1}{x} = 1$$

⊘ Notice that when using l'Hospital's Rule we differentiate the numerator and denominator *separately*. We do *not* use the Quotient Rule.

EXAMPLE 2 Calculate $\displaystyle\lim_{x \to \infty} \frac{e^x}{x^2}$.

SOLUTION We have $\lim_{x \to \infty} e^x = \infty$ and $\lim_{x \to \infty} x^2 = \infty$, so the limit is an indeterminate form of type ∞/∞, and l'Hospital's Rule gives

$$\lim_{x \to \infty} \frac{e^x}{x^2} = \lim_{x \to \infty} \frac{\dfrac{d}{dx}(e^x)}{\dfrac{d}{dx}(x^2)} = \lim_{x \to \infty} \frac{e^x}{2x}$$

Since $e^x \to \infty$ and $2x \to \infty$ as $x \to \infty$, the limit on the right side is also indeterminate, but a second application of l'Hospital's Rule gives

$$\lim_{x \to \infty} \frac{e^x}{x^2} = \lim_{x \to \infty} \frac{e^x}{2x} = \lim_{x \to \infty} \frac{e^x}{2} = \infty$$

The graph of the function of Example 2 is shown in Figure 2. We have noticed previously that exponential functions grow far more rapidly than power functions, so the result of Example 2 is not unexpected. See also Exercise 73.

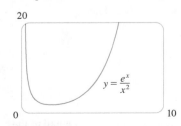

FIGURE 2

The graph of the function of Example 3 is shown in Figure 3. We have discussed previously the slow growth of logarithms, so it isn't surprising that this ratio approaches 0 as $x \to \infty$. See also Exercise 74.

FIGURE 3

EXAMPLE 3 Calculate $\displaystyle\lim_{x \to \infty} \frac{\ln x}{\sqrt{x}}$.

SOLUTION Since $\ln x \to \infty$ and $\sqrt{x} \to \infty$ as $x \to \infty$, l'Hospital's Rule applies:

$$\lim_{x \to \infty} \frac{\ln x}{\sqrt{x}} = \lim_{x \to \infty} \frac{1/x}{\frac{1}{2}x^{-1/2}} = \lim_{x \to \infty} \frac{1/x}{1/(2\sqrt{x})}$$

Notice that the limit on the right side is now indeterminate of type $\frac{0}{0}$. But instead of applying l'Hospital's Rule a second time as we did in Example 2, we simplify the expression and see that a second application is unnecessary:

$$\lim_{x \to \infty} \frac{\ln x}{\sqrt{x}} = \lim_{x \to \infty} \frac{1/x}{1/(2\sqrt{x})} = \lim_{x \to \infty} \frac{2}{\sqrt{x}} = 0$$

In both Examples 2 and 3 we evaluated limits of type ∞/∞, but we got two different results. In Example 2, the infinite limit tells us that the numerator e^x increases significantly faster than the denominator x^2, resulting in larger and larger ratios. In fact, $y = e^x$ grows more quickly than all the power functions $y = x^n$ (see Exercise 73). In Example 3 we have the opposite situation; the limit of 0 means that the denominator outpaces the numerator, and the ratio eventually approaches 0.

EXAMPLE 4 Find $\displaystyle\lim_{x \to 0} \frac{\tan x - x}{x^3}$. (See Exercise 2.2.50.)

SOLUTION Noting that both $\tan x - x \to 0$ and $x^3 \to 0$ as $x \to 0$, we use l'Hospital's Rule:

$$\lim_{x \to 0} \frac{\tan x - x}{x^3} = \lim_{x \to 0} \frac{\sec^2 x - 1}{3x^2}$$

Since the limit on the right side is still indeterminate of type $\frac{0}{0}$, we apply l'Hospital's Rule again:

$$\lim_{x \to 0} \frac{\sec^2 x - 1}{3x^2} = \lim_{x \to 0} \frac{2\sec^2 x \tan x}{6x}$$

The graph in Figure 4 gives visual confirmation of the result of Example 4. If we were to zoom in too far, however, we would get an inaccurate graph because $\tan x$ is close to x when x is small. See Exercise 2.2.50(d).

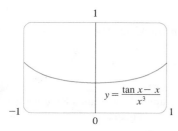

FIGURE 4

Because $\lim_{x \to 0} \sec^2 x = 1$, we simplify the calculation by writing

$$\lim_{x \to 0} \frac{2\sec^2 x \tan x}{6x} = \frac{1}{3} \lim_{x \to 0} \sec^2 x \cdot \lim_{x \to 0} \frac{\tan x}{x} = \frac{1}{3} \lim_{x \to 0} \frac{\tan x}{x}$$

We can evaluate this last limit either by using l'Hospital's Rule a third time or by writing $\tan x$ as $(\sin x)/(\cos x)$ and making use of our knowledge of trigonometric limits. Putting together all the steps, we get

$$\lim_{x \to 0} \frac{\tan x - x}{x^3} = \lim_{x \to 0} \frac{\sec^2 x - 1}{3x^2} = \lim_{x \to 0} \frac{2\sec^2 x \tan x}{6x}$$

$$= \frac{1}{3} \lim_{x \to 0} \frac{\tan x}{x} = \frac{1}{3} \lim_{x \to 0} \frac{\sec^2 x}{1} = \frac{1}{3}$$

EXAMPLE 5 Find $\lim\limits_{x \to \pi^-} \dfrac{\sin x}{1 - \cos x}$.

SOLUTION If we blindly attempted to use l'Hospital's Rule, we would get

$$\lim_{x \to \pi^-} \frac{\sin x}{1 - \cos x} = \lim_{x \to \pi^-} \frac{\cos x}{\sin x} = -\infty$$

This is wrong! Although the numerator $\sin x \to 0$ as $x \to \pi^-$, notice that the denominator $(1 - \cos x)$ does not approach 0, so l'Hospital's Rule can't be applied here.

The required limit is, in fact, easy to find because the function is continuous at π and the denominator is nonzero there:

$$\lim_{x \to \pi^-} \frac{\sin x}{1 - \cos x} = \frac{\sin \pi}{1 - \cos \pi} = \frac{0}{1 - (-1)} = 0 \qquad \blacksquare$$

Example 5 shows what can go wrong if you use l'Hospital's Rule without thinking. Other limits *can* be found using l'Hospital's Rule but are more easily found by other methods. (See Examples 2.3.3, 2.3.5, and 2.6.3, and the discussion at the beginning of this section.) So when evaluating any limit, you should consider other methods before using l'Hospital's Rule.

■ Indeterminate Products

If $\lim_{x \to a} f(x) = 0$ and $\lim_{x \to a} g(x) = \infty$ (or $-\infty$), then it isn't clear what the value of $\lim_{x \to a} [f(x) g(x)]$, if any, will be. There is a struggle between f and g. If f wins, the answer will be 0; if g wins, the answer will be ∞ (or $-\infty$). Or there may be a compromise where the answer is a finite nonzero number. This kind of limit is called an **indeterminate form of type $0 \cdot \infty$**. We can deal with it by writing the product fg as a quotient:

$$fg = \frac{f}{1/g} \qquad \text{or} \qquad fg = \frac{g}{1/f}$$

This converts the given limit into an indeterminate form of type $\frac{0}{0}$ or ∞/∞ so that we can use l'Hospital's Rule.

Figure 5 shows the graph of the function in Example 6. Notice that the function is undefined at $x = 0$; the graph approaches the origin but never quite reaches it.

EXAMPLE 6 Evaluate $\lim\limits_{x \to 0^+} x \ln x$.

SOLUTION The given limit is indeterminate because, as $x \to 0^+$, the first factor (x) approaches 0 while the second factor $(\ln x)$ approaches $-\infty$. Writing $x = 1/(1/x)$, we have $1/x \to \infty$ as $x \to 0^+$, so l'Hospital's Rule gives

$$\lim_{x \to 0^+} x \ln x = \lim_{x \to 0^+} \frac{\ln x}{1/x} = \lim_{x \to 0^+} \frac{1/x}{-1/x^2} = \lim_{x \to 0^+} (-x) = 0 \qquad \blacksquare$$

NOTE In solving Example 6 another possible option would have been to write

$$\lim_{x \to 0^+} x \ln x = \lim_{x \to 0^+} \frac{x}{1/\ln x}$$

This gives an indeterminate form of the type $\frac{0}{0}$, but if we apply l'Hospital's Rule we get a more complicated expression than the one we started with. In general, when we rewrite an indeterminate product, we try to choose the option that leads to the simpler limit.

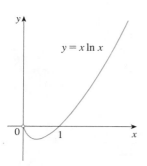

$y = x \ln x$

FIGURE 5

Indeterminate Differences

If $\lim_{x\to a} f(x) = \infty$ and $\lim_{x\to a} g(x) = \infty$, then the limit

$$\lim_{x\to a} [f(x) - g(x)]$$

is called an **indeterminate form of type $\infty - \infty$**. Again there is a contest between f and g. Will the answer be ∞ (f wins) or will it be $-\infty$ (g wins) or will they compromise on a finite number? To find out, we try to convert the difference into a quotient (for instance, by using a common denominator, or rationalization, or factoring out a common factor) so that we have an indeterminate form of type $\frac{0}{0}$ or ∞/∞.

EXAMPLE 7 Compute $\displaystyle\lim_{x\to 1^+} \left(\frac{1}{\ln x} - \frac{1}{x-1} \right)$.

SOLUTION First notice that $1/(\ln x) \to \infty$ and $1/(x-1) \to \infty$ as $x \to 1^+$, so the limit is indeterminate of type $\infty - \infty$. Here we can start with a common denominator:

$$\lim_{x\to 1^+} \left(\frac{1}{\ln x} - \frac{1}{x-1} \right) = \lim_{x\to 1^+} \frac{x - 1 - \ln x}{(x-1)\ln x}$$

Both numerator and denominator have a limit of 0, so l'Hospital's Rule applies, giving

$$\lim_{x\to 1^+} \frac{x - 1 - \ln x}{(x-1)\ln x} = \lim_{x\to 1^+} \frac{1 - \dfrac{1}{x}}{(x-1)\cdot\dfrac{1}{x} + \ln x} = \lim_{x\to 1^+} \frac{x-1}{x - 1 + x\ln x}$$

Again we have an indeterminate limit of type $\frac{0}{0}$, so we apply l'Hospital's Rule a second time:

$$\lim_{x\to 1^+} \frac{x-1}{x - 1 + x\ln x} = \lim_{x\to 1^+} \frac{1}{1 + x\cdot\dfrac{1}{x} + \ln x}$$

$$= \lim_{x\to 1^+} \frac{1}{2 + \ln x} = \frac{1}{2} \qquad \blacksquare$$

EXAMPLE 8 Calculate $\displaystyle\lim_{x\to\infty} (e^x - x)$.

SOLUTION This is an indeterminate difference because both e^x and x approach infinity. We would expect the limit to be infinity because $e^x \to \infty$ much faster than x. But we can verify this by factoring out x:

$$e^x - x = x\left(\frac{e^x}{x} - 1 \right)$$

The term $e^x/x \to \infty$ as $x \to \infty$ by l'Hospital's Rule and so we now have a product in which both factors grow large:

$$\lim_{x\to\infty} (e^x - x) = \lim_{x\to\infty} \left[x\left(\frac{e^x}{x} - 1 \right) \right] = \infty \qquad \blacksquare$$

■ Indeterminate Powers

Several indeterminate forms arise from the limit

$$\lim_{x \to a} = [f(x)]^{g(x)}$$

1. $\lim\limits_{x \to a} f(x) = 0$ and $\lim\limits_{x \to a} g(x) = 0$ type 0^0

2. $\lim\limits_{x \to a} f(x) = \infty$ and $\lim\limits_{x \to a} g(x) = 0$ type ∞^0

3. $\lim\limits_{x \to a} f(x) = 1$ and $\lim\limits_{x \to a} g(x) = \pm\infty$ type 1^∞

Each of these three cases can be treated either by taking the natural logarithm:

$$\text{let} \quad y = [f(x)]^{g(x)}, \quad \text{then} \quad \ln y = g(x) \ln f(x)$$

or by writing the function as an exponential:

$$[f(x)]^{g(x)} = e^{g(x)\ln f(x)}$$

(Recall that both of these methods were used in differentiating such functions.) In either method we are led to the indeterminate product $g(x) \ln f(x)$, which is of type $0 \cdot \infty$.

EXAMPLE 9 Calculate $\lim\limits_{x \to 0^+} (1 + \sin 4x)^{\cot x}$.

SOLUTION First notice that as $x \to 0^+$, we have $1 + \sin 4x \to 1$ and $\cot x \to \infty$, so the given limit is indeterminate (type 1^∞). Let

$$y = (1 + \sin 4x)^{\cot x}$$

Then $\ln y = \ln[(1 + \sin 4x)^{\cot x}] = \cot x \ln(1 + \sin 4x) = \dfrac{\ln(1 + \sin 4x)}{\tan x}$

so l'Hospital's Rule gives

$$\lim_{x \to 0^+} \ln y = \lim_{x \to 0^+} \frac{\ln(1 + \sin 4x)}{\tan x} = \lim_{x \to 0^+} \frac{\dfrac{4 \cos 4x}{1 + \sin 4x}}{\sec^2 x} = 4$$

So far we have computed the limit of $\ln y$, but what we want is the limit of y. To find this we use the fact that $y = e^{\ln y}$:

$$\lim_{x \to 0^+} (1 + \sin 4x)^{\cot x} = \lim_{x \to 0^+} y = \lim_{x \to 0^+} e^{\ln y} = e^4 \qquad ■$$

EXAMPLE 10 Find $\lim\limits_{x \to 0^+} x^x$.

SOLUTION Notice that this limit is indeterminate since $0^x = 0$ for any $x > 0$ but $x^0 = 1$ for any $x \neq 0$. (Recall that 0^0 is undefined.) We could proceed as in Example 9 or by writing the function as an exponential:

$$x^x = (e^{\ln x})^x = e^{x \ln x}$$

In Example 6 we used l'Hospital's Rule to show that

$$\lim_{x \to 0^+} x \ln x = 0$$

Although forms of the type 0^0, ∞^0, and 1^∞ are indeterminate, the form 0^∞ is not indeterminate. (See Exercise 86.)

The graph of the function $y = x^x$, $x > 0$, is shown in Figure 6. Notice that although 0^0 is not defined, the values of the function approach 1 as $x \to 0^+$. This confirms the result of Example 10.

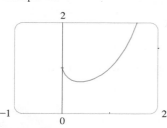

FIGURE 6

Therefore

$$\lim_{x \to 0^+} x^x = \lim_{x \to 0^+} e^{x \ln x} = e^0 = 1$$

4.4 EXERCISES

1–4 Given that

$$\lim_{x \to a} f(x) = 0 \qquad \lim_{x \to a} g(x) = 0 \qquad \lim_{x \to a} h(x) = 1$$

$$\lim_{x \to a} p(x) = \infty \qquad \lim_{x \to a} q(x) = \infty$$

which of the following limits are indeterminate forms? For those that are not an indeterminate form, evaluate the limit where possible.

1. (a) $\displaystyle\lim_{x \to a} \frac{f(x)}{g(x)}$ (b) $\displaystyle\lim_{x \to a} \frac{f(x)}{p(x)}$

(c) $\displaystyle\lim_{x \to a} \frac{h(x)}{p(x)}$ (d) $\displaystyle\lim_{x \to a} \frac{p(x)}{f(x)}$

(e) $\displaystyle\lim_{x \to a} \frac{p(x)}{q(x)}$

2. (a) $\displaystyle\lim_{x \to a} [f(x)p(x)]$ (b) $\displaystyle\lim_{x \to a} [h(x)p(x)]$

(c) $\displaystyle\lim_{x \to a} [p(x)q(x)]$

3. (a) $\displaystyle\lim_{x \to a} [f(x) - p(x)]$ (b) $\displaystyle\lim_{x \to a} [p(x) - q(x)]$

(c) $\displaystyle\lim_{x \to a} [p(x) + q(x)]$

4. (a) $\displaystyle\lim_{x \to a} [f(x)]^{g(x)}$ (b) $\displaystyle\lim_{x \to a} [f(x)]^{p(x)}$

(c) $\displaystyle\lim_{x \to a} [h(x)]^{p(x)}$ (d) $\displaystyle\lim_{x \to a} [p(x)]^{f(x)}$

(e) $\displaystyle\lim_{x \to a} [p(x)]^{q(x)}$ (f) $\displaystyle\lim_{x \to a} \sqrt[q(x)]{p(x)}$

5–6 Use the graphs of f and g and their tangent lines at $(2, 0)$ to find $\displaystyle\lim_{x \to 2} \frac{f(x)}{g(x)}$.

5.

6.
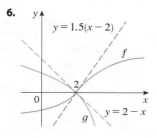

7. The graph of a function f and its tangent line at 0 are shown. What is the value of $\displaystyle\lim_{x \to 0} \frac{f(x)}{e^x - 1}$?

8–68 Find the limit. Use l'Hospital's Rule where appropriate. If there is a more elementary method, consider using it. If l'Hospital's Rule doesn't apply, explain why.

8. $\displaystyle\lim_{x \to 3} \frac{x - 3}{x^2 - 9}$

9. $\displaystyle\lim_{x \to 4} \frac{x^2 - 2x - 8}{x - 4}$ **10.** $\displaystyle\lim_{x \to -2} \frac{x^3 + 8}{x + 2}$

11. $\displaystyle\lim_{x \to 1} \frac{x^3 - 2x^2 + 1}{x^3 - 1}$ **12.** $\displaystyle\lim_{x \to 1/2} \frac{6x^2 + 5x - 4}{4x^2 + 16x - 9}$

13. $\displaystyle\lim_{x \to (\pi/2)^+} \frac{\cos x}{1 - \sin x}$ **14.** $\displaystyle\lim_{x \to 0} \frac{\tan 3x}{\sin 2x}$

15. $\displaystyle\lim_{t \to 0} \frac{e^{2t} - 1}{\sin t}$ **16.** $\displaystyle\lim_{x \to 0} \frac{x^2}{1 - \cos x}$

17. $\displaystyle\lim_{\theta \to \pi/2} \frac{1 - \sin \theta}{1 + \cos 2\theta}$ **18.** $\displaystyle\lim_{\theta \to \pi} \frac{1 + \cos \theta}{1 - \cos \theta}$

19. $\displaystyle\lim_{x \to \infty} \frac{\ln x}{\sqrt{x}}$ **20.** $\displaystyle\lim_{x \to \infty} \frac{x + x^2}{1 - 2x^2}$

21. $\displaystyle\lim_{x \to 0^+} \frac{\ln x}{x}$ **22.** $\displaystyle\lim_{x \to \infty} \frac{\ln \sqrt{x}}{x^2}$

23. $\displaystyle\lim_{t \to 1} \frac{t^8 - 1}{t^5 - 1}$ **24.** $\displaystyle\lim_{t \to 0} \frac{8^t - 5^t}{t}$

25. $\displaystyle\lim_{x \to 0} \frac{\sqrt{1 + 2x} - \sqrt{1 - 4x}}{x}$ **26.** $\displaystyle\lim_{u \to \infty} \frac{e^{u/10}}{u^3}$

27. $\displaystyle\lim_{x \to 0} \frac{e^x - 1 - x}{x^2}$ **28.** $\displaystyle\lim_{x \to 0} \frac{\sinh x - x}{x^3}$

29. $\displaystyle\lim_{x\to 0}\frac{\tanh x}{\tan x}$

30. $\displaystyle\lim_{x\to 0}\frac{x-\sin x}{x-\tan x}$

31. $\displaystyle\lim_{x\to 0}\frac{\sin^{-1}x}{x}$

32. $\displaystyle\lim_{x\to\infty}\frac{(\ln x)^2}{x}$

33. $\displaystyle\lim_{x\to 0}\frac{x3^x}{3^x-1}$

34. $\displaystyle\lim_{x\to 0}\frac{\cos mx-\cos nx}{x^2}$

35. $\displaystyle\lim_{x\to 0}\frac{\ln(1+x)}{\cos x+e^x-1}$

36. $\displaystyle\lim_{x\to 1}\frac{x\sin(x-1)}{2x^2-x-1}$

37. $\displaystyle\lim_{x\to 0^+}\frac{\arctan(2x)}{\ln x}$

38. $\displaystyle\lim_{x\to 0^+}\frac{x^x-1}{\ln x+x-1}$

39. $\displaystyle\lim_{x\to 1}\frac{x^a-1}{x^b-1},\ b\neq 0$

40. $\displaystyle\lim_{x\to 0}\frac{e^x-e^{-x}-2x}{x-\sin x}$

41. $\displaystyle\lim_{x\to 0}\frac{\cos x-1+\frac{1}{2}x^2}{x^4}$

42. $\displaystyle\lim_{x\to a^+}\frac{\cos x\ln(x-a)}{\ln(e^x-e^a)}$

43. $\displaystyle\lim_{x\to\infty}x\sin(\pi/x)$

44. $\displaystyle\lim_{x\to\infty}\sqrt{x}\,e^{-x/2}$

45. $\displaystyle\lim_{x\to 0}\sin 5x\csc 3x$

46. $\displaystyle\lim_{x\to-\infty}x\ln\left(1-\frac{1}{x}\right)$

47. $\displaystyle\lim_{x\to\infty}x^3e^{-x^2}$

48. $\displaystyle\lim_{x\to\infty}x^{3/2}\sin(1/x)$

49. $\displaystyle\lim_{x\to 1^+}\ln x\tan(\pi x/2)$

50. $\displaystyle\lim_{x\to(\pi/2)^-}\cos x\sec 5x$

51. $\displaystyle\lim_{x\to 1}\left(\frac{x}{x-1}-\frac{1}{\ln x}\right)$

52. $\displaystyle\lim_{x\to 0}(\csc x-\cot x)$

53. $\displaystyle\lim_{x\to 0^+}\left(\frac{1}{x}-\frac{1}{e^x-1}\right)$

54. $\displaystyle\lim_{x\to 0^+}\left(\frac{1}{x}-\frac{1}{\tan^{-1}x}\right)$

55. $\displaystyle\lim_{x\to\infty}(x-\ln x)$

56. $\displaystyle\lim_{x\to 1^+}\left[\ln(x^7-1)-\ln(x^5-1)\right]$

57. $\displaystyle\lim_{x\to 0^+}x^{\sqrt{x}}$

58. $\displaystyle\lim_{x\to 0^+}(\tan 2x)^x$

59. $\displaystyle\lim_{x\to 0}(1-2x)^{1/x}$

60. $\displaystyle\lim_{x\to\infty}\left(1+\frac{a}{x}\right)^{bx}$

61. $\displaystyle\lim_{x\to 1^+}x^{1/(1-x)}$

62. $\displaystyle\lim_{x\to\infty}x^{(\ln 2)/(1+\ln x)}$

63. $\displaystyle\lim_{x\to\infty}x^{1/x}$

64. $\displaystyle\lim_{x\to\infty}x^{e^{-x}}$

65. $\displaystyle\lim_{x\to 0^+}(4x+1)^{\cot x}$

66. $\displaystyle\lim_{x\to 1}(2-x)^{\tan(\pi x/2)}$

67. $\displaystyle\lim_{x\to 0^+}(1+\sin 3x)^{1/x}$

68. $\displaystyle\lim_{x\to\infty}\left(\frac{2x-3}{2x+5}\right)^{2x+1}$

69–70 Use a graph to estimate the value of the limit. Then use l'Hospital's Rule to find the exact value.

69. $\displaystyle\lim_{x\to\infty}\left(1+\frac{2}{x}\right)^x$

70. $\displaystyle\lim_{x\to 0}\frac{5^x-4^x}{3^x-2^x}$

71–72 Illustrate l'Hospital's Rule by graphing both $f(x)/g(x)$ and $f'(x)/g'(x)$ near $x=0$ to see that these ratios have the same limit as $x\to 0$. Also, calculate the exact value of the limit.

71. $f(x)=e^x-1,\quad g(x)=x^3+4x$

72. $f(x)=2x\sin x,\quad g(x)=\sec x-1$

73. Prove that

$$\lim_{x\to\infty}\frac{e^x}{x^n}=\infty$$

for any positive integer n. This shows that the exponential function approaches infinity faster than any power of x.

74. Prove that

$$\lim_{x\to\infty}\frac{\ln x}{x^p}=0$$

for any number $p>0$. This shows that the logarithmic function approaches infinity more slowly than any power of x.

75–76 What happens if you try to use l'Hospital's Rule to find the limit? Evaluate the limit using another method.

75. $\displaystyle\lim_{x\to\infty}\frac{x}{\sqrt{x^2+1}}$

76. $\displaystyle\lim_{x\to(\pi/2)^-}\frac{\sec x}{\tan x}$

77. Investigate the family of curves $f(x)=e^x-cx$. In particular, find the limits as $x\to\pm\infty$ and determine the values of c for which f has an absolute minimum. What happens to the minimum points as c increases?

78. If an object with mass m is dropped from rest, one model for its speed v after t seconds, taking air resistance into account, is

$$v=\frac{mg}{c}\left(1-e^{-ct/m}\right)$$

where g is the acceleration due to gravity and c is a positive constant. (In Chapter 9 we will be able to deduce this equation from the assumption that the air resistance is proportional to the speed of the object; c is the proportionality constant.)

(a) Calculate $\lim_{t\to\infty}v$. What is the meaning of this limit?

(b) For fixed t, use l'Hospital's Rule to calculate $\lim_{c\to 0^+}v$. What can you conclude about the velocity of a falling object in a vacuum?

79. If an initial amount A_0 of money is invested at an interest rate r compounded n times a year, the value of the investment after t years is

$$A = A_0 \left(1 + \frac{r}{n} \right)^{nt}$$

If we let $n \to \infty$, we refer to the *continuous compounding* of interest. Use l'Hospital's Rule to show that if interest is compounded continuously, then the amount after t years is

$$A = A_0 e^{rt}$$

80. Light enters the eye through the pupil and strikes the retina, where photoreceptor cells sense light and color. W. Stanley Stiles and B. H. Crawford studied the phenomenon in which measured brightness decreases as light enters farther from the center of the pupil. (See the figure.)

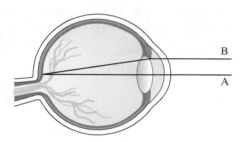

A light beam A that enters through the center of the pupil measures brighter than a beam B entering near the edge of the pupil.

They detailed their findings of this phenomenon, known as the *Stiles–Crawford effect of the first kind*, in an important paper published in 1933. In particular, they observed that the amount of luminance sensed was *not* proportional to the area of the pupil as they expected. The percentage P of the total luminance entering a pupil of radius r mm that is sensed at the retina can be described by

$$P = \frac{1 - 10^{-\rho r^2}}{\rho r^2 \ln 10}$$

where ρ is an experimentally determined constant, typically about 0.05.
(a) What is the percentage of luminance sensed by a pupil of radius 3 mm? Use $\rho = 0.05$.
(b) Compute the percentage of luminance sensed by a pupil of radius 2 mm. Does it make sense that it is larger than the answer to part (a)?
(c) Compute $\lim_{r \to 0^+} P$. Is the result what you would expect? Is this result physically possible?

Source: Adapted from W. Stiles and B. Crawford, "The Luminous Efficiency of Rays Entering the Eye Pupil at Different Points." *Proceedings of the Royal Society of London, Series B: Biological Sciences* 112 (1933): 428–50.

81. Some populations initally grow exponentially but eventually level off. Equations of the form

$$P(t) = \frac{M}{1 + Ae^{-kt}}$$

where M, A, and k are positive constants, are called *logistic equations* and are often used to model such populations. (We will investigate these in detail in Chapter 9.) Here M is called the *carrying capacity* and represents the maximum population size that can be supported, and $A = \dfrac{M - P_0}{P_0}$, where P_0 is the initial population.
(a) Compute $\lim_{t \to \infty} P(t)$. Explain why your answer is to be expected.
(b) Compute $\lim_{M \to \infty} P(t)$. (Note that A is defined in terms of M.) What kind of function is your result?

82. A metal cable has radius r and is covered by insulation so that the distance from the center of the cable to the exterior of the insulation is R. The velocity v of an electrical impulse in the cable is

$$v = -c \left(\frac{r}{R} \right)^2 \ln \left(\frac{r}{R} \right)$$

where c is a positive constant. Find the following limits and interpret your answers.
(a) $\lim_{R \to r^+} v$ (b) $\lim_{r \to 0^+} v$

83. The first appearance in print of l'Hospital's Rule was in the book *Analyse des Infiniment Petits* published by the Marquis de l'Hospital in 1696. This was the first calculus textbook ever published and the example that the Marquis used in that book to illustrate his rule was to find the limit of the function

$$y = \frac{\sqrt{2a^3 x - x^4} - a\sqrt[3]{aax}}{a - \sqrt[4]{ax^3}}$$

as x approaches a, where $a > 0$. (At that time it was common to write aa instead of a^2.) Solve this problem.

84. The figure shows a sector of a circle with central angle θ. Let $A(\theta)$ be the area of the segment between the chord PR and the arc PR. Let $B(\theta)$ be the area of the triangle PQR. Find $\lim_{\theta \to 0^+} A(\theta)/B(\theta)$.

85. Evaluate

$$\lim_{x \to \infty} \left[x - x^2 \ln \left(\frac{1 + x}{x} \right) \right].$$

86. Suppose f is a positive function. If $\lim_{x \to a} f(x) = 0$ and $\lim_{x \to a} g(x) = \infty$, show that

$$\lim_{x \to a} [f(x)]^{g(x)} = 0$$

This shows that 0^∞ is not an indeterminate form.

87. If f' is continuous, $f(2) = 0$, and $f'(2) = 7$, evaluate

$$\lim_{x \to 0} \frac{f(2 + 3x) + f(2 + 5x)}{x}$$

88. For what values of a and b is the following equation true?

$$\lim_{x \to 0} \left(\frac{\sin 2x}{x^3} + a + \frac{b}{x^2} \right) = 0$$

89. If f' is continuous, use l'Hospital's Rule to show that

$$\lim_{h \to 0} \frac{f(x + h) - f(x - h)}{2h} = f'(x)$$

Explain the meaning of this equation with the aid of a diagram.

90. If f'' is continuous, show that

$$\lim_{h \to 0} \frac{f(x + h) - 2f(x) + f(x - h)}{h^2} = f''(x)$$

91. Let

$$f(x) = \begin{cases} e^{-1/x^2} & \text{if } x \neq 0 \\ 0 & \text{if } x = 0 \end{cases}$$

(a) Use the definition of derivative to compute $f'(0)$.
(b) Show that f has derivatives of all orders that are defined on $\mathbb{R}$. [*Hint:* First show by induction that there is a polynomial $p_n(x)$ and a nonnegative integer k_n such that $f^{(n)}(x) = p_n(x)f(x)/x^{k_n}$ for $x \neq 0$.]

92. Let

$$f(x) = \begin{cases} |x|^x & \text{if } x \neq 0 \\ 1 & \text{if } x = 0 \end{cases}$$

(a) Show that f is continuous at 0.
(b) Investigate graphically whether f is differentiable at 0 by zooming in several times toward the point $(0, 1)$ on the graph of f.
(c) Show that f is not differentiable at 0. How can you reconcile this fact with the appearance of the graphs in part (b)?

WRITING PROJECT   THE ORIGINS OF L'HOSPITAL'S RULE

ANALYSE
DES
INFINIMENT PETITS,
POUR
L'INTELLIGENCE DES LIGNES COURBES.
Par M^r le Marquis DE L'HOSPITAL.
SECONDE EDITION.

A PARIS,
Chez ETIENNE PAPILLON, rue S. Jacques,
aux Armes d'Angleterre.
MDCCXVI.
AVEC APPROBATION ET PRIVILEGE DU ROY.

www.stewartcalculus.com
The Internet is another source of information for this project. Click on *History of Mathematics* for a list of reliable websites.

L'Hospital's Rule was first published in 1696 in the Marquis de l'Hospital's calculus textbook *Analyse des Infiniment Petits*, but the rule was discovered in 1694 by the Swiss mathematician John (Johann) Bernoulli. The explanation is that these two mathematicians had entered into a curious business arrangement whereby the Marquis de l'Hospital bought the rights to Bernoulli's mathematical discoveries. The details, including a translation of l'Hospital's letter to Bernoulli proposing the arrangement, can be found in the book by Eves [1].

Write a report on the historical and mathematical origins of l'Hospital's Rule. Start by providing brief biographical details of both men (the dictionary edited by Gillispie [2] is a good source) and outline the business deal between them. Then give l'Hospital's statement of his rule, which is found in Struik's sourcebook [4] and more briefly in the book of Katz [3]. Notice that l'Hospital and Bernoulli formulated the rule geometrically and gave the answer in terms of differentials. Compare their statement with the version of l'Hospital's Rule given in Section 4.4 and show that the two statements are essentially the same.

1. Howard Eves, *In Mathematical Circles (Volume 2: Quadrants III and IV)* (Boston: Prindle, Weber and Schmidt, 1969), pp. 20–22.

2. C. C. Gillispie, ed., *Dictionary of Scientific Biography* (New York: Scribner's, 1974). See the article on Johann Bernoulli by E. A. Fellmann and J. O. Fleckenstein in Volume II and the article on the Marquis de l'Hospital by Abraham Robinson in Volume VIII.

3. Victor Katz, *A History of Mathematics: An Introduction* (New York: HarperCollins, 1993), p. 484.

4. D. J. Struik, ed., *A Sourcebook in Mathematics, 1200–1800* (Princeton, NJ: Princeton University Press, 1969), pp. 315–16.

4.5 Summary of Curve Sketching

So far we have been concerned with some particular aspects of curve sketching: domain, range, and symmetry in Chapter 1; limits, continuity, and asymptotes in Chapter 2; derivatives and tangents in Chapters 2 and 3; and extreme values, intervals of increase and decrease, concavity, points of inflection, and l'Hospital's Rule in this chapter. It is now time to put all of this information together to sketch graphs that reveal the important features of functions.

You might ask: Why don't we just use a graphing calculator or computer to graph a curve? Why do we need to use calculus?

It's true that modern technology is capable of producing very accurate graphs. But even the best graphing devices have to be used intelligently. It is easy to arrive at a misleading graph, or to miss important details of a curve, when relying solely on technology. (See "Graphing Calculators and Computers" at *www.stewartcalculus.com*, especially Examples 1, 3, 4, and 5. See also Section 4.6.) The use of calculus enables us to discover the most interesting aspects of graphs and in many cases to calculate maximum and minimum points and inflection points *exactly* instead of approximately.

For instance, Figure 1 shows the graph of $f(x) = 8x^3 - 21x^2 + 18x + 2$. At first glance it seems reasonable: It has the same shape as cubic curves like $y = x^3$, and it appears to have no maximum or minimum point. But if you compute the derivative, you will see that there is a maximum when $x = 0.75$ and a minimum when $x = 1$. Indeed, if we zoom in to this portion of the graph, we see that behavior exhibited in Figure 2. Without calculus, we could easily have overlooked it.

In the next section we will graph functions by using the interaction between calculus and graphing devices. In this section we draw graphs by first considering the following information. We don't assume that you have a graphing device, but if you do have one you should use it as a check on your work.

FIGURE 1

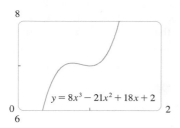

FIGURE 2

▮ Guidelines for Sketching a Curve

The following checklist is intended as a guide to sketching a curve $y = f(x)$ by hand. Not every item is relevant to every function. (For instance, a given curve might not have an asymptote or possess symmetry.) But the guidelines provide all the information you need to make a sketch that displays the most important aspects of the function.

A. Domain It's often useful to start by determining the domain D of f, that is, the set of values of x for which $f(x)$ is defined.

B. Intercepts The y-intercept is $f(0)$ and this tells us where the curve intersects the y-axis. To find the x-intercepts, we set $y = 0$ and solve for x. (You can omit this step if the equation is difficult to solve.)

C. Symmetry

  (i) If $f(-x) = f(x)$ for all x in D, that is, the equation of the curve is unchanged when x is replaced by $-x$, then f is an **even function** and the curve is symmetric about the y-axis. This means that our work is cut in half. If we know what the curve looks like for $x \geq 0$, then we need only reflect about the y-axis to obtain the complete curve [see Figure 3(a)]. Here are some examples: $y = x^2$, $y = x^4$, $y = |x|$, and $y = \cos x$.

  (ii) If $f(-x) = -f(x)$ for all x in D, then f is an **odd function** and the curve is symmetric about the origin. Again we can obtain the complete curve if we know what it looks like for $x \geq 0$. [Rotate 180° about the origin; see Figure 3(b).] Some simple examples of odd functions are $y = x$, $y = x^3$, $y = x^5$, and $y = \sin x$.

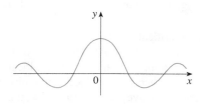

(a) Even function: reflectional symmetry

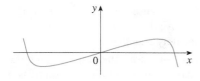

(b) Odd function: rotational symmetry

FIGURE 3

(iii) If $f(x + p) = f(x)$ for all x in D, where p is a positive constant, then f is called a **periodic function** and the smallest such number p is called the **period**. For instance, $y = \sin x$ has period 2π and $y = \tan x$ has period π. If we know what the graph looks like in an interval of length p, then we can use translation to sketch the entire graph (see Figure 4).

FIGURE 4

Periodic function:
translational symmetry

D. Asymptotes

(i) *Horizontal Asymptotes.* Recall from Section 2.6 that if either $\lim_{x \to \infty} f(x) = L$ or $\lim_{x \to -\infty} f(x) = L$, then the line $y = L$ is a horizontal asymptote of the curve $y = f(x)$. If it turns out that $\lim_{x \to \infty} f(x) = \infty$ (or $-\infty$), then we do not have an asymptote to the right, but this fact is still useful information for sketching the curve.

(ii) *Vertical Asymptotes.* Recall from Section 2.2 that the line $x = a$ is a vertical asymptote if at least one of the following statements is true:

$$\boxed{1} \qquad \lim_{x \to a^+} f(x) = \infty \qquad\qquad \lim_{x \to a^-} f(x) = \infty$$

$$\lim_{x \to a^+} f(x) = -\infty \qquad\qquad \lim_{x \to a^-} f(x) = -\infty$$

(For rational functions you can locate the vertical asymptotes by equating the denominator to 0 after canceling any common factors. But for other functions this method does not apply.) Furthermore, in sketching the curve it is very useful to know exactly which of the statements in (1) is true. If $f(a)$ is not defined but a is an endpoint of the domain of f, then you should compute $\lim_{x \to a^-} f(x)$ or $\lim_{x \to a^+} f(x)$, whether or not this limit is infinite.

(iii) *Slant Asymptotes.* These are discussed at the end of this section.

E. Intervals of Increase or Decrease
Use the I/D Test. Compute $f'(x)$ and find the intervals on which $f'(x)$ is positive (f is increasing) and the intervals on which $f'(x)$ is negative (f is decreasing).

F. Local Maximum and Minimum Values
Find the critical numbers of f [the numbers c where $f'(c) = 0$ or $f'(c)$ does not exist]. Then use the First Derivative Test. If f' changes from positive to negative at a critical number c, then $f(c)$ is a local maximum. If f' changes from negative to positive at c, then $f(c)$ is a local minimum. Although it is usually preferable to use the First Derivative Test, you can use the Second Derivative Test if $f'(c) = 0$ and $f''(c) \neq 0$. Then $f''(c) > 0$ implies that $f(c)$ is a local minimum, whereas $f''(c) < 0$ implies that $f(c)$ is a local maximum.

G. Concavity and Points of Inflection
Compute $f''(x)$ and use the Concavity Test. The curve is concave upward where $f''(x) > 0$ and concave downward where $f''(x) < 0$. Inflection points occur where the direction of concavity changes.

H. Sketch the Curve
Using the information in items A–G, draw the graph. Sketch the asymptotes as dashed lines. Plot the intercepts, maximum and minimum points, and inflection points. Then make the curve pass through these points, rising and falling according to E, with concavity according to G, and approaching the asymptotes.

If additional accuracy is desired near any point, you can compute the value of the derivative there. The tangent indicates the direction in which the curve proceeds.

EXAMPLE 1 Use the guidelines to sketch the curve $y = \dfrac{2x^2}{x^2 - 1}$.

A. The domain is

$$\{x \mid x^2 - 1 \neq 0\} = \{x \mid x \neq \pm 1\} = (-\infty, -1) \cup (-1, 1) \cup (1, \infty)$$

B. The x- and y-intercepts are both 0.

C. Since $f(-x) = f(x)$, the function f is even. The curve is symmetric about the y-axis.

D.
$$\lim_{x \to \pm\infty} \frac{2x^2}{x^2 - 1} = \lim_{x \to \pm\infty} \frac{2}{1 - 1/x^2} = 2$$

Therefore the line $y = 2$ is a horizontal asymptote.

Since the denominator is 0 when $x = \pm 1$, we compute the following limits:

$$\lim_{x \to 1^+} \frac{2x^2}{x^2 - 1} = \infty \qquad \lim_{x \to 1^-} \frac{2x^2}{x^2 - 1} = -\infty$$

$$\lim_{x \to -1^+} \frac{2x^2}{x^2 - 1} = -\infty \qquad \lim_{x \to -1^-} \frac{2x^2}{x^2 - 1} = \infty$$

Therefore the lines $x = 1$ and $x = -1$ are vertical asymptotes. This information about limits and asymptotes enables us to draw the preliminary sketch in Figure 5, showing the parts of the curve near the asymptotes.

FIGURE 5
Preliminary sketch

We have shown the curve approaching its horizontal asymptote from above in Figure 5. This is confirmed by the intervals of increase and decrease.

E.
$$f'(x) = \frac{(x^2 - 1)(4x) - 2x^2 \cdot 2x}{(x^2 - 1)^2} = \frac{-4x}{(x^2 - 1)^2}$$

Since $f'(x) > 0$ when $x < 0$ ($x \neq -1$) and $f'(x) < 0$ when $x > 0$ ($x \neq 1$), f is increasing on $(-\infty, -1)$ and $(-1, 0)$ and decreasing on $(0, 1)$ and $(1, \infty)$.

F. The only critical number is $x = 0$. Since f' changes from positive to negative at 0, $f(0) = 0$ is a local maximum by the First Derivative Test.

G.
$$f''(x) = \frac{(x^2 - 1)^2(-4) + 4x \cdot 2(x^2 - 1)2x}{(x^2 - 1)^4} = \frac{12x^2 + 4}{(x^2 - 1)^3}$$

Since $12x^2 + 4 > 0$ for all x, we have

$$f''(x) > 0 \iff x^2 - 1 > 0 \iff |x| > 1$$

FIGURE 6
Finished sketch of $y = \dfrac{2x^2}{x^2 - 1}$

and $f''(x) < 0 \iff |x| < 1$. Thus the curve is concave upward on the intervals $(-\infty, -1)$ and $(1, \infty)$ and concave downward on $(-1, 1)$. It has no point of inflection since 1 and -1 are not in the domain of f.

H. Using the information in E–G, we finish the sketch in Figure 6. ∎

EXAMPLE 2 Sketch the graph of $f(x) = \dfrac{x^2}{\sqrt{x + 1}}$.

A. Domain $= \{x \mid x + 1 > 0\} = \{x \mid x > -1\} = (-1, \infty)$

B. The x- and y-intercepts are both 0.

C. Symmetry: None

D. Since

$$\lim_{x \to \infty} \frac{x^2}{\sqrt{x + 1}} = \infty$$

there is no horizontal asymptote. Since $\sqrt{x + 1} \to 0$ as $x \to -1^+$ and $f(x)$ is always positive, we have

$$\lim_{x \to -1^+} \frac{x^2}{\sqrt{x + 1}} = \infty$$

and so the line $x = -1$ is a vertical asymptote.

E. $$f'(x) = \frac{\sqrt{x + 1}\,(2x) - x^2 \cdot 1/(2\sqrt{x + 1})}{x + 1} = \frac{3x^2 + 4x}{2(x + 1)^{3/2}} = \frac{x(3x + 4)}{2(x + 1)^{3/2}}$$

We see that $f'(x) = 0$ when $x = 0$ (notice that $-\frac{4}{3}$ is not in the domain of f), so the only critical number is 0. Since $f'(x) < 0$ when $-1 < x < 0$ and $f'(x) > 0$ when $x > 0$, f is decreasing on $(-1, 0)$ and increasing on $(0, \infty)$.

F. Since $f'(0) = 0$ and f' changes from negative to positive at 0, $f(0) = 0$ is a local (and absolute) minimum by the First Derivative Test.

G. $$f''(x) = \frac{2(x + 1)^{3/2}(6x + 4) - (3x^2 + 4x)3(x + 1)^{1/2}}{4(x + 1)^3} = \frac{3x^2 + 8x + 8}{4(x + 1)^{5/2}}$$

Note that the denominator is always positive. The numerator is the quadratic $3x^2 + 8x + 8$, which is always positive because its discriminant is $b^2 - 4ac = -32$, which is negative, and the coefficient of x^2 is positive. Thus $f''(x) > 0$ for all x in the domain of f, which means that f is concave upward on $(-1, \infty)$ and there is no point of inflection.

H. The curve is sketched in Figure 7.

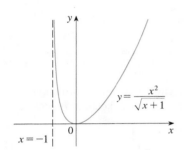

FIGURE 7

EXAMPLE 3 Sketch the graph of $f(x) = xe^x$.

A. The domain is $\mathbb{R}$.

B. The x- and y-intercepts are both 0.

C. Symmetry: None

D. Because both x and e^x become large as $x \to \infty$, we have $\lim_{x \to \infty} xe^x = \infty$. As $x \to -\infty$, however, $e^x \to 0$ and so we have an indeterminate product that requires the use of l'Hospital's Rule:

$$\lim_{x \to -\infty} xe^x = \lim_{x \to -\infty} \frac{x}{e^{-x}} = \lim_{x \to -\infty} \frac{1}{-e^{-x}} = \lim_{x \to -\infty} (-e^x) = 0$$

Thus the x-axis is a horizontal asymptote.

E. $$f'(x) = xe^x + e^x = (x + 1)e^x$$

Since e^x is always positive, we see that $f'(x) > 0$ when $x + 1 > 0$, and $f'(x) < 0$ when $x + 1 < 0$. So f is increasing on $(-1, \infty)$ and decreasing on $(-\infty, -1)$.

F. Because $f'(-1) = 0$ and f' changes from negative to positive at $x = -1$, $f(-1) = -e^{-1} \approx -0.37$ is a local (and absolute) minimum.

G. $$f''(x) = (x + 1)e^x + e^x = (x + 2)e^x$$

Since $f''(x) > 0$ if $x > -2$ and $f''(x) < 0$ if $x < -2$, f is concave upward on

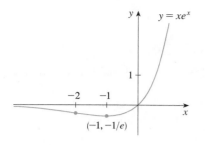

FIGURE 8

$(-2, \infty)$ and concave downward on $(-\infty, -2)$. The inflection point is $(-2, -2e^{-2}) \approx (-2, -0.27)$.

H. We use this information to sketch the curve in Figure 8.

EXAMPLE 4 Sketch the graph of $f(x) = \dfrac{\cos x}{2 + \sin x}$.

A. The domain is $\mathbb{R}$.

B. The y-intercept is $f(0) = \frac{1}{2}$. The x-intercepts occur when $\cos x = 0$, that is, $x = (\pi/2) + n\pi$, where n is an integer.

C. f is neither even nor odd, but $f(x + 2\pi) = f(x)$ for all x and so f is periodic and has period 2π. Thus, in what follows, we need to consider only $0 \leqslant x \leqslant 2\pi$ and then extend the curve by translation in part H.

D. Asymptotes: None

E.
$$f'(x) = \frac{(2 + \sin x)(-\sin x) - \cos x \,(\cos x)}{(2 + \sin x)^2} = -\frac{2\sin x + 1}{(2 + \sin x)^2}$$

The denominator is always positive, so $f'(x) > 0$ when $2\sin x + 1 < 0 \iff$ $\sin x < -\frac{1}{2} \iff 7\pi/6 < x < 11\pi/6$. So f is increasing on $(7\pi/6, 11\pi/6)$ and decreasing on $(0, 7\pi/6)$ and $(11\pi/6, 2\pi)$.

F. From part E and the First Derivative Test, we see that the local minimum value is $f(7\pi/6) = -1/\sqrt{3}$ and the local maximum value is $f(11\pi/6) = 1/\sqrt{3}$.

G. If we use the Quotient Rule again and simplify, we get

$$f''(x) = -\frac{2\cos x \,(1 - \sin x)}{(2 + \sin x)^3}$$

Because $(2 + \sin x)^3 > 0$ and $1 - \sin x \geqslant 0$ for all x, we know that $f''(x) > 0$ when $\cos x < 0$, that is, $\pi/2 < x < 3\pi/2$. So f is concave upward on $(\pi/2, 3\pi/2)$ and concave downward on $(0, \pi/2)$ and $(3\pi/2, 2\pi)$. The inflection points are $(\pi/2, 0)$ and $(3\pi/2, 0)$.

H. The graph of the function restricted to $0 \leqslant x \leqslant 2\pi$ is shown in Figure 9. Then we extend it, using periodicity, to the complete graph in Figure 10.

FIGURE 9

FIGURE 10

EXAMPLE 5 Sketch the graph of $y = \ln(4 - x^2)$.

A. The domain is

$$\{x \mid 4 - x^2 > 0\} = \{x \mid x^2 < 4\} = \{x \mid |x| < 2\} = (-2, 2)$$

B. The y-intercept is $f(0) = \ln 4$. To find the x-intercept we set

$$y = \ln(4 - x^2) = 0$$

We know that $\ln 1 = 0$, so we have $4 - x^2 = 1 \implies x^2 = 3$ and therefore the x-intercepts are $\pm\sqrt{3}$.

C. Since $f(-x) = f(x)$, f is even and the curve is symmetric about the y-axis.

D. We look for vertical asymptotes at the endpoints of the domain. Since $4 - x^2 \to 0^+$ as $x \to 2^-$ and also as $x \to -2^+$, we have

$$\lim_{x \to 2^-} \ln(4 - x^2) = -\infty \qquad \lim_{x \to -2^+} \ln(4 - x^2) = -\infty$$

Thus the lines $x = 2$ and $x = -2$ are vertical asymptotes.

E.
$$f'(x) = \frac{-2x}{4 - x^2}$$

Since $f'(x) > 0$ when $-2 < x < 0$ and $f'(x) < 0$ when $0 < x < 2$, f is increasing on $(-2, 0)$ and decreasing on $(0, 2)$.

F. The only critical number is $x = 0$. Since f' changes from positive to negative at 0, $f(0) = \ln 4$ is a local maximum by the First Derivative Test.

G.
$$f''(x) = \frac{(4 - x^2)(-2) + 2x(-2x)}{(4 - x^2)^2} = \frac{-8 - 2x^2}{(4 - x^2)^2}$$

Since $f''(x) < 0$ for all x, the curve is concave downward on $(-2, 2)$ and has no inflection point.

H. Using this information, we sketch the curve in Figure 11.

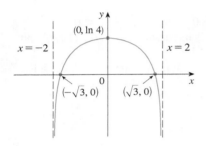

FIGURE 11
$y = \ln(4 - x^2)$

Slant Asymptotes

Some curves have asymptotes that are *oblique*, that is, neither horizontal nor vertical. If

$$\lim_{x \to \infty} [f(x) - (mx + b)] = 0$$

where $m \neq 0$, then the line $y = mx + b$ is called a **slant asymptote** because the vertical distance between the curve $y = f(x)$ and the line $y = mx + b$ approaches 0, as in Figure 12. (A similar situation exists if we let $x \to -\infty$.) For rational functions, slant asymptotes occur when the degree of the numerator is one more than the degree of the denominator. In such a case the equation of the slant asymptote can be found by long division as in the following example.

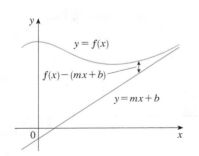

FIGURE 12

EXAMPLE 6 Sketch the graph of $f(x) = \dfrac{x^3}{x^2 + 1}$.

A. The domain is $\mathbb{R} = (-\infty, \infty)$.

B. The x- and y-intercepts are both 0.

C. Since $f(-x) = -f(x)$, f is odd and its graph is symmetric about the origin.

D. Since $x^2 + 1$ is never 0, there is no vertical asymptote. Since $f(x) \to \infty$ as $x \to \infty$ and $f(x) \to -\infty$ as $x \to -\infty$, there is no horizontal asymptote. But long division gives

$$f(x) = \frac{x^3}{x^2 + 1} = x - \frac{x}{x^2 + 1}$$

This equation suggests that $y = x$ is a candidate for a slant asymptote. In fact,

$$f(x) - x = -\frac{x}{x^2 + 1} = -\frac{\dfrac{1}{x}}{1 + \dfrac{1}{x^2}} \to 0 \quad \text{as} \quad x \to \pm\infty$$

So the line $y = x$ is a slant asymptote.

E.
$$f'(x) = \frac{(x^2 + 1)(3x^2) - x^3 \cdot 2x}{(x^2 + 1)^2} = \frac{x^2(x^2 + 3)}{(x^2 + 1)^2}$$

Since $f'(x) > 0$ for all x (except 0), f is increasing on $(-\infty, \infty)$.

F. Although $f'(0) = 0$, f' does not change sign at 0, so there is no local maximum or minimum.

G. $f''(x) = \dfrac{(x^2 + 1)^2(4x^3 + 6x) - (x^4 + 3x^2) \cdot 2(x^2 + 1)2x}{(x^2 + 1)^4} = \dfrac{2x(3 - x^2)}{(x^2 + 1)^3}$

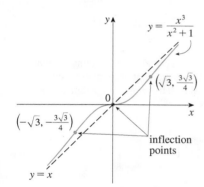

$y = \dfrac{x^3}{x^2 + 1}$

$\left(\sqrt{3}, \dfrac{3\sqrt{3}}{4}\right)$

$\left(-\sqrt{3}, -\dfrac{3\sqrt{3}}{4}\right)$

inflection points

$y = x$

FIGURE 13

Since $f''(x) = 0$ when $x = 0$ or $x = \pm\sqrt{3}$, we set up the following chart:

Interval	x	$3 - x^2$	$(x^2 + 1)^3$	$f''(x)$	f
$x < -\sqrt{3}$	$-$	$-$	$+$	$+$	CU on $(-\infty, -\sqrt{3})$
$-\sqrt{3} < x < 0$	$-$	$+$	$+$	$-$	CD on $(-\sqrt{3}, 0)$
$0 < x < \sqrt{3}$	$+$	$+$	$+$	$+$	CU on $(0, \sqrt{3})$
$x > \sqrt{3}$	$+$	$-$	$+$	$-$	CD on $(\sqrt{3}, \infty)$

The points of inflection are $\left(-\sqrt{3}, -\frac{3}{4}\sqrt{3}\right)$, $(0, 0)$, and $\left(\sqrt{3}, \frac{3}{4}\sqrt{3}\right)$.

H. The graph of f is sketched in Figure 13. ∎

4.5 EXERCISES

1–54 Use the guidelines of this section to sketch the curve.

1. $y = x^3 + 3x^2$

2. $y = 2 + 3x^2 - x^3$

3. $y = x^4 - 4x$

4. $y = x^4 - 8x^2 + 8$

5. $y = x(x - 4)^3$

6. $y = x^5 - 5x$

7. $y = \frac{1}{5}x^5 - \frac{8}{3}x^3 + 16x$

8. $y = (4 - x^2)^5$

9. $y = \dfrac{x}{x - 1}$

10. $y = \dfrac{x^2 + 5x}{25 - x^2}$

11. $y = \dfrac{x - x^2}{2 - 3x + x^2}$

12. $y = 1 + \dfrac{1}{x} + \dfrac{1}{x^2}$

13. $y = \dfrac{x}{x^2 - 4}$

14. $y = \dfrac{1}{x^2 - 4}$

15. $y = \dfrac{x^2}{x^2 + 3}$

16. $y = \dfrac{(x - 1)^2}{x^2 + 1}$

17. $y = \dfrac{x - 1}{x^2}$

18. $y = \dfrac{x}{x^3 - 1}$

19. $y = \dfrac{x^3}{x^3 + 1}$

20. $y = \dfrac{x^3}{x - 2}$

21. $y = (x - 3)\sqrt{x}$

22. $y = (x - 4)\sqrt[3]{x}$

23. $y = \sqrt{x^2 + x - 2}$

24. $y = \sqrt{x^2 + x} - x$

25. $y = \dfrac{x}{\sqrt{x^2 + 1}}$

26. $y = x\sqrt{2 - x^2}$

27. $y = \dfrac{\sqrt{1 - x^2}}{x}$

28. $y = \dfrac{x}{\sqrt{x^2 - 1}}$

29. $y = x - 3x^{1/3}$

30. $y = x^{5/3} - 5x^{2/3}$

31. $y = \sqrt[3]{x^2 - 1}$

32. $y = \sqrt[3]{x^3 + 1}$

33. $y = \sin^3 x$

34. $y = x + \cos x$

35. $y = x \tan x, \quad -\pi/2 < x < \pi/2$

36. $y = 2x - \tan x, \quad -\pi/2 < x < \pi/2$

37. $y = \sin x + \sqrt{3} \cos x, \quad -2\pi \leqslant x \leqslant 2\pi$

38. $y = \csc x - 2\sin x, \quad 0 < x < \pi$

39. $y = \dfrac{\sin x}{1 + \cos x}$

40. $y = \dfrac{\sin x}{2 + \cos x}$

41. $y = \arctan(e^x)$

42. $y = (1 - x)e^x$

43. $y = 1/(1 + e^{-x})$

44. $y = e^{-x} \sin x, \quad 0 \leqslant x \leqslant 2\pi$

45. $y = \dfrac{1}{x} + \ln x$

46. $y = e^{2x} - e^x$

47. $y = (1 + e^x)^{-2}$

48. $y = e^x/x^2$

49. $y = \ln(\sin x)$

50. $y = \ln(1 + x^3)$

51. $y = xe^{-1/x}$

52. $y = \dfrac{\ln x}{x^2}$

53. $y = e^{\arctan x}$

54. $y = \tan^{-1}\left(\dfrac{x - 1}{x + 1}\right)$

---

55. In the theory of relativity, the mass of a particle is

$$m = \frac{m_0}{\sqrt{1 - v^2/c^2}}$$

where m_0 is the rest mass of the particle, m is the mass when the particle moves with speed v relative to the observer, and c is the speed of light. Sketch the graph of m as a function of v.

56. In the theory of relativity, the energy of a particle is

$$E = \sqrt{m_0^2 c^4 + h^2 c^2/\lambda^2}$$

where m_0 is the rest mass of the particle, λ is its wave

length, and h is Planck's constant. Sketch the graph of E as a function of λ. What does the graph say about the energy?

57. A model for the spread of a rumor is given by the equation

$$p(t) = \frac{1}{1 + ae^{-kt}}$$

where $p(t)$ is the proportion of the population that knows the rumor at time t and a and k are positive constants.
(a) When will half the population have heard the rumor?
(b) When is the rate of spread of the rumor greatest?
(c) Sketch the graph of p.

58. A model for the concentration at time t of a drug injected into the bloodstream is

$$C(t) = K(e^{-at} - e^{-bt})$$

where a, b, and K are positive constants and $b > a$. Sketch the graph of the concentration function. What does the graph tell us about how the concentration varies as time passes?

59. The figure shows a beam of length L embedded in concrete walls. If a constant load W is distributed evenly along its length, the beam takes the shape of the deflection curve

$$y = -\frac{W}{24EI}x^4 + \frac{WL}{12EI}x^3 - \frac{WL^2}{24EI}x^2$$

where E and I are positive constants. (E is Young's modulus of elasticity and I is the moment of inertia of a cross-section of the beam.) Sketch the graph of the deflection curve.

60. Coulomb's Law states that the force of attraction between two charged particles is directly proportional to the product of the charges and inversely proportional to the square of the distance between them. The figure shows particles with charge 1 located at positions 0 and 2 on a coordinate line and a particle with charge -1 at a position x between them. It follows from Coulomb's Law that the net force acting on the middle particle is

$$F(x) = -\frac{k}{x^2} + \frac{k}{(x - 2)^2} \qquad 0 < x < 2$$

where k is a positive constant. Sketch the graph of the net force function. What does the graph say about the force?

61–64 Find an equation of the slant asymptote. Do not sketch the curve.

61. $y = \dfrac{x^2 + 1}{x + 1}$

62. $y = \dfrac{4x^3 - 10x^2 - 11x + 1}{x^2 - 3x}$

63. $y = \dfrac{2x^3 - 5x^2 + 3x}{x^2 - x - 2}$

64. $y = \dfrac{-6x^4 + 2x^3 + 3}{2x^3 - x}$

65–70 Use the guidelines of this section to sketch the curve. In guideline D find an equation of the slant asymptote.

65. $y = \dfrac{x^2}{x - 1}$

66. $y = \dfrac{1 + 5x - 2x^2}{x - 2}$

67. $y = \dfrac{x^3 + 4}{x^2}$

68. $y = \dfrac{x^3}{(x + 1)^2}$

69. $y = 1 + \frac{1}{2}x + e^{-x}$

70. $y = 1 - x + e^{1 + x/3}$

---

71. Show that the curve $y = x - \tan^{-1}x$ has two slant asymptotes: $y = x + \pi/2$ and $y = x - \pi/2$. Use this fact to help sketch the curve.

72. Show that the curve $y = \sqrt{x^2 + 4x}$ has two slant asymptotes: $y = x + 2$ and $y = -x - 2$. Use this fact to help sketch the curve.

73. Show that the lines $y = (b/a)x$ and $y = -(b/a)x$ are slant asymptotes of the hyperbola $(x^2/a^2) - (y^2/b^2) = 1$.

74. Let $f(x) = (x^3 + 1)/x$. Show that

$$\lim_{x \to \pm\infty} [f(x) - x^2] = 0$$

This shows that the graph of f approaches the graph of $y = x^2$, and we say that the curve $y = f(x)$ is *asymptotic* to the parabola $y = x^2$. Use this fact to help sketch the graph of f.

75. Discuss the asymptotic behavior of $f(x) = (x^4 + 1)/x$ in the same manner as in Exercise 74. Then use your results to help sketch the graph of f.

76. Use the asymptotic behavior of $f(x) = \sin x + e^{-x}$ to sketch its graph without going through the curve-sketching procedure of this section.

4.6 Graphing with Calculus *and* Calculators

You may want to read "Graphing Calculators and Computers" at www.stewartcalculus.com if you haven't already. In particular, it explains how to avoid some of the pitfalls of graphing devices by choosing appropriate viewing rectangles.

The method we used to sketch curves in the preceding section was a culmination of much of our study of differential calculus. The graph was the final object that we produced. In this section our point of view is completely different. Here we *start* with a graph produced by a graphing calculator or computer and then we refine it. We use calculus to make sure that we reveal all the important aspects of the curve. And with the use of graphing devices we can tackle curves that would be far too complicated to consider without technology. The theme is the *interaction* between calculus and calculators.

EXAMPLE 1 Graph the polynomial $f(x) = 2x^6 + 3x^5 + 3x^3 - 2x^2$. Use the graphs of f' and f'' to estimate all maximum and minimum points and intervals of concavity.

SOLUTION If we specify a domain but not a range, many graphing devices will deduce a suitable range from the values computed. Figure 1 shows the plot from one such device if we specify that $-5 \le x \le 5$. Although this viewing rectangle is useful for showing that the asymptotic behavior (or end behavior) is the same as for $y = 2x^6$, it is obviously hiding some finer detail. So we change to the viewing rectangle $[-3, 2]$ by $[-50, 100]$ shown in Figure 2.

FIGURE 1

FIGURE 2

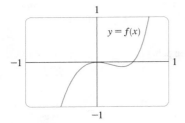

FIGURE 3

From this graph it appears that there is an absolute minimum value of about -15.33 when $x \approx -1.62$ (by using the cursor) and f is decreasing on $(-\infty, -1.62)$ and increasing on $(-1.62, \infty)$. Also there appears to be a horizontal tangent at the origin and inflection points when $x = 0$ and when x is somewhere between -2 and -1.

Now let's try to confirm these impressions using calculus. We differentiate and get

$$f'(x) = 12x^5 + 15x^4 + 9x^2 - 4x$$

$$f''(x) = 60x^4 + 60x^3 + 18x - 4$$

When we graph f' in Figure 3 we see that $f'(x)$ changes from negative to positive when $x \approx -1.62$; this confirms (by the First Derivative Test) the minimum value that we found earlier. But, perhaps to our surprise, we also notice that $f'(x)$ changes from positive to negative when $x = 0$ and from negative to positive when $x \approx 0.35$. This means that f has a local maximum at 0 and a local minimum when $x \approx 0.35$, but these were hidden in Figure 2. Indeed, if we now zoom in toward the origin in Figure 4, we see what we missed before: a local maximum value of 0 when $x = 0$ and a local minimum value of about -0.1 when $x \approx 0.35$.

What about concavity and inflection points? From Figures 2 and 4 there appear to be inflection points when x is a little to the left of -1 and when x is a little to the right of 0. But it's difficult to determine inflection points from the graph of f, so we graph the second derivative f'' in Figure 5. We see that f'' changes from positive to negative when $x \approx -1.23$ and from negative to positive when $x \approx 0.19$. So, correct to two decimal places, f is concave upward on $(-\infty, -1.23)$ and $(0.19, \infty)$ and concave downward on $(-1.23, 0.19)$. The inflection points are $(-1.23, -10.18)$ and $(0.19, -0.05)$.

We have discovered that no single graph reveals all the important features of this polynomial. But Figures 2 and 4, when taken together, do provide an accurate picture. ∎

FIGURE 5

EXAMPLE 2 Draw the graph of the function

$$f(x) = \frac{x^2 + 7x + 3}{x^2}$$

in a viewing rectangle that contains all the important features of the function. Estimate the maximum and minimum values and the intervals of concavity. Then use calculus to find these quantities exactly.

SOLUTION Figure 6, produced by a computer with automatic scaling, is a disaster. Some graphing calculators use $[-10, 10]$ by $[-10, 10]$ as the default viewing rectangle, so let's try it. We get the graph shown in Figure 7; it's a major improvement.

FIGURE 6

FIGURE 7

FIGURE 8

FIGURE 9

The y-axis appears to be a vertical asymptote and indeed it is because

$$\lim_{x \to 0} \frac{x^2 + 7x + 3}{x^2} = \infty$$

Figure 7 also allows us to estimate the x-intercepts: about -0.5 and -6.5. The exact values are obtained by using the quadratic formula to solve the equation $x^2 + 7x + 3 = 0$; we get $x = \left(-7 \pm \sqrt{37}\right)/2$.

To get a better look at horizontal asymptotes, we change to the viewing rectangle $[-20, 20]$ by $[-5, 10]$ in Figure 8. It appears that $y = 1$ is the horizontal asymptote and this is easily confirmed:

$$\lim_{x \to \pm\infty} \frac{x^2 + 7x + 3}{x^2} = \lim_{x \to \pm\infty} \left(1 + \frac{7}{x} + \frac{3}{x^2}\right) = 1$$

To estimate the minimum value we zoom in to the viewing rectangle $[-3, 0]$ by $[-4, 2]$ in Figure 9. The cursor indicates that the absolute minimum value is about -3.1 when $x \approx -0.9$, and we see that the function decreases on $(-\infty, -0.9)$ and $(0, \infty)$ and increases on $(-0.9, 0)$. The exact values are obtained by differentiating:

$$f'(x) = -\frac{7}{x^2} - \frac{6}{x^3} = -\frac{7x + 6}{x^3}$$

This shows that $f'(x) > 0$ when $-\frac{6}{7} < x < 0$ and $f'(x) < 0$ when $x < -\frac{6}{7}$ and when $x > 0$. The exact minimum value is $f\left(-\frac{6}{7}\right) = -\frac{37}{12} \approx -3.08$.

Figure 9 also shows that an inflection point occurs somewhere between $x = -1$ and $x = -2$. We could estimate it much more accurately using the graph of the second derivative, but in this case it's just as easy to find exact values. Since

$$f''(x) = \frac{14}{x^3} + \frac{18}{x^4} = \frac{2(7x + 9)}{x^4}$$

we see that $f''(x) > 0$ when $x > -\frac{9}{7}$ $(x \neq 0)$. So f is concave upward on $\left(-\frac{9}{7}, 0\right)$ and $(0, \infty)$ and concave downward on $\left(-\infty, -\frac{9}{7}\right)$. The inflection point is $\left(-\frac{9}{7}, -\frac{71}{27}\right)$.

The analysis using the first two derivatives shows that Figure 8 displays all the major aspects of the curve. ∎

EXAMPLE 3 Graph the function $f(x) = \dfrac{x^2(x + 1)^3}{(x - 2)^2(x - 4)^4}$.

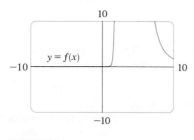

FIGURE 10

SOLUTION Drawing on our experience with a rational function in Example 2, let's start by graphing f in the viewing rectangle $[-10, 10]$ by $[-10, 10]$. From Figure 10 we have the feeling that we are going to have to zoom in to see some finer detail and also zoom out to see the larger picture. But, as a guide to intelligent zooming, let's first take a close look at the expression for $f(x)$. Because of the factors $(x - 2)^2$ and $(x - 4)^4$ in the denominator, we expect $x = 2$ and $x = 4$ to be the vertical asymptotes. Indeed

$$\lim_{x \to 2} \frac{x^2(x + 1)^3}{(x - 2)^2(x - 4)^4} = \infty \quad \text{and} \quad \lim_{x \to 4} \frac{x^2(x + 1)^3}{(x - 2)^2(x - 4)^4} = \infty$$

To find the horizontal asymptotes, we divide numerator and denominator by x^6:

$$\frac{x^2(x+1)^3}{(x-2)^2(x-4)^4} = \frac{\dfrac{x^2}{x^3} \cdot \dfrac{(x+1)^3}{x^3}}{\dfrac{(x-2)^2}{x^2} \cdot \dfrac{(x-4)^4}{x^4}} = \frac{\dfrac{1}{x}\left(1+\dfrac{1}{x}\right)^3}{\left(1-\dfrac{2}{x}\right)^2\left(1-\dfrac{4}{x}\right)^4}$$

This shows that $f(x) \to 0$ as $x \to \pm\infty$, so the x-axis is a horizontal asymptote.

It is also very useful to consider the behavior of the graph near the x-intercepts using an analysis like that in Example 2.6.12. Since x^2 is positive, $f(x)$ does not change sign at 0 and so its graph doesn't cross the x-axis at 0. But, because of the factor $(x+1)^3$, the graph does cross the x-axis at -1 and has a horizontal tangent there. Putting all this information together, but without using derivatives, we see that the curve has to look something like the one in Figure 11.

Now that we know what to look for, we zoom in (several times) to produce the graphs in Figures 12 and 13 and zoom out (several times) to get Figure 14.

FIGURE 11

FIGURE 12

FIGURE 13

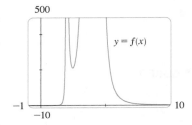

FIGURE 14

We can read from these graphs that the absolute minimum is about -0.02 and occurs when $x \approx -20$. There is also a local maximum ≈ 0.00002 when $x \approx -0.3$ and a local minimum ≈ 211 when $x \approx 2.5$. These graphs also show three inflection points near -35, -5, and -1 and two between -1 and 0. To estimate the inflection points closely we would need to graph f'', but to compute f'' by hand is an unreasonable chore. If you have a computer algebra system, then it's easy to do (see Exercise 15).

We have seen that, for this particular function, *three* graphs (Figures 12, 13, and 14) are necessary to convey all the useful information. The only way to display all these features of the function on a single graph is to draw it by hand. Despite the exaggerations and distortions, Figure 11 does manage to summarize the essential nature of the function. ∎

EXAMPLE 4 Graph the function $f(x) = \sin(x + \sin 2x)$. For $0 \leqslant x \leqslant \pi$, estimate all maximum and minimum values, intervals of increase and decrease, and inflection points.

SOLUTION We first note that f is periodic with period 2π. Also, f is odd and $|f(x)| \leqslant 1$ for all x. So the choice of a viewing rectangle is not a problem for this function: We start with $[0, \pi]$ by $[-1.1, 1.1]$. (See Figure 15.) It appears that there are three local maximum values and two local minimum values in that window. To confirm

FIGURE 15

The family of functions

$$f(x) = \sin(x + \sin cx)$$

where c is a constant, occurs in applications to frequency modulation (FM) synthesis. A sine wave is modulated by a wave with a different frequency ($\sin cx$). The case where $c = 2$ is studied in Example 4. Exercise 27 explores another special case.

FIGURE 16

FIGURE 17

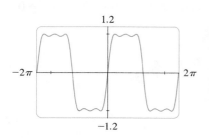

FIGURE 18

this and locate them more accurately, we calculate that

$$f'(x) = \cos(x + \sin 2x) \cdot (1 + 2\cos 2x)$$

and graph both f and f' in Figure 16.

Using zoom-in and the First Derivative Test, we find the following approximate values:

Intervals of increase: $(0, 0.6)$, $(1.0, 1.6)$, $(2.1, 2.5)$

Intervals of decrease: $(0.6, 1.0)$, $(1.6, 2.1)$, $(2.5, \pi)$

Local maximum values: $f(0.6) \approx 1$, $f(1.6) \approx 1$, $f(2.5) \approx 1$

Local minimum values: $f(1.0) \approx 0.94$, $f(2.1) \approx 0.94$

The second derivative is

$$f''(x) = -(1 + 2\cos 2x)^2 \sin(x + \sin 2x) - 4\sin 2x \cos(x + \sin 2x)$$

Graphing both f and f'' in Figure 17, we obtain the following approximate values:

Concave upward on: $(0.8, 1.3)$, $(1.8, 2.3)$

Concave downward on: $(0, 0.8)$, $(1.3, 1.8)$, $(2.3, \pi)$

Inflection points: $(0, 0)$, $(0.8, 0.97)$, $(1.3, 0.97)$, $(1.8, 0.97)$, $(2.3, 0.97)$

Having checked that Figure 15 does indeed represent f accurately for $0 \leqslant x \leqslant \pi$, we can state that the extended graph in Figure 18 represents f accurately for $-2\pi \leqslant x \leqslant 2\pi$. ∎

Our final example is concerned with *families* of functions. This means that the functions in the family are related to each other by a formula that contains one or more arbitrary constants. Each value of the constant gives rise to a member of the family and the idea is to see how the graph of the function changes as the constant changes.

EXAMPLE 5 How does the graph of $f(x) = 1/(x^2 + 2x + c)$ vary as c varies?

SOLUTION The graphs in Figures 19 and 20 (the special cases $c = 2$ and $c = -2$) show two very different-looking curves.

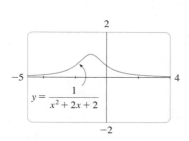

FIGURE 19
$c = 2$

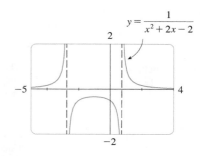

FIGURE 20
$c = -2$

Before drawing any more graphs, let's see what members of this family have in common. Since

$$\lim_{x \to \pm\infty} \frac{1}{x^2 + 2x + c} = 0$$

for any value of c, they all have the x-axis as a horizontal asymptote. A vertical asymptote will occur when $x^2 + 2x + c = 0$. Solving this quadratic equation, we get $x = -1 \pm \sqrt{1 - c}$. When $c > 1$, there is no vertical asymptote (as in Figure 19). When $c = 1$, the graph has a single vertical asymptote $x = -1$ because

$$\lim_{x \to -1} \frac{1}{x^2 + 2x + 1} = \lim_{x \to -1} \frac{1}{(x + 1)^2} = \infty$$

When $c < 1$, there are two vertical asymptotes: $x = -1 \pm \sqrt{1 - c}$ (as in Figure 20)

Now we compute the derivative:

$$f'(x) = -\frac{2x + 2}{(x^2 + 2x + c)^2}$$

This shows that $f'(x) = 0$ when $x = -1$ (if $c \neq 1$), $f'(x) > 0$ when $x < -1$, and $f'(x) < 0$ when $x > -1$. For $c \geq 1$, this means that f increases on $(-\infty, -1)$ and decreases on $(-1, \infty)$. For $c > 1$, there is an absolute maximum value $f(-1) = 1/(c - 1)$. For $c < 1$, $f(-1) = 1/(c - 1)$ is a local maximum value and the intervals of increase and decrease are interrupted at the vertical asymptotes.

Figure 21 is a "slide show" displaying five members of the family, all graphed in the viewing rectangle $[-5, 4]$ by $[-2, 2]$. As predicted, a transition takes place from two vertical asymptotes to one at $c = 1$, and then to none for $c > 1$. As c increases from 1, we see that the maximum point becomes lower; this is explained by the fact that $1/(c - 1) \to 0$ as $c \to \infty$. As c decreases from 1, the vertical asymptotes become more widely separated because the distance between them is $2\sqrt{1 - c}$, which becomes large as $c \to -\infty$. Again, the maximum point approaches the x-axis because $1/(c - 1) \to 0$ as $c \to -\infty$.

TEC See an animation of Figure 21 in Visual 4.6.

   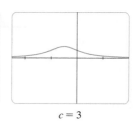

$c = -1$ $c = 0$ $c = 1$ $c = 2$ $c = 3$

FIGURE 21
The family of functions
$f(x) = 1/(x^2 + 2x + c)$

There is clearly no inflection point when $c \leq 1$. For $c > 1$ we calculate that

$$f''(x) = \frac{2(3x^2 + 6x + 4 - c)}{(x^2 + 2x + c)^3}$$

and deduce that inflection points occur when $x = -1 \pm \sqrt{3(c - 1)}/3$. So the inflection points become more spread out as c increases and this seems plausible from the last two parts of Figure 21.

4.6 ⊞ EXERCISES

1–8 Produce graphs of f that reveal all the important aspects of the curve. In particular, you should use graphs of f' and f'' to estimate the intervals of increase and decrease, extreme values, intervals of concavity, and inflection points.

1. $f(x) = x^5 - 5x^4 - x^3 - 28x^2 - 2x$

2. $f(x) = -2x^6 + 5x^5 + 140x^3 - 110x^2$

3. $f(x) = x^6 - 5x^5 + 25x^3 - 6x^2 - 48x$

4. $f(x) = \dfrac{x^4 - x^3 - 8}{x^2 - x - 6}$ **5.** $f(x) = \dfrac{x}{x^3 + x^2 + 1}$

6. $f(x) = 6 \sin x - x^2$, $-5 \le x \le 3$

7. $f(x) = 6 \sin x + \cot x$, $-\pi \le x \le \pi$

8. $f(x) = e^x - 0.186x^4$

9–10 Produce graphs of f that reveal all the important aspects of the curve. Estimate the intervals of increase and decrease and intervals of concavity, and use calculus to find these intervals exactly.

9. $f(x) = 1 + \dfrac{1}{x} + \dfrac{8}{x^2} + \dfrac{1}{x^3}$ **10.** $f(x) = \dfrac{1}{x^8} - \dfrac{2 \times 10^8}{x^4}$

11–12
(a) Graph the function.
(b) Use l'Hospital's Rule to explain the behavior as $x \to 0$.
(c) Estimate the minimum value and intervals of concavity. Then use calculus to find the exact values.

11. $f(x) = x^2 \ln x$ **12.** $f(x) = xe^{1/x}$

13–14 Sketch the graph by hand using asymptotes and intercepts, but not derivatives. Then use your sketch as a guide to producing graphs (with a graphing device) that display the major features of the curve. Use these graphs to estimate the maximum and minimum values.

13. $f(x) = \dfrac{(x + 4)(x - 3)^2}{x^4(x - 1)}$

14. $f(x) = \dfrac{(2x + 3)^2(x - 2)^5}{x^3(x - 5)^2}$

CAS **15.** If f is the function considered in Example 3, use a computer algebra system to calculate f' and then graph it to confirm that all the maximum and minimum values are as given in the example. Calculate f'' and use it to estimate the intervals of concavity and inflection points.

CAS **16.** If f is the function of Exercise 14, find f' and f'' and use their graphs to estimate the intervals of increase and decrease and concavity of f.

CAS **17–22** Use a computer algebra system to graph f and to find f' and f''. Use graphs of these derivatives to estimate the intervals of increase and decrease, extreme values, intervals of concavity, and inflection points of f.

17. $f(x) = \dfrac{x^3 + 5x^2 + 1}{x^4 + x^3 - x^2 + 2}$

18. $f(x) = \dfrac{x^{2/3}}{1 + x + x^4}$

19. $f(x) = \sqrt{x + 5 \sin x}$, $x \le 20$

20. $f(x) = x - \tan^{-1}(x^2)$

21. $f(x) = \dfrac{1 - e^{1/x}}{1 + e^{1/x}}$

22. $f(x) = \dfrac{3}{3 + 2 \sin x}$

CAS **23–24** Graph the function using as many viewing rectangles as you need to depict the true nature of the function.

23. $f(x) = \dfrac{1 - \cos(x^4)}{x^8}$

24. $f(x) = e^x + \ln|x - 4|$

CAS **25–26**
(a) Graph the function.
(b) Explain the shape of the graph by computing the limit as $x \to 0^+$ or as $x \to \infty$.
(c) Estimate the maximum and minimum values and then use calculus to find the exact values.
(d) Use a graph of f'' to estimate the x-coordinates of the inflection points.

25. $f(x) = x^{1/x}$ **26.** $f(x) = (\sin x)^{\sin x}$

---

27. In Example 4 we considered a member of the family of functions $f(x) = \sin(x + \sin cx)$ that occur in FM synthesis. Here we investigate the function with $c = 3$. Start by graphing f in the viewing rectangle $[0, \pi]$ by $[-1.2, 1.2]$. How many local maximum points do you see? The graph has more than are visible to the naked eye. To discover the hidden maximum and minimum points you will need to examine the graph of f' very carefully. In fact, it helps to look at the graph of f'' at the same time. Find all the maximum and minimum values and inflection points. Then graph f in the viewing rectangle $[-2\pi, 2\pi]$ by $[-1.2, 1.2]$ and comment on symmetry.

28–35 Describe how the graph of f varies as c varies. Graph several members of the family to illustrate the trends that you discover. In particular, you should investigate how maximum and minimum points and inflection points move when c changes. You should also identify any transitional values of c at which the basic shape of the curve changes.

28. $f(x) = x^3 + cx$

29. $f(x) = x^2 + 6x + c/x$ (Trident of Newton)

30. $f(x) = x\sqrt{c^2 - x^2}$ **31.** $f(x) = e^x + ce^{-x}$

32. $f(x) = \ln(x^2 + c)$ **33.** $f(x) = \dfrac{cx}{1 + c^2x^2}$

34. $f(x) = \dfrac{\sin x}{c + \cos x}$ **35.** $f(x) = cx + \sin x$

36. The family of functions $f(t) = C(e^{-at} - e^{-bt})$, where a, b, and C are positive numbers and $b > a$, has been used to model the concentration of a drug injected into the bloodstream at time $t = 0$. Graph several members of this family. What do they have in common? For fixed values of C and a, discover graphically what happens as b increases. Then use calculus to prove what you have discovered.

37. Investigate the family of curves given by $f(x) = xe^{-cx}$, where c is a real number. Start by computing the limits as

$x \to \pm\infty$. Identify any transitional values of c where the basic shape changes. What happens to the maximum or minimum points and inflection points as c changes? Illustrate by graphing several members of the family.

38. Investigate the family of curves given by the equation $f(x) = x^4 + cx^2 + x$. Start by determining the transitional value of c at which the number of inflection points changes. Then graph several members of the family to see what shapes are possible. There is another transitional value of c at which the number of critical numbers changes. Try to discover it graphically. Then prove what you have discovered.

39. (a) Investigate the family of polynomials given by the equation $f(x) = cx^4 - 2x^2 + 1$. For what values of c does the curve have minimum points?
   (b) Show that the minimum and maximum points of every curve in the family lie on the parabola $y = 1 - x^2$. Illustrate by graphing this parabola and several members of the family.

40. (a) Investigate the family of polynomials given by the equation $f(x) = 2x^3 + cx^2 + 2x$. For what values of c does the curve have maximum and minimum points?
   (b) Show that the minimum and maximum points of every curve in the family lie on the curve $y = x - x^3$. Illustrate by graphing this curve and several members of the family.

4.7 Optimization Problems

The methods we have learned in this chapter for finding extreme values have practical applications in many areas of life. A businessperson wants to minimize costs and maximize profits. A traveler wants to minimize transportation time. Fermat's Principle in optics states that light follows the path that takes the least time. In this section we solve such problems as maximizing areas, volumes, and profits and minimizing distances, times, and costs.

In solving such practical problems the greatest challenge is often to convert the word problem into a mathematical optimization problem by setting up the function that is to be maximized or minimized. Let's recall the problem-solving principles discussed on page 71 and adapt them to this situation:

Steps In Solving Optimization Problems

1. **Understand the Problem** The first step is to read the problem carefully until it is clearly understood. Ask yourself: What is the unknown? What are the given quantities? What are the given conditions?

2. **Draw a Diagram** In most problems it is useful to draw a diagram and identify the given and required quantities on the diagram.

3. **Introduce Notation** Assign a symbol to the quantity that is to be maximized or minimized (let's call it Q for now). Also select symbols $(a, b, c, \ldots, x, y)$ for other unknown quantities and label the diagram with these symbols. It may help to use initials as suggestive symbols—for example, A for area, h for height, t for time.

4. Express Q in terms of some of the other symbols from Step 3.

5. If Q has been expressed as a function of more than one variable in Step 4, use the given information to find relationships (in the form of equations) among these variables. Then use these equations to eliminate all but one of the variables in the expression for Q. Thus Q will be expressed as a function of *one* variable x, say, $Q = f(x)$. Write the domain of this function in the given context.

6. Use the methods of Sections 4.1 and 4.3 to find the *absolute* maximum or minimum value of f. In particular, if the domain of f is a closed interval, then the Closed Interval Method in Section 4.1 can be used.

EXAMPLE 1 A farmer has 2400 ft of fencing and wants to fence off a rectangular field that borders a straight river. He needs no fence along the river. What are the dimensions of the field that has the largest area?

PS Understand the problem
PS Analogy: Try special cases
PS Draw diagrams

SOLUTION In order to get a feeling for what is happening in this problem, let's experiment with some specific cases. Figure 1 (not to scale) shows three possible ways of laying out the 2400 ft of fencing.

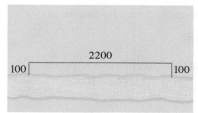

Area = 100 · 2200 = 220,000 ft²

Area = 700 · 1000 = 700,000 ft²

Area = 1000 · 400 = 400,000 ft²

FIGURE 1

We see that when we try shallow, wide fields or deep, narrow fields, we get relatively small areas. It seems plausible that there is some intermediate configuration that produces the largest area.

Figure 2 illustrates the general case. We wish to maximize the area A of the rectangle. Let x and y be the depth and width of the rectangle (in feet). Then we express A in terms of x and y:

PS Introduce notation

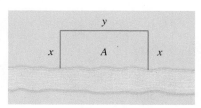

FIGURE 2

$$A = xy$$

We want to express A as a function of just one variable, so we eliminate y by expressing it in terms of x. To do this we use the given information that the total length of the fencing is 2400 ft. Thus

$$2x + y = 2400$$

From this equation we have $y = 2400 - 2x$, which gives

$$A = xy = x(2400 - 2x) = 2400x - 2x^2$$

Note that the largest x can be is 1200 (this uses all the fence for the depth and none for the width) and x can't be negative, so the function that we wish to maximize is

$$A(x) = 2400x - 2x^2 \qquad 0 \leqslant x \leqslant 1200$$

The derivative is $A'(x) = 2400 - 4x$, so to find the critical numbers we solve the equation

$$2400 - 4x = 0$$

which gives $x = 600$. The maximum value of A must occur either at this critical number or at an endpoint of the interval. Since $A(0) = 0$, $A(600) = 720{,}000$, and $A(1200) = 0$, the Closed Interval Method gives the maximum value as $A(600) = 720{,}000$.

[Alternatively, we could have observed that $A''(x) = -4 < 0$ for all x, so A is always concave downward and the local maximum at $x = 600$ must be an absolute maximum.]

The corresponding y-value is $y = 2400 - 2(600) = 1200$; so the rectangular field should be 600 ft deep and 1200 ft wide.   ∎

EXAMPLE 2 A cylindrical can is to be made to hold 1 L of oil. Find the dimensions that will minimize the cost of the metal to manufacture the can.

SOLUTION Draw the diagram as in Figure 3, where r is the radius and h the height (both in centimeters). In order to minimize the cost of the metal, we minimize the total surface area of the cylinder (top, bottom, and sides). From Figure 4 we see that the sides are made from a rectangular sheet with dimensions $2\pi r$ and h. So the surface area is

$$A = 2\pi r^2 + 2\pi rh$$

We would like to express A in terms of one variable, r. To eliminate h we use the fact that the volume is given as 1 L, which is equivalent to 1000 cm³. Thus

$$\pi r^2 h = 1000$$

which gives $h = 1000/(\pi r^2)$. Substitution of this into the expression for A gives

$$A = 2\pi r^2 + 2\pi r\left(\frac{1000}{\pi r^2}\right) = 2\pi r^2 + \frac{2000}{r}$$

We know r must be positive, and there are no limitations on how large r can be. Therefore the function that we want to minimize is

$$A(r) = 2\pi r^2 + \frac{2000}{r} \qquad r > 0$$

To find the critical numbers, we differentiate:

$$A'(r) = 4\pi r - \frac{2000}{r^2} = \frac{4(\pi r^3 - 500)}{r^2}$$

Then $A'(r) = 0$ when $\pi r^3 = 500$, so the only critical number is $r = \sqrt[3]{500/\pi}$.

Since the domain of A is $(0, \infty)$, we can't use the argument of Example 1 concerning endpoints. But we can observe that $A'(r) < 0$ for $r < \sqrt[3]{500/\pi}$ and $A'(r) > 0$ for $r > \sqrt[3]{500/\pi}$, so A is decreasing for *all r* to the left of the critical number and increasing for *all r* to the right. Thus $r = \sqrt[3]{500/\pi}$ must give rise to an *absolute* minimum.

[Alternatively, we could argue that $A(r) \to \infty$ as $r \to 0^+$ and $A(r) \to \infty$ as $r \to \infty$, so there must be a minimum value of $A(r)$, which must occur at the critical number. See Figure 5.]

FIGURE 3

Area $2(\pi r^2)$ Area $(2\pi r)h$

FIGURE 4

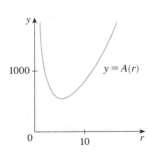

FIGURE 5

In the Applied Project on page 343 we investigate the most economical shape for a can by taking into account other manufacturing costs.

The value of h corresponding to $r = \sqrt[3]{500/\pi}$ is

$$h = \frac{1000}{\pi r^2} = \frac{1000}{\pi(500/\pi)^{2/3}} = 2\sqrt[3]{\frac{500}{\pi}} = 2r$$

Thus, to minimize the cost of the can, the radius should be $\sqrt[3]{500/\pi}$ cm and the height should be equal to twice the radius, namely, the diameter. ∎

NOTE 1 The argument used in Example 2 to justify the absolute minimum is a variant of the First Derivative Test (which applies only to *local* maximum or minimum values) and is stated here for future reference.

> **First Derivative Test for Absolute Extreme Values** Suppose that c is a critical number of a continuous function f defined on an interval.
> (a) If $f'(x) > 0$ for all $x < c$ and $f'(x) < 0$ for all $x > c$, then $f(c)$ is the absolute maximum value of f.
> (b) If $f'(x) < 0$ for all $x < c$ and $f'(x) > 0$ for all $x > c$, then $f(c)$ is the absolute minimum value of f.

TEC Module 4.7 takes you through six additional optimization problems, including animations of the physical situations.

NOTE 2 An alternative method for solving optimization problems is to use implicit differentiation. Let's look at Example 2 again to illustrate the method. We work with the same equations

$$A = 2\pi r^2 + 2\pi rh \qquad \pi r^2 h = 1000$$

but instead of eliminating h, we differentiate both equations implicitly with respect to r:

$$A' = 4\pi r + 2\pi rh' + 2\pi h \qquad \pi r^2 h' + 2\pi rh = 0$$

The minimum occurs at a critical number, so we set $A' = 0$, simplify, and arrive at the equations

$$2r + rh' + h = 0 \qquad rh' + 2h = 0$$

and subtraction gives $2r - h = 0$, or $h = 2r$.

EXAMPLE 3 Find the point on the parabola $y^2 = 2x$ that is closest to the point $(1, 4)$.

SOLUTION The distance between the point $(1, 4)$ and the point (x, y) is

$$d = \sqrt{(x-1)^2 + (y-4)^2}$$

(See Figure 6.) But if (x, y) lies on the parabola, then $x = \frac{1}{2}y^2$, so the expression for d becomes

$$d = \sqrt{\left(\tfrac{1}{2}y^2 - 1\right)^2 + (y-4)^2}$$

(Alternatively, we could have substituted $y = \sqrt{2x}$ to get d in terms of x alone.) Instead of minimizing d, we minimize its square:

$$d^2 = f(y) = \left(\tfrac{1}{2}y^2 - 1\right)^2 + (y-4)^2$$

FIGURE 6

(You should convince yourself that the minimum of d occurs at the same point as the minimum of d^2, but d^2 is easier to work with.) Note that there are no restrictions on y, so the domain is all real numbers. Differentiating, we obtain

$$f'(y) = 2\left(\tfrac{1}{2}y^2 - 1\right)y + 2(y - 4) = y^3 - 8$$

so $f'(y) = 0$ when $y = 2$. Observe that $f'(y) < 0$ when $y < 2$ and $f'(y) > 0$ when $y > 2$, so by the First Derivative Test for Absolute Extreme Values, the absolute minimum occurs when $y = 2$. (Or we could simply say that because of the geometric nature of the problem, it's obvious that there is a closest point but not a farthest point.) The corresponding value of x is $x = \tfrac{1}{2}y^2 = 2$. Thus the point on $y^2 = 2x$ closest to $(1, 4)$ is $(2, 2)$. [The distance between the points is $d = \sqrt{f(2)} = \sqrt{5}$.] ∎

EXAMPLE 4 A man launches his boat from point A on a bank of a straight river, 3 km wide, and wants to reach point B, 8 km downstream on the opposite bank, as quickly as possible (see Figure 7). He could row his boat directly across the river to point C and then run to B, or he could row directly to B, or he could row to some point D between C and B and then run to B. If he can row 6 km/h and run 8 km/h, where should he land to reach B as soon as possible? (We assume that the speed of the water is negligible compared with the speed at which the man rows.)

SOLUTION If we let x be the distance from C to D, then the running distance is $|DB| = 8 - x$ and the Pythagorean Theorem gives the rowing distance as $|AD| = \sqrt{x^2 + 9}$. We use the equation

$$\text{time} = \frac{\text{distance}}{\text{rate}}$$

Then the rowing time is $\sqrt{x^2 + 9}/6$ and the running time is $(8 - x)/8$, so the total time T as a function of x is

$$T(x) = \frac{\sqrt{x^2 + 9}}{6} + \frac{8 - x}{8}$$

The domain of this function T is $[0, 8]$. Notice that if $x = 0$, he rows to C and if $x = 8$, he rows directly to B. The derivative of T is

$$T'(x) = \frac{x}{6\sqrt{x^2 + 9}} - \frac{1}{8}$$

Thus, using the fact that $x \geq 0$, we have

$$T'(x) = 0 \quad \Longleftrightarrow \quad \frac{x}{6\sqrt{x^2 + 9}} = \frac{1}{8} \quad \Longleftrightarrow \quad 4x = 3\sqrt{x^2 + 9}$$

$$\Longleftrightarrow \quad 16x^2 = 9(x^2 + 9) \quad \Longleftrightarrow \quad 7x^2 = 81$$

$$\Longleftrightarrow \quad x = \frac{9}{\sqrt{7}}$$

The only critical number is $x = 9/\sqrt{7}$. To see whether the minimum occurs at this critical number or at an endpoint of the domain $[0, 8]$, we follow the Closed Interval

3 km

A   C

x

D

8 km

B

FIGURE 7

FIGURE 8

Method by evaluating T at all three points:

$$T(0) = 1.5 \qquad T\left(\frac{9}{\sqrt{7}}\right) = 1 + \frac{\sqrt{7}}{8} \approx 1.33 \qquad T(8) = \frac{\sqrt{73}}{6} \approx 1.42$$

Since the smallest of these values of T occurs when $x = 9/\sqrt{7}$, the absolute minimum value of T must occur there. Figure 8 illustrates this calculation by showing the graph of T.

Thus the man should land the boat at a point $9/\sqrt{7}$ km (≈ 3.4 km) downstream from his starting point. ∎

EXAMPLE 5 Find the area of the largest rectangle that can be inscribed in a semicircle of radius r.

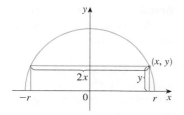

FIGURE 9

SOLUTION 1 Let's take the semicircle to be the upper half of the circle $x^2 + y^2 = r^2$ with center the origin. Then the word *inscribed* means that the rectangle has two vertices on the semicircle and two vertices on the x-axis as shown in Figure 9.

Let (x, y) be the vertex that lies in the first quadrant. Then the rectangle has sides of lengths $2x$ and y, so its area is

$$A = 2xy$$

To eliminate y we use the fact that (x, y) lies on the circle $x^2 + y^2 = r^2$ and so $y = \sqrt{r^2 - x^2}$. Thus

$$A = 2x\sqrt{r^2 - x^2}$$

The domain of this function is $0 \le x \le r$. Its derivative is

$$A' = 2\sqrt{r^2 - x^2} - \frac{2x^2}{\sqrt{r^2 - x^2}} = \frac{2(r^2 - 2x^2)}{\sqrt{r^2 - x^2}}$$

which is 0 when $2x^2 = r^2$, that is, $x = r/\sqrt{2}$ (since $x \ge 0$). This value of x gives a maximum value of A since $A(0) = 0$ and $A(r) = 0$. Therefore the area of the largest inscribed rectangle is

$$A\left(\frac{r}{\sqrt{2}}\right) = 2\frac{r}{\sqrt{2}}\sqrt{r^2 - \frac{r^2}{2}} = r^2$$

FIGURE 10

SOLUTION 2 A simpler solution is possible if we think of using an angle as a variable. Let θ be the angle shown in Figure 10. Then the area of the rectangle is

$$A(\theta) = (2r\cos\theta)(r\sin\theta) = r^2(2\sin\theta\cos\theta) = r^2\sin 2\theta$$

We know that $\sin 2\theta$ has a maximum value of 1 and it occurs when $2\theta = \pi/2$. So $A(\theta)$ has a maximum value of r^2 and it occurs when $\theta = \pi/4$.

Notice that this trigonometric solution doesn't involve differentiation. In fact, we didn't need to use calculus at all. ∎

■ Applications to Business and Economics

In Section 3.7 we introduced the idea of marginal cost. Recall that if $C(x)$, the **cost function**, is the cost of producing x units of a certain product, then the **marginal cost** is the rate of change of C with respect to x. In other words, the marginal cost function is the derivative, $C'(x)$, of the cost function.

Now let's consider marketing. Let $p(x)$ be the price per unit that the company can charge if it sells x units. Then p is called the **demand function** (or **price function**) and we would expect it to be a decreasing function of x. (More units sold corresponds to a lower price.) If x units are sold and the price per unit is $p(x)$, then the total revenue is

$$R(x) = \text{quantity} \times \text{price} = xp(x)$$

and R is called the **revenue function**. The derivative R' of the revenue function is called the **marginal revenue function** and is the rate of change of revenue with respect to the number of units sold.

If x units are sold, then the total profit is

$$P(x) = R(x) - C(x)$$

and P is called the **profit function**. The **marginal profit function** is P', the derivative of the profit function. In Exercises 59–63 you are asked to use the marginal cost, revenue, and profit functions to minimize costs and maximize revenues and profits.

EXAMPLE 6 A store has been selling 200 flat-screen TVs a week at \$350 each. A market survey indicates that for each \$10 rebate offered to buyers, the number of TVs sold will increase by 20 a week. Find the demand function and the revenue function. How large a rebate should the store offer to maximize its revenue?

SOLUTION If x is the number of TVs sold per week, then the weekly increase in sales is $x - 200$. For each increase of 20 units sold, the price is decreased by \$10. So for each additional unit sold, the decrease in price will be $\frac{1}{20} \times 10$ and the demand function is

$$p(x) = 350 - \tfrac{10}{20}(x - 200) = 450 - \tfrac{1}{2}x$$

The revenue function is

$$R(x) = xp(x) = 450x - \tfrac{1}{2}x^2$$

Since $R'(x) = 450 - x$, we see that $R'(x) = 0$ when $x = 450$. This value of x gives an absolute maximum by the First Derivative Test (or simply by observing that the graph of R is a parabola that opens downward). The corresponding price is

$$p(450) = 450 - \tfrac{1}{2}(450) = 225$$

and the rebate is $350 - 225 = 125$. Therefore, to maximize revenue, the store should offer a rebate of \$125. ∎

4.7 EXERCISES

1. Consider the following problem: Find two numbers whose sum is 23 and whose product is a maximum.
   (a) Make a table of values, like the one at the right, so that the sum of the numbers in the first two columns is always 23. On the basis of the evidence in your table, estimate the answer to the problem.
   (b) Use calculus to solve the problem and compare with your answer to part (a).

First number	Second number	Product
1	22	22
2	21	42
3	20	60
⋮	⋮	⋮

2. Find two numbers whose difference is 100 and whose product is a minimum.

3. Find two positive numbers whose product is 100 and whose sum is a minimum.

4. The sum of two positive numbers is 16. What is the smallest possible value of the sum of their squares?

5. What is the maximum vertical distance between the line $y = x + 2$ and the parabola $y = x^2$ for $-1 \leqslant x \leqslant 2$?

6. What is the minimum vertical distance between the parabolas $y = x^2 + 1$ and $y = x - x^2$?

7. Find the dimensions of a rectangle with perimeter 100 m whose area is as large as possible.

8. Find the dimensions of a rectangle with area 1000 m² whose perimeter is as small as possible.

9. A model used for the yield Y of an agricultural crop as a function of the nitrogen level N in the soil (measured in appropriate units) is

$$Y = \frac{kN}{1 + N^2}$$

where k is a positive constant. What nitrogen level gives the best yield?

10. The rate (in mg carbon/m³/h) at which photosynthesis takes place for a species of phytoplankton is modeled by the function

$$P = \frac{100I}{I^2 + I + 4}$$

where I is the light intensity (measured in thousands of foot-candles). For what light intensity is P a maximum?

11. Consider the following problem: A farmer with 750 ft of fencing wants to enclose a rectangular area and then divide it into four pens with fencing parallel to one side of the rectangle. What is the largest possible total area of the four pens?
   (a) Draw several diagrams illustrating the situation, some with shallow, wide pens and some with deep, narrow pens. Find the total areas of these configurations. Does it appear that there is a maximum area? If so, estimate it.
   (b) Draw a diagram illustrating the general situation. Introduce notation and label the diagram with your symbols.
   (c) Write an expression for the total area.
   (d) Use the given information to write an equation that relates the variables.
   (e) Use part (d) to write the total area as a function of one variable.
   (f) Finish solving the problem and compare the answer with your estimate in part (a).

12. Consider the following problem: A box with an open top is to be constructed from a square piece of cardboard, 3 ft wide, by cutting out a square from each of the four corners and bend-

ing up the sides. Find the largest volume that such a box can have.
   (a) Draw several diagrams to illustrate the situation, some short boxes with large bases and some tall boxes with small bases. Find the volumes of several such boxes. Does it appear that there is a maximum volume? If so, estimate it.
   (b) Draw a diagram illustrating the general situation. Introduce notation and label the diagram with your symbols.
   (c) Write an expression for the volume.
   (d) Use the given information to write an equation that relates the variables.
   (e) Use part (d) to write the volume as a function of one variable.
   (f) Finish solving the problem and compare the answer with your estimate in part (a).

13. A farmer wants to fence in an area of 1.5 million square feet in a rectangular field and then divide it in half with a fence parallel to one of the sides of the rectangle. How can he do this so as to minimize the cost of the fence?

14. A box with a square base and open top must have a volume of 32,000 cm³. Find the dimensions of the box that minimize the amount of material used.

15. If 1200 cm² of material is available to make a box with a square base and an open top, find the largest possible volume of the box.

16. A rectangular storage container with an open top is to have a volume of 10 m³. The length of its base is twice the width. Material for the base costs $10 per square meter. Material for the sides costs $6 per square meter. Find the cost of materials for the cheapest such container.

17. Do Exercise 16 assuming the container has a lid that is made from the same material as the sides.

18. A farmer wants to fence in a rectangular plot of land adjacent to the north wall of his barn. No fencing is needed along the barn, and the fencing along the west side of the plot is shared with a neighbor who will split the cost of that portion of the fence. If the fencing costs $20 per linear foot to install and the farmer is not willing to spend more than $5000, find the dimensions for the plot that would enclose the most area.

19. If the farmer in Exercise 18 wants to enclose 8000 square feet of land, what dimensions will minimize the cost of the fence?

20. (a) Show that of all the rectangles with a given area, the one with smallest perimeter is a square.
   (b) Show that of all the rectangles with a given perimeter, the one with greatest area is a square.

21. Find the point on the line $y = 2x + 3$ that is closest to the origin.

22. Find the point on the curve $y = \sqrt{x}$ that is closest to the point (3, 0).

23. Find the points on the ellipse $4x^2 + y^2 = 4$ that are farthest away from the point $(1, 0)$.

24. Find, correct to two decimal places, the coordinates of the point on the curve $y = \sin x$ that is closest to the point $(4, 2)$.

25. Find the dimensions of the rectangle of largest area that can be inscribed in a circle of radius r.

26. Find the area of the largest rectangle that can be inscribed in the ellipse $x^2/a^2 + y^2/b^2 = 1$.

27. Find the dimensions of the rectangle of largest area that can be inscribed in an equilateral triangle of side L if one side of the rectangle lies on the base of the triangle.

28. Find the area of the largest trapezoid that can be inscribed in a circle of radius 1 and whose base is a diameter of the circle.

29. Find the dimensions of the isosceles triangle of largest area that can be inscribed in a circle of radius r.

30. If the two equal sides of an isosceles triangle have length a, find the length of the third side that maximizes the area of the triangle.

31. A right circular cylinder is inscribed in a sphere of radius r. Find the largest possible volume of such a cylinder.

32. A right circular cylinder is inscribed in a cone with height h and base radius r. Find the largest possible volume of such a cylinder.

33. A right circular cylinder is inscribed in a sphere of radius r. Find the largest possible surface area of such a cylinder.

34. A Norman window has the shape of a rectangle surmounted by a semicircle. (Thus the diameter of the semicircle is equal to the width of the rectangle. See Exercise 1.1.62.) If the perimeter of the window is 30 ft, find the dimensions of the window so that the greatest possible amount of light is admitted.

35. The top and bottom margins of a poster are each 6 cm and the side margins are each 4 cm. If the area of printed material on the poster is fixed at 384 cm², find the dimensions of the poster with the smallest area.

36. A poster is to have an area of 180 in² with 1-inch margins at the bottom and sides and a 2-inch margin at the top. What dimensions will give the largest printed area?

37. A piece of wire 10 m long is cut into two pieces. One piece is bent into a square and the other is bent into an equilateral triangle. How should the wire be cut so that the total area enclosed is (a) a maximum? (b) A minimum?

38. Answer Exercise 37 if one piece is bent into a square and the other into a circle.

39. If you are offered one slice from a round pizza (in other words, a sector of a circle) and the slice must have a perimeter of 32 inches, what diameter pizza will reward you with the largest slice?

40. A fence 8 ft tall runs parallel to a tall building at a distance of 4 ft from the building. What is the length of the shortest ladder that will reach from the ground over the fence to the wall of the building?

41. A cone-shaped drinking cup is made from a circular piece of paper of radius R by cutting out a sector and joining the edges CA and CB. Find the maximum capacity of such a cup.

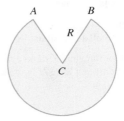

42. A cone-shaped paper drinking cup is to be made to hold 27 cm³ of water. Find the height and radius of the cup that will use the smallest amount of paper.

43. A cone with height h is inscribed in a larger cone with height H so that its vertex is at the center of the base of the larger cone. Show that the inner cone has maximum volume when $h = \frac{1}{3}H$.

44. An object with weight W is dragged along a horizontal plane by a force acting along a rope attached to the object. If the rope makes an angle θ with a plane, then the magnitude of the force is

$$F = \frac{\mu W}{\mu \sin \theta + \cos \theta}$$

where μ is a constant called the coefficient of friction. For what value of θ is F smallest?

45. If a resistor of R ohms is connected across a battery of E volts with internal resistance r ohms, then the power (in watts) in the external resistor is

$$P = \frac{E^2 R}{(R + r)^2}$$

If E and r are fixed but R varies, what is the maximum value of the power?

46. For a fish swimming at a speed v relative to the water, the energy expenditure per unit time is proportional to v^3. It is believed that migrating fish try to minimize the total energy required to swim a fixed distance. If the fish are swimming against a current u ($u < v$), then the time required to swim a distance L is $L/(v - u)$ and the total energy E required to swim the distance is given by

$$E(v) = av^3 \cdot \frac{L}{v - u}$$

where a is the proportionality constant.
(a) Determine the value of v that minimizes E.
(b) Sketch the graph of E.

Note: This result has been verified experimentally; migrating fish swim against a current at a speed 50% greater than the current speed.

47. In a beehive, each cell is a regular hexagonal prism, open at one end with a trihedral angle at the other end as in the figure. It is believed that bees form their cells in such a way as to minimize the surface area for a given side length and height, thus using the least amount of wax in cell construction. Examination of these cells has shown that the measure of the apex angle θ is amazingly consistent. Based on the geometry of the cell, it can be shown that the surface area S is given by

$$S = 6sh - \tfrac{3}{2}s^2 \cot \theta + \left(3s^2\sqrt{3}/2\right) \csc \theta$$

where s, the length of the sides of the hexagon, and h, the height, are constants.
(a) Calculate $dS/d\theta$.
(b) What angle should the bees prefer?
(c) Determine the minimum surface area of the cell (in terms of s and h).

Note: Actual measurements of the angle θ in beehives have been made, and the measures of these angles seldom differ from the calculated value by more than $2°$.

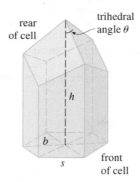

48. A boat leaves a dock at 2:00 PM and travels due south at a speed of 20 km/h. Another boat has been heading due east at 15 km/h and reaches the same dock at 3:00 PM. At what time were the two boats closest together?

49. Solve the problem in Example 4 if the river is 5 km wide and point B is only 5 km downstream from A.

50. A woman at a point A on the shore of a circular lake with radius 2 mi wants to arrive at the point C diametrically

opposite A on the other side of the lake in the shortest possible time (see the figure). She can walk at the rate of 4 mi/h and row a boat at 2 mi/h. How should she proceed?

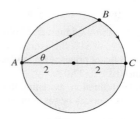

51. An oil refinery is located on the north bank of a straight river that is 2 km wide. A pipeline is to be constructed from the refinery to storage tanks located on the south bank of the river 6 km east of the refinery. The cost of laying pipe is $400,000/km over land to a point P on the north bank and $800,000/km under the river to the tanks. To minimize the cost of the pipeline, where should P be located?

52. Suppose the refinery in Exercise 51 is located 1 km north of the river. Where should P be located?

53. The illumination of an object by a light source is directly proportional to the strength of the source and inversely proportional to the square of the distance from the source. If two light sources, one three times as strong as the other, are placed 10 ft apart, where should an object be placed on the line between the sources so as to receive the least illumination?

54. Find an equation of the line through the point $(3, 5)$ that cuts off the least area from the first quadrant.

55. Let a and b be positive numbers. Find the length of the shortest line segment that is cut off by the first quadrant and passes through the point (a, b).

56. At which points on the curve $y = 1 + 40x^3 - 3x^5$ does the tangent line have the largest slope?

57. What is the shortest possible length of the line segment that is cut off by the first quadrant and is tangent to the curve $y = 3/x$ at some point?

58. What is the smallest possible area of the triangle that is cut off by the first quadrant and whose hypotenuse is tangent to the parabola $y = 4 - x^2$ at some point?

59. (a) If $C(x)$ is the cost of producing x units of a commodity, then the **average cost** per unit is $c(x) = C(x)/x$. Show that if the average cost is a minimum, then the marginal cost equals the average cost.
(b) If $C(x) = 16{,}000 + 200x + 4x^{3/2}$, in dollars, find (i) the cost, average cost, and marginal cost at a production level of 1000 units; (ii) the production level that will minimize the average cost; and (iii) the minimum average cost.

60. (a) Show that if the profit $P(x)$ is a maximum, then the marginal revenue equals the marginal cost.
(b) If $C(x) = 16{,}000 + 500x - 1.6x^2 + 0.004x^3$ is the cost function and $p(x) = 1700 - 7x$ is the demand function, find the production level that will maximize profit.

61. A baseball team plays in a stadium that holds 55,000 spectators. With ticket prices at $10, the average attendance had been 27,000. When ticket prices were lowered to $8, the average attendance rose to 33,000.
(a) Find the demand function, assuming that it is linear.
(b) How should ticket prices be set to maximize revenue?

62. During the summer months Terry makes and sells necklaces on the beach. Last summer he sold the necklaces for $10 each and his sales averaged 20 per day. When he increased the price by $1, he found that the average decreased by two sales per day.
(a) Find the demand function, assuming that it is linear.
(b) If the material for each necklace costs Terry $6, what should the selling price be to maximize his profit?

63. A retailer has been selling 1200 tablet computers a week at $350 each. The marketing department estimates that an additional 80 tablets will sell each week for every $10 that the price is lowered.
(a) Find the demand function.
(b) What should the price be set at in order to maximize revenue?
(c) If the retailer's weekly cost function is

$$C(x) = 35{,}000 + 120x$$

what price should it choose in order to maximize its profit?

64. A company operates 16 oil wells in a designated area. Each pump, on average, extracts 240 barrels of oil daily. The company can add more wells but every added well reduces the average daily ouput of each of the wells by 8 barrels. How many wells should the company add in order to maximize daily production?

65. Show that of all the isosceles triangles with a given perimeter, the one with the greatest area is equilateral.

66. Consider the situation in Exercise 51 if the cost of laying pipe under the river is considerably higher than the cost of laying pipe over land ($400,000/km). You may suspect that in some instances, the minimum distance possible under the river should be used, and P should be located 6 km from the refinery, directly across from the storage tanks. Show that this is *never* the case, no matter what the "under river" cost is.

67. Consider the tangent line to the ellipse $\dfrac{x^2}{a^2} + \dfrac{y^2}{b^2} = 1$ at a point (p, q) in the first quadrant.
(a) Show that the tangent line has x-intercept a^2/p and y-intercept b^2/q.
(b) Show that the portion of the tangent line cut off by the coordinate axes has minimum length $a + b$.

(c) Show that the triangle formed by the tangent line and the coordinate axes has minimum area ab.

68. The frame for a kite is to be made from six pieces of wood. The four exterior pieces have been cut with the lengths indicated in the figure. To maximize the area of the kite, how long should the diagonal pieces be?

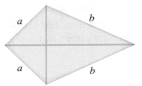

69. A point P needs to be located somewhere on the line AD so that the total length L of cables linking P to the points A, B. and C is minimized (see the figure). Express L as a function of $x = |AP|$ and use the graphs of L and dL/dx to estimate the minimum value of L.

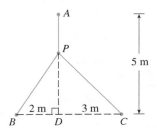

70. The graph shows the fuel consumption c of a car (measured in gallons per hour) as a function of the speed v of the car. At very low speeds the engine runs inefficiently, so initially c decreases as the speed increases. But at high speeds the fuel consumption increases. You can see that $c(v)$ is minimized for this car when $v \approx 30$ mi/h. However, for fuel efficiency, what must be minimized is not the consumption in gallons per hour but rather the fuel consumption in gallons *per mile*. Let's call this consumption G. Using the graph, estimate the speed at which G has its minimum value.

71. Let v_1 be the velocity of light in air and v_2 the velocity of light in water. According to Fermat's Principle, a ray of light will travel from a point A in the air to a point B in the water by a path ACB that minimizes the time taken. Show that

$$\frac{\sin\theta_1}{\sin\theta_2} = \frac{v_1}{v_2}$$

where θ_1 (the angle of incidence) and θ_2 (the angle of refraction) are as shown. This equation is known as Snell's Law.

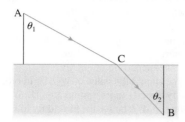

72. Two vertical poles PQ and ST are secured by a rope PRS going from the top of the first pole to a point R on the ground between the poles and then to the top of the second pole as in the figure. Show that the shortest length of such a rope occurs when $\theta_1 = \theta_2$.

73. The upper right-hand corner of a piece of paper, 12 in. by 8 in., as in the figure, is folded over to the bottom edge. How would you fold it so as to minimize the length of the fold? In other words, how would you choose x to minimize y?

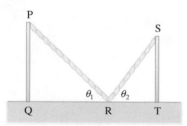

74. A steel pipe is being carried down a hallway 9 ft wide. At the end of the hall there is a right-angled turn into a narrower hallway 6 ft wide. What is the length of the longest pipe that can be carried horizontally around the corner?

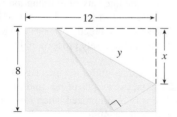

75. An observer stands at a point P, one unit away from a track. Two runners start at the point S in the figure and run along the track. One runner runs three times as fast as the other. Find the maximum value of the observer's angle of sight θ between the runners.

76. A rain gutter is to be constructed from a metal sheet of width 30 cm by bending up one-third of the sheet on each side through an angle θ. How should θ be chosen so that the gutter will carry the maximum amount of water?

77. Where should the point P be chosen on the line segment AB so as to maximize the angle θ?

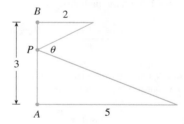

78. A painting in an art gallery has height h and is hung so that its lower edge is a distance d above the eye of an observer (as in the figure). How far from the wall should the observer stand to get the best view? (In other words, where should the observer stand so as to maximize the angle θ subtended at his eye by the painting?)

79. Find the maximum area of a rectangle that can be circumscribed about a given rectangle with length L and width W. [*Hint:* Express the area as a function of an angle θ.]

80. The blood vascular system consists of blood vessels (arteries, arterioles, capillaries, and veins) that convey blood from the heart to the organs and back to the heart. This system should work so as to minimize the energy expended by the heart in pumping the blood. In particular, this energy is reduced when the resistance of the blood is lowered. One of Poiseuille's Laws gives the resistance R of the blood as

$$R = C\frac{L}{r^4}$$

where L is the length of the blood vessel, r is the radius, and C is a positive constant determined by the viscosity of the blood. (Poiseuille established this law experimentally, but it also follows from Equation 8.4.2.) The figure shows a main blood vessel with radius r_1 branching at an angle θ into a smaller vessel with radius r_2.

(a) Use Poiseuille's Law to show that the total resistance of the blood along the path ABC is

$$R = C\left(\frac{a - b\cot\theta}{r_1^4} + \frac{b\csc\theta}{r_2^4}\right)$$

where a and b are the distances shown in the figure.

(b) Prove that this resistance is minimized when

$$\cos\theta = \frac{r_2^4}{r_1^4}$$

(c) Find the optimal branching angle (correct to the nearest degree) when the radius of the smaller blood vessel is two-thirds the radius of the larger vessel.

81. Ornithologists have determined that some species of birds tend to avoid flights over large bodies of water during daylight hours. It is believed that more energy is required to fly over water than over land because air generally rises over land and falls over water during the day. A bird with these tendencies is released from an island that is 5 km from the nearest point B on a straight shoreline, flies to a point C on the shoreline, and then flies along the shoreline to its nesting area D. Assume that the bird instinctively chooses a path that will minimize its energy expenditure. Points B and D are 13 km apart.

(a) In general, if it takes 1.4 times as much energy to fly over water as it does over land, to what point C should the bird

fly in order to minimize the total energy expended in returning to its nesting area?

(b) Let W and L denote the energy (in joules) per kilometer flown over water and land, respectively. What would a large value of the ratio W/L mean in terms of the bird's flight? What would a small value mean? Determine the ratio W/L corresponding to the minimum expenditure of energy.

(c) What should the value of W/L be in order for the bird to fly directly to its nesting area D? What should the value of W/L be for the bird to fly to B and then along the shore to D?

(d) If the ornithologists observe that birds of a certain species reach the shore at a point 4 km from B, how many times more energy does it take a bird to fly over water than over land?

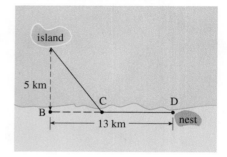

82. Two light sources of identical strength are placed 10 m apart. An object is to be placed at a point P on a line ℓ, parallel to the line joining the light sources and at a distance d meters from it (see the figure). We want to locate P on ℓ so that the intensity of illumination is minimized. We need to use the fact that the intensity of illumination for a single source is directly proportional to the strength of the source and inversely proportional to the square of the distance from the source.

(a) Find an expression for the intensity $I(x)$ at the point P.

(b) If $d = 5$ m, use graphs of $I(x)$ and $I'(x)$ to show that the intensity is minimized when $x = 5$ m, that is, when P is at the midpoint of ℓ.

(c) If $d = 10$ m, show that the intensity (perhaps surprisingly) is *not* minimized at the midpoint.

(d) Somewhere between $d = 5$ m and $d = 10$ m there is a transitional value of d at which the point of minimal illumination abruptly changes. Estimate this value of d by graphical methods. Then find the exact value of d.

APPLIED PROJECT   **THE SHAPE OF A CAN**

Discs cut from squares

Discs cut from hexagons

In this project we investigate the most economical shape for a can. We first interpret this to mean that the volume V of a cylindrical can is given and we need to find the height h and radius r that minimize the cost of the metal to make the can (see the figure). If we disregard any waste metal in the manufacturing process, then the problem is to minimize the surface area of the cylinder. We solved this problem in Example 4.7.2 and we found that $h = 2r$; that is, the height should be the same as the diameter. But if you go to your cupboard or your supermarket with a ruler, you will discover that the height is usually greater than the diameter and the ratio h/r varies from 2 up to about 3.8. Let's see if we can explain this phenomenon.

1. The material for the cans is cut from sheets of metal. The cylindrical sides are formed by bending rectangles; these rectangles are cut from the sheet with little or no waste. But if the top and bottom discs are cut from squares of side $2r$ (as in the figure), this leaves considerable waste metal, which may be recycled but has little or no value to the can makers. If this is the case, show that the amount of metal used is minimized when

$$\frac{h}{r} = \frac{8}{\pi} \approx 2.55$$

2. A more efficient packing of the discs is obtained by dividing the metal sheet into hexagons and cutting the circular lids and bases from the hexagons (see the figure). Show that if this strategy is adopted, then

$$\frac{h}{r} = \frac{4\sqrt{3}}{\pi} \approx 2.21$$

3. The values of h/r that we found in Problems 1 and 2 are a little closer to the ones that actually occur on supermarket shelves, but they still don't account for everything. If we look more closely at some real cans, we see that the lid and the base are formed from discs with radius larger than r that are bent over the ends of the can. If we allow for this we would increase h/r. More significantly, in addition to the cost of the metal we need to incorporate the manufacturing of the can into the cost. Let's assume that most of the expense is incurred in joining the sides to the rims of the cans. If we cut the discs from hexagons as in Problem 2, then the total cost is proportional to

$$4\sqrt{3}\,r^2 + 2\pi rh + k(4\pi r + h)$$

where k is the reciprocal of the length that can be joined for the cost of one unit area of metal. Show that this expression is minimized when

$$\frac{\sqrt[3]{V}}{k} = \sqrt[3]{\frac{\pi h}{r}} \cdot \frac{2\pi - h/r}{\pi h/r - 4\sqrt{3}}$$

4. Plot $\sqrt[3]{V}/k$ as a function of $x = h/r$ and use your graph to argue that when a can is large or joining is cheap, we should make h/r approximately 2.21 (as in Problem 2). But when the can is small or joining is costly, h/r should be substantially larger.

5. Our analysis shows that large cans should be almost square but small cans should be tall and thin. Take a look at the relative shapes of the cans in a supermarket. Is our conclusion usually true in practice? Are there exceptions? Can you suggest reasons why small cans are not always tall and thin?

APPLIED PROJECT   **PLANES AND BIRDS: MINIMIZING ENERGY**

Small birds like finches alternate between flapping their wings and keeping them folded while gliding (see Figure 1). In this project we analyze this phenomenon and try to determine how frequently a bird should flap its wings. Some of the principles are the same as for fixed-wing aircraft and so we begin by considering how required power and energy depend on the speed of airplanes.[1]

Path of wing tip

FIGURE 1

1. The power needed to propel an airplane forward at velocity v is

$$P = Av^3 + \frac{BL^2}{v}$$

where A and B are positive constants specific to the particular aircraft and L is the lift; the upward force supporting the weight of the plane. Find the speed that minimizes the required power.

2. The speed found in Problem 1 minimizes power but a faster speed might use less fuel. The energy needed to propel the airplane a unit distance is $E = P/v$. At what speed is energy minimized?

3. Hows much faster is the speed for minimum energy than the speed for minimum power?

4. In applying the equation of Problem 1 to bird flight we split the term Av^3 into two parts: $A_b v^3$ for the bird's body and $A_w v^3$ for its wings. Let x be the fraction of flying time spent in flapping mode. If m is the bird's mass and all the lift occurs during flapping, then the lift is mg/x and so the power needed during flapping is

$$P_{\text{flap}} = (A_b + A_w)v^3 + \frac{B(mg/x)^2}{v}$$

The power while wings are folded is $P_{\text{fold}} = A_b v^3$. Show that the average power over an entire flight cycle is

$$\overline{P} = xP_{\text{flap}} + (1 - x)P_{\text{fold}} = A_b v^3 + xA_w v^3 + \frac{Bm^2 g^2}{xv}$$

5. For what value of x is the average power a minimum? What can you conclude if the bird flies slowly? What can you conclude if the bird flies faster and faster?

6. The average energy over a cycle is $\overline{E} = \overline{P}/v$. What value of x minimizes $\overline{E}$?

---

1. Adapted from R. McNeill Alexander, *Optima for Animals* (Princeton, NJ: Princeton University Press, 1996.)

4.8   Newton's Method

Suppose that a car dealer offers to sell you a car for $18,000 or for payments of $375 per month for five years. You would like to know what monthly interest rate the dealer is, in effect, charging you. To find the answer, you have to solve the equation

$$\boxed{1} \qquad 48x(1 + x)^{60} - (1 + x)^{60} + 1 = 0$$

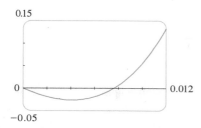

0.15

0

0.012

−0.05

FIGURE 1

Try to solve Equation 1 numerically using your calculator or computer. Some machines are not able to solve it. Others are successful but require you to specify a starting point for the search.

(The details are explained in Exercise 41.) How would you solve such an equation?

For a quadratic equation $ax^2 + bx + c = 0$ there is a well-known formula for the solutions. For third- and fourth-degree equations there are also formulas for the solutions, but they are extremely complicated. If f is a polynomial of degree 5 or higher, there is no such formula (see the note on page 211). Likewise, there is no formula that will enable us to find the exact roots of a transcendental equation such as $\cos x = x$.

We can find an *approximate* solution to Equation 1 by plotting the left side of the equation. Using a graphing device, and after experimenting with viewing rectangles, we produce the graph in Figure 1.

We see that in addition to the solution $x = 0$, which doesn't interest us, there is a solution between 0.007 and 0.008. Zooming in shows that the root is approximately 0.0076. If we need more accuracy we could zoom in repeatedly, but that becomes tiresome. A faster alternative is to use a calculator or computer algebra system to solve the equation numerically. If we do so, we find that the root, correct to nine decimal places, is 0.007628603.

How do these devices solve equations? They use a variety of methods, but most of them make some use of **Newton's method**, also called the **Newton-Raphson method**. We will explain how this method works, partly to show what happens inside a calculator or computer, and partly as an application of the idea of linear approximation.

The geometry behind Newton's method is shown in Figure 2. We wish to solve an equation of the form $f(x) = 0$, so the roots of the equation correspond to the x-intercepts of the graph of f. The root that we are trying to find is labeled r in the figure. We start with a first approximation x_1, which is obtained by guessing, or from a rough sketch of the graph of f, or from a computer-generated graph of f. Consider the tangent line L to the curve $y = f(x)$ at the point $(x_1, f(x_1))$ and look at the x-intercept of L, labeled x_2. The idea behind Newton's method is that the tangent line is close to the curve and so its x-intercept, x_2, is close to the x-intercept of the curve (namely, the root r that we are seeking). Because the tangent is a line, we can easily find its x-intercept.

To find a formula for x_2 in terms of x_1 we use the fact that the slope of L is $f'(x_1)$, so its equation is

$$y - f(x_1) = f'(x_1)(x - x_1)$$

Since the x-intercept of L is x_2, we know that the point $(x_2, 0)$ is on the line, and so

$$0 - f(x_1) = f'(x_1)(x_2 - x_1)$$

If $f'(x_1) \neq 0$, we can solve this equation for x_2:

$$x_2 = x_1 - \frac{f(x_1)}{f'(x_1)}$$

We use x_2 as a second approximation to r.

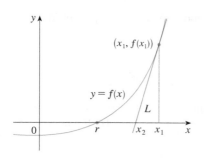

y

$(x_1, f(x_1))$

$y = f(x)$

L

0 r x_2 x_1 x

FIGURE 2

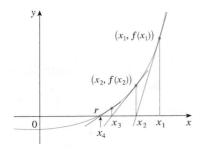

FIGURE 3

Sequences were briefly introduced in *A Preview of Calculus* on page 5. A more thorough discussion starts in Section 11.1.

FIGURE 4

TEC In Module 4.8 you can investigate how Newton's method works for several functions and what happens when you change x_1.

Figure 5 shows the geometry behind the first step in Newton's method in Example 1. Since $f'(2) = 10$, the tangent line to $y = x^3 - 2x - 5$ at $(2, -1)$ has equation $y = 10x - 21$ so its x-intercept is $x_2 = 2.1$.

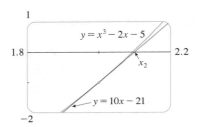

FIGURE 5

Next we repeat this procedure with x_1 replaced by the second approximation x_2, using the tangent line at $(x_2, f(x_2))$. This gives a third approximation:

$$x_3 = x_2 - \frac{f(x_2)}{f'(x_2)}$$

If we keep repeating this process, we obtain a sequence of approximations $x_1, x_2, x_3, x_4, \ldots$ as shown in Figure 3. In general, if the nth approximation is x_n and $f'(x_n) \neq 0$, then the next approximation is given by

$$\boxed{2} \qquad \boxed{x_{n+1} = x_n - \frac{f(x_n)}{f'(x_n)}}$$

If the numbers x_n become closer and closer to r as n becomes large, then we say that the sequence *converges* to r and we write

$$\lim_{n \to \infty} x_n = r$$

⊘ Although the sequence of successive approximations converges to the desired root for functions of the type illustrated in Figure 3, in certain circumstances the sequence may not converge. For example, consider the situation shown in Figure 4. You can see that x_2 is a worse approximation than x_1. This is likely to be the case when $f'(x_1)$ is close to 0. It might even happen that an approximation (such as x_3 in Figure 4) falls outside the domain of f. Then Newton's method fails and a better initial approximation x_1 should be chosen. See Exercises 31–34 for specific examples in which Newton's method works very slowly or does not work at all.

EXAMPLE 1 Starting with $x_1 = 2$, find the third approximation x_3 to the root of the equation $x^3 - 2x - 5 = 0$.

SOLUTION We apply Newton's method with

$$f(x) = x^3 - 2x - 5 \qquad \text{and} \qquad f'(x) = 3x^2 - 2$$

Newton himself used this equation to illustrate his method and he chose $x_1 = 2$ after some experimentation because $f(1) = -6$, $f(2) = -1$, and $f(3) = 16$. Equation 2 becomes

$$x_{n+1} = x_n - \frac{f(x_n)}{f'(x_n)} = x_n - \frac{x_n^3 - 2x_n - 5}{3x_n^2 - 2}$$

With $n = 1$ we have

$$x_2 = x_1 - \frac{f(x_1)}{f'(x_1)} = x_1 - \frac{x_1^3 - 2x_1 - 5}{3x_1^2 - 2}$$

$$= 2 - \frac{2^3 - 2(2) - 5}{3(2)^2 - 2} = 2.1$$

Then with $n = 2$ we obtain

$$x_3 = x_2 - \frac{x_2^3 - 2x_2 - 5}{3x_2^2 - 2} = 2.1 - \frac{(2.1)^3 - 2(2.1) - 5}{3(2.1)^2 - 2} \approx 2.0946$$

It turns out that this third approximation $x_3 \approx 2.0946$ is accurate to four decimal places.   ∎

Suppose that we want to achieve a given accuracy, say to eight decimal places, using Newton's method. How do we know when to stop? The rule of thumb that is generally used is that we can stop when successive approximations x_n and x_{n+1} agree to eight decimal places. (A precise statement concerning accuracy in Newton's method will be given in Exercise 11.11.39.)

Notice that the procedure in going from n to $n + 1$ is the same for all values of n. (It is called an *iterative* process.) This means that Newton's method is particularly convenient for use with a programmable calculator or a computer.

EXAMPLE 2 Use Newton's method to find $\sqrt[6]{2}$ correct to eight decimal places.

SOLUTION First we observe that finding $\sqrt[6]{2}$ is equivalent to finding the positive root of the equation

$$x^6 - 2 = 0$$

so we take $f(x) = x^6 - 2$. Then $f'(x) = 6x^5$ and Formula 2 (Newton's method) becomes

$$x_{n+1} = x_n - \frac{f(x_n)}{f'(x_n)} = x_n - \frac{x_n^6 - 2}{6x_n^5}$$

If we choose $x_1 = 1$ as the initial approximation, then we obtain

$$x_2 \approx 1.16666667$$
$$x_3 \approx 1.12644368$$
$$x_4 \approx 1.12249707$$
$$x_5 \approx 1.12246205$$
$$x_6 \approx 1.12246205$$

Since x_5 and x_6 agree to eight decimal places, we conclude that

$$\sqrt[6]{2} \approx 1.12246205$$

to eight decimal places.   ∎

EXAMPLE 3 Find, correct to six decimal places, the root of the equation $\cos x = x$.

SOLUTION We first rewrite the equation in standard form:

$$\cos x - x = 0$$

Therefore we let $f(x) = \cos x - x$. Then $f'(x) = -\sin x - 1$, so Formula 2 becomes

$$x_{n+1} = x_n - \frac{\cos x_n - x_n}{-\sin x_n - 1} = x_n + \frac{\cos x_n - x_n}{\sin x_n + 1}$$

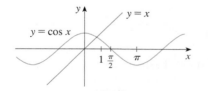

FIGURE 6

In order to guess a suitable value for x_1 we sketch the graphs of $y = \cos x$ and $y = x$ in Figure 6. It appears that they intersect at a point whose x-coordinate is somewhat less than 1, so let's take $x_1 = 1$ as a convenient first approximation. Then, remembering to put our calculator in radian mode, we get

$$x_2 \approx 0.75036387$$

$$x_3 \approx 0.73911289$$

$$x_4 \approx 0.73908513$$

$$x_5 \approx 0.73908513$$

Since x_4 and x_5 agree to six decimal places (eight, in fact), we conclude that the root of the equation, correct to six decimal places, is 0.739085.   ■

Instead of using the rough sketch in Figure 6 to get a starting approximation for Newton's method in Example 3, we could have used the more accurate graph that a calculator or computer provides. Figure 7 suggests that we use $x_1 = 0.75$ as the initial approximation. Then Newton's method gives

$$x_2 \approx 0.73911114$$

$$x_3 \approx 0.73908513$$

$$x_4 \approx 0.73908513$$

and so we obtain the same answer as before, but with one fewer step.

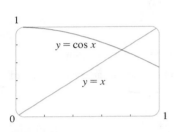

FIGURE 7

4.8   EXERCISES

1. The figure shows the graph of a function f. Suppose that Newton's method is used to approximate the root s of the equation $f(x) = 0$ with initial approximation $x_1 = 6$.
   (a) Draw the tangent lines that are used to find x_2 and x_3, and estimate the numerical values of x_2 and x_3.
   (b) Would $x_1 = 8$ be a better first approximation? Explain.

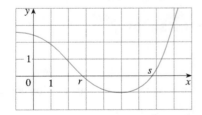

2. Follow the instructions for Exercise 1(a) but use $x_1 = 1$ as the starting approximation for finding the root r.

3. Suppose the tangent line to the curve $y = f(x)$ at the point $(2, 5)$ has the equation $y = 9 - 2x$. If Newton's method is used to locate a root of the equation $f(x) = 0$ and the initial approximation is $x_1 = 2$, find the second approximation x_2.

4. For each initial approximation, determine graphically what happens if Newton's method is used for the function whose graph is shown.

(a) $x_1 = 0$ (b) $x_1 = 1$ (c) $x_1 = 3$
(d) $x_1 = 4$ (e) $x_1 = 5$

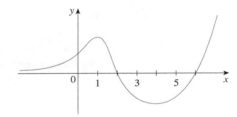

5. For which of the initial approximations $x_1 = a, b, c,$ and d do you think Newton's method will work and lead to the root of the equation $f(x) = 0$?

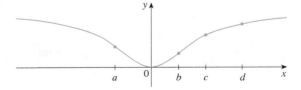

6–8 Use Newton's method with the specified initial approximation x_1 to find x_3, the third approximation to the root of the given equation. (Give your answer to four decimal places.)

6. $2x^3 - 3x^2 + 2 = 0$, $x_1 = -1$

7. $\dfrac{2}{x} - x^2 + 1 = 0$, $x_1 = 2$ **8.** $x^7 + 4 = 0$, $x_1 = -1$

9. Use Newton's method with initial approximation $x_1 = -1$ to find x_2, the second approximation to the root of the equation $x^3 + x + 3 = 0$. Explain how the method works by first graphing the function and its tangent line at $(-1, 1)$.

10. Use Newton's method with initial approximation $x_1 = 1$ to find x_2, the second approximation to the root of the equation $x^4 - x - 1 = 0$. Explain how the method works by first graphing the function and its tangent line at $(1, -1)$.

11–12 Use Newton's method to approximate the given number correct to eight decimal places.

11. $\sqrt[4]{75}$ **12.** $\sqrt[8]{500}$

13–14 (a) Explain how we know that the given equation must have a root in the given interval. (b) Use Newton's method to approximate the root correct to six decimal places.

13. $3x^4 - 8x^3 + 2 = 0$, $[2, 3]$

14. $-2x^5 + 9x^4 - 7x^3 - 11x = 0$, $[3, 4]$

15–16 Use Newton's method to approximate the indicated root of the equation correct to six decimal places.

15. The negative root of $e^x = 4 - x^2$

16. The positive root of $3 \sin x = x$

17–22 Use Newton's method to find all solutions of the equation correct to six decimal places.

17. $3 \cos x = x + 1$ **18.** $\sqrt{x + 1} = x^2 - x$

19. $2^x = 2 - x^2$ **20.** $\ln x = \dfrac{1}{x - 3}$

21. $x^3 = \tan^{-1}x$ **22.** $\sin x = x^2 - 2$

23–28 Use Newton's method to find all the solutions of the equation correct to eight decimal places. Start by drawing a graph to find initial approximations.

23. $-2x^7 - 5x^4 + 9x^3 + 5 = 0$

24. $x^5 - 3x^4 + x^3 - x^2 - x + 6 = 0$

25. $\dfrac{x}{x^2 + 1} = \sqrt{1 - x}$

26. $\cos(x^2 - x) = x^4$

27. $4e^{-x^2} \sin x = x^2 - x + 1$

28. $\ln(x^2 + 2) = \dfrac{3x}{\sqrt{x^2 + 1}}$

29. (a) Apply Newton's method to the equation $x^2 - a = 0$ to derive the following square-root algorithm (used by the ancient Babylonians to compute $\sqrt{a}$):

$$x_{n+1} = \frac{1}{2}\left(x_n + \frac{a}{x_n}\right)$$

(b) Use part (a) to compute $\sqrt{1000}$ correct to six decimal places.

30. (a) Apply Newton's method to the equation $1/x - a = 0$ to derive the following reciprocal algorithm:

$$x_{n+1} = 2x_n - ax_n^2$$

(This algorithm enables a computer to find reciprocals without actually dividing.)

(b) Use part (a) to compute $1/1.6984$ correct to six decimal places.

31. Explain why Newton's method doesn't work for finding the root of the equation

$$x^3 - 3x + 6 = 0$$

if the initial approximation is chosen to be $x_1 = 1$.

32. (a) Use Newton's method with $x_1 = 1$ to find the root of the equation $x^3 - x = 1$ correct to six decimal places.
(b) Solve the equation in part (a) using $x_1 = 0.6$ as the initial approximation.
(c) Solve the equation in part (a) using $x_1 = 0.57$. (You definitely need a programmable calculator for this part.)
(d) Graph $f(x) = x^3 - x - 1$ and its tangent lines at $x_1 = 1, 0.6$, and 0.57 to explain why Newton's method is so sensitive to the value of the initial approximation.

33. Explain why Newton's method fails when applied to the equation $\sqrt[3]{x} = 0$ with any initial approximation $x_1 \neq 0$. Illustrate your explanation with a sketch.

34. If

$$f(x) = \begin{cases} \sqrt{x} & \text{if } x \geq 0 \\ -\sqrt{-x} & \text{if } x < 0 \end{cases}$$

then the root of the equation $f(x) = 0$ is $x = 0$. Explain why Newton's method fails to find the root no matter which initial approximation $x_1 \neq 0$ is used. Illustrate your explanation with a sketch.

35. (a) Use Newton's method to find the critical numbers of the function

$$f(x) = x^6 - x^4 + 3x^3 - 2x$$

correct to six decimal places.
(b) Find the absolute minimum value of f correct to four decimal places.

36. Use Newton's method to find the absolute maximum value of the function $f(x) = x \cos x$, $0 \leqslant x \leqslant \pi$, correct to six decimal places.

37. Use Newton's method to find the coordinates of the inflection point of the curve $y = x^2 \sin x$, $0 \leqslant x \leqslant \pi$, correct to six decimal places.

38. Of the infinitely many lines that are tangent to the curve $y = -\sin x$ and pass through the origin, there is one that has the largest slope. Use Newton's method to find the slope of that line correct to six decimal places.

39. Use Newton's method to find the coordinates, correct to six decimal places, of the point on the parabola $y = (x - 1)^2$ that is closest to the origin.

40. In the figure, the length of the chord AB is 4 cm and the length of the arc AB is 5 cm. Find the central angle θ, in radians, correct to four decimal places. Then give the answer to the nearest degree.

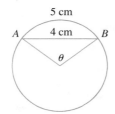

41. A car dealer sells a new car for $18,000. He also offers to sell the same car for payments of $375 per month for five years. What monthly interest rate is this dealer charging?

To solve this problem you will need to use the formula for the present value A of an annuity consisting of n equal payments of size R with interest rate i per time period:

$$A = \frac{R}{i}\left[1 - (1 + i)^{-n}\right]$$

Replacing i by x, show that

$$48x(1 + x)^{60} - (1 + x)^{60} + 1 = 0$$

Use Newton's method to solve this equation.

42. The figure shows the sun located at the origin and the earth at the point $(1, 0)$. (The unit here is the distance between the centers of the earth and the sun, called an *astronomical unit:* 1 AU $\approx$ 1.496 $\times$ 10^8 km.) There are five locations L_1, L_2, L_3, L_4, and L_5 in this plane of rotation of the earth about the sun where a satellite remains motionless with respect to the earth because the forces acting on the satellite (including the gravitational attractions of the earth and the sun) balance each other. These locations are called *libration points.* (A solar research satellite has been placed at one of these libration points.) If m_1 is the mass of the sun, m_2 is the mass of the earth, and $r = m_2/(m_1 + m_2)$, it turns out that the x-coordinate of L_1 is the unique root of the fifth-degree equation

$$p(x) = x^5 - (2 + r)x^4 + (1 + 2r)x^3 - (1 - r)x^2$$
$$+ 2(1 - r)x + r - 1 = 0$$

and the x-coordinate of L_2 is the root of the equation

$$p(x) - 2rx^2 = 0$$

Using the value $r \approx 3.04042 \times 10^{-6}$, find the locations of the libration points (a) L_1 and (b) L_2.

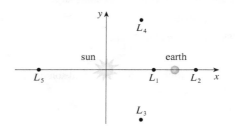

4.9   Antiderivatives

A physicist who knows the velocity of a particle might wish to know its position at a given time. An engineer who can measure the variable rate at which water is leaking from a tank wants to know the amount leaked over a certain time period. A biologist who knows the rate at which a bacteria population is increasing might want to deduce what the size of the population will be at some future time. In each case, the problem is to find a function F whose derivative is a known function f. If such a function F exists, it is called an *antiderivative* of f.

> **Definition** A function F is called an **antiderivative** of f on an interval I if $F'(x) = f(x)$ for all x in I.

For instance, let $f(x) = x^2$. It isn't difficult to discover an antiderivative of f if we keep the Power Rule in mind. In fact, if $F(x) = \frac{1}{3}x^3$, then $F'(x) = x^2 = f(x)$. But the function $G(x) = \frac{1}{3}x^3 + 100$ also satisfies $G'(x) = x^2$. Therefore both F and G are antiderivatives of f. Indeed, any function of the form $H(x) = \frac{1}{3}x^3 + C$, where C is a constant, is an antiderivative of f. The question arises: Are there any others?

To answer this question, recall that in Section 4.2 we used the Mean Value Theorem to prove that if two functions have identical derivatives on an interval, then they must differ by a constant (Corollary 4.2.7). Thus if F and G are any two antiderivatives of f, then

$$F'(x) = f(x) = G'(x)$$

so $G(x) - F(x) = C$, where C is a constant. We can write this as $G(x) = F(x) + C$, so we have the following result.

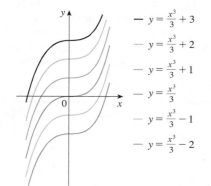

- $y = \frac{x^3}{3} + 3$
- $y = \frac{x^3}{3} + 2$
- $y = \frac{x^3}{3} + 1$
- $y = \frac{x^3}{3}$
- $y = \frac{x^3}{3} - 1$
- $y = \frac{x^3}{3} - 2$

FIGURE 1
Members of the family of antiderivatives of $f(x) = x^2$

> **1 Theorem** If F is an antiderivative of f on an interval I, then the most general antiderivative of f on I is
>
> $$F(x) + C$$
>
> where C is an arbitrary constant.

Going back to the function $f(x) = x^2$, we see that the general antiderivative of f is $\frac{1}{3}x^3 + C$. By assigning specific values to the constant C, we obtain a family of functions whose graphs are vertical translates of one another (see Figure 1). This makes sense because each curve must have the same slope at any given value of x.

EXAMPLE 1 Find the most general antiderivative of each of the following functions.
(a) $f(x) = \sin x$ (b) $f(x) = 1/x$ (c) $f(x) = x^n$, $n \neq -1$

SOLUTION
(a) If $F(x) = -\cos x$, then $F'(x) = \sin x$, so an antiderivative of $\sin x$ is $-\cos x$. By Theorem 1, the most general antiderivative is $G(x) = -\cos x + C$.
(b) Recall from Section 3.6 that

$$\frac{d}{dx}(\ln x) = \frac{1}{x}$$

So on the interval $(0, \infty)$ the general antiderivative of $1/x$ is $\ln x + C$. We also learned that

$$\frac{d}{dx}(\ln |x|) = \frac{1}{x}$$

for all $x \neq 0$. Theorem 1 then tells us that the general antiderivative of $f(x) = 1/x$ is $\ln |x| + C$ on any interval that doesn't contain 0. In particular, this is true on each of the intervals $(-\infty, 0)$ and $(0, \infty)$. So the general antiderivative of f is

$$F(x) = \begin{cases} \ln x + C_1 & \text{if } x > 0 \\ \ln(-x) + C_2 & \text{if } x < 0 \end{cases}$$

(c) We use the Power Rule to discover an antiderivative of x^n. In fact, if $n \neq -1$, then

$$\frac{d}{dx}\left(\frac{x^{n+1}}{n+1} \right) = \frac{(n+1)x^n}{n+1} = x^n$$

Therefore the general antiderivative of $f(x) = x^n$ is

$$F(x) = \frac{x^{n+1}}{n+1} + C$$

This is valid for $n \geq 0$ since then $f(x) = x^n$ is defined on an interval. If n is negative (but $n \neq -1$), it is valid on any interval that doesn't contain 0. ∎

As in Example 1, every differentiation formula, when read from right to left, gives rise to an antidifferentiation formula. In Table 2 we list some particular antiderivatives. Each formula in the table is true because the derivative of the function in the right column appears in the left column. In particular, the first formula says that the antiderivative of a constant times a function is the constant times the antiderivative of the function. The second formula says that the antiderivative of a sum is the sum of the antiderivatives. (We use the notation $F' = f$, $G' = g$.)

2 Table of Antidifferentiation Formulas

To obtain the most general antiderivative from the particular ones in Table 2, we have to add a constant (or constants), as in Example 1.

Function	Particular antiderivative	Function	Particular antiderivative
$cf(x)$	$cF(x)$	$\sin x$	$-\cos x$
$f(x) + g(x)$	$F(x) + G(x)$	$\sec^2 x$	$\tan x$
x^n $(n \neq -1)$	$\dfrac{x^{n+1}}{n+1}$	$\sec x \tan x$	$\sec x$
$\dfrac{1}{x}$	$\ln|x|$	$\dfrac{1}{\sqrt{1-x^2}}$	$\sin^{-1}x$
e^x	e^x	$\dfrac{1}{1+x^2}$	$\tan^{-1}x$
b^x	$\dfrac{b^x}{\ln b}$	$\cosh x$	$\sinh x$
$\cos x$	$\sin x$	$\sinh x$	$\cosh x$

EXAMPLE 2 Find all functions g such that

$$g'(x) = 4\sin x + \frac{2x^5 - \sqrt{x}}{x}$$

SOLUTION We first rewrite the given function as follows:

$$g'(x) = 4\sin x + \frac{2x^5}{x} - \frac{\sqrt{x}}{x} = 4\sin x + 2x^4 - \frac{1}{\sqrt{x}}$$

Thus we want to find an antiderivative of

$$g'(x) = 4\sin x + 2x^4 - x^{-1/2}$$

Using the formulas in Table 2 together with Theorem 1, we obtain

We often use a capital letter F to represent an antiderivative of a function f. If we begin with derivative notation, f', an antiderivative is f, of course.

$$g(x) = 4(-\cos x) + 2\frac{x^5}{5} - \frac{x^{1/2}}{\frac{1}{2}} + C$$

$$= -4\cos x + \tfrac{2}{5}x^5 - 2\sqrt{x} + C \qquad ∎$$

In applications of calculus it is very common to have a situation as in Example 2, where it is required to find a function, given knowledge about its derivatives. An equation that involves the derivatives of a function is called a **differential equation**. These will be studied in some detail in Chapter 9, but for the present we can solve some elementary differential equations. The general solution of a differential equation involves an arbitrary constant (or constants) as in Example 2. However, there may be some extra conditions given that will determine the constants and therefore uniquely specify the solution.

EXAMPLE 3 Find f if $f'(x) = e^x + 20(1 + x^2)^{-1}$ and $f(0) = -2$.

SOLUTION The general antiderivative of

$$f'(x) = e^x + \frac{20}{1 + x^2}$$

Figure 2 shows the graphs of the function f' in Example 3 and its antiderivative f. Notice that $f'(x) > 0$, so f is always increasing. Also notice that when f' has a maximum or minimum, f appears to have an inflection point. So the graph serves as a check on our calculation.

is

$$f(x) = e^x + 20 \tan^{-1}x + C$$

To determine C we use the fact that $f(0) = -2$:

$$f(0) = e^0 + 20 \tan^{-1} 0 + C = -2$$

Thus we have $C = -2 - 1 = -3$, so the particular solution is

$$f(x) = e^x + 20 \tan^{-1}x - 3$$ ∎

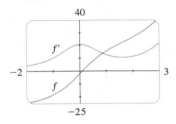

FIGURE 2

EXAMPLE 4 Find f if $f''(x) = 12x^2 + 6x - 4$, $f(0) = 4$, and $f(1) = 1$.

SOLUTION The general antiderivative of $f''(x) = 12x^2 + 6x - 4$ is

$$f'(x) = 12\frac{x^3}{3} + 6\frac{x^2}{2} - 4x + C = 4x^3 + 3x^2 - 4x + C$$

Using the antidifferentiation rules once more, we find that

$$f(x) = 4\frac{x^4}{4} + 3\frac{x^3}{3} - 4\frac{x^2}{2} + Cx + D = x^4 + x^3 - 2x^2 + Cx + D$$

To determine C and D we use the given conditions that $f(0) = 4$ and $f(1) = 1$. Since $f(0) = 0 + D = 4$, we have $D = 4$. Since

$$f(1) = 1 + 1 - 2 + C + 4 = 1$$

we have $C = -3$. Therefore the required function is

$$f(x) = x^4 + x^3 - 2x^2 - 3x + 4$$ ∎

If we are given the graph of a function f, it seems reasonable that we should be able to sketch the graph of an antiderivative F. Suppose, for instance, that we are given that $F(0) = 1$. Then we have a place to start, the point $(0, 1)$, and the direction in which we move our pencil is given at each stage by the derivative $F'(x) = f(x)$. In the next example we use the principles of this chapter to show how to graph F even when we don't have a formula for f. This would be the case, for instance, when $f(x)$ is determined by experimental data.

FIGURE 3

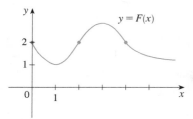

FIGURE 4

EXAMPLE 5 The graph of a function f is given in Figure 3. Make a rough sketch of an antiderivative F, given that $F(0) = 2$.

SOLUTION We are guided by the fact that the slope of $y = F(x)$ is $f(x)$. We start at the point $(0, 2)$ and draw F as an initially decreasing function since $f(x)$ is negative when $0 < x < 1$. Notice that $f(1) = f(3) = 0$, so F has horizontal tangents when $x = 1$ and $x = 3$. For $1 < x < 3$, $f(x)$ is positive and so F is increasing. We see that F has a local minimum when $x = 1$ and a local maximum when $x = 3$. For $x > 3$, $f(x)$ is negative and so F is decreasing on $(3, \infty)$. Since $f(x) \to 0$ as $x \to \infty$, the graph of F becomes flatter as $x \to \infty$. Also notice that $F''(x) = f'(x)$ changes from positive to negative at $x = 2$ and from negative to positive at $x = 4$, so F has inflection points when $x = 2$ and $x = 4$. We use this information to sketch the graph of the antiderivative in Figure 4.   ■

■ Rectilinear Motion

Antidifferentiation is particularly useful in analyzing the motion of an object moving in a straight line. Recall that if the object has position function $s = f(t)$, then the velocity function is $v(t) = s'(t)$. This means that the position function is an antiderivative of the velocity function. Likewise, the acceleration function is $a(t) = v'(t)$, so the velocity function is an antiderivative of the acceleration. If the acceleration and the initial values $s(0)$ and $v(0)$ are known, then the position function can be found by antidifferentiating twice.

EXAMPLE 6 A particle moves in a straight line and has acceleration given by $a(t) = 6t + 4$. Its initial velocity is $v(0) = -6$ cm/s and its initial displacement is $s(0) = 9$ cm. Find its position function $s(t)$.

SOLUTION Since $v'(t) = a(t) = 6t + 4$, antidifferentiation gives

$$v(t) = 6\frac{t^2}{2} + 4t + C = 3t^2 + 4t + C$$

Note that $v(0) = C$. But we are given that $v(0) = -6$, so $C = -6$ and

$$v(t) = 3t^2 + 4t - 6$$

Since $v(t) = s'(t)$, s is the antiderivative of v:

$$s(t) = 3\frac{t^3}{3} + 4\frac{t^2}{2} - 6t + D = t^3 + 2t^2 - 6t + D$$

This gives $s(0) = D$. We are given that $s(0) = 9$, so $D = 9$ and the required position function is

$$s(t) = t^3 + 2t^2 - 6t + 9$$ ■

An object near the surface of the earth is subject to a gravitational force that produces a downward acceleration denoted by g. For motion close to the ground we may assume that g is constant, its value being about 9.8 m/s² (or 32 ft/s²).

EXAMPLE 7 A ball is thrown upward with a speed of 48 ft/s from the edge of a cliff 432 ft above the ground. Find its height above the ground t seconds later. When does it reach its maximum height? When does it hit the ground?

SOLUTION The motion is vertical and we choose the positive direction to be upward. At time t the distance above the ground is $s(t)$ and the velocity $v(t)$ is decreasing. Therefore the acceleration must be negative and we have

$$a(t) = \frac{dv}{dt} = -32$$

Taking antiderivatives, we have

$$v(t) = -32t + C$$

To determine C we use the given information that $v(0) = 48$. This gives $48 = 0 + C$, so

$$v(t) = -32t + 48$$

The maximum height is reached when $v(t) = 0$, that is, after 1.5 seconds. Since $s'(t) = v(t)$, we antidifferentiate again and obtain

$$s(t) = -16t^2 + 48t + D$$

Using the fact that $s(0) = 432$, we have $432 = 0 + D$ and so

$$s(t) = -16t^2 + 48t + 432$$

Figure 5 shows the position function of the ball in Example 7. The graph corroborates the conclusions we reached: The ball reaches its maximum height after 1.5 seconds and hits the ground after about 6.9 seconds.

The expression for $s(t)$ is valid until the ball hits the ground. This happens when $s(t) = 0$, that is, when

$$-16t^2 + 48t + 432 = 0$$

or, equivalently,

$$t^2 - 3t - 27 = 0$$

Using the quadratic formula to solve this equation, we get

$$t = \frac{3 \pm 3\sqrt{13}}{2}$$

We reject the solution with the minus sign since it gives a negative value for t. Therefore the ball hits the ground after $3(1 + \sqrt{13})/2 \approx 6.9$ seconds. ∎

500

0   8

FIGURE 5

4.9 EXERCISES

1–22 Find the most general antiderivative of the function. (Check your answer by differentiation.)

1. $f(x) = 4x + 7$

2. $f(x) = x^2 - 3x + 2$

3. $f(x) = 2x^3 - \frac{2}{3}x^2 + 5x$

4. $f(x) = 6x^5 - 8x^4 - 9x^2$

5. $f(x) = x(12x + 8)$

6. $f(x) = (x - 5)^2$

7. $f(x) = 7x^{2/5} + 8x^{-4/5}$

8. $f(x) = x^{3.4} - 2x^{\sqrt{2}-1}$

9. $f(x) = \sqrt{2}$

10. $f(x) = e^2$

11. $f(x) = 3\sqrt{x} - 2\sqrt[3]{x}$

12. $f(x) = \sqrt[3]{x^2} + x\sqrt{x}$

13. $f(x) = \dfrac{1}{5} - \dfrac{2}{x}$

14. $f(t) = \dfrac{3t^4 - t^3 + 6t^2}{t^4}$

15. $g(t) = \dfrac{1 + t + t^2}{\sqrt{t}}$

16. $r(\theta) = \sec\theta \tan\theta - 2e^\theta$

17. $h(\theta) = 2\sin\theta - \sec^2\theta$

18. $g(v) = 2\cos v - \dfrac{3}{\sqrt{1 - v^2}}$

19. $f(x) = 2^x + 4\sinh x$

20. $f(x) = 1 + 2\sin x + 3/\sqrt{x}$

21. $f(x) = \dfrac{2x^4 + 4x^3 - x}{x^3}$, $x > 0$

22. $f(x) = \dfrac{2x^2 + 5}{x^2 + 1}$

23–24 Find the antiderivative F of f that satisfies the given condition. Check your answer by comparing the graphs of f and F.

23. $f(x) = 5x^4 - 2x^5$, $F(0) = 4$

24. $f(x) = 4 - 3(1 + x^2)^{-1}$, $F(1) = 0$

25–48 Find f.

25. $f''(x) = 20x^3 - 12x^2 + 6x$

26. $f''(x) = x^6 - 4x^4 + x + 1$

27. $f''(x) = 2x + 3e^x$ **28.** $f''(x) = 1/x^2$

29. $f'''(t) = 12 + \sin t$ **30.** $f'''(t) = \sqrt{t} - 2\cos t$

31. $f'(x) = 1 + 3\sqrt{x}$, $f(4) = 25$

32. $f'(x) = 5x^4 - 3x^2 + 4$, $f(-1) = 2$

33. $f'(t) = 4/(1 + t^2)$, $f(1) = 0$

34. $f'(t) = t + 1/t^3$, $t > 0$, $f(1) = 6$

35. $f'(x) = 5x^{2/3}$, $f(8) = 21$

36. $f'(x) = (x + 1)/\sqrt{x}$, $f(1) = 5$

37. $f'(t) = \sec t\,(\sec t + \tan t)$, $-\pi/2 < t < \pi/2$, $f(\pi/4) = -1$

38. $f'(t) = 3^t - 3/t$, $f(1) = 2$, $f(-1) = 1$

39. $f''(x) = -2 + 12x - 12x^2$, $f(0) = 4$, $f'(0) = 12$

40. $f''(x) = 8x^3 + 5$, $f(1) = 0$, $f'(1) = 8$

41. $f''(\theta) = \sin\theta + \cos\theta$, $f(0) = 3$, $f'(0) = 4$

42. $f''(t) = t^2 + 1/t^2$, $t > 0$, $f(2) = 3$, $f'(1) = 2$

43. $f''(x) = 4 + 6x + 24x^2$, $f(0) = 3$, $f(1) = 10$

44. $f''(x) = x^3 + \sinh x$, $f(0) = 1$, $f(2) = 2.6$

45. $f''(x) = e^x - 2\sin x$, $f(0) = 3$, $f(\pi/2) = 0$

46. $f''(t) = \sqrt[3]{t} - \cos t$, $f(0) = 2$, $f(1) = 2$

47. $f''(x) = x^{-2}$, $x > 0$, $f(1) = 0$, $f(2) = 0$

48. $f'''(x) = \cos x$, $f(0) = 1$, $f'(0) = 2$, $f''(0) = 3$

---

49. Given that the graph of f passes through the point $(2, 5)$ and that the slope of its tangent line at $(x, f(x))$ is $3 - 4x$, find $f(1)$.

50. Find a function f such that $f'(x) = x^3$ and the line $x + y = 0$ is tangent to the graph of f.

51–52 The graph of a function f is shown. Which graph is an antiderivative of f and why?

51.

52.

53. The graph of a function is shown in the figure. Make a rough sketch of an antiderivative F, given that $F(0) = 1$.

54. The graph of the velocity function of a particle is shown in the figure. Sketch the graph of a position function.

55. The graph of f' is shown in the figure. Sketch the graph of f if f is continuous and $f(0) = -1$.

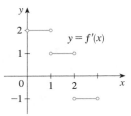

56. (a) Use a graphing device to graph $f(x) = 2x - 3\sqrt{x}$.
 (b) Starting with the graph in part (a), sketch a rough graph of the antiderivative F that satisfies $F(0) = 1$.
 (c) Use the rules of this section to find an expression for $F(x)$.
 (d) Graph F using the expression in part (c). Compare with your sketch in part (b).

57–58 Draw a graph of f and use it to make a rough sketch of the antiderivative that passes through the origin.

57. $f(x) = \dfrac{\sin x}{1 + x^2}$, $-2\pi \le x \le 2\pi$

58. $f(x) = \sqrt{x^4 - 2x^2 + 2} - 2$, $-3 \le x \le 3$

59–64 A particle is moving with the given data. Find the position of the particle.

59. $v(t) = \sin t - \cos t$, $\quad s(0) = 0$

60. $v(t) = t^2 - 3\sqrt{t}$, $\quad s(4) = 8$

61. $a(t) = 2t + 1$, $\quad s(0) = 3$, $\quad v(0) = -2$

62. $a(t) = 3\cos t - 2\sin t$, $\quad s(0) = 0$, $\quad v(0) = 4$

63. $a(t) = 10\sin t + 3\cos t$, $\quad s(0) = 0$, $\quad s(2\pi) = 12$

64. $a(t) = t^2 - 4t + 6$, $\quad s(0) = 0$, $\quad s(1) = 20$

---

65. A stone is dropped from the upper observation deck (the Space Deck) of the CN Tower, 450 m above the ground.
 (a) Find the distance of the stone above ground level at time t.
 (b) How long does it take the stone to reach the ground?
 (c) With what velocity does it strike the ground?
 (d) If the stone is thrown downward with a speed of 5 m/s, how long does it take to reach the ground?

66. Show that for motion in a straight line with constant acceleration a, initial velocity v_0, and initial displacement s_0, the displacement after time t is
$$s = \tfrac{1}{2}at^2 + v_0 t + s_0$$

67. An object is projected upward with initial velocity v_0 meters per second from a point s_0 meters above the ground. Show that
$$[v(t)]^2 = v_0^2 - 19.6[s(t) - s_0]$$

68. Two balls are thrown upward from the edge of the cliff in Example 7. The first is thrown with a speed of 48 ft/s and the other is thrown a second later with a speed of 24 ft/s. Do the balls ever pass each other?

69. A stone was dropped off a cliff and hit the ground with a speed of 120 ft/s. What is the height of the cliff?

70. If a diver of mass m stands at the end of a diving board with length L and linear density ρ, then the board takes on the shape of a curve $y = f(x)$, where
$$EIy'' = mg(L - x) + \tfrac{1}{2}\rho g(L - x)^2$$
E and I are positive constants that depend on the material of the board and g (< 0) is the acceleration due to gravity.
 (a) Find an expression for the shape of the curve.
 (b) Use $f(L)$ to estimate the distance below the horizontal at the end of the board.

71. A company estimates that the marginal cost (in dollars per item) of producing x items is $1.92 - 0.002x$. If the cost of

producing one item is \$562, find the cost of producing 100 items.

72. The linear density of a rod of length 1 m is given by $\rho(x) = 1/\sqrt{x}$, in grams per centimeter, where x is measured in centimeters from one end of the rod. Find the mass of the rod.

73. Since raindrops grow as they fall, their surface area increases and therefore the resistance to their falling increases. A raindrop has an initial downward velocity of 10 m/s and its downward acceleration is
$$a = \begin{cases} 9 - 0.9t & \text{if } 0 \le t \le 10 \\ 0 & \text{if } t > 10 \end{cases}$$
If the raindrop is initially 500 m above the ground, how long does it take to fall?

74. A car is traveling at 50 mi/h when the brakes are fully applied, producing a constant deceleration of 22 ft/s². What is the distance traveled before the car comes to a stop?

75. What constant acceleration is required to increase the speed of a car from 30 mi/h to 50 mi/h in 5 seconds?

76. A car braked with a constant deceleration of 16 ft/s², producing skid marks measuring 200 ft before coming to a stop. How fast was the car traveling when the brakes were first applied?

77. A car is traveling at 100 km/h when the driver sees an accident 80 m ahead and slams on the brakes. What constant deceleration is required to stop the car in time to avoid a pileup?

78. A model rocket is fired vertically upward from rest. Its acceleration for the first three seconds is $a(t) = 60t$, at which time the fuel is exhausted and it becomes a freely "falling" body. Fourteen seconds later, the rocket's parachute opens, and the (downward) velocity slows linearly to -18 ft/s in 5 seconds. The rocket then "floats" to the ground at that rate.
 (a) Determine the position function s and the velocity function v (for all times t). Sketch the graphs of s and v.
 (b) At what time does the rocket reach its maximum height, and what is that height?
 (c) At what time does the rocket land?

79. A high-speed bullet train accelerates and decelerates at the rate of 4 ft/s². Its maximum cruising speed is 90 mi/h.
 (a) What is the maximum distance the train can travel if it accelerates from rest until it reaches its cruising speed and then runs at that speed for 15 minutes?
 (b) Suppose that the train starts from rest and must come to a complete stop in 15 minutes. What is the maximum distance it can travel under these conditions?
 (c) Find the minimum time that the train takes to travel between two consecutive stations that are 45 miles apart.
 (d) The trip from one station to the next takes 37.5 minutes. How far apart are the stations?

4   REVIEW

1. Explain the difference between an absolute maximum and a local maximum. Illustrate with a sketch.

2. (a) What does the Extreme Value Theorem say?
    (b) Explain how the Closed Interval Method works.

3. (a) State Fermat's Theorem.
    (b) Define a critical number of f.

4. (a) State Rolle's Theorem.
    (b) State the Mean Value Theorem and give a geometric interpretation.

5. (a) State the Increasing/Decreasing Test.
    (b) What does it mean to say that f is concave upward on an interval I?
    (c) State the Concavity Test.
    (d) What are inflection points? How do you find them?

6. (a) State the First Derivative Test.
    (b) State the Second Derivative Test.
    (c) What are the relative advantages and disadvantages of these tests?

7. (a) What does l'Hospital's Rule say?
    (b) How can you use l'Hospital's Rule if you have a product $f(x)g(x)$ where $f(x) \to 0$ and $g(x) \to \infty$ as $x \to a$?
    (c) How can you use l'Hospital's Rule if you have a difference $f(x) - g(x)$ where $f(x) \to \infty$ and $g(x) \to \infty$ as $x \to a$?

(d) How can you use l'Hospital's Rule if you have a power $[f(x)]^{g(x)}$ where $f(x) \to 0$ and $g(x) \to 0$ as $x \to a$?

8. State whether each of the following limit forms is indeterminate. Where possible, state the limit.

(a) $\dfrac{0}{0}$ (b) $\dfrac{\infty}{\infty}$ (c) $\dfrac{0}{\infty}$ (d) $\dfrac{\infty}{0}$

(e) $\infty + \infty$ (f) $\infty - \infty$ (g) $\infty \cdot \infty$ (h) $\infty \cdot 0$

(i) 0^0 (j) 0^∞ (k) ∞^0 (l) 1^∞

9. If you have a graphing calculator or computer, why do you need calculus to graph a function?

10. (a) Given an initial approximation x_1 to a root of the equation $f(x) = 0$, explain geometrically, with a diagram, how the second approximation x_2 in Newton's method is obtained.
    (b) Write an expression for x_2 in terms of x_1, $f(x_1)$, and $f'(x_1)$.
    (c) Write an expression for x_{n+1} in terms of x_n, $f(x_n)$, and $f'(x_n)$.
    (d) Under what circumstances is Newton's method likely to fail or to work very slowly?

11. (a) What is an antiderivative of a function f?
    (b) Suppose F_1 and F_2 are both antiderivatives of f on an interval I. How are F_1 and F_2 related?

Determine whether the statement is true or false. If it is true, explain why. If it is false, explain why or give an example that disproves the statement.

1. If $f'(c) = 0$, then f has a local maximum or minimum at c.

2. If f has an absolute minimum value at c, then $f'(c) = 0$.

3. If f is continuous on (a, b), then f attains an absolute maximum value $f(c)$ and an absolute minimum value $f(d)$ at some numbers c and d in (a, b).

4. If f is differentiable and $f(-1) = f(1)$, then there is a number c such that $|c| < 1$ and $f'(c) = 0$.

5. If $f'(x) < 0$ for $1 < x < 6$, then f is decreasing on $(1, 6)$.

6. If $f''(2) = 0$, then $(2, f(2))$ is an inflection point of the curve $y = f(x)$.

7. If $f'(x) = g'(x)$ for $0 < x < 1$, then $f(x) = g(x)$ for $0 < x < 1$.

8. There exists a function f such that $f(1) = -2$, $f(3) = 0$, and $f'(x) > 1$ for all x.

9. There exists a function f such that $f(x) > 0$, $f'(x) < 0$, and $f''(x) > 0$ for all x.

10. There exists a function f such that $f(x) < 0$, $f'(x) < 0$, and $f''(x) > 0$ for all x.

11. If f and g are increasing on an interval I, then $f + g$ is increasing on I.

12. If f and g are increasing on an interval I, then $f - g$ is increasing on I.

13. If f and g are increasing on an interval I, then fg is increasing on I.

14. If f and g are positive increasing functions on an interval I, then fg is increasing on I.

15. If f is increasing and $f(x) > 0$ on I, then $g(x) = 1/f(x)$ is decreasing on I.

16. If f is even, then f' is even.

17. If f is periodic, then f' is periodic.

18. The most general antiderivative of $f(x) = x^{-2}$ is

$$F(x) = -\frac{1}{x} + C$$

19. If $f'(x)$ exists and is nonzero for all x, then $f(1) \neq f(0)$.

20. If $\lim\limits_{x \to \infty} f(x) = 1$ and $\lim\limits_{x \to \infty} g(x) = \infty$, then

$$\lim_{x \to \infty} [f(x)]^{g(x)} = 1$$

21. $\lim\limits_{x \to 0} \dfrac{x}{e^x} = 1$

EXERCISES

1–6 Find the local and absolute extreme values of the function on the given interval.

1. $f(x) = x^3 - 9x^2 + 24x - 2, \quad [0, 5]$

2. $f(x) = x\sqrt{1 - x}, \quad [-1, 1]$

3. $f(x) = \dfrac{3x - 4}{x^2 + 1}, \quad [-2, 2]$

4. $f(x) = \sqrt{x^2 + x + 1}, \quad [-2, 1]$

5. $f(x) = x + 2\cos x, \quad [-\pi, \pi]$

6. $f(x) = x^2 e^{-x}, \quad [-1, 3]$

7–14 Evaluate the limit.

7. $\lim\limits_{x \to 0} \dfrac{e^x - 1}{\tan x}$

8. $\lim\limits_{x \to 0} \dfrac{\tan 4x}{x + \sin 2x}$

9. $\lim\limits_{x \to 0} \dfrac{e^{2x} - e^{-2x}}{\ln(x + 1)}$

10. $\lim\limits_{x \to \infty} \dfrac{e^{2x} - e^{-2x}}{\ln(x + 1)}$

11. $\lim\limits_{x \to -\infty} (x^2 - x^3)e^{2x}$

12. $\lim\limits_{x \to \pi^-} (x - \pi)\csc x$

13. $\lim\limits_{x \to 1^+} \left(\dfrac{x}{x - 1} - \dfrac{1}{\ln x} \right)$

14. $\lim\limits_{x \to (\pi/2)^-} (\tan x)^{\cos x}$

15–17 Sketch the graph of a function that satisfies the given conditions.

15. $f(0) = 0, \quad f'(-2) = f'(1) = f'(9) = 0,$

$\lim\limits_{x \to \infty} f(x) = 0, \quad \lim\limits_{x \to 6} f(x) = -\infty,$

$f'(x) < 0$ on $(-\infty, -2)$, $(1, 6)$, and $(9, \infty)$,

$f'(x) > 0$ on $(-2, 1)$ and $(6, 9)$,

$f''(x) > 0$ on $(-\infty, 0)$ and $(12, \infty)$,

$f''(x) < 0$ on $(0, 6)$ and $(6, 12)$

16. $f(0) = 0, \quad f$ is continuous and even,

$f'(x) = 2x$ if $0 < x < 1, \quad f'(x) = -1$ if $1 < x < 3,$

$f'(x) = 1$ if $x > 3$

17. f is odd, $\quad f'(x) < 0$ for $0 < x < 2,$

$f'(x) > 0$ for $x > 2, \quad f''(x) > 0$ for $0 < x < 3,$

$f''(x) < 0$ for $x > 3, \quad \lim\limits_{x \to \infty} f(x) = -2$

18. The figure shows the graph of the *derivative* f' of a function f.
   (a) On what intervals is f increasing or decreasing?
   (b) For what values of x does f have a local maximum or minimum?
   (c) Sketch the graph of f''.
   (d) Sketch a possible graph of f.

19–34 Use the guidelines of Section 4.5 to sketch the curve.

19. $y = 2 - 2x - x^3$

20. $y = -2x^3 - 3x^2 + 12x + 5$

21. $y = 3x^4 - 4x^3 + 2$

22. $y = \dfrac{x}{1 - x^2}$

23. $y = \dfrac{1}{x(x - 3)^2}$

24. $y = \dfrac{1}{x^2} - \dfrac{1}{(x - 2)^2}$

25. $y = \dfrac{(x - 1)^3}{x^2}$

26. $y = \sqrt{1 - x} + \sqrt{1 + x}$

27. $y = x\sqrt{2 + x}$

28. $y = x^{2/3}(x - 3)^2$

29. $y = e^x \sin x, \quad -\pi \leqslant x \leqslant \pi$

30. $y = 4x - \tan x, \quad -\pi/2 < x < \pi/2$

31. $y = \sin^{-1}(1/x)$ **32.** $y = e^{2x - x^2}$

33. $y = (x - 2)e^{-x}$ **34.** $y = x + \ln(x^2 + 1)$

35–38 Produce graphs of f that reveal all the important aspects of the curve. Use graphs of f' and f'' to estimate the intervals of increase and decrease, extreme values, intervals of concavity, and inflection points. In Exercise 35 use calculus to find these quantities exactly.

35. $f(x) = \dfrac{x^2 - 1}{x^3}$

36. $f(x) = \dfrac{x^3 + 1}{x^6 + 1}$

37. $f(x) = 3x^6 - 5x^5 + x^4 - 5x^3 - 2x^2 + 2$

38. $f(x) = x^2 + 6.5 \sin x, \quad -5 \leqslant x \leqslant 5$

---

39. Graph $f(x) = e^{-1/x^2}$ in a viewing rectangle that shows all the main aspects of this function. Estimate the inflection points. Then use calculus to find them exactly.

40. (a) Graph the function $f(x) = 1/(1 + e^{1/x})$.
  (b) Explain the shape of the graph by computing the limits of $f(x)$ as x approaches ∞, $-\infty$, 0^+, and 0^-.
  (c) Use the graph of f to estimate the coordinates of the inflection points.
  (d) Use your CAS to compute and graph f''.
  (e) Use the graph in part (d) to estimate the inflection points more accurately.

41–42 Use the graphs of f, f', and f'' to estimate the x-coordinates of the maximum and minimum points and inflection points of f.

41. $f(x) = \dfrac{\cos^2 x}{\sqrt{x^2 + x + 1}}, \quad -\pi \leqslant x \leqslant \pi$

42. $f(x) = e^{-0.1x} \ln(x^2 - 1)$

---

43. Investigate the family of functions $f(x) = \ln(\sin x + C)$. What features do the members of this family have in common? How do they differ? For which values of C is f continuous on $(-\infty, \infty)$? For which values of C does f have no graph at all? What happens as $C \to \infty$?

44. Investigate the family of functions $f(x) = cxe^{-cx^2}$. What happens to the maximum and minimum points and the inflection points as c changes? Illustrate your conclusions by graphing several members of the family.

45. Show that the equation $3x + 2 \cos x + 5 = 0$ has exactly one real root.

46. Suppose that f is continuous on $[0, 4]$, $f(0) = 1$, and $2 \leqslant f'(x) \leqslant 5$ for all x in $(0, 4)$. Show that $9 \leqslant f(4) \leqslant 21$.

47. By applying the Mean Value Theorem to the function $f(x) = x^{1/5}$ on the interval $[32, 33]$, show that

$$2 < \sqrt[5]{33} < 2.0125$$

48. For what values of the constants a and b is $(1, 3)$ a point of inflection of the curve $y = ax^3 + bx^2$?

49. Let $g(x) = f(x^2)$, where f is twice differentiable for all x, $f'(x) > 0$ for all $x \neq 0$, and f is concave downward on $(-\infty, 0)$ and concave upward on $(0, \infty)$.
  (a) At what numbers does g have an extreme value?
  (b) Discuss the concavity of g.

50. Find two positive integers such that the sum of the first number and four times the second number is 1000 and the product of the numbers is as large as possible.

51. Show that the shortest distance from the point (x_1, y_1) to the straight line $Ax + By + C = 0$ is

$$\frac{|Ax_1 + By_1 + C|}{\sqrt{A^2 + B^2}}$$

52. Find the point on the hyperbola $xy = 8$ that is closest to the point $(3, 0)$.

53. Find the smallest possible area of an isosceles triangle that is circumscribed about a circle of radius r.

54. Find the volume of the largest circular cone that can be inscribed in a sphere of radius r.

55. In $\triangle ABC$, D lies on AB, $CD \perp AB$, $|AD| = |BD| = 4$ cm, and $|CD| = 5$ cm. Where should a point P be chosen on CD so that the sum $|PA| + |PB| + |PC|$ is a minimum?

56. Solve Exercise 55 when $|CD| = 2$ cm.

57. The velocity of a wave of length L in deep water is

$$v = K\sqrt{\frac{L}{C} + \frac{C}{L}}$$

where K and C are known positive constants. What is the length of the wave that gives the minimum velocity?

58. A metal storage tank with volume V is to be constructed in the shape of a right circular cylinder surmounted by a hemisphere. What dimensions will require the least amount of metal?

59. A hockey team plays in an arena with a seating capacity of 15,000 spectators. With the ticket price set at $12, average attendance at a game has been 11,000. A market survey indicates that for each dollar the ticket price is lowered, average attendance will increase by 1000. How should the owners of the team set the ticket price to maximize their revenue from ticket sales?

60. A manufacturer determines that the cost of making x units of a commodity is

$$C(x) = 1800 + 25x - 0.2x^2 + 0.001x^3$$

and the demand function is $p(x) = 48.2 - 0.03x$.
   (a) Graph the cost and revenue functions and use the graphs to estimate the production level for maximum profit.
   (b) Use calculus to find the production level for maximum profit.
   (c) Estimate the production level that minimizes the average cost.

61. Use Newton's method to find the root of the equation

$$x^5 - x^4 + 3x^2 - 3x - 2 = 0$$

in the interval $[1, 2]$ correct to six decimal places.

62. Use Newton's method to find all solutions of the equation $\sin x = x^2 - 3x + 1$ correct to six decimal places.

63. Use Newton's method to find the absolute maximum value of the function $f(t) = \cos t + t - t^2$ correct to eight decimal places.

64. Use the guidelines in Section 4.5 to sketch the curve $y = x \sin x$, $0 \le x \le 2\pi$. Use Newton's method when necessary.

65–68 Find the most general antiderivative of the function.

65. $f(x) = 4\sqrt{x} - 6x^2 + 3$

66. $g(x) = \dfrac{1}{x} + \dfrac{1}{x^2 + 1}$

67. $f(t) = 2 \sin t - 3e^t$

68. $f(x) = x^{-3} + \cosh x$

69–72 Find f.

69. $f'(t) = 2t - 3 \sin t$, $f(0) = 5$

70. $f'(u) = \dfrac{u^2 + \sqrt{u}}{u}$, $f(1) = 3$

71. $f''(x) = 1 - 6x + 48x^2$, $f(0) = 1$, $f'(0) = 2$

72. $f''(x) = 5x^3 + 6x^2 + 2$, $f(0) = 3$, $f(1) = -2$

73–74 A particle is moving with the given data. Find the position of the particle.

73. $v(t) = 2t - 1/(1 + t^2)$, $s(0) = 1$

74. $a(t) = \sin t + 3 \cos t$, $s(0) = 0$, $v(0) = 2$

75. (a) If $f(x) = 0.1e^x + \sin x$, $-4 \le x \le 4$, use a graph of f to sketch a rough graph of the antiderivative F of f that satisfies $F(0) = 0$.
   (b) Find an expression for $F(x)$.
   (c) Graph F using the expression in part (b). Compare with your sketch in part (a).

76. Investigate the family of curves given by

$$f(x) = x^4 + x^3 + cx^2$$

In particular you should determine the transitional value of c at which the number of critical numbers changes and the transitional value at which the number of inflection points changes. Illustrate the various possible shapes with graphs.

77. A canister is dropped from a helicopter 500 m above the ground. Its parachute does not open, but the canister has been designed to withstand an impact velocity of 100 m/s. Will it burst?

78. In an automobile race along a straight road, car A passed car B twice. Prove that at some time during the race their accelerations were equal. State the assumptions that you make.

79. A rectangular beam will be cut from a cylindrical log of radius 10 inches.
   (a) Show that the beam of maximal cross-sectional area is a square.
   (b) Four rectangular planks will be cut from the four sections of the log that remain after cutting the square

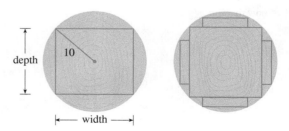

beam. Determine the dimensions of the planks that will have maximal cross-sectional area.

(c) Suppose that the strength of a rectangular beam is proportional to the product of its width and the square of its depth. Find the dimensions of the strongest beam that can be cut from the cylindrical log.

80. If a projectile is fired with an initial velocity v at an angle of inclination θ from the horizontal, then its trajectory, neglecting air resistance, is the parabola

$$y = (\tan\theta)x - \frac{g}{2v^2\cos^2\theta}x^2 \qquad 0 < \theta < \frac{\pi}{2}$$

(a) Suppose the projectile is fired from the base of a plane that is inclined at an angle α, $\alpha > 0$, from the horizontal, as shown in the figure. Show that the range of the projectile, measured up the slope, is given by

$$R(\theta) = \frac{2v^2\cos\theta\,\sin(\theta - \alpha)}{g\cos^2\alpha}$$

(b) Determine θ so that R is a maximum.

(c) Suppose the plane is at an angle α *below* the horizontal. Determine the range R in this case, and determine the angle at which the projectile should be fired to maximize R.

81. If an electrostatic field E acts on a liquid or a gaseous polar dielectric, the net dipole moment P per unit volume is

$$P(E) = \frac{e^E + e^{-E}}{e^E - e^{-E}} - \frac{1}{E}$$

Show that $\lim_{E \to 0^+} P(E) = 0$.

82. If a metal ball with mass m is projected in water and the force of resistance is proportional to the square of the velocity, then the distance the ball travels in time t is

$$s(t) = \frac{m}{c}\ln\cosh\sqrt{\frac{gc}{mt}}$$

where c is a positive constant. Find $\lim_{c \to 0^+} s(t)$.

83. Show that, for $x > 0$,

$$\frac{x}{1 + x^2} < \tan^{-1}x < x$$

84. Sketch the graph of a function f such that $f'(x) < 0$ for all x, $f''(x) > 0$ for $|x| > 1$, $f''(x) < 0$ for $|x| < 1$, and $\lim_{x \to \pm\infty}[f(x) + x] = 0$.

85. A light is to be placed atop a pole of height h feet to illuminate a busy traffic circle, which has a radius of 40 ft. The intensity of illumination I at any point P on the circle is directly proportional to the cosine of the angle θ (see the figure) and inversely proportional to the square of the distance d from the source.

(a) How tall should the light pole be to maximize I?

(b) Suppose that the light pole is h feet tall and that a woman is walking away from the base of the pole at the rate of 4 ft/s. At what rate is the intensity of the light at the point on her back 4 ft above the ground decreasing when she reaches the outer edge of the traffic circle?

86. Water is flowing at a constant rate into a spherical tank. Let $V(t)$ be the volume of water in the tank and $H(t)$ be the height of the water in the tank at time t.

(a) What are the meanings of $V'(t)$ and $H'(t)$? Are these derivatives positive, negative, or zero?

(b) Is $V''(t)$ positive, negative, or zero? Explain.

(c) Let t_1, t_2, and t_3 be the times when the tank is one-quarter full, half full, and three-quarters full, respectively. Are the values $H''(t_1)$, $H''(t_2)$, and $H''(t_3)$ positive, negative, or zero? Why?

Problems Plus

1. If a rectangle has its base on the x-axis and two vertices on the curve $y = e^{-x^2}$, show that the rectangle has the largest possible area when the two vertices are at the points of inflection of the curve.

2. Show that $|\sin x - \cos x| \leq \sqrt{2}$ for all x.

3. Does the function $f(x) = e^{10|x-2|-x^2}$ have an absolute maximum? If so, find it. What about an absolute minimum?

4. Show that $x^2 y^2 (4 - x^2)(4 - y^2) \leq 16$ for all numbers x and y such that $|x| \leq 2$ and $|y| \leq 2$.

5. Show that the inflection points of the curve $y = (\sin x)/x$ lie on the curve $y^2(x^4 + 4) = 4$.

6. Find the point on the parabola $y = 1 - x^2$ at which the tangent line cuts from the first quadrant the triangle with the smallest area.

7. If a, b, c, and d are constants such that

$$\lim_{x \to 0} \frac{ax^2 + \sin bx + \sin cx + \sin dx}{3x^2 + 5x^4 + 7x^6} = 8$$

find the value of the sum $a + b + c + d$.

8. Evaluate

$$\lim_{x \to \infty} \frac{(x + 2)^{1/x} - x^{1/x}}{(x + 3)^{1/x} - x^{1/x}}$$

9. Find the highest and lowest points on the curve $x^2 + xy + y^2 = 12$.

10. Sketch the set of all points (x, y) such that $|x + y| \leq e^x$.

11. If $P(a, a^2)$ is any point on the parabola $y = x^2$, except for the origin, let Q be the point where the normal line at P intersects the parabola again (see the figure).
    (a) Show that the y-coordinate of Q is smallest when $a = 1/\sqrt{2}$.
    (b) Show that the line segment PQ has the shortest possible length when $a = 1/\sqrt{2}$.

FIGURE FOR PROBLEM 11

12. For what values of c does the curve $y = cx^3 + e^x$ have inflection points?

13. An isosceles triangle is circumscribed about the unit circle so that the equal sides meet at the point $(0, a)$ on the y-axis (see the figure). Find the value of a that minimizes the lengths of the equal sides. (You may be surprised that the result does not give an equilateral triangle.).

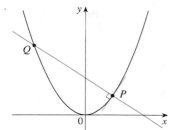

FIGURE FOR PROBLEM 13

14. Sketch the region in the plane consisting of all points (x, y) such that

$$2xy \leq |x - y| \leq x^2 + y^2$$

15. The line $y = mx + b$ intersects the parabola $y = x^2$ in points A and B. (See the figure.) Find the point P on the arc AOB of the parabola that maximizes the area of the triangle PAB.

16. $ABCD$ is a square piece of paper with sides of length 1 m. A quarter-circle is drawn from B to D with center A. The piece of paper is folded along EF, with E on AB and F on AD, so that A falls on the quarter-circle. Determine the maximum and minimum areas that the triangle AEF can have.

17. For which positive numbers a does the curve $y = a^x$ intersect the line $y = x$?

18. For what value of a is the following equation true?

$$\lim_{x \to \infty} \left(\frac{x + a}{x - a} \right)^x = e$$

FIGURE FOR PROBLEM 15

FIGURE FOR PROBLEM 20

FIGURE FOR PROBLEM 21

FIGURE FOR PROBLEM 23

FIGURE FOR PROBLEM 26

19. Let $f(x) = a_1 \sin x + a_2 \sin 2x + \cdots + a_n \sin nx$, where $a_1, a_2, \ldots, a_n$ are real numbers and n is a positive integer. If it is given that $|f(x)| \leq |\sin x|$ for all x, show that

$$|a_1 + 2a_2 + \cdots + na_n| \leq 1$$

20. An arc PQ of a circle subtends a central angle θ as in the figure. Let $A(\theta)$ be the area between the chord PQ and the arc PQ. Let $B(\theta)$ be the area between the tangent lines PR, QR, and the arc. Find

$$\lim_{\theta \to 0^+} \frac{A(\theta)}{B(\theta)}$$

21. The speeds of sound c_1 in an upper layer and c_2 in a lower layer of rock and the thickness h of the upper layer can be determined by seismic exploration if the speed of sound in the lower layer is greater than the speed in the upper layer. A dynamite charge is detonated at a point P and the transmitted signals are recorded at a point Q, which is a distance D from P. The first signal to arrive at Q travels along the surface and takes T_1 seconds. The next signal travels from P to a point R, from R to S in the lower layer, and then to Q, taking T_2 seconds. The third signal is reflected off the lower layer at the midpoint O of RS and takes T_3 seconds to reach Q. (See the figure.)

(a) Express T_1, T_2, and T_3 in terms of D, h, c_1, c_2, and θ.

(b) Show that T_2 is a minimum when $\sin \theta = c_1/c_2$.

(c) Suppose that $D = 1$ km, $T_1 = 0.26$ s, $T_2 = 0.32$ s, and $T_3 = 0.34$ s. Find c_1, c_2, and h.

Note: Geophysicists use this technique when studying the structure of the earth's crust, whether searching for oil or examining fault lines.

22. For what values of c is there a straight line that intersects the curve

$$y = x^4 + cx^3 + 12x^2 - 5x + 2$$

in four distinct points?

23. One of the problems posed by the Marquis de l'Hospital in his calculus textbook *Analyse des Infiniment Petits* concerns a pulley that is attached to the ceiling of a room at a point C by a rope of length r. At another point B on the ceiling, at a distance d from C (where $d > r$), a rope of length ℓ is attached and passed through the pulley at F and connected to a weight W. The weight is released and comes to rest at its equilibrium position D. (See the figure.) As l'Hospital argued, this happens when the distance $|ED|$ is maximized. Show that when the system reaches equilibrium, the value of x is

$$\frac{r}{4d}\left(r + \sqrt{r^2 + 8d^2}\right)$$

Notice that this expression is independent of both W and ℓ.

24. Given a sphere with radius r, find the height of a pyramid of minimum volume whose base is a square and whose base and triangular faces are all tangent to the sphere. What if the base of the pyramid is a regular n-gon? (A regular n-gon is a polygon with n equal sides and angles.) (Use the fact that the volume of a pyramid is $\frac{1}{3}Ah$, where A is the area of the base.)

25. Assume that a snowball melts so that its volume decreases at a rate proportional to its surface area. If it takes three hours for the snowball to decrease to half its original volume, how much longer will it take for the snowball to melt completely?

26. A hemispherical bubble is placed on a spherical bubble of radius 1. A smaller hemispherical bubble is then placed on the first one. This process is continued until n chambers, including the sphere, are formed. (The figure shows the case $n = 4$.) Use mathematical induction to prove that the maximum height of any bubble tower with n chambers is $1 + \sqrt{n}$.

5 Integrals

JRC, Inc. / Alamy

The photo shows Lake Lanier, which is a reservoir in Georgia, USA. In Exercise 70 in Section 5.4 you will estimate the amount of water that flowed into Lake Lanier during a certain time period.

IN CHAPTER 2 we used the tangent and velocity problems to introduce the derivative, which is the central idea in differential calculus. In much the same way, this chapter starts with the area and distance problems and uses them to formulate the idea of a definite integral, which is the basic concept of integral calculus. We will see in Chapters 6 and 8 how to use the integral to solve problems concerning volumes, lengths of curves, population predictions, cardiac output, forces on a dam, work, consumer surplus, and baseball, among many others.

There is a connection between integral calculus and differential calculus. The Fundamental Theorem of Calculus relates the integral to the derivative, and we will see in this chapter that it greatly simplifies the solution of many problems.

5.1 Areas and Distances

Now is a good time to read (or reread) *A Preview of Calculus* (see page 1). It discusses the unifying ideas of calculus and helps put in perspective where we have been and where we are going.

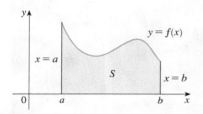

FIGURE 1
$S = \{(x, y) \mid a \leqslant x \leqslant b, 0 \leqslant y \leqslant f(x)\}$

In this section we discover that in trying to find the area under a curve or the distance traveled by a car, we end up with the same special type of limit.

■ The Area Problem

We begin by attempting to solve the *area problem:* Find the area of the region S that lies under the curve $y = f(x)$ from a to b. This means that S, illustrated in Figure 1, is bounded by the graph of a continuous function f [where $f(x) \geqslant 0$], the vertical lines $x = a$ and $x = b$, and the x-axis.

In trying to solve the area problem we have to ask ourselves: What is the meaning of the word *area*? This question is easy to answer for regions with straight sides. For a rectangle, the area is defined as the product of the length and the width. The area of a triangle is half the base times the height. The area of a polygon is found by dividing it into triangles (as in Figure 2) and adding the areas of the triangles.

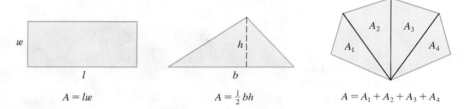

FIGURE 2

$$A = lw \qquad A = \tfrac{1}{2} bh \qquad A = A_1 + A_2 + A_3 + A_4$$

However, it isn't so easy to find the area of a region with curved sides. We all have an intuitive idea of what the area of a region is. But part of the area problem is to make this intuitive idea precise by giving an exact definition of area.

Recall that in defining a tangent we first approximated the slope of the tangent line by slopes of secant lines and then we took the limit of these approximations. We pursue a similar idea for areas. We first approximate the region S by rectangles and then we take the limit of the areas of these rectangles as we increase the number of rectangles. The following example illustrates the procedure.

EXAMPLE 1 Use rectangles to estimate the area under the parabola $y = x^2$ from 0 to 1 (the parabolic region S illustrated in Figure 3).

SOLUTION We first notice that the area of S must be somewhere between 0 and 1 because S is contained in a square with side length 1, but we can certainly do better than that. Suppose we divide S into four strips S_1, S_2, S_3, and S_4 by drawing the vertical lines $x = \frac{1}{4}$, $x = \frac{1}{2}$, and $x = \frac{3}{4}$ as in Figure 4(a).

FIGURE 3

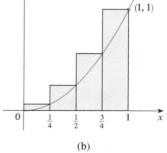

FIGURE 4   (a)   (b)

We can approximate each strip by a rectangle that has the same base as the strip and whose height is the same as the right edge of the strip [see Figure 4(b)]. In other words, the heights of these rectangles are the values of the function $f(x) = x^2$ at the *right* endpoints of the subintervals $\left[0, \frac{1}{4}\right], \left[\frac{1}{4}, \frac{1}{2}\right], \left[\frac{1}{2}, \frac{3}{4}\right]$, and $\left[\frac{3}{4}, 1\right]$.

Each rectangle has width $\frac{1}{4}$ and the heights are $\left(\frac{1}{4}\right)^2, \left(\frac{1}{2}\right)^2, \left(\frac{3}{4}\right)^2$, and 1^2. If we let R_4 be the sum of the areas of these approximating rectangles, we get

$$R_4 = \tfrac{1}{4} \cdot \left(\tfrac{1}{4}\right)^2 + \tfrac{1}{4} \cdot \left(\tfrac{1}{2}\right)^2 + \tfrac{1}{4} \cdot \left(\tfrac{3}{4}\right)^2 + \tfrac{1}{4} \cdot 1^2 = \tfrac{15}{32} = 0.46875$$

From Figure 4(b) we see that the area A of S is less than R_4, so

$$A < 0.46875$$

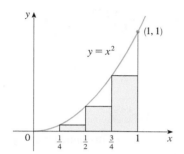

FIGURE 5

Instead of using the rectangles in Figure 4(b) we could use the smaller rectangles in Figure 5 whose heights are the values of f at the *left* endpoints of the subintervals. (The leftmost rectangle has collapsed because its height is 0.) The sum of the areas of these approximating rectangles is

$$L_4 = \tfrac{1}{4} \cdot 0^2 + \tfrac{1}{4} \cdot \left(\tfrac{1}{4}\right)^2 + \tfrac{1}{4} \cdot \left(\tfrac{1}{2}\right)^2 + \tfrac{1}{4} \cdot \left(\tfrac{3}{4}\right)^2 = \tfrac{7}{32} = 0.21875$$

We see that the area of S is larger than L_4, so we have lower and upper estimates for A:

$$0.21875 < A < 0.46875$$

We can repeat this procedure with a larger number of strips. Figure 6 shows what happens when we divide the region S into eight strips of equal width.

 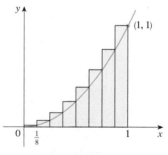

FIGURE 6
Approximating S with eight rectangles

(a) Using left endpoints   (b) Using right endpoints

By computing the sum of the areas of the smaller rectangles (L_8) and the sum of the areas of the larger rectangles (R_8), we obtain better lower and upper estimates for A:

$$0.2734375 < A < 0.3984375$$

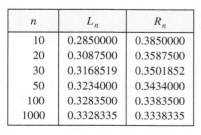

n	L_n	R_n
10	0.2850000	0.3850000
20	0.3087500	0.3587500
30	0.3168519	0.3501852
50	0.3234000	0.3434000
100	0.3283500	0.3383500
1000	0.3328335	0.3338335

So one possible answer to the question is to say that the true area of S lies somewhere between 0.2734375 and 0.3984375.

We could obtain better estimates by increasing the number of strips. The table at the left shows the results of similar calculations (with a computer) using n rectangles whose heights are found with left endpoints (L_n) or right endpoints (R_n). In particular, we see by using 50 strips that the area lies between 0.3234 and 0.3434. With 1000 strips we narrow it down even more: A lies between 0.3328335 and 0.3338335. A good estimate is obtained by averaging these numbers: $A \approx 0.3333335$.

From the values in the table in Example 1, it looks as if R_n is approaching $\frac{1}{3}$ as n increases. We confirm this in the next example.

EXAMPLE 2 For the region S in Example 1, show that the sum of the areas of the upper approximating rectangles approaches $\frac{1}{3}$, that is,

$$\lim_{n \to \infty} R_n = \tfrac{1}{3}$$

SOLUTION R_n is the sum of the areas of the n rectangles in Figure 7. Each rectangle has width $1/n$ and the heights are the values of the function $f(x) = x^2$ at the points $1/n, 2/n, 3/n, \ldots, n/n$; that is, the heights are $(1/n)^2, (2/n)^2, (3/n)^2, \ldots, (n/n)^2$. Thus

$$R_n = \frac{1}{n}\left(\frac{1}{n}\right)^2 + \frac{1}{n}\left(\frac{2}{n}\right)^2 + \frac{1}{n}\left(\frac{3}{n}\right)^2 + \cdots + \frac{1}{n}\left(\frac{n}{n}\right)^2$$

$$= \frac{1}{n} \cdot \frac{1}{n^2} (1^2 + 2^2 + 3^2 + \cdots + n^2)$$

$$= \frac{1}{n^3} (1^2 + 2^2 + 3^2 + \cdots + n^2)$$

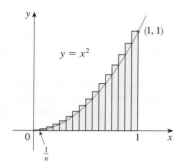

FIGURE 7

Here we need the formula for the sum of the squares of the first n positive integers:

$$\boxed{1} \qquad\qquad 1^2 + 2^2 + 3^2 + \cdots + n^2 = \frac{n(n+1)(2n+1)}{6}$$

Perhaps you have seen this formula before. It is proved in Example 5 in Appendix E. Putting Formula 1 into our expression for R_n, we get

$$R_n = \frac{1}{n^3} \cdot \frac{n(n+1)(2n+1)}{6} = \frac{(n+1)(2n+1)}{6n^2}$$

Here we are computing the limit of the sequence $\{R_n\}$. Sequences and their limits were discussed in *A Preview of Calculus* and will be studied in detail in Section 11.1. The idea is very similar to a limit at infinity (Section 2.6) except that in writing $\lim_{n \to \infty}$ we restrict n to be a positive integer. In particular, we know that

$$\lim_{n \to \infty} \frac{1}{n} = 0$$

When we write $\lim_{n \to \infty} R_n = \frac{1}{3}$ we mean that we can make R_n as close to $\frac{1}{3}$ as we like by taking n sufficiently large.

Thus we have

$$\lim_{n \to \infty} R_n = \lim_{n \to \infty} \frac{(n+1)(2n+1)}{6n^2}$$

$$= \lim_{n \to \infty} \frac{1}{6}\left(\frac{n+1}{n}\right)\left(\frac{2n+1}{n}\right)$$

$$= \lim_{n \to \infty} \frac{1}{6}\left(1 + \frac{1}{n}\right)\left(2 + \frac{1}{n}\right)$$

$$= \frac{1}{6} \cdot 1 \cdot 2 = \frac{1}{3}$$

It can be shown that the lower approximating sums also approach $\frac{1}{3}$, that is,

$$\lim_{n \to \infty} L_n = \tfrac{1}{3}$$

From Figures 8 and 9 it appears that, as n increases, both L_n and R_n become better and better approximations to the area of S. Therefore we *define* the area A to be the limit of the sums of the areas of the approximating rectangles, that is,

TEC In Visual 5.1 you can create pictures like those in Figures 8 and 9 for other values of n.

$$A = \lim_{n \to \infty} R_n = \lim_{n \to \infty} L_n = \tfrac{1}{3}$$

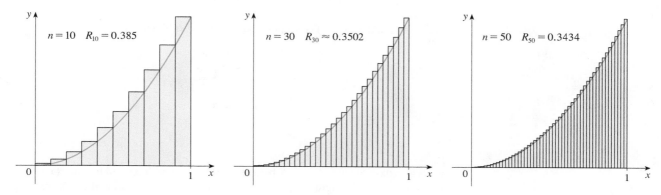

FIGURE 8 Right endpoints produce upper sums because $f(x) = x^2$ is increasing.

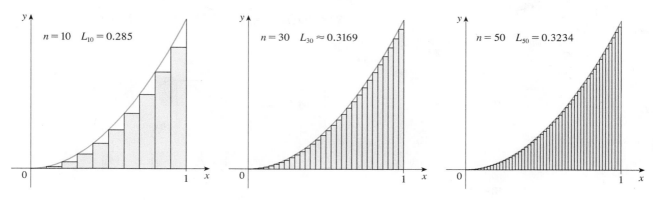

FIGURE 9 Left endpoints produce lower sums because $f(x) = x^2$ is increasing.

Let's apply the idea of Examples 1 and 2 to the more general region S of Figure 1. We start by subdividing S into n strips $S_1, S_2, \ldots, S_n$ of equal width as in Figure 10.

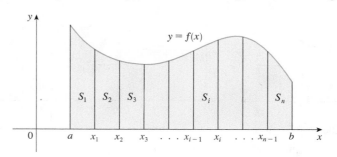

FIGURE 10

The width of the interval $[a, b]$ is $b - a$, so the width of each of the n strips is

$$\Delta x = \frac{b - a}{n}$$

These strips divide the interval $[a, b]$ into n subintervals

$$[x_0, x_1], \quad [x_1, x_2], \quad [x_2, x_3], \quad \ldots, \quad [x_{n-1}, x_n]$$

where $x_0 = a$ and $x_n = b$. The right endpoints of the subintervals are

$$x_1 = a + \Delta x,$$

$$x_2 = a + 2\,\Delta x,$$

$$x_3 = a + 3\,\Delta x,$$

$$\vdots$$

Let's approximate the ith strip S_i by a rectangle with width Δx and height $f(x_i)$, which is the value of f at the right endpoint (see Figure 11). Then the area of the ith rectangle is $f(x_i)\,\Delta x$. What we think of intuitively as the area of S is approximated by the sum of the areas of these rectangles, which is

$$R_n = f(x_1)\,\Delta x + f(x_2)\,\Delta x + \cdots + f(x_n)\,\Delta x$$

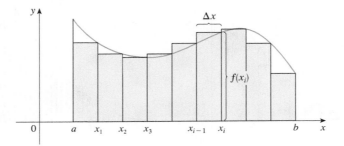

FIGURE 11

Figure 12 shows this approximation for $n = 2, 4, 8,$ and 12. Notice that this approximation appears to become better and better as the number of strips increases, that is, as $n \to \infty$. Therefore we define the area A of the region S in the following way.

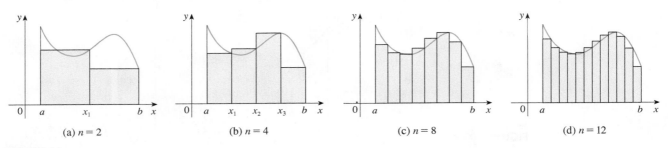

(a) $n = 2$ (b) $n = 4$ (c) $n = 8$ (d) $n = 12$

FIGURE 12

> **2** **Definition** The **area** A of the region S that lies under the graph of the continuous function f is the limit of the sum of the areas of approximating rectangles:
>
> $$A = \lim_{n \to \infty} R_n = \lim_{n \to \infty} \left[f(x_1)\,\Delta x + f(x_2)\,\Delta x + \cdots + f(x_n)\,\Delta x \right]$$

It can be proved that the limit in Definition 2 always exists, since we are assuming that f is continuous. It can also be shown that we get the same value if we use left endpoints:

$$\boxed{3} \qquad A = \lim_{n \to \infty} L_n = \lim_{n \to \infty} \left[f(x_0)\,\Delta x + f(x_1)\,\Delta x + \cdots + f(x_{n-1})\,\Delta x \right]$$

In fact, instead of using left endpoints or right endpoints, we could take the height of the ith rectangle to be the value of f at *any* number x_i^* in the ith subinterval $[x_{i-1}, x_i]$. We call the numbers $x_1^*, x_2^*, \ldots, x_n^*$ the **sample points**. Figure 13 shows approximating rectangles when the sample points are not chosen to be endpoints. So a more general expression for the area of S is

$$\boxed{4} \qquad A = \lim_{n \to \infty} \left[f(x_1^*)\,\Delta x + f(x_2^*)\,\Delta x + \cdots + f(x_n^*)\,\Delta x \right]$$

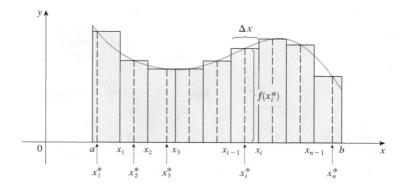

FIGURE 13

NOTE It can be shown that an equivalent definition of area is the following: *A is the unique number that is smaller than all the upper sums and bigger than all the lower sums.* We saw in Examples 1 and 2, for instance, that the area $\left(A = \frac{1}{3} \right)$ is trapped between all the left approximating sums L_n and all the right approximating sums R_n. The function in those examples, $f(x) = x^2$, happens to be increasing on $[0, 1]$ and so the lower sums arise from left endpoints and the upper sums from right endpoints. (See Figures 8 and 9.) In general, we form **lower** (and **upper**) **sums** by choosing the sample points x_i^* so that $f(x_i^*)$ is the minimum (and maximum) value of f on the ith subinterval. (See Figure 14 and Exercises 7–8.)

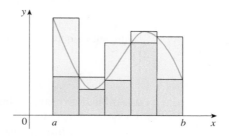

FIGURE 14
Lower sums (short rectangles) and
upper sums (tall rectangles)

This tells us to
end with $i = n$.

This tells us
to add. $\displaystyle\sum_{i=m}^{n} f(x_i)\,\Delta x$

This tells us to
start with $i = m$.

If you need practice with sigma notation, look at the examples and try some of the exercises in Appendix E.

We often use **sigma notation** to write sums with many terms more compactly. For instance,

$$\sum_{i=1}^{n} f(x_i)\,\Delta x = f(x_1)\,\Delta x + f(x_2)\,\Delta x + \cdots + f(x_n)\,\Delta x$$

So the expressions for area in Equations 2, 3, and 4 can be written as follows:

$$A = \lim_{n\to\infty} \sum_{i=1}^{n} f(x_i)\,\Delta x$$

$$A = \lim_{n\to\infty} \sum_{i=1}^{n} f(x_{i-1})\,\Delta x$$

$$A = \lim_{n\to\infty} \sum_{i=1}^{n} f(x_i^*)\,\Delta x$$

We can also rewrite Formula 1 in the following way:

$$\sum_{i=1}^{n} i^2 = \frac{n(n+1)(2n+1)}{6}$$

EXAMPLE 3 Let A be the area of the region that lies under the graph of $f(x) = e^{-x}$ between $x = 0$ and $x = 2$.
(a) Using right endpoints, find an expression for A as a limit. Do not evaluate the limit.
(b) Estimate the area by taking the sample points to be midpoints and using four subintervals and then ten subintervals.

SOLUTION
(a) Since $a = 0$ and $b = 2$, the width of a subinterval is

$$\Delta x = \frac{2-0}{n} = \frac{2}{n}$$

So $x_1 = 2/n$, $x_2 = 4/n$, $x_3 = 6/n$, $x_i = 2i/n$, and $x_n = 2n/n$. The sum of the areas of the approximating rectangles is

$$R_n = f(x_1)\,\Delta x + f(x_2)\,\Delta x + \cdots + f(x_n)\,\Delta x$$

$$= e^{-x_1}\,\Delta x + e^{-x_2}\,\Delta x + \cdots + e^{-x_n}\,\Delta x$$

$$= e^{-2/n}\left(\frac{2}{n}\right) + e^{-4/n}\left(\frac{2}{n}\right) + \cdots + e^{-2n/n}\left(\frac{2}{n}\right)$$

According to Definition 2, the area is

$$A = \lim_{n\to\infty} R_n = \lim_{n\to\infty} \frac{2}{n}\left(e^{-2/n} + e^{-4/n} + e^{-6/n} + \cdots + e^{-2n/n}\right)$$

Using sigma notation we could write

$$A = \lim_{n\to\infty} \frac{2}{n} \sum_{i=1}^{n} e^{-2i/n}$$

It is difficult to evaluate this limit directly by hand, but with the aid of a computer algebra system it isn't hard (see Exercise 30). In Section 5.3 we will be able to find A more easily using a different method.

FIGURE 15

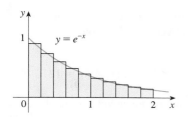

FIGURE 16

(b) With $n = 4$ the subintervals of equal width $\Delta x = 0.5$ are $[0, 0.5]$, $[0.5, 1]$, $[1, 1.5]$, and $[1.5, 2]$. The midpoints of these subintervals are $x_1^* = 0.25$, $x_2^* = 0.75$, $x_3^* = 1.25$, and $x_4^* = 1.75$, and the sum of the areas of the four approximating rectangles (see Figure 15) is

$$M_4 = \sum_{i=1}^{4} f(x_i^*)\, \Delta x$$

$$= f(0.25)\, \Delta x + f(0.75)\, \Delta x + f(1.25)\, \Delta x + f(1.75)\, \Delta x$$

$$= e^{-0.25}(0.5) + e^{-0.75}(0.5) + e^{-1.25}(0.5) + e^{-1.75}(0.5)$$

$$= \tfrac{1}{2}(e^{-0.25} + e^{-0.75} + e^{-1.25} + e^{-1.75}) \approx 0.8557$$

So an estimate for the area is

$$A \approx 0.8557$$

With $n = 10$ the subintervals are $[0, 0.2]$, $[0.2, 0.4]$, ..., $[1.8, 2]$ and the midpoints are $x_1^* = 0.1$, $x_2^* = 0.3$, $x_3^* = 0.5$, ..., $x_{10}^* = 1.9$. Thus

$$A \approx M_{10} = f(0.1)\, \Delta x + f(0.3)\, \Delta x + f(0.5)\, \Delta x + \cdots + f(1.9)\, \Delta x$$

$$= 0.2(e^{-0.1} + e^{-0.3} + e^{-0.5} + \cdots + e^{-1.9}) \approx 0.8632$$

From Figure 16 it appears that this estimate is better than the estimate with $n = 4$. ∎

■ The Distance Problem

Now let's consider the *distance problem:* Find the distance traveled by an object during a certain time period if the velocity of the object is known at all times. (In a sense this is the inverse problem of the velocity problem that we discussed in Section 2.1.) If the velocity remains constant, then the distance problem is easy to solve by means of the formula

$$\text{distance} = \text{velocity} \times \text{time}$$

But if the velocity varies, it's not so easy to find the distance traveled. We investigate the problem in the following example.

EXAMPLE 4 Suppose the odometer on our car is broken and we want to estimate the distance driven over a 30-second time interval. We take speedometer readings every five seconds and record them in the following table:

Time (s)	0	5	10	15	20	25	30
Velocity (mi/h)	17	21	24	29	32	31	28

In order to have the time and the velocity in consistent units, let's convert the velocity readings to feet per second (1 mi/h $= 5280/3600$ ft/s):

Time (s)	0	5	10	15	20	25	30
Velocity (ft/s)	25	31	35	43	47	45	41

During the first five seconds the velocity doesn't change very much, so we can estimate the distance traveled during that time by assuming that the velocity is constant. If we take the velocity during that time interval to be the initial velocity (25 ft/s), then we obtain the approximate distance traveled during the first five seconds:

$$25 \text{ ft/s} \times 5 \text{ s} = 125 \text{ ft}$$

Similarly, during the second time interval the velocity is approximately constant and we take it to be the velocity when $t = 5$ s. So our estimate for the distance traveled from $t = 5$ s to $t = 10$ s is

$$31 \text{ ft/s} \times 5 \text{ s} = 155 \text{ ft}$$

If we add similar estimates for the other time intervals, we obtain an estimate for the total distance traveled:

$$(25 \times 5) + (31 \times 5) + (35 \times 5) + (43 \times 5) + (47 \times 5) + (45 \times 5) = 1130 \text{ ft}$$

We could just as well have used the velocity at the *end* of each time period instead of the velocity at the beginning as our assumed constant velocity. Then our estimate becomes

$$(31 \times 5) + (35 \times 5) + (43 \times 5) + (47 \times 5) + (45 \times 5) + (41 \times 5) = 1210 \text{ ft}$$

If we had wanted a more accurate estimate, we could have taken velocity readings every two seconds, or even every second. ∎

Perhaps the calculations in Example 4 remind you of the sums we used earlier to estimate areas. The similarity is explained when we sketch a graph of the velocity function of the car in Figure 17 and draw rectangles whose heights are the initial velocities for each time interval. The area of the first rectangle is $25 \times 5 = 125$, which is also our estimate for the distance traveled in the first five seconds. In fact, the area of each rectangle can be interpreted as a distance because the height represents velocity and the width represents time. The sum of the areas of the rectangles in Figure 17 is $L_6 = 1130$, which is our initial estimate for the total distance traveled.

In general, suppose an object moves with velocity $v = f(t)$, where $a \leq t \leq b$ and $f(t) \geq 0$ (so the object always moves in the positive direction). We take velocity readings at times $t_0 \, (= a), t_1, t_2, \ldots, t_n \, (= b)$ so that the velocity is approximately constant on each subinterval. If these times are equally spaced, then the time between consecutive readings is $\Delta t = (b - a)/n$. During the first time interval the velocity is approximately $f(t_0)$ and so the distance traveled is approximately $f(t_0) \, \Delta t$. Similarly, the distance traveled during the second time interval is about $f(t_1) \, \Delta t$ and the total distance traveled during the time interval $[a, b]$ is approximately

$$f(t_0) \, \Delta t + f(t_1) \, \Delta t + \cdots + f(t_{n-1}) \, \Delta t = \sum_{i=1}^{n} f(t_{i-1}) \, \Delta t$$

If we use the velocity at right endpoints instead of left endpoints, our estimate for the total distance becomes

$$f(t_1) \, \Delta t + f(t_2) \, \Delta t + \cdots + f(t_n) \, \Delta t = \sum_{i=1}^{n} f(t_i) \, \Delta t$$

The more frequently we measure the velocity, the more accurate our estimates become,

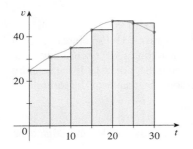

FIGURE 17

so it seems plausible that the *exact* distance d traveled is the *limit* of such expressions:

$$\boxed{5} \qquad d = \lim_{n \to \infty} \sum_{i=1}^{n} f(t_{i-1}) \, \Delta t = \lim_{n \to \infty} \sum_{i=1}^{n} f(t_i) \, \Delta t$$

We will see in Section 5.4 that this is indeed true.

Because Equation 5 has the same form as our expressions for area in Equations 2 and 3, it follows that the distance traveled is equal to the area under the graph of the velocity function. In Chapter 6 we will see that other quantities of interest in the natural and social sciences—such as the work done by a variable force or the cardiac output of the heart—can also be interpreted as the area under a curve. So when we compute areas in this chapter, bear in mind that they can be interpreted in a variety of practical ways.

5.1   EXERCISES

1. (a) By reading values from the given graph of f, use five rectangles to find a lower estimate and an upper estimate for the area under the given graph of f from $x = 0$ to $x = 10$. In each case sketch the rectangles that you use.

(b) Find new estimates using ten rectangles in each case.

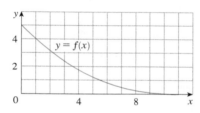

2. (a) Use six rectangles to find estimates of each type for the area under the given graph of f from $x = 0$ to $x = 12$.

(i) L_6 (sample points are left endpoints)
(ii) R_6 (sample points are right endpoints)
(iii) M_6 (sample points are midpoints)

(b) Is L_6 an underestimate or overestimate of the true area?

(c) Is R_6 an underestimate or overestimate of the true area?

(d) Which of the numbers L_6, R_6, or M_6 gives the best estimate? Explain.

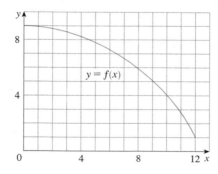

3. (a) Estimate the area under the graph of $f(x) = 1/x$ from $x = 1$ to $x = 2$ using four approximating rectangles and right endpoints. Sketch the graph and the rectangles. Is your estimate an underestimate or an overestimate?

(b) Repeat part (a) using left endpoints.

4. (a) Estimate the area under the graph of $f(x) = \sin x$ from $x = 0$ to $x = \pi/2$ using four approximating rectangles and right endpoints. Sketch the graph and the rectangles. Is your estimate an underestimate or an overestimate?

(b) Repeat part (a) using left endpoints.

5. (a) Estimate the area under the graph of $f(x) = 1 + x^2$ from $x = -1$ to $x = 2$ using three rectangles and right endpoints. Then improve your estimate by using six rectangles. Sketch the curve and the approximating rectangles.

(b) Repeat part (a) using left endpoints.

(c) Repeat part (a) using midpoints.

(d) From your sketches in parts (a)–(c), which appears to be the best estimate?

6. (a) Graph the function

$$f(x) = x - 2 \ln x \qquad 1 \leqslant x \leqslant 5$$

(b) Estimate the area under the graph of f using four approximating rectangles and taking the sample points to be (i) right endpoints and (ii) midpoints. In each case sketch the curve and the rectangles.

(c) Improve your estimates in part (b) by using eight rectangles.

7. Evaluate the upper and lower sums for $f(x) = 2 + \sin x$, $0 \leqslant x \leqslant \pi$, with $n = 2$, 4, and 8. Illustrate with diagrams like Figure 14.

8. Evaluate the upper and lower sums for $f(x) = 1 + x^2$, $-1 \leqslant x \leqslant 1$, with $n = 3$ and 4. Illustrate with diagrams like Figure 14.

9–10 With a programmable calculator (or a computer), it is possible to evaluate the expressions for the sums of areas of approximating rectangles, even for large values of n, using looping. (On a TI use the Is> command or a For-EndFor loop, on a Casio use Isz, on an HP or in BASIC use a FOR-NEXT loop.) Compute the sum of the areas of approximating rectangles using equal subintervals and right endpoints for $n = 10$, 30, 50, and 100. Then guess the value of the exact area.

9. The region under $y = x^4$ from 0 to 1

10. The region under $y = \cos x$ from 0 to $\pi/2$

CAS 11. Some computer algebra systems have commands that will draw approximating rectangles and evaluate the sums of their areas, at least if x_i^* is a left or right endpoint. (For instance, in Maple use `leftbox`, `rightbox`, `leftsum`, and `rightsum`.)
  (a) If $f(x) = 1/(x^2 + 1)$, $0 \le x \le 1$, find the left and right sums for $n = 10$, 30, and 50.
  (b) Illustrate by graphing the rectangles in part (a).
  (c) Show that the exact area under f lies between 0.780 and 0.791.

CAS 12. (a) If $f(x) = \ln x$, $1 \le x \le 4$, use the commands discussed in Exercise 11 to find the left and right sums for $n = 10$, 30, and 50.
  (b) Illustrate by graphing the rectangles in part (a).
  (c) Show that the exact area under f lies between 2.50 and 2.59.

13. The speed of a runner increased steadily during the first three seconds of a race. Her speed at half-second intervals is given in the table. Find lower and upper estimates for the distance that she traveled during these three seconds.

t (s)	0	0.5	1.0	1.5	2.0	2.5	3.0
v (ft/s)	0	6.2	10.8	14.9	18.1	19.4	20.2

14. The table shows speedometer readings at 10-second intervals during a 1-minute period for a car racing at the Daytona International Speedway in Florida.
  (a) Estimate the distance the race car traveled during this

Time (s)	Velocity (mi/h)
0	182.9
10	168.0
20	106.6
30	99.8
40	124.5
50	176.1
60	175.6

time period using the velocities at the beginning of the time intervals.
  (b) Give another estimate using the velocities at the end of the time periods.
  (c) Are your estimates in parts (a) and (b) upper and lower estimates? Explain.

15. Oil leaked from a tank at a rate of $r(t)$ liters per hour. The rate decreased as time passed and values of the rate at two-hour time intervals are shown in the table. Find lower and upper estimates for the total amount of oil that leaked out.

t (h)	0	2	4	6	8	10
$r(t)$ (L/h)	8.7	7.6	6.8	6.2	5.7	5.3

16. When we estimate distances from velocity data, it is sometimes necessary to use times $t_0, t_1, t_2, t_3, \ldots$ that are not equally spaced. We can still estimate distances using the time periods $\Delta t_i = t_i - t_{i-1}$. For example, on May 7, 1992, the space shuttle *Endeavour* was launched on mission STS-49, the purpose of which was to install a new perigee kick motor in an Intelsat communications satellite. The table, provided by NASA, gives the velocity data for the shuttle between liftoff and the jettisoning of the solid rocket boosters. Use these data to estimate the height above the earth's surface of the *Endeavour*, 62 seconds after liftoff.

Event	Time (s)	Velocity (ft/s)
Launch	0	0
Begin roll maneuver	10	185
End roll maneuver	15	319
Throttle to 89%	20	447
Throttle to 67%	32	742
Throttle to 104%	59	1325
Maximum dynamic pressure	62	1445
Solid rocket booster separation	125	4151

17. The velocity graph of a braking car is shown. Use it to estimate the distance traveled by the car while the brakes are applied.

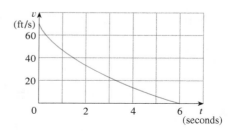

18. The velocity graph of a car accelerating from rest to a speed of 120 km/h over a period of 30 seconds is shown. Estimate the distance traveled during this period.

19. In someone infected with measles, the virus level N (measured in number of infected cells per mL of blood plasma) reaches a peak density at about $t = 12$ days (when a rash appears) and then decreases fairly rapidly as a result of immune response. The area under the graph of $N(t)$ from $t = 0$ to $t = 12$ (as shown in the figure) is equal to the total amount of infection needed to develop symptoms (measured in density of infected cells $\times$ time). The function N has been modeled by the function

$$f(t) = -t(t - 21)(t + 1)$$

Use this model with six subintervals and their midpoints to estimate the total amount of infection needed to develop symptoms of measles.

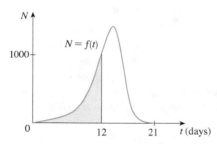

Source: J. M. Heffernan et al., "An In-Host Model of Acute Infection: Measles as a Case Study," *Theoretical Population Biology* 73 (2006): 134–47.

20. The table shows the number of people per day who died from SARS in Singapore at two-week intervals beginning on March 1, 2003.

Date	Deaths per day	Date	Deaths per day
March 1	0.0079	April 26	0.5620
March 15	0.0638	May 10	0.4630
March 29	0.1944	May 24	0.2897
April 12	0.4435		

(a) By using an argument similar to that in Example 4, estimate the number of people who died of SARS in Singapore between March 1 and May 24, 2003, using both left endpoints and right endpoints.

(b) How would you interpret the number of SARS deaths as an area under a curve?

Source: A. Gumel et al., "Modelling Strategies for Controlling SARS Outbreaks," *Proceedings of the Royal Society of London: Series B* 271 (2004): 2223–32.

21–23 Use Definition 2 to find an expression for the area under the graph of f as a limit. Do not evaluate the limit.

21. $f(x) = \dfrac{2x}{x^2 + 1}$, $1 \leqslant x \leqslant 3$

22. $f(x) = x^2 + \sqrt{1 + 2x}$, $4 \leqslant x \leqslant 7$

23. $f(x) = \sqrt{\sin x}$, $0 \leqslant x \leqslant \pi$

24–25 Determine a region whose area is equal to the given limit. Do not evaluate the limit.

24. $\displaystyle \lim_{n \to \infty} \sum_{i=1}^{n} \frac{3}{n} \sqrt{1 + \frac{3i}{n}}$ **25.** $\displaystyle \lim_{n \to \infty} \sum_{i=1}^{n} \frac{\pi}{4n} \tan \frac{i\pi}{4n}$

26. (a) Use Definition 2 to find an expression for the area under the curve $y = x^3$ from 0 to 1 as a limit.

(b) The following formula for the sum of the cubes of the first n integers is proved in Appendix E. Use it to evaluate the limit in part (a).

$$1^3 + 2^3 + 3^3 + \cdots + n^3 = \left[\frac{n(n + 1)}{2} \right]^2$$

27. Let A be the area under the graph of an increasing continuous function f from a to b, and let L_n and R_n be the approximations to A with n subintervals using left and right endpoints, respectively.

(a) How are A, L_n, and R_n related?

(b) Show that

$$R_n - L_n = \frac{b - a}{n} [f(b) - f(a)]$$

Then draw a diagram to illustrate this equation by showing that the n rectangles representing $R_n - L_n$ can be reassembled to form a single rectangle whose area is the right side of the equation.

(c) Deduce that

$$R_n - A < \frac{b - a}{n} [f(b) - f(a)]$$

28. If A is the area under the curve $y = e^x$ from 1 to 3, use Exercise 27 to find a value of n such that $R_n - A < 0.0001$.

CAS **29.** (a) Express the area under the curve $y = x^5$ from 0 to 2 as a limit.
   (b) Use a computer algebra system to find the sum in your expression from part (a).
   (c) Evaluate the limit in part (a).

CAS **30.** Find the exact area of the region under the graph of $y = e^{-x}$ from 0 to 2 by using a computer algebra system to evaluate the sum and then the limit in Example 3(a). Compare your answer with the estimate obtained in Example 3(b).

CAS **31.** Find the exact area under the cosine curve $y = \cos x$ from $x = 0$ to $x = b$, where $0 \leqslant b \leqslant \pi/2$. (Use a computer algebra system both to evaluate the sum and compute the limit.) In particular, what is the area if $b = \pi/2$?

32. (a) Let A_n be the area of a polygon with n equal sides inscribed in a circle with radius r. By dividing the polygon into n congruent triangles with central angle $2\pi/n$, show that

$$A_n = \tfrac{1}{2}nr^2 \sin\left(\frac{2\pi}{n}\right)$$

   (b) Show that $\lim_{n \to \infty} A_n = \pi r^2$. [*Hint:* Use Equation 3.3.2 on page 191.]

5.2   The Definite Integral

We saw in Section 5.1 that a limit of the form

$$\boxed{1} \quad \lim_{n \to \infty} \sum_{i=1}^{n} f(x_i^*)\,\Delta x = \lim_{n \to \infty}\left[f(x_1^*)\,\Delta x + f(x_2^*)\,\Delta x + \cdots + f(x_n^*)\,\Delta x\right]$$

arises when we compute an area. We also saw that it arises when we try to find the distance traveled by an object. It turns out that this same type of limit occurs in a wide variety of situations even when f is not necessarily a positive function. In Chapters 6 and 8 we will see that limits of the form (1) also arise in finding lengths of curves, volumes of solids, centers of mass, force due to water pressure, and work, as well as other quantities. We therefore give this type of limit a special name and notation.

$\boxed{2}$ **Definition of a Definite Integral** If f is a function defined for $a \leqslant x \leqslant b$, we divide the interval $[a, b]$ into n subintervals of equal width $\Delta x = (b - a)/n$. We let $x_0\, (= a),\ x_1, x_2, \ldots, x_n\, (= b)$ be the endpoints of these subintervals and we let $x_1^*, x_2^*, \ldots, x_n^*$ be any **sample points** in these subintervals, so x_i^* lies in the ith subinterval $[x_{i-1}, x_i]$. Then the **definite integral of f from a to b** is

$$\int_a^b f(x)\,dx = \lim_{n \to \infty} \sum_{i=1}^{n} f(x_i^*)\,\Delta x$$

provided that this limit exists and gives the same value for all possible choices of sample points. If it does exist, we say that f is **integrable** on $[a, b]$.

The precise meaning of the limit that defines the integral is as follows:

For every number $\varepsilon > 0$ there is an integer N such that

$$\left| \int_a^b f(x)\,dx - \sum_{i=1}^{n} f(x_i^*)\,\Delta x \right| < \varepsilon$$

for every integer $n > N$ and for every choice of x_i^* in $[x_{i-1}, x_i]$.

NOTE 1 The symbol $\int$ was introduced by Leibniz and is called an **integral sign**. It is an elongated S and was chosen because an integral is a limit of sums. In the notation

$\int_a^b f(x)\,dx$, $f(x)$ is called the **integrand** and a and b are called the **limits of integration**; a is the **lower limit** and b is the **upper limit**. For now, the symbol dx has no meaning by itself; $\int_a^b f(x)\,dx$ is all one symbol. The dx simply indicates that the independent variable is x. The procedure of calculating an integral is called **integration**.

NOTE 2 The definite integral $\int_a^b f(x)\,dx$ is a number; it does not depend on x. In fact, we could use any letter in place of x without changing the value of the integral:

$$\int_a^b f(x)\,dx = \int_a^b f(t)\,dt = \int_a^b f(r)\,dr$$

NOTE 3 The sum

$$\sum_{i=1}^{n} f(x_i^*)\,\Delta x$$

Riemann

Bernhard Riemann received his Ph.D. under the direction of the legendary Gauss at the University of Göttingen and remained there to teach. Gauss, who was not in the habit of praising other mathematicians, spoke of Riemann's "creative, active, truly mathematical mind and gloriously fertile originality." The definition (2) of an integral that we use is due to Riemann. He also made major contributions to the theory of functions of a complex variable, mathematical physics, number theory, and the foundations of geometry. Riemann's broad concept of space and geometry turned out to be the right setting, 50 years later, for Einstein's general relativity theory. Riemann's health was poor throughout his life, and he died of tuberculosis at the age of 39.

that occurs in Definition 2 is called a **Riemann sum** after the German mathematician Bernhard Riemann (1826–1866). So Definition 2 says that the definite integral of an integrable function can be approximated to within any desired degree of accuracy by a Riemann sum.

We know that if f happens to be positive, then the Riemann sum can be interpreted as a sum of areas of approximating rectangles (see Figure 1). By comparing Definition 2 with the definition of area in Section 5.1, we see that the definite integral $\int_a^b f(x)\,dx$ can be interpreted as the area under the curve $y = f(x)$ from a to b. (See Figure 2.)

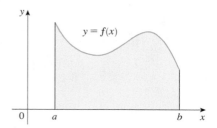

FIGURE 1
If $f(x) \geqslant 0$, the Riemann sum $\sum f(x_i^*)\,\Delta x$ is the sum of areas of rectangles.

FIGURE 2
If $f(x) \geqslant 0$, the integral $\int_a^b f(x)\,dx$ is the area under the curve $y = f(x)$ from a to b.

FIGURE 3

$\sum f(x_i^*)\,\Delta x$ is an approximation to the net area.

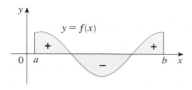

FIGURE 4

$\int_a^b f(x)\,dx$ is the net area.

If f takes on both positive and negative values, as in Figure 3, then the Riemann sum is the sum of the areas of the rectangles that lie above the x-axis and the *negatives* of the areas of the rectangles that lie below the x-axis (the areas of the blue rectangles *minus* the areas of the gold rectangles). When we take the limit of such Riemann sums, we get the situation illustrated in Figure 4. A definite integral can be interpreted as a **net area**, that is, a difference of areas:

$$\int_a^b f(x)\,dx = A_1 - A_2$$

where A_1 is the area of the region above the x-axis and below the graph of f, and A_2 is the area of the region below the x-axis and above the graph of f.

NOTE 4 Although we have defined $\int_a^b f(x)\,dx$ by dividing $[a, b]$ into subintervals of equal width, there are situations in which it is advantageous to work with subintervals of unequal width. For instance, in Exercise 5.1.16, NASA provided velocity data at times that were not equally spaced, but we were still able to estimate the distance traveled. And there are methods for numerical integration that take advantage of unequal subintervals.

If the subinterval widths are $\Delta x_1, \Delta x_2, \ldots, \Delta x_n$, we have to ensure that all these widths approach 0 in the limiting process. This happens if the largest width, max Δx_i, approaches 0. So in this case the definition of a definite integral becomes

$$\int_a^b f(x)\,dx = \lim_{\max \Delta x_i \to 0} \sum_{i=1}^n f(x_i^*)\,\Delta x_i$$

NOTE 5 We have defined the definite integral for an integrable function, but not all functions are integrable (see Exercises 71–72). The following theorem shows that the most commonly occurring functions are in fact integrable. The theorem is proved in more advanced courses.

---

3   **Theorem**   If f is continuous on $[a, b]$, or if f has only a finite number of jump discontinuities, then f is integrable on $[a, b]$; that is, the definite integral $\int_a^b f(x)\,dx$ exists.

If f is integrable on $[a, b]$, then the limit in Definition 2 exists and gives the same value no matter how we choose the sample points x_i^*. To simplify the calculation of the integral we often take the sample points to be right endpoints. Then $x_i^* = x_i$ and the definition of an integral simplifies as follows.

---

4   **Theorem**   If f is integrable on $[a, b]$, then

$$\int_a^b f(x)\,dx = \lim_{n \to \infty} \sum_{i=1}^n f(x_i)\,\Delta x$$

where $$\Delta x = \frac{b - a}{n} \qquad \text{and} \qquad x_i = a + i\,\Delta x$$

EXAMPLE 1   Express

$$\lim_{n \to \infty} \sum_{i=1}^n (x_i^3 + x_i \sin x_i)\,\Delta x$$

as an integral on the interval $[0, \pi]$.

SOLUTION Comparing the given limit with the limit in Theorem 4, we see that they will be identical if we choose $f(x) = x^3 + x \sin x$. We are given that $a = 0$ and $b = \pi$. Therefore, by Theorem 4, we have

$$\lim_{n \to \infty} \sum_{i=1}^n (x_i^3 + x_i \sin x_i)\,\Delta x = \int_0^\pi (x^3 \, 1 \, x \sin x)\,dx \qquad \blacksquare$$

Later, when we apply the definite integral to physical situations, it will be important to recognize limits of sums as integrals, as we did in Example 1. When Leibniz chose the notation for an integral, he chose the ingredients as reminders of the limiting process. In general, when we write

$$\lim_{n \to \infty} \sum_{i=1}^{n} f(x_i^*) \, \Delta x = \int_a^b f(x) \, dx$$

we replace $\lim \Sigma$ by $\int$, x_i^* by x, and Δx by dx.

■ Evaluating Integrals

When we use a limit to evaluate a definite integral, we need to know how to work with sums. The following three equations give formulas for sums of powers of positive integers. Equation 5 may be familiar to you from a course in algebra. Equations 6 and 7 were discussed in Section 5.1 and are proved in Appendix E.

$\boxed{5}$

$$\sum_{i=1}^{n} i = \frac{n(n+1)}{2}$$

$\boxed{6}$

$$\sum_{i=1}^{n} i^2 = \frac{n(n+1)(2n+1)}{6}$$

$\boxed{7}$

$$\sum_{i=1}^{n} i^3 = \left[\frac{n(n+1)}{2} \right]^2$$

The remaining formulas are simple rules for working with sigma notation:

Formulas 8–11 are proved by writing out each side in expanded form. The left side of Equation 9 is

$$ca_1 + ca_2 + \cdots + ca_n$$

The right side is

$$c(a_1 + a_2 + \cdots + a_n)$$

These are equal by the distributive property. The other formulas are discussed in Appendix E.

$\boxed{8}$

$$\sum_{i=1}^{n} c = nc$$

$\boxed{9}$

$$\sum_{i=1}^{n} ca_i = c \sum_{i=1}^{n} a_i$$

$\boxed{10}$

$$\sum_{i=1}^{n} (a_i + b_i) = \sum_{i=1}^{n} a_i + \sum_{i=1}^{n} b_i$$

$\boxed{11}$

$$\sum_{i=1}^{n} (a_i - b_i) = \sum_{i=1}^{n} a_i - \sum_{i=1}^{n} b_i$$

EXAMPLE 2

(a) Evaluate the Riemann sum for $f(x) = x^3 - 6x$, taking the sample points to be right endpoints and $a = 0$, $b = 3$, and $n = 6$.

(b) Evaluate $\int_0^3 (x^3 - 6x) \, dx$.

SOLUTION

(a) With $n = 6$ the interval width is

$$\Delta x = \frac{b - a}{n} = \frac{3 - 0}{6} = \frac{1}{2}$$

and the right endpoints are $x_1 = 0.5$, $x_2 = 1.0$, $x_3 = 1.5$, $x_4 = 2.0$, $x_5 = 2.5$, and

$x_6 = 3.0$. So the Riemann sum is

$$R_6 = \sum_{i=1}^{6} f(x_i)\, \Delta x$$

$$= f(0.5)\, \Delta x + f(1.0)\, \Delta x + f(1.5)\, \Delta x + f(2.0)\, \Delta x + f(2.5)\, \Delta x + f(3.0)\, \Delta x$$

$$= \tfrac{1}{2}(-2.875 - 5 - 5.625 - 4 + 0.625 + 9)$$

$$= -3.9375$$

Notice that f is not a positive function and so the Riemann sum does not represent a sum of areas of rectangles. But it does represent the sum of the areas of the blue rectangles (above the x-axis) minus the sum of the areas of the gold rectangles (below the x-axis) in Figure 5.

(b) With n subintervals we have

$$\Delta x = \frac{b - a}{n} = \frac{3}{n}$$

So $x_0 = 0$, $x_1 = 3/n$, $x_2 = 6/n$, $x_3 = 9/n$, and, in general, $x_i = 3i/n$. Since we are using right endpoints, we can use Theorem 4:

$$\int_0^3 (x^3 - 6x)\, dx = \lim_{n \to \infty} \sum_{i=1}^{n} f(x_i)\, \Delta x = \lim_{n \to \infty} \sum_{i=1}^{n} f\!\left(\frac{3i}{n}\right) \frac{3}{n}$$

In the sum, n is a constant (unlike i), so we can move $3/n$ in front of the Σ sign.

$$= \lim_{n \to \infty} \frac{3}{n} \sum_{i=1}^{n} \left[\left(\frac{3i}{n}\right)^3 - 6\left(\frac{3i}{n}\right) \right] \qquad \text{(Equation 9 with } c = 3/n)$$

$$= \lim_{n \to \infty} \frac{3}{n} \sum_{i=1}^{n} \left[\frac{27}{n^3} i^3 - \frac{18}{n} i \right]$$

$$= \lim_{n \to \infty} \left[\frac{81}{n^4} \sum_{i=1}^{n} i^3 - \frac{54}{n^2} \sum_{i=1}^{n} i \right] \qquad \text{(Equations 11 and 9)}$$

$$= \lim_{n \to \infty} \left\{ \frac{81}{n^4} \left[\frac{n(n + 1)}{2} \right]^2 - \frac{54}{n^2}\, \frac{n(n + 1)}{2} \right\} \qquad \text{(Equations 7 and 5)}$$

$$= \lim_{n \to \infty} \left[\frac{81}{4} \left(1 + \frac{1}{n}\right)^2 - 27\left(1 + \frac{1}{n}\right) \right]$$

$$= \frac{81}{4} - 27 = -\frac{27}{4} = -6.75$$

This integral can't be interpreted as an area because f takes on both positive and negative values. But it can be interpreted as the difference of areas $A_1 - A_2$, where A_1 and A_2 are shown in Figure 6.

FIGURE 5

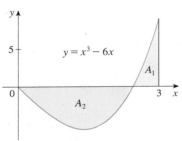

FIGURE 6

$$\int_0^3 (x^3 - 6x)\, dx = A_1 - A_2 = -6.75$$

Figure 7 illustrates the calculation by showing the positive and negative terms in the right Riemann sum R_n for $n = 40$. The values in the table show the Riemann sums approaching the exact value of the integral, -6.75, as $n \to \infty$.

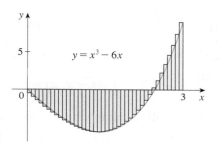

n	R_n
40	-6.3998
100	-6.6130
500	-6.7229
1000	-6.7365
5000	-6.7473

FIGURE 7
$R_{40} \approx -6.3998$

A much simpler method for evaluating the integral in Example 2 will be given in Section 5.4.

EXAMPLE 3

(a) Set up an expression for $\int_1^3 e^x\,dx$ as a limit of sums.

(b) Use a computer algebra system to evaluate the expression.

Because $f(x) = e^x$ is positive, the integral in Example 3 represents the area shown in Figure 8.

FIGURE 8

SOLUTION

(a) Here we have $f(x) = e^x$, $a = 1$, $b = 3$, and

$$\Delta x = \frac{b-a}{n} = \frac{2}{n}$$

So $x_0 = 1$, $x_1 = 1 + 2/n$, $x_2 = 1 + 4/n$, $x_3 = 1 + 6/n$, and

$$x_i = 1 + \frac{2i}{n}$$

From Theorem 4, we get

$$\int_1^3 e^x\,dx = \lim_{n\to\infty} \sum_{i=1}^n f(x_i)\,\Delta x$$

$$= \lim_{n\to\infty} \sum_{i=1}^n f\left(1 + \frac{2i}{n}\right)\frac{2}{n}$$

$$= \lim_{n\to\infty} \frac{2}{n}\sum_{i=1}^n e^{1+2i/n}$$

A computer algebra system is able to find an explicit expression for this sum because it is a geometric series. The limit could be found using l'Hospital's Rule.

(b) If we ask a computer algebra system to evaluate the sum and simplify, we obtain

$$\sum_{i=1}^n e^{1+2i/n} = \frac{e^{(3n+2)/n} - e^{(n+2)/n}}{e^{2/n} - 1}$$

Now we ask the computer algebra system to evaluate the limit:

$$\int_1^3 e^x\,dx = \lim_{n\to\infty} \frac{2}{n} \cdot \frac{e^{(3n+2)/n} - e^{(n+2)/n}}{e^{2/n} - 1} = e^3 - e$$

We will learn a much easier method for the evaluation of integrals in the next section.

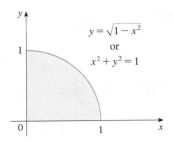

$y = \sqrt{1 - x^2}$
or
$x^2 + y^2 = 1$

FIGURE 9

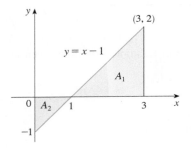

$(3, 2)$

$y = x - 1$

A_1

A_2

FIGURE 10

TEC Module 5.2/7.7 shows how the Midpoint Rule estimates improve as n increases.

EXAMPLE 4 Evaluate the following integrals by interpreting each in terms of areas.

(a) $\displaystyle\int_0^1 \sqrt{1 - x^2}\, dx$ (b) $\displaystyle\int_0^3 (x - 1)\, dx$

SOLUTION

(a) Since $f(x) = \sqrt{1 - x^2} \geqslant 0$, we can interpret this integral as the area under the curve $y = \sqrt{1 - x^2}$ from 0 to 1. But, since $y^2 = 1 - x^2$, we get $x^2 + y^2 = 1$, which shows that the graph of f is the quarter-circle with radius 1 in Figure 9. Therefore

$$\int_0^1 \sqrt{1 - x^2}\, dx = \tfrac{1}{4}\pi(1)^2 = \frac{\pi}{4}$$

(In Section 7.3 we will be able to *prove* that the area of a circle of radius r is πr^2.)

(b) The graph of $y = x - 1$ is the line with slope 1 shown in Figure 10. We compute the integral as the difference of the areas of the two triangles:

$$\int_0^3 (x - 1)\, dx = A_1 - A_2 = \tfrac{1}{2}(2 \cdot 2) - \tfrac{1}{2}(1 \cdot 1) = 1.5 \qquad \blacksquare$$

▪ The Midpoint Rule

We often choose the sample point x_i^* to be the right endpoint of the ith subinterval because it is convenient for computing the limit. But if the purpose is to find an *approximation* to an integral, it is usually better to choose x_i^* to be the midpoint of the interval, which we denote by $\bar{x}_i$. Any Riemann sum is an approximation to an integral, but if we use midpoints we get the following approximation.

Midpoint Rule

$$\int_a^b f(x)\, dx \approx \sum_{i=1}^{n} f(\bar{x}_i)\, \Delta x = \Delta x\, [f(\bar{x}_1) + \cdots + f(\bar{x}_n)]$$

where

$$\Delta x = \frac{b - a}{n}$$

and

$$\bar{x}_i = \tfrac{1}{2}(x_{i-1} + x_i) = \text{midpoint of } [x_{i-1}, x_i]$$

EXAMPLE 5 Use the Midpoint Rule with $n = 5$ to approximate $\displaystyle\int_1^2 \frac{1}{x}\, dx$.

SOLUTION The endpoints of the five subintervals are 1, 1.2, 1.4, 1.6, 1.8, and 2.0, so the midpoints are 1.1, 1.3, 1.5, 1.7, and 1.9. The width of the subintervals is $\Delta x = (2 - 1)/5 = \tfrac{1}{5}$, so the Midpoint Rule gives

$$\int_1^2 \frac{1}{x}\, dx \approx \Delta x\, [f(1.1) + f(1.3) + f(1.5) + f(1.7) + f(1.9)]$$

$$= \frac{1}{5} \left(\frac{1}{1.1} + \frac{1}{1.3} + \frac{1}{1.5} + \frac{1}{1.7} + \frac{1}{1.9} \right)$$

$$\approx 0.691908$$

Since $f(x) = 1/x > 0$ for $1 \leqslant x \leqslant 2$, the integral represents an area, and the approxi-

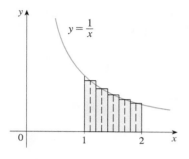

$y = \dfrac{1}{x}$

FIGURE 11

TEC In Visual 5.2 you can compare left, right, and midpoint approximations to the integral in Example 2 for different values of n.

FIGURE 12
$M_{40} \approx -6.7563$

mation given by the Midpoint Rule is the sum of the areas of the rectangles shown in Figure 11.

At the moment we don't know how accurate the approximation in Example 5 is, but in Section 7.7 we will learn a method for estimating the error involved in using the Midpoint Rule. At that time we will discuss other methods for approximating definite integrals.

If we apply the Midpoint Rule to the integral in Example 2, we get the picture in Figure 12. The approximation $M_{40} \approx -6.7563$ is much closer to the true value -6.75 than the right endpoint approximation, $R_{40} \approx -6.3998$, shown in Figure 7.

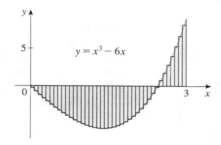

$y = x^3 - 6x$

Properties of the Definite Integral

When we defined the definite integral $\int_a^b f(x)\,dx$, we implicitly assumed that $a < b$. But the definition as a limit of Riemann sums makes sense even if $a > b$. Notice that if we reverse a and b, then Δx changes from $(b - a)/n$ to $(a - b)/n$. Therefore

$$\int_b^a f(x)\,dx = -\int_a^b f(x)\,dx$$

If $a = b$, then $\Delta x = 0$ and so

$$\int_a^a f(x)\,dx = 0$$

We now develop some basic properties of integrals that will help us to evaluate integrals in a simple manner. We assume that f and g are continuous functions.

Properties of the Integral

1. $\displaystyle\int_a^b c\,dx = c(b - a)$,   where c is any constant

2. $\displaystyle\int_a^b [f(x) + g(x)]\,dx = \int_a^b f(x)\,dx + \int_a^b g(x)\,dx$

3. $\displaystyle\int_a^b cf(x)\,dx = c\int_a^b f(x)\,dx$,   where c is any constant

4. $\displaystyle\int_a^b [f(x) - g(x)]\,dx = \int_a^b f(x)\,dx - \int_a^b g(x)\,dx$

Property 1 says that the integral of a constant function $f(x) = c$ is the constant times the length of the interval. If $c > 0$ and $a < b$, this is to be expected because $c(b - a)$ is the area of the shaded rectangle in Figure 13.

FIGURE 13

$$\int_a^b c \, dx = c(b - a)$$

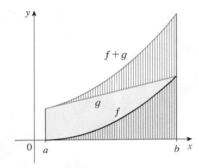

FIGURE 14

$$\int_a^b [f(x) + g(x)] \, dx =$$
$$\int_a^b f(x) \, dx + \int_a^b g(x) \, dx$$

Property 3 seems intuitively reasonable because we know that multiplying a function by a positive number c stretches or shrinks its graph vertically by a factor of c. So it stretches or shrinks each approximating rectangle by a factor c and therefore it has the effect of multiplying the area by c.

Property 2 says that the integral of a sum is the sum of the integrals. For positive functions it says that the area under $f + g$ is the area under f plus the area under g. Figure 14 helps us understand why this is true: In view of how graphical addition works, the corresponding vertical line segments have equal height.

In general, Property 2 follows from Theorem 4 and the fact that the limit of a sum is the sum of the limits:

$$\int_a^b [f(x) + g(x)] \, dx = \lim_{n \to \infty} \sum_{i=1}^n [f(x_i) + g(x_i)] \, \Delta x$$

$$= \lim_{n \to \infty} \left[\sum_{i=1}^n f(x_i) \, \Delta x + \sum_{i=1}^n g(x_i) \, \Delta x \right]$$

$$= \lim_{n \to \infty} \sum_{i=1}^n f(x_i) \, \Delta x + \lim_{n \to \infty} \sum_{i=1}^n g(x_i) \, \Delta x$$

$$= \int_a^b f(x) \, dx + \int_a^b g(x) \, dx$$

Property 3 can be proved in a similar manner and says that the integral of a constant times a function is the constant times the integral of the function. In other words, a constant (but *only* a constant) can be taken in front of an integral sign. Property 4 is proved by writing $f - g = f + (-g)$ and using Properties 2 and 3 with $c = -1$.

EXAMPLE 6 Use the properties of integrals to evaluate $\int_0^1 (4 + 3x^2) \, dx$.

SOLUTION Using Properties 2 and 3 of integrals, we have

$$\int_0^1 (4 + 3x^2) \, dx = \int_0^1 4 \, dx + \int_0^1 3x^2 \, dx = \int_0^1 4 \, dx + 3 \int_0^1 x^2 \, dx$$

We know from Property 1 that

$$\int_0^1 4 \, dx = 4(1 - 0) = 4$$

and we found in Example 5.1.2 that $\int_0^1 x^2 \, dx = \frac{1}{3}$. So

$$\int_0^1 (4 + 3x^2) \, dx = \int_0^1 4 \, dx + 3 \int_0^1 x^2 \, dx$$

$$= 4 + 3 \cdot \frac{1}{3} = 5 \qquad \blacksquare$$

The next property tells us how to combine integrals of the same function over adjacent intervals.

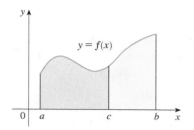

$y = f(x)$

FIGURE 15

5.
$$\int_a^c f(x)\, dx + \int_c^b f(x)\, dx = \int_a^b f(x)\, dx$$

This is not easy to prove in general, but for the case where $f(x) \geqslant 0$ and $a < c < b$ Property 5 can be seen from the geometric interpretation in Figure 15: the area under $y = f(x)$ from a to c plus the area from c to b is equal to the total area from a to b.

EXAMPLE 7 If it is known that $\int_0^{10} f(x)\, dx = 17$ and $\int_0^8 f(x)\, dx = 12$, find $\int_8^{10} f(x)\, dx$.

SOLUTION By Property 5, we have

$$\int_0^8 f(x)\, dx + \int_8^{10} f(x)\, dx = \int_0^{10} f(x)\, dx$$

so

$$\int_8^{10} f(x)\, dx = \int_0^{10} f(x)\, dx - \int_0^8 f(x)\, dx = 17 - 12 = 5 \qquad \blacksquare$$

Properties 1–5 are true whether $a < b$, $a = b$, or $a > b$. The following properties, in which we compare sizes of functions and sizes of integrals, are true only if $a \leqslant b$.

Comparison Properties of the Integral

6. If $f(x) \geqslant 0$ for $a \leqslant x \leqslant b$, then $\int_a^b f(x)\, dx \geqslant 0$.

7. If $f(x) \geqslant g(x)$ for $a \leqslant x \leqslant b$, then $\int_a^b f(x)\, dx \geqslant \int_a^b g(x)\, dx$.

8. If $m \leqslant f(x) \leqslant M$ for $a \leqslant x \leqslant b$, then

$$m(b - a) \leqslant \int_a^b f(x)\, dx \leqslant M(b - a)$$

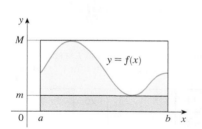

$y = f(x)$

FIGURE 16

If $f(x) \geqslant 0$, then $\int_a^b f(x)\, dx$ represents the area under the graph of f, so the geometric interpretation of Property 6 is simply that areas are positive. (It also follows directly from the definition because all the quantities involved are positive.) Property 7 says that a bigger function has a bigger integral. It follows from Properties 6 and 4 because $f - g \geqslant 0$.

Property 8 is illustrated by Figure 16 for the case where $f(x) \geqslant 0$. If f is continuous, we could take m and M to be the absolute minimum and maximum values of f on the interval $[a, b]$. In this case Property 8 says that the area under the graph of f is greater than the area of the rectangle with height m and less than the area of the rectangle with height M.

PROOF OF PROPERTY 8 Since $m \leqslant f(x) \leqslant M$, Property 7 gives

$$\int_a^b m\, dx \leqslant \int_a^b f(x)\, dx \leqslant \int_a^b M\, dx$$

Using Property 1 to evaluate the integrals on the left and right sides, we obtain

$$m(b - a) \leqslant \int_a^b f(x)\, dx \leqslant M(b - a) \qquad \blacksquare$$

Property 8 is useful when all we want is a rough estimate of the size of an integral without going to the bother of using the Midpoint Rule.

EXAMPLE 8 Use Property 8 to estimate $\int_0^1 e^{-x^2}\,dx$.

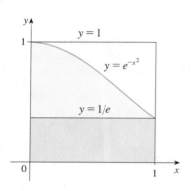

y = 1

$y = e^{-x^2}$

y = 1/e

SOLUTION Because $f(x) = e^{-x^2}$ is a decreasing function on $[0, 1]$, its absolute maximum value is $M = f(0) = 1$ and its absolute minimum value is $m = f(1) = e^{-1}$. Thus, by Property 8,

$$e^{-1}(1 - 0) \leqslant \int_0^1 e^{-x^2}\,dx \leqslant 1(1 - 0)$$

or

$$e^{-1} \leqslant \int_0^1 e^{-x^2}\,dx \leqslant 1$$

Since $e^{-1} \approx 0.3679$, we can write

$$0.367 \leqslant \int_0^1 e^{-x^2}\,dx \leqslant 1$$ ■

FIGURE 17

The result of Example 8 is illustrated in Figure 17. The integral is greater than the area of the lower rectangle and less than the area of the square.

5.2 EXERCISES

1. Evaluate the Riemann sum for $f(x) = x - 1$, $-6 \leqslant x \leqslant 4$, with five subintervals, taking the sample points to be right endpoints. Explain, with the aid of a diagram, what the Riemann sum represents.

2. If

$$f(x) = \cos x \qquad 0 \leqslant x \leqslant 3\pi/4$$

evaluate the Riemann sum with $n = 6$, taking the sample points to be left endpoints. (Give your answer correct to six decimal places.) What does the Riemann sum represent? Illustrate with a diagram.

3. If $f(x) = x^2 - 4$, $0 \leqslant x \leqslant 3$, find the Riemann sum with $n = 6$, taking the sample points to be midpoints. What does the Riemann sum represent? Illustrate with a diagram.

4. (a) Find the Riemann sum for $f(x) = 1/x$, $1 \leqslant x \leqslant 2$, with four terms, taking the sample points to be right endpoints. (Give your answer correct to six decimal places.) Explain what the Riemann sum represents with the aid of a sketch.
   (b) Repeat part (a) with midpoints as the sample points.

5. The graph of a function f is given. Estimate $\int_0^{10} f(x)\,dx$ using five subintervals with (a) right endpoints, (b) left endpoints, and (c) midpoints.

6. The graph of g is shown. Estimate $\int_{-2}^{4} g(x)\,dx$ with six subintervals using (a) right endpoints, (b) left endpoints, and (c) midpoints.

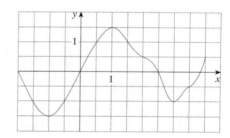

7. A table of values of an increasing function f is shown. Use the table to find lower and upper estimates for $\int_{10}^{30} f(x)\,dx$.

x	10	14	18	22	26	30
$f(x)$	-12	-6	-2	1	3	8

8. The table gives the values of a function obtained from an experiment. Use them to estimate $\int_3^9 f(x)\,dx$ using three equal subintervals with (a) right endpoints, (b) left endpoints, and (c) midpoints. If the function is known to be an increasing function, can you say whether your estimates are less than or greater than the exact value of the integral?

x	3	4	5	6	7	8	9
$f(x)$	-3.4	-2.1	-0.6	0.3	0.9	1.4	1.8

9–12 Use the Midpoint Rule with the given value of n to approximate the integral. Round the answer to four decimal places.

9. $\displaystyle\int_0^8 \sin\sqrt{x}\, dx, \quad n = 4$

10. $\displaystyle\int_0^1 \sqrt{x^3 + 1}\, dx, \quad n = 5$

11. $\displaystyle\int_0^2 \frac{x}{x+1}\, dx, \quad n = 5$

12. $\displaystyle\int_0^\pi x \sin^2 x\, dx, \quad n = 4$

CAS 13. If you have a CAS that evaluates midpoint approximations and graphs the corresponding rectangles (use RiemannSum or middlesum and middlebox commands in Maple), check the answer to Exercise 11 and illustrate with a graph. Then repeat with $n = 10$ and $n = 20$.

14. With a programmable calculator or computer (see the instructions for Exercise 5.1.9), compute the left and right Riemann sums for the function $f(x) = x/(x + 1)$ on the interval $[0, 2]$ with $n = 100$. Explain why these estimates show that

$$0.8946 < \int_0^2 \frac{x}{x+1}\, dx < 0.9081$$

15. Use a calculator or computer to make a table of values of right Riemann sums R_n for the integral $\int_0^\pi \sin x\, dx$ with $n = 5, 10, 50,$ and 100. What value do these numbers appear to be approaching?

16. Use a calculator or computer to make a table of values of left and right Riemann sums L_n and R_n for the integral $\int_0^2 e^{-x^2}\, dx$ with $n = 5, 10, 50,$ and 100. Between what two numbers must the value of the integral lie? Can you make a similar statement for the integral $\int_{-1}^2 e^{-x^2}\, dx$? Explain.

17–20 Express the limit as a definite integral on the given interval.

17. $\displaystyle\lim_{n\to\infty} \sum_{i=1}^n \frac{e^{x_i}}{1 + x_i} \Delta x, \quad [0, 1]$

18. $\displaystyle\lim_{n\to\infty} \sum_{i=1}^n x_i \sqrt{1 + x_i^3}\, \Delta x, \quad [2, 5]$

19. $\displaystyle\lim_{n\to\infty} \sum_{i=1}^n [5(x_i^*)^3 - 4x_i^*] \Delta x, \quad [2, 7]$

20. $\displaystyle\lim_{n\to\infty} \sum_{i=1}^n \frac{x_i^*}{(x_i^*)^2 + 4} \Delta x, \quad [1, 3]$

21–25 Use the form of the definition of the integral given in Theorem 4 to evaluate the integral.

21. $\displaystyle\int_2^5 (4 - 2x)\, dx$

22. $\displaystyle\int_1^4 (x^2 - 4x + 2)\, dx$

23. $\displaystyle\int_{-2}^0 (x^2 + x)\, dx$

24. $\displaystyle\int_0^2 (2x - x^3)\, dx$

25. $\displaystyle\int_0^1 (x^3 - 3x^2)\, dx$

26. (a) Find an approximation to the integral $\int_0^4 (x^2 - 3x)\, dx$ using a Riemann sum with right endpoints and $n = 8$.
(b) Draw a diagram like Figure 3 to illustrate the approximation in part (a).
(c) Use Theorem 4 to evaluate $\int_0^4 (x^2 - 3x)\, dx$.
(d) Interpret the integral in part (c) as a difference of areas and illustrate with a diagram like Figure 4.

27. Prove that $\displaystyle\int_a^b x\, dx = \frac{b^2 - a^2}{2}$.

28. Prove that $\displaystyle\int_a^b x^2\, dx = \frac{b^3 - a^3}{3}$.

29–30 Express the integral as a limit of Riemann sums. Do not evaluate the limit.

29. $\displaystyle\int_1^3 \sqrt{4 + x^2}\, dx$

30. $\displaystyle\int_2^5 \left(x^2 + \frac{1}{x}\right) dx$

CAS 31–32 Express the integral as a limit of sums. Then evaluate, using a computer algebra system to find both the sum and the limit.

31. $\displaystyle\int_0^\pi \sin 5x\, dx$

32. $\displaystyle\int_2^{10} x^6\, dx$

---

33. The graph of f is shown. Evaluate each integral by interpreting it in terms of areas.

(a) $\displaystyle\int_0^2 f(x)\, dx$

(b) $\displaystyle\int_0^5 f(x)\, dx$

(c) $\displaystyle\int_5^7 f(x)\, dx$

(d) $\displaystyle\int_0^9 f(x)\, dx$

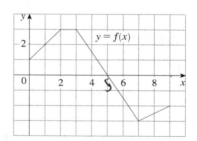

34. The graph of g consists of two straight lines and a semicircle. Use it to evaluate each integral.

(a) $\displaystyle\int_0^2 g(x)\, dx$

(b) $\displaystyle\int_2^6 g(x)\, dx$

(c) $\displaystyle\int_0^7 g(x)\, dx$

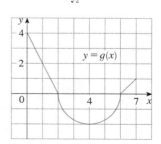

35–40 Evaluate the integral by interpreting it in terms of areas.

35. $\int_{-1}^{2} (1 - x)\, dx$

36. $\int_{0}^{9} \left(\frac{1}{3}x - 2\right) dx$

37. $\int_{-3}^{0} \left(1 + \sqrt{9 - x^2}\right) dx$

38. $\int_{-5}^{5} \left(x - \sqrt{25 - x^2}\right) dx$

39. $\int_{-4}^{3} \left|\frac{1}{2}x\right| dx$

40. $\int_{0}^{1} |2x - 1|\, dx$

---

41. Evaluate $\int_{1}^{1} \sqrt{1 + x^4}\, dx$.

42. Given that $\int_{0}^{\pi} \sin^4 x\, dx = \frac{3}{8}\pi$, what is $\int_{\pi}^{0} \sin^4 \theta\, d\theta$?

43. In Example 5.1.2 we showed that $\int_{0}^{1} x^2\, dx = \frac{1}{3}$. Use this fact and the properties of integrals to evaluate $\int_{0}^{1} (5 - 6x^2)\, dx$.

44. Use the properties of integrals and the result of Example 3 to evaluate $\int_{1}^{3} (2e^x - 1)\, dx$.

45. Use the result of Example 3 to evaluate $\int_{1}^{3} e^{x+2}\, dx$.

46. Use the result of Exercise 27 and the fact that $\int_{0}^{\pi/2} \cos x\, dx = 1$ (from Exercise 5.1.31), together with the properties of integrals, to evaluate $\int_{0}^{\pi/2} (2 \cos x - 5x)\, dx$.

47. Write as a single integral in the form $\int_{a}^{b} f(x)\, dx$:

$$\int_{-2}^{2} f(x)\, dx + \int_{2}^{5} f(x)\, dx - \int_{-2}^{-1} f(x)\, dx$$

48. If $\int_{2}^{8} f(x)\, dx = 7.3$ and $\int_{2}^{4} f(x)\, dx = 5.9$, find $\int_{4}^{8} f(x)\, dx$.

49. If $\int_{0}^{9} f(x)\, dx = 37$ and $\int_{0}^{9} g(x)\, dx = 16$, find

$$\int_{0}^{9} [2f(x) + 3g(x)]\, dx$$

50. Find $\int_{0}^{5} f(x)\, dx$ if

$$f(x) = \begin{cases} 3 & \text{for } x < 3 \\ x & \text{for } x \geqslant 3 \end{cases}$$

51. For the function f whose graph is shown, list the following quantities in increasing order, from smallest to largest, and explain your reasoning.

(A) $\int_{0}^{8} f(x)\, dx$ (B) $\int_{0}^{3} f(x)\, dx$ (C) $\int_{3}^{8} f(x)\, dx$

(D) $\int_{4}^{8} f(x)\, dx$ (E) $f'(1)$

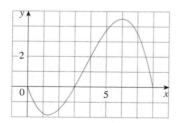

52. If $F(x) = \int_{2}^{x} f(t)\, dt$, where f is the function whose graph is given, which of the following values is largest?

(A) $F(0)$ (B) $F(1)$ (C) $F(2)$

(D) $F(3)$ (E) $F(4)$

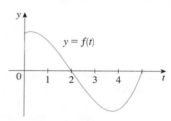

53. Each of the regions A, B, and C bounded by the graph of f and the x-axis has area 3. Find the value of

$$\int_{-4}^{2} [f(x) + 2x + 5]\, dx$$

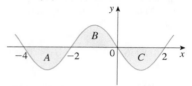

54. Suppose f has absolute minimum value m and absolute maximum value M. Between what two values must $\int_{0}^{2} f(x)\, dx$ lie? Which property of integrals allows you to make your conclusion?

55–58 Use the properties of integrals to verify the inequality without evaluating the integrals.

55. $\int_{0}^{4} (x^2 - 4x + 4)\, dx \geqslant 0$

56. $\int_{0}^{1} \sqrt{1 + x^2}\, dx \leqslant \int_{0}^{1} \sqrt{1 + x}\, dx$

57. $2 \leqslant \int_{-1}^{1} \sqrt{1 + x^2}\, dx \leqslant 2\sqrt{2}$

58. $\dfrac{\pi}{12} \leqslant \int_{\pi/6}^{\pi/3} \sin x\, dx \leqslant \dfrac{\sqrt{3}\,\pi}{12}$

59–64 Use Property 8 to estimate the value of the integral.

59. $\int_{0}^{1} x^3\, dx$

60. $\int_{0}^{3} \dfrac{1}{x + 4}\, dx$

61. $\int_{\pi/4}^{\pi/3} \tan x\, dx$

62. $\int_{0}^{2} (x^3 - 3x + 3)\, dx$

63. $\int_{0}^{2} x e^{-x}\, dx$

64. $\int_{\pi}^{2\pi} (x - 2 \sin x)\, dx$

65–66 Use properties of integrals, together with Exercises 27 and 28, to prove the inequality.

65. $\int_{1}^{3} \sqrt{x^4 + 1}\, dx \geqslant \dfrac{26}{3}$

66. $\int_{0}^{\pi/2} x \sin x\, dx \leqslant \dfrac{\pi^2}{8}$

67. Which of the integrals $\int_1^2 \arctan x \, dx$, $\int_1^2 \arctan \sqrt{x} \, dx$, and $\int_1^2 \arctan(\sin x) \, dx$ has the largest value? Why?

68. Which of the integrals $\int_0^{0.5} \cos(x^2) \, dx$, $\int_0^{0.5} \cos \sqrt{x} \, dx$ is larger? Why?

69. Prove Property 3 of integrals.

70. (a) If f is continuous on $[a, b]$, show that

$$\left| \int_a^b f(x) \, dx \right| \le \int_a^b |f(x)| \, dx$$

[*Hint:* $-|f(x)| \le f(x) \le |f(x)|$.]

(b) Use the result of part (a) to show that

$$\left| \int_0^{2\pi} f(x) \sin 2x \, dx \right| \le \int_0^{2\pi} |f(x)| \, dx$$

71. Let $f(x) = 0$ if x is any rational number and $f(x) = 1$ if x is any irrational number. Show that f is not integrable on $[0, 1]$.

72. Let $f(0) = 0$ and $f(x) = 1/x$ if $0 < x \le 1$. Show that f is not integrable on $[0, 1]$. [*Hint:* Show that the first term in the Riemann sum, $f(x_1^*) \, \Delta x$, can be made arbitrarily large.]

73–74 Express the limit as a definite integral.

73. $\displaystyle \lim_{n \to \infty} \sum_{i=1}^n \frac{i^4}{n^5}$ [*Hint:* Consider $f(x) = x^4$.]

74. $\displaystyle \lim_{n \to \infty} \frac{1}{n} \sum_{i=1}^n \frac{1}{1 + (i/n)^2}$

75. Find $\int_1^2 x^{-2} \, dx$. *Hint:* Choose x_i^* to be the geometric mean of x_{i-1} and x_i (that is, $x_i^* = \sqrt{x_{i-1} x_i}$) and use the identity

$$\frac{1}{m(m+1)} = \frac{1}{m} - \frac{1}{m+1}$$

DISCOVERY PROJECT   AREA FUNCTIONS

1. (a) Draw the line $y = 2t + 1$ and use geometry to find the area under this line, above the t-axis, and between the vertical lines $t = 1$ and $t = 3$.

(b) If $x > 1$, let $A(x)$ be the area of the region that lies under the line $y = 2t + 1$ between $t = 1$ and $t = x$. Sketch this region and use geometry to find an expression for $A(x)$.

(c) Differentiate the area function $A(x)$. What do you notice?

2. (a) If $x \ge -1$, let

$$A(x) = \int_{-1}^x (1 + t^2) \, dt$$

$A(x)$ represents the area of a region. Sketch that region.

(b) Use the result of Exercise 5.2.28 to find an expression for $A(x)$.

(c) Find $A'(x)$. What do you notice?

(d) If $x \ge -1$ and h is a small positive number, then $A(x + h) - A(x)$ represents the area of a region. Describe and sketch the region.

(e) Draw a rectangle that approximates the region in part (d). By comparing the areas of these two regions, show that

$$\frac{A(x + h) - A(x)}{h} \approx 1 + x^2$$

(f) Use part (e) to give an intuitive explanation for the result of part (c).

3. (a) Draw the graph of the function $f(x) = \cos(x^2)$ in the viewing rectangle $[0, 2]$ by $[-1.25, 1.25]$.

(b) If we define a new function g by

$$g(x) = \int_0^x \cos(t^2) \, dt$$

then $g(x)$ is the area under the graph of f from 0 to x [until $f(x)$ becomes negative, at which point $g(x)$ becomes a difference of areas]. Use part (a) to determine the value of x at which $g(x)$ starts to decrease. [Unlike the integral in Problem 2, it is impossible to evaluate the integral defining g to obtain an explicit expression for $g(x)$.]

(c) Use the integration command on your calculator or computer to estimate $g(0.2)$, $g(0.4)$, $g(0.6)$, ..., $g(1.8)$, $g(2)$. Then use these values to sketch a graph of g.

(d) Use your graph of g from part (c) to sketch the graph of g' using the interpretation of $g'(x)$ as the slope of a tangent line. How does the graph of g' compare with the graph of f?

4. Suppose f is a continuous function on the interval $[a, b]$ and we define a new function g by the equation

$$g(x) = \int_a^x f(t)\, dt$$

Based on your results in Problems 1–3, conjecture an expression for $g'(x)$.

5.3   The Fundamental Theorem of Calculus

The Fundamental Theorem of Calculus is appropriately named because it establishes a connection between the two branches of calculus: differential calculus and integral calculus. Differential calculus arose from the tangent problem, whereas integral calculus arose from a seemingly unrelated problem, the area problem. Newton's mentor at Cambridge, Isaac Barrow (1630–1677), discovered that these two problems are actually closely related. In fact, he realized that differentiation and integration are inverse processes. The Fundamental Theorem of Calculus gives the precise inverse relationship between the derivative and the integral. It was Newton and Leibniz who exploited this relationship and used it to develop calculus into a systematic mathematical method. In particular, they saw that the Fundamental Theorem enabled them to compute areas and integrals very easily without having to compute them as limits of sums as we did in Sections 5.1 and 5.2.

The first part of the Fundamental Theorem deals with functions defined by an equation of the form

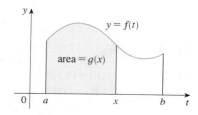

$$\boxed{1} \qquad g(x) = \int_a^x f(t)\, dt$$

where f is a continuous function on $[a, b]$ and x varies between a and b. Observe that g depends only on x, which appears as the variable upper limit in the integral. If x is a fixed number, then the integral $\int_a^x f(t)\, dt$ is a definite number. If we then let x vary, the number $\int_a^x f(t)\, dt$ also varies and defines a function of x denoted by $g(x)$.

If f happens to be a positive function, then $g(x)$ can be interpreted as the area under the graph of f from a to x, where x can vary from a to b. (Think of g as the "area so far" function; see Figure 1.)

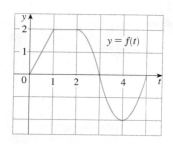

FIGURE 1

FIGURE 2

EXAMPLE 1 If f is the function whose graph is shown in Figure 2 and $g(x) = \int_0^x f(t)\, dt$, find the values of $g(0)$, $g(1)$, $g(2)$, $g(3)$, $g(4)$, and $g(5)$. Then sketch a rough graph of g.

SOLUTION First we notice that $g(0) = \int_0^0 f(t)\, dt = 0$. From Figure 3 we see that $g(1)$ is the area of a triangle:

$$g(1) = \int_0^1 f(t)\, dt = \tfrac{1}{2}\,(1 \cdot 2) = 1$$

To find $g(2)$ we add to $g(1)$ the area of a rectangle:

$$g(2) = \int_0^2 f(t)\, dt = \int_0^1 f(t)\, dt + \int_1^2 f(t)\, dt = 1 + (1 \cdot 2) = 3$$

We estimate that the area under f from 2 to 3 is about 1.3, so

$$g(3) = g(2) + \int_2^3 f(t)\, dt \approx 3 + 1.3 = 4.3$$

$g(1) = 1$

$g(2) = 3$

$g(3) \approx 4.3$

$g(4) \approx 3$

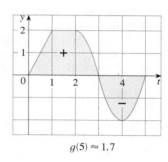

$g(5) \approx 1.7$

FIGURE 3

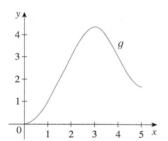

FIGURE 4

$$g(x) = \int_0^x f(t)\, dt$$

For $t > 3$, $f(t)$ is negative and so we start subtracting areas:

$$g(4) = g(3) + \int_3^4 f(t)\, dt \approx 4.3 + (-1.3) = 3.0$$

$$g(5) = g(4) + \int_4^5 f(t)\, dt \approx 3 + (-1.3) = 1.7$$

We use these values to sketch the graph of g in Figure 4. Notice that, because $f(t)$ is positive for $t < 3$, we keep adding area for $t < 3$ and so g is increasing up to $x = 3$, where it attains a maximum value. For $x > 3$, g decreases because $f(t)$ is negative. ∎

If we take $f(t) = t$ and $a = 0$, then, using Exercise 5.2.27, we have

$$g(x) = \int_0^x t\, dt = \frac{x^2}{2}$$

Notice that $g'(x) = x$, that is, $g' = f$. In other words, if g is defined as the integral of f by Equation 1, then g turns out to be an antiderivative of f, at least in this case. And if we sketch the derivative of the function g shown in Figure 4 by estimating slopes of tangents, we get a graph like that of f in Figure 2. So we suspect that $g' = f$ in Example 1 too.

To see why this might be generally true we consider any continuous function f with $f(x) \geqslant 0$. Then $g(x) = \int_a^x f(t)\, dt$ can be interpreted as the area under the graph of f from a to x, as in Figure 1.

In order to compute $g'(x)$ from the definition of a derivative we first observe that, for $h > 0$, $g(x + h) - g(x)$ is obtained by subtracting areas, so it is the area under the graph of f from x to $x + h$ (the blue area in Figure 5). For small h you can see from the figure that this area is approximately equal to the area of the rectangle with height $f(x)$ and width h:

$$g(x + h) - g(x) \approx h f(x)$$

so

$$\frac{g(x + h) - g(x)}{h} \approx f(x)$$

FIGURE 5

Intuitively, we therefore expect that

$$g'(x) = \lim_{h \to 0} \frac{g(x + h) - g(x)}{h} = f(x)$$

The fact that this is true, even when f is not necessarily positive, is the first part of the Fundamental Theorem of Calculus.

We abbreviate the name of this theorem as FTC1. In words, it says that the derivative of a definite integral with respect to its upper limit is the integrand evaluated at the upper limit.

> **The Fundamental Theorem of Calculus, Part 1**   If f is continuous on $[a, b]$, then the function g defined by
>
> $$g(x) = \int_a^x f(t)\, dt \qquad a \leqslant x \leqslant b$$
>
> is continuous on $[a, b]$ and differentiable on (a, b), and $g'(x) = f(x)$.

PROOF   If x and $x + h$ are in (a, b), then

$$g(x + h) - g(x) = \int_a^{x+h} f(t)\, dt - \int_a^x f(t)\, dt$$

$$= \left(\int_a^x f(t)\, dt + \int_x^{x+h} f(t)\, dt \right) - \int_a^x f(t)\, dt \qquad \text{(by Property 5)}$$

$$= \int_x^{x+h} f(t)\, dt$$

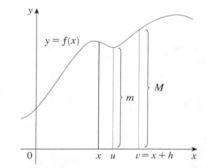

FIGURE 6

and so, for $h \neq 0$,

$$\boxed{2} \qquad \frac{g(x + h) - g(x)}{h} = \frac{1}{h} \int_x^{x+h} f(t)\, dt$$

For now let's assume that $h > 0$. Since f is continuous on $[x, x + h]$, the Extreme Value Theorem says that there are numbers u and v in $[x, x + h]$ such that $f(u) = m$ and $f(v) = M$, where m and M are the absolute minimum and maximum values of f on $[x, x + h]$. (See Figure 6.)

By Property 8 of integrals, we have

$$mh \leqslant \int_x^{x+h} f(t)\, dt \leqslant Mh$$

that is,

$$f(u)h \leqslant \int_x^{x+h} f(t)\, dt \leqslant f(v)h$$

Since $h > 0$, we can divide this inequality by h:

$$f(u) \leqslant \frac{1}{h} \int_x^{x+h} f(t)\, dt \leqslant f(v)$$

Now we use Equation 2 to replace the middle part of this inequality:

$$\boxed{3} \qquad f(u) \leqslant \frac{g(x + h) - g(x)}{h} \leqslant f(v)$$

TEC   Module 5.3 provides visual evidence for FTC1.

Inequality 3 can be proved in a similar manner for the case where $h < 0$. (See Exercise 77.)

Now we let $h \to 0$. Then $u \to x$ and $v \to x$, since u and v lie between x and $x + h$. Therefore

$$\lim_{h \to 0} f(u) = \lim_{u \to x} f(u) = f(x) \qquad \text{and} \qquad \lim_{h \to 0} f(v) = \lim_{v \to x} f(v) = f(x)$$

because f is continuous at x. We conclude, from (3) and the Squeeze Theorem, that

$$\boxed{4} \qquad\qquad g'(x) = \lim_{h \to 0} \frac{g(x + h) - g(x)}{h} = f(x)$$

If $x = a$ or b, then Equation 4 can be interpreted as a one-sided limit. Then Theorem 2.8.4 (modified for one-sided limits) shows that g is continuous on $[a, b]$. ■

Using Leibniz notation for derivatives, we can write FTC1 as

$$\boxed{5} \qquad\qquad \frac{d}{dx} \int_a^x f(t)\, dt = f(x)$$

when f is continuous. Roughly speaking, Equation 5 says that if we first integrate f and then differentiate the result, we get back to the original function f.

EXAMPLE 2 Find the derivative of the function $g(x) = \displaystyle\int_0^x \sqrt{1 + t^2}\, dt$.

SOLUTION Since $f(t) = \sqrt{1 + t^2}$ is continuous, Part 1 of the Fundamental Theorem of Calculus gives

$$g'(x) = \sqrt{1 + x^2}$$

■

EXAMPLE 3 Although a formula of the form $g(x) = \int_a^x f(t)\, dt$ may seem like a strange way of defining a function, books on physics, chemistry, and statistics are full of such functions. For instance, the **Fresnel function**

$$S(x) = \int_0^x \sin(\pi t^2/2)\, dt$$

is named after the French physicist Augustin Fresnel (1788–1827), who is famous for his works in optics. This function first appeared in Fresnel's theory of the diffraction of light waves, but more recently it has been applied to the design of highways.

Part 1 of the Fundamental Theorem tells us how to differentiate the Fresnel function:

$$S'(x) = \sin(\pi x^2/2)$$

This means that we can apply all the methods of differential calculus to analyze S (see Exercise 71).

Figure 7 shows the graphs of $f(x) = \sin(\pi x^2/2)$ and the Fresnel function $S(x) = \int_0^x f(t)\, dt$. A computer was used to graph S by computing the value of this integral for many values of x. It does indeed look as if $S(x)$ is the area under the graph of f from 0 to x [until $x \approx 1.4$ when $S(x)$ becomes a difference of areas]. Figure 8 shows a larger part of the graph of S.

If we now start with the graph of S in Figure 7 and think about what its derivative should look like, it seems reasonable that $S'(x) = f(x)$. [For instance, S is increasing when $f(x) > 0$ and decreasing when $f(x) < 0$.] So this gives a visual confirmation of Part 1 of the Fundamental Theorem of Calculus. ■

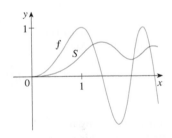

FIGURE 7
$f(x) = \sin(\pi x^2/2)$
$S(x) = \displaystyle\int_0^x \sin(\pi t^2/2)\, dt$

FIGURE 8
The Fresnel function
$S(x) = \displaystyle\int_0^x \sin(\pi t^2/2)\, dt$

EXAMPLE 4 Find $\dfrac{d}{dx} \displaystyle\int_1^{x^4} \sec t \, dt$.

SOLUTION Here we have to be careful to use the Chain Rule in conjunction with FTC1. Let $u = x^4$. Then

$$\frac{d}{dx} \int_1^{x^4} \sec t \, dt = \frac{d}{dx} \int_1^{u} \sec t \, dt$$

$$= \frac{d}{du} \left[\int_1^{u} \sec t \, dt \right] \frac{du}{dx} \qquad \text{(by the Chain Rule)}$$

$$= \sec u \, \frac{du}{dx} \qquad \text{(by FTC1)}$$

$$= \sec(x^4) \cdot 4x^3 \qquad\qquad\qquad \blacksquare$$

In Section 5.2 we computed integrals from the definition as a limit of Riemann sums and we saw that this procedure is sometimes long and difficult. The second part of the Fundamental Theorem of Calculus, which follows easily from the first part, provides us with a much simpler method for the evaluation of integrals.

We abbreviate this theorem as FTC2.

> **The Fundamental Theorem of Calculus, Part 2** If f is continuous on $[a, b]$, then
>
> $$\int_a^b f(x) \, dx = F(b) - F(a)$$
>
> where F is any antiderivative of f, that is, a function such that $F' = f$.

PROOF Let $g(x) = \int_a^x f(t) \, dt$. We know from Part 1 that $g'(x) = f(x)$; that is, g is an antiderivative of f. If F is any other antiderivative of f on $[a, b]$, then we know from Corollary 4.2.7 that F and g differ by a constant:

$$\boxed{6} \qquad\qquad F(x) = g(x) + C$$

for $a < x < b$. But both F and g are continuous on $[a, b]$ and so, by taking limits of both sides of Equation 6 (as $x \to a^+$ and $x \to b^-$), we see that it also holds when $x = a$ and $x = b$. So $F(x) = g(x) + C$ for all x in $[a, b]$.

If we put $x = a$ in the formula for $g(x)$, we get

$$g(a) = \int_a^a f(t) \, dt = 0$$

So, using Equation 6 with $x = b$ and $x = a$, we have

$$F(b) - F(a) = [g(b) + C] - [g(a) + C]$$

$$= g(b) - g(a) = g(b) = \int_a^b f(t) \, dt \qquad\qquad \blacksquare$$

Part 2 of the Fundamental Theorem states that if we know an antiderivative F of f, then we can evaluate $\int_a^b f(x) \, dx$ simply by subtracting the values of F at the endpoints of the interval $[a, b]$. It's very surprising that $\int_a^b f(x) \, dx$, which was defined by a complicated procedure involving all of the values of $f(x)$ for $a \leq x \leq b$, can be found by knowing the values of $F(x)$ at only two points, a and b.

Although the theorem may be surprising at first glance, it becomes plausible if we interpret it in physical terms. If $v(t)$ is the velocity of an object and $s(t)$ is its position at time t, then $v(t) = s'(t)$, so s is an antiderivative of v. In Section 5.1 we considered an object that always moves in the positive direction and made the guess that the area under the velocity curve is equal to the distance traveled. In symbols:

$$\int_a^b v(t)\, dt = s(b) - s(a)$$

That is exactly what FTC2 says in this context.

EXAMPLE 5 Evaluate the integral $\int_1^3 e^x\, dx$.

SOLUTION The function $f(x) = e^x$ is continuous everywhere and we know that an anti-derivative is $F(x) = e^x$, so Part 2 of the Fundamental Theorem gives

$$\int_1^3 e^x\, dx = F(3) - F(1) = e^3 - e$$

Compare the calculation in Example 5 with the much harder one in Example 5.2.3.

Notice that FTC2 says we can use *any* antiderivative F of f. So we may as well use the simplest one, namely $F(x) = e^x$, instead of $e^x + 7$ or $e^x + C$. ∎

We often use the notation

$$F(x)\Big]_a^b = F(b) - F(a)$$

So the equation of FTC2 can be written as

$$\int_a^b f(x)\, dx = F(x)\Big]_a^b \qquad \text{where} \qquad F' = f$$

Other common notations are $F(x)\big|_a^b$ and $[F(x)]_a^b$.

EXAMPLE 6 Find the area under the parabola $y = x^2$ from 0 to 1.

SOLUTION An antiderivative of $f(x) = x^2$ is $F(x) = \frac{1}{3}x^3$. The required area A is found using Part 2 of the Fundamental Theorem:

In applying the Fundamental Theorem we use a particular antiderivative F of f. It is not necessary to use the most general antiderivative.

$$A = \int_0^1 x^2\, dx = \frac{x^3}{3}\Big]_0^1 = \frac{1^3}{3} - \frac{0^3}{3} = \frac{1}{3}$$ ∎

If you compare the calculation in Example 6 with the one in Example 5.1.2, you will see that the Fundamental Theorem gives a *much* shorter method.

EXAMPLE 7 Evaluate $\int_3^6 \dfrac{dx}{x}$.

SOLUTION The given integral is an abbreviation for

$$\int_3^6 \frac{1}{x}\, dx$$

An antiderivative of $f(x) = 1/x$ is $F(x) = \ln|x|$ and, because $3 \leq x \leq 6$, we can write $F(x) = \ln x$. So

$$\int_3^6 \frac{1}{x}\, dx = \ln x\Big]_3^6 = \ln 6 - \ln 3 = \ln \frac{6}{3} = \ln 2$$ ∎

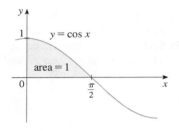

FIGURE 9

EXAMPLE 8 Find the area under the cosine curve from 0 to b, where $0 \le b \le \pi/2$.

SOLUTION Since an antiderivative of $f(x) = \cos x$ is $F(x) = \sin x$, we have

$$A = \int_0^b \cos x \, dx = \sin x \Big]_0^b = \sin b - \sin 0 = \sin b$$

In particular, taking $b = \pi/2$, we have proved that the area under the cosine curve from 0 to $\pi/2$ is $\sin(\pi/2) = 1$. (See Figure 9.) ∎

When the French mathematician Gilles de Roberval first found the area under the sine and cosine curves in 1635, this was a very challenging problem that required a great deal of ingenuity. If we didn't have the benefit of the Fundamental Theorem, we would have to compute a difficult limit of sums using obscure trigonometric identities (or a computer algebra system as in Exercise 5.1.31). It was even more difficult for Roberval because the apparatus of limits had not been invented in 1635. But in the 1660s and 1670s, when the Fundamental Theorem was discovered by Barrow and exploited by Newton and Leibniz, such problems became very easy, as you can see from Example 8.

EXAMPLE 9 What is wrong with the following calculation?

$$\int_{-1}^{3} \frac{1}{x^2} \, dx = \frac{x^{-1}}{-1} \Bigg]_{-1}^{3} = -\frac{1}{3} - 1 = -\frac{4}{3}$$

SOLUTION To start, we notice that this calculation must be wrong because the answer is negative but $f(x) = 1/x^2 \ge 0$ and Property 6 of integrals says that $\int_a^b f(x) \, dx \ge 0$ when $f \ge 0$. The Fundamental Theorem of Calculus applies to continuous functions. It can't be applied here because $f(x) = 1/x^2$ is not continuous on $[-1, 3]$. In fact, f has an infinite discontinuity at $x = 0$, so

$$\int_{-1}^{3} \frac{1}{x^2} \, dx \qquad \text{does not exist.} \qquad ∎$$

■ Differentiation and Integration as Inverse Processes

We end this section by bringing together the two parts of the Fundamental Theorem.

The Fundamental Theorem of Calculus Suppose f is continuous on $[a, b]$.

1. If $g(x) = \int_a^x f(t) \, dt$, then $g'(x) = f(x)$.

2. $\int_a^b f(x) \, dx = F(b) - F(a)$, where F is any antiderivative of f, that is, $F' = f$.

We noted that Part 1 can be rewritten as

$$\frac{d}{dx} \int_a^x f(t) \, dt = f(x)$$

which says that if f is integrated and then the result is differentiated, we arrive back at the original function f. Since $F'(x) = f(x)$, Part 2 can be rewritten as

$$\int_a^b F'(x) \, dx = F(b) - F(a)$$

This version says that if we take a function F, first differentiate it, and then integrate the result, we arrive back at the original function F, but in the form $F(b) - F(a)$. Taken together, the two parts of the Fundamental Theorem of Calculus say that differentiation and integration are inverse processes. Each undoes what the other does.

The Fundamental Theorem of Calculus is unquestionably the most important theorem in calculus and, indeed, it ranks as one of the great accomplishments of the human mind. Before it was discovered, from the time of Eudoxus and Archimedes to the time of Galileo and Fermat, problems of finding areas, volumes, and lengths of curves were so difficult that only a genius could meet the challenge. But now, armed with the systematic method that Newton and Leibniz fashioned out of the Fundamental Theorem, we will see in the chapters to come that these challenging problems are accessible to all of us.

5.3 EXERCISES

1. Explain exactly what is meant by the statement that "differentiation and integration are inverse processes."

2. Let $g(x) = \int_0^x f(t)\, dt$, where f is the function whose graph is shown.
   (a) Evaluate $g(x)$ for $x = 0, 1, 2, 3, 4, 5,$ and 6.
   (b) Estimate $g(7)$.
   (c) Where does g have a maximum value? Where does it have a minimum value?
   (d) Sketch a rough graph of g.

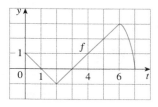

3. Let $g(x) = \int_0^x f(t)\, dt$, where f is the function whose graph is shown.
   (a) Evaluate $g(0), g(1), g(2), g(3),$ and $g(6)$.
   (b) On what interval is g increasing?
   (c) Where does g have a maximum value?
   (d) Sketch a rough graph of g.

4. Let $g(x) = \int_0^x f(t)\, dt$, where f is the function whose graph is shown.
   (a) Evaluate $g(0)$ and $g(6)$.
   (b) Estimate $g(x)$ for $x = 1, 2, 3, 4,$ and 5.
   (c) On what interval is g increasing?

(d) Where does g have a maximum value?
(e) Sketch a rough graph of g.
(f) Use the graph in part (e) to sketch the graph of $g'(x)$. Compare with the graph of f.

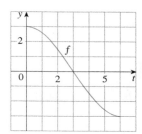

5–6 Sketch the area represented by $g(x)$. Then find $g'(x)$ in two ways: (a) by using Part 1 of the Fundamental Theorem and (b) by evaluating the integral using Part 2 and then differentiating.

5. $g(x) = \int_1^x t^2\, dt$

6. $g(x) = \int_0^x (2 + \sin t)\, dt$

7–18 Use Part 1 of the Fundamental Theorem of Calculus to find the derivative of the function.

7. $g(x) = \int_0^x \sqrt{t + t^3}\, dt$

8. $g(x) = \int_1^x \ln(1 + t^2)\, dt$

9. $g(s) = \int_5^s (t - t^2)^8\, dt$

10. $h(u) = \int_0^u \dfrac{\sqrt{t}}{t + 1}\, dt$

11. $F(x) = \int_x^0 \sqrt{1 + \sec t}\, dt$

$$\left[\text{Hint: } \int_x^0 \sqrt{1 + \sec t}\, dt = -\int_0^x \sqrt{1 + \sec t}\, dt \right]$$

12. $R(y) = \int_y^2 t^3 \sin t\, dt$

13. $h(x) = \int_1^{e^x} \ln t\, dt$

14. $h(x) = \int_1^{\sqrt{x}} \dfrac{z^2}{z^4 + 1}\, dz$

15. $y = \int_1^{3x+2} \dfrac{t}{1+t^3}\,dt$

16. $y = \int_0^{x^4} \cos^2\theta\,d\theta$

17. $y = \int_{\sqrt{x}}^{\pi/4} \theta\tan\theta\,d\theta$

18. $y = \int_{\sin x}^{1} \sqrt{1+t^2}\,dt$

19–44 Evaluate the integral.

19. $\int_1^3 (x^2 + 2x - 4)\,dx$

20. $\int_{-1}^{1} x^{100}\,dx$

21. $\int_0^2 \left(\tfrac{4}{5}t^3 - \tfrac{3}{4}t^2 + \tfrac{2}{5}t\right) dt$

22. $\int_0^1 (1 - 8v^3 + 16v^7)\,dv$

23. $\int_1^9 \sqrt{x}\,dx$

24. $\int_1^8 x^{-2/3}\,dx$

25. $\int_{\pi/6}^{\pi} \sin\theta\,d\theta$

26. $\int_{-5}^{5} e\,dx$

27. $\int_0^1 (u+2)(u-3)\,du$

28. $\int_0^4 (4-t)\sqrt{t}\,dt$

29. $\int_1^4 \dfrac{2+x^2}{\sqrt{x}}\,dx$

30. $\int_{-1}^{2} (3u-2)(u+1)\,du$

31. $\int_{\pi/6}^{\pi/2} \csc t \cot t\,dt$

32. $\int_{\pi/4}^{\pi/3} \csc^2\theta\,d\theta$

33. $\int_0^1 (1+r)^3\,dr$

34. $\int_0^3 (2\sin x - e^x)\,dx$

35. $\int_1^2 \dfrac{v^3 + 3v^6}{v^4}\,dv$

36. $\int_1^{18} \sqrt{\dfrac{3}{z}}\,dz$

37. $\int_0^1 (x^e + e^x)\,dx$

38. $\int_0^1 \cosh t\,dt$

39. $\int_{1/\sqrt{3}}^{\sqrt{3}} \dfrac{8}{1+x^2}\,dx$

40. $\int_1^3 \dfrac{y^3 - 2y^2 - y}{y^2}\,dy$

41. $\int_0^4 2^s\,ds$

42. $\int_{1/2}^{1/\sqrt{2}} \dfrac{4}{\sqrt{1-x^2}}\,dx$

43. $\int_0^{\pi} f(x)\,dx$ where $f(x) = \begin{cases} \sin x & \text{if } 0 \le x < \pi/2 \\ \cos x & \text{if } \pi/2 \le x \le \pi \end{cases}$

44. $\int_{-2}^{2} f(x)\,dx$ where $f(x) = \begin{cases} 2 & \text{if } -2 \le x \le 0 \\ 4 - x^2 & \text{if } 0 < x \le 2 \end{cases}$

45–48 Sketch the region enclosed by the given curves and calculate its area.

45. $y = \sqrt{x}$, $y = 0$, $x = 4$

46. $y = x^3$, $y = 0$, $x = 1$

47. $y = 4 - x^2$, $y = 0$

48. $y = 2x - x^2$, $y = 0$

49–52 Use a graph to give a rough estimate of the area of the region that lies beneath the given curve. Then find the exact area.

49. $y = \sqrt[3]{x}$, $0 \le x \le 27$

50. $y = x^{-4}$, $1 \le x \le 6$

51. $y = \sin x$, $0 \le x \le \pi$

52. $y = \sec^2 x$, $0 \le x \le \pi/3$

53–54 Evaluate the integral and interpret it as a difference of areas. Illustrate with a sketch.

53. $\int_{-1}^{2} x^3\,dx$

54. $\int_{\pi/6}^{2\pi} \cos x\,dx$

55–58 What is wrong with the equation?

55. $\int_{-2}^{1} x^{-4}\,dx = \dfrac{x^{-3}}{-3}\Big]_{-2}^{1} = -\dfrac{3}{8}$

56. $\int_{-1}^{2} \dfrac{4}{x^3}\,dx = -\dfrac{2}{x^2}\Big]_{-1}^{2} = \dfrac{3}{2}$

57. $\int_{\pi/3}^{\pi} \sec\theta\tan\theta\,d\theta = \sec\theta\Big]_{\pi/3}^{\pi} = -3$

58. $\int_0^{\pi} \sec^2 x\,dx = \tan x\Big]_0^{\pi} = 0$

59–63 Find the derivative of the function.

59. $g(x) = \int_{2x}^{3x} \dfrac{u^2 - 1}{u^2 + 1}\,du$

$\left[\text{Hint: } \int_{2x}^{3x} f(u)\,du = \int_{2x}^{0} f(u)\,du + \int_0^{3x} f(u)\,du \right]$

60. $g(x) = \int_{1-2x}^{1+2x} t\sin t\,dt$

61. $F(x) = \int_x^{x^2} e^{t^2}\,dt$

62. $F(x) = \int_{\sqrt{x}}^{2x} \arctan t\,dt$

63. $y = \int_{\cos x}^{\sin x} \ln(1 + 2v)\,dv$

---

64. If $f(x) = \int_0^x (1 - t^2)e^{t^2}\,dt$, on what interval is f increasing?

65. On what interval is the curve

$$y = \int_0^x \dfrac{t^2}{t^2 + t + 2}\,dt$$

concave downward?

66. Let $F(x) = \int_1^x f(t)\, dt$, where f is the function whose graph is shown. Where is F concave downward?

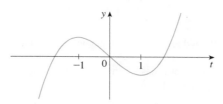

67. Let $F(x) = \int_2^x e^{t^2}\, dt$. Find an equation of the tangent line to the curve $y = F(x)$ at the point with x-coordinate 2.

68. If $f(x) = \int_0^{\sin x} \sqrt{1 + t^2}\, dt$ and $g(y) = \int_3^y f(x)\, dx$, find $g''(\pi/6)$.

69. If $f(1) = 12$, f' is continuous, and $\int_1^4 f'(x)\, dx = 17$, what is the value of $f(4)$?

70. The **error function**

$$\text{erf}(x) = \frac{2}{\sqrt{\pi}} \int_0^x e^{-t^2}\, dt$$

is used in probability, statistics, and engineering.

(a) Show that $\int_a^b e^{-t^2}\, dt = \frac{1}{2}\sqrt{\pi}\, [\text{erf}(b) - \text{erf}(a)]$.

(b) Show that the function $y = e^{x^2}\text{erf}(x)$ satisfies the differential equation $y' = 2xy + 2/\sqrt{\pi}$.

71. The Fresnel function S was defined in Example 3 and graphed in Figures 7 and 8.

(a) At what values of x does this function have local maximum values?

(b) On what intervals is the function concave upward?

(c) Use a graph to solve the following equation correct to two decimal places:

$$\int_0^x \sin(\pi t^2/2)\, dt = 0.2$$

72. The **sine integral function**

$$\text{Si}(x) = \int_0^x \frac{\sin t}{t}\, dt$$

is important in electrical engineering. [The integrand $f(t) = (\sin t)/t$ is not defined when $t = 0$, but we know that its limit is 1 when $t \to 0$. So we define $f(0) = 1$ and this makes f a continuous function everywhere.]

(a) Draw the graph of Si.

(b) At what values of x does this function have local maximum values?

(c) Find the coordinates of the first inflection point to the right of the origin.

(d) Does this function have horizontal asymptotes?

(e) Solve the following equation correct to one decimal place:

$$\int_0^x \frac{\sin t}{t}\, dt = 1$$

73–74 Let $g(x) = \int_0^x f(t)\, dt$, where f is the function whose graph is shown.

(a) At what values of x do the local maximum and minimum values of g occur?

(b) Where does g attain its absolute maximum value?

(c) On what intervals is g concave downward?

(d) Sketch the graph of g.

73.

74.

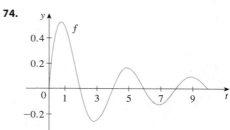

75–76 Evaluate the limit by first recognizing the sum as a Riemann sum for a function defined on $[0, 1]$.

75. $\displaystyle\lim_{n \to \infty} \sum_{i=1}^{n} \left(\frac{i^4}{n^5} + \frac{i}{n^2} \right)$

76. $\displaystyle\lim_{n \to \infty} \frac{1}{n} \left(\sqrt{\frac{1}{n}} + \sqrt{\frac{2}{n}} + \sqrt{\frac{3}{n}} + \cdots + \sqrt{\frac{n}{n}} \right)$

77. Justify (3) for the case $h < 0$.

78. If f is continuous and g and h are differentiable functions, find a formula for

$$\frac{d}{dx} \int_{g(x)}^{h(x)} f(t)\, dt$$

79. (a) Show that $1 \le \sqrt{1 + x^3} \le 1 + x^3$ for $x \ge 0$.

(b) Show that $1 \le \int_0^1 \sqrt{1 + x^3}\, dx \le 1.25$.

80. (a) Show that $\cos(x^2) \ge \cos x$ for $0 \le x \le 1$.

(b) Deduce that $\int_0^{\pi/6} \cos(x^2)\, dx \ge \frac{1}{2}$.

81. Show that

$$0 \le \int_5^{10} \frac{x^2}{x^4 + x^2 + 1}\, dx \le 0.1$$

by comparing the integrand to a simpler function.

82. Let

$$f(x) = \begin{cases} 0 & \text{if } x < 0 \\ x & \text{if } 0 \leq x \leq 1 \\ 2 - x & \text{if } 1 < x \leq 2 \\ 0 & \text{if } x > 2 \end{cases}$$

and

$$g(x) = \int_0^x f(t)\, dt$$

(a) Find an expression for $g(x)$ similar to the one for $f(x)$.

(b) Sketch the graphs of f and g.

(c) Where is f differentiable? Where is g differentiable?

83. Find a function f and a number a such that

$$6 + \int_a^x \frac{f(t)}{t^2}\, dt = 2\sqrt{x} \qquad \text{for all } x > 0$$

84. The area labeled B is three times the area labeled A. Express b in terms of a.

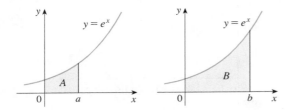

85. A manufacturing company owns a major piece of equipment that depreciates at the (continuous) rate $f = f(t)$, where t is the time measured in months since its last overhaul. Because a fixed cost A is incurred each time the machine is overhauled, the company wants to determine the optimal time T (in months) between overhauls.

(a) Explain why $\int_0^t f(s)\, ds$ represents the loss in value of the machine over the period of time t since the last overhaul.

(b) Let $C = C(t)$ be given by

$$C(t) = \frac{1}{t}\left[A + \int_0^t f(s)\, ds \right]$$

What does C represent and why would the company want to minimize C?

(c) Show that C has a minimum value at the numbers $t = T$ where $C(T) = f(T)$.

86. A high-tech company purchases a new computing system whose initial value is V. The system will depreciate at the rate $f = f(t)$ and will accumulate maintenance costs at the rate $g = g(t)$, where t is the time measured in months. The company wants to determine the optimal time to replace the system.

(a) Let

$$C(t) = \frac{1}{t} \int_0^t [f(s) + g(s)]\, ds$$

Show that the critical numbers of C occur at the numbers t where $C(t) = f(t) + g(t)$.

(b) Suppose that

$$f(t) = \begin{cases} \dfrac{V}{15} - \dfrac{V}{450}t & \text{if } 0 < t \leq 30 \\ 0 & \text{if } t > 30 \end{cases}$$

and

$$g(t) = \frac{Vt^2}{12{,}900} \qquad t > 0$$

Determine the length of time T for the total depreciation $D(t) = \int_0^t f(s)\, ds$ to equal the initial value V.

(c) Determine the absolute minimum of C on $(0, T]$.

(d) Sketch the graphs of C and $f + g$ in the same coordinate system, and verify the result in part (a) in this case.

5.4 Indefinite Integrals and the Net Change Theorem

We saw in Section 5.3 that the second part of the Fundamental Theorem of Calculus provides a very powerful method for evaluating the definite integral of a function, assuming that we can find an antiderivative of the function. In this section we introduce a notation for antiderivatives, review the formulas for antiderivatives, and use them to evaluate definite integrals. We also reformulate FTC2 in a way that makes it easier to apply to science and engineering problems.

■ Indefinite Integrals

Both parts of the Fundamental Theorem establish connections between antiderivatives and definite integrals. Part 1 says that if f is continuous, then $\int_a^x f(t)\, dt$ is an antiderivative of f. Part 2 says that $\int_a^b f(x)\, dx$ can be found by evaluating $F(b) - F(a)$, where F is an antiderivative of f.

We need a convenient notation for antiderivatives that makes them easy to work with. Because of the relation between antiderivatives and integrals given by the Fundamental

Theorem, the notation $\int f(x)\,dx$ is traditionally used for an antiderivative of f and is called an **indefinite integral**. Thus

$$\int f(x)\,dx = F(x) \qquad \text{means} \qquad F'(x) = f(x)$$

For example, we can write

$$\int x^2\,dx = \frac{x^3}{3} + C \qquad \text{because} \qquad \frac{d}{dx}\left(\frac{x^3}{3} + C\right) = x^2$$

So we can regard an indefinite integral as representing an entire *family* of functions (one antiderivative for each value of the constant C).

You should distinguish carefully between definite and indefinite integrals. A definite integral $\int_a^b f(x)\,dx$ is a *number*, whereas an indefinite integral $\int f(x)\,dx$ is a *function* (or family of functions). The connection between them is given by Part 2 of the Fundamental Theorem: If f is continuous on $[a, b]$, then

$$\int_a^b f(x)\,dx = \int f(x)\,dx \Big]_a^b$$

The effectiveness of the Fundamental Theorem depends on having a supply of antiderivatives of functions. We therefore restate the Table of Antidifferentiation Formulas from Section 4.9, together with a few others, in the notation of indefinite integrals. Any formula can be verified by differentiating the function on the right side and obtaining the integrand. For instance,

$$\int \sec^2 x\,dx = \tan x + C \qquad \text{because} \qquad \frac{d}{dx}(\tan x + C) = \sec^2 x$$

1   **Table of Indefinite Integrals**

$$\int cf(x)\,dx = c\int f(x)\,dx \qquad\qquad \int [f(x) + g(x)]\,dx = \int f(x)\,dx + \int g(x)\,dx$$

$$\int k\,dx = kx + C$$

$$\int x^n\,dx = \frac{x^{n+1}}{n+1} + C \quad (n \neq -1) \qquad \int \frac{1}{x}\,dx = \ln|x| + C$$

$$\int e^x\,dx = e^x + C \qquad\qquad \int b^x\,dx = \frac{b^x}{\ln b} + C$$

$$\int \sin x\,dx = -\cos x + C \qquad\qquad \int \cos x\,dx = \sin x + C$$

$$\int \sec^2 x\,dx = \tan x + C \qquad\qquad \int \csc^2 x\,dx = -\cot x + C$$

$$\int \sec x \tan x\,dx = \sec x + C \qquad\qquad \int \csc x \cot x\,dx = -\csc x + C$$

$$\int \frac{1}{x^2 + 1}\,dx = \tan^{-1}x + C \qquad\qquad \int \frac{1}{\sqrt{1 - x^2}}\,dx = \sin^{-1}x + C$$

$$\int \sinh x\,dx = \cosh x + C \qquad\qquad \int \cosh x\,dx = \sinh x + C$$

Recall from Theorem 4.9.1 that the most general antiderivative *on a given interval* is obtained by adding a constant to a particular antiderivative. **We adopt the convention** that when a formula for a general indefinite integral is given, it is valid only on an interval. Thus we write

$$\int \frac{1}{x^2}\, dx = -\frac{1}{x} + C$$

with the understanding that it is valid on the interval $(0, \infty)$ or on the interval $(-\infty, 0)$. This is true despite the fact that the general antiderivative of the function $f(x) = 1/x^2$, $x \neq 0$, is

$$F(x) = \begin{cases} -\dfrac{1}{x} + C_1 & \text{if } x < 0 \\[2mm] -\dfrac{1}{x} + C_2 & \text{if } x > 0 \end{cases}$$

FIGURE 1

The indefinite integral in Example 1 is graphed in Figure 1 for several values of C. Here the value of C is the y-intercept.

EXAMPLE 1 Find the general indefinite integral

$$\int (10x^4 - 2\sec^2 x)\, dx$$

SOLUTION Using our convention and Table 1, we have

$$\int (10x^4 - 2\sec^2 x)\, dx = 10 \int x^4\, dx - 2 \int \sec^2 x\, dx$$

$$= 10\frac{x^5}{5} - 2\tan x + C$$

$$= 2x^5 - 2\tan x + C$$

You should check this answer by differentiating it.   ∎

EXAMPLE 2 Evaluate $\displaystyle\int \frac{\cos\theta}{\sin^2\theta}\, d\theta$.

SOLUTION This indefinite integral isn't immediately apparent in Table 1, so we use trigonometric identities to rewrite the function before integrating:

$$\int \frac{\cos\theta}{\sin^2\theta}\, d\theta = \int \left(\frac{1}{\sin\theta}\right)\left(\frac{\cos\theta}{\sin\theta}\right) d\theta$$

$$= \int \csc\theta \cot\theta\, d\theta = -\csc\theta + C$$ ∎

EXAMPLE 3 Evaluate $\displaystyle\int_0^3 (x^3 - 6x)\, dx$.

SOLUTION Using FTC2 and Table 1, we have

$$\int_0^3 (x^3 - 6x)\, dx = \frac{x^4}{4} - 6\frac{x^2}{2}\Bigg]_0^3$$

$$= \left(\tfrac{1}{4}\cdot 3^4 - 3\cdot 3^2\right) - \left(\tfrac{1}{4}\cdot 0^4 - 3\cdot 0^2\right)$$

$$= \tfrac{81}{4} - 27 - 0 + 0 = -6.75$$

Compare this calculation with Example 5.2.2(b).   ∎

Figure 2 shows the graph of the integrand in Example 4. We know from Section 5.2 that the value of the integral can be interpreted as a net area: the sum of the areas labeled with a plus sign minus the area labeled with a minus sign.

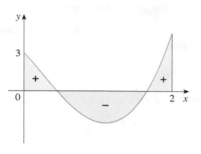

FIGURE 2

EXAMPLE 4 Find $\displaystyle\int_0^2 \left(2x^3 - 6x + \frac{3}{x^2 + 1}\right) dx$ and interpret the result in terms of areas.

SOLUTION The Fundamental Theorem gives

$$\int_0^2 \left(2x^3 - 6x + \frac{3}{x^2 + 1}\right) dx = 2\frac{x^4}{4} - 6\frac{x^2}{2} + 3\tan^{-1}x\Big]_0^2$$

$$= \tfrac{1}{2}x^4 - 3x^2 + 3\tan^{-1}x\Big]_0^2$$

$$= \tfrac{1}{2}(2^4) - 3(2^2) + 3\tan^{-1}2 - 0$$

$$= -4 + 3\tan^{-1}2$$

This is the exact value of the integral. If a decimal approximation is desired, we can use a calculator to approximate $\tan^{-1}2$. Doing so, we get

$$\int_0^2 \left(2x^3 - 6x + \frac{3}{x^2 + 1}\right) dx \approx -0.67855 \qquad \blacksquare$$

EXAMPLE 5 Evaluate $\displaystyle\int_1^9 \frac{2t^2 + t^2\sqrt{t} - 1}{t^2} dt.$

SOLUTION First we need to write the integrand in a simpler form by carrying out the division:

$$\int_1^9 \frac{2t^2 + t^2\sqrt{t} - 1}{t^2} dt = \int_1^9 (2 + t^{1/2} - t^{-2}) dt$$

$$= 2t + \frac{t^{3/2}}{\frac{3}{2}} - \frac{t^{-1}}{-1}\Big]_1^9 = 2t + \tfrac{2}{3}t^{3/2} + \frac{1}{t}\Big]_1^9$$

$$= \left(2 \cdot 9 + \tfrac{2}{3} \cdot 9^{3/2} + \tfrac{1}{9}\right) - \left(2 \cdot 1 + \tfrac{2}{3} \cdot 1^{3/2} + \tfrac{1}{1}\right)$$

$$= 18 + 18 + \tfrac{1}{9} - 2 - \tfrac{2}{3} - 1 = 32\tfrac{4}{9} \qquad \blacksquare$$

■ **Applications**

Part 2 of the Fundamental Theorem says that if f is continuous on $[a, b]$, then

$$\int_a^b f(x)\, dx = F(b) - F(a)$$

where F is any antiderivative of f. This means that $F' = f$, so the equation can be rewritten as

$$\int_a^b F'(x)\, dx = F(b) - F(a)$$

We know that $F'(x)$ represents the rate of change of $y = F(x)$ with respect to x and $F(b) - F(a)$ is the change in y when x changes from a to b. [Note that y could, for instance, increase, then decrease, then increase again. Although y might change in both directions, $F(b) - F(a)$ represents the *net* change in *y*.] So we can reformulate FTC2 in words as follows.

Net Change Theorem   The integral of a rate of change is the net change:

$$\int_a^b F'(x)\,dx = F(b) - F(a)$$

This principle can be applied to all of the rates of change in the natural and social sciences that we discussed in Section 3.7. Here are a few instances of this idea:

- If $V(t)$ is the volume of water in a reservoir at time t, then its derivative $V'(t)$ is the rate at which water flows into the reservoir at time t. So

$$\int_{t_1}^{t_2} V'(t)\,dt = V(t_2) - V(t_1)$$

is the change in the amount of water in the reservoir between time t_1 and time t_2.

- If $[C](t)$ is the concentration of the product of a chemical reaction at time t, then the rate of reaction is the derivative $d[C]/dt$. So

$$\int_{t_1}^{t_2} \frac{d[C]}{dt}\,dt = [C](t_2) - [C](t_1)$$

is the change in the concentration of C from time t_1 to time t_2.

- If the mass of a rod measured from the left end to a point x is $m(x)$, then the linear density is $\rho(x) = m'(x)$. So

$$\int_a^b \rho(x)\,dx = m(b) - m(a)$$

is the mass of the segment of the rod that lies between $x = a$ and $x = b$.

- If the rate of growth of a population is dn/dt, then

$$\int_{t_1}^{t_2} \frac{dn}{dt}\,dt = n(t_2) - n(t_1)$$

is the net change in population during the time period from t_1 to t_2. (The population increases when births happen and decreases when deaths occur. The net change takes into account both births and deaths.)

- If $C(x)$ is the cost of producing x units of a commodity, then the marginal cost is the derivative $C'(x)$. So

$$\int_{x_1}^{x_2} C'(x)\,dx = C(x_2) - C(x_1)$$

is the increase in cost when production is increased from x_1 units to x_2 units.

- If an object moves along a straight line with position function $s(t)$, then its velocity is $v(t) = s'(t)$, so

$$\boxed{2} \qquad \int_{t_1}^{t_2} v(t)\,dt = s(t_2) - s(t_1)$$

is the net change of position, or *displacement*, of the particle during the time period from t_1 to t_2. In Section 5.1 we guessed that this was true for the case where the object moves in the positive direction, but now we have proved that it is always true.

- If we want to calculate the distance the object travels during the time interval, we have to consider the intervals when $v(t) \geq 0$ (the particle moves to the right) and also the intervals when $v(t) \leq 0$ (the particle moves to the left). In both cases the distance is computed by integrating $|v(t)|$, the speed. Therefore

3
$$\int_{t_1}^{t_2} |v(t)| \, dt = \text{total distance traveled}$$

Figure 3 shows how both displacement and distance traveled can be interpreted in terms of areas under a velocity curve.

$$\text{displacement} = \int_{t_1}^{t_2} v(t) \, dt = A_1 - A_2 + A_3$$

$$\text{distance} = \int_{t_1}^{t_2} |v(t)| \, dt = A_1 + A_2 + A_3$$

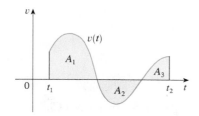

FIGURE 3

- The acceleration of the object is $a(t) = v'(t)$, so

$$\int_{t_1}^{t_2} a(t) \, dt = v(t_2) - v(t_1)$$

is the change in velocity from time t_1 to time t_2.

EXAMPLE 6 A particle moves along a line so that its velocity at time t is $v(t) = t^2 - t - 6$ (measured in meters per second).
(a) Find the displacement of the particle during the time period $1 \leq t \leq 4$.
(b) Find the distance traveled during this time period.

SOLUTION
(a) By Equation 2, the displacement is

$$s(4) - s(1) = \int_1^4 v(t) \, dt = \int_1^4 (t^2 - t - 6) \, dt$$

$$= \left[\frac{t^3}{3} - \frac{t^2}{2} - 6t \right]_1^4 = -\frac{9}{2}$$

This means that the particle moved 4.5 m toward the left.

(b) Note that $v(t) = t^2 - t - 6 = (t - 3)(t + 2)$ and so $v(t) \leq 0$ on the interval $[1, 3]$ and $v(t) \geq 0$ on $[3, 4]$. Thus, from Equation 3, the distance traveled is

To integrate the absolute value of $v(t)$, we use Property 5 of integrals from Section 5.2 to split the integral into two parts, one where $v(t) \leq 0$ and one where $v(t) \geq 0$.

$$\int_1^4 |v(t)| \, dt = \int_1^3 [-v(t)] \, dt + \int_3^4 v(t) \, dt$$

$$= \int_1^3 (-t^2 + t + 6) \, dt + \int_3^4 (t^2 - t - 6) \, dt$$

$$= \left[-\frac{t^3}{3} + \frac{t^2}{2} + 6t \right]_1^3 + \left[\frac{t^3}{3} - \frac{t^2}{2} - 6t \right]_3^4$$

$$= \frac{61}{6} \approx 10.17 \text{ m}$$

∎

EXAMPLE 7 Figure 4 shows the power consumption in the city of San Francisco for a day in September (P is measured in megawatts; t is measured in hours starting at midnight). Estimate the energy used on that day.

FIGURE 4

Pacific Gas & Electric

SOLUTION Power is the rate of change of energy: $P(t) = E'(t)$. So, by the Net Change Theorem,

$$\int_0^{24} P(t)\,dt = \int_0^{24} E'(t)\,dt = E(24) - E(0)$$

is the total amount of energy used on that day. We approximate the value of the integral using the Midpoint Rule with 12 subintervals and $\Delta t = 2$:

$$\int_0^{24} P(t)\,dt \approx [P(1) + P(3) + P(5) + \cdots + P(21) + P(23)]\,\Delta t$$

$$\approx (440 + 400 + 420 + 620 + 790 + 840 + 850$$

$$+ 840 + 810 + 690 + 670 + 550)(2)$$

$$= 15{,}840$$

The energy used was approximately 15,840 megawatt-hours. ∎

A note on units

How did we know what units to use for energy in Example 7? The integral $\int_0^{24} P(t)\,dt$ is defined as the limit of sums of terms of the form $P(t_i^*)\,\Delta t$. Now $P(t_i^*)$ is measured in megawatts and Δt is measured in hours, so their product is measured in megawatt-hours. The same is true of the limit. In general, the unit of measurement for $\int_a^b f(x)\,dx$ is the product of the unit for $f(x)$ and the unit for x.

5.4   EXERCISES

1–4 Verify by differentiation that the formula is correct.

1. $\displaystyle\int \frac{1}{x^2\sqrt{1+x^2}}\,dx = -\frac{\sqrt{1+x^2}}{x} + C$

2. $\displaystyle\int \cos^2 x\,dx = \tfrac{1}{2}x + \tfrac{1}{4}\sin 2x + C$

3. $\displaystyle\int \tan^2 x\,dx = \tan x - x + C$

4. $\displaystyle\int x\sqrt{a+bx}\,dx = \frac{2}{15b^2}(3bx - 2a)(a+bx)^{3/2} + C$

5–18 Find the general indefinite integral.

5. $\displaystyle\int (x^{1.3} + 7x^{2.5})\,dx$

6. $\displaystyle\int \sqrt[4]{x^5}\,dx$

7. $\int \left(5 + \frac{2}{3}x^2 + \frac{3}{4}x^3\right) dx$

8. $\int \left(u^6 - 2u^5 - u^3 + \frac{2}{7}\right) du$

9. $\int (u + 4)(2u + 1)\, du$

10. $\int \sqrt{t}\,(t^2 + 3t + 2)\, dt$

11. $\int \dfrac{1 + \sqrt{x} + x}{x}\, dx$

12. $\int \left(x^2 + 1 + \dfrac{1}{x^2 + 1}\right) dx$

13. $\int (\sin x + \sinh x)\, dx$

14. $\int \left(\dfrac{1 + r}{r}\right)^2 dr$

15. $\int (2 + \tan^2 \theta)\, d\theta$

16. $\int \sec t\,(\sec t + \tan t)\, dt$

17. $\int 2^t(1 + 5^t)\, dt$

18. $\int \dfrac{\sin 2x}{\sin x}\, dx$

19–20 Find the general indefinite integral. Illustrate by graphing several members of the family on the same screen.

19. $\int \left(\cos x + \frac{1}{2}x\right) dx$

20. $\int (e^x - 2x^2)\, dx$

21–46 Evaluate the integral.

21. $\int_{-2}^{3} (x^2 - 3)\, dx$

22. $\int_{1}^{2} (4x^3 - 3x^2 + 2x)\, dx$

23. $\int_{-2}^{0} \left(\frac{1}{2}t^4 + \frac{1}{4}t^3 - t\right) dt$

24. $\int_{0}^{3} (1 + 6w^2 - 10w^4)\, dw$

25. $\int_{0}^{2} (2x - 3)(4x^2 + 1)\, dx$

26. $\int_{-1}^{1} t(1 - t)^2\, dt$

27. $\int_{0}^{\pi} (5e^x + 3\sin x)\, dx$

28. $\int_{1}^{2} \left(\dfrac{1}{x^2} - \dfrac{4}{x^3}\right) dx$

29. $\int_{1}^{4} \left(\dfrac{4 + 6u}{\sqrt{u}}\right) du$

30. $\int_{0}^{1} \dfrac{4}{1 + p^2}\, dp$

31. $\int_{0}^{1} x(\sqrt[3]{x} + \sqrt[4]{x})\, dx$

32. $\int_{1}^{4} \dfrac{\sqrt{y} - y}{y^2}\, dy$

33. $\int_{1}^{2} \left(\dfrac{x}{2} - \dfrac{2}{x}\right) dx$

34. $\int_{0}^{1} (5x - 5^x)\, dx$

35. $\int_{0}^{1} (x^{10} + 10^x)\, dx$

36. $\int_{0}^{\pi/4} \sec\theta \tan\theta\, d\theta$

37. $\int_{0}^{\pi/4} \dfrac{1 + \cos^2\theta}{\cos^2\theta}\, d\theta$

38. $\int_{0}^{\pi/3} \dfrac{\sin\theta + \sin\theta \tan^2\theta}{\sec^2\theta}\, d\theta$

39. $\int_{1}^{8} \dfrac{2 + t}{\sqrt[3]{t^2}}\, dt$

40. $\int_{-10}^{10} \dfrac{2e^x}{\sinh x + \cosh x}\, dx$

41. $\int_{0}^{\sqrt{3}/2} \dfrac{dr}{\sqrt{1 - r^2}}$

42. $\int_{1}^{2} \dfrac{(x - 1)^3}{x^2}\, dx$

43. $\int_{0}^{1/\sqrt{3}} \dfrac{t^2 - 1}{t^4 - 1}\, dt$

44. $\int_{0}^{2} |2x - 1|\, dx$

45. $\int_{-1}^{2} (x - 2|x|)\, dx$

46. $\int_{0}^{3\pi/2} |\sin x|\, dx$

---

47. Use a graph to estimate the x-intercepts of the curve $y = 1 - 2x - 5x^4$. Then use this information to estimate the area of the region that lies under the curve and above the x-axis.

48. Repeat Exercise 47 for the curve $y = (x^2 + 1)^{-1} - x^4$.

49. The area of the region that lies to the right of the y-axis and to the left of the parabola $x = 2y - y^2$ (the shaded region in the figure) is given by the integral $\int_{0}^{2} (2y - y^2)\, dy$. (Turn your head clockwise and think of the region as lying below the curve $x = 2y - y^2$ from $y = 0$ to $y = 2$.) Find the area of the region.

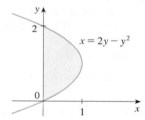

50. The boundaries of the shaded region are the y-axis, the line $y = 1$, and the curve $y = \sqrt[4]{x}$. Find the area of this region by writing x as a function of y and integrating with respect to y (as in Exercise 49).

51. If $w'(t)$ is the rate of growth of a child in pounds per year, what does $\int_{5}^{10} w'(t)\, dt$ represent?

52. The current in a wire is defined as the derivative of the charge: $I(t) = Q'(t)$. (See Example 3.7.3.) What does $\int_{a}^{b} I(t)\, dt$ represent?

53. If oil leaks from a tank at a rate of $r(t)$ gallons per minute at time t, what does $\int_0^{120} r(t)\, dt$ represent?

54. A honeybee population starts with 100 bees and increases at a rate of $n'(t)$ bees per week. What does $100 + \int_0^{15} n'(t)\, dt$ represent?

55. In Section 4.7 we defined the marginal revenue function $R'(x)$ as the derivative of the revenue function $R(x)$, where x is the number of units sold. What does $\int_{1000}^{5000} R'(x)\, dx$ represent?

56. If $f(x)$ is the slope of a trail at a distance of x miles from the start of the trail, what does $\int_3^5 f(x)\, dx$ represent?

57. If x is measured in meters and $f(x)$ is measured in newtons, what are the units for $\int_0^{100} f(x)\, dx$?

58. If the units for x are feet and the units for $a(x)$ are pounds per foot, what are the units for da/dx? What units does $\int_2^8 a(x)\, dx$ have?

59–60 The velocity function (in meters per second) is given for a particle moving along a line. Find (a) the displacement and (b) the distance traveled by the particle during the given time interval.

59. $v(t) = 3t - 5, \quad 0 \le t \le 3$

60. $v(t) = t^2 - 2t - 3, \quad 2 \le t \le 4$

61–62 The acceleration function (in m/s^2) and the initial velocity are given for a particle moving along a line. Find (a) the velocity at time t and (b) the distance traveled during the given time interval.

61. $a(t) = t + 4, \quad v(0) = 5, \quad 0 \le t \le 10$

62. $a(t) = 2t + 3, \quad v(0) = -4, \quad 0 \le t \le 3$

---

63. The linear density of a rod of length 4 m is given by $\rho(x) = 9 + 2\sqrt{x}$ measured in kilograms per meter, where x is measured in meters from one end of the rod. Find the total mass of the rod.

64. Water flows from the bottom of a storage tank at a rate of $r(t) = 200 - 4t$ liters per minute, where $0 \le t \le 50$. Find the amount of water that flows from the tank during the first 10 minutes.

65. The velocity of a car was read from its speedometer at 10-second intervals and recorded in the table. Use the Midpoint Rule to estimate the distance traveled by the car.

t (s)	v (mi/h)	t (s)	v (mi/h)
0	0	60	56
10	38	70	53
20	52	80	50
30	58	90	47
40	55	100	45
50	51		

66. Suppose that a volcano is erupting and readings of the rate $r(t)$ at which solid materials are spewed into the atmosphere are given in the table. The time t is measured in seconds and the units for $r(t)$ are tonnes (metric tons) per second.

t	0	1	2	3	4	5	6
$r(t)$	2	10	24	36	46	54	60

(a) Give upper and lower estimates for the total quantity $Q(6)$ of erupted materials after six seconds.

(b) Use the Midpoint Rule to estimate $Q(6)$.

67. The marginal cost of manufacturing x yards of a certain fabric is

$$C'(x) = 3 - 0.01x + 0.000006x^2$$

(in dollars per yard). Find the increase in cost if the production level is raised from 2000 yards to 4000 yards.

68. Water flows into and out of a storage tank. A graph of the rate of change $r(t)$ of the volume of water in the tank, in liters per day, is shown. If the amount of water in the tank at time $t = 0$ is 25,000 L, use the Midpoint Rule to estimate the amount of water in the tank four days later.

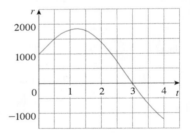

69. The graph of the acceleration $a(t)$ of a car measured in ft/s^2 is shown. Use the Midpoint Rule to estimate the increase in the velocity of the car during the six-second time interval.

70. Lake Lanier in Georgia, USA, is a reservoir created by Buford Dam on the Chattahoochee River. The table shows the rate of inflow of water, in cubic feet per second, as measured every morning at 7:30 AM by the US Army Corps of Engineers. Use the Midpoint Rule to estimate the amount of water that flowed into Lake Lanier from July 18th, 2013, at 7:30 AM to July 26th at 7:30 AM.

Day	Inflow rate (ft³/s)
July 18	5275
July 19	6401
July 20	2554
July 21	4249
July 22	3016
July 23	3821
July 24	2462
July 25	2628
July 26	3003

71. A bacteria population is 4000 at time $t = 0$ and its rate of growth is $1000 \cdot 2^t$ bacteria per hour after t hours. What is the population after one hour?

72. Shown is the graph of traffic on an Internet service provider's T1 data line from midnight to 8:00 AM. D is the data throughput, measured in megabits per second. Use the Midpoint Rule to estimate the total amount of data transmitted during that time period.

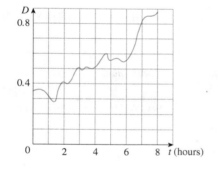

73. Shown is the power consumption in the province of Ontario, Canada, for December 9, 2004 (P is measured in megawatts; t is measured in hours starting at midnight). Using the fact that power is the rate of change of energy, estimate the energy used on that day.

Independent Electricity Market Operator

74. On May 7, 1992, the space shuttle *Endeavour* was launched on mission STS-49, the purpose of which was to install a new perigee kick motor in an Intelsat communications satellite. The table gives the velocity data for the shuttle between liftoff and the jettisoning of the solid rocket boosters.
   (a) Use a graphing calculator or computer to model these data by a third-degree polynomial.
   (b) Use the model in part (a) to estimate the height reached by the *Endeavour*, 125 seconds after liftoff.

Event	Time (s)	Velocity (ft/s)
Launch	0	0
Begin roll maneuver	10	185
End roll maneuver	15	319
Throttle to 89%	20	447
Throttle to 67%	32	742
Throttle to 104%	59	1325
Maximum dynamic pressure	62	1445
Solid rocket booster separation	125	4151

WRITING PROJECT   NEWTON, LEIBNIZ, AND THE INVENTION OF CALCULUS

We sometimes read that the inventors of calculus were Sir Isaac Newton (1642–1727) and Gottfried Wilhelm Leibniz (1646–1716). But we know that the basic ideas behind integration were investigated 2500 years ago by ancient Greeks such as Eudoxus and Archimedes, and methods for finding tangents were pioneered by Pierre Fermat (1601–1665), Isaac Barrow (1630–1677), and others. Barrow—who taught at Cambridge and was a major influence on Newton—was the first to understand the inverse relationship between differentiation and integration. What Newton and Leibniz did was to use this relationship, in the form of the Fundamental Theorem of Calculus, in order to develop calculus into a systematic mathematical discipline. It is in this sense that Newton and Leibniz are credited with the invention of calculus.

Read about the contributions of these men in one or more of the given references and write a report on one of the following three topics. You can include biographical details, but the main

412   CHAPTER 5   Integrals

thrust of your report should be a description, in some detail, of their methods and notations. In particular, you should consult one of the sourcebooks, which give excerpts from the original publications of Newton and Leibniz, translated from Latin to English.

- The Role of Newton in the Development of Calculus
- The Role of Leibniz in the Development of Calculus
- The Controversy between the Followers of Newton and Leibniz over Priority in the Invention of Calculus

References

1. Carl Boyer and Uta Merzbach, *A History of Mathematics* (New York: Wiley, 1987), Chapter 19.

2. Carl Boyer, *The History of the Calculus and Its Conceptual Development* (New York: Dover, 1959), Chapter V.

3. C. H. Edwards, *The Historical Development of the Calculus* (New York: Springer-Verlag, 1979), Chapters 8 and 9.

4. Howard Eves, *An Introduction to the History of Mathematics,* 6th ed. (New York: Saunders, 1990), Chapter 11.

5. C. C. Gillispie, ed., *Dictionary of Scientific Biography* (New York: Scribner's, 1974). See the article on Leibniz by Joseph Hofmann in Volume VIII and the article on Newton by I. B. Cohen in Volume X.

6. Victor Katz, *A History of Mathematics: An Introduction* (New York: HarperCollins, 1993), Chapter 12.

7. Morris Kline, *Mathematical Thought from Ancient to Modern Times* (New York: Oxford University Press, 1972), Chapter 17.

Sourcebooks

1. John Fauvel and Jeremy Gray, eds., *The History of Mathematics: A Reader* (London: MacMillan Press, 1987), Chapters 12 and 13.

2. D. E. Smith, ed., *A Sourcebook in Mathematics* (New York: Dover, 1959), Chapter V.

3. D. J. Struik, ed., *A Sourcebook in Mathematics,* 1200–1800 (Princeton, NJ: Princeton University Press, 1969), Chapter V.

5.5   The Substitution Rule

Because of the Fundamental Theorem, it's important to be able to find antiderivatives. But our antidifferentiation formulas don't tell us how to evaluate integrals such as

$$\boxed{1} \qquad \int 2x\sqrt{1+x^2}\,dx$$

To find this integral we use the problem-solving strategy of *introducing something extra.* Here the "something extra" is a new variable; we change from the variable x to a new variable u. Suppose that we let u be the quantity under the root sign in (1), $u = 1 + x^2$. Then the differential of u is $du = 2x\,dx$. Notice that if the dx in the notation for an inte-

Differentials were defined in Section 3.10. If $u = f(x)$, then

$$du = f'(x)\,dx$$

gral were to be interpreted as a differential, then the differential $2x\,dx$ would occur in (1) and so, formally, without justifying our calculation, we could write

$$\boxed{2} \qquad \int 2x\sqrt{1 + x^2}\,dx = \int \sqrt{1 + x^2}\,2x\,dx = \int \sqrt{u}\,du$$

$$= \tfrac{2}{3}u^{3/2} + C = \tfrac{2}{3}(1 + x^2)^{3/2} + C$$

But now we can check that we have the correct answer by using the Chain Rule to differentiate the final function of Equation 2:

$$\frac{d}{dx}\left[\tfrac{2}{3}(1 + x^2)^{3/2} + C\right] = \tfrac{2}{3} \cdot \tfrac{3}{2}(1 + x^2)^{1/2} \cdot 2x = 2x\sqrt{1 + x^2}$$

In general, this method works whenever we have an integral that we can write in the form $\int f(g(x))\,g'(x)\,dx$. Observe that if $F' = f$, then

$$\boxed{3} \qquad \int F'(g(x))\,g'(x)\,dx = F(g(x)) + C$$

because, by the Chain Rule,

$$\frac{d}{dx}[F(g(x))] = F'(g(x))\,g'(x)$$

If we make the "change of variable" or "substitution" $u = g(x)$, then from Equation 3 we have

$$\int F'(g(x))\,g'(x)\,dx = F(g(x)) + C = F(u) + C = \int F'(u)\,du$$

or, writing $F' = f$, we get

$$\int f(g(x))\,g'(x)\,dx = \int f(u)\,du$$

Thus we have proved the following rule.

$\boxed{4}$ **The Substitution Rule** If $u = g(x)$ is a differentiable function whose range is an interval I and f is continuous on I, then

$$\int f(g(x))\,g'(x)\,dx = \int f(u)\,du$$

Notice that the Substitution Rule for integration was proved using the Chain Rule for differentiation. Notice also that if $u = g(x)$, then $du = g'(x)\,dx$, so a way to remember the Substitution Rule is to think of dx and du in (4) as differentials.

Thus the Substitution Rule says: **It is permissible to operate with dx and du after integral signs as if they were differentials.**

EXAMPLE 1 Find $\int x^3 \cos(x^4 + 2)\,dx$.

SOLUTION We make the substitution $u = x^4 + 2$ because its differential is $du = 4x^3\,dx$, which, apart from the constant factor 4, occurs in the integral. Thus, using

$x^3 \, dx = \frac{1}{4} \, du$ and the Substitution Rule, we have

$$\int x^3 \cos(x^4 + 2) \, dx = \int \cos u \cdot \tfrac{1}{4} \, du = \tfrac{1}{4} \int \cos u \, du$$

$$= \tfrac{1}{4} \sin u + C$$

$$= \tfrac{1}{4} \sin(x^4 + 2) + C$$

Check the answer by differentiating it.   Notice that at the final stage we had to return to the original variable x.   ■

The idea behind the Substitution Rule is to replace a relatively complicated integral by a simpler integral. This is accomplished by changing from the original variable x to a new variable u that is a function of x. Thus in Example 1 we replaced the integral $\int x^3 \cos(x^4 + 2) \, dx$ by the simpler integral $\frac{1}{4} \int \cos u \, du$.

The main challenge in using the Substitution Rule is to think of an appropriate substitution. You should try to choose u to be some function in the integrand whose differential also occurs (except for a constant factor). This was the case in Example 1. If that is not possible, try choosing u to be some complicated part of the integrand (perhaps the inner function in a composite function). Finding the right substitution is a bit of an art. It's not unusual to guess wrong; if your first guess doesn't work, try another substitution.

EXAMPLE 2 Evaluate $\int \sqrt{2x + 1} \, dx$.

SOLUTION 1 Let $u = 2x + 1$. Then $du = 2 \, dx$, so $dx = \frac{1}{2} \, du$. Thus the Substitution Rule gives

$$\int \sqrt{2x + 1} \, dx = \int \sqrt{u} \cdot \tfrac{1}{2} \, du = \tfrac{1}{2} \int u^{1/2} \, du$$

$$= \frac{1}{2} \cdot \frac{u^{3/2}}{3/2} + C = \tfrac{1}{3} u^{3/2} + C$$

$$= \tfrac{1}{3}(2x + 1)^{3/2} + C$$

SOLUTION 2 Another possible substitution is $u = \sqrt{2x + 1}$. Then

$$du = \frac{dx}{\sqrt{2x + 1}} \qquad \text{so} \qquad dx = \sqrt{2x + 1} \, du = u \, du$$

(Or observe that $u^2 = 2x + 1$, so $2u \, du = 2 \, dx$.) Therefore

$$\int \sqrt{2x + 1} \, dx = \int u \cdot u \, du = \int u^2 \, du$$

$$= \frac{u^3}{3} + C = \tfrac{1}{3}(2x + 1)^{3/2} + C$$ ■

EXAMPLE 3 Find $\int \dfrac{x}{\sqrt{1 - 4x^2}} \, dx$.

SOLUTION Let $u = 1 - 4x^2$. Then $du = -8x \, dx$, so $x \, dx = -\frac{1}{8} \, du$ and

$$\int \frac{x}{\sqrt{1 - 4x^2}} \, dx = -\tfrac{1}{8} \int \frac{1}{\sqrt{u}} \, du = -\tfrac{1}{8} \int u^{-1/2} \, du$$

$$= -\tfrac{1}{8}(2\sqrt{u}) + C = -\tfrac{1}{4}\sqrt{1 - 4x^2} + C$$ ■

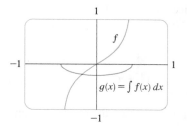

FIGURE 1

$$f(x) = \frac{x}{\sqrt{1 - 4x^2}}$$

$$g(x) = \int f(x)\, dx = -\tfrac{1}{4}\sqrt{1 - 4x^2}$$

The answer to Example 3 could be checked by differentiation, but instead let's check it with a graph. In Figure 1 we have used a computer to graph both the integrand $f(x) = x/\sqrt{1 - 4x^2}$ and its indefinite integral $g(x) = -\tfrac{1}{4}\sqrt{1 - 4x^2}$ (we take the case $C = 0$). Notice that $g(x)$ decreases when $f(x)$ is negative, increases when $f(x)$ is positive, and has its minimum value when $f(x) = 0$. So it seems reasonable, from the graphical evidence, that g is an antiderivative of f.

EXAMPLE 4 Calculate $\int e^{5x}\, dx$.

SOLUTION If we let $u = 5x$, then $du = 5\, dx$, so $dx = \tfrac{1}{5}\, du$. Therefore

$$\int e^{5x}\, dx = \tfrac{1}{5} \int e^u\, du = \tfrac{1}{5} e^u + C = \tfrac{1}{5} e^{5x} + C \qquad \blacksquare$$

NOTE With some experience, you might be able to evaluate integrals like those in Examples 1–4 without going to the trouble of making an explicit substitution. By recognizing the pattern in Equation 3, where the integrand on the left side is the product of the derivative of an outer function and the derivative of the inner function, we could work Example 1 as follows:

$$\int x^3 \cos(x^4 + 2)\, dx = \int \cos(x^4 + 2) \cdot x^3\, dx = \tfrac{1}{4} \int \cos(x^4 + 2) \cdot (4x^3)\, dx$$

$$= \tfrac{1}{4} \int \cos(x^4 + 2) \cdot \frac{d}{dx}(x^4 + 2)\, dx = \tfrac{1}{4} \sin(x^4 + 2) + C$$

Similarly, the solution to Example 4 could be written like this:

$$\int e^{5x}\, dx = \tfrac{1}{5} \int 5e^{5x}\, dx = \tfrac{1}{5} \int \frac{d}{dx}(e^{5x})\, dx = \tfrac{1}{5} e^{5x} + C$$

The following example, however, is more complicated and so an explicit substitution is advisable.

EXAMPLE 5 Find $\int \sqrt{1 + x^2}\; x^5\, dx$.

SOLUTION An appropriate substitution becomes more obvious if we factor x^5 as $x^4 \cdot x$. Let $u = 1 + x^2$. Then $du = 2x\, dx$, so $x\, dx = \tfrac{1}{2}\, du$. Also $x^2 = u - 1$, so $x^4 = (u - 1)^2$:

$$\int \sqrt{1 + x^2}\; x^5\, dx = \int \sqrt{1 + x^2}\; x^4 \cdot x\, dx$$

$$= \int \sqrt{u}\; (u - 1)^2 \cdot \tfrac{1}{2}\, du = \tfrac{1}{2} \int \sqrt{u}\; (u^2 - 2u + 1)\, du$$

$$= \tfrac{1}{2} \int (u^{5/2} - 2u^{3/2} + u^{1/2})\, du$$

$$= \tfrac{1}{2}\left(\tfrac{2}{7} u^{7/2} - 2 \cdot \tfrac{2}{5} u^{5/2} + \tfrac{2}{3} u^{3/2}\right) + C$$

$$= \tfrac{1}{7}(1 + x^2)^{7/2} - \tfrac{2}{5}(1 + x^2)^{5/2} + \tfrac{1}{3}(1 + x^2)^{3/2} + C \qquad \blacksquare$$

EXAMPLE 6 Calculate $\int \tan x\, dx$.

SOLUTION First we write tangent in terms of sine and cosine:

$$\int \tan x\, dx = \int \frac{\sin x}{\cos x}\, dx$$

This suggests that we should substitute $u = \cos x$, since then $du = -\sin x \, dx$ and so $\sin x \, dx = -du$:

$$\int \tan x \, dx = \int \frac{\sin x}{\cos x} \, dx = -\int \frac{1}{u} \, du$$

$$= -\ln|u| + C = -\ln|\cos x| + C \qquad \blacksquare$$

Since $-\ln|\cos x| = \ln(|\cos x|^{-1}) = \ln(1/|\cos x|) = \ln|\sec x|$, the result of Example 6 can also be written as

$$\boxed{5} \qquad \boxed{\int \tan x \, dx = \ln|\sec x| + C}$$

Definite Integrals

When evaluating a *definite* integral by substitution, two methods are possible. One method is to evaluate the indefinite integral first and then use the Fundamental Theorem. For instance, using the result of Example 2, we have

$$\int_0^4 \sqrt{2x + 1} \, dx = \int \sqrt{2x + 1} \, dx \Big]_0^4$$

$$= \tfrac{1}{3}(2x + 1)^{3/2} \Big]_0^4 = \tfrac{1}{3}(9)^{3/2} - \tfrac{1}{3}(1)^{3/2}$$

$$= \tfrac{1}{3}(27 - 1) = \tfrac{26}{3}$$

Another method, which is usually preferable, is to change the limits of integration when the variable is changed.

This rule says that when using a substitution in a definite integral, we must put everything in terms of the new variable u, not only x and dx but also the limits of integration. The new limits of integration are the values of u that correspond to $x = a$ and $x = b$.

> $\boxed{6}$ **The Substitution Rule for Definite Integrals** If g' is continuous on $[a, b]$ and f is continuous on the range of $u = g(x)$, then
>
> $$\int_a^b f(g(x)) g'(x) \, dx = \int_{g(a)}^{g(b)} f(u) \, du$$

PROOF   Let F be an antiderivative of f. Then, by (3), $F(g(x))$ is an antiderivative of $f(g(x)) g'(x)$, so by Part 2 of the Fundamental Theorem, we have

$$\int_a^b f(g(x)) g'(x) \, dx = F(g(x)) \Big]_a^b = F(g(b)) - F(g(a))$$

But, applying FTC2 a second time, we also have

$$\int_{g(a)}^{g(b)} f(u) \, du = F(u) \Big]_{g(a)}^{g(b)} = F(g(b)) - F(g(a)) \qquad \blacksquare$$

EXAMPLE 7   Evaluate $\displaystyle\int_0^4 \sqrt{2x + 1} \, dx$ using (6).

SOLUTION   Using the substitution from Solution 1 of Example 2, we have $u = 2x + 1$ and $dx = \tfrac{1}{2} \, du$. To find the new limits of integration we note that

$$\text{when } x = 0, \; u = 2(0) + 1 = 1 \qquad \text{and} \qquad \text{when } x = 4, \; u = 2(4) + 1 = 9$$

Therefore

$$\int_0^4 \sqrt{2x + 1}\, dx = \int_1^9 \tfrac{1}{2}\sqrt{u}\, du$$

$$= \tfrac{1}{2} \cdot \tfrac{2}{3} u^{3/2} \Big]_1^9$$

$$= \tfrac{1}{3}(9^{3/2} - 1^{3/2}) = \tfrac{26}{3}$$

Observe that when using (6) we do *not* return to the variable x after integrating. We simply evaluate the expression in u between the appropriate values of u. ■

The integral given in Example 8 is an abbreviation for

$$\int_1^2 \frac{1}{(3 - 5x)^2}\, dx$$

EXAMPLE 8 Evaluate $\displaystyle\int_1^2 \frac{dx}{(3 - 5x)^2}$.

SOLUTION Let $u = 3 - 5x$. Then $du = -5\, dx$, so $dx = -\tfrac{1}{5}\, du$. When $x = 1$, $u = -2$ and when $x = 2$, $u = -7$. Thus

$$\int_1^2 \frac{dx}{(3 - 5x)^2} = -\frac{1}{5}\int_{-2}^{-7} \frac{du}{u^2}$$

$$= -\frac{1}{5}\left[-\frac{1}{u} \right]_{-2}^{-7} = \frac{1}{5u}\Big]_{-2}^{-7}$$

$$= \frac{1}{5}\left(-\frac{1}{7} + \frac{1}{2} \right) = \frac{1}{14}$$ ■

Since the function $f(x) = (\ln x)/x$ in Example 9 is positive for $x > 1$, the integral represents the area of the shaded region in Figure 2.

EXAMPLE 9 Calculate $\displaystyle\int_1^e \frac{\ln x}{x}\, dx$.

SOLUTION We let $u = \ln x$ because its differential $du = dx/x$ occurs in the integral. When $x = 1$, $u = \ln 1 = 0$; when $x = e$, $u = \ln e = 1$. Thus

$$\int_1^e \frac{\ln x}{x}\, dx = \int_0^1 u\, du = \frac{u^2}{2}\Big]_0^1 = \frac{1}{2}$$ ■

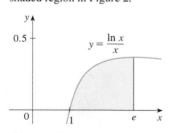

FIGURE 2

■ Symmetry

The next theorem uses the Substitution Rule for Definite Integrals (6) to simplify the calculation of integrals of functions that possess symmetry properties.

---

7 **Integrals of Symmetric Functions** Suppose f is continuous on $[-a, a]$.

(a) If f is even $[f(-x) = f(x)]$, then $\displaystyle\int_{-a}^a f(x)\, dx = 2\int_0^a f(x)\, dx$.

(b) If f is odd $[f(-x) = -f(x)]$, then $\displaystyle\int_{-a}^a f(x)\, dx = 0$.

PROOF We split the integral in two:

$$\boxed{8}\qquad \int_{-a}^a f(x)\, dx = \int_{-a}^0 f(x)\, dx + \int_0^a f(x)\, dx = -\int_0^{-a} f(x)\, dx + \int_0^a f(x)\, dx$$

In the first integral on the far right side we make the substitution $u = -x$. Then $du = -dx$ and when $x = -a$, $u = a$. Therefore

$$-\int_0^{-a} f(x)\, dx = -\int_0^a f(-u)\,(-du) = \int_0^a f(-u)\, du$$

and so Equation 8 becomes

$$\boxed{9} \qquad \int_{-a}^a f(x)\, dx = \int_0^a f(-u)\, du + \int_0^a f(x)\, dx$$

(a) If f is even, then $f(-u) = f(u)$ so Equation 9 gives

$$\int_{-a}^a f(x)\, dx = \int_0^a f(u)\, du + \int_0^a f(x)\, dx = 2\int_0^a f(x)\, dx$$

(b) If f is odd, then $f(-u) = -f(u)$ and so Equation 9 gives

$$\int_{-a}^a f(x)\, dx = -\int_0^a f(u)\, du + \int_0^a f(x)\, dx = 0 \qquad \blacksquare$$

(a) f even, $\int_{-a}^a f(x)\, dx = 2\int_0^a f(x)\, dx$

(b) f odd, $\int_{-a}^a f(x)\, dx = 0$

FIGURE 3

Theorem 7 is illustrated by Figure 3. For the case where f is positive and even, part (a) says that the area under $y = f(x)$ from $-a$ to a is twice the area from 0 to a because of symmetry. Recall that an integral $\int_a^b f(x)\, dx$ can be expressed as the area above the x-axis and below $y = f(x)$ minus the area below the axis and above the curve. Thus part (b) says the integral is 0 because the areas cancel.

EXAMPLE 10 Since $f(x) = x^6 + 1$ satisfies $f(-x) = f(x)$, it is even and so

$$\int_{-2}^2 (x^6 + 1)\, dx = 2\int_0^2 (x^6 + 1)\, dx$$

$$= 2\left[\tfrac{1}{7}x^7 + x\right]_0^2 = 2\left(\tfrac{128}{7} + 2\right) = \tfrac{284}{7} \qquad \blacksquare$$

EXAMPLE 11 Since $f(x) = (\tan x)/(1 + x^2 + x^4)$ satisfies $f(-x) = -f(x)$, it is odd and so

$$\int_{-1}^1 \frac{\tan x}{1 + x^2 + x^4}\, dx = 0 \qquad \blacksquare$$

5.5   EXERCISES

1–6 Evaluate the integral by making the given substitution.

1. $\displaystyle\int \cos 2x\, dx, \quad u = 2x$

2. $\displaystyle\int xe^{-x^2}\, dx, \quad u = -x^2$

3. $\displaystyle\int x^2\sqrt{x^3 + 1}\, dx, \quad u = x^3 + 1$

4. $\displaystyle\int \sin^2\theta \cos\theta\, d\theta, \quad u = \sin\theta$

5. $\displaystyle\int \frac{x^3}{x^4 - 5}\, dx, \quad u = x^4 - 5$

6. $\displaystyle\int \sqrt{2t + 1}\, dt, \quad u = 2t + 1$

7–48 Evaluate the indefinite integral.

7. $\displaystyle\int x\sqrt{1 - x^2}\, dx$

8. $\displaystyle\int x^2 e^{x^3}\, dx$

9. $\displaystyle\int (1 - 2x)^9\, dx$

10. $\displaystyle\int \sin t\sqrt{1 + \cos t}\, dt$

11. $\displaystyle\int \cos(\pi t/2)\, dt$

12. $\displaystyle\int \sec^2 2\theta\, d\theta$

13. $\int \dfrac{dx}{5 - 3x}$

14. $\int y^2(4 - y^3)^{2/3}\, dy$

15. $\int \cos^3\theta \sin\theta\, d\theta$

16. $\int e^{-5r}\, dr$

17. $\int \dfrac{e^u}{(1 - e^u)^2}\, du$

18. $\int \dfrac{\sin\sqrt{x}}{\sqrt{x}}\, dx$

19. $\int \dfrac{a + bx^2}{\sqrt{3ax + bx^3}}\, dx$

20. $\int \dfrac{z^2}{z^3 + 1}\, dz$

21. $\int \dfrac{(\ln x)^2}{x}\, dx$

22. $\int \sin x \sin(\cos x)\, dx$

23. $\int \sec^2\theta \tan^3\theta\, d\theta$

24. $\int x\sqrt{x + 2}\, dx$

25. $\int e^x\sqrt{1 + e^x}\, dx$

26. $\int \dfrac{dx}{ax + b}\ (a \neq 0)$

27. $\int (x^2 + 1)(x^3 + 3x)^4\, dx$

28. $\int e^{\cos t} \sin t\, dt$

29. $\int 5^t \sin(5^t)\, dt$

30. $\int \dfrac{\sec^2 x}{\tan^2 x}\, dx$

31. $\int \dfrac{(\arctan x)^2}{x^2 + 1}\, dx$

32. $\int \dfrac{x}{x^2 + 4}\, dx$

33. $\int \cos(1 + 5t)\, dt$

34. $\int \dfrac{\cos(\pi/x)}{x^2}\, dx$

35. $\int \sqrt{\cot x}\ \csc^2 x\, dx$

36. $\int \dfrac{2^t}{2^t + 3}\, dt$

37. $\int \sinh^2 x \cosh x\, dx$

38. $\int \dfrac{dt}{\cos^2 t\sqrt{1 + \tan t}}$

39. $\int \dfrac{\sin 2x}{1 + \cos^2 x}\, dx$

40. $\int \dfrac{\sin x}{1 + \cos^2 x}\, dx$

41. $\int \cot x\, dx$

42. $\int \dfrac{\cos(\ln t)}{t}\, dt$

43. $\int \dfrac{dx}{\sqrt{1 - x^2}\ \sin^{-1} x}$

44. $\int \dfrac{x}{1 + x^4}\, dx$

45. $\int \dfrac{1 + x}{1 + x^2}\, dx$

46. $\int x^2\sqrt{2 + x}\, dx$

47. $\int x(2x + 5)^8\, dx$

48. $\int x^3\sqrt{x^2 + 1}\, dx$

49–52 Evaluate the indefinite integral. Illustrate and check that your answer is reasonable by graphing both the function and its antiderivative (take $C = 0$).

49. $\int x(x^2 - 1)^3\, dx$

50. $\int \tan^2\theta \sec^2\theta\, d\theta$

51. $\int e^{\cos x} \sin x\, dx$

52. $\int \sin x \cos^4 x\, dx$

53–73 Evaluate the definite integral.

53. $\int_0^1 \cos(\pi t/2)\, dt$

54. $\int_0^1 (3t - 1)^{50}\, dt$

55. $\int_0^1 \sqrt[3]{1 + 7x}\, dx$

56. $\int_0^3 \dfrac{dx}{5x + 1}$

57. $\int_0^{\pi/6} \dfrac{\sin t}{\cos^2 t}\, dt$

58. $\int_{\pi/3}^{2\pi/3} \csc^2\left(\tfrac{1}{2}t\right)\, dt$

59. $\int_1^2 \dfrac{e^{1/x}}{x^2}\, dx$

60. $\int_0^1 xe^{-x^2}\, dx$

61. $\int_{-\pi/4}^{\pi/4} (x^3 + x^4 \tan x)\, dx$

62. $\int_0^{\pi/2} \cos x \sin(\sin x)\, dx$

63. $\int_0^{13} \dfrac{dx}{\sqrt[3]{(1 + 2x)^2}}$

64. $\int_0^a x\sqrt{a^2 - x^2}\, dx$

65. $\int_0^a x\sqrt{x^2 + a^2}\, dx\ \ (a > 0)$

66. $\int_{-\pi/3}^{\pi/3} x^4 \sin x\, dx$

67. $\int_1^2 x\sqrt{x - 1}\, dx$

68. $\int_0^4 \dfrac{x}{\sqrt{1 + 2x}}\, dx$

69. $\int_e^{e^4} \dfrac{dx}{x\sqrt{\ln x}}$

70. $\int_0^2 (x - 1)e^{(x-1)^2}\, dx$

71. $\int_0^1 \dfrac{e^z + 1}{e^z + z}\, dz$

72. $\int_0^{T/2} \sin(2\pi t/T - \alpha)\, dt$

73. $\int_0^1 \dfrac{dx}{(1 + \sqrt{x})^4}$

---

74. Verify that $f(x) = \sin\sqrt[3]{x}$ is an odd function and use that fact to show that
$$0 \le \int_{-2}^3 \sin\sqrt[3]{x}\, dx \le 1$$

75–76 Use a graph to give a rough estimate of the area of the region that lies under the given curve. Then find the exact area.

75. $y = \sqrt{2x + 1},\ \ 0 \le x \le 1$

76. $y = 2\sin x - \sin 2x,\ \ 0 \le x \le \pi$

77. Evaluate $\int_{-2}^2 (x + 3)\sqrt{4 - x^2}\, dx$ by writing it as a sum of two integrals and interpreting one of those integrals in terms of an area.

78. Evaluate $\int_0^1 x\sqrt{1 - x^4}\, dx$ by making a substitution and interpreting the resulting integral in terms of an area.

79. Which of the following areas are equal? Why?

80. A model for the basal metabolism rate, in kcal/h, of a young man is $R(t) = 85 - 0.18 \cos(\pi t/12)$, where t is the time in hours measured from 5:00 AM. What is the total basal metabolism of this man, $\int_0^{24} R(t)\, dt$, over a 24-hour time period?

81. An oil storage tank ruptures at time $t = 0$ and oil leaks from the tank at a rate of $r(t) = 100e^{-0.01t}$ liters per minute. How much oil leaks out during the first hour?

82. A bacteria population starts with 400 bacteria and grows at a rate of $r(t) = (450.268)e^{1.12567t}$ bacteria per hour. How many bacteria will there be after three hours?

83. Breathing is cyclic and a full respiratory cycle from the beginning of inhalation to the end of exhalation takes about 5 s. The maximum rate of air flow into the lungs is about 0.5 L/s. This explains, in part, why the function $f(t) = \frac{1}{2} \sin(2\pi t/5)$ has often been used to model the rate of air flow into the lungs. Use this model to find the volume of inhaled air in the lungs at time t.

84. The rate of growth of a fish population was modeled by the equation

$$G(t) = \frac{60{,}000e^{-0.6t}}{(1 + 5e^{-0.6t})^2}$$

where t is measured in years and G in kilograms per year. If the biomass was 25,000 kg in the year 2000, what is the predicted biomass for the year 2020?

85. Dialysis treatment removes urea and other waste products from a patient's blood by diverting some of the bloodflow externally through a machine called a dialyzer. The rate at which urea is removed from the blood (in mg/min) is often well described by the equation

$$u(t) = \frac{r}{V} C_0 e^{-rt/V}$$

where r is the rate of flow of blood through the dialyzer (in mL/min), V is the volume of the patient's blood (in mL), and C_0 is the amount of urea in the blood (in mg) at time $t = 0$. Evaluate the integral $\int_0^{30} u(t)\, dt$ and interpret it.

86. Alabama Instruments Company has set up a production line to manufacture a new calculator. The rate of production of these calculators after t weeks is

$$\frac{dx}{dt} = 5000\left(1 - \frac{100}{(t + 10)^2}\right) \text{ calculators/week}$$

(Notice that production approaches 5000 per week as time goes on, but the initial production is lower because of the workers' unfamiliarity with the new techniques.) Find the number of calculators produced from the beginning of the third week to the end of the fourth week.

87. If f is continuous and $\int_0^4 f(x)\, dx = 10$, find $\int_0^2 f(2x)\, dx$.

88. If f is continuous and $\int_0^9 f(x)\, dx = 4$, find $\int_0^3 xf(x^2)\, dx$.

89. If f is continuous on $\mathbb{R}$, prove that

$$\int_a^b f(-x)\, dx = \int_{-b}^{-a} f(x)\, dx$$

For the case where $f(x) \geqslant 0$ and $0 < a < b$, draw a diagram to interpret this equation geometrically as an equality of areas.

90. If f is continuous on $\mathbb{R}$, prove that

$$\int_a^b f(x + c)\, dx = \int_{a+c}^{b+c} f(x)\, dx$$

For the case where $f(x) \geqslant 0$, draw a diagram to interpret this equation geometrically as an equality of areas.

91. If a and b are positive numbers, show that

$$\int_0^1 x^a(1 - x)^b\, dx = \int_0^1 x^b(1 - x)^a\, dx$$

92. If f is continuous on $[0, \pi]$, use the substitution $u = \pi - x$ to show that

$$\int_0^\pi xf(\sin x)\, dx = \frac{\pi}{2} \int_0^\pi f(\sin x)\, dx$$

93. Use Exercise 92 to evaluate the integral

$$\int_0^\pi \frac{x \sin x}{1 + \cos^2 x}\, dx$$

94. (a) If f is continuous, prove that

$$\int_0^{\pi/2} f(\cos x)\, dx = \int_0^{\pi/2} f(\sin x)\, dx$$

(b) Use part (a) to evaluate $\int_0^{\pi/2} \cos^2 x\, dx$ and $\int_0^{\pi/2} \sin^2 x\, dx$.

5   REVIEW

CONCEPT CHECK

Answers to the Concept Check can be found on the back endpapers.

1. (a) Write an expression for a Riemann sum of a function f. Explain the meaning of the notation that you use.

  (b) If $f(x) \geq 0$, what is the geometric interpretation of a Riemann sum? Illustrate with a diagram.

  (c) If $f(x)$ takes on both positive and negative values, what is the geometric interpretation of a Riemann sum? Illustrate with a diagram.

2. (a) Write the definition of the definite integral of a continuous function from a to b.

  (b) What is the geometric interpretation of $\int_a^b f(x)\,dx$ if $f(x) \geq 0$?

  (c) What is the geometric interpretation of $\int_a^b f(x)\,dx$ if $f(x)$ takes on both positive and negative values? Illustrate with a diagram.

3. State the Midpoint Rule.

4. State both parts of the Fundamental Theorem of Calculus.

5. (a) State the Net Change Theorem.

  (b) If $r(t)$ is the rate at which water flows into a reservoir, what does $\int_{t_1}^{t_2} r(t)\,dt$ represent?

6. Suppose a particle moves back and forth along a straight line with velocity $v(t)$, measured in feet per second, and acceleration $a(t)$.

  (a) What is the meaning of $\int_{60}^{120} v(t)\,dt$?

  (b) What is the meaning of $\int_{60}^{120} |v(t)|\,dt$?

  (c) What is the meaning of $\int_{60}^{120} a(t)\,dt$?

7. (a) Explain the meaning of the indefinite integral $\int f(x)\,dx$.

  (b) What is the connection between the definite integral $\int_a^b f(x)\,dx$ and the indefinite integral $\int f(x)\,dx$?

8. Explain exactly what is meant by the statement that "differentiation and integration are inverse processes."

9. State the Substitution Rule. In practice, how do you use it?

TRUE–FALSE QUIZ

Determine whether the statement is true or false. If it is true, explain why. If it is false, explain why or give an example that disproves the statement.

1. If f and g are continuous on $[a, b]$, then

$$\int_a^b [f(x) + g(x)]\,dx = \int_a^b f(x)\,dx + \int_a^b g(x)\,dx$$

2. If f and g are continuous on $[a, b]$, then

$$\int_a^b [f(x)g(x)]\,dx = \left(\int_a^b f(x)\,dx\right)\left(\int_a^b g(x)\,dx\right)$$

3. If f is continuous on $[a, b]$, then

$$\int_a^b 5f(x)\,dx = 5\int_a^b f(x)\,dx$$

4. If f is continuous on $[a, b]$, then

$$\int_a^b xf(x)\,dx = x\int_a^b f(x)\,dx$$

5. If f is continuous on $[a, b]$ and $f(x) \geq 0$, then

$$\int_a^b \sqrt{f(x)}\,dx = \sqrt{\int_a^b f(x)\,dx}$$

6. If f' is continuous on $[1, 3]$, then $\int_1^3 f'(v)\,dv = f(3) - f(1)$.

7. If f and g are continuous and $f(x) \geq g(x)$ for $a \leq x \leq b$, then

$$\int_a^b f(x)\,dx \geq \int_a^b g(x)\,dx$$

8. If f and g are differentiable and $f(x) \geq g(x)$ for $a < x < b$, then $f'(x) \geq g'(x)$ for $a < x < b$.

9. $\displaystyle\int_{-1}^1 \left(x^5 - 6x^9 + \frac{\sin x}{(1 + x^4)^2}\right) dx = 0$

10. $\displaystyle\int_{-5}^5 (ax^2 + bx + c)\,dx = 2\int_0^5 (ax^2 + c)\,dx$

11. All continuous functions have derivatives.

12. All continuous functions have antiderivatives.

13. $\displaystyle\int_0^3 e^{x^2}\,dx = \int_0^5 e^{x^2}\,dx + \int_5^3 e^{x^2}\,dx$

14. If $\int_0^1 f(x)\,dx = 0$, then $f(x) = 0$ for $0 \leq x \leq 1$.

15. If f is continuous on $[a, b]$, then

$$\frac{d}{dx}\left(\int_a^b f(x)\,dx\right) = f(x)$$

16. $\int_0^2 (x - x^3)\,dx$ represents the area under the curve $y = x - x^3$ from 0 to 2.

17. $\displaystyle\int_{-2}^1 \frac{1}{x^4}\,dx = -\frac{3}{8}$

18. If f has a discontinuity at 0, then $\int_{-1}^1 f(x)\,dx$ does not exist.

EXERCISES

1. Use the given graph of f to find the Riemann sum with six subintervals. Take the sample points to be (a) left endpoints and (b) midpoints. In each case draw a diagram and explain what the Riemann sum represents.

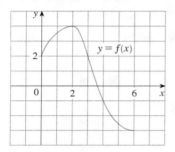

2. (a) Evaluate the Riemann sum for
$$f(x) = x^2 - x \qquad 0 \le x \le 2$$
with four subintervals, taking the sample points to be right endpoints. Explain, with the aid of a diagram, what the Riemann sum represents.

(b) Use the definition of a definite integral (with right endpoints) to calculate the value of the integral
$$\int_0^2 (x^2 - x)\, dx$$

(c) Use the Fundamental Theorem to check your answer to part (b).

(d) Draw a diagram to explain the geometric meaning of the integral in part (b).

3. Evaluate
$$\int_0^1 \left(x + \sqrt{1 - x^2}\right) dx$$
by interpreting it in terms of areas.

4. Express
$$\lim_{n \to \infty} \sum_{i=1}^{n} \sin x_i \, \Delta x$$
as a definite integral on the interval $[0, \pi]$ and then evaluate the integral.

5. If $\int_0^6 f(x)\, dx = 10$ and $\int_0^4 f(x)\, dx = 7$, find $\int_4^6 f(x)\, dx$.

CAS **6.** (a) Write $\int_1^5 (x + 2x^5)\, dx$ as a limit of Riemann sums, taking the sample points to be right endpoints. Use a computer algebra system to evaluate the sum and to compute the limit.

(b) Use the Fundamental Theorem to check your answer to part (a).

7. The figure shows the graphs of f, f', and $\int_0^x f(t)\, dt$. Identify each graph, and explain your choices.

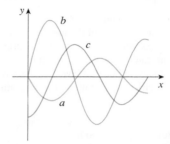

8. Evaluate:

(a) $\displaystyle\int_0^1 \frac{d}{dx}\left(e^{\arctan x}\right) dx$ (b) $\displaystyle\frac{d}{dx}\int_0^1 e^{\arctan x}\, dx$

(c) $\displaystyle\frac{d}{dx}\int_0^x e^{\arctan t}\, dt$

9. The graph of f consists of the three line segments shown. If $g(x) = \int_0^x f(t)\, dt$, find $g(4)$ and $g'(4)$.

10. If f is the function in Exercise 9, find $g''(4)$.

11–40 Evaluate the integral, if it exists.

11. $\displaystyle\int_1^2 (8x^3 + 3x^2)\, dx$ **12.** $\displaystyle\int_0^T (x^4 - 8x + 7)\, dx$

13. $\displaystyle\int_0^1 (1 - x^9)\, dx$ **14.** $\displaystyle\int_0^1 (1 - x)^9\, dx$

15. $\displaystyle\int_1^9 \frac{\sqrt{u} - 2u^2}{u}\, du$ **16.** $\displaystyle\int_0^1 \left(\sqrt[4]{u} + 1\right)^2 du$

17. $\displaystyle\int_0^1 y(y^2 + 1)^5\, dy$ **18.** $\displaystyle\int_0^2 y^2\sqrt{1 + y^3}\, dy$

19. $\displaystyle\int_1^5 \frac{dt}{(t - 4)^2}$ **20.** $\displaystyle\int_0^1 \sin(3\pi t)\, dt$

21. $\displaystyle\int_0^1 v^2 \cos(v^3)\, dv$ **22.** $\displaystyle\int_{-1}^1 \frac{\sin x}{1 + x^2}\, dx$

23. $\displaystyle\int_{-\pi/4}^{\pi/4} \frac{t^4 \tan t}{2 + \cos t}\, dt$ **24.** $\displaystyle\int_0^1 \frac{e^x}{1 + e^{2x}}\, dx$

25. $\displaystyle\int \left(\frac{1 - x}{x}\right)^2 dx$ **26.** $\displaystyle\int_1^{10} \frac{x}{x^2 - 4}\, dx$

27. $\displaystyle\int \frac{x + 2}{\sqrt{x^2 + 4x}}\, dx$

28. $\displaystyle\int \frac{\csc^2 x}{1 + \cot x}\, dx$

29. $\displaystyle\int \sin \pi t \cos \pi t\, dt$

30. $\displaystyle\int \sin x \cos(\cos x)\, dx$

31. $\displaystyle\int \frac{e^{\sqrt{x}}}{\sqrt{x}}\, dx$

32. $\displaystyle\int \frac{\sin(\ln x)}{x}\, dx$

33. $\displaystyle\int \tan x \ln(\cos x)\, dx$

34. $\displaystyle\int \frac{x}{\sqrt{1 - x^4}}\, dx$

35. $\displaystyle\int \frac{x^3}{1 + x^4}\, dx$

36. $\displaystyle\int \sinh(1 + 4x)\, dx$

37. $\displaystyle\int \frac{\sec \theta \tan \theta}{1 + \sec \theta}\, d\theta$

38. $\displaystyle\int_0^{\pi/4} (1 + \tan t)^3 \sec^2 t\, dt$

39. $\displaystyle\int_0^3 |x^2 - 4|\, dx$

40. $\displaystyle\int_0^4 |\sqrt{x} - 1|\, dx$

41–42 Evaluate the indefinite integral. Illustrate and check that your answer is reasonable by graphing both the function and its antiderivative (take $C = 0$).

41. $\displaystyle\int \frac{\cos x}{\sqrt{1 + \sin x}}\, dx$

42. $\displaystyle\int \frac{x^3}{\sqrt{x^2 + 1}}\, dx$

---

43. Use a graph to give a rough estimate of the area of the region that lies under the curve $y = x\sqrt{x}$, $0 \leqslant x \leqslant 4$. Then find the exact area.

44. Graph the function $f(x) = \cos^2 x \sin x$ and use the graph to guess the value of the integral $\int_0^{2\pi} f(x)\, dx$. Then evaluate the integral to confirm your guess.

45–50 Find the derivative of the function.

45. $\displaystyle F(x) = \int_0^x \frac{t^2}{1 + t^3}\, dt$

46. $\displaystyle F(x) = \int_x^1 \sqrt{t + \sin t}\, dt$

47. $\displaystyle g(x) = \int_0^{x^4} \cos(t^2)\, dt$

48. $\displaystyle g(x) = \int_1^{\sin x} \frac{1 - t^2}{1 + t^4}\, dt$

49. $\displaystyle y = \int_{\sqrt{x}}^x \frac{e^t}{t}\, dt$

50. $\displaystyle y = \int_{2x}^{3x+1} \sin(t^4)\, dt$

51–52 Use Property 8 of integrals to estimate the value of the integral.

51. $\displaystyle\int_1^3 \sqrt{x^2 + 3}\, dx$

52. $\displaystyle\int_3^5 \frac{1}{x + 1}\, dx$

53–56 Use the properties of integrals to verify the inequality.

53. $\displaystyle\int_0^1 x^2 \cos x\, dx \leqslant \frac{1}{3}$

54. $\displaystyle\int_{\pi/4}^{\pi/2} \frac{\sin x}{x}\, dx \leqslant \frac{\sqrt{2}}{2}$

55. $\displaystyle\int_0^1 e^x \cos x\, dx \leqslant e - 1$

56. $\displaystyle\int_0^1 x \sin^{-1} x\, dx \leqslant \pi/4$

---

57. Use the Midpoint Rule with $n = 6$ to approximate $\int_0^3 \sin(x^3)\, dx$.

58. A particle moves along a line with velocity function $v(t) = t^2 - t$, where v is measured in meters per second. Find (a) the displacement and (b) the distance traveled by the particle during the time interval $[0, 5]$.

59. Let $r(t)$ be the rate at which the world's oil is consumed, where t is measured in years starting at $t = 0$ on January 1, 2000, and $r(t)$ is measured in barrels per year. What does $\int_0^8 r(t)\, dt$ represent?

60. A radar gun was used to record the speed of a runner at the times given in the table. Use the Midpoint Rule to estimate the distance the runner covered during those 5 seconds.

t (s)	v (m/s)	t (s)	v (m/s)
0	0	3.0	10.51
0.5	4.67	3.5	10.67
1.0	7.34	4.0	10.76
1.5	8.86	4.5	10.81
2.0	9.73	5.0	10.81
2.5	10.22		

61. A population of honeybees increased at a rate of $r(t)$ bees per week, where the graph of r is as shown. Use the Midpoint Rule with six subintervals to estimate the increase in the bee population during the first 24 weeks.

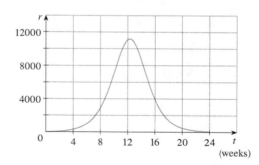

62. Let

$$f(x) = \begin{cases} -x - 1 & \text{if } -3 \leqslant x \leqslant 0 \\ -\sqrt{1 - x^2} & \text{if } 0 \leqslant x \leqslant 1 \end{cases}$$

Evaluate $\int_{-3}^1 f(x)\, dx$ by interpreting the integral as a difference of areas.

63. If f is continuous and $\int_0^2 f(x)\, dx = 6$, evaluate $\int_0^{\pi/2} f(2 \sin \theta) \cos \theta\, d\theta$.

64. The Fresnel function $S(x) = \int_0^x \sin\left(\frac{1}{2}\pi t^2\right) dt$ was introduced in Section 5.3. Fresnel also used the function

$$C(x) = \int_0^x \cos\left(\frac{1}{2}\pi t^2\right) dt$$

in his theory of the diffraction of light waves.
(a) On what intervals is C increasing?
(b) On what intervals is C concave upward?
[CAS] (c) Use a graph to solve the following equation correct to two decimal places:

$$\int_0^x \cos\left(\frac{1}{2}\pi t^2\right) dt = 0.7$$

[CAS] (d) Plot the graphs of C and S on the same screen. How are these graphs related?

65. Estimate the value of the number c such that the area under the curve $y = \sinh cx$ between $x = 0$ and $x = 1$ is equal to 1.

66. Suppose that the temperature in a long, thin rod placed along the x-axis is initially $C/(2a)$ if $|x| \le a$ and 0 if $|x| > a$. It can be shown that if the heat diffusivity of the rod is k, then the temperature of the rod at the point x at time t is

$$T(x, t) = \frac{C}{a\sqrt{4\pi kt}} \int_0^a e^{-(x-u)^2/(4kt)} du$$

To find the temperature distribution that results from an initial hot spot concentrated at the origin, we need to compute $\lim_{a \to 0} T(x, t)$. Use l'Hospital's Rule to find this limit.

67. If f is a continuous function such that

$$\int_1^x f(t) dt = (x - 1)e^{2x} + \int_1^x e^{-t} f(t) dt$$

for all x, find an explicit formula for $f(x)$.

68. Suppose h is a function such that $h(1) = -2$, $h'(1) = 2$, $h''(1) = 3$, $h(2) = 6$, $h'(2) = 5$, $h''(2) = 13$, and h'' is continuous everywhere. Evaluate $\int_1^2 h''(u) du$.

69. If f' is continuous on $[a, b]$, show that

$$2 \int_a^b f(x) f'(x) dx = [f(b)]^2 - [f(a)]^2$$

70. Find

$$\lim_{h \to 0} \frac{1}{h} \int_2^{2+h} \sqrt{1 + t^3} \, dt$$

71. If f is continuous on $[0, 1]$, prove that

$$\int_0^1 f(x) dx = \int_0^1 f(1 - x) dx$$

72. Evaluate

$$\lim_{n \to \infty} \frac{1}{n} \left[\left(\frac{1}{n}\right)^9 + \left(\frac{2}{n}\right)^9 + \left(\frac{3}{n}\right)^9 + \cdots + \left(\frac{n}{n}\right)^9 \right]$$

73. Suppose f is continuous, $f(0) = 0$, $f(1) = 1$, $f'(x) > 0$, and $\int_0^1 f(x) dx = \frac{1}{3}$. Find the value of the integral $\int_0^1 f^{-1}(y) dy$.

Problems Plus

Before you look at the solution of the following example, cover it up and first try to solve the problem yourself.

EXAMPLE Evaluate $\lim\limits_{x \to 3} \left(\dfrac{x}{x-3} \displaystyle\int_3^x \dfrac{\sin t}{t}\, dt \right)$.

SOLUTION Let's start by having a preliminary look at the ingredients of the function. What happens to the first factor, $x/(x-3)$, when x approaches 3? The numerator approaches 3 and the denominator approaches 0, so we have

$$\frac{x}{x-3} \to \infty \quad \text{as} \quad x \to 3^+ \qquad \text{and} \qquad \frac{x}{x-3} \to -\infty \quad \text{as} \quad x \to 3^-$$

The second factor approaches $\int_3^3 (\sin t)/t\, dt$, which is 0. It's not clear what happens to the function as a whole. (One factor is becoming large while the other is becoming small.) So how do we proceed?

PS The principles of problem solving are discussed on page 71.

One of the principles of problem solving is *recognizing something familiar.* Is there a part of the function that reminds us of something we've seen before? Well, the integral

$$\int_3^x \frac{\sin t}{t}\, dt$$

has x as its upper limit of integration and that type of integral occurs in Part 1 of the Fundamental Theorem of Calculus:

$$\frac{d}{dx} \int_a^x f(t)\, dt = f(x)$$

This suggests that differentiation might be involved.

Once we start thinking about differentiation, the denominator $(x - 3)$ reminds us of something else that should be familiar: One of the forms of the definition of the derivative in Chapter 2 is

$$F'(a) = \lim_{x \to a} \frac{F(x) - F(a)}{x - a}$$

and with $a = 3$ this becomes

$$F'(3) = \lim_{x \to 3} \frac{F(x) - F(3)}{x - 3}$$

So what is the function F in our situation? Notice that if we define

$$F(x) = \int_3^x \frac{\sin t}{t}\, dt$$

then $F(3) = 0$. What about the factor x in the numerator? That's just a red herring, so let's factor it out and put together the calculation:

Another approach is to use l'Hospital's Rule

$$\lim_{x \to 3} \left(\frac{x}{x-3} \int_3^x \frac{\sin t}{t}\, dt \right) = \lim_{x \to 3} x \cdot \lim_{x \to 3} \frac{\displaystyle\int_3^x \frac{\sin t}{t}\, dt}{x-3} = 3 \lim_{x \to 3} \frac{F(x) - F(3)}{x - 3}$$

$$= 3F'(3) = 3\, \frac{\sin 3}{3} = \sin 3 \qquad \text{(FTC1)} \qquad \blacksquare$$

Problems

1. If $x \sin \pi x = \displaystyle\int_0^{x^2} f(t)\, dt$, where f is a continuous function, find $f(4)$.

2. Find the minimum value of the area of the region under the curve $y = x + 1/x$ from $x = a$ to $x = a + 1.5$, for all $a > 0$.

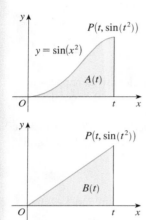

$P(t, \sin(t^2))$

$y = \sin(x^2)$

$A(t)$

$P(t, \sin(t^2))$

$B(t)$

FIGURE FOR PROBLEM 8

FIGURE FOR PROBLEM 16

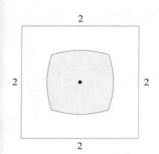

FIGURE FOR PROBLEM 18

3. If $\int_0^4 e^{(x-2)^4} dx = k$, find the value of $\int_0^4 xe^{(x-2)^4} dx$.

4. (a) Graph several members of the family of functions $f(x) = (2cx - x^2)/c^3$ for $c > 0$ and look at the regions enclosed by these curves and the x-axis. Make a conjecture about how the areas of these regions are related.

(b) Prove your conjecture in part (a).

(c) Take another look at the graphs in part (a) and use them to sketch the curve traced out by the vertices (highest points) of the family of functions. Can you guess what kind of curve this is?

(d) Find an equation of the curve you sketched in part (c).

5. If $f(x) = \int_0^{g(x)} \dfrac{1}{\sqrt{1 + t^3}} \, dt$, where $g(x) = \int_0^{\cos x} [1 + \sin(t^2)] \, dt$, find $f'(\pi/2)$.

6. If $f(x) = \int_0^x x^2 \sin(t^2) \, dt$, find $f'(x)$.

7. Evaluate $\displaystyle\lim_{x \to 0} \dfrac{1}{x} \int_0^x (1 - \tan 2t)^{1/t} \, dt$.

8. The figure shows two regions in the first quadrant: $A(t)$ is the area under the curve $y = \sin(x^2)$ from 0 to t, and $B(t)$ is the area of the triangle with vertices O, P, and $(t, 0)$. Find $\displaystyle\lim_{t \to 0^+} A(t)/B(t)$.

9. Find the interval $[a, b]$ for which the value of the integral $\int_a^b (2 + x - x^2) \, dx$ is a maximum.

10. Use an integral to estimate the sum $\displaystyle\sum_{i=1}^{10000} \sqrt{i}$.

11. (a) Evaluate $\int_0^n [\![x]\!] \, dx$, where n is a positive integer.

(b) Evaluate $\int_a^b [\![x]\!] \, dx$, where a and b are real numbers with $0 \le a < b$.

12. Find $\dfrac{d^2}{dx^2} \displaystyle\int_0^x \left(\int_1^{\sin t} \sqrt{1 + u^4} \, du \right) dt$.

13. Suppose the coefficients of the cubic polynomial $P(x) = a + bx + cx^2 + dx^3$ satisfy the equation

$$a + \frac{b}{2} + \frac{c}{3} + \frac{d}{4} = 0$$

Show that the equation $P(x) = 0$ has a root between 0 and 1. Can you generalize this result for an nth-degree polynomial?

14. A circular disk of radius r is used in an evaporator and is rotated in a vertical plane. If it is to be partially submerged in the liquid so as to maximize the exposed wetted area of the disk, show that the center of the disk should be positioned at a height $r/\sqrt{1 + \pi^2}$ above the surface of the liquid.

15. Prove that if f is continuous, then $\displaystyle\int_0^x f(u)(x - u) \, du = \int_0^x \left(\int_0^u f(t) \, dt \right) du$.

16. The figure shows a parabolic segment, that is, a portion of a parabola cut off by a chord AB. It also shows a point C on the parabola with the property that the tangent line at C is parallel to the chord AB. Archimedes proved that the area of the parabolic segment is $\frac{4}{3}$ times the area of the inscribed triangle ABC. Verify Archimedes' result for the parabola $y = 4 - x^2$ and the line $y = x + 2$.

17. Given the point (a, b) in the first quadrant, find the downward-opening parabola that passes through the point (a, b) and the origin such that the area under the parabola is a minimum.

18. The figure shows a region consisting of all points inside a square that are closer to the center than to the sides of the square. Find the area of the region.

19. Evaluate $\displaystyle\lim_{n \to \infty} \left(\dfrac{1}{\sqrt{n} \sqrt{n + 1}} + \dfrac{1}{\sqrt{n} \sqrt{n + 2}} + \cdots + \dfrac{1}{\sqrt{n} \sqrt{n + n}} \right)$.

20. For any number c, we let $f_c(x)$ be the smaller of the two numbers $(x - c)^2$ and $(x - c - 2)^2$. Then we define $g(c) = \int_0^1 f_c(x) \, dx$. Find the maximum and minimum values of $g(c)$ if $-2 \le c \le 2$.

Appendixes

A   Numbers, Inequalities, and Absolute Values

Calculus is based on the real number system. We start with the **integers**:

$$\ldots, \quad -3, \quad -2, \quad -1, \quad 0, \quad 1, \quad 2, \quad 3, \quad 4, \quad \ldots$$

Then we construct the **rational numbers**, which are ratios of integers. Thus any rational number r can be expressed as

$$r = \frac{m}{n} \qquad \text{where } m \text{ and } n \text{ are integers and } n \neq 0$$

Examples are

$$\tfrac{1}{2} \qquad -\tfrac{3}{7} \qquad 46 = \tfrac{46}{1} \qquad 0.17 = \tfrac{17}{100}$$

(Recall that division by 0 is always ruled out, so expressions like $\frac{3}{0}$ and $\frac{0}{0}$ are undefined.) Some real numbers, such as $\sqrt{2}$, can't be expressed as a ratio of integers and are therefore called **irrational numbers**. It can be shown, with varying degrees of difficulty, that the following are also irrational numbers:

$$\sqrt{3} \qquad \sqrt{5} \qquad \sqrt[3]{2} \qquad \pi \qquad \sin 1° \qquad \log_{10} 2$$

The set of all real numbers is usually denoted by the symbol $\mathbb{R}$. When we use the word *number* without qualification, we mean "real number."

Every number has a decimal representation. If the number is rational, then the corresponding decimal is repeating. For example,

$$\tfrac{1}{2} = 0.5000\ldots = 0.5\overline{0} \qquad\qquad \tfrac{2}{3} = 0.66666\ldots = 0.\overline{6}$$

$$\tfrac{157}{495} = 0.317171717\ldots = 0.3\overline{17} \qquad\qquad \tfrac{9}{7} = 1.285714285714\ldots = 1.\overline{285714}$$

(The bar indicates that the sequence of digits repeats forever.) On the other hand, if the number is irrational, the decimal is nonrepeating:

$$\sqrt{2} = 1.414213562373095\ldots \qquad\qquad \pi = 3.141592653589793\ldots$$

If we stop the decimal expansion of any number at a certain place, we get an approximation to the number. For instance, we can write

$$\pi \approx 3.14159265$$

where the symbol $\approx$ is read "is approximately equal to." The more decimal places we retain, the better the approximation we get.

The real numbers can be represented by points on a line as in Figure 1. The positive direction (to the right) is indicated by an arrow. We choose an arbitrary reference point O, called the **origin**, which corresponds to the real number 0. Given any convenient unit of measurement, each positive number x is represented by the point on the line a distance of x units to the right of the origin, and each negative number $-x$ is represented by the point x units to the left of the origin. Thus every real number is represented by a point on the line, and every point P on the line corresponds to exactly one real number. The number associated with the point P is called the **coordinate** of P and the line is then called a

coordinate line, or a **real number line**, or simply a **real line**. Often we identify the point with its coordinate and think of a number as being a point on the real line.

FIGURE 1

The real numbers are ordered. We say *a is less than b* and write $a < b$ if $b - a$ is a positive number. Geometrically this means that a lies to the left of b on the number line. (Equivalently, we say *b is greater than a* and write $b > a$.) The symbol $a \leq b$ (or $b \geq a$) means that either $a < b$ or $a = b$ and is read "*a* is less than or equal to *b*." For instance, the following are true inequalities:

$$7 < 7.4 < 7.5 \qquad -3 > -\pi \qquad \sqrt{2} < 2 \qquad \sqrt{2} \leq 2 \qquad 2 \leq 2$$

In what follows we need to use *set notation*. A **set** is a collection of objects, and these objects are called the **elements** of the set. If S is a set, the notation $a \in S$ means that a is an element of S, and $a \notin S$ means that a is not an element of S. For example, if Z represents the set of integers, then $-3 \in Z$ but $\pi \notin Z$. If S and T are sets, then their **union** $S \cup T$ is the set consisting of all elements that are in S or T (or in both S and T). The **intersection** of S and T is the set $S \cap T$ consisting of all elements that are in both S and T. In other words, $S \cap T$ is the common part of S and T. The empty set, denoted by $\varnothing$, is the set that contains no element.

Some sets can be described by listing their elements between braces. For instance, the set A consisting of all positive integers less than 7 can be written as

$$A = \{1, 2, 3, 4, 5, 6\}$$

We could also write A in *set-builder notation* as

$$A = \{x \mid x \text{ is an integer and } 0 < x < 7\}$$

which is read "A is the set of x such that x is an integer and $0 < x < 7$."

■ Intervals

Certain sets of real numbers, called **intervals**, occur frequently in calculus and correspond geometrically to line segments. For example, if $a < b$, the **open interval** from a to b consists of all numbers between a and b and is denoted by the symbol (a, b). Using set-builder notation, we can write

$$(a, b) = \{x \mid a < x < b\}$$

FIGURE 2
Open interval (a, b)

Notice that the endpoints of the interval—namely, a and b—are excluded. This is indicated by the round brackets ( ) and by the open dots in Figure 2. The **closed interval** from a to b is the set

$$[a, b] = \{x \mid a \leq x \leq b\}$$

FIGURE 3
Closed interval $[a, b]$

Here the endpoints of the interval are included. This is indicated by the square brackets [ ] and by the solid dots in Figure 3. It is also possible to include only one endpoint in an interval, as shown in Table 1.

We also need to consider infinite intervals such as

$$(a, \infty) = \{x \mid x > a\}$$

This does not mean that ∞ ("infinity") is a number. The notation (a, ∞) stands for the set of all numbers that are greater than a, so the symbol ∞ simply indicates that the interval extends indefinitely far in the positive direction.

1 **Table of Intervals**

Notation	Set description	Picture
(a, b)	$\{x \mid a < x < b\}$	
$[a, b]$	$\{x \mid a \leq x \leq b\}$	
$[a, b)$	$\{x \mid a \leq x < b\}$	
$(a, b]$	$\{x \mid a < x \leq b\}$	
(a, ∞)	$\{x \mid x > a\}$	
$[a, \infty)$	$\{x \mid x \geq a\}$	
$(-\infty, b)$	$\{x \mid x < b\}$	
$(-\infty, b]$	$\{x \mid x \leq b\}$	
$(-\infty, \infty)$	$\mathbb{R}$ (set of all real numbers)	

Table 1 lists the nine possible types of intervals. When these intervals are discussed, it is always assumed that $a < b$.

Inequalities

When working with inequalities, note the following rules.

2 **Rules for Inequalities**

1. If $a < b$, then $a + c < b + c$.
2. If $a < b$ and $c < d$, then $a + c < b + d$.
3. If $a < b$ and $c > 0$, then $ac < bc$.
4. If $a < b$ and $c < 0$, then $ac > bc$.
5. If $0 < a < b$, then $1/a > 1/b$.

Rule 1 says that we can add any number to both sides of an inequality, and Rule 2 says that two inequalities can be added. However, we have to be careful with multiplication. Rule 3 says that we can multiply both sides of an inequality by a *positive* number, but Rule 4 says that if we multiply both sides of an inequality by a negative number, then we reverse the direction of the inequality. For example, if we take the inequality $3 < 5$ and multiply by 2, we get $6 < 10$, but if we multiply by -2, we get $-6 > -10$. Finally, Rule 5 says that if we take reciprocals, then we reverse the direction of an inequality (provided the numbers are positive).

EXAMPLE 1 Solve the inequality $1 + x < 7x + 5$.

SOLUTION The given inequality is satisfied by some values of x but not by others. To *solve* an inequality means to determine the set of numbers x for which the inequality is true. This is called the *solution set*.

First we subtract 1 from each side of the inequality (using Rule 1 with $c = -1$):

$$x < 7x + 4$$

Then we subtract $7x$ from both sides (Rule 1 with $c = -7x$):

$$-6x < 4$$

Now we divide both sides by -6 $\left(\text{Rule 4 with } c = -\frac{1}{6}\right)$:

$$x > -\tfrac{4}{6} = -\tfrac{2}{3}$$

These steps can all be reversed, so the solution set consists of all numbers greater than $-\frac{2}{3}$. In other words, the solution of the inequality is the interval $\left(-\frac{2}{3}, \infty\right)$. ∎

EXAMPLE 2 Solve the inequalities $4 \leqslant 3x - 2 < 13$.

SOLUTION Here the solution set consists of all values of x that satisfy both inequalities. Using the rules given in (2), we see that the following inequalities are equivalent:

$$4 \leqslant 3x - 2 < 13$$

$$6 \leqslant 3x < 15 \qquad \text{(add 2)}$$

$$2 \leqslant x < 5 \qquad \text{(divide by 3)}$$

Therefore the solution set is $[2, 5)$. ∎

EXAMPLE 3 Solve the inequality $x^2 - 5x + 6 \leqslant 0$.

SOLUTION First we factor the left side:

$$(x - 2)(x - 3) \leqslant 0$$

We know that the corresponding equation $(x - 2)(x - 3) = 0$ has the solutions 2 and 3. The numbers 2 and 3 divide the real line into three intervals:

$$(-\infty, 2) \qquad (2, 3) \qquad (3, \infty)$$

A visual method for solving Example 3 is to use a graphing device to graph the parabola $y = x^2 - 5x + 6$ (as in Figure 4) and observe that the curve lies on or below the x-axis when $2 \leqslant x \leqslant 3$.

On each of these intervals we determine the signs of the factors. For instance,

$$x \in (-\infty, 2) \quad \Rightarrow \quad x < 2 \quad \Rightarrow \quad x - 2 < 0$$

Then we record these signs in the following chart:

Interval	$x - 2$	$x - 3$	$(x - 2)(x - 3)$
$x < 2$	−	−	+
$2 < x < 3$	+	−	−
$x > 3$	+	+	+

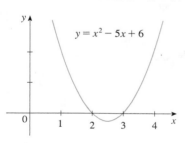

FIGURE 4

Another method for obtaining the information in the chart is to use *test values*. For instance, if we use the test value $x = 1$ for the interval $(-\infty, 2)$, then substitution in $x^2 - 5x + 6$ gives

$$1^2 - 5(1) + 6 = 2$$

The polynomial $x^2 - 5x + 6$ doesn't change sign inside any of the three intervals, so we conclude that it is positive on $(-\infty, 2)$.

Then we read from the chart that $(x - 2)(x - 3)$ is negative when $2 < x < 3$. Thus the solution of the inequality $(x - 2)(x - 3) \leqslant 0$ is

$$\{x \mid 2 \leqslant x \leqslant 3\} = [2, 3]$$

FIGURE 5

Notice that we have included the endpoints 2 and 3 because we are looking for values of x such that the product is either negative or zero. The solution is illustrated in Figure 5. ∎

EXAMPLE 4 Solve $x^3 + 3x^2 > 4x$.

SOLUTION First we take all nonzero terms to one side of the inequality sign and factor the resulting expression:

$$x^3 + 3x^2 - 4x > 0 \qquad \text{or} \qquad x(x - 1)(x + 4) > 0$$

As in Example 3 we solve the corresponding equation $x(x - 1)(x + 4) = 0$ and use the solutions $x = -4$, $x = 0$, and $x = 1$ to divide the real line into four intervals $(-\infty, -4)$, $(-4, 0)$, $(0, 1)$, and $(1, \infty)$. On each interval the product keeps a constant sign, which we list in the following chart:

Interval	x	$x - 1$	$x + 4$	$x(x - 1)(x + 4)$
$x < -4$	$-$	$-$	$-$	$-$
$-4 < x < 0$	$-$	$-$	$+$	$+$
$0 < x < 1$	$+$	$-$	$+$	$-$
$x > 1$	$+$	$+$	$+$	$+$

Then we read from the chart that the solution set is

$$\{x \mid -4 < x < 0 \text{ or } x > 1\} = (-4, 0) \cup (1, \infty)$$

The solution is illustrated in Figure 6. ∎

FIGURE 6

Absolute Value

The **absolute value** of a number a, denoted by $|a|$, is the distance from a to 0 on the real number line. Distances are always positive or 0, so we have

$$|a| \geqslant 0 \qquad \text{for every number } a$$

For example,

$$|3| = 3 \qquad |-3| = 3 \qquad |0| = 0 \qquad |\sqrt{2} - 1| = \sqrt{2} - 1 \qquad |3 - \pi| = \pi - 3$$

In general, we have

Remember that if a is negative, then $-a$ is positive.

$\boxed{3}$

$$|a| = a \qquad \text{if } a \geqslant 0$$

$$|a| = -a \qquad \text{if } a < 0$$

EXAMPLE 5 Express $|3x - 2|$ without using the absolute-value symbol.

SOLUTION

$$|3x - 2| = \begin{cases} 3x - 2 & \text{if } 3x - 2 \geq 0 \\ -(3x - 2) & \text{if } 3x - 2 < 0 \end{cases}$$

$$= \begin{cases} 3x - 2 & \text{if } x \geq \frac{2}{3} \\ 2 - 3x & \text{if } x < \frac{2}{3} \end{cases} \qquad \blacksquare$$

Recall that the symbol $\sqrt{}$ means "the positive square root of." Thus $\sqrt{r} = s$ means $s^2 = r$ and $s \geq 0$. Therefore the equation $\sqrt{a^2} = a$ is not always true. It is true only when $a \geq 0$. If $a < 0$, then $-a > 0$, so we have $\sqrt{a^2} = -a$. In view of (3), we then have the equation

4

$$\boxed{\sqrt{a^2} = |a|}$$

which is true for all values of a.

Hints for the proofs of the following properties are given in the exercises.

---

5 **Properties of Absolute Values** Suppose a and b are any real numbers and n is an integer. Then

1. $|ab| = |a||b|$ **2.** $\left|\dfrac{a}{b}\right| = \dfrac{|a|}{|b|}$ $(b \neq 0)$ **3.** $|a^n| = |a|^n$

For solving equations or inequalities involving absolute values, it's often very helpful to use the following statements.

---

6 Suppose $a > 0$. Then

4. $|x| = a$ if and only if $x = \pm a$

5. $|x| < a$ if and only if $-a < x < a$

6. $|x| > a$ if and only if $x > a$ or $x < -a$

For instance, the inequality $|x| < a$ says that the distance from x to the origin is less than a, and you can see from Figure 7 that this is true if and only if x lies between $-a$ and a.

If a and b are any real numbers, then the distance between a and b is the absolute value of the difference, namely, $|a - b|$, which is also equal to $|b - a|$. (See Figure 8.)

EXAMPLE 6 Solve $|2x - 5| = 3$.

SOLUTION By Property 4 of (6), $|2x - 5| = 3$ is equivalent to

$$2x - 5 = 3 \qquad \text{or} \qquad 2x - 5 = -3$$

So $2x = 8$ or $2x = 2$. Thus $x = 4$ or $x = 1$. $\blacksquare$

FIGURE 7

FIGURE 8
Length of a line segment $= |a - b|$

EXAMPLE 7 Solve $|x - 5| < 2$.

SOLUTION 1 By Property 5 of (6), $|x - 5| < 2$ is equivalent to

$$-2 < x - 5 < 2$$

Therefore, adding 5 to each side, we have

$$3 < x < 7$$

and the solution set is the open interval $(3, 7)$.

FIGURE 9

SOLUTION 2 Geometrically the solution set consists of all numbers x whose distance from 5 is less than 2. From Figure 9 we see that this is the interval $(3, 7)$. ∎

EXAMPLE 8 Solve $|3x + 2| \geq 4$.

SOLUTION By Properties 4 and 6 of (6), $|3x + 2| \geq 4$ is equivalent to

$$3x + 2 \geq 4 \qquad \text{or} \qquad 3x + 2 \leq -4$$

In the first case $3x \geq 2$, which gives $x \geq \frac{2}{3}$. In the second case $3x \leq -6$, which gives $x \leq -2$. So the solution set is

$$\left\{x \mid x \leq -2 \ \text{or} \ x \geq \tfrac{2}{3}\right\} = (-\infty, -2] \cup \left[\tfrac{2}{3}, \infty\right)$$ ∎

Another important property of absolute value, called the Triangle Inequality, is used frequently not only in calculus but throughout mathematics in general.

> **7** **The Triangle Inequality** If a and b are any real numbers, then
>
> $$|a + b| \leq |a| + |b|$$

Observe that if the numbers a and b are both positive or both negative, then the two sides in the Triangle Inequality are actually equal. But if a and b have opposite signs, the left side involves a subtraction and the right side does not. This makes the Triangle Inequality seem reasonable, but we can prove it as follows.

Notice that

$$-|a| \leq a \leq |a|$$

is always true because a equals either $|a|$ or $-|a|$. The corresponding statement for b is

$$-|b| \leq b \leq |b|$$

Adding these inequalities, we get

$$-(|a| + |b|) \leq a + b \leq |a| + |b|$$

If we now apply Properties 4 and 5 $\big($with x replaced by $a + b$ and a by $|a| + |b|\big)$, we obtain

$$|a + b| \leq |a| + |b|$$

which is what we wanted to show.

EXAMPLE 9 If $|x - 4| < 0.1$ and $|y - 7| < 0.2$, use the Triangle Inequality to estimate $|(x + y) - 11|$.

SOLUTION In order to use the given information, we use the Triangle Inequality with $a = x - 4$ and $b = y - 7$:

$$|(x + y) - 11| = |(x - 4) + (y - 7)|$$
$$\leq |x - 4| + |y - 7|$$
$$< 0.1 + 0.2 = 0.3$$

Thus $$|(x + y) - 11| < 0.3$$ ∎

A   EXERCISES

1–12 Rewrite the expression without using the absolute-value symbol.

1. $|5 - 23|$

2. $|5| - |-23|$

3. $|-\pi|$

4. $|\pi - 2|$

5. $|\sqrt{5} - 5|$

6. $||-2| - |-3||$

7. $|x - 2|$ if $x < 2$

8. $|x - 2|$ if $x > 2$

9. $|x + 1|$

10. $|2x - 1|$

11. $|x^2 + 1|$

12. $|1 - 2x^2|$

13–38 Solve the inequality in terms of intervals and illustrate the solution set on the real number line.

13. $2x + 7 > 3$

14. $3x - 11 < 4$

15. $1 - x \leq 2$

16. $4 - 3x \geq 6$

17. $2x + 1 < 5x - 8$

18. $1 + 5x > 5 - 3x$

19. $-1 < 2x - 5 < 7$

20. $1 < 3x + 4 \leq 16$

21. $0 \leq 1 - x < 1$

22. $-5 \leq 3 - 2x \leq 9$

23. $4x < 2x + 1 \leq 3x + 2$

24. $2x - 3 < x + 4 < 3x - 2$

25. $(x - 1)(x - 2) > 0$

26. $(2x + 3)(x - 1) \geq 0$

27. $2x^2 + x \leq 1$

28. $x^2 < 2x + 8$

29. $x^2 + x + 1 > 0$

30. $x^2 + x > 1$

31. $x^2 < 3$

32. $x^2 \geq 5$

33. $x^3 - x^2 \leq 0$

34. $(x + 1)(x - 2)(x + 3) \geq 0$

35. $x^3 > x$

36. $x^3 + 3x < 4x^2$

37. $\dfrac{1}{x} < 4$

38. $-3 < \dfrac{1}{x} \leq 1$

39. The relationship between the Celsius and Fahrenheit temperature scales is given by $C = \frac{5}{9}(F - 32)$, where C is the temperature in degrees Celsius and F is the temperature in degrees Fahrenheit. What interval on the Celsius scale corresponds to the temperature range $50 \leq F \leq 95$?

40. Use the relationship between C and F given in Exercise 39 to find the interval on the Fahrenheit scale corresponding to the temperature range $20 \leq C \leq 30$.

41. As dry air moves upward, it expands and in so doing cools at a rate of about 1°C for each 100-m rise, up to about 12 km.
(a) If the ground temperature is 20°C, write a formula for the temperature at height h.
(b) What range of temperature can be expected if a plane takes off and reaches a maximum height of 5 km?

42. If a ball is thrown upward from the top of a building 128 ft high with an initial velocity of 16 ft/s, then the height h above the ground t seconds later will be

$$h = 128 + 16t - 16t^2$$

During what time interval will the ball be at least 32 ft above the ground?

43–46 Solve the equation for x.

43. $|2x| = 3$

44. $|3x + 5| = 1$

45. $|x + 3| = |2x + 1|$

46. $\left|\dfrac{2x - 1}{x + 1}\right| = 3$

47–56 Solve the inequality.

47. $|x| < 3$

48. $|x| \geq 3$

49. $|x - 4| < 1$

50. $|x - 6| < 0.1$

51. $|x + 5| \geq 2$

52. $|x + 1| \geq 3$

53. $|2x - 3| \leq 0.4$

54. $|5x - 2| < 6$

55. $1 \leq |x| \leq 4$

56. $0 < |x - 5| < \frac{1}{2}$

57–58 Solve for x, assuming a, b, and c are positive constants.

57. $a(bx - c) \geqslant bc$ **58.** $a \leqslant bx + c < 2a$

59–60 Solve for x, assuming a, b, and c are negative constants.

59. $ax + b < c$ **60.** $\dfrac{ax + b}{c} \leqslant b$

---

61. Suppose that $|x - 2| < 0.01$ and $|y - 3| < 0.04$. Use the Triangle Inequality to show that $|(x + y) - 5| < 0.05$.

62. Show that if $|x + 3| < \frac{1}{2}$, then $|4x + 13| < 3$.

63. Show that if $a < b$, then $a < \dfrac{a + b}{2} < b$.

64. Use Rule 3 to prove Rule 5 of (2).

65. Prove that $|ab| = |a||b|$. [*Hint:* Use Equation 4.]

66. Prove that $\left|\dfrac{a}{b}\right| = \dfrac{|a|}{|b|}$.

67. Show that if $0 < a < b$, then $a^2 < b^2$.

68. Prove that $|x - y| \geqslant |x| - |y|$. [*Hint:* Use the Triangle Inequality with $a = x - y$ and $b = y$.]

69. Show that the sum, difference, and product of rational numbers are rational numbers.

70. (a) Is the sum of two irrational numbers always an irrational number?
    (b) Is the product of two irrational numbers always an irrational number?

B   Coordinate Geometry and Lines

Just as the points on a line can be identified with real numbers by assigning them coordinates, as described in Appendix A, so the points in a plane can be identified with ordered pairs of real numbers. We start by drawing two perpendicular coordinate lines that intersect at the origin O on each line. Usually one line is horizontal with positive direction to the right and is called the *x*-axis; the other line is vertical with positive direction upward and is called the *y*-axis.

Any point P in the plane can be located by a unique ordered pair of numbers as follows. Draw lines through P perpendicular to the *x*- and *y*-axes. These lines intersect the axes in points with coordinates a and b as shown in Figure 1. Then the point P is assigned the ordered pair (a, b). The first number a is called the **x-coordinate** of P; the second number b is called the **y-coordinate** of P. We say that P is the point with coordinates (a, b), and we denote the point by the symbol $P(a, b)$. Several points are labeled with their coordinates in Figure 2.

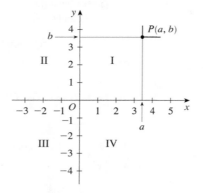

FIGURE 1   **FIGURE 2**

By reversing the preceding process we can start with an ordered pair (a, b) and arrive at the corresponding point P. Often we identify the point P with the ordered pair (a, b) and refer to "the point (a, b)." [Although the notation used for an open interval (a, b) is the same as the notation used for a point (a, b), you will be able to tell from the context which meaning is intended.]

This coordinate system is called the **rectangular coordinate system** or the **Cartesian coordinate system** in honor of the French mathematician René Descartes (1596–1650), even though another Frenchman, Pierre Fermat (1601–1665), invented the principles of analytic geometry at about the same time as Descartes. The plane supplied with this coordinate system is called the **coordinate plane** or the **Cartesian plane** and is denoted by $\mathbb{R}^2$.

The x- and y-axes are called the **coordinate axes** and divide the Cartesian plane into four quadrants, which are labeled I, II, III, and IV in Figure 1. Notice that the first quadrant consists of those points whose x- and y-coordinates are both positive.

EXAMPLE 1 Describe and sketch the regions given by the following sets.

(a) $\{(x, y) \mid x \geqslant 0\}$ (b) $\{(x, y) \mid y = 1\}$ (c) $\{(x, y) \mid |y| < 1\}$

SOLUTION

(a) The points whose x-coordinates are 0 or positive lie on the y-axis or to the right of it as indicated by the shaded region in Figure 3(a).

FIGURE 3   (a) $x \geqslant 0$ (b) $y = 1$ (c) $|y| < 1$

(b) The set of all points with y-coordinate 1 is a horizontal line one unit above the x-axis [see Figure 3(b)].

(c) Recall from Appendix A that

$$|y| < 1 \qquad \text{if and only if} \qquad -1 < y < 1$$

The given region consists of those points in the plane whose y-coordinates lie between -1 and 1. Thus the region consists of all points that lie between (but not on) the horizontal lines $y = 1$ and $y = -1$. [These lines are shown as dashed lines in Figure 3(c) to indicate that the points on these lines don't lie in the set.] ∎

Recall from Appendix A that the distance between points a and b on a number line is $|a - b| = |b - a|$. Thus the distance between points $P_1(x_1, y_1)$ and $P_3(x_2, y_1)$ on a horizontal line must be $|x_2 - x_1|$ and the distance between $P_2(x_2, y_2)$ and $P_3(x_2, y_1)$ on a vertical line must be $|y_2 - y_1|$. (See Figure 4.)

To find the distance $|P_1P_2|$ between any two points $P_1(x_1, y_1)$ and $P_2(x_2, y_2)$, we note that triangle $P_1P_2P_3$ in Figure 4 is a right triangle, and so by the Pythagorean Theorem we have

$$|P_1P_2| = \sqrt{|P_1P_3|^2 + |P_2P_3|^2} = \sqrt{|x_2 - x_1|^2 + |y_2 - y_1|^2}$$

$$= \sqrt{(x_2 - x_1)^2 + (y_2 - y_1)^2}$$

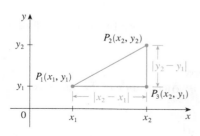

FIGURE 4

> **1 Distance Formula** The distance between the points $P_1(x_1, y_1)$ and $P_2(x_2, y_2)$ is
>
> $$|P_1P_2| = \sqrt{(x_2 - x_1)^2 + (y_2 - y_1)^2}$$

EXAMPLE 2 The distance between $(1, -2)$ and $(5, 3)$ is

$$\sqrt{(5 - 1)^2 + [3 - (-2)]^2} = \sqrt{4^2 + 5^2} = \sqrt{41}$$ ∎

Lines

We want to find an equation of a given line L; such an equation is satisfied by the coordinates of the points on L and by no other point. To find the equation of L we use its *slope*, which is a measure of the steepness of the line.

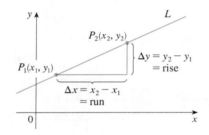

> **2 Definition** The **slope** of a nonvertical line that passes through the points $P_1(x_1, y_1)$ and $P_2(x_2, y_2)$ is
>
> $$m = \frac{\Delta y}{\Delta x} = \frac{y_2 - y_1}{x_2 - x_1}$$
>
> The slope of a vertical line is not defined.

FIGURE 5

Thus the slope of a line is the ratio of the change in y, Δy, to the change in x, Δx. (See Figure 5.) The slope is therefore the rate of change of y with respect to x. The fact that the line is straight means that the rate of change is constant.

Figure 6 shows several lines labeled with their slopes. Notice that lines with positive slope slant upward to the right, whereas lines with negative slope slant downward to the right. Notice also that the steepest lines are the ones for which the absolute value of the slope is largest, and a horizontal line has slope 0.

Now let's find an equation of the line that passes through a given point $P_1(x_1, y_1)$ and has slope m. A point $P(x, y)$ with $x \neq x_1$ lies on this line if and only if the slope of the line through P_1 and P is equal to m; that is,

$$\frac{y - y_1}{x - x_1} = m$$

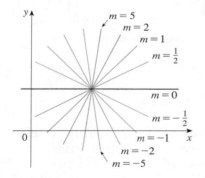

FIGURE 6

This equation can be rewritten in the form

$$y - y_1 = m(x - x_1)$$

and we observe that this equation is also satisfied when $x = x_1$ and $y = y_1$. Therefore it is an equation of the given line.

> **3 Point-Slope Form of the Equation of a Line** An equation of the line passing through the point $P_1(x_1, y_1)$ and having slope m is
>
> $$y - y_1 = m(x - x_1)$$

EXAMPLE 3 Find an equation of the line through $(1, -7)$ with slope $-\frac{1}{2}$.

SOLUTION Using (3) with $m = -\frac{1}{2}$, $x_1 = 1$, and $y_1 = -7$, we obtain an equation of the line as

$$y + 7 = -\tfrac{1}{2}(x - 1)$$

which we can rewrite as

$$2y + 14 = -x + 1 \qquad \text{or} \qquad x + 2y + 13 = 0 \qquad\blacksquare$$

EXAMPLE 4 Find an equation of the line through the points $(-1, 2)$ and $(3, -4)$.

SOLUTION By Definition 2 the slope of the line is

$$m = \frac{-4 - 2}{3 - (-1)} = -\frac{3}{2}$$

Using the point-slope form with $x_1 = -1$ and $y_1 = 2$, we obtain

$$y - 2 = -\tfrac{3}{2}(x + 1)$$

which simplifies to $\qquad\qquad\qquad\qquad 3x + 2y = 1 \qquad\blacksquare$

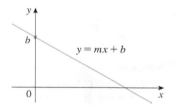

FIGURE 7

Suppose a nonvertical line has slope m and y-intercept b. (See Figure 7.) This means it intersects the y-axis at the point $(0, b)$, so the point-slope form of the equation of the line, with $x_1 = 0$ and $y_1 = b$, becomes

$$y - b = m(x - 0)$$

This simplifies as follows.

---

4 Slope-Intercept Form of the Equation of a Line An equation of the line with slope m and y-intercept b is

$$y = mx + b$$

FIGURE 8

In particular, if a line is horizontal, its slope is $m = 0$, so its equation is $y = b$, where b is the y-intercept (see Figure 8). A vertical line does not have a slope, but we can write its equation as $x = a$, where a is the x-intercept, because the x-coordinate of every point on the line is a.

Observe that the equation of every line can be written in the form

5 $$Ax + By + C = 0$$

because a vertical line has the equation $x = a$ or $x - a = 0$ ($A = 1$, $B = 0$, $C = -a$) and a nonvertical line has the equation $y = mx + b$ or $-mx + y - b = 0$ ($A = -m$, $B = 1$, $C = -b$). Conversely, if we start with a general first-degree equation, that is, an equation of the form (5), where A, B, and C are constants and A and B are not both 0, then we can show that it is the equation of a line. If $B = 0$, the equation becomes $Ax + C = 0$ or $x = -C/A$, which represents a vertical line with x-intercept $-C/A$. If $B \neq 0$, the equation can be rewritten by solving for y:

$$y = -\frac{A}{B}x - \frac{C}{B}$$

and we recognize this as being the slope-intercept form of the equation of a line ($m = -A/B$, $b = -C/B$). Therefore an equation of the form (5) is called a **linear equation** or the **general equation of a line**. For brevity, we often refer to "the line $Ax + By + C = 0$" instead of "the line whose equation is $Ax + By + C = 0$."

EXAMPLE 5 Sketch the graph of the equation $3x - 5y = 15$.

SOLUTION Since the equation is linear, its graph is a line. To draw the graph, we can simply find two points on the line. It's easiest to find the intercepts. Substituting $y = 0$ (the equation of the x-axis) in the given equation, we get $3x = 15$, so $x = 5$ is the x-intercept. Substituting $x = 0$ in the equation, we see that the y-intercept is -3. This allows us to sketch the graph as in Figure 9.   ∎

FIGURE 9

EXAMPLE 6 Graph the inequality $x + 2y > 5$.

SOLUTION We are asked to sketch the graph of the set $\{(x, y) \mid x + 2y > 5\}$ and we begin by solving the inequality for y:

$$x + 2y > 5$$

$$2y > -x + 5$$

$$y > -\tfrac{1}{2}x + \tfrac{5}{2}$$

Compare this inequality with the equation $y = -\frac{1}{2}x + \frac{5}{2}$, which represents a line with slope $-\frac{1}{2}$ and y-intercept $\frac{5}{2}$. We see that the given graph consists of points whose y-coordinates are *larger* than those on the line $y = -\frac{1}{2}x + \frac{5}{2}$. Thus the graph is the region that lies *above* the line, as illustrated in Figure 10.   ∎

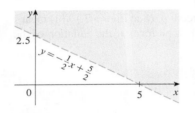

FIGURE 10

▪ Parallel and Perpendicular Lines

Slopes can be used to show that lines are parallel or perpendicular. The following facts are proved, for instance, in *Precalculus: Mathematics for Calculus,* Seventh Edition, by Stewart, Redlin, and Watson (Belmont, CA, 2016).

---

6 **Parallel and Perpendicular Lines**

1. Two nonvertical lines are parallel if and only if they have the same slope.
2. Two lines with slopes m_1 and m_2 are perpendicular if and only if $m_1 m_2 = -1$; that is, their slopes are negative reciprocals:

$$m_2 = -\frac{1}{m_1}$$

EXAMPLE 7 Find an equation of the line through the point $(5, 2)$ that is parallel to the line $4x + 6y + 5 = 0$.

SOLUTION The given line can be written in the form

$$y = -\tfrac{2}{3}x - \tfrac{5}{6}$$

which is in slope-intercept form with $m = -\frac{2}{3}$. Parallel lines have the same slope, so the required line has slope $-\frac{2}{3}$ and its equation in point-slope form is

$$y - 2 = -\tfrac{2}{3}(x - 5)$$

We can write this equation as $2x + 3y = 16$. ■

EXAMPLE 8 Show that the lines $2x + 3y = 1$ and $6x - 4y - 1 = 0$ are perpendicular.

SOLUTION The equations can be written as

$$y = -\tfrac{2}{3}x + \tfrac{1}{3} \qquad \text{and} \qquad y = \tfrac{3}{2}x - \tfrac{1}{4}$$

from which we see that the slopes are

$$m_1 = -\tfrac{2}{3} \qquad \text{and} \qquad m_2 = \tfrac{3}{2}$$

Since $m_1 m_2 = -1$, the lines are perpendicular. ■

B EXERCISES

1–6 Find the distance between the points.

1. $(1, 1)$, $(4, 5)$

2. $(1, -3)$, $(5, 7)$

3. $(6, -2)$, $(-1, 3)$

4. $(1, -6)$, $(-1, -3)$

5. $(2, 5)$, $(4, -7)$

6. (a, b), (b, a)

7–10 Find the slope of the line through P and Q.

7. $P(1, 5)$, $Q(4, 11)$

8. $P(-1, 6)$, $Q(4, -3)$

9. $P(-3, 3)$, $Q(-1, -6)$

10. $P(-1, -4)$, $Q(6, 0)$

11. Show that the triangle with vertices $A(0, 2)$, $B(-3, -1)$, and $C(-4, 3)$ is isosceles.

12. (a) Show that the triangle with vertices $A(6, -7)$, $B(11, -3)$, and $C(2, -2)$ is a right triangle using the converse of the Pythagorean Theorem.
(b) Use slopes to show that ABC is a right triangle.
(c) Find the area of the triangle.

13. Show that the points $(-2, 9)$, $(4, 6)$, $(1, 0)$, and $(-5, 3)$ are the vertices of a square.

14. (a) Show that the points $A(-1, 3)$, $B(3, 11)$, and $C(5, 15)$ are collinear (lie on the same line) by showing that $|AB| + |BC| = |AC|$.
(b) Use slopes to show that A, B, and C are collinear.

15. Show that $A(1, 1)$, $B(7, 4)$, $C(5, 10)$, and $D(-1, 7)$ are vertices of a parallelogram.

16. Show that $A(1, 1)$, $B(11, 3)$, $C(10, 8)$, and $D(0, 6)$ are vertices of a rectangle.

17–20 Sketch the graph of the equation.

17. $x = 3$

18. $y = -2$

19. $xy = 0$

20. $|y| = 1$

21–36 Find an equation of the line that satisfies the given conditions.

21. Through $(2, -3)$, slope 6

22. Through $(-1, 4)$, slope -3

23. Through $(1, 7)$, slope $\frac{2}{3}$

24. Through $(-3, -5)$, slope $-\frac{7}{2}$

25. Through $(2, 1)$ and $(1, 6)$

26. Through $(-1, -2)$ and $(4, 3)$

27. Slope 3, y-intercept -2

28. Slope $\frac{2}{5}$, y-intercept 4

29. x-intercept 1, y-intercept -3

30. x-intercept -8, y-intercept 6

31. Through $(4, 5)$, parallel to the x-axis

32. Through $(4, 5)$, parallel to the y-axis

33. Through $(1, -6)$, parallel to the line $x + 2y = 6$

34. y-intercept 6, parallel to the line $2x + 3y + 4 = 0$

35. Through $(-1, -2)$, perpendicular to the line $2x + 5y + 8 = 0$

36. Through $\left(\frac{1}{2}, -\frac{2}{3}\right)$, perpendicular to the line $4x - 8y = 1$

37–42 Find the slope and y-intercept of the line and draw its graph.

37. $x + 3y = 0$

38. $2x - 5y = 0$

39. $y = -2$

40. $2x - 3y + 6 = 0$

41. $3x - 4y = 12$

42. $4x + 5y = 10$

43–52 Sketch the region in the xy-plane.

43. $\{(x, y) \mid x < 0\}$

44. $\{(x, y) \mid y > 0\}$

45. $\{(x, y) \mid xy < 0\}$

46. $\{(x, y) \mid x \geqslant 1 \text{ and } y < 3\}$

47. $\{(x, y) \mid |x| \leqslant 2\}$

48. $\{(x, y) \mid |x| < 3 \text{ and } |y| < 2\}$

49. $\{(x, y) \mid 0 \leqslant y \leqslant 4 \text{ and } x \leqslant 2\}$

50. $\{(x, y) \mid y > 2x - 1\}$

51. $\{(x, y) \mid 1 + x \leqslant y \leqslant 1 - 2x\}$

52. $\{(x, y) \mid -x \leqslant y < \frac{1}{2}(x + 3)\}$

53. Find a point on the y-axis that is equidistant from $(5, -5)$ and $(1, 1)$.

54. Show that the midpoint of the line segment from $P_1(x_1, y_1)$ to $P_2(x_2, y_2)$ is
$$\left(\frac{x_1 + x_2}{2}, \frac{y_1 + y_2}{2} \right)$$

55. Find the midpoint of the line segment joining the given points.
(a) $(1, 3)$ and $(7, 15)$ (b) $(-1, 6)$ and $(8, -12)$

56. Find the lengths of the medians of the triangle with vertices $A(1, 0)$, $B(3, 6)$, and $C(8, 2)$. (A median is a line segment from a vertex to the midpoint of the opposite side.)

57. Show that the lines $2x - y = 4$ and $6x - 2y = 10$ are not parallel and find their point of intersection.

58. Show that the lines $3x - 5y + 19 = 0$ and $10x + 6y - 50 = 0$ are perpendicular and find their point of intersection.

59. Find an equation of the perpendicular bisector of the line segment joining the points $A(1, 4)$ and $B(7, -2)$.

60. (a) Find equations for the sides of the triangle with vertices $P(1, 0)$, $Q(3, 4)$, and $R(-1, 6)$.
(b) Find equations for the medians of this triangle. Where do they intersect?

61. (a) Show that if the x- and y-intercepts of a line are nonzero numbers a and b, then the equation of the line can be put in the form
$$\frac{x}{a} + \frac{y}{b} = 1$$

This equation is called the **two-intercept form** of an equation of a line.
(b) Use part (a) to find an equation of the line whose x-intercept is 6 and whose y-intercept is -8.

62. A car leaves Detroit at 2:00 PM, traveling at a constant speed west along I-96. It passes Ann Arbor, 40 mi from Detroit, at 2:50 PM.
(a) Express the distance traveled in terms of the time elapsed.
(b) Draw the graph of the equation in part (a).
(c) What is the slope of this line? What does it represent?

C   Graphs of Second-Degree Equations

In Appendix B we saw that a first-degree, or linear, equation $Ax + By + C = 0$ represents a line. In this section we discuss second-degree equations such as

$$x^2 + y^2 = 1 \qquad y = x^2 + 1 \qquad \frac{x^2}{9} + \frac{y^2}{4} = 1 \qquad x^2 - y^2 = 1$$

which represent a circle, a parabola, an ellipse, and a hyperbola, respectively.

The graph of such an equation in x and y is the set of all points (x, y) that satisfy the equation; it gives a visual representation of the equation. Conversely, given a curve in the xy-plane, we may have to find an equation that represents it, that is, an equation satisfied by the coordinates of the points on the curve and by no other point. This is the other half of the basic principle of analytic geometry as formulated by Descartes and Fermat. The idea is that if a geometric curve can be represented by an algebraic equation, then the rules of algebra can be used to analyze the geometric problem.

▮ Circles

As an example of this type of problem, let's find an equation of the circle with radius r and center (h, k). By definition, the circle is the set of all points $P(x, y)$ whose distance

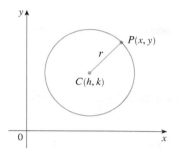

FIGURE 1

from the center $C(h, k)$ is r. (See Figure 1.) Thus P is on the circle if and only if $|PC| = r$. From the distance formula, we have

$$\sqrt{(x - h)^2 + (y - k)^2} = r$$

or equivalently, squaring both sides, we get

$$(x - h)^2 + (y - k)^2 = r^2$$

This is the desired equation.

> **1** **Equation of a Circle** An equation of the circle with center (h, k) and radius r is
>
> $$(x - h)^2 + (y - k)^2 = r^2$$
>
> In particular, if the center is the origin $(0, 0)$, the equation is
>
> $$x^2 + y^2 = r^2$$

EXAMPLE 1 Find an equation of the circle with radius 3 and center $(2, -5)$.

SOLUTION From Equation 1 with $r = 3$, $h = 2$, and $k = -5$, we obtain

$$(x - 2)^2 + (y + 5)^2 = 9$$ ∎

EXAMPLE 2 Sketch the graph of the equation $x^2 + y^2 + 2x - 6y + 7 = 0$ by first showing that it represents a circle and then finding its center and radius.

SOLUTION We first group the x-terms and y-terms as follows:

$$(x^2 + 2x) + (y^2 - 6y) = -7$$

Then we complete the square within each grouping, adding the appropriate constants (the squares of half the coefficients of x and y) to both sides of the equation:

$$(x^2 + 2x + 1) + (y^2 - 6y + 9) = -7 + 1 + 9$$

or $$(x + 1)^2 + (y - 3)^2 = 3$$

Comparing this equation with the standard equation of a circle (1), we see that $h = -1$, $k = 3$, and $r = \sqrt{3}$, so the given equation represents a circle with center $(-1, 3)$ and radius $\sqrt{3}$. It is sketched in Figure 2.

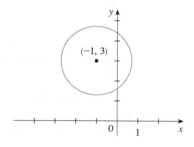

FIGURE 2
$x^2 + y^2 + 2x - 6y + 7 = 0$

▉ Parabolas

The geometric properties of parabolas are reviewed in Section 10.5. Here we regard a parabola as a graph of an equation of the form $y = ax^2 + bx + c$.

EXAMPLE 3 Draw the graph of the parabola $y = x^2$.

SOLUTION We set up a table of values, plot points, and join them by a smooth curve to obtain the graph in Figure 3.

x	$y = x^2$
0	0
$\pm\frac{1}{2}$	$\frac{1}{4}$
± 1	1
± 2	4
± 3	9

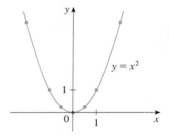

FIGURE 3   ▉

Figure 4 shows the graphs of several parabolas with equations of the form $y = ax^2$ for various values of the number a. In each case the *vertex,* the point where the parabola changes direction, is the origin. We see that the parabola $y = ax^2$ opens upward if $a > 0$ and downward if $a < 0$ (as in Figure 5).

FIGURE 4

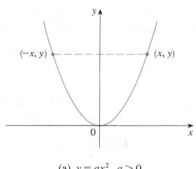

(a) $y = ax^2$, $a > 0$

FIGURE 5

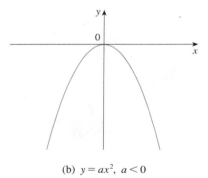

(b) $y = ax^2$, $a < 0$

Notice that if (x, y) satisfies $y = ax^2$, then so does $(-x, y)$. This corresponds to the geometric fact that if the right half of the graph is reflected about the y-axis, then the left half of the graph is obtained. We say that the graph is **symmetric with respect to the y-axis**.

> The graph of an equation is symmetric with respect to the y-axis if the equation is unchanged when x is replaced by $-x$.

If we interchange x and y in the equation $y = ax^2$, the result is $x = ay^2$, which also represents a parabola. (Interchanging x and y amounts to reflecting about the diagonal line $y = x$.) The parabola $x = ay^2$ opens to the right if $a > 0$ and to the left if $a < 0$. (See

Figure 6.) This time the parabola is symmetric with respect to the x-axis because if (x, y) satisfies $x = ay^2$, then so does $(x, -y)$.

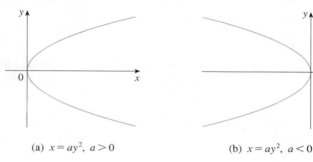

FIGURE 6

(a) $x = ay^2$, $a > 0$ (b) $x = ay^2$, $a < 0$

> The graph of an equation is symmetric with respect to the x-axis if the equation is unchanged when y is replaced by $-y$.

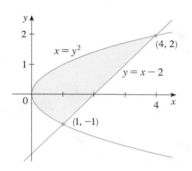

FIGURE 7

EXAMPLE 4 Sketch the region bounded by the parabola $x = y^2$ and the line $y = x - 2$.

SOLUTION First we find the points of intersection by solving the two equations. Substituting $x = y + 2$ into the equation $x = y^2$, we get $y + 2 = y^2$, which gives

$$0 = y^2 - y - 2 = (y - 2)(y + 1)$$

so $y = 2$ or -1. Thus the points of intersection are $(4, 2)$ and $(1, -1)$, and we draw the line $y = x - 2$ passing through these points. We then sketch the parabola $x = y^2$ by referring to Figure 6(a) and having the parabola pass through $(4, 2)$ and $(1, -1)$. The region bounded by $x = y^2$ and $y = x - 2$ means the finite region whose boundaries are these curves. It is sketched in Figure 7. ∎

Ellipses

The curve with equation

$$\boxed{2} \qquad \frac{x^2}{a^2} + \frac{y^2}{b^2} = 1$$

where a and b are positive numbers, is called an **ellipse** in standard position. (Geometric properties of ellipses are discussed in Section 10.5.) Observe that Equation 2 is unchanged if x is replaced by $-x$ or y is replaced by $-y$, so the ellipse is symmetric with respect to both axes. As a further aid to sketching the ellipse, we find its intercepts.

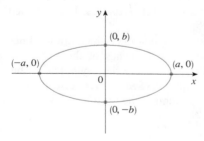

FIGURE 8
$$\frac{x^2}{a^2} + \frac{y^2}{b^2} = 1$$

> The **x-intercepts** of a graph are the x-coordinates of the points where the graph intersects the x-axis. They are found by setting $y = 0$ in the equation of the graph.
>
> The **y-intercepts** are the y-coordinates of the points where the graph intersects the y-axis. They are found by setting $x = 0$ in its equation.

If we set $y = 0$ in Equation 2, we get $x^2 = a^2$ and so the x-intercepts are $\pm a$. Setting $x = 0$, we get $y^2 = b^2$, so the y-intercepts are $\pm b$. Using this information, together with symmetry, we sketch the ellipse in Figure 8. If $a = b$, the ellipse is a circle with radius a.

EXAMPLE 5 Sketch the graph of $9x^2 + 16y^2 = 144$.

SOLUTION We divide both sides of the equation by 144:

$$\frac{x^2}{16} + \frac{y^2}{9} = 1$$

The equation is now in the standard form for an ellipse (2), so we have $a^2 = 16$, $b^2 = 9$, $a = 4$, and $b = 3$. The x-intercepts are ± 4; the y-intercepts are ± 3. The graph is sketched in Figure 9.

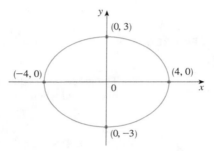

FIGURE 9

$9x^2 + 16y^2 = 144$

Hyperbolas

The curve with equation

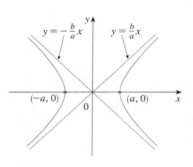

FIGURE 10

The hyperbola $\dfrac{x^2}{a^2} - \dfrac{y^2}{b^2} = 1$

$$\boxed{3} \qquad \frac{x^2}{a^2} - \frac{y^2}{b^2} = 1$$

is called a **hyperbola** in standard position. Again, Equation 3 is unchanged when x is replaced by $-x$ or y is replaced by $-y$, so the hyperbola is symmetric with respect to both axes. To find the x-intercepts we set $y = 0$ and obtain $x^2 = a^2$ and $x = \pm a$. However, if we put $x = 0$ in Equation 3, we get $y^2 = -b^2$, which is impossible, so there is no y-intercept. In fact, from Equation 3 we obtain

$$\frac{x^2}{a^2} = 1 + \frac{y^2}{b^2} \geq 1$$

which shows that $x^2 \geq a^2$ and so $|x| = \sqrt{x^2} \geq a$. Therefore we have $x \geq a$ or $x \leq -a$. This means that the hyperbola consists of two parts, called its *branches*. It is sketched in Figure 10.

In drawing a hyperbola it is useful to draw first its *asymptotes*, which are the lines $y = (b/a)x$ and $y = -(b/a)x$ shown in Figure 10. Both branches of the hyperbola approach the asymptotes; that is, they come arbitrarily close to the asymptotes. This involves the idea of a limit, which is discussed in Chapter 2. (See also Exercise 4.5.73.)

By interchanging the roles of x and y we get an equation of the form

$$\frac{y^2}{a^2} - \frac{x^2}{b^2} = 1$$

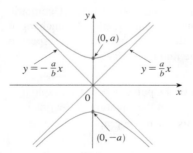

FIGURE 11

The hyperbola $\dfrac{y^2}{a^2} - \dfrac{x^2}{b^2} = 1$

which also represents a hyperbola and is sketched in Figure 11.

EXAMPLE 6 Sketch the curve $9x^2 - 4y^2 = 36$.

SOLUTION Dividing both sides by 36, we obtain

$$\frac{x^2}{4} - \frac{y^2}{9} = 1$$

which is the standard form of the equation of a hyperbola (Equation 3). Since $a^2 = 4$, the x-intercepts are ± 2. Since $b^2 = 9$, we have $b = 3$ and the asymptotes are $y = \pm\frac{3}{2}x$. The hyperbola is sketched in Figure 12.

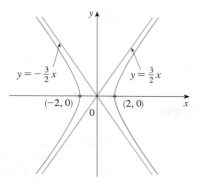

FIGURE 12
The hyperbola $9x^2 - 4y^2 = 36$

If $b = a$, a hyperbola has the equation $x^2 - y^2 = a^2$ (or $y^2 - x^2 = a^2$) and is called an *equilateral hyperbola* [see Figure 13(a)]. Its asymptotes are $y = \pm x$, which are perpendicular. If an equilateral hyperbola is rotated by 45°, the asymptotes become the x- and y-axes, and it can be shown that the new equation of the hyperbola is $xy = k$, where k is a constant [see Figure 13(b)].

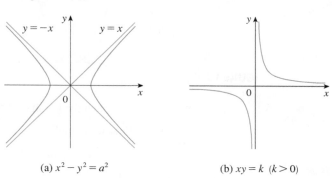

FIGURE 13
Equilateral hyperbolas

(a) $x^2 - y^2 = a^2$

(b) $xy = k$ $(k > 0)$

Shifted Conics

Recall that an equation of the circle with center the origin and radius r is $x^2 + y^2 = r^2$, but if the center is the point (h, k), then the equation of the circle becomes

$$(x - h)^2 + (y - k)^2 = r^2$$

Similarly, if we take the ellipse with equation

4

$$\frac{x^2}{a^2} + \frac{y^2}{b^2} = 1$$

and translate it (shift it) so that its center is the point (h, k), then its equation becomes

$$\boxed{5} \qquad \frac{(x - h)^2}{a^2} + \frac{(y - k)^2}{b^2} = 1$$

(See Figure 14.)

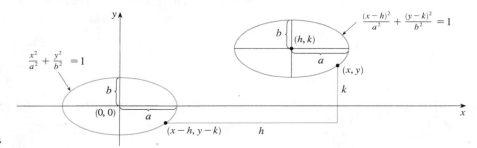

FIGURE 14

Notice that in shifting the ellipse, we replaced x by $x - h$ and y by $y - k$ in Equation 4 to obtain Equation 5. We use the same procedure to shift the parabola $y = ax^2$ so that its vertex (the origin) becomes the point (h, k) as in Figure 15. Replacing x by $x - h$ and y by $y - k$, we see that the new equation is

$$y - k = a(x - h)^2 \qquad \text{or} \qquad y = a(x - h)^2 + k$$

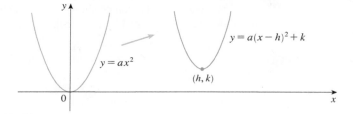

FIGURE 15

EXAMPLE 7 Sketch the graph of the equation $y = 2x^2 - 4x + 1$.

SOLUTION First we complete the square:

$$y = 2(x^2 - 2x) + 1 = 2(x - 1)^2 - 1$$

In this form we see that the equation represents the parabola obtained by shifting $y = 2x^2$ so that its vertex is at the point $(1, -1)$. The graph is sketched in Figure 16.

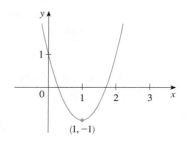

FIGURE 16
$y = 2x^2 - 4x + 1$

EXAMPLE 8 Sketch the curve $x = 1 - y^2$.

SOLUTION This time we start with the parabola $x = -y^2$ (as in Figure 6 with $a = -1$) and shift one unit to the right to get the graph of $x = 1 - y^2$. (See Figure 17.)

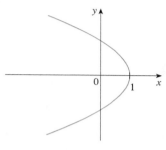

FIGURE 17

(a) $x = -y^2$

(b) $x = 1 - y^2$

C   EXERCISES

1–4 Find an equation of a circle that satisfies the given conditions.

1. Center $(3, -1)$, radius 5

2. Center $(-2, -8)$, radius 10

3. Center at the origin, passes through $(4, 7)$

4. Center $(-1, 5)$, passes through $(-4, -6)$

5–9 Show that the equation represents a circle and find the center and radius.

5. $x^2 + y^2 - 4x + 10y + 13 = 0$

6. $x^2 + y^2 + 6y + 2 = 0$

7. $x^2 + y^2 + x = 0$

8. $16x^2 + 16y^2 + 8x + 32y + 1 = 0$

9. $2x^2 + 2y^2 - x + y = 1$

10. Under what condition on the coefficients a, b, and c does the equation $x^2 + y^2 + ax + by + c = 0$ represent a circle? When that condition is satisfied, find the center and radius of the circle.

11–32 Identify the type of curve and sketch the graph. Do not plot points. Just use the standard graphs given in Figures 5, 6, 8, 10, and 11 and shift if necessary.

11. $y = -x^2$

12. $y^2 - x^2 = 1$

13. $x^2 + 4y^2 = 16$

14. $x = -2y^2$

15. $16x^2 - 25y^2 = 400$

16. $25x^2 + 4y^2 = 100$

17. $4x^2 + y^2 = 1$

18. $y = x^2 + 2$

19. $x = y^2 - 1$

20. $9x^2 - 25y^2 = 225$

21. $9y^2 - x^2 = 9$

22. $2x^2 + 5y^2 = 10$

23. $xy = 4$

24. $y = x^2 + 2x$

25. $9(x - 1)^2 + 4(y - 2)^2 = 36$

26. $16x^2 + 9y^2 - 36y = 108$

27. $y = x^2 - 6x + 13$

28. $x^2 - y^2 - 4x + 3 = 0$

29. $x = 4 - y^2$

30. $y^2 - 2x + 6y + 5 = 0$

31. $x^2 + 4y^2 - 6x + 5 = 0$

32. $4x^2 + 9y^2 - 16x + 54y + 61 = 0$

33–34 Sketch the region bounded by the curves.

33. $y = 3x$, $y = x^2$

34. $y = 4 - x^2$, $x - 2y = 2$

35. Find an equation of the parabola with vertex $(1, -1)$ that passes through the points $(-1, 3)$ and $(3, 3)$.

36. Find an equation of the ellipse with center at the origin that passes through the points $\left(1, -10\sqrt{2}/3\right)$ and $\left(-2, 5\sqrt{5}/3\right)$.

37–40 Sketch the graph of the set.

37. $\{(x, y) \mid x^2 + y^2 \le 1\}$

38. $\{(x, y) \mid x^2 + y^2 > 4\}$

39. $\{(x, y) \mid y \ge x^2 - 1\}$

40. $\{(x, y) \mid x^2 + 4y^2 \le 4\}$

D Trigonometry

Angles

Angles can be measured in degrees or in radians (abbreviated as rad). The angle given by a complete revolution contains 360°, which is the same as 2π rad. Therefore

$$\boxed{1} \qquad \pi \text{ rad} = 180°$$

and

$$\boxed{2} \qquad 1 \text{ rad} = \left(\frac{180}{\pi}\right)° \approx 57.3° \qquad 1° = \frac{\pi}{180} \text{ rad} \approx 0.017 \text{ rad}$$

EXAMPLE 1

(a) Find the radian measure of 60°. \qquad (b) Express $5\pi/4$ rad in degrees.

SOLUTION

(a) From Equation 1 or 2 we see that to convert from degrees to radians we multiply by $\pi/180$. Therefore

$$60° = 60\left(\frac{\pi}{180}\right) = \frac{\pi}{3} \text{ rad}$$

(b) To convert from radians to degrees we multiply by $180/\pi$. Thus

$$\frac{5\pi}{4} \text{ rad} = \frac{5\pi}{4}\left(\frac{180}{\pi}\right) = 225°$$

In calculus we use radians to measure angles except when otherwise indicated. The following table gives the correspondence between degree and radian measures of some common angles.

Degrees	0°	30°	45°	60°	90°	120°	135°	150°	180°	270°	360°
Radians	0	$\dfrac{\pi}{6}$	$\dfrac{\pi}{4}$	$\dfrac{\pi}{3}$	$\dfrac{\pi}{2}$	$\dfrac{2\pi}{3}$	$\dfrac{3\pi}{4}$	$\dfrac{5\pi}{6}$	π	$\dfrac{3\pi}{2}$	2π

Figure 1 shows a sector of a circle with central angle θ and radius r subtending an arc with length a. Since the length of the arc is proportional to the size of the angle, and since the entire circle has circumference $2\pi r$ and central angle 2π, we have

$$\frac{\theta}{2\pi} = \frac{a}{2\pi r}$$

Solving this equation for θ and for a, we obtain

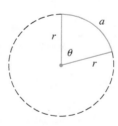

FIGURE 1

$$\boxed{3} \qquad \theta = \frac{a}{r} \qquad\qquad a = r\theta$$

Remember that Equations 3 are valid only when θ is measured in radians.

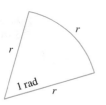

FIGURE 2

In particular, putting $a = r$ in Equation 3, we see that an angle of 1 rad is the angle subtended at the center of a circle by an arc equal in length to the radius of the circle (see Figure 2).

EXAMPLE 2

(a) If the radius of a circle is 5 cm, what angle is subtended by an arc of 6 cm?

(b) If a circle has radius 3 cm, what is the length of an arc subtended by a central angle of $3\pi/8$ rad?

SOLUTION

(a) Using Equation 3 with $a = 6$ and $r = 5$, we see that the angle is

$$\theta = \tfrac{6}{5} = 1.2 \text{ rad}$$

(b) With $r = 3$ cm and $\theta = 3\pi/8$ rad, the arc length is

$$a = r\theta = 3\left(\frac{3\pi}{8}\right) = \frac{9\pi}{8} \text{ cm}$$

The **standard position** of an angle occurs when we place its vertex at the origin of a coordinate system and its initial side on the positive x-axis as in Figure 3. A **positive** angle is obtained by rotating the initial side counterclockwise until it coincides with the terminal side. Likewise, **negative** angles are obtained by clockwise rotation as in Figure 4.

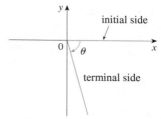

FIGURE 3 $\theta \geqslant 0$ **FIGURE 4** $\theta < 0$

Figure 5 shows several examples of angles in standard position. Notice that different angles can have the same terminal side. For instance, the angles $3\pi/4$, $-5\pi/4$, and $11\pi/4$ have the same initial and terminal sides because

$$\frac{3\pi}{4} - 2\pi = -\frac{5\pi}{4} \qquad \frac{3\pi}{4} + 2\pi = \frac{11\pi}{4}$$

and 2π rad represents a complete revolution.

FIGURE 5
Angles in standard position

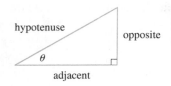

FIGURE 6

■ The Trigonometric Functions

For an acute angle θ the six trigonometric functions are defined as ratios of lengths of sides of a right triangle as follows (see Figure 6).

4

$$\sin \theta = \frac{\text{opp}}{\text{hyp}} \qquad \csc \theta = \frac{\text{hyp}}{\text{opp}}$$

$$\cos \theta = \frac{\text{adj}}{\text{hyp}} \qquad \sec \theta = \frac{\text{hyp}}{\text{adj}}$$

$$\tan \theta = \frac{\text{opp}}{\text{adj}} \qquad \cot \theta = \frac{\text{adj}}{\text{opp}}$$

FIGURE 7

This definition doesn't apply to obtuse or negative angles, so for a general angle θ in standard position we let $P(x, y)$ be any point on the terminal side of θ and we let r be the distance $|OP|$ as in Figure 7. Then we define

5

$$\sin \theta = \frac{y}{r} \qquad \csc \theta = \frac{r}{y}$$

$$\cos \theta = \frac{x}{r} \qquad \sec \theta = \frac{r}{x}$$

$$\tan \theta = \frac{y}{x} \qquad \cot \theta = \frac{x}{y}$$

If we put $r = 1$ in Definition 5 and draw a unit circle with center the origin and label θ as in Figure 8, then the coordinates of P are $(\cos \theta, \sin \theta)$.

Since division by 0 is not defined, $\tan \theta$ and $\sec \theta$ are undefined when $x = 0$ and $\csc \theta$ and $\cot \theta$ are undefined when $y = 0$. Notice that the definitions in (4) and (5) are consistent when θ is an acute angle.

If θ is a number, the convention is that $\sin \theta$ means the sine of the angle whose *radian* measure is θ. For example, the expression $\sin 3$ implies that we are dealing with an angle of 3 rad. When finding a calculator approximation to this number, we must remember to set our calculator in radian mode, and then we obtain

$$\sin 3 \approx 0.14112$$

If we want to know the sine of the angle 3° we would write $\sin 3°$ and, with our calculator in degree mode, we find that

$$\sin 3° \approx 0.05234$$

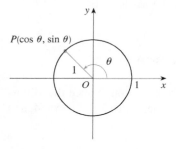

FIGURE 8

The exact trigonometric ratios for certain angles can be read from the triangles in Figure 9. For instance,

$$\sin \frac{\pi}{4} = \frac{1}{\sqrt{2}} \qquad \sin \frac{\pi}{6} = \frac{1}{2} \qquad \sin \frac{\pi}{3} = \frac{\sqrt{3}}{2}$$

$$\cos \frac{\pi}{4} = \frac{1}{\sqrt{2}} \qquad \cos \frac{\pi}{6} = \frac{\sqrt{3}}{2} \qquad \cos \frac{\pi}{3} = \frac{1}{2}$$

$$\tan \frac{\pi}{4} = 1 \qquad \tan \frac{\pi}{6} = \frac{1}{\sqrt{3}} \qquad \tan \frac{\pi}{3} = \sqrt{3}$$

FIGURE 9

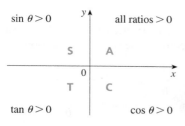

sin θ > 0   all ratios > 0

S   A

T   C

tan θ > 0   cos θ > 0

FIGURE 10

The signs of the trigonometric functions for angles in each of the four quadrants can be remembered by means of the rule "All Students Take Calculus" shown in Figure 10.

EXAMPLE 3 Find the exact trigonometric ratios for $\theta = 2\pi/3$.

SOLUTION From Figure 11 we see that a point on the terminal line for $\theta = 2\pi/3$ is $P\left(-1, \sqrt{3}\right)$. Therefore, taking

$$x = -1 \qquad y = \sqrt{3} \qquad r = 2$$

in the definitions of the trigonometric ratios, we have

$$\sin \frac{2\pi}{3} = \frac{\sqrt{3}}{2} \qquad \cos \frac{2\pi}{3} = -\frac{1}{2} \qquad \tan \frac{2\pi}{3} = -\sqrt{3}$$

$$\csc \frac{2\pi}{3} = \frac{2}{\sqrt{3}} \qquad \sec \frac{2\pi}{3} = -2 \qquad \cot \frac{2\pi}{3} = -\frac{1}{\sqrt{3}} \qquad \blacksquare$$

$P(-1, \sqrt{3})$

FIGURE 11

The following table gives some values of $\sin \theta$ and $\cos \theta$ found by the method of Example 3.

θ	0	$\dfrac{\pi}{6}$	$\dfrac{\pi}{4}$	$\dfrac{\pi}{3}$	$\dfrac{\pi}{2}$	$\dfrac{2\pi}{3}$	$\dfrac{3\pi}{4}$	$\dfrac{5\pi}{6}$	π	$\dfrac{3\pi}{2}$	2π
$\sin \theta$	0	$\dfrac{1}{2}$	$\dfrac{1}{\sqrt{2}}$	$\dfrac{\sqrt{3}}{2}$	1	$\dfrac{\sqrt{3}}{2}$	$\dfrac{1}{\sqrt{2}}$	$\dfrac{1}{2}$	0	-1	0
$\cos \theta$	1	$\dfrac{\sqrt{3}}{2}$	$\dfrac{1}{\sqrt{2}}$	$\dfrac{1}{2}$	0	$-\dfrac{1}{2}$	$-\dfrac{1}{\sqrt{2}}$	$-\dfrac{\sqrt{3}}{2}$	-1	0	1

EXAMPLE 4 If $\cos \theta = \frac{2}{5}$ and $0 < \theta < \pi/2$, find the other five trigonometric functions of θ.

SOLUTION Since $\cos \theta = \frac{2}{5}$, we can label the hypotenuse as having length 5 and the adjacent side as having length 2 in Figure 12. If the opposite side has length x, then the Pythagorean Theorem gives $x^2 + 4 = 25$ and so $x^2 = 21$, $x = \sqrt{21}$. We can now use the diagram to write the other five trigonometric functions:

$$\sin \theta = \frac{\sqrt{21}}{5} \qquad \tan \theta = \frac{\sqrt{21}}{2}$$

$$\csc \theta = \frac{5}{\sqrt{21}} \qquad \sec \theta = \frac{5}{2} \qquad \cot \theta = \frac{2}{\sqrt{21}} \qquad \blacksquare$$

$x = \sqrt{21}$

FIGURE 12

FIGURE 13

EXAMPLE 5 Use a calculator to approximate the value of x in Figure 13.

SOLUTION From the diagram we see that

$$\tan 40° = \frac{16}{x}$$

Therefore

$$x = \frac{16}{\tan 40°} \approx 19.07 \qquad \blacksquare$$

Trigonometric Identities

A trigonometric identity is a relationship among the trigonometric functions. The most elementary are the following, which are immediate consequences of the definitions of the trigonometric functions.

$$\boxed{6} \qquad \csc\theta = \frac{1}{\sin\theta} \qquad\qquad \sec\theta = \frac{1}{\cos\theta} \qquad\qquad \cot\theta = \frac{1}{\tan\theta}$$

$$\tan\theta = \frac{\sin\theta}{\cos\theta} \qquad\qquad \cot\theta = \frac{\cos\theta}{\sin\theta}$$

For the next identity we refer back to Figure 7. The distance formula (or, equivalently, the Pythagorean Theorem) tells us that $x^2 + y^2 = r^2$. Therefore

$$\sin^2\theta + \cos^2\theta = \frac{y^2}{r^2} + \frac{x^2}{r^2} = \frac{x^2 + y^2}{r^2} = \frac{r^2}{r^2} = 1$$

We have therefore proved one of the most useful of all trigonometric identities:

$$\boxed{7} \qquad\qquad \sin^2\theta + \cos^2\theta = 1$$

If we now divide both sides of Equation 7 by $\cos^2\theta$ and use Equations 6, we get

$$\boxed{8} \qquad\qquad \tan^2\theta + 1 = \sec^2\theta$$

Similarly, if we divide both sides of Equation 7 by $\sin^2\theta$, we get

$$\boxed{9} \qquad\qquad 1 + \cot^2\theta = \csc^2\theta$$

The identities

$$\boxed{10a} \qquad\qquad \sin(-\theta) = -\sin\theta$$

$$\boxed{10b} \qquad\qquad \cos(-\theta) = \cos\theta$$

Odd functions and even functions are discussed in Section 1.1.

show that sine is an odd function and cosine is an even function. They are easily proved by drawing a diagram showing θ and $-\theta$ in standard position (see Exercise 39).

Since the angles θ and $\theta + 2\pi$ have the same terminal side, we have

$$\boxed{11} \qquad\qquad \sin(\theta + 2\pi) = \sin\theta \qquad\qquad \cos(\theta + 2\pi) = \cos\theta$$

These identities show that the sine and cosine functions are periodic with period 2π.

The remaining trigonometric identities are all consequences of two basic identities called the **addition formulas**:

12a
$$\sin(x + y) = \sin x \cos y + \cos x \sin y$$

12b
$$\cos(x + y) = \cos x \cos y - \sin x \sin y$$

The proofs of these addition formulas are outlined in Exercises 85, 86, and 87.

By substituting $-y$ for y in Equations 12a and 12b and using Equations 10a and 10b, we obtain the following **subtraction formulas**:

13a
$$\sin(x - y) = \sin x \cos y - \cos x \sin y$$

13b
$$\cos(x - y) = \cos x \cos y + \sin x \sin y$$

Then, by dividing the formulas in Equations 12 or Equations 13, we obtain the corresponding formulas for $\tan(x \pm y)$:

14a
$$\tan(x + y) = \frac{\tan x + \tan y}{1 - \tan x \tan y}$$

14b
$$\tan(x - y) = \frac{\tan x - \tan y}{1 + \tan x \tan y}$$

If we put $y = x$ in the addition formulas (12), we get the **double-angle formulas**:

15a
$$\sin 2x = 2 \sin x \cos x$$

15b
$$\cos 2x = \cos^2 x - \sin^2 x$$

Then, by using the identity $\sin^2 x + \cos^2 x = 1$, we obtain the following alternate forms of the double-angle formulas for $\cos 2x$:

16a
$$\cos 2x = 2 \cos^2 x - 1$$

16b
$$\cos 2x = 1 - 2 \sin^2 x$$

If we now solve these equations for $\cos^2 x$ and $\sin^2 x$, we get the following **half-angle formulas**, which are useful in integral calculus:

17a
$$\cos^2 x = \frac{1 + \cos 2x}{2}$$

17b
$$\sin^2 x = \frac{1 - \cos 2x}{2}$$

Finally, we state the **product formulas**, which can be deduced from Equations 12 and 13:

18a $$\sin x \cos y = \tfrac{1}{2}[\sin(x + y) + \sin(x - y)]$$

18b $$\cos x \cos y = \tfrac{1}{2}[\cos(x + y) + \cos(x - y)]$$

18c $$\sin x \sin y = \tfrac{1}{2}[\cos(x - y) - \cos(x + y)]$$

There are many other trigonometric identities, but those we have stated are the ones used most often in calculus. If you forget any of the identities 13–18, remember that they can all be deduced from Equations 12a and 12b.

EXAMPLE 6 Find all values of x in the interval $[0, 2\pi]$ such that $\sin x = \sin 2x$.

SOLUTION Using the double-angle formula (15a), we rewrite the given equation as

$$\sin x = 2 \sin x \cos x \qquad \text{or} \qquad \sin x(1 - 2 \cos x) = 0$$

Therefore there are two possibilities:

$$\sin x = 0 \qquad \text{or} \qquad 1 - 2 \cos x = 0$$

$$x = 0, \pi, 2\pi \qquad\qquad \cos x = \tfrac{1}{2}$$

$$x = \frac{\pi}{3}, \frac{5\pi}{3}$$

The given equation has five solutions: $0, \pi/3, \pi, 5\pi/3$, and 2π. ▪

■ Graphs of the Trigonometric Functions

The graph of the function $f(x) = \sin x$, shown in Figure 14(a), is obtained by plotting points for $0 \le x \le 2\pi$ and then using the periodic nature of the function (from Equation 11) to complete the graph. Notice that the zeros of the sine function occur at the

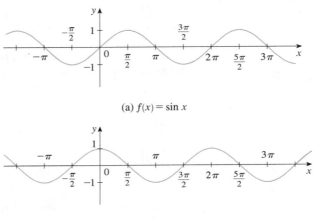

(a) $f(x) = \sin x$

(b) $g(x) = \cos x$

FIGURE 14

integer multiples of π, that is,

$$\sin x = 0 \qquad \text{whenever } x = n\pi, \quad n \text{ an integer}$$

Because of the identity

$$\cos x = \sin\left(x + \frac{\pi}{2}\right)$$

(which can be verified using Equation 12a), the graph of cosine is obtained by shifting the graph of sine by an amount $\pi/2$ to the left [see Figure 14(b)]. Note that for both the sine and cosine functions the domain is $(-\infty, \infty)$ and the range is the closed interval $[-1, 1]$. Thus, for all values of x, we have

$$-1 \leqslant \sin x \leqslant 1 \qquad -1 \leqslant \cos x \leqslant 1$$

The graphs of the remaining four trigonometric functions are shown in Figure 15 and their domains are indicated there. Notice that tangent and cotangent have range $(-\infty, \infty)$, whereas cosecant and secant have range $(-\infty, -1] \cup [1, \infty)$. All four functions are periodic: tangent and cotangent have period π, whereas cosecant and secant have period 2π.

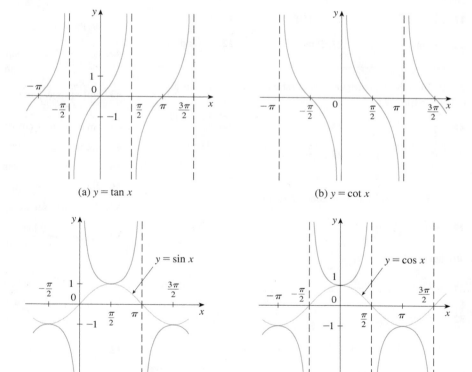

(a) $y = \tan x$

(b) $y = \cot x$

(c) $y = \csc x$

(d) $y = \sec x$

FIGURE 15

D EXERCISES

1–6 Convert from degrees to radians.

1. $210°$

2. $300°$

3. $9°$

4. $-315°$

5. $900°$

6. $36°$

7–12 Convert from radians to degrees.

7. 4π

8. $-\dfrac{7\pi}{2}$

9. $\dfrac{5\pi}{12}$

10. $\dfrac{8\pi}{3}$

11. $-\dfrac{3\pi}{8}$

12. 5

13. Find the length of a circular arc subtended by an angle of $\pi/12$ rad if the radius of the circle is 36 cm.

14. If a circle has radius 10 cm, find the length of the arc subtended by a central angle of $72°$.

15. A circle has radius 1.5 m. What angle is subtended at the center of the circle by an arc 1 m long?

16. Find the radius of a circular sector with angle $3\pi/4$ and arc length 6 cm.

17–22 Draw, in standard position, the angle whose measure is given.

17. $315°$

18. $-150°$

19. $-\dfrac{3\pi}{4}$ rad

20. $\dfrac{7\pi}{3}$ rad

21. 2 rad

22. -3 rad

23–28 Find the exact trigonometric ratios for the angle whose radian measure is given.

23. $\dfrac{3\pi}{4}$

24. $\dfrac{4\pi}{3}$

25. $\dfrac{9\pi}{2}$

26. -5π

27. $\dfrac{5\pi}{6}$

28. $\dfrac{11\pi}{4}$

29–34 Find the remaining trigonometric ratios.

29. $\sin\theta = \dfrac{3}{5}, \quad 0 < \theta < \dfrac{\pi}{2}$

30. $\tan\alpha = 2, \quad 0 < \alpha < \dfrac{\pi}{2}$

31. $\sec\phi = -1.5, \quad \dfrac{\pi}{2} < \phi < \pi$

32. $\cos x = -\dfrac{1}{3}, \quad \pi < x < \dfrac{3\pi}{2}$

33. $\cot\beta = 3, \quad \pi < \beta < 2\pi$

34. $\csc\theta = -\dfrac{4}{3}, \quad \dfrac{3\pi}{2} < \theta < 2\pi$

35–38 Find, correct to five decimal places, the length of the side labeled x.

35.

36.

37.

38.

39–41 Prove each equation.

39. (a) Equation 10a   (b) Equation 10b

40. (a) Equation 14a   (b) Equation 14b

41. (a) Equation 18a   (b) Equation 18b
    (c) Equation 18c

42–58 Prove the identity.

42. $\cos\left(\dfrac{\pi}{2} - x\right) = \sin x$

43. $\sin\left(\dfrac{\pi}{2} + x\right) = \cos x$ **44.** $\sin(\pi - x) = \sin x$

45. $\sin\theta\,\cot\theta = \cos\theta$ **46.** $(\sin x + \cos x)^2 = 1 + \sin 2x$

47. $\sec y - \cos y = \tan y \sin y$

48. $\tan^2\alpha - \sin^2\alpha = \tan^2\alpha\,\sin^2\alpha$

49. $\cot^2\theta + \sec^2\theta = \tan^2\theta + \csc^2\theta$

50. $2\csc 2t = \sec t\,\csc t$

51. $\tan 2\theta = \dfrac{2\tan\theta}{1 - \tan^2\theta}$

52. $\dfrac{1}{1 - \sin\theta} + \dfrac{1}{1 + \sin\theta} = 2\sec^2\theta$

53. $\sin x \sin 2x + \cos x \cos 2x = \cos x$

54. $\sin^2 x - \sin^2 y = \sin(x + y)\sin(x - y)$

55. $\dfrac{\sin\phi}{1 - \cos\phi} = \csc\phi + \cot\phi$

56. $\tan x + \tan y = \dfrac{\sin(x + y)}{\cos x \cos y}$

57. $\sin 3\theta + \sin \theta = 2 \sin 2\theta \cos \theta$

58. $\cos 3\theta = 4 \cos^3\theta - 3 \cos \theta$

59–64 If $\sin x = \frac{1}{3}$ and $\sec y = \frac{5}{4}$, where x and y lie between 0 and $\pi/2$, evaluate the expression.

59. $\sin(x + y)$ **60.** $\cos(x + y)$

61. $\cos(x - y)$ **62.** $\sin(x - y)$

63. $\sin 2y$ **64.** $\cos 2y$

65–72 Find all values of x in the interval $[0, 2\pi]$ that satisfy the equation.

65. $2 \cos x - 1 = 0$ **66.** $3 \cot^2 x = 1$

67. $2 \sin^2 x = 1$ **68.** $|\tan x| = 1$

69. $\sin 2x = \cos x$ **70.** $2 \cos x + \sin 2x = 0$

71. $\sin x = \tan x$ **72.** $2 + \cos 2x = 3 \cos x$

73–76 Find all values of x in the interval $[0, 2\pi]$ that satisfy the inequality.

73. $\sin x \leqslant \frac{1}{2}$ **74.** $2 \cos x + 1 > 0$

75. $-1 < \tan x < 1$ **76.** $\sin x > \cos x$

77–82 Graph the function by starting with the graphs in Figures 14 and 15 and applying the transformations of Section 1.3 where appropriate.

77. $y = \cos\left(x - \dfrac{\pi}{3}\right)$ **78.** $y = \tan 2x$

79. $y = \dfrac{1}{3} \tan\left(x - \dfrac{\pi}{2}\right)$ **80.** $y = 1 + \sec x$

81. $y = |\sin x|$ **82.** $y = 2 + \sin\left(x + \dfrac{\pi}{4}\right)$

---

83. Prove the **Law of Cosines**: If a triangle has sides with lengths a, b, and c, and θ is the angle between the sides with lengths a and b, then
$$c^2 = a^2 + b^2 - 2ab \cos\theta$$

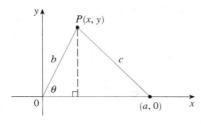

[*Hint:* Introduce a coordinate system so that θ is in standard

position, as in the figure. Express x and y in terms of θ and then use the distance formula to compute c.]

84. In order to find the distance $|AB|$ across a small inlet, a point C was located as in the figure and the following measurements were recorded:

$$\angle C = 103° \qquad |AC| = 820 \text{ m} \qquad |BC| = 910 \text{ m}$$

Use the Law of Cosines from Exercise 83 to find the required distance.

85. Use the figure to prove the subtraction formula
$$\cos(\alpha - \beta) = \cos \alpha \cos \beta + \sin \alpha \sin \beta$$

[*Hint:* Compute c^2 in two ways (using the Law of Cosines from Exercise 83 and also using the distance formula) and compare the two expressions.]

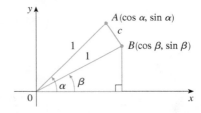

86. Use the formula in Exercise 85 to prove the addition formula for cosine (12b).

87. Use the addition formula for cosine and the identities
$$\cos\left(\dfrac{\pi}{2} - \theta\right) = \sin \theta \qquad \sin\left(\dfrac{\pi}{2} - \theta\right) = \cos \theta$$

to prove the subtraction formula (13a) for the sine function.

88. Show that the area of a triangle with sides of lengths a and b and with included angle θ is
$$A = \tfrac{1}{2}ab \sin \theta$$

89. Find the area of triangle ABC, correct to five decimal places, if
$$|AB| = 10 \text{ cm} \qquad |BC| = 3 \text{ cm} \qquad \angle ABC = 107°$$

E   Sigma Notation

A convenient way of writing sums uses the Greek letter Σ (capital sigma, corresponding to our letter S) and is called **sigma notation**.

This tells us to end with $i = n$.

This tells us to add.

$\displaystyle\sum_{i=m}^{n} a_i$

This tells us to start with $i = m$.

---

1 Definition If $a_m, a_{m+1}, \ldots, a_n$ are real numbers and m and n are integers such that $m \leq n$, then

$$\sum_{i=m}^{n} a_i = a_m + a_{m+1} + a_{m+2} + \cdots + a_{n-1} + a_n$$

With function notation, Definition 1 can be written as

$$\sum_{i=m}^{n} f(i) = f(m) + f(m + 1) + f(m + 2) + \cdots + f(n - 1) + f(n)$$

Thus the symbol $\sum_{i=m}^{n}$ indicates a summation in which the letter i (called the **index of summation**) takes on consecutive integer values beginning with m and ending with n, that is, $m, m + 1, \ldots, n$. Other letters can also be used as the index of summation.

EXAMPLE 1

(a) $\displaystyle\sum_{i=1}^{4} i^2 = 1^2 + 2^2 + 3^2 + 4^2 = 30$

(b) $\displaystyle\sum_{i=3}^{n} i = 3 + 4 + 5 + \cdots + (n - 1) + n$

(c) $\displaystyle\sum_{j=0}^{5} 2^j = 2^0 + 2^1 + 2^2 + 2^3 + 2^4 + 2^5 = 63$

(d) $\displaystyle\sum_{k=1}^{n} \frac{1}{k} = 1 + \frac{1}{2} + \frac{1}{3} + \cdots + \frac{1}{n}$

(e) $\displaystyle\sum_{i=1}^{3} \frac{i - 1}{i^2 + 3} = \frac{1 - 1}{1^2 + 3} + \frac{2 - 1}{2^2 + 3} + \frac{3 - 1}{3^2 + 3} = 0 + \frac{1}{7} + \frac{1}{6} = \frac{13}{42}$

(f) $\displaystyle\sum_{i=1}^{4} 2 = 2 + 2 + 2 + 2 = 8$

EXAMPLE 2 Write the sum $2^3 + 3^3 + \cdots + n^3$ in sigma notation.

SOLUTION There is no unique way of writing a sum in sigma notation. We could write

$$2^3 + 3^3 + \cdots + n^3 = \sum_{i=2}^{n} i^3$$

or

$$2^3 + 3^3 + \cdots + n^3 = \sum_{j=1}^{n-1} (j + 1)^3$$

or

$$2^3 + 3^3 + \cdots + n^3 = \sum_{k=0}^{n-2} (k + 2)^3$$

The following theorem gives three simple rules for working with sigma notation.

2 **Theorem** If c is any constant (that is, it does not depend on i), then

(a) $\displaystyle\sum_{i=m}^{n} ca_i = c \sum_{i=m}^{n} a_i$ (b) $\displaystyle\sum_{i=m}^{n} (a_i + b_i) = \sum_{i=m}^{n} a_i + \sum_{i=m}^{n} b_i$

(c) $\displaystyle\sum_{i=m}^{n} (a_i - b_i) = \sum_{i=m}^{n} a_i - \sum_{i=m}^{n} b_i$

PROOF To see why these rules are true, all we have to do is write both sides in expanded form. Rule (a) is just the distributive property of real numbers:

$$ca_m + ca_{m+1} + \cdots + ca_n = c(a_m + a_{m+1} + \cdots + a_n)$$

Rule (b) follows from the associative and commutative properties:

$$(a_m + b_m) + (a_{m+1} + b_{m+1}) + \cdots + (a_n + b_n)$$
$$= (a_m + a_{m+1} + \cdots + a_n) + (b_m + b_{m+1} + \cdots + b_n)$$

Rule (c) is proved similarly. ∎

EXAMPLE 3 Find $\displaystyle\sum_{i=1}^{n} 1$.

SOLUTION
$$\sum_{i=1}^{n} 1 = \underbrace{1 + 1 + \cdots + 1}_{n \text{ terms}} = n$$

EXAMPLE 4 Prove the formula for the sum of the first n positive integers:

$$\sum_{i=1}^{n} i = 1 + 2 + 3 + \cdots + n = \frac{n(n+1)}{2}$$

SOLUTION This formula can be proved by mathematical induction (see page 72) or by the following method used by the German mathematician Karl Friedrich Gauss (1777–1855) when he was ten years old.

Write the sum S twice, once in the usual order and once in reverse order:

$$S = 1 + \quad 2 \quad + \quad 3 \quad + \cdots + (n-1) + n$$
$$S = n + (n-1) + (n-2) + \cdots + \quad 2 \quad + 1$$

Adding all columns vertically, we get

$$2S = (n+1) + (n+1) + (n+1) + \cdots + (n+1) + (n+1)$$

On the right side there are n terms, each of which is $n+1$, so

$$2S = n(n+1) \qquad \text{or} \qquad S = \frac{n(n+1)}{2}$$

EXAMPLE 5 Prove the formula for the sum of the squares of the first n positive integers:

$$\sum_{i=1}^{n} i^2 = 1^2 + 2^2 + 3^2 + \cdots + n^2 = \frac{n(n+1)(2n+1)}{6}$$

SOLUTION 1 Let S be the desired sum. We start with the *telescoping sum* (or collapsing sum):

Most terms cancel in pairs.

$$\sum_{i=1}^{n} [(1 + i)^3 - i^3] = (2^3 - 1^3) + (3^3 - 2^3) + (4^3 - 3^3) + \cdots + [(n + 1)^3 - n^3]$$

$$= (n + 1)^3 - 1^3 = n^3 + 3n^2 + 3n$$

On the other hand, using Theorem 2 and Examples 3 and 4, we have

$$\sum_{i=1}^{n} [(1 + i)^3 - i^3] = \sum_{i=1}^{n} [3i^2 + 3i + 1] = 3\sum_{i=1}^{n} i^2 + 3\sum_{i=1}^{n} i + \sum_{i=1}^{n} 1$$

$$= 3S + 3\frac{n(n + 1)}{2} + n = 3S + \tfrac{3}{2}n^2 + \tfrac{5}{2}n$$

Thus we have

$$n^3 + 3n^2 + 3n = 3S + \tfrac{3}{2}n^2 + \tfrac{5}{2}n$$

Solving this equation for S, we obtain

$$3S = n^3 + \tfrac{3}{2}n^2 + \tfrac{1}{2}n$$

or

$$S = \frac{2n^3 + 3n^2 + n}{6} = \frac{n(n + 1)(2n + 1)}{6}$$

Principle of Mathematical Induction
Let S_n be a statement involving the positive integer n. Suppose that
   1. S_1 is true.
   2. If S_k is true, then S_{k+1} is true.
Then S_n is true for all positive integers n.

See pages 72 and 74 for a more thorough discussion of mathematical induction.

SOLUTION 2 Let S_n be the given formula.

1. S_1 is true because $1^2 = \dfrac{1(1 + 1)(2 \cdot 1 + 1)}{6}$

2. Assume that S_k is true; that is,

$$1^2 + 2^2 + 3^2 + \cdots + k^2 = \frac{k(k + 1)(2k + 1)}{6}$$

Then

$$1^2 + 2^2 + 3^2 + \cdots + (k + 1)^2 = (1^2 + 2^2 + 3^2 + \cdots + k^2) + (k + 1)^2$$

$$= \frac{k(k + 1)(2k + 1)}{6} + (k + 1)^2$$

$$= (k + 1)\frac{k(2k + 1) + 6(k + 1)}{6}$$

$$= (k + 1)\frac{2k^2 + 7k + 6}{6}$$

$$= \frac{(k + 1)(k + 2)(2k + 3)}{6}$$

$$= \frac{(k + 1)[(k + 1) + 1][2(k + 1) + 1]}{6}$$

So S_{k+1} is true.

By the Principle of Mathematical Induction, S_n is true for all n.

We list the results of Examples 3, 4, and 5 together with a similar result for cubes (see Exercises 37–40) as Theorem 3. These formulas are needed for finding areas and evaluating integrals in Chapter 5.

---

3 **Theorem** Let c be a constant and n a positive integer. Then

(a) $\displaystyle\sum_{i=1}^{n} 1 = n$ (b) $\displaystyle\sum_{i=1}^{n} c = nc$

(c) $\displaystyle\sum_{i=1}^{n} i = \frac{n(n+1)}{2}$ (d) $\displaystyle\sum_{i=1}^{n} i^2 = \frac{n(n+1)(2n+1)}{6}$

(e) $\displaystyle\sum_{i=1}^{n} i^3 = \left[\frac{n(n+1)}{2}\right]^2$

EXAMPLE 6 Evaluate $\displaystyle\sum_{i=1}^{n} i(4i^2 - 3)$.

SOLUTION Using Theorems 2 and 3, we have

$$\sum_{i=1}^{n} i(4i^2 - 3) = \sum_{i=1}^{n} (4i^3 - 3i) = 4\sum_{i=1}^{n} i^3 - 3\sum_{i=1}^{n} i$$

$$= 4\left[\frac{n(n+1)}{2}\right]^2 - 3\,\frac{n(n+1)}{2}$$

$$= \frac{n(n+1)[2n(n+1) - 3]}{2}$$

$$= \frac{n(n+1)(2n^2 + 2n - 3)}{2}$$

EXAMPLE 7 Find $\displaystyle\lim_{n\to\infty} \sum_{i=1}^{n} \frac{3}{n}\left[\left(\frac{i}{n}\right)^2 + 1\right]$.

The type of calculation in Example 7 arises in Chapter 5 when we compute areas.

SOLUTION

$$\lim_{n\to\infty} \sum_{i=1}^{n} \frac{3}{n}\left[\left(\frac{i}{n}\right)^2 + 1\right] = \lim_{n\to\infty} \sum_{i=1}^{n} \left[\frac{3}{n^3} i^2 + \frac{3}{n}\right]$$

$$= \lim_{n\to\infty} \left[\frac{3}{n^3} \sum_{i=1}^{n} i^2 + \frac{3}{n} \sum_{i=1}^{n} 1\right]$$

$$= \lim_{n\to\infty} \left[\frac{3}{n^3}\, \frac{n(n+1)(2n+1)}{6} + \frac{3}{n} \cdot n\right]$$

$$= \lim_{n\to\infty} \left[\frac{1}{2} \cdot \frac{n}{n} \cdot \left(\frac{n+1}{n}\right)\left(\frac{2n+1}{n}\right) + 3\right]$$

$$= \lim_{n\to\infty} \left[\frac{1}{2} \cdot 1\left(1 + \frac{1}{n}\right)\left(2 + \frac{1}{n}\right) + 3\right]$$

$$= \tfrac{1}{2} \cdot 1 \cdot 1 \cdot 2 + 3 = 4$$

E   EXERCISES

1–10 Write the sum in expanded form.

1. $\displaystyle\sum_{i=1}^{5} \sqrt{i}$

2. $\displaystyle\sum_{i=1}^{6} \frac{1}{i+1}$

3. $\displaystyle\sum_{i=4}^{6} 3^i$

4. $\displaystyle\sum_{i=4}^{6} i^3$

5. $\displaystyle\sum_{k=0}^{4} \frac{2k-1}{2k+1}$

6. $\displaystyle\sum_{k=5}^{8} x^k$

7. $\displaystyle\sum_{i=1}^{n} i^{10}$

8. $\displaystyle\sum_{j=n}^{n+3} j^2$

9. $\displaystyle\sum_{j=0}^{n-1} (-1)^j$

10. $\displaystyle\sum_{i=1}^{n} f(x_i)\, \Delta x_i$

11–20 Write the sum in sigma notation.

11. $1 + 2 + 3 + 4 + \cdots + 10$

12. $\sqrt{3} + \sqrt{4} + \sqrt{5} + \sqrt{6} + \sqrt{7}$

13. $\frac{1}{2} + \frac{2}{3} + \frac{3}{4} + \frac{4}{5} + \cdots + \frac{19}{20}$

14. $\frac{3}{7} + \frac{4}{8} + \frac{5}{9} + \frac{6}{10} + \cdots + \frac{23}{27}$

15. $2 + 4 + 6 + 8 + \cdots + 2n$

16. $1 + 3 + 5 + 7 + \cdots + (2n-1)$

17. $1 + 2 + 4 + 8 + 16 + 32$

18. $\frac{1}{1} + \frac{1}{4} + \frac{1}{9} + \frac{1}{16} + \frac{1}{25} + \frac{1}{36}$

19. $x + x^2 + x^3 + \cdots + x^n$

20. $1 - x + x^2 - x^3 + \cdots + (-1)^n x^n$

21–35 Find the value of the sum.

21. $\displaystyle\sum_{i=4}^{8} (3i-2)$

22. $\displaystyle\sum_{i=3}^{6} i(i+2)$

23. $\displaystyle\sum_{j=1}^{6} 3^{j+1}$

24. $\displaystyle\sum_{k=0}^{8} \cos k\pi$

25. $\displaystyle\sum_{n=1}^{20} (-1)^n$

26. $\displaystyle\sum_{i=1}^{100} 4$

27. $\displaystyle\sum_{i=0}^{4} (2^i + i^2)$

28. $\displaystyle\sum_{i=-2}^{4} 2^{3-i}$

29. $\displaystyle\sum_{i=1}^{n} 2i$

30. $\displaystyle\sum_{i=1}^{n} (2-5i)$

31. $\displaystyle\sum_{i=1}^{n} (i^2 + 3i + 4)$

32. $\displaystyle\sum_{i=1}^{n} (3+2i)^2$

33. $\displaystyle\sum_{i=1}^{n} (i+1)(i+2)$

34. $\displaystyle\sum_{i=1}^{n} i(i+1)(i+2)$

35. $\displaystyle\sum_{i=1}^{n} (i^3 - i - 2)$

---

36. Find the number n such that $\displaystyle\sum_{i=1}^{n} i = 78$.

37. Prove formula (b) of Theorem 3.

38. Prove formula (e) of Theorem 3 using mathematical induction.

39. Prove formula (e) of Theorem 3 using a method similar to that of Example 5, Solution 1 [start with $(1+i)^4 - i^4$].

40. Prove formula (e) of Theorem 3 using the following method published by Abu Bekr Mohammed ibn Alhusain Alkarchi in about AD 1010. The figure shows a square $ABCD$ in which sides AB and AD have been divided into segments of lengths $1, 2, 3, \ldots, n$. Thus the side of the square has length $n(n+1)/2$ so the area is $[n(n+1)/2]^2$. But the area is also the sum of the areas of the n "gnomons" $G_1, G_2, \ldots, G_n$ shown in the figure. Show that the area of G_i is i^3 and conclude that formula (e) is true.

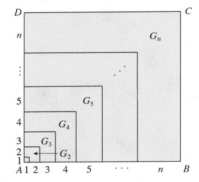

41. Evaluate each telescoping sum.

(a) $\displaystyle\sum_{i=1}^{n} [i^4 - (i-1)^4]$

(b) $\displaystyle\sum_{i=1}^{100} (5^i - 5^{i-1})$

(c) $\displaystyle\sum_{i=3}^{99} \left(\frac{1}{i} - \frac{1}{i+1}\right)$

(d) $\displaystyle\sum_{i=1}^{n} (a_i - a_{i-1})$

42. Prove the generalized triangle inequality:

$$\left|\sum_{i=1}^{n} a_i\right| \leq \sum_{i=1}^{n} |a_i|$$

43–46 Find the limit.

43. $\displaystyle\lim_{n \to \infty} \sum_{i=1}^{n} \frac{1}{n}\left(\frac{i}{n}\right)^2$

44. $\displaystyle\lim_{n \to \infty} \sum_{i=1}^{n} \frac{1}{n}\left[\left(\frac{i}{n}\right)^3 + 1\right]$

45. $\displaystyle\lim_{n \to \infty} \sum_{i=1}^{n} \frac{2}{n}\left[\left(\frac{2i}{n}\right)^3 + 5\left(\frac{2i}{n}\right)\right]$

46. $\lim\limits_{n \to \infty} \sum\limits_{i=1}^{n} \dfrac{3}{n} \left[\left(1 + \dfrac{3i}{n}\right)^3 - 2\left(1 + \dfrac{3i}{n}\right) \right]$

48. Evaluate $\sum\limits_{i=1}^{n} \dfrac{3}{2^{i-1}}$.

---

47. Prove the formula for the sum of a finite geometric series with first term a and common ratio $r \neq 1$:

$$\sum_{i=1}^{n} ar^{i-1} = a + ar + ar^2 + \cdots + ar^{n-1} = \frac{a(r^n - 1)}{r - 1}$$

49. Evaluate $\sum\limits_{i=1}^{n} (2i + 2^i)$.

50. Evaluate $\sum\limits_{i=1}^{m} \left[\sum\limits_{j=1}^{n} (i + j) \right]$.

F   Proofs of Theorems

In this appendix we present proofs of several theorems that are stated in the main body of the text. The sections in which they occur are indicated in the margin.

Section 2.3

> **Limit Laws** Suppose that c is a constant and the limits
>
> $$\lim_{x \to a} f(x) = L \qquad \text{and} \qquad \lim_{x \to a} g(x) = M$$
>
> exist. Then
>
> **1.** $\lim\limits_{x \to a} [f(x) + g(x)] = L + M$ **2.** $\lim\limits_{x \to a} [f(x) - g(x)] = L - M$
>
> **3.** $\lim\limits_{x \to a} [cf(x)] = cL$ **4.** $\lim\limits_{x \to a} [f(x)g(x)] = LM$
>
> **5.** $\lim\limits_{x \to a} \dfrac{f(x)}{g(x)} = \dfrac{L}{M}$ if $M \neq 0$

PROOF OF LAW 4 Let $\varepsilon > 0$ be given. We want to find $\delta > 0$ such that

$$\text{if} \qquad 0 < |x - a| < \delta \qquad \text{then} \qquad |f(x)g(x) - LM| < \varepsilon$$

In order to get terms that contain $|f(x) - L|$ and $|g(x) - M|$, we add and subtract $Lg(x)$ as follows:

$$
\begin{aligned}
|f(x)g(x) - LM| &= |f(x)g(x) - Lg(x) + Lg(x) - LM| \\
&= |[f(x) - L]g(x) + L[g(x) - M]| \\
&\leq |[f(x) - L]g(x)| + |L[g(x) - M]| \qquad \text{(Triangle Inequality)} \\
&= |f(x) - L||g(x)| + |L||g(x) - M|
\end{aligned}
$$

We want to make each of these terms less than $\varepsilon/2$.

Since $\lim_{x \to a} g(x) = M$, there is a number $\delta_1 > 0$ such that

$$\text{if} \qquad 0 < |x - a| < \delta_1 \qquad \text{then} \qquad |g(x) - M| < \frac{\varepsilon}{2(1 + |L|)}$$

Also, there is a number $\delta_2 > 0$ such that if $0 < |x - a| < \delta_2$, then

$$|g(x) - M| < 1$$

and therefore

$$|g(x)| = |g(x) - M + M| \leq |g(x) - M| + |M| < 1 + |M|$$

Since $\lim_{x \to a} f(x) = L$, there is a number $\delta_3 > 0$ such that

$$\text{if} \qquad 0 < |x - a| < \delta_3 \qquad \text{then} \qquad |f(x) - L| < \frac{\varepsilon}{2(1 + |M|)}$$

Let $\delta = \min\{\delta_1, \delta_2, \delta_3\}$. If $0 < |x - a| < \delta$, then we have $0 < |x - a| < \delta_1$, $0 < |x - a| < \delta_2$, and $0 < |x - a| < \delta_3$, so we can combine the inequalities to obtain

$$|f(x)g(x) - LM| \leq |f(x) - L||g(x)| + |L||g(x) - M|$$

$$< \frac{\varepsilon}{2(1 + |M|)}(1 + |M|) + |L|\frac{\varepsilon}{2(1 + |L|)}$$

$$< \frac{\varepsilon}{2} + \frac{\varepsilon}{2} = \varepsilon$$

This shows that $\lim_{x \to a} [f(x)g(x)] = LM$.   ■

PROOF OF LAW 3 If we take $g(x) = c$ in Law 4, we get

$$\lim_{x \to a} [cf(x)] = \lim_{x \to a} [g(x)f(x)] = \lim_{x \to a} g(x) \cdot \lim_{x \to a} f(x)$$

$$= \lim_{x \to a} c \cdot \lim_{x \to a} f(x)$$

$$= c \lim_{x \to a} f(x) \qquad \text{(by Law 7)}$$ ■

PROOF OF LAW 2 Using Law 1 and Law 3 with $c = -1$, we have

$$\lim_{x \to a} [f(x) - g(x)] = \lim_{x \to a} [f(x) + (-1)g(x)] = \lim_{x \to a} f(x) + \lim_{x \to a} (-1)g(x)$$

$$= \lim_{x \to a} f(x) + (-1) \lim_{x \to a} g(x) = \lim_{x \to a} f(x) - \lim_{x \to a} g(x)$$ ■

PROOF OF LAW 5 First let us show that

$$\lim_{x \to a} \frac{1}{g(x)} = \frac{1}{M}$$

To do this we must show that, given $\varepsilon > 0$, there exists $\delta > 0$ such that

$$\text{if} \qquad 0 < |x - a| < \delta \qquad \text{then} \qquad \left| \frac{1}{g(x)} - \frac{1}{M} \right| < \varepsilon$$

Observe that

$$\left| \frac{1}{g(x)} - \frac{1}{M} \right| = \frac{|M - g(x)|}{|Mg(x)|}$$

We know that we can make the numerator small. But we also need to know that the denominator is not small when x is near a. Since $\lim_{x \to a} g(x) = M$, there is a number $\delta_1 > 0$ such that, whenever $0 < |x - a| < \delta_1$, we have

$$|g(x) - M| < \frac{|M|}{2}$$

and therefore $$|M| = |M - g(x) + g(x)| \leq |M - g(x)| + |g(x)|$$

$$< \frac{|M|}{2} + |g(x)|$$

This shows that

$$\text{if} \qquad 0 < |x - a| < \delta_1 \qquad \text{then} \qquad |g(x)| > \frac{|M|}{2}$$

and so, for these values of x,

$$\frac{1}{|Mg(x)|} = \frac{1}{|M||g(x)|} < \frac{1}{|M|} \cdot \frac{2}{|M|} = \frac{2}{M^2}$$

Also, there exists $\delta_2 > 0$ such that

$$\text{if} \qquad 0 < |x - a| < \delta_2 \qquad \text{then} \qquad |g(x) - M| < \frac{M^2}{2} \varepsilon$$

Let $\delta = \min\{\delta_1, \delta_2\}$. Then, for $0 < |x - a| < \delta$, we have

$$\left| \frac{1}{g(x)} - \frac{1}{M} \right| = \frac{|M - g(x)|}{|Mg(x)|} < \frac{2}{M^2} \frac{M^2}{2} \varepsilon = \varepsilon$$

It follows that $\lim_{x \to a} 1/g(x) = 1/M$. Finally, using Law 4, we obtain

$$\lim_{x \to a} \frac{f(x)}{g(x)} = \lim_{x \to a} \left(f(x) \cdot \frac{1}{g(x)} \right) = \lim_{x \to a} f(x) \lim_{x \to a} \frac{1}{g(x)} = L \cdot \frac{1}{M} = \frac{L}{M} \qquad \blacksquare$$

---

2 Theorem If $f(x) \leq g(x)$ for all x in an open interval that contains a (except possibly at a) and

$$\lim_{x \to a} f(x) = L \qquad \text{and} \qquad \lim_{x \to a} g(x) = M$$

then $L \leq M$.

PROOF We use the method of proof by contradiction. Suppose, if possible, that $L > M$. Law 2 of limits says that

$$\lim_{x \to a} [g(x) - f(x)] = M - L$$

Therefore, for any $\varepsilon > 0$, there exists $\delta > 0$ such that

$$\text{if} \qquad 0 < |x - a| < \delta \qquad \text{then} \qquad |[g(x) - f(x)] - (M - L)| < \varepsilon$$

In particular, taking $\varepsilon = L - M$ (noting that $L - M > 0$ by hypothesis), we have a number $\delta > 0$ such that

$$\text{if} \qquad 0 < |x - a| < \delta \qquad \text{then} \qquad |[g(x) - f(x)] - (M - L)| < L - M$$

Since $b \leq |b|$ for any number b, we have

$$\text{if} \qquad 0 < |x - a| < \delta \qquad \text{then} \qquad [g(x) - f(x)] - (M - L) < L - M$$

which simplifies to

$$\text{if} \qquad 0 < |x - a| < \delta \qquad \text{then} \qquad g(x) < f(x)$$

But this contradicts $f(x) \leq g(x)$. Thus the inequality $L > M$ must be false. Therefore $L \leq M$. $\qquad \blacksquare$

> **3** **The Squeeze Theorem** If $f(x) \leqslant g(x) \leqslant h(x)$ for all x in an open interval that contains a (except possibly at a) and
>
> $$\lim_{x \to a} f(x) = \lim_{x \to a} h(x) = L$$
>
> then
> $$\lim_{x \to a} g(x) = L$$

PROOF Let $\varepsilon > 0$ be given. Since $\lim_{x \to a} f(x) = L$, there is a number $\delta_1 > 0$ such that

$$\text{if} \quad 0 < |x - a| < \delta_1 \quad \text{then} \quad |f(x) - L| < \varepsilon$$

that is,

$$\text{if} \quad 0 < |x - a| < \delta_1 \quad \text{then} \quad L - \varepsilon < f(x) < L + \varepsilon$$

Since $\lim_{x \to a} h(x) = L$, there is a number $\delta_2 > 0$ such that

$$\text{if} \quad 0 < |x - a| < \delta_2 \quad \text{then} \quad |h(x) - L| < \varepsilon$$

that is,

$$\text{if} \quad 0 < |x - a| < \delta_2 \quad \text{then} \quad L - \varepsilon < h(x) < L + \varepsilon$$

Let $\delta = \min\{\delta_1, \delta_2\}$. If $0 < |x - a| < \delta$, then $0 < |x - a| < \delta_1$ and $0 < |x - a| < \delta_2$, so

$$L - \varepsilon < f(x) \leqslant g(x) \leqslant h(x) < L + \varepsilon$$

In particular, $$L - \varepsilon < g(x) < L + \varepsilon$$

and so $|g(x) - L| < \varepsilon$. Therefore $\lim_{x \to a} g(x) = L$. ∎

Section 2.5

> **Theorem** If f is a one-to-one continuous function defined on an interval (a, b), then its inverse function f^{-1} is also continuous.

PROOF First we show that if f is both one-to-one and continuous on (a, b), then it must be either increasing or decreasing on (a, b). If it were neither increasing nor decreasing, then there would exist numbers x_1, x_2, and x_3 in (a, b) with $x_1 < x_2 < x_3$ such that $f(x_2)$ does not lie between $f(x_1)$ and $f(x_3)$. There are two possibilities: either (1) $f(x_3)$ lies between $f(x_1)$ and $f(x_2)$ or (2) $f(x_1)$ lies between $f(x_2)$ and $f(x_3)$. (Draw a picture.) In case (1) we apply the Intermediate Value Theorem to the continuous function f to get a number c between x_1 and x_2 such that $f(c) = f(x_3)$. In case (2) the Intermediate Value Theorem gives a number c between x_2 and x_3 such that $f(c) = f(x_1)$. In either case we have contradicted the fact that f is one-to-one.

Let us assume, for the sake of definiteness, that f is increasing on (a, b). We take any number y_0 in the domain of f^{-1} and we let $f^{-1}(y_0) = x_0$; that is, x_0 is the number in (a, b) such that $f(x_0) = y_0$. To show that f^{-1} is continuous at y_0 we take any $\varepsilon > 0$ such that the interval $(x_0 - \varepsilon, x_0 + \varepsilon)$ is contained in the interval (a, b). Since f is increasing, it maps the numbers in the interval $(x_0 - \varepsilon, x_0 + \varepsilon)$ onto the numbers in the interval $(f(x_0 - \varepsilon), f(x_0 + \varepsilon))$ and f^{-1} reverses the correspondence. If we let δ denote the smaller of the numbers $\delta_1 = y_0 - f(x_0 - \varepsilon)$ and $\delta_2 = f(x_0 + \varepsilon) - y_0$, then the interval $(y_0 - \delta, y_0 + \delta)$ is contained in the interval $(f(x_0 - \varepsilon), f(x_0 + \varepsilon))$ and so is mapped into the interval $(x_0 - \varepsilon, x_0 + \varepsilon)$ by f^{-1}. (See the arrow diagram in Figure 1.) We have therefore found a number $\delta > 0$ such that

$$\text{if} \quad |y - y_0| < \delta \quad \text{then} \quad |f^{-1}(y) - f^{-1}(y_0)| < \varepsilon$$

FIGURE 1

This shows that $\lim_{y \to y_0} f^{-1}(y) = f^{-1}(y_0)$ and so f^{-1} is continuous at any number y_0 in its domain. ∎

---

8 Theorem If f is continuous at b and $\lim_{x \to a} g(x) = b$, then

$$\lim_{x \to a} f(g(x)) = f(b)$$

PROOF Let $\varepsilon > 0$ be given. We want to find a number $\delta > 0$ such that

$$\text{if} \quad 0 < |x - a| < \delta \quad \text{then} \quad |f(g(x)) - f(b)| < \varepsilon$$

Since f is continuous at b, we have

$$\lim_{y \to b} f(y) = f(b)$$

and so there exists $\delta_1 > 0$ such that

$$\text{if} \quad 0 < |y - b| < \delta_1 \quad \text{then} \quad |f(y) - f(b)| < \varepsilon$$

Since $\lim_{x \to a} g(x) = b$, there exists $\delta > 0$ such that

$$\text{if} \quad 0 < |x - a| < \delta \quad \text{then} \quad |g(x) - b| < \delta_1$$

Combining these two statements, we see that whenever $0 < |x - a| < \delta$ we have $|g(x) - b| < \delta_1$, which implies that $|f(g(x)) - f(b)| < \varepsilon$. Therefore we have proved that $\lim_{x \to a} f(g(x)) = f(b)$. ∎

Section 3.3

The proof of the following result was promised when we proved that $\lim_{\theta \to 0} \dfrac{\sin \theta}{\theta} = 1$.

Theorem If $0 < \theta < \pi/2$, then $\theta \leqslant \tan \theta$.

PROOF Figure 2 shows a sector of a circle with center O, central angle θ, and radius 1. Then

$$|AD| = |OA| \tan \theta = \tan \theta$$

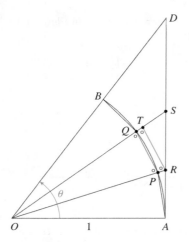

FIGURE 2

We approximate the arc AB by an inscribed polygon consisting of n equal line segments and we look at a typical segment PQ. We extend the lines OP and OQ to meet AD in the points R and S. Then we draw $RT \parallel PQ$ as in Figure 2. Observe that

$$\angle RTO = \angle PQO < 90°$$

and so $\angle RTS > 90°$. Therefore we have

$$|PQ| < |RT| < |RS|$$

If we add n such inequalities, we get

$$L_n < |AD| = \tan\theta$$

where L_n is the length of the inscribed polygon. Thus, by Theorem 2.3.2, we have

$$\lim_{n\to\infty} L_n \leqslant \tan\theta$$

But the arc length is defined in Equation 8.1.1 as the limit of the lengths of inscribed polygons, so

$$\theta = \lim_{n\to\infty} L_n \leqslant \tan\theta \qquad\blacksquare$$

Section 4.3

> **Concavity Test**
> (a) If $f''(x) > 0$ for all x in I, then the graph of f is concave upward on I.
> (b) If $f''(x) < 0$ for all x in I, then the graph of f is concave downward on I.

PROOF OF (a) Let a be any number in I. We must show that the curve $y = f(x)$ lies above the tangent line at the point $(a, f(a))$. The equation of this tangent is

$$y = f(a) + f'(a)(x - a)$$

So we must show that

$$f(x) > f(a) + f'(a)(x - a)$$

whenever $x \in I$ ($x \neq a$). (See Figure 3.)

First let us take the case where $x > a$. Applying the Mean Value Theorem to f on the interval $[a, x]$, we get a number c, with $a < c < x$, such that

$$\boxed{1} \qquad f(x) - f(a) = f'(c)(x - a)$$

Since $f'' > 0$ on I, we know from the Increasing/Decreasing Test that f' is increasing on I. Thus, since $a < c$, we have

$$f'(a) < f'(c)$$

and so, multiplying this inequality by the positive number $x - a$, we get

$$\boxed{2} \qquad f'(a)(x - a) < f'(c)(x - a)$$

FIGURE 3

Now we add $f(a)$ to both sides of this inequality:

$$f(a) + f'(a)(x - a) < f(a) + f'(c)(x - a)$$

But from Equation 1 we have $f(x) = f(a) + f'(c)(x - a)$. So this inequality becomes

$$\boxed{3} \qquad\qquad f(x) > f(a) + f'(a)(x - a)$$

which is what we wanted to prove.

For the case where $x < a$ we have $f'(c) < f'(a)$, but multiplication by the negative number $x - a$ reverses the inequality, so we get (2) and (3) as before.   ∎

Section 4.4

In order to give the promised proof of l'Hospital's Rule, we first need a generalization of the Mean Value Theorem. The following theorem is named after another French mathematician, Augustin-Louis Cauchy (1789–1857).

See the biographical sketch of Cauchy on page 109.

---

1 Cauchy's Mean Value Theorem Suppose that the functions f and g are continuous on $[a, b]$ and differentiable on (a, b), and $g'(x) \neq 0$ for all x in (a, b). Then there is a number c in (a, b) such that

$$\frac{f'(c)}{g'(c)} = \frac{f(b) - f(a)}{g(b) - g(a)}$$

Notice that if we take the special case in which $g(x) = x$, then $g'(c) = 1$ and Theorem 1 is just the ordinary Mean Value Theorem. Furthermore, Theorem 1 can be proved in a similar manner. You can verify that all we have to do is change the function h given by Equation 4.2.4 to the function

$$h(x) = f(x) - f(a) - \frac{f(b) - f(a)}{g(b) - g(a)}[g(x) - g(a)]$$

and apply Rolle's Theorem as before.

L'Hospital's Rule Suppose f and g are differentiable and $g'(x) \neq 0$ on an open interval I that contains a (except possibly at a). Suppose that

$$\lim_{x \to a} f(x) = 0 \qquad \text{and} \qquad \lim_{x \to a} g(x) = 0$$

or that

$$\lim_{x \to a} f(x) = \pm\infty \qquad \text{and} \qquad \lim_{x \to a} g(x) = \pm\infty$$

(In other words, we have an indeterminate form of type $\frac{0}{0}$ or ∞/∞.) Then

$$\lim_{x \to a} \frac{f(x)}{g(x)} = \lim_{x \to a} \frac{f'(x)}{g'(x)}$$

if the limit on the right side exists (or is ∞ or $-\infty$).

PROOF OF L'HOSPITAL'S RULE We are assuming that $\lim_{x \to a} f(x) = 0$ and $\lim_{x \to a} g(x) = 0$. Let

$$L = \lim_{x \to a} \frac{f'(x)}{g'(x)}$$

We must show that $\lim_{x \to a} f(x)/g(x) = L$. Define

$$F(x) = \begin{cases} f(x) & \text{if } x \neq a \\ 0 & \text{if } x = a \end{cases} \qquad G(x) = \begin{cases} g(x) & \text{if } x \neq a \\ 0 & \text{if } x = a \end{cases}$$

Then F is continuous on I since f is continuous on $\{x \in I \mid x \neq a\}$ and

$$\lim_{x \to a} F(x) = \lim_{x \to a} f(x) = 0 = F(a)$$

Likewise, G is continuous on I. Let $x \in I$ and $x > a$. Then F and G are continuous on $[a, x]$ and differentiable on (a, x) and $G' \neq 0$ there (since $F' = f'$ and $G' = g'$). Therefore, by Cauchy's Mean Value Theorem, there is a number y such that $a < y < x$ and

$$\frac{F'(y)}{G'(y)} = \frac{F(x) - F(a)}{G(x) - G(a)} = \frac{F(x)}{G(x)}$$

Here we have used the fact that, by definition, $F(a) = 0$ and $G(a) = 0$. Now, if we let $x \to a^+$, then $y \to a^+$ (since $a < y < x$), so

$$\lim_{x \to a^+} \frac{f(x)}{g(x)} = \lim_{x \to a^+} \frac{F(x)}{G(x)} = \lim_{y \to a^+} \frac{F'(y)}{G'(y)} = \lim_{y \to a^+} \frac{f'(y)}{g'(y)} = L$$

A similar argument shows that the left-hand limit is also L. Therefore

$$\lim_{x \to a} \frac{f(x)}{g(x)} = L$$

This proves l'Hospital's Rule for the case where a is finite.

If a is infinite, we let $t = 1/x$. Then $t \to 0^+$ as $x \to \infty$, so we have

$$\lim_{x \to \infty} \frac{f(x)}{g(x)} = \lim_{t \to 0^+} \frac{f(1/t)}{g(1/t)}$$

$$= \lim_{t \to 0^+} \frac{f'(1/t)(-1/t^2)}{g'(1/t)(-1/t^2)} \qquad \text{(by l'Hospital's Rule for finite a)}$$

$$= \lim_{t \to 0^+} \frac{f'(1/t)}{g'(1/t)} = \lim_{x \to \infty} \frac{f'(x)}{g'(x)}$$

Section 11.8

In order to prove Theorem 11.8.4, we first need the following results.

Theorem

1. If a power series $\sum c_n x^n$ converges when $x = b$ (where $b \neq 0$), then it converges whenever $|x| < |b|$.

2. If a power series $\sum c_n x^n$ diverges when $x = d$ (where $d \neq 0$), then it diverges whenever $|x| > |d|$.

PROOF OF 1 Suppose that $\Sigma\, c_n b^n$ converges. Then, by Theorem 11.2.6, we have $\lim_{n\to\infty} c_n b^n = 0$. According to Definition 11.1.2 with $\varepsilon = 1$, there is a positive integer N such that $|c_n b^n| < 1$ whenever $n \geqslant N$. Thus, for $n \geqslant N$, we have

$$|c_n x^n| = \left| \frac{c_n b^n x^n}{b^n} \right| = |c_n b^n| \left| \frac{x}{b} \right|^n < \left| \frac{x}{b} \right|^n$$

If $|x| < |b|$, then $|x/b| < 1$, so $\Sigma\, |x/b|^n$ is a convergent geometric series. Therefore, by the Comparison Test, the series $\Sigma_{n=N}^{\infty}\, |c_n x^n|$ is convergent. Thus the series $\Sigma\, c_n x^n$ is absolutely convergent and therefore convergent.   ■

PROOF OF 2 Suppose that $\Sigma\, c_n d^n$ diverges. If x is any number such that $|x| > |d|$, then $\Sigma\, c_n x^n$ cannot converge because, by part 1, the convergence of $\Sigma\, c_n x^n$ would imply the convergence of $\Sigma\, c_n d^n$. Therefore $\Sigma\, c_n x^n$ diverges whenever $|x| > |d|$.   ■

Theorem For a power series $\Sigma\, c_n x^n$ there are only three possibilities:

1. The series converges only when $x = 0$.

2. The series converges for all x.

3. There is a positive number R such that the series converges if $|x| < R$ and diverges if $|x| > R$.

PROOF Suppose that neither case 1 nor case 2 is true. Then there are nonzero numbers b and d such that $\Sigma\, c_n x^n$ converges for $x = b$ and diverges for $x = d$. Therefore the set $S = \{x \mid \Sigma\, c_n x^n \text{ converges}\}$ is not empty. By the preceding theorem, the series diverges if $|x| > |d|$, so $|x| \leqslant |d|$ for all $x \in S$. This says that $|d|$ is an upper bound for the set S. Thus, by the Completeness Axiom (see Section 11.1), S has a least upper bound R. If $|x| > R$, then $x \notin S$, so $\Sigma\, c_n x^n$ diverges. If $|x| < R$, then $|x|$ is not an upper bound for S and so there exists $b \in S$ such that $b > |x|$. Since $b \in S$, $\Sigma\, c_n x^n$ converges, so by the preceding theorem $\Sigma\, c_n x^n$ converges.   ■

---

4 Theorem For a power series $\Sigma\, c_n(x - a)^n$ there are only three possibilities:

1. The series converges only when $x = a$.

2. The series converges for all x.

3. There is a positive number R such that the series converges if $|x - a| < R$ and diverges if $|x - a| > R$.

PROOF If we make the change of variable $u = x - a$, then the power series becomes $\Sigma\, c_n u^n$ and we can apply the preceding theorem to this series. In case 3 we have convergence for $|u| < R$ and divergence for $|u| > R$. Thus we have convergence for $|x - a| < R$ and divergence for $|x - a| > R$.   ■

> **Clairaut's Theorem** Suppose f is defined on a disk D that contains the point (a, b). If the functions f_{xy} and f_{yx} are both continuous on D, then $f_{xy}(a, b) = f_{yx}(a, b)$.

PROOF For small values of h, $h \neq 0$, consider the difference

$$\Delta(h) = [f(a + h, b + h) - f(a + h, b)] - [f(a, b + h) - f(a, b)]$$

Notice that if we let $g(x) = f(x, b + h) - f(x, b)$, then

$$\Delta(h) = g(a + h) - g(a)$$

By the Mean Value Theorem, there is a number c between a and $a + h$ such that

$$g(a + h) - g(a) = g'(c)h = h[f_x(c, b + h) - f_x(c, b)]$$

Applying the Mean Value Theorem again, this time to f_x, we get a number d between b and $b + h$ such that

$$f_x(c, b + h) - f_x(c, b) = f_{xy}(c, d)h$$

Combining these equations, we obtain

$$\Delta(h) = h^2 f_{xy}(c, d)$$

If $h \to 0$, then $(c, d) \to (a, b)$, so the continuity of f_{xy} at (a, b) gives

$$\lim_{h \to 0} \frac{\Delta(h)}{h^2} = \lim_{(c, d) \to (a, b)} f_{xy}(c, d) = f_{xy}(a, b)$$

Similarly, by writing

$$\Delta(h) = [f(a + h, b + h) - f(a, b + h)] - [f(a + h, b) - f(a, b)]$$

and using the Mean Value Theorem twice and the continuity of f_{yx} at (a, b), we obtain

$$\lim_{h \to 0} \frac{\Delta(h)}{h^2} = f_{yx}(a, b)$$

It follows that $f_{xy}(a, b) = f_{yx}(a, b)$. ■

> **8** **Theorem** If the partial derivatives f_x and f_y exist near (a, b) and are continuous at (a, b), then f is differentiable at (a, b).

PROOF Let

$$\Delta z = f(a + \Delta x, b + \Delta y) - f(a, b)$$

According to (14.4.7), to prove that f is differentiable at (a, b) we have to show that we can write Δz in the form

$$\Delta z = f_x(a, b) \, \Delta x + f_y(a, b) \, \Delta y + \varepsilon_1 \, \Delta x + \varepsilon_2 \, \Delta y$$

where ε_1 and $\varepsilon_2 \to 0$ as $(\Delta x, \Delta y) \to (0, 0)$.

Referring to Figure 4, we write

$$\boxed{1} \quad \Delta z = [f(a + \Delta x, b + \Delta y) - f(a, b + \Delta y)] + [f(a, b + \Delta y) - f(a, b)]$$

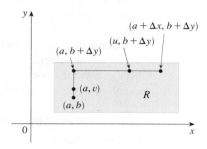

FIGURE 4

Observe that the function of a single variable

$$g(x) = f(x, b + \Delta y)$$

is defined on the interval $[a, a + \Delta x]$ and $g'(x) = f_x(x, b + \Delta y)$. If we apply the Mean Value Theorem to g, we get

$$g(a + \Delta x) - g(a) = g'(u) \, \Delta x$$

where u is some number between a and $a + \Delta x$. In terms of f, this equation becomes

$$f(a + \Delta x, b + \Delta y) - f(a, b + \Delta y) = f_x(u, b + \Delta y) \, \Delta x$$

This gives us an expression for the first part of the right side of Equation 1. For the second part we let $h(y) = f(a, y)$. Then h is a function of a single variable defined on the interval $[b, b + \Delta y]$ and $h'(y) = f_y(a, y)$. A second application of the Mean Value Theorem then gives

$$h(b + \Delta y) - h(b) = h'(v) \, \Delta y$$

where v is some number between b and $b + \Delta y$. In terms of f, this becomes

$$f(a, b + \Delta y) - f(a, b) = f_y(a, v) \, \Delta y$$

We now substitute these expressions into Equation 1 and obtain

$$\Delta z = f_x(u, b + \Delta y) \, \Delta x + f_y(a, v) \, \Delta y$$

$$= f_x(a, b) \, \Delta x + [f_x(u, b + \Delta y) - f_x(a, b)] \, \Delta x + f_y(a, b) \, \Delta y$$

$$+ [f_y(a, v) - f_y(a, b)] \, \Delta y$$

$$= f_x(a, b) \, \Delta x + f_y(a, b) \, \Delta y + \varepsilon_1 \, \Delta x + \varepsilon_2 \, \Delta y$$

where

$$\varepsilon_1 = f_x(u, b + \Delta y) - f_x(a, b)$$

$$\varepsilon_2 = f_y(a, v) - f_y(a, b)$$

Since $(u, b + \Delta y) \to (a, b)$ and $(a, v) \to (a, b)$ as $(\Delta x, \Delta y) \to (0, 0)$ and since f_x and f_y are continuous at (a, b), we see that $\varepsilon_1 \to 0$ and $\varepsilon_2 \to 0$ as $(\Delta x, \Delta y) \to (0, 0)$.

Therefore f is differentiable at (a, b). ∎

G   The Logarithm Defined as an Integral

Our treatment of exponential and logarithmic functions until now has relied on our intuition, which is based on numerical and visual evidence. (See Sections 1.4, 1.5, and 3.1.) Here we use the Fundamental Theorem of Calculus to give an alternative treatment that provides a surer footing for these functions.

Instead of starting with b^x and defining $\log_b x$ as its inverse, this time we start by defining $\ln x$ as an integral and then define the exponential function as its inverse. You should bear in mind that we do not use any of our previous definitions and results concerning exponential and logarithmic functions.

◼ The Natural Logarithm

We first define $\ln x$ as an integral.

> **1 Definition** The **natural logarithmic function** is the function defined by
> $$\ln x = \int_1^x \frac{1}{t}\,dt \qquad x > 0$$

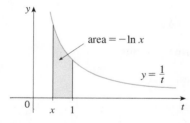

FIGURE 1

The existence of this function depends on the fact that the integral of a continuous function always exists. If $x > 1$, then $\ln x$ can be interpreted geometrically as the area under the hyperbola $y = 1/t$ from $t = 1$ to $t = x$. (See Figure 1.) For $x = 1$, we have

$$\ln 1 = \int_1^1 \frac{1}{t}\,dt = 0$$

For $0 < x < 1$, $\qquad \ln x = \int_1^x \frac{1}{t}\,dt = -\int_x^1 \frac{1}{t}\,dt < 0$

and so $\ln x$ is the negative of the area shown in Figure 2.

FIGURE 2

EXAMPLE 1

(a) By comparing areas, show that $\frac{1}{2} < \ln 2 < \frac{3}{4}$.

(b) Use the Midpoint Rule with $n = 10$ to estimate the value of $\ln 2$.

SOLUTION

(a) We can interpret $\ln 2$ as the area under the curve $y = 1/t$ from 1 to 2. From Figure 3 we see that this area is larger than the area of rectangle $BCDE$ and smaller than the area of trapezoid $ABCD$. Thus we have

$$\tfrac{1}{2} \cdot 1 < \ln 2 < 1 \cdot \tfrac{1}{2}\left(1 + \tfrac{1}{2}\right)$$

$$\tfrac{1}{2} < \ln 2 < \tfrac{3}{4}$$

(b) If we use the Midpoint Rule with $f(t) = 1/t$, $n = 10$, and $\Delta t = 0.1$, we get

$$\ln 2 = \int_1^2 \frac{1}{t}\,dt \approx (0.1)[f(1.05) + f(1.15) + \cdots + f(1.95)]$$

$$= (0.1)\left(\frac{1}{1.05} + \frac{1}{1.15} + \cdots + \frac{1}{1.95}\right) \approx 0.693 \qquad ◼$$

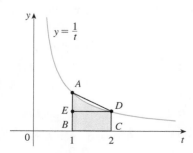

FIGURE 3

Notice that the integral that defines $\ln x$ is exactly the type of integral discussed in Part 1 of the Fundamental Theorem of Calculus (see Section 5.3). In fact, using that theorem, we have

$$\frac{d}{dx} \int_1^x \frac{1}{t}\, dt = \frac{1}{x}$$

and so

2
$$\frac{d}{dx}(\ln x) = \frac{1}{x}$$

We now use this differentiation rule to prove the following properties of the logarithm function.

---

3   Laws of Logarithms If x and y are positive numbers and r is a rational number, then

1. $\ln(xy) = \ln x + \ln y$ 2. $\ln\left(\dfrac{x}{y}\right) = \ln x - \ln y$ 3. $\ln(x^r) = r \ln x$

PROOF

1. Let $f(x) = \ln(ax)$, where a is a positive constant. Then, using Equation 2 and the Chain Rule, we have

$$f'(x) = \frac{1}{ax}\frac{d}{dx}(ax) = \frac{1}{ax}\cdot a = \frac{1}{x}$$

Therefore $f(x)$ and $\ln x$ have the same derivative and so they must differ by a constant:

$$\ln(ax) = \ln x + C$$

Putting $x = 1$ in this equation, we get $\ln a = \ln 1 + C = 0 + C = C$. Thus

$$\ln(ax) = \ln x + \ln a$$

If we now replace the constant a by any number y, we have

$$\ln(xy) = \ln x + \ln y$$

2. Using Law 1 with $x = 1/y$, we have

$$\ln\frac{1}{y} + \ln y = \ln\left(\frac{1}{y}\cdot y\right) = \ln 1 = 0$$

and so

$$\ln\frac{1}{y} = -\ln y$$

Using Law 1 again, we have

$$\ln\left(\frac{x}{y}\right) = \ln\left(x\cdot\frac{1}{y}\right) = \ln x + \ln\frac{1}{y} = \ln x - \ln y$$

The proof of Law 3 is left as an exercise. ■

In order to graph $y = \ln x$, we first determine its limits:

$\boxed{4}$

$$\text{(a) } \lim_{x \to \infty} \ln x = \infty \qquad \text{(b) } \lim_{x \to 0^+} \ln x = -\infty$$

PROOF

(a) Using Law 3 with $x = 2$ and $r = n$ (where n is any positive integer), we have $\ln(2^n) = n \ln 2$. Now $\ln 2 > 0$, so this shows that $\ln(2^n) \to \infty$ as $n \to \infty$. But $\ln x$ is an increasing function since its derivative $1/x > 0$. Therefore $\ln x \to \infty$ as $x \to \infty$.

(b) If we let $t = 1/x$, then $t \to \infty$ as $x \to 0^+$. Thus, using (a), we have

$$\lim_{x \to 0^+} \ln x = \lim_{t \to \infty} \ln\left(\frac{1}{t}\right) = \lim_{t \to \infty} (-\ln t) = -\infty \qquad \blacksquare$$

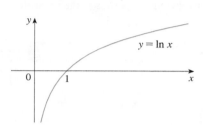

FIGURE 4

If $y = \ln x$, $x > 0$, then

$$\frac{dy}{dx} = \frac{1}{x} > 0 \qquad \text{and} \qquad \frac{d^2 y}{dx^2} = -\frac{1}{x^2} < 0$$

which shows that $\ln x$ is increasing and concave downward on $(0, \infty)$. Putting this information together with (4), we draw the graph of $y = \ln x$ in Figure 4.

Since $\ln 1 = 0$ and $\ln x$ is an increasing continuous function that takes on arbitrarily large values, the Intermediate Value Theorem shows that there is a number where $\ln x$ takes on the value 1. (See Figure 5.) This important number is denoted by e.

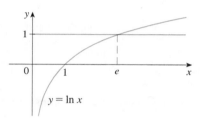

FIGURE 5

$\boxed{5}$ **Definition** e is the number such that $\ln e = 1$.

We will show (in Theorem 19) that this definition is consistent with our previous definition of e.

■ The Natural Exponential Function

Since $\ln$ is an increasing function, it is one-to-one and therefore has an inverse function, which we denote by exp. Thus, according to the definition of an inverse function,

$f^{-1}(x) = y \iff f(y) = x$

$\boxed{6}$

$$\exp(x) = y \iff \ln y = x$$

and the cancellation equations are

$f^{-1}(f(x)) = x$
$f(f^{-1}(x)) = x$

$\boxed{7}$

$$\exp(\ln x) = x \qquad \text{and} \qquad \ln(\exp x) = x$$

In particular, we have

$$\exp(0) = 1 \quad \text{since} \quad \ln 1 = 0$$

$$\exp(1) = e \quad \text{since} \quad \ln e = 1$$

We obtain the graph of $y = \exp x$ by reflecting the graph of $y = \ln x$ about the line

FIGURE 6

FIGURE 7
The natural exponential function

$y = x$. (See Figure 6.) The domain of exp is the range of ln, that is, $(-\infty, \infty)$; the range of exp is the domain of ln, that is, $(0, \infty)$.

If r is any rational number, then the third law of logarithms gives

$$\ln(e^r) = r \ln e = r$$

Therefore, by (6),
$$\exp(r) = e^r$$

Thus $\exp(x) = e^x$ whenever x is a rational number. This leads us to define e^x, even for irrational values of x, by the equation

$$e^x = \exp(x)$$

In other words, for the reasons given, we define e^x to be the inverse of the function $\ln x$. In this notation (6) becomes

8
$$e^x = y \iff \ln y = x$$

and the cancellation equations (7) become

9
$$e^{\ln x} = x \qquad x > 0$$

10
$$\ln(e^x) = x \qquad \text{for all } x$$

The natural exponential function $f(x) = e^x$ is one of the most frequently occurring functions in calculus and its applications, so it is important to be familiar with its graph (Figure 7) and its properties (which follow from the fact that it is the inverse of the natural logarithmic function).

Properties of the Exponential Function The exponential function $f(x) = e^x$ is an increasing continuous function with domain $\mathbb{R}$ and range $(0, \infty)$. Thus $e^x > 0$ for all x. Also
$$\lim_{x \to -\infty} e^x = 0 \qquad \lim_{x \to \infty} e^x = \infty$$
So the x-axis is a horizontal asymptote of $f(x) = e^x$.

We now verify that f has the other properties expected of an exponential function.

11 **Laws of Exponents** If x and y are real numbers and r is rational, then

1. $e^{x+y} = e^x e^y$ **2.** $e^{x-y} = \dfrac{e^x}{e^y}$ **3.** $(e^x)^r = e^{rx}$

PROOF OF LAW 1 Using the first law of logarithms and Equation 10, we have

$$\ln(e^x e^y) = \ln(e^x) + \ln(e^y) = x + y = \ln(e^{x+y})$$

Since ln is a one-to-one function, it follows that $e^x e^y = e^{x+y}$.

Laws 2 and 3 are proved similarly (see Exercises 6 and 7). As we will soon see, Law 3 actually holds when r is any real number. ∎

We now prove the differentiation formula for e^x.

$$\boxed{\frac{d}{dx}(e^x) = e^x}$$

[12]

PROOF The function $y = e^x$ is differentiable because it is the inverse function of $y = \ln x$, which we know is differentiable with nonzero derivative. To find its derivative, we use the inverse function method. Let $y = e^x$. Then $\ln y = x$ and, differentiating this latter equation implicitly with respect to x, we get

$$\frac{1}{y}\frac{dy}{dx} = 1$$

$$\frac{dy}{dx} = y = e^x$$

∎

▧ General Exponential Functions

If $b > 0$ and r is any rational number, then by (9) and (11),

$$b^r = (e^{\ln b})^r = e^{r \ln b}$$

Therefore, even for irrational numbers x, we *define*

$$\boxed{b^x = e^{x \ln b}}$$

[13]

Thus, for instance,

$$2^{\sqrt{3}} = e^{\sqrt{3} \ln 2} \approx e^{1.20} \approx 3.32$$

The function $f(x) = b^x$ is called the **exponential function with base b**. Notice that b^x is positive for all x because e^x is positive for all x.

Definition 13 allows us to extend one of the laws of logarithms. We already know that $\ln(b^r) = r \ln b$ when r is rational. But if we now let r be *any* real number we have, from Definition 13,

$$\ln b^r = \ln(e^{r \ln b}) = r \ln b$$

Thus

[14] $$\ln b^r = r \ln b \qquad \text{for any real number } r$$

The general laws of exponents follow from Definition 13 together with the laws of exponents for e^x.

---

15   Laws of Exponents If x and y are real numbers and $a, b > 0$, then

1. $b^{x+y} = b^x b^y$ **2.** $b^{x-y} = b^x/b^y$ **3.** $(b^x)^y = b^{xy}$ **4.** $(ab)^x = a^x b^x$

PROOF

1. Using Definition 13 and the laws of exponents for e^x, we have

$$b^{x+y} = e^{(x+y)\ln b} = e^{x\ln b + y\ln b}$$

$$= e^{x\ln b}e^{y\ln b} = b^x b^y$$

3. Using Equation 14 we obtain

$$(b^x)^y = e^{y\ln(b^x)} = e^{yx\ln b} = e^{xy\ln b} = b^{xy}$$

The remaining proofs are left as exercises. ∎

The differentiation formula for exponential functions is also a consequence of Definition 13:

16
$$\frac{d}{dx}(b^x) = b^x \ln b$$

PROOF

$$\frac{d}{dx}(b^x) = \frac{d}{dx}(e^{x\ln b}) = e^{x\ln b}\frac{d}{dx}(x\ln b) = b^x \ln b$$ ∎

If $b > 1$, then $\ln b > 0$, so $(d/dx)\,b^x = b^x \ln b > 0$, which shows that $y = b^x$ is increasing (see Figure 8). If $0 < b < 1$, then $\ln b < 0$ and so $y = b^x$ is decreasing (see Figure 9).

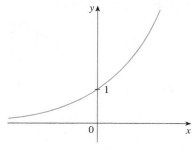

$$\lim_{x\to-\infty} b^x = 0, \ \lim_{x\to\infty} b^x = \infty$$

FIGURE 8 $y = b^x, \ b > 1$

$$\lim_{x\to-\infty} b^x = \infty, \ \lim_{x\to\infty} b^x = 0$$

FIGURE 9 $y = b^x, \ 0 < b < 1$

■ General Logarithmic Functions

If $b > 0$ and $b \neq 1$, then $f(x) = b^x$ is a one-to-one function. Its inverse function is called the **logarithmic function with base b** and is denoted by $\log_b$. Thus

17
$$\log_b x = y \iff b^y = x$$

In particular, we see that

$$\log_e x = \ln x$$

The laws of logarithms are similar to those for the natural logarithm and can be deduced from the laws of exponents (see Exercise 10).

To differentiate $y = \log_b x$, we write the equation as $b^y = x$. From Equation 14 we have $y \ln b = \ln x$, so

$$\log_b x = y = \frac{\ln x}{\ln b}$$

Since $\ln b$ is a constant, we can differentiate as follows:

$$\frac{d}{dx}(\log_b x) = \frac{d}{dx}\frac{\ln x}{\ln b} = \frac{1}{\ln b}\frac{d}{dx}(\ln x) = \frac{1}{x \ln b}$$

[18]
$$\frac{d}{dx}(\log_b x) = \frac{1}{x \ln b}$$

■ The Number e Expressed as a Limit

In this section we defined e as the number such that $\ln e = 1$. The next theorem shows that this is the same as the number e defined in Section 3.1 (see Equation 3.6.5).

[19]
$$e = \lim_{x \to 0}(1 + x)^{1/x}$$

PROOF Let $f(x) = \ln x$. Then $f'(x) = 1/x$, so $f'(1) = 1$. But, by the definition of derivative,

$$f'(1) = \lim_{h \to 0}\frac{f(1 + h) - f(1)}{h} = \lim_{x \to 0}\frac{f(1 + x) - f(1)}{x}$$

$$= \lim_{x \to 0}\frac{\ln(1 + x) - \ln 1}{x} = \lim_{x \to 0}\frac{1}{x}\ln(1 + x) = \lim_{x \to 0}\ln(1 + x)^{1/x}$$

Because $f'(1) = 1$, we have

$$\lim_{x \to 0}\ln(1 + x)^{1/x} = 1$$

Then, by Theorem 2.5.8 and the continuity of the exponential function, we have

$$e = e^1 = e^{\lim_{x \to 0}\ln(1+x)^{1/x}} = \lim_{x \to 0}e^{\ln(1+x)^{1/x}} = \lim_{x \to 0}(1 + x)^{1/x}$$ ■

G   EXERCISES

1. (a) By comparing areas, show that

$$\tfrac{1}{3} < \ln 1.5 < \tfrac{5}{12}$$

(b) Use the Midpoint Rule with $n = 10$ to estimate $\ln 1.5$.

2. Refer to Example 1.
(a) Find the equation of the tangent line to the curve $y = 1/t$ that is parallel to the secant line AD.

(b) Use part (a) to show that $\ln 2 > 0.66$.

3. By comparing areas, show that

$$\frac{1}{2} + \frac{1}{3} + \cdots + \frac{1}{n} < \ln n < 1 + \frac{1}{2} + \frac{1}{3} + \cdots + \frac{1}{n - 1}$$

4. (a) By comparing areas, show that $\ln 2 < 1 < \ln 3$.
(b) Deduce that $2 < e < 3$.

5. Prove the third law of logarithms. [*Hint:* Start by showing that both sides of the equation have the same derivative.]

6. Prove the second law of exponents for e^x [see (11)].

7. Prove the third law of exponents for e^x [see (11)].

8. Prove the second law of exponents [see (15)].

9. Prove the fourth law of exponents [see (15)].

10. Deduce the following laws of logarithms from (15):
    (a) $\log_b(xy) = \log_b x + \log_b y$
    (b) $\log_b(x/y) = \log_b x - \log_b y$
    (c) $\log_b(x^y) = y \log_b x$

H   Complex Numbers

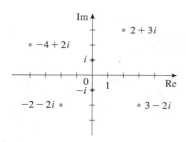

FIGURE 1
Complex numbers as points in the Argand plane

A **complex number** can be represented by an expression of the form $a + bi$, where a and b are real numbers and i is a symbol with the property that $i^2 = -1$. The complex number $a + bi$ can also be represented by the ordered pair (a, b) and plotted as a point in a plane (called the Argand plane) as in Figure 1. Thus the complex number $i = 0 + 1 \cdot i$ is identified with the point $(0, 1)$.

The **real part** of the complex number $a + bi$ is the real number a and the **imaginary part** is the real number b. Thus the real part of $4 - 3i$ is 4 and the imaginary part is -3. Two complex numbers $a + bi$ and $c + di$ are **equal** if $a = c$ and $b = d$, that is, their real parts are equal and their imaginary parts are equal. In the Argand plane the horizontal axis is called the real axis and the vertical axis is called the imaginary axis.

The sum and difference of two complex numbers are defined by adding or subtracting their real parts and their imaginary parts:

$$(a + bi) + (c + di) = (a + c) + (b + d)i$$
$$(a + bi) - (c + di) = (a - c) + (b - d)i$$

For instance,
$$(1 - i) + (4 + 7i) = (1 + 4) + (-1 + 7)i = 5 + 6i$$

The product of complex numbers is defined so that the usual commutative and distributive laws hold:

$$(a + bi)(c + di) = a(c + di) + (bi)(c + di)$$
$$= ac + adi + bci + bdi^2$$

Since $i^2 = -1$, this becomes

$$(a + bi)(c + di) = (ac - bd) + (ad + bc)i$$

EXAMPLE 1

$$(-1 + 3i)(2 - 5i) = (-1)(2 - 5i) + 3i(2 - 5i)$$
$$= -2 + 5i + 6i - 15(-1) = 13 + 11i \qquad \blacksquare$$

Division of complex numbers is much like rationalizing the denominator of a rational expression. For the complex number $z = a + bi$, we define its **complex conjugate** to be $\bar{z} = a - bi$. To find the quotient of two complex numbers we multiply numerator and denominator by the complex conjugate of the denominator.

EXAMPLE 2 Express the number $\dfrac{-1 + 3i}{2 + 5i}$ in the form $a + bi$.

SOLUTION We multiply numerator and denominator by the complex conjugate of $2 + 5i$, namely, $2 - 5i$, and we take advantage of the result of Example 1:

$$\frac{-1 + 3i}{2 + 5i} = \frac{-1 + 3i}{2 + 5i} \cdot \frac{2 - 5i}{2 - 5i} = \frac{13 + 11i}{2^2 + 5^2} = \frac{13}{29} + \frac{11}{29} i$$

The geometric interpretation of the complex conjugate is shown in Figure 2: $\bar{z}$ is the reflection of z in the real axis. We list some of the properties of the complex conjugate in the following box. The proofs follow from the definition and are requested in Exercise 18.

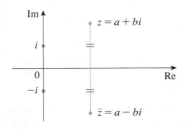

FIGURE 2

Properties of Conjugates

$$\overline{z + w} = \bar{z} + \bar{w} \qquad \overline{zw} = \bar{z}\,\bar{w} \qquad \overline{z^n} = \bar{z}^n$$

The **modulus**, or **absolute value**, $|z|$ of a complex number $z = a + bi$ is its distance from the origin. From Figure 3 we see that if $z = a + bi$, then

$$|z| = \sqrt{a^2 + b^2}$$

FIGURE 3

Notice that

$$z\bar{z} = (a + bi)(a - bi) = a^2 + abi - abi - b^2i^2 = a^2 + b^2$$

and so

$$z\bar{z} = |z|^2$$

This explains why the division procedure in Example 2 works in general:

$$\frac{z}{w} = \frac{z\bar{w}}{w\bar{w}} = \frac{z\bar{w}}{|w|^2}$$

Since $i^2 = -1$, we can think of i as a square root of -1. But notice that we also have $(-i)^2 = i^2 = -1$ and so $-i$ is also a square root of -1. We say that i is the **principal square root** of -1 and write $\sqrt{-1} = i$. In general, if c is any positive number, we write

$$\sqrt{-c} = \sqrt{c}\, i$$

With this convention, the usual derivation and formula for the roots of the quadratic equation $ax^2 + bx + c = 0$ are valid even when $b^2 - 4ac < 0$:

$$x = \frac{-b \pm \sqrt{b^2 - 4ac}}{2a}$$

EXAMPLE 3 Find the roots of the equation $x^2 + x + 1 = 0$.

SOLUTION Using the quadratic formula, we have

$$x = \frac{-1 \pm \sqrt{1^2 - 4 \cdot 1}}{2} = \frac{-1 \pm \sqrt{-3}}{2} = \frac{-1 \pm \sqrt{3}\, i}{2}$$

We observe that the solutions of the equation in Example 3 are complex conjugates of each other. In general, the solutions of any quadratic equation $ax^2 + bx + c = 0$ with real coefficients a, b, and c are always complex conjugates. (If z is real, $\bar{z} = z$, so z is its own conjugate.)

We have seen that if we allow complex numbers as solutions, then every quadratic equation has a solution. More generally, it is true that every polynomial equation

$$a_n x^n + a_{n-1} x^{n-1} + \cdots + a_1 x + a_0 = 0$$

of degree at least one has a solution among the complex numbers. This fact is known as the Fundamental Theorem of Algebra and was proved by Gauss.

■ Polar Form

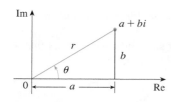

FIGURE 4

We know that any complex number $z = a + bi$ can be considered as a point (a, b) and that any such point can be represented by polar coordinates (r, θ) with $r \geqslant 0$. In fact,

$$a = r \cos \theta \qquad b = r \sin \theta$$

as in Figure 4. Therefore we have

$$z = a + bi = (r \cos \theta) + (r \sin \theta)i$$

Thus we can write any complex number z in the form

$$z = r(\cos \theta + i \sin \theta)$$

where

$$r = |z| = \sqrt{a^2 + b^2} \qquad \text{and} \qquad \tan \theta = \frac{b}{a}$$

The angle θ is called the **argument** of z and we write $\theta = \arg(z)$. Note that $\arg(z)$ is not unique; any two arguments of z differ by an integer multiple of 2π.

EXAMPLE 4 Write the following numbers in polar form.

(a) $z = 1 + i$ (b) $w = \sqrt{3} - i$

SOLUTION

(a) We have $r = |z| = \sqrt{1^2 + 1^2} = \sqrt{2}$ and $\tan \theta = 1$, so we can take $\theta = \pi/4$. Therefore the polar form is

$$z = \sqrt{2}\left(\cos \frac{\pi}{4} + i \sin \frac{\pi}{4}\right)$$

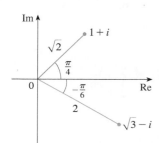

FIGURE 5

(b) Here we have $r = |w| = \sqrt{3 + 1} = 2$ and $\tan \theta = -1/\sqrt{3}$. Since w lies in the fourth quadrant, we take $\theta = -\pi/6$ and

$$w = 2\left[\cos\left(-\frac{\pi}{6}\right) + i \sin\left(-\frac{\pi}{6}\right)\right]$$

The numbers z and w are shown in Figure 5. ∎

The polar form of complex numbers gives insight into multiplication and division. Let

$$z_1 = r_1(\cos\theta_1 + i\sin\theta_1) \qquad z_2 = r_2(\cos\theta_2 + i\sin\theta_2)$$

be two complex numbers written in polar form. Then

$$z_1 z_2 = r_1 r_2(\cos\theta_1 + i\sin\theta_1)(\cos\theta_2 + i\sin\theta_2)$$
$$= r_1 r_2[(\cos\theta_1\cos\theta_2 - \sin\theta_1\sin\theta_2) + i(\sin\theta_1\cos\theta_2 + \cos\theta_1\sin\theta_2)]$$

Therefore, using the addition formulas for cosine and sine, we have

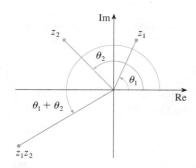

FIGURE 6

$$\boxed{1} \qquad z_1 z_2 = r_1 r_2[\cos(\theta_1 + \theta_2) + i\sin(\theta_1 + \theta_2)]$$

This formula says that *to multiply two complex numbers we multiply the moduli and add the arguments.* (See Figure 6.)

A similar argument using the subtraction formulas for sine and cosine shows that *to divide two complex numbers we divide the moduli and subtract the arguments.*

$$\frac{z_1}{z_2} = \frac{r_1}{r_2}[\cos(\theta_1 - \theta_2) + i\sin(\theta_1 - \theta_2)] \qquad z_2 \neq 0$$

In particular, taking $z_1 = 1$ and $z_2 = z$ (and therefore $\theta_1 = 0$ and $\theta_2 = \theta$), we have the following, which is illustrated in Figure 7.

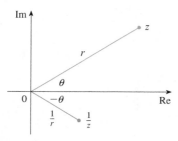

FIGURE 7

$$\text{If}\quad z = r(\cos\theta + i\sin\theta), \quad\text{then}\quad \frac{1}{z} = \frac{1}{r}(\cos\theta - i\sin\theta).$$

EXAMPLE 5 Find the product of the complex numbers $1 + i$ and $\sqrt{3} - i$ in polar form.

SOLUTION From Example 4 we have

$$1 + i = \sqrt{2}\left(\cos\frac{\pi}{4} + i\sin\frac{\pi}{4}\right)$$

and

$$\sqrt{3} - i = 2\left[\cos\left(-\frac{\pi}{6}\right) + i\sin\left(-\frac{\pi}{6}\right)\right]$$

So, by Equation 1,

$$(1 + i)(\sqrt{3} - i) = 2\sqrt{2}\left[\cos\left(\frac{\pi}{4} - \frac{\pi}{6}\right) + i\sin\left(\frac{\pi}{4} - \frac{\pi}{6}\right)\right]$$

$$= 2\sqrt{2}\left(\cos\frac{\pi}{12} + i\sin\frac{\pi}{12}\right)$$

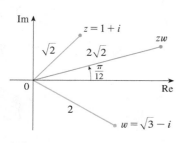

FIGURE 8

This is illustrated in Figure 8.

Repeated use of Formula 1 shows how to compute powers of a complex number. If

$$z = r(\cos\theta + i\sin\theta)$$

then $\qquad\qquad z^2 = r^2(\cos 2\theta + i\sin 2\theta)$

and $\qquad\qquad z^3 = zz^2 = r^3(\cos 3\theta + i\sin 3\theta)$

In general, we obtain the following result, which is named after the French mathematician Abraham De Moivre (1667–1754).

> **2 De Moivre's Theorem** If $z = r(\cos\theta + i\sin\theta)$ and n is a positive integer, then
>
> $$z^n = [r(\cos\theta + i\sin\theta)]^n = r^n(\cos n\theta + i\sin n\theta)$$

This says that *to take the nth power of a complex number we take the nth power of the modulus and multiply the argument by n.*

EXAMPLE 6 Find $\left(\frac{1}{2} + \frac{1}{2}i\right)^{10}$.

SOLUTION Since $\frac{1}{2} + \frac{1}{2}i = \frac{1}{2}(1 + i)$, it follows from Example 4(a) that $\frac{1}{2} + \frac{1}{2}i$ has the polar form

$$\frac{1}{2} + \frac{1}{2}i = \frac{\sqrt{2}}{2}\left(\cos\frac{\pi}{4} + i\sin\frac{\pi}{4}\right)$$

So by De Moivre's Theorem,

$$\left(\frac{1}{2} + \frac{1}{2}i\right)^{10} = \left(\frac{\sqrt{2}}{2}\right)^{10}\left(\cos\frac{10\pi}{4} + i\sin\frac{10\pi}{4}\right)$$

$$= \frac{2^5}{2^{10}}\left(\cos\frac{5\pi}{2} + i\sin\frac{5\pi}{2}\right) = \frac{1}{32}i \qquad\blacksquare$$

De Moivre's Theorem can also be used to find the nth roots of complex numbers. An nth root of the complex number z is a complex number w such that

$$w^n = z$$

Writing these two numbers in trigonometric form as

$$w = s(\cos\phi + i\sin\phi) \qquad \text{and} \qquad z = r(\cos\theta + i\sin\theta)$$

and using De Moivre's Theorem, we get

$$s^n(\cos n\phi + i\sin n\phi) = r(\cos\theta + i\sin\theta)$$

The equality of these two complex numbers shows that

$$s^n = r \qquad \text{or} \qquad s = r^{1/n}$$

and $\qquad\qquad\qquad \cos n\phi = \cos\theta \qquad \text{and} \qquad \sin n\phi = \sin\theta$

From the fact that sine and cosine have period 2π, it follows that

$$n\phi = \theta + 2k\pi \quad \text{or} \quad \phi = \frac{\theta + 2k\pi}{n}$$

Thus

$$w = r^{1/n}\left[\cos\left(\frac{\theta + 2k\pi}{n}\right) + i\sin\left(\frac{\theta + 2k\pi}{n}\right)\right]$$

Since this expression gives a different value of w for $k = 0, 1, 2, \ldots, n - 1$, we have the following.

---

3   Roots of a Complex Number Let $z = r(\cos\theta + i\sin\theta)$ and let n be a positive integer. Then z has the n distinct nth roots

$$w_k = r^{1/n}\left[\cos\left(\frac{\theta + 2k\pi}{n}\right) + i\sin\left(\frac{\theta + 2k\pi}{n}\right)\right]$$

where $k = 0, 1, 2, \ldots, n - 1$.

Notice that each of the nth roots of z has modulus $|w_k| = r^{1/n}$. Thus all the nth roots of z lie on the circle of radius $r^{1/n}$ in the complex plane. Also, since the argument of each successive nth root exceeds the argument of the previous root by $2\pi/n$, we see that the nth roots of z are equally spaced on this circle.

EXAMPLE 7 Find the six sixth roots of $z = -8$ and graph these roots in the complex plane.

SOLUTION In trigonometric form, $z = 8(\cos\pi + i\sin\pi)$. Applying Equation 3 with $n = 6$, we get

$$w_k = 8^{1/6}\left(\cos\frac{\pi + 2k\pi}{6} + i\sin\frac{\pi + 2k\pi}{6}\right)$$

We get the six sixth roots of -8 by taking $k = 0, 1, 2, 3, 4, 5$ in this formula:

$$w_0 = 8^{1/6}\left(\cos\frac{\pi}{6} + i\sin\frac{\pi}{6}\right) = \sqrt{2}\left(\frac{\sqrt{3}}{2} + \frac{1}{2}i\right)$$

$$w_1 = 8^{1/6}\left(\cos\frac{\pi}{2} + i\sin\frac{\pi}{2}\right) = \sqrt{2}\,i$$

$$w_2 = 8^{1/6}\left(\cos\frac{5\pi}{6} + i\sin\frac{5\pi}{6}\right) = \sqrt{2}\left(-\frac{\sqrt{3}}{2} + \frac{1}{2}i\right)$$

$$w_3 = 8^{1/6}\left(\cos\frac{7\pi}{6} + i\sin\frac{7\pi}{6}\right) = \sqrt{2}\left(-\frac{\sqrt{3}}{2} - \frac{1}{2}i\right)$$

$$w_4 = 8^{1/6}\left(\cos\frac{3\pi}{2} + i\sin\frac{3\pi}{2}\right) = -\sqrt{2}\,i$$

$$w_5 = 8^{1/6}\left(\cos\frac{11\pi}{6} + i\sin\frac{11\pi}{6}\right) = \sqrt{2}\left(\frac{\sqrt{3}}{2} - \frac{1}{2}i\right)$$

All these points lie on the circle of radius $\sqrt{2}$ as shown in Figure 9.

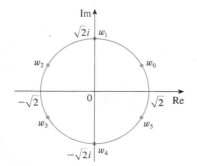

FIGURE 9
The six sixth roots of $z = -8$

■ Complex Exponentials

We also need to give a meaning to the expression e^z when $z = x + iy$ is a complex number. The theory of infinite series as developed in Chapter 11 can be extended to the case where the terms are complex numbers. Using the Taylor series for e^x (11.10.11) as our guide, we define

$$\boxed{4} \qquad e^z = \sum_{n=0}^{\infty} \frac{z^n}{n!} = 1 + z + \frac{z^2}{2!} + \frac{z^3}{3!} + \cdots$$

and it turns out that this complex exponential function has the same properties as the real exponential function. In particular, it is true that

$$\boxed{5} \qquad e^{z_1 + z_2} = e^{z_1} e^{z_2}$$

If we put $z = iy$, where y is a real number, in Equation 4, and use the facts that

$$i^2 = -1, \quad i^3 = i^2 i = -i, \quad i^4 = 1, \quad i^5 = i, \quad \ldots$$

we get
$$e^{iy} = 1 + iy + \frac{(iy)^2}{2!} + \frac{(iy)^3}{3!} + \frac{(iy)^4}{4!} + \frac{(iy)^5}{5!} + \cdots$$

$$= 1 + iy - \frac{y^2}{2!} - i\frac{y^3}{3!} + \frac{y^4}{4!} + i\frac{y^5}{5!} + \cdots$$

$$= \left(1 - \frac{y^2}{2!} + \frac{y^4}{4!} - \frac{y^6}{6!} + \cdots \right) + i\left(y - \frac{y^3}{3!} + \frac{y^5}{5!} - \cdots \right)$$

$$= \cos y + i \sin y$$

Here we have used the Taylor series for $\cos y$ and $\sin y$ (Equations 11.10.16 and 11.10.15). The result is a famous formula called **Euler's formula**:

$$\boxed{6} \qquad \boxed{e^{iy} = \cos y + i \sin y}$$

Combining Euler's formula with Equation 5, we get

$$\boxed{7} \qquad e^{x+iy} = e^x e^{iy} = e^x(\cos y + i \sin y)$$

EXAMPLE 8 Evaluate: (a) $e^{i\pi}$ (b) $e^{-1+i\pi/2}$

SOLUTION

(a) From Euler's equation (6) we have

$$e^{i\pi} = \cos \pi + i \sin \pi = -1 + i(0) = -1$$

(b) Using Equation 7 we get

$$e^{-1+i\pi/2} = e^{-1}\left(\cos \frac{\pi}{2} + i \sin \frac{\pi}{2} \right) = \frac{1}{e}[0 + i(1)] = \frac{i}{e}$$ ■

We could write the result of Example 8(a) as

$$e^{i\pi} + 1 = 0$$

This equation relates the five most famous numbers in all of mathematics: 0, 1, e, i, and π.

Finally, we note that Euler's equation provides us with an easier method of proving De Moivre's Theorem:

$$[r(\cos \theta + i \sin \theta)]^n = (re^{i\theta})^n = r^n e^{in\theta} = r^n(\cos n\theta + i \sin n\theta)$$

H   EXERCISES

1–14 Evaluate the expression and write your answer in the form $a + bi$.

1. $(5 - 6i) + (3 + 2i)$

2. $\left(4 - \frac{1}{2}i\right) - \left(9 + \frac{5}{2}i\right)$

3. $(2 + 5i)(4 - i)$

4. $(1 - 2i)(8 - 3i)$

5. $\overline{12 + 7i}$

6. $\overline{2i\left(\frac{1}{2} - i\right)}$

7. $\dfrac{1 + 4i}{3 + 2i}$

8. $\dfrac{3 + 2i}{1 - 4i}$

9. $\dfrac{1}{1 + i}$

10. $\dfrac{3}{4 - 3i}$

11. i^3

12. i^{100}

13. $\sqrt{-25}$

14. $\sqrt{-3}\,\sqrt{-12}$

15–17 Find the complex conjugate and the modulus of the number.

15. $12 - 5i$

16. $-1 + 2\sqrt{2}\,i$

17. $-4i$

18. Prove the following properties of complex numbers.
   (a) $\overline{z + w} = \bar{z} + \bar{w}$ (b) $\overline{zw} = \bar{z}\,\bar{w}$
   (c) $\overline{z^n} = \bar{z}^n$, where n is a positive integer
   [*Hint:* Write $z = a + bi$, $w = c + di$.]

19–24 Find all solutions of the equation.

19. $4x^2 + 9 = 0$

20. $x^4 = 1$

21. $x^2 + 2x + 5 = 0$

22. $2x^2 - 2x + 1 = 0$

23. $z^2 + z + 2 = 0$

24. $z^2 + \frac{1}{2}z + \frac{1}{4} = 0$

25–28 Write the number in polar form with argument between 0 and 2π.

25. $-3 + 3i$

26. $1 - \sqrt{3}\,i$

27. $3 + 4i$

28. $8i$

29–32 Find polar forms for zw, z/w, and $1/z$ by first putting z and w into polar form.

29. $z = \sqrt{3} + i$, $w = 1 + \sqrt{3}\,i$

30. $z = 4\sqrt{3} - 4i$, $w = 8i$

31. $z = 2\sqrt{3} - 2i$, $w = -1 + i$

32. $z = 4\left(\sqrt{3} + i\right)$, $w = -3 - 3i$

33–36 Find the indicated power using De Moivre's Theorem.

33. $(1 + i)^{20}$

34. $\left(1 - \sqrt{3}\,i\right)^5$

35. $\left(2\sqrt{3} + 2i\right)^5$

36. $(1 - i)^8$

37–40 Find the indicated roots. Sketch the roots in the complex plane.

37. The eighth roots of 1

38. The fifth roots of 32

39. The cube roots of i

40. The cube roots of $1 + i$

41–46 Write the number in the form $a + bi$.

41. $e^{i\pi/2}$

42. $e^{2\pi i}$

43. $e^{i\pi/3}$

44. $e^{-i\pi}$

45. $e^{2+i\pi}$

46. $e^{\pi+i}$

47. Use De Moivre's Theorem with $n = 3$ to express $\cos 3\theta$ and $\sin 3\theta$ in terms of $\cos\theta$ and $\sin\theta$.

48. Use Euler's formula to prove the following formulas for $\cos x$ and $\sin x$:

$$\cos x = \frac{e^{ix} + e^{-ix}}{2} \qquad \sin x = \frac{e^{ix} - e^{-ix}}{2i}$$

49. If $u(x) = f(x) + ig(x)$ is a complex-valued function of a real variable x and the real and imaginary parts $f(x)$ and $g(x)$ are differentiable functions of x, then the derivative of u is defined to be $u'(x) = f'(x) + ig'(x)$. Use this together with Equation 7 to prove that if $F(x) = e^{rx}$, then $F'(x) = re^{rx}$ when $r = a + bi$ is a complex number.

50. (a) If u is a complex-valued function of a real variable, its indefinite integral $\int u(x)\,dx$ is an antiderivative of u. Evaluate

$$\int e^{(1+i)x}\,dx$$

   (b) By considering the real and imaginary parts of the integral in part (a), evaluate the real integrals

$$\int e^x \cos x\,dx \qquad \text{and} \qquad \int e^x \sin x\,dx$$

   (c) Compare with the method used in Example 7.1.4.

Answers to Odd-Numbered Exercises

CHAPTER 1

1. Yes

3. (a) 3   (b) -0.2 (c) 0, 3   (d) -0.8
(e) $[-2, 4], [-1, 3]$ (f) $[-2, 1]$

5. $[-85, 115]$ **7.** No

9. Yes, $[-3, 2], [-3, -2) \cup [-1, 3]$

11. (a) $13.8°C$ (b) 1990   (c) 1910, 2005   (d) $[13.5, 14.5]$

13.

15. (a) 500 MW; 730 MW   (b) 4 AM; noon; yes

17.

19.

21.

23. (a)   (b) $74°F$

25. $12, 16, 3a^2 - a + 2, 3a^2 + a + 2, 3a^2 + 5a + 4,$
$6a^2 - 2a + 4, 12a^2 - 2a + 2, 3a^4 - a^2 + 2,$
$9a^4 - 6a^3 + 13a^2 - 4a + 4, 3a^2 + 6ah + 3h^2 - a - h + 2$

27. $-3 - h$ **29.** $-1/(ax)$

31. $(-\infty, -3) \cup (-3, 3) \cup (3, \infty)$ **33.** $(-\infty, \infty)$

35. $(-\infty, 0) \cup (5, \infty)$ **37.** $[0, 4]$

39. $(-\infty, \infty)$

41. $-1, 1, -1$

43. $-2, 0, 4$

45.

47.

49.

51. $f(x) = \frac{5}{2}x - \frac{11}{2}, 1 \le x \le 5$ **53.** $f(x) = 1 - \sqrt{-x}$

55. $f(x) = \begin{cases} -x + 3 & \text{if } 0 \le x \le 3 \\ 2x - 6 & \text{if } 3 < x \le 5 \end{cases}$

57. $A(L) = 10L - L^2, 0 < L < 10$

59. $A(x) = \sqrt{3}x^2/4, x > 0$ **61.** $S(x) = x^2 + (8/x), x > 0$

63. $V(x) = 4x^3 - 64x^2 + 240x, 0 < x < 6$

65. $F(x) = \begin{cases} 15(40 - x) & \text{if } 0 \le x < 40 \\ 0 & \text{if } 40 \le x \le 65 \\ 15(x - 65) & \text{if } x > 65 \end{cases}$

67. (a)

R (%)

(b) $400, $1900

(c)

T (in dollars)

69. f is odd, g is even   **71.** (a) $(-5, 3)$ (b) $(-5, -3)$
73. Odd   **75.** Neither   **77.** Even
79. Even; odd; neither (unless $f = 0$ or $g = 0$)

EXERCISES 1.2 ▪ **PAGE 33**

1. (a) Logarithmic   (b) Root   (c) Rational
(d) Polynomial, degree 2   (e) Exponential   (f) Trigonometric
3. (a) h (b) f (c) g
5. $\{x \mid x \neq \pi/2 + 2n\pi\}$, n an integer

7. (a) $y = 2x + b$,
where b is the y-intercept.

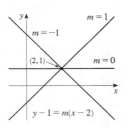

(b) $y = mx + 1 - 2m$,
where m is the slope.

(c) $y = 2x - 3$

9. Their graphs have slope -1.

11. $f(x) = -3x(x + 1)(x - 2)$
13. (a) 8.34, change in mg for every 1 year change
(b) 8.34 mg
15. (a)

$F = \frac{9}{5}C + 32$
(100, 212)
32
$(-40, -40)$

(b) $\frac{9}{5}$, change in °F for every 1°C change; 32, Fahrenheit temperature corresponding to 0°C
17. (a) $T = \frac{1}{6}N + \frac{307}{6}$ (b) $\frac{1}{6}$, change in °F for every chirp per
minute change   (c) 76°F
19. (a) $P = 0.434d + 15$ (b) 196 ft
21. (a) Cosine   (b) Linear
23. (a)

15   61,000

A linear model is
appropriate.

(b) $y = -0.000105x + 14.521$

15
(b)
(c)
0   61,000

(c) $y = -0.00009979x + 13.951$
(d) About 11.5 per 100 population
(e) About 6%   (f) No

25. (a) See graph in part (b).
(b) $y = 1.88074x + 82.64974$

180

Height (cm)

150
35   55
Femur length (cm)

(c) 182.3 cm

27. (a) A linear model is appropriate. See graph in part (b).
(b) $y = 1116.64x + 60{,}188.33$

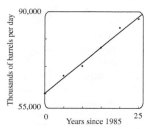

(c) In thousands of barrels per day: 79,171 and 90,338
29. Four times as bright
31. (a) $N = 3.1046A^{0.308}$ (b) 18

EXERCISES 1.3 ■ PAGE 42

1. (a) $y = f(x) + 3$ (b) $y = f(x) - 3$ (c) $y = f(x - 3)$
(d) $y = f(x + 3)$ (e) $y = -f(x)$ (f) $y = f(-x)$
(g) $y = 3f(x)$ (h) $y = \frac{1}{3}f(x)$
3. (a) 3 (b) 1 (c) 4 (d) 5 (e) 2
5. (a) (b)

(c) (d)

7. $y = -\sqrt{-x^2 - 5x - 4} - 1$
9.

11.

13.

15.

17.

19.

21.

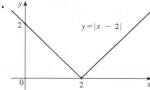

23.

25. $L(t) = 12 + 2 \sin\left[\dfrac{2\pi}{365}(t - 80)\right]$

27. $D(t) = 5 \cos[(\pi/6)(t - 6.75)] + 7$
29. (a) The portion of the graph of $y = f(x)$ to the right of the y-axis is reflected about the y-axis.
(b) (c)

31. (a) $(f + g)(x) = x^3 + 5x^2 - 1, (-\infty, \infty)$
(b) $(f - g)(x) = x^3 - x^2 + 1, (-\infty, \infty)$
(c) $(fg)(x) = 3x^5 + 6x^4 - x^3 - 2x^2, (-\infty, \infty)$
(d) $(f/g)(x) = \dfrac{x^3 + 2x^2}{3x^2 - 1}, \left\{x \mid x \neq \pm\dfrac{1}{\sqrt{3}}\right\}$

33. (a) $(f \circ g)(x) = 3x^2 + 3x + 5, (-\infty, \infty)$
(b) $(g \circ f)(x) = 9x^2 + 33x + 30, (-\infty, \infty)$
(c) $(f \circ f)(x) = 9x + 20, (-\infty, \infty)$
(d) $(g \circ g)(x) = x^4 + 2x^3 + 2x^2 + x, (-\infty, \infty)$

35. (a) $(f \circ g)(x) = \sqrt{4x - 2}, \left[\tfrac{1}{2}, \infty\right)$
(b) $(g \circ f)(x) = 4\sqrt{x + 1} - 3, [-1, \infty)$
(c) $(f \circ f)(x) = \sqrt{\sqrt{x + 1} + 1}, [-1, \infty)$
(d) $(g \circ g)(x) = 16x - 15, (-\infty, \infty)$

37. (a) $(f \circ g)(x) = \dfrac{2x^2 + 6x + 5}{(x + 2)(x + 1)}, \{x \mid x \neq -2, -1\}$
(b) $(g \circ f)(x) = \dfrac{x^2 + x + 1}{(x + 1)^2}, \{x \mid x \neq -1, 0\}$
(c) $(f \circ f)(x) = \dfrac{x^4 + 3x^2 + 1}{x(x^2 + 1)}, \{x \mid x \neq 0\}$
(d) $(g \circ g)(x) = \dfrac{2x + 3}{3x + 5}, \left\{x \mid x \neq -2, -\tfrac{5}{3}\right\}$

39. $(f \circ g \circ h)(x) = 3\sin(x^2) - 2$
41. $(f \circ g \circ h)(x) = \sqrt{x^6 + 4x^3 + 1}$
43. $g(x) = 2x + x^2, f(x) = x^4$
45. $g(x) = \sqrt[3]{x}, f(x) = x/(1 + x)$
47. $g(t) = t^2, f(t) = \sec t \tan t$
49. $h(x) = \sqrt{x}, g(x) = x - 1, f(x) = \sqrt{x}$
51. $h(t) = \cos t, g(t) = \sin t, f(t) = t^2$
53. (a) 4   (b) 3   (c) 0   (d) Does not exist; $f(6) = 6$ is not in the domain of g.   (e) 4   (f) -2
55. (a) $r(t) = 60t$ (b) $(A \circ r)(t) = 3600\pi t^2$; the area of the circle as a function of time
57. (a) $s = \sqrt{d^2 + 36}$ (b) $d = 30t$
(c) $(f \circ g)(t) = \sqrt{900t^2 + 36}$; the distance between the lighthouse and the ship as a function of the time elapsed since noon
59. (a)   (b)

$V(t) = 120H(t)$

(c)

$V(t) = 240H(t - 5)$

61. Yes; $m_1 m_2$
63. (a) $f(x) = x^2 + 6$ (b) $g(x) = x^2 + x - 1$
65. Yes

EXERCISES 1.4 ■ **PAGE 53**

1. (a) 4   (b) $x^{-4/3}$
3. (a) $16b^{12}$ (b) $648y^7$
5. (a) $f(x) = b^x, b > 0$ (b) $\mathbb{R}$ (c) $(0, \infty)$
(d) See Figures 4(c), 4(b), and 4(a), respectively.
7.

All approach 0 as $x \to -\infty$, all pass through $(0, 1)$, and all are increasing. The larger the base, the faster the rate of increase.

9.

The functions with base greater than 1 are increasing and those with base less than 1 are decreasing. The latter are reflections of the former about the y-axis.

11.   **13.**

15.

$y = 20^x$ $y = 5^x$ $y = e^x$ $y = 2^x$

17. (a) $y = e^x - 2$ (b) $y = e^{x-2}$ (c) $y = -e^x$
(d) $y = e^{-x}$ (e) $y = -e^{-x}$
19. (a) $(-\infty, -1) \cup (-1, 1) \cup (1, \infty)$ (b) $(-\infty, \infty)$
21. $f(x) = 3 \cdot 2^x$ **27.** At $x \approx 35.8$
29. (a) See graph in part (c).
(b) $f(t) = 36.89301(1.06614)^t$
(c)

About 10.87 h

31. (a) 25 mg   (b) $200 \cdot 2^{-t/5}$ mg
(c) 10.9 mg   (d) 38.2 days

33. 3.5 days
35. $P = 2614.086(1.01693)^t$; 5381 million; 8466 million

EXERCISES 1.5 ■ PAGE 66

1. (a) See Definition 1.
(b) It must pass the Horizontal Line Test.
3. No   **5.** No   **7.** Yes   **9.** Yes   **11.** No   **13.** No
15. (a) 6   (b) 3   **17.** 0
19. $F = \frac{9}{5}C + 32$; the Fahrenheit temperature as a function of the Celsius temperature; $[-273.15, \infty)$
21. $f^{-1}(x) = \frac{1}{3}(x-1)^2 - \frac{2}{3}, x \ge 1$
23. $f^{-1}(x) = \frac{1}{2}(1 + \ln x)$ **25.** $y = e^x - 3$
27. $f^{-1}(x) = \frac{1}{4}(x^2 - 3), x \ge 0$

29.

31. (a) $f^{-1}(x) = \sqrt{1 - x^2}, 0 \le x \le 1$; f^{-1} and f are the same function.   (b) Quarter-circle in the first quadrant
33. (a) It's defined as the inverse of the exponential function with base b, that is, $\log_b x = y \iff b^y = x$.
(b) $(0, \infty)$ (c) $\mathbb{R}$ (d) See Figure 11.
35. (a) 5   (b) $\frac{1}{3}$ **37.** (a) 2   (b) $\frac{2}{3}$
39. $\ln 250$ **41.** $\ln \dfrac{\sqrt{x}}{x+1}$

43.

All graphs approach $-\infty$ as $x \to 0^+$, all pass through (1, 0), and all are increasing. The larger the base, the slower the rate of increase.

45. About 1,084,588 mi

47. (a)

(b)

49. (a) $(0, \infty)$; $(-\infty, \infty)$ (b) e^{-2}
(c)

51. (a) $\frac{1}{4}(7 - \ln 6)$ (b) $\frac{1}{3}(e^2 + 10)$
53. (a) $5 + \log_2 3$ or $5 + (\ln 3)/\ln 2$ (b) $\frac{1}{2}(1 + \sqrt{1 + 4e})$
55. (a) $0 < x < 1$ (b) $x > \ln 5$
57. (a) $(\ln 3, \infty)$ (b) $f^{-1}(x) = \ln(e^x + 3)$; $\mathbb{R}$
59. The graph passes the Horizontal Line Test.

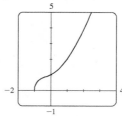

$f^{-1}(x) = -\frac{1}{6}\sqrt[3]{4}\left(\sqrt[3]{D - 27x^2 + 20} - \sqrt[3]{D + 27x^2 - 20} + \sqrt[3]{2}\right)$, where $D = 3\sqrt{3}\sqrt{27x^4 - 40x^2 + 16}$; two of the expressions are complex.
61. (a) $f^{-1}(n) = (3/\ln 2)\ln(n/100)$; the time elapsed when there are n bacteria   (b) After about 26.9 hours
63. (a) π (b) $\pi/6$
65. (a) $\pi/4$ (b) $\pi/2$
67. (a) $5\pi/6$ (b) $\pi/3$
71. $x/\sqrt{1 + x^2}$
73.

The second graph is the reflection of the first graph about the line $y = x$

75. $\left[-\frac{2}{3}, 0\right]$, $[-\pi/2, \pi/2]$
77. (a) $g^{-1}(x) = f^{-1}(x) - c$ (b) $h^{-1}(x) = (1/c)f^{-1}(x)$

CHAPTER 1 REVIEW ■ PAGE 69

True-False Quiz
1. False   **3.** False   **5.** True   **7.** False   **9.** True
11. False   **13.** False

Exercises
1. (a) 2.7   (b) 2.3, 5.6   (c) $[-6, 6]$ (d) $[-4, 4]$
(e) $[-4, 4]$ (f) No; it fails the Horizontal Line Test.
(g) Odd; its graph is symmetric about the origin.
3. $2a + h - 2$ **5.** $\left(-\infty, \frac{1}{3}\right) \cup \left(\frac{1}{3}, \infty\right)$, $(-\infty, 0) \cup (0, \infty)$
7. $(-6, \infty)$, $\mathbb{R}$

9. (a) Shift the graph 8 units upward.
(b) Shift the graph 8 units to the left.
(c) Stretch the graph vertically by a factor of 2, then shift it 1 unit upward.
(d) Shift the graph 2 units to the right and 2 units downward.
(e) Reflect the graph about the x-axis.
(f) Reflect the graph about the line $y = x$ (assuming f is one-to-one).

11.

$y = (x - 2)^3$

13.

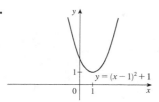

$y = (x - 1)^2 + 1$

15.

$y = -\cos 2x$

17. (a) Neither   (b) Odd   (c) Even   (d) Neither
19. (a) $(f \circ g)(x) = \ln(x^2 - 9), (-\infty, -3) \cup (3, \infty)$
(b) $(g \circ f)(x) = (\ln x)^2 - 9, (0, \infty)$
(c) $(f \circ f)(x) = \ln \ln x, (1, \infty)$
(d) $(g \circ g)(x) = (x^2 - 9)^2 - 9, (-\infty, \infty)$

21. $y = 0.2493x - 423.4818$; about 77.6 years
23. 1   **25.** (a) 9   (b) 2   (c) $1/\sqrt{3}$ (d) $\frac{3}{5}$
27. (a) $\frac{1}{16}$ g   (b) $m(t) = 2^{-t/4}$
(c) $t(m) = -4 \log_2 m$; the time elapsed when there are m grams of ^{100}Pd
(d) About 26.6 days

PRINCIPLES OF PROBLEM SOLVING ■ **PAGE 76**

1. $a = 4\sqrt{h^2 - 16}/h$, where a is the length of the altitude and h is the length of the hypotenuse
3. $-\frac{7}{3}, 9$
5.

7.

9. (a)

$f(x) = \max\{x, 1/x\}$

(b)

$f(x) = \max\{\sin x, \cos x\}$

$-\sqrt{2}/2$

(c)

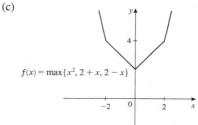

$f(x) = \max\{x^2, 2 + x, 2 - x\}$

11. 5   **13.** $x \in \left[-1, 1 - \sqrt{3}\right) \cup \left(1 + \sqrt{3}, 3\right]$
15. 40 mi/h   **19.** $f_n(x) = x^{2^{n+1}}$

CHAPTER 2

EXERCISES 2.1 ■ **PAGE 82**

1. (a) $-44.4, -38.8, -27.8, -22.2, -16.\overline{6}$
(b) -33.3 (c) $-33\frac{1}{3}$
3. (a) (i) 2   (ii) 1.111111   (iii) 1.010101   (iv) 1.001001
(v) 0.666667   (vi) 0.909091   (vii) 0.990099
(viii) 0.999001   (b) 1   (c) $y = x - 3$
5. (a) (i) -32 ft/s   (ii) -25.6 ft/s   (iii) -24.8 ft/s
(iv) -24.16 ft/s   (b) -24 ft/s
7. (a) (i) 29.3 ft/s   (ii) 32.7 ft/s   (iii) 45.6 ft/s
(iv) 48.75 ft/s   (b) 29.7 ft/s
9. (a) 0, 1.7321, $-1.0847, -2.7433, 4.3301, -2.8173, 0,$
$-2.1651, -2.6061, -5, 3.4202$; no   (c) -31.4

EXERCISES 2.2 ■ **PAGE 92**

1. Yes
3. (a) $\lim_{x \to -3} f(x) = \infty$ means that the values of $f(x)$ can be made arbitrarily large (as large as we please) by taking x sufficiently close to -3 (but not equal to -3).
(b) $\lim_{x \to 4^+} f(x) = -\infty$ means that the values of $f(x)$ can be made arbitrarily large negative by taking x sufficiently close to 4 through values larger than 4.

5. (a) 2   (b) 1   (c) 4   (d) Does not exist   (e) 3

7. (a) −1   (b) −2   (c) Does not exist   (d) 2   (e) 0
(f) Does not exist   (g) 1   (h) 3

9. (a) −∞   (b) ∞   (c) ∞   (d) −∞   (e) ∞
(f) $x = -7, x = -3, x = 0, x = 6$

11. $\lim\limits_{x \to a} f(x)$ exists for all a except $a = -1$.

13. (a) 1   (b) 0   (c) Does not exist

15.   **17.**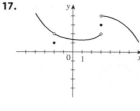

19. $\frac{1}{2}$ **21.** 5   **23.** 0.25   **25.** 1.5   **27.** 1

29. (a) −1.5   **31.** ∞   **33.** ∞   **35.** −∞   **37.** −∞

39. −∞   **41.** ∞   **43.** −∞   **45.** −∞; ∞

47. (a) 2.71828   (b)

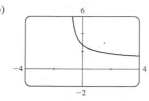

49. (a) 0.998000, 0.638259, 0.358484, 0.158680, 0.038851, 0.008928, 0.001465; 0
(b) 0.000572, −0.000614, −0.000907, −0.000978, −0.000993, −0.001000; −0.001

51. No matter how many times we zoom in toward the origin, the graph appears to consist of almost-vertical lines. This indicates more and more frequent oscillations as $x \to 0$.

53. $x \approx \pm 0.90, \pm 2.24; x = \pm \sin^{-1}(\pi/4), \pm(\pi - \sin^{-1}(\pi/4))$

55. (a) 6   (b) Within 0.0649 of 1

EXERCISES 2.3 ■ PAGE 102

1. (a) −6   (b) −8   (c) 2   (d) −6
(e) Does not exist   (f) 0

3. 105   **5.** $\frac{7}{8}$ **7.** 390   **9.** $\frac{3}{2}$ **11.** 4

13. Does not exist   **15.** $\frac{6}{5}$ **17.** −10   **19.** $\frac{1}{12}$

21. $\frac{1}{6}$ **23.** $-\frac{1}{9}$ **25.** 1   **27.** $\frac{1}{128}$ **29.** $-\frac{1}{2}$

31. $3x^2$ **33.** (a), (b) $\frac{2}{3}$ **37.** 7   **41.** 6   **43.** −4

45. Does not exist

47. (a)   (b) (i) 1
          (ii) −1
          (iii) Does not exist
          (iv) 1

49. (a) (i) 5   (ii) −5   (b) Does not exist
(c)

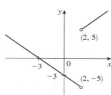

51. 7

53. (a) (i) −2   (ii) Does not exist   (iii) −3
(b) (i) $n - 1$ (ii) n (c) a is not an integer.

59. 8   **65.** 15; −1

EXERCISES 2.4 ■ PAGE 113

1. 0.1 (or any smaller positive number)

3. 1.44 (or any smaller positive number)

5. 0.0906 (or any smaller positive number)

7. 0.0219 (or any positive number);
0.011 (or any smaller positive number)

9. (a) 0.01 (or any smaller positive number)

(b) $\lim\limits_{x \to 2^+} \dfrac{1}{\ln(x - 1)} = \infty$

11. (a) $\sqrt{1000/\pi}$ cm   (b) Within approximately 0.0445 cm
(c) Radius; area; $\sqrt{1000/\pi}$; 1000; 5; ≈0.0445

13. (a) 0.025   (b) 0.0025

35. (a) 0.093   (b) $\delta = (B^{2/3} - 12)/(6B^{1/3}) - 1$, where
$B = 216 + 108\varepsilon + 12\sqrt{336 + 324\varepsilon + 81\varepsilon^2}$

41. Within 0.1

EXERCISES 2.5 ■ PAGE 124

1. $\lim_{x \to 4} f(x) = f(4)$

3. (a) −4, −2, 2, 4; $f(-4)$ is not defined and $\lim\limits_{x \to a} f(x)$ does not exist for $a = -2, 2,$ and 4
(b) −4, neither; −2, left; 2, right; 4, right

5.   **7.**

9. (a)

17. $f(-2)$ is undefined.

19. $\lim\limits_{x \to -1} f(x)$ does not exist.   **21.** $\lim\limits_{x \to 0} f(x) \neq f(0)$

23. Define $f(2) = 3$.   **25.** $(-\infty, \infty)$

27. $\left(-\infty, \sqrt[3]{2}\right) \cup \left(\sqrt[3]{2}, \infty\right)$ **29.** $[-1, 0]$

31. $(-\infty, -1] \cup (0, \infty)$

33. $x = 0$

35. 8   **37.** ln 2

41. -1, right   **43.** 0, right; 1, left

    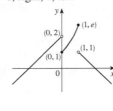

45. $\frac{2}{3}$ **47.** 4

49. (a) $g(x) = x^3 + x^2 + x + 1$ (b) $g(x) = x^2 + x$

57. (b) $(0.86, 0.87)$ **59.** (b) 70.347 **67.** None

69. Yes

1. (a) As x becomes large, $f(x)$ approaches 5.
(b) As x becomes large negative, $f(x)$ approaches 3.
3. (a) -2 (b) 2   (c) ∞ (d) $-\infty$
(e) $x = 1, x = 3, y = -2, y = 2$

5.

7.

9.

11. 0   **13.** $\frac{2}{5}$ **15.** $\frac{3}{2}$ **17.** 0   **19.** -1 **21.** 4

23. -2 **25.** $\frac{\sqrt{3}}{4}$ **27.** $\frac{1}{6}$ **29.** $\frac{1}{2}(a - b)$ **31.** ∞

33. $-\infty$ **35.** $\pi/2$ **37.** $-\frac{1}{2}$ **39.** 0   **41.** ∞

43. (a) (i) 0   (ii) $-\infty$ (iii) ∞ (b) ∞
(c)

45. (a), (b) $-\frac{1}{2}$ **47.** $y = 4, x = -3$

49. $y = 2; x = -2, x = 1$ **51.** $x = 5$ **53.** $y = 3$

55. (a) 0   (b) $\pm\infty$

57. $f(x) = \dfrac{2 - x}{x^2(x - 3)}$ **59.** (a) $\frac{5}{4}$ (b) 5

61. $-\infty, -\infty$ **63.** $-\infty, \infty$

65. (a) 0   (b) An infinite number of times

67. 5

69. (a) v^* (b) 1.2 ≈ 0.47 s

71. $N \geqslant 15$ **73.** $N \leqslant -9, N \leqslant -19$
75. (a) $x > 100$

EXERCISES 2.7 ■ PAGE 148

1. (a) $\dfrac{f(x) - f(3)}{x - 3}$ (b) $\lim\limits_{x \to 3} \dfrac{f(x) - f(3)}{x - 3}$
3. (a) 2 (b) $y = 2x + 1$ (c)

5. $y = -8x + 12$ **7.** $y = \frac{1}{2}x + \frac{1}{2}$
9. (a) $8a - 6a^2$ (b) $y = 2x + 3, y = -8x + 19$
(c)

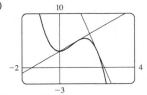

11. (a) Right: $0 < t < 1$ and $4 < t < 6$; left: $2 < t < 3$;
standing still: $1 < t < 2$ and $3 < t < 4$

(b)

13. -24 ft/s
15. $-2/a^3$ m/s; -2 m/s; $-\frac{1}{4}$ m/s; $-\frac{2}{27}$ m/s
17. $g'(0), 0, g'(4), g'(2), g'(-2)$
19. (a) 26 (b) No (c) Yes
21. $f(2) = 3; f'(2) = 4$
23.

25.

27. $y = 3x - 1$

29. (a) $-\frac{3}{5}; y = -\frac{3}{5}x + \frac{16}{5}$ (b)

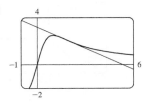

31. $6a - 4$ **33.** $\dfrac{5}{(a + 3)^2}$ **35.** $-\dfrac{1}{\sqrt{1 - 2a}}$
37. $f(x) = \sqrt{x}, a = 9$ **39.** $f(x) = x^6, a = 2$
41. $f(x) = \cos x, a = \pi$ or $f(x) = \cos(\pi + x), a = 0$
43. 32 m/s; 32 m/s
45.

Greater (in magnitude)

47. (a) In (mg/mL)/h: (i) -0.15 (ii) -0.12
(iii) -0.12 (iv) -0.11 (b) -0.12 (mg/mL)/h; After
2 hours, the BAC is decreasing at a rate of 0.12 (mg/mL)/h.

49. (a) 1169.6 thousands of barrels of oil per day per year; oil
consumption rose by an average of 1169.6 thousands of
oil per day each year from 1990 to 2005.
(b) 1397.8 thousands of barrels of oil per day per year

51. (a) (i) \$20.25/unit (ii) \$20.05/unit (b) \$20/unit

53. (a) The rate at which the cost is changing per ounce of gold
produced; dollars per ounce
(b) When the 800th ounce of gold is produced, the cost of
production is \$17/oz.
(c) Decrease in the short term; increase in the long term

55. (a) The rate at which daily heating costs change with respect
to temperature when the temperature is 58°F; dollars/°F
(b) Negative; If the outside temperature increases, the building
should require less heating.

57. (a) The rate at which the oxygen solubility changes with
respect to the water temperature; (mg/L)/°C
(b) $S'(16) \approx -0.25$; as the temperature increases past 16°C,
the oxygen solubility is decreasing at a rate of 0.25 (mg/L)/°C.

59. Does not exist

61. (a)

Slope appears to be 1.

(b)

Yes

(c)

Yes; 0

EXERCISES 2.8 ■ PAGE 160

1. (a) −0.2   (b) 0   (c) 1   (d) 2
(e) 1   (f) 0   (g) −0.2

3. (a) II   (b) IV   (c) I   (d) III
5.

7.

9.

11.

13. (a) The instantaneous rate of change of percentage of full
capacity with respect to elapsed time in hours

(b)

The rate of change of percentage of
full capacity is decreasing and
approaching 0.

15.

1963 to 1971

17.

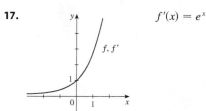

$f'(x) = e^x$

19. (a) 0, 1, 2, 4   (b) −1, −2, −4   (c) $f'(x) = 2x$
21. $f'(x) = 3$, $\mathbb{R}$, $\mathbb{R}$ **23.** $f'(t) = 5t + 6$, $\mathbb{R}$, $\mathbb{R}$
25. $f'(x) = 2x - 6x^2$, $\mathbb{R}$, $\mathbb{R}$

27. $g'(x) = -\dfrac{1}{2\sqrt{9-x}}$, $(-\infty, 9]$, $(-\infty, 9)$

29. $G'(t) = \dfrac{-7}{(3+t)^2}$, $(-\infty, -3) \cup (-3, \infty)$, $(-\infty, -3) \cup (-3, \infty)$

31. $f'(x) = 4x^3$, $\mathbb{R}$, $\mathbb{R}$ **33.** (a) $f'(x) = 4x^3 + 2$
35. (a) The rate at which the unemployment rate is changing, in
percent unemployed per year
(b)

t	$U'(t)$	t	$U'(t)$
2003	−0.50	2008	2.35
2004	−0.45	2009	1.90
2005	−0.45	2010	−0.20
2006	−0.25	2011	−0.75
2007	0.60	2012	−0.80

37.

t	14	21	28	35	42	49
$H'(t)$	$\frac{13}{7}$	$\frac{23}{14}$	$\frac{9}{7}$	1	$\frac{11}{14}$	$\frac{5}{7}$

39. (a) The rate at which the percentage of electrical power produced by solar panels is changing, in percentage points per year.
(b) On January 1, 2002, the percentage of electrical power produced by solar panels was increasing at a rate of 3.5 percentage points per year.
41. -4 (corner); 0 (discontinuity)
43. 1 (not defined); 5 (vertical tangent)
45.

Differentiable at -1; not differentiable at 0

47. $f''(1)$ **49.** $a = f, b = f', c = f''$
51. a = acceleration, b = velocity, c = position
53. $6x + 2$; 6

55.

$f'(x) = 4x - 3x^2$,
$f''(x) = 4 - 6x$,
$f'''(x) = -6$,
$f^{(4)}(x) = 0$

57. (a) $\frac{1}{3}a^{-2/3}$

59. $f'(x) = \begin{cases} -1 & \text{if } x < 6 \\ 1 & \text{if } x > 6 \end{cases}$

or $f'(x) = \dfrac{x - 6}{|x - 6|}$

61. (a)

(b) All x
(c) $f'(x) = 2|x|$

65. (a)

(b)

67. $63°$

CHAPTER 2 REVIEW ■ **PAGE 166**

True-False Quiz
1. False   **3.** True   **5.** True   **7.** False   **9.** True
11. True   **13.** True   **15.** False   **17.** True
19. True   **21.** False   **23.** False   **25.** True

Exercises
1. (a) (i) 3   (ii) 0   (iii) Does not exist   (iv) 2
(v) ∞ (vi) $-\infty$ (vii) 4   (viii) -1
(b) $y = 4, y = -1$ (c) $x = 0, x = 2$ (d) $-3, 0, 2, 4$
3. 1   **5.** $\frac{3}{2}$ **7.** 3   **9.** ∞ **11.** $\frac{4}{7}$ **13.** $\frac{1}{2}$
15. $-\infty$ **17.** 2   **19.** $\pi/2$ **21.** $x = 0, y = 0$ **23.** 1
29. (a) (i) 3   (ii) 0   (iii) Does not exist
(iv) 0   (v) 0   (vi) 0
(b) At 0 and 3   (c)

31. $\mathbb{R}$ **35.** (a) -8 (b) $y = -8x + 17$
37. (a) (i) 3 m/s   (ii) 2.75 m/s   (iii) 2.625 m/s
(iv) 2.525 m/s   (b) 2.5 m/s
39. (a) 10   (b) $y = 10x - 16$
(c)

41. (a) The rate at which the cost changes with respect to the interest rate; dollars/(percent per year)
(b) As the interest rate increases past 10%, the cost is increasing at a rate of $1200/(percent per year).
(c) Always positive
43.

45. (a) $f'(x) = -\frac{5}{2}(3 - 5x)^{-1/2}$ (b) $\left(-\infty, \frac{3}{5}\right], \left(-\infty, \frac{3}{5}\right)$
(c)

47. -4 (discontinuity), -1 (corner), 2 (discontinuity), 5 (vertical tangent)

49.

51. The rate at which the number of US $20 bills in circulation is changing with respect to time; 0.156 billion bills per year
53. 0

PROBLEMS PLUS ■ **PAGE 169**

1. $\frac{2}{3}$ **3.** -4 **5.** (a) Does not exist   (b) 1
7. $a = \frac{1}{2} \pm \frac{1}{2}\sqrt{5}$ **9.** $\frac{3}{4}$ **11.** (b) Yes   (c) Yes; no
13. (a) 0   (b) 1   (c) $f'(x) = x^2 + 1$

CHAPTER 3

EXERCISES 3.1 ■ **PAGE 180**

1. (a) e is the number such that $\lim\limits_{h \to 0} \dfrac{e^h - 1}{h} = 1$.
(b) 0.99, 1.03; $2.7 < e < 2.8$
3. $f'(x) = 0$ **5.** $f'(x) = 5.2$ **7.** $f'(t) = 6t^2 - 6t - 4$
9. $g'(x) = 2x - 6x^2$ **11.** $g'(t) = -\frac{3}{2}t^{-7/4}$
13. $F'(r) = -15/r^4$ **15.** $R'(a) = 18a + 6$
17. $S'(p) = \frac{1}{2}p^{-1/2} - 1$ **19.** $y' = 3e^x - \frac{4}{3}x^{-4/3}$
21. $h'(u) = 3Au^2 + 2Bu + C$
23. $y' = \frac{3}{2}\sqrt{x} + \dfrac{2}{\sqrt{x}} - \dfrac{3}{2x\sqrt{x}}$ **25.** $j'(x) = 2.4x^{1.4}$
27. $G'(q) = -2q^{-2} - 2q^{-3}$ **29.** $f'(v) = -\frac{2}{3}v^{-5/3} - 2e^v$
31. $z' = -10A/y^{11} + Be^y$ **33.** $y = 4x - 1$
35. $y = \frac{1}{2}x + 2$
37. Tangent: $y = 2x + 2$; normal: $y = -\frac{1}{2}x + 2$
39. $y = 3x - 1$ **41.** $f'(x) = 4x^3 - 6x^2 + 2x$
43. (a)   (c) $4x^3 - 9x^2 - 12x + 7$

45. $f'(x) = 0.005x^4 - 0.06x^2$, $f''(x) = 0.02x^3 - 0.12x$
47. $f'(x) = 2 - \frac{15}{4}x^{-1/4}$, $f''(x) = \frac{15}{16}x^{-5/4}$
49. (a) $v(t) = 3t^2 - 3$, $a(t) = 6t$ (b) 12 m/s^2
(c) $a(1) = 6 \text{ m/s}^2$
51. 1.718; instantaneous rate of change of the length with respect to the age at 12 yr
53. (a) $V = 5.3/P$
(b) -0.00212; instantaneous rate of change of the volume with respect to the pressure at 25°C; m^3/kPa
55. $(-2, 21), (1, -6)$ **59.** $y = 3x - 3, y = 3x - 7$
61. $y = -2x + 3$
63. $(\pm 2, 4)$ **67.** $P(x) = x^2 - x + 3$

69. $y = \frac{3}{16}x^3 - \frac{9}{4}x + 3$
71. No

73. (a) Not differentiable at 3 or -3
$$f'(x) = \begin{cases} 2x & \text{if } |x| > 3 \\ -2x & \text{if } |x| < 3 \end{cases}$$
(b)

75. $y = 2x^2 - x$ **77.** $a = -\frac{1}{2}, b = 2$ **79.** $-\frac{1}{3}$
81. $m = 4, b = -4$ **83.** 1000   **85.** 3; 1

EXERCISES 3.2 ■ **PAGE 188**

1. $1 - 2x + 6x^2 - 8x^3$ **3.** $f'(x) = e^x(3x^2 + x - 5)$
5. $y' = \dfrac{1 - x}{e^x}$ **7.** $g'(x) = \dfrac{10}{(3 - 4x)^2}$ **9.** $H'(u) = 2u - 1$
11. $F'(y) = 5 + \dfrac{14}{y^2} + \dfrac{9}{y^4}$ **13.** $y' = \dfrac{x(-x^3 - 3x - 2)}{(x^3 - 1)^2}$
15. $y' = \dfrac{t^4 - 8t^3 + 6t^2 + 9}{(t^2 - 4t + 3)^2}$
17. $y' = e^p\left(1 + \frac{3}{2}\sqrt{p} + p + p\sqrt{p}\right)$ **19.** $y' = \dfrac{3 - 2\sqrt{s}}{2s^{5/2}}$
21. $f'(t) = \dfrac{-2t - 3}{3t^{2/3}(t - 3)^2}$ **23.** $f'(x) = \dfrac{xe^x(x^3 + 2e^x)}{(x^2 + e^x)^2}$
25. $f'(x) = \dfrac{2cx}{(x^2 + c)^2}$
27. $(x^3 + 3x^2 + 1)e^x; (x^3 + 6x^2 + 6x + 1)e^x$
29. $\dfrac{x(2 + 2e^x - xe^x)}{(1 + e^x)^2}$;
$\dfrac{2 + 4e^x - 4xe^x - x^2e^x + 2e^{2x} - 4xe^{2x} + x^2e^{2x}}{(1 + e^x)^3}$
31. $y = \frac{2}{3}x - \frac{2}{3}$ **33.** $y = 2x; y = -\frac{1}{2}x$
35. (a) $y = \frac{1}{2}x + 1$ (b)

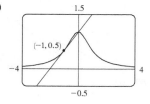

37. (a) $e^x(x^3 + 3x^2 - x - 1)$
(b)

39. (a) $f'(x) = \dfrac{4x}{(x^2 + 1)^2}$; $f''(x) = \dfrac{4(1 - 3x^2)}{(x^2 + 1)^3}$
(b)

41. $\frac{1}{4}$ **43.** (a) -16 (b) $-\frac{20}{9}$ (c) 20   **45.** 7
47. $y = -2x + 18$ **49.** (a) 0   (b) $-\frac{2}{3}$
51. (a) $y' = xg'(x) + g(x)$ (b) $y' = \dfrac{g(x) - xg'(x)}{[g(x)]^2}$
(c) $y' = \dfrac{xg'(x) - g(x)}{x^2}$
53. Two, $\left(-2 \pm \sqrt{3}, \frac{1}{2}(1 \mp \sqrt{3})\right)$ **55.** 1
57. \$1.627 billion/year
59. $\dfrac{0.0021}{(0.015 + [S])^2}$;
The rate of change of the rate of an enzymatic reaction with respect to the concentration of a substrate S.
61. (c) $3e^{3x}$
63. $f'(x) = (x^2 + 2x)e^x$, $f''(x) = (x^2 + 4x + 2)e^x$,
$f'''(x) = (x^2 + 6x + 6)e^x$, $f^{(4)}(x) = (x^2 + 8x + 12)e^x$,
$f^{(5)}(x) = (x^2 + 10x + 20)e^x$; $f^{(n)}(x) = [x^2 + 2nx + n(n-1)]e^x$

EXERCISES 3.3 ■ PAGE 196

1. $f'(x) = x^2 \cos x + 2x \sin x$ **3.** $f'(x) = e^x(\cos x - \sin x)$
5. $y' = \sec\theta\,(\sec^2\theta + \tan^2\theta)$
7. $y' = -c \sin t + t(t \cos t + 2 \sin t)$
9. $y' = \dfrac{2 - \tan x + x \sec^2 x}{(2 - \tan x)^2}$ **11.** $f'(\theta) = \dfrac{1}{1 + \cos\theta}$
13. $y' = \dfrac{(t^2 + t)\cos t + \sin t}{(1 + t)^2}$
15. $f'(\theta) = \frac{1}{2}\sin 2\theta + \theta \cos 2\theta$
21. $y = x + 1$ **23.** $y = x - \pi - 1$
25. (a) $y = 2x$ (b)

27. (a) $\sec x \tan x - 1$
29. $\theta \cos\theta + \sin\theta$; $2 \cos\theta - \theta \sin\theta$

31. (a) $f'(x) = (1 + \tan x)/\sec x$ (b) $f'(x) = \cos x + \sin x$
33. $(2n + 1)\pi \pm \frac{1}{3}\pi$, n an integer
35. (a) $v(t) = 8 \cos t$, $a(t) = -8 \sin t$
(b) $4\sqrt{3}, -4, -4\sqrt{3}$; to the left
37. 5 ft/rad   **39.** $\frac{5}{3}$ **41.** 3   **43.** $-\frac{3}{4}$
45. $\frac{1}{2}$ **47.** $-\frac{1}{4}$ **49.** $-\sqrt{2}$ **51.** $-\cos x$
53. $A = -\frac{3}{10}$, $B = -\frac{1}{10}$
55. (a) $\sec^2 x = \dfrac{1}{\cos^2 x}$ (b) $\sec x \tan x = \dfrac{\sin x}{\cos^2 x}$
(c) $\cos x - \sin x = \dfrac{\cot x - 1}{\csc x}$
57. 1

EXERCISES 3.4 ■ PAGE 204

1. $\dfrac{4}{3\sqrt[3]{(1 + 4x)^2}}$ **3.** $\pi \sec^2 \pi x$ **5.** $\dfrac{e^{\sqrt{x}}}{2\sqrt{x}}$
7. $F'(x) = 24x^{11}(5x^3 + 2)^3(5x^3 + 1)$
9. $f'(x) = \dfrac{5}{2\sqrt{5x + 1}}$ **11.** $f'(\theta) = -2\theta \sin(\theta^2)$
13. $y' = xe^{-3x}(2 - 3x)$ **15.** $f'(t) = e^{at}(b \cos bt + a \sin bt)$
17. $f'(x) = (2x - 3)^3(x^2 + x + 1)^4(28x^2 - 12x - 7)$
19. $h'(t) = \frac{2}{3}(t + 1)^{-1/3}(2t^2 - 1)^2(20t^2 + 18t - 1)$
21. $y' = \dfrac{1}{2\sqrt{x}\,(x + 1)^{3/2}}$ **23.** $y' = (\sec^2\theta)\,e^{\tan\theta}$
25. $g'(u) = \dfrac{48u^2(u^3 - 1)^7}{(u^3 + 1)^9}$ **27.** $r'(t) = \dfrac{(\ln 10)10^{2\sqrt{t}}}{\sqrt{t}}$
29. $H'(r) = \dfrac{2(r^2 - 1)^2(r^2 + 3r + 5)}{(2r + 1)^6}$
31. $F'(t) = e^{t \sin 2t}(2t \cos 2t + \sin 2t)$
33. $G'(x) = -C(\ln 4)\dfrac{4^{C/x}}{x^2}$
35. $y' = \dfrac{4e^{2x}}{(1 + e^{2x})^2}\sin\dfrac{1 - e^{2x}}{1 + e^{2x}}$
37. $y' = -2 \cos\theta \cot(\sin\theta) \csc^2(\sin\theta)$
39. $f'(t) = -\sec^2(\sec(\cos t))\sec(\cos t)\tan(\cos t)\sin t$
41. $f'(t) = 4 \sin(e^{\sin^2 t})\cos(e^{\sin^2 t})e^{\sin^2 t}\sin t \cos t$
43. $g'(x) = 2r^2 p(\ln a)(2ra^{rx} + n)^{p-1}a^{rx}$
45. $y' = \dfrac{-\pi \cos(\tan \pi x)\sec^2(\pi x)\sin\sqrt{\sin(\tan \pi x)}}{2\sqrt{\sin(\tan \pi x)}}$
47. $y' = -3 \cos 3\theta \sin(\sin 3\theta)$;
$y'' = -9 \cos^2(3\theta)\cos(\sin 3\theta) + 9(\sin 3\theta)\sin(\sin 3\theta)$
49. $y' = \dfrac{-\sec t \tan t}{2\sqrt{1 - \sec t}}$;
$y'' = \dfrac{\sec t\,(3 \sec^3 t - 4 \sec^2 t - \sec t + 2)}{4(1 - \sec t)^{3/2}}$
51. $y = (\ln 2)x + 1$ **53.** $y = -x + \pi$
55. (a) $y = \frac{1}{2}x + 1$ (b)

57. (a) $f'(x) = \dfrac{2 - 2x^2}{\sqrt{2 - x^2}}$

59. $((\pi/2) + 2n\pi, 3), ((3\pi/2) + 2n\pi, -1)$, n an integer

61. 24   **63.** (a) 30   (b) 36

65. (a) $\frac{3}{4}$ (b) Does not exist   (c) -2 **67.** $-\frac{1}{6}\sqrt{2}$

69. (a) $F'(x) = e^x f'(e^x)$ (b) $G'(x) = e^{f(x)} f'(x)$

71. 120   **73.** 96

77. $-2^{50} \cos 2x$ **79.** $v(t) = \frac{5}{2}\pi \cos(10\pi t)$ cm/s

81. (a) $\dfrac{dB}{dt} = \dfrac{7\pi}{54} \cos \dfrac{2\pi t}{5.4}$ (b) 0.16

83. $v(t) = 2e^{-1.5t}(2\pi \cos 2\pi t - 1.5 \sin 2\pi t)$

85. (a) 0.0075 (mg/mL)/min   (b) 0.0030 (mg/mL)/min

87. dv/dt is the rate of change of velocity with respect to time; dv/ds is the rate of change of velocity with respect to displacement

89. (a) $Q = ab^t$ where $a \approx 100.01244$ and $b \approx 0.000045146$

(b) -670.63 μA

91. (b) The factored form   **95.** (b) $-n \cos^{n-1}x \sin[(n + 1)x]$

EXERCISES 3.5 ■ PAGE 215

1. (a) $y' = 9x/y$ (b) $y = \pm\sqrt{9x^2 - 1}$, $y' = \pm 9x/\sqrt{9x^2 - 1}$

3. (a) $y' = -\sqrt{y}/\sqrt{x}$ (b) $y = (1 - \sqrt{x})^2$, $y' = 1 - 1/\sqrt{x}$

5. $y' = \dfrac{2y - x}{y - 2x}$ **7.** $y' = -\dfrac{2x(2x^2 + y^2)}{y(2x^2 + 3y)}$

9. $y' = \dfrac{x(x + 2y)}{2x^2y + 4xy^2 + 2y^3 + x^2}$ **11.** $y' = \dfrac{2x + y\sin x}{\cos x - 2y}$

13. $y' = \dfrac{1 - 8x^3\sqrt{x + y}}{8y^3\sqrt{x + y} - 1}$ **15.** $y' = \dfrac{y(y - e^{x/y})}{y^2 - xe^{x/y}}$

17. $y' = \dfrac{1 + x^4y^2 + y^2 + x^4y^4 - 2xy}{x^2 - 2xy - 2x^5y^3}$

19. $y' = -\dfrac{y\cos(xy) + \sin(x + y)}{x\cos(xy) + \sin(x + y)}$ **21.** $-\frac{16}{13}$

23. $x' = \dfrac{-2x^4y + x^3 - 6xy^2}{4x^3y^2 - 3x^2y + 2y^3}$ **25.** $y = \frac{1}{2}x$

27. $y = \frac{3}{4}x - \frac{1}{2}$ **29.** $y = x + \frac{1}{2}$ **31.** $y = -\frac{9}{13}x + \frac{40}{13}$

33. (a) $y = \frac{9}{2}x - \frac{5}{2}$ (b)

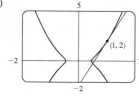

35. $-1/(4y^3)$ **37.** $\dfrac{\cos^2 y \cos x + \sin^2 x \sin y}{\cos^3 y}$ **39.** $1/e^2$

41. (a)

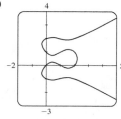

Eight; $x \approx 0.42$, 1.58

(b) $y = -x + 1$, $y = \frac{1}{3}x + 2$ (c) $1 \mp \frac{1}{3}\sqrt{3}$

43. $\left(\pm\frac{5}{4}\sqrt{3}, \pm\frac{5}{4}\right)$ **45.** $(x_0 x/a^2) - (y_0 y/b^2) = 1$

49. $y' = \dfrac{2\tan^{-1}x}{1 + x^2}$ **51.** $y' = \dfrac{1}{\sqrt{-x^2 - x}}$

53. $F'(x) = \dfrac{3}{\sqrt{x^6 - 1}} + \sec^{-1}(x^3)$ **55.** $h'(t) = 0$

57. $y' = \sin^{-1}x$ **59.** $y' = \dfrac{\sqrt{a^2 - b^2}}{a + b\cos x}$

61. $1 - \dfrac{x\arcsin x}{\sqrt{1 - x^2}}$

65.   **67.**

71. (a) $\dfrac{V^3(nb - V)}{PV^3 - n^2aV + 2n^3ab}$ (b) -4.04 L/atm

73. $(\pm\sqrt{3}, 0)$ **75.** $(-1, -1), (1, 1)$ **77.** (b) $\frac{3}{2}$

79. (a) 0   (b) $-\frac{1}{2}$

EXERCISES 3.6 ■ PAGE 223

1. The differentiation formula is simplest.

3. $f'(x) = \dfrac{\cos(\ln x)}{x}$ **5.** $f'(x) = -\dfrac{1}{x}$

7. $f'(x) = \dfrac{-\sin x}{(1 + \cos x)\ln 10}$ **9.** $g'(x) = \dfrac{1}{x} - 2$

11. $F'(t) = \ln t\left(\ln t \cos t + \dfrac{2\sin t}{t}\right)$

13. $G'(y) = \dfrac{10}{2y + 1} - \dfrac{y}{y^2 + 1}$ **15.** $F'(s) = \dfrac{1}{s\ln s}$

17. $T'(z) = 2^z\left(\dfrac{1}{z\ln 2} + \ln z\right)$

19. $y' = \dfrac{-x}{1 + x}$ **21.** $y' = \sec^2[\ln(ax + b)]\dfrac{a}{ax + b}$

23. $y' = (2 + \ln x)/(2\sqrt{x})$; $y'' = -\ln x/(4x\sqrt{x})$

25. $y' = \tan x$; $y'' = \sec^2 x$

27. $f'(x) = \dfrac{2x - 1 - (x - 1)\ln(x - 1)}{(x - 1)[1 - \ln(x - 1)]^2}$;

$(1, 1 + e) \cup (1 + e, \infty)$

29. $f'(x) = \dfrac{2(x - 1)}{x(x - 2)}$; $(-\infty, 0) \cup (2, \infty)$ **31.** 2

33. $y = 3x - 9$ **35.** $\cos x + 1/x$ **37.** 7

39. $y' = (x^2 + 2)^2 (x^4 + 4)^4 \left(\dfrac{4x}{x^2 + 2} + \dfrac{16x^3}{x^4 + 4} \right)$

41. $y' = \sqrt{\dfrac{x - 1}{x^4 + 1}} \left(\dfrac{1}{2x - 2} - \dfrac{2x^3}{x^4 + 1} \right)$

43. $y' = x^x (1 + \ln x)$

45. $y' = x^{\sin x} \left(\dfrac{\sin x}{x} + \cos x \ln x \right)$

47. $y' = (\cos x)^x (-x \tan x + \ln \cos x)$

49. $y' = (\tan x)^{1/x} \left(\dfrac{\sec^2 x}{x \tan x} - \dfrac{\ln \tan x}{x^2} \right)$

51. $y' = \dfrac{2x}{x^2 + y^2 - 2y}$ **53.** $f^{(n)}(x) = \dfrac{(-1)^{n-1}(n - 1)!}{(x - 1)^n}$

EXERCISES 3.7 ■ PAGE 233

1. (a) $3t^2 - 16t + 24$ (b) 11 ft/s   (c) Never   (d) Always
(e) 72 ft
(f)

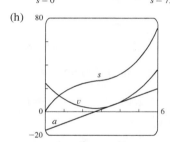

$t = 0$ $t = 6$ s
$s = 0$ $s = 72$

(g) $6t - 16$; -10 ft/s²

(h)

(i) Speeding up when $t > \frac{8}{3}$; slowing down when $0 \le t < \frac{8}{3}$

3. (a) $(\pi/2) \cos(\pi t/2)$ (b) 0 ft/s
(c) $t = 2n + 1$, t a nonnegative integer
(d) $0 < t < 1$, $3 < t < 5$, $7 < t < 9$, and so on   (e) 6 ft
(f)

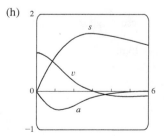

$t = 6$ (g) $(-\pi^2/4) \sin(\pi t/2)$;
$s = 1.2$ $-\pi^2/4$ ft/s²

$t = 0$ $t = 3$ s
$s = 0$ $s = 1.5$

(h)

(i) Speeding up when $1 < t < 2$, $3 < t < 4$, and $5 < t < 6$;
slowing down when $0 < t < 1$, $2 < t < 3$, and $4 < t < 5$
5. (a) Speeding up when $0 < t < 1$ or $2 < t < 3$;
slowing down when $1 < t < 2$

(b) Speeding up when $1 < t < 2$ or $3 < t < 4$;
slowing down when $0 < t < 1$ or $2 < t < 3$
7. (a) 4.9 m/s; -14.7 m/s   (b) After 2.5 s   (c) $32\frac{5}{8}$ m
(d) ≈ 5.08 s   (e) ≈ -25.3 m/s
9. (a) 7.56 m/s   (b) 6.24 m/s; -6.24 m/s
11. (a) 30 mm²/mm; the rate at which the area is increasing
with respect to side length as x reaches 15 mm
(b) $\Delta A \approx 2x\,\Delta x$
13. (a) (i) 5π (ii) 4.5π (iii) 4.1π
(b) 4π (c) $\Delta A \approx 2\pi r\,\Delta r$
15. (a) 8π ft²/ft   (b) 16π ft²/ft   (c) 24π ft²/ft
The rate increases as the radius increases.
17. (a) 6 kg/m   (b) 12 kg/m   (c) 18 kg/m
At the right end; at the left end
19. (a) 4.75 A   (b) 5 A; $t = \frac{2}{3}$ s
23. (a) $dV/dP = -C/P^2$ (b) At the beginning
25. $400(3^t) \ln 3$; ≈ 6850 bacteria/h
27. (a) 16 million/year; 78.5 million/year
(b) $P(t) = at^3 + bt^2 + ct + d$, where $a \approx -0.0002849$,
$b \approx 0.5224331$, $c \approx -6.395641$, $d \approx 1720.586$
(c) $P'(t) = 3at^2 + 2bt + c$
(d) 14.16 million/year (smaller); 71.72 million/year (smaller)
(e) $f'(t) = (1.43653 \times 10^9) \cdot (1.01395)^t \ln 1.01395$
(f) 26.25 million/year (larger); 60.28 million/year (smaller)
(g) $P'(85) \approx 76.24$ million/year, $f'(85) = 64.61$ million/year
29. (a) 0.926 cm/s; 0.694 cm/s; 0
(b) 0; -92.6 (cm/s)/cm; -185.2 (cm/s)/cm
(c) At the center; at the edge
31. (a) $C'(x) = 3 + 0.02x + 0.0006x^2$
(b) $11/pair; the rate at which the cost is changing as the 100th
pair of jeans is being produced; the cost of the 101st pair
(c) $11.07
33. (a) $[xp'(x) - p(x)]/x^2$; the average productivity increases as
new workers are added.

35. $\dfrac{dt}{dc} = \dfrac{3\sqrt{9c^2 - 8c} + 9c - 4}{\sqrt{9c^2 - 8c}\,\left(3c + \sqrt{9c^2 - 8c}\,\right)}$; the rate of change

of duration of dialysis required with respect to the initial urea
concentration
37. -0.2436 K/min
39. (a) 0 and 0   (b) $C = 0$
(c) $(0, 0)$, $(500, 50)$; it is possible for the species to coexist.

EXERCISES 3.8 ■ PAGE 242

1. About 235
3. (a) $100(4.2)^t$ (b) ≈ 7409 (c) $\approx 10{,}632$ bacteria/h
(d) $(\ln 100)/(\ln 4.2) \approx 3.2$ h
5. (a) 1508 million, 1871 million   (b) 2161 million
(c) 3972 million; wars in the first half of century, increased life
expectancy in second half
7. (a) $Ce^{-0.0005t}$ (b) $-2000 \ln 0.9 \approx 211$ s
9. (a) $100 \times 2^{-t/30}$ mg   (b) ≈ 9.92 mg   (c) ≈ 199.3 years
11. ≈ 2500 years   **13.** Yes; 12.5 billion years
15. (a) $\approx 137°$F   (b) ≈ 116 min
17. (a) $13.\overline{3}°$C   (b) ≈ 67.74 min
19. (a) ≈ 64.5 kPa   (b) ≈ 39.9 kPa

21. (a) (i) \$3828.84   (ii) \$3840.25   (iii) \$3850.08
(iv) \$3851.61   (v) \$3852.01   (vi) \$3852.08
(b) $dA/dt = 0.05A$, $A(0) = 3000$

EXERCISES 3.9 ■ PAGE 249

1. $dV/dt = 3x^2 \, dx/dt$ **3.** $48 \text{ cm}^2/\text{s}$ **5.** $3/(25\pi) \text{ m/min}$
7. $128\pi \text{ cm}^2/\text{min}$ **9.** (a) 1   (b) 25   **11.** -18
13. (a) The plane's altitude is 1 mi and its speed is 500 mi/h.
(b) The rate at which the distance from the plane to the station is
increasing when the plane is 2 mi from the station
(c)   (d) $y^2 = x^2 + 1$
(e) $250\sqrt{3} \text{ mi/h}$

15. (a) The height of the pole (15 ft), the height of the man (6 ft),
and the speed of the man (5 ft/s)
(b) The rate at which the tip of the man's shadow is moving when
he is 40 ft from the pole
(c)   (d) $\dfrac{15}{6} = \dfrac{x + y}{y}$ (e) $\frac{25}{3} \text{ ft/s}$

17. 65 mi/h   **19.** $837/\sqrt{8674} \approx 8.99 \text{ ft/s}$
21. -1.6 cm/min **23.** $\frac{720}{13} \approx 55.4 \text{ km/h}$
25. $(10{,}000 + 800{,}000\pi/9) \approx 2.89 \times 10^5 \text{ cm}^3/\text{min}$
27. $\frac{10}{3} \text{ cm/min}$ **29.** $6/(5\pi) \approx 0.38 \text{ ft/min}$
31. $150\sqrt{3} \text{ cm}^2/\text{min}$ **33.** 5 m   **35.** $\pi r^2/60 \text{ cm}^2/\text{min}$
37. $80 \text{ cm}^3/\text{min}$ **39.** $\frac{107}{810} \approx 0.132 \, \Omega/\text{s}$
41. $\sqrt{7}\,\pi/21 \approx 0.396 \text{ m/min}$
43. (a) 360 ft/s   (b) 0.096 rad/s
45. $\frac{10}{9}\pi \text{ km/min}$ **47.** $1650/\sqrt{31} \approx 296 \text{ km/h}$
49. $\frac{7}{4}\sqrt{15} \approx 6.78 \text{ m/s}$

EXERCISES 3.10 ■ PAGE 256

1. $L(x) = 16x + 23$ **3.** $L(x) = \frac{1}{4}x + 1$
5. $\sqrt{1 - x} \approx 1 - \frac{1}{2}x$;
$\sqrt{0.9} \approx 0.95$,
$\sqrt{0.99} \approx 0.995$

7. $-0.383 < x < 0.516$ **9.** $-0.368 < x < 0.677$
11. (a) $dy = (1 - 4x)e^{-4x}\, dx$ (b) $dy = -\dfrac{2t^3}{\sqrt{1 - t^4}}\, dt$
13. (a) $dy = \dfrac{\sec^2\sqrt{t}}{2\sqrt{t}}\, dt$ (b) $dy = \dfrac{-4v}{(1 + v^2)^2}\, dv$
15. (a) $dy = \frac{1}{10}e^{x/10}\, dx$ (b) 0.01
17. (a) $dy = \dfrac{x}{\sqrt{3 + x^2}}\, dx$ (b) -0.05

19. $\Delta y = 1.25$, $dy = 1$

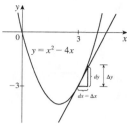

21. $\Delta y \approx 0.34$, $dy = 0.4$

23. 15.968   **25.** $10.00\overline{3}$ **27.** 1.1
33. (a) 270 cm^3, 0.01, 1%   (b) 36 cm^2, $0.00\overline{6}$, $0.\overline{6}\%$
35. (a) $84/\pi \approx 27 \text{ cm}^2$; $\frac{1}{84} \approx 0.012 = 1.2\%$
(b) $1764/\pi^2 \approx 179 \text{ cm}^3$; $\frac{1}{56} \approx 0.018 = 1.8\%$
37. (a) $2\pi rh\, \Delta r$ (b) $\pi(\Delta r)^2 h$
43. (a) 4.8, 5.2   (b) Too large

EXERCISES 3.11 ■ PAGE 264

1. (a) 0   (b) 1   **3.** (a) $\frac{13}{5}$ (b) $\frac{1}{2}(e^5 + e^{-5}) \approx 74.20995$
5. (a) 1   (b) 0
21. $\operatorname{sech} x = \frac{3}{5}$, $\sinh x = \frac{4}{3}$, $\operatorname{csch} x = \frac{3}{4}$, $\tanh x = \frac{4}{5}$, $\coth x = \frac{5}{4}$
23. (a) 1   (b) -1 (c) ∞ (d) $-\infty$ (e) 0   (f) 1
(g) ∞ (h) $-\infty$ (i) 0   (j) $\frac{1}{2}$
31. $f'(x) = \dfrac{\operatorname{sech}^2\sqrt{x}}{2\sqrt{x}}$ **33.** $h'(x) = 2x\cosh(x^2)$
35. $G'(t) = \dfrac{t^2 + 1}{2t^2}$
37. $y' = 3e^{\cosh 3x}\sinh 3x$
39. $g'(t) = \coth\sqrt{t^2 + 1} - \dfrac{t^2}{\sqrt{t^2 + 1}}\operatorname{csch}^2\sqrt{t^2 + 1}$
41. $y' = \dfrac{1}{2\sqrt{x(x - 1)}}$
43. $y' = \sinh^{-1}(x/3)$ **45.** $y' = -\csc x$
51. (a) 0.3572   (b) 70.34°
53. (a) 164.50 m   (b) 120 m; 164.13 m
55. (b) $y = 2\sinh 3x - 4\cosh 3x$
57. $\left(\ln\left(1 + \sqrt{2}\right), \sqrt{2}\right)$

CHAPTER 3 REVIEW ■ PAGE 266

True-False Quiz
1. True   **3.** True   **5.** False   **7.** False   **9.** True
11. True   **13.** True   **15.** True

Exercises

1. $4x^7(x+1)^3(3x+2)$ **3.** $\frac{3}{2}\sqrt{x} - \frac{1}{2\sqrt{x}} - \frac{1}{\sqrt{x^3}}$

5. $x(\pi x \cos \pi x + 2 \sin \pi x)$

7. $\frac{8t^3}{(t^4+1)^2}$ **9.** $\frac{1+\ln x}{x \ln x}$ **11.** $\frac{\cos \sqrt{x} - \sqrt{x} \sin \sqrt{x}}{2\sqrt{x}}$

13. $-\frac{e^{1/x}(1+2x)}{x^4}$ **15.** $\frac{2xy - \cos y}{1 - x \sin y - x^2}$

17. $\frac{1}{2\sqrt{\arctan x}\,(1+x^2)}$ **19.** $\frac{1-t^2}{(1+t^2)^2}\sec^2\left(\frac{t}{1+t^2}\right)$

21. $3^{x \ln x}(\ln 3)(1 + \ln x)$ **23.** $-(x-1)^{-2}$

25. $\frac{2x - y \cos(xy)}{x \cos(xy) + 1}$ **27.** $\frac{2}{(1+2x)\ln 5}$

29. $\cot x - \sin x \cos x$ **31.** $\frac{4x}{1+16x^2} + \tan^{-1}(4x)$

33. $5 \sec 5x$ **35.** $-6x \csc^2(3x^2+5)$

37. $\cos(\tan \sqrt{1+x^3})(\sec^2 \sqrt{1+x^3}) \frac{3x^2}{2\sqrt{1+x^3}}$

39. $2 \cos \theta \tan(\sin \theta) \sec^2(\sin \theta)$

41. $\frac{(2-x)^4(3x^2 - 55x - 52)}{2\sqrt{x+1}\,(x+3)^8}$ **43.** $2x^2 \cosh(x^2) + \sinh(x^2)$

45. $3 \tanh 3x$ **47.** $\frac{\cosh x}{\sqrt{\sinh^2 x - 1}}$

49. $\frac{-3\sin(e^{\sqrt{\tan 3x}})e^{\sqrt{\tan 3x}}\sec^2(3x)}{2\sqrt{\tan 3x}}$ **51.** $-\frac{4}{27}$

53. $-5x^4/y^{11}$ **57.** $y = 2\sqrt{3}x + 1 - \pi\sqrt{3}/3$

59. $y = 2x + 1$ **61.** $y = -x + 2; y = x + 2$

63. (a) $\frac{10-3x}{2\sqrt{5-x}}$ (b) $y = \frac{7}{4}x + \frac{1}{4}, y = -x + 8$

(c)

65. $(\pi/4, \sqrt{2}), (5\pi/4, -\sqrt{2})$

69. (a) 4   (b) 6   (c) $\frac{7}{9}$ (d) 12

71. $2xg(x) + x^2g'(x)$ **73.** $2g(x)g'(x)$

75. $g'(e^x)e^x$ **77.** $g'(x)/g(x)$

79. $\frac{f'(x)[g(x)]^2 + g'(x)[f(x)]^2}{[f(x)+g(x)]^2}$

81. $f'(g(\sin 4x))g'(\sin 4x)(\cos 4x)(4)$

83. $(-3, 0)$ **85.** $y = -\frac{2}{3}x^2 + \frac{14}{3}x$

87. $v(t) = -Ae^{-ct}[c \cos(\omega t + \delta) + \omega \sin(\omega t + \delta)],$
$a(t) = Ae^{-ct}[(c^2 - \omega^2) \cos(\omega t + \delta) + 2c\omega \sin(\omega t + \delta)]$

89. (a) $v(t) = 3t^2 - 12; a(t) = 6t$ (b) $t > 2; 0 \le t < 2$
(c) 23   (d)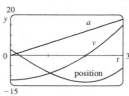

(e) $t > 2; 0 < t < 2$
91. 4 kg/m
93. (a) $200(3.24)^t$ (b) $\approx 22{,}040$
(c) $\approx 25{,}910$ bacteria/h   (d) $(\ln 50)/(\ln 3.24) \approx 3.33$ h
95. (a) $C_0 e^{-kt}$ (b) ≈ 100 h   **97.** $\frac{4}{3}$ cm²/min
99. 13 ft/s   **101.** 400 ft/h
103. (a) $L(x) = 1 + x; \sqrt[3]{1+3x} \approx 1 + x; \sqrt[3]{1.03} \approx 1.01$
(b) $-0.23 < x < 0.40$
105. $12 + \frac{3}{2}\pi \approx 16.7$ cm²   **107.** $\frac{1}{32}$ **109.** $\frac{1}{4}$ **111.** $\frac{1}{8}x^2$

PROBLEMS PLUS ■ PAGE 271

1. $(\pm\sqrt{3}/2, \frac{1}{4})$ **5.** $3\sqrt{2}$ **11.** $(0, \frac{5}{4})$

13. 3 lines: $(0, 2)$, $(\frac{4}{3}\sqrt{2}, \frac{2}{3})$ and $(\frac{2}{3}\sqrt{2}, \frac{10}{3})$, $(-\frac{4}{3}\sqrt{2}, \frac{2}{3})$ and $(-\frac{2}{3}\sqrt{2}, \frac{10}{3})$

15. (a) $4\pi\sqrt{3}/\sqrt{11}$ rad/s   (b) $40(\cos\theta + \sqrt{8 + \cos^2\theta})$ cm
(c) $-480\pi \sin\theta \left(1 + \cos\theta/\sqrt{8 + \cos^2\theta}\right)$ cm/s

19. $x_T \in (3, \infty), y_T \in (2, \infty), x_N \in (0, \frac{5}{3}), y_N \in (-\frac{5}{2}, 0)$

21. (b) (i) 53° (or 127°)   (ii) 63° (or 117°)
23. R approaches the midpoint of the radius AO.

25. $-\sin a$ **27.** $2\sqrt{e}$ **31.** $(1, -2), (-1, 0)$

33. $\sqrt{29}/58$ **35.** $2 + \frac{375}{128}\pi \approx 11.204$ cm³/min

CHAPTER 4

EXERCISES 4.1 ■ PAGE 283

Abbreviations: abs, absolute; loc, local; max, maximum; min, minimum
1. Abs min: smallest function value on the entire domain of the function; loc min at c: smallest function value when x is near c
3. Abs max at s, abs min at r, loc max at c, loc min at b and r, neither a max nor a min at a and d
5. Abs max $f(4) = 5$, loc max $f(4) = 5$ and $f(6) = 4$, loc min $f(2) = 2$ and $f(1) = f(5) = 3$

7.   **9.**

11. (a)   (b)

(c)

13. (a)   (b)

15. Abs max $f(3) = 4$ **17.** Abs max $f(1) = 1$
19. Abs min $f(0) = 0$
21. Abs max $f(\pi/2) = 1$; abs min $f(-\pi/2) = -1$
23. Abs max $f(2) = \ln 2$ **25.** Abs max $f(0) = 1$
27. Abs min $f(1) = -1$ **29.** $\frac{1}{3}$ **31.** $-2, 3$ **33.** 0
35. $0, 2$ **37.** $0, \frac{4}{9}$ **39.** $0, \frac{8}{7}, 4$ **41.** $n\pi$ (n an integer)
43. $0, \frac{2}{3}$ **45.** 10   **47.** $f(2) = 16, f(5) = 7$
49. $f(-1) = 8, f(2) = -19$ **51.** $f(-2) = 33, f(2) = -31$
53. $f(0.2) = 5.2, f(1) = 2$
55. $f(4) = 4 - \sqrt[3]{4}, f(\sqrt{3}/9) = -2\sqrt{3}/9$
57. $f(\pi/6) = \frac{3}{2}\sqrt{3}, f(\pi/2) = 0$
59. $f(e^{1/2}) = 1/(2e), f(\frac{1}{2}) = -4\ln 2$
61. $f(1) = \ln 3, f(-\frac{1}{2}) = \ln\frac{3}{4}$
63. $f\left(\dfrac{a}{a+b}\right) = \dfrac{a^a b^b}{(a+b)^{a+b}}$
65. (a) $2.19, 1.81$ (b) $\frac{6}{25}\sqrt{\frac{3}{5}} + 2, -\frac{6}{25}\sqrt{\frac{3}{5}} + 2$
67. (a) $0.32, 0.00$ (b) $\frac{3}{16}\sqrt{3}, 0$
69. 0.177 mg/mL; 21.4 min   **71.** $\approx 3.9665°C$
73. About 4.1 months after Jan. 1
75. (a) $r = \frac{2}{3}r_0$ (b) $v = \frac{4}{27}kr_0^3$

(c)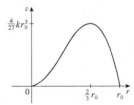

1. $1, 5$
3. (a) g is continuous on $[0, 8]$ and differentiable on $(0, 8)$.
(b) $2.1, 6.3$ (c) $3.6, 5.6$
5. 1   **7.** π
9. f is not differentiable on $(-1, 1)$ **11.** 1
13. $3/\ln 4$ **15.** 1; yes   **17.** f is not continous at 3
25. 16   **27.** No   **33.** No

1. (a) $(1, 3), (4, 6)$ (b) $(0, 1), (3, 4)$ (c) $(0, 2)$
(d) $(2, 4), (4, 6)$ (e) $(2, 3)$
3. (a) I/D Test   (b) Concavity Test
(c) Find points at which the concavity changes.
5. (a) Inc on $(1, 5)$; dec on $(0, 1)$ and $(5, 6)$
(b) Loc max at $x = 5$, loc min at $x = 1$
7. (a) $3, 5$ (b) $2, 4, 6$ (c) $1, 7$
9. (a) Inc on $(-\infty, -1), (3, \infty)$; dec on $(-1, 3)$
(b) Loc max $f(-1) = 9$; loc min $f(3) = -23$
(c) CU on $(1, \infty)$, CD on $(-\infty, 1)$; IP $(1, -7)$
11. (a) Inc on $(-1, 0), (1, \infty)$; dec on $(-\infty, -1), (0, 1)$
(b) Loc max $f(0) = 3$; loc min $f(\pm 1) = 2$
(c) CU on $\left(-\infty, -\sqrt{3}/3\right), \left(\sqrt{3}/3, \infty\right)$;
CD on $\left(-\sqrt{3}/3, \sqrt{3}/3\right)$; IP $\left(\pm\sqrt{3}/3, \frac{22}{9}\right)$
13. (a) Inc on $(0, \pi/4), (5\pi/4, 2\pi)$; dec on $(\pi/4, 5\pi/4)$
(b) Loc max $f(\pi/4) = \sqrt{2}$; loc min $f(5\pi/4) = -\sqrt{2}$
(c) CU on $(3\pi/4, 7\pi/4)$; CD on $(0, 3\pi/4), (7\pi/4, 2\pi)$;
IP $(3\pi/4, 0), (7\pi/4, 0)$
15. (a) Inc on $\left(-\frac{1}{3}\ln 2, \infty\right)$; dec on $\left(-\infty, -\frac{1}{3}\ln 2\right)$
(b) Loc min $f\left(-\frac{1}{3}\ln 2\right) = 2^{-2/3} + 2^{1/3}$ (c) CU on $(-\infty, \infty)$
17. (a) Inc on $(1, \infty)$; dec on $(0, 1)$ (b) Loc min $f(1) = 0$
(c) CU on $(0, \infty)$; No IP
19. Loc max $f(1) = 2$; loc min $f(0) = 1$
21. Loc min $f\left(\frac{1}{16}\right) = -\frac{1}{4}$
23. (a) f has a local maximum at 2.
(b) f has a horizontal tangent at 6.
25. (a)   (b)

27.   **29.**

31.

33. (a) No
(b) Yes

(c)

35. (a) Inc on $(0, 2)$, $(4, 6)$, $(8, \infty)$;
dec on $(2, 4)$, $(6, 8)$
(b) Loc max at $x = 2, 6$;
loc min at $x = 4, 8$
(c) CU on $(3, 6)$, $(6, \infty)$;
CD on $(0, 3)$ (d) 3
(e) See graph at right.

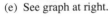

37. (a) Inc on $(-\infty, -2)$, $(2, \infty)$; dec on $(-2, 2)$
(b) Loc max $f(-2) = 18$; loc min $f(2) = -14$
(c) CU on $(0, \infty)$, CD on $(-\infty, 0)$; IP $(0, 2)$
(d)

39. (a) Inc on $(-2, 0)$, $(2, \infty)$; dec on $(-\infty, -2)$, $(0, 2)$
(b) Loc max $f(0) = 3$; loc min $f(\pm 2) = -5$

(c) CU on $\left(-\infty, -\dfrac{2}{\sqrt{3}}\right)$, $\left(\dfrac{2}{\sqrt{3}}, \infty\right)$; CD on $\left(-\dfrac{2}{\sqrt{3}}, \dfrac{2}{\sqrt{3}}\right)$;

IPs $\left(\pm\dfrac{2}{\sqrt{3}}, -\dfrac{13}{9}\right)$

(d)

41. (a) Inc on $(-\infty, -2)$, $(0, \infty)$;
dec on $(-2, 0)$
(b) Loc max $h(-2) = 7$;
loc min $h(0) = -1$
(c) CU on $(-1, \infty)$;
CD on $(-\infty, -1)$; IP $(-1, 3)$
(d) See graph at right.

43. (a) Inc on $(-\infty, 4)$;
dec on $(4, 6)$
(b) Loc max $F(4) = 4\sqrt{2}$
(c) CD on $(-\infty, 6)$; No IP
(d) See graph at right.

45. (a) Inc on $(-1, \infty)$;
dec on $(-\infty, -1)$
(b) Loc min $C(-1) = -3$
(c) CU on $(-\infty, 0)$, $(2, \infty)$;
CD on $(0, 2)$;
IPs $(0, 0)$, $\left(2, 6\sqrt[3]{2}\right)$
(d) See graph at right.

47. (a) Inc on $(\pi, 2\pi)$;
dec on $(0, \pi)$
(b) Loc min $f(\pi) = -1$
(c) CU on $(\pi/3, 5\pi/3)$;
CD on $(0, \pi/3)$, $(5\pi/3, 2\pi)$;
IPs $\left(\pi/3, \frac{5}{4}\right)$, $\left(5\pi/3, \frac{5}{4}\right)$
(d) See graph at right.

49. (a) VA $x = 0$; HA $y = 1$
(b) Inc on $(0, 2)$;
dec on $(-\infty, 0)$, $(2, \infty)$
(c) Loc max $f(2) = \frac{5}{4}$
(d) CU on $(3, \infty)$;
CD on $(-\infty, 0)$, $(0, 3)$; IP $\left(3, \frac{11}{9}\right)$
(e) See graph at right.

51. (a) HA $y = 0$
(b) Dec on $(-\infty, \infty)$
(c) None
(d) CU on $(-\infty, \infty)$
(e) See graph at right.

53. (a) HA $y = 0$
(b) Inc on $(-\infty, 0)$,
dec on $(0, \infty)$
(c) Loc max $f(0) = 1$
(d) CU on $\left(-\infty, -1\sqrt{2}\,\right)$,
$\left(1/\sqrt{2}, \infty\right)$; CD on $\left(-1/\sqrt{2}, 1/\sqrt{2}\,\right)$; IPs $\left(\pm 1/\sqrt{2}, e^{-1/2}\right)$
(e) See graph at right.

55. (a) VA $x = 0, x = e$
(b) Dec on $(0, e)$
(c) None
(d) CU on $(0, 1)$; CD on $(1, e)$;
IP $(1, 0)$
(e) See graph at right.

57. $(3, \infty)$

59. (a) Loc and abs max $f(1) = \sqrt{2}$, no min   (b) $\frac{1}{4}\left(3 - \sqrt{17}\right)$

61. (b) CU on $(0, 0.85)$, $(1.57, 2.29)$; CD on $(0.85, 1.57)$, $(2.29, \pi)$; IP $(0.85, 0.74)$, $(1.57, 0)$, $(2.29, -0.74)$

63. CU on $(-\infty, -0.6)$, $(0.0, \infty)$; CD on $(-0.6, 0.0)$

65. (a) The rate of increase is initially very small, increases to a maximum at $t \approx 8$ h, then decreases toward 0.
(b) When $t = 8$ (c) CU on $(0, 8)$; CD on $(8, 18)$
(d) $(8, 350)$

67. If $D(t)$ is the size of the deficit as a function of time, then at the time of the speech $D'(t) > 0$, but $D''(t) < 0$.

69. $K(3) - K(2)$; CD

71. 28.57 min, when the rate of increase of drug level in the bloodstream is greatest; 85.71 min, when rate of decrease is greatest

73. $f(x) = \frac{1}{9}(2x^3 + 3x^2 - 12x + 7)$

75. (a) $a = 0, b = -1$ (b) $y = -x$ at $(0, 0)$

EXERCISES 4.4 ▪ **PAGE 311**

1. (a) Indeterminate   (b) 0   (c) 0
(d) ∞, $-\infty$, or does not exist   (e) Indeterminate
3. (a) $-\infty$ (b) Indeterminate   (c) ∞
5. $\frac{9}{4}$ **7.** 1   **9.** 6   **11.** $-\frac{1}{3}$
13. $-\infty$ **15.** 2   **17.** $\frac{1}{4}$ **19.** 0   **21.** $-\infty$
23. $\frac{8}{5}$ **25.** 3   **27.** $\frac{1}{2}$ **29.** 1   **31.** 1
33. $1/\ln 3$ **35.** 0   **37.** 0   **39.** a/b
41. $\frac{1}{24}$ **43.** π **45.** $\frac{5}{3}$ **47.** 0   **49.** $-2/\pi$
51. $\frac{1}{2}$ **53.** $\frac{1}{2}$ **55.** ∞ **57.** 1   **59.** e^{-2}
61. $1/e$ **63.** 1   **65.** e^4 **67.** e^3 **69.** e^2
71. $\frac{1}{4}$ **75.** 1
77. f has an absolute minimum for $c > 0$. As c increases, the minimum points get farther away from the origin.
81. (a) M; the population should approach its maximum size as time increases   (b) $P_0 e^{kt}$; exponential
83. $\frac{16}{9}a$ **85.** $\frac{1}{2}$ **87.** 56   **91.** (a) 0

EXERCISES 4.5 ▪ **PAGE 321**

Abbreviation: int, intercept; SA, slant asymptote
1. A. $\mathbb{R}$ B. y-int 0; x-int $-3, 0$
C. None   D. None
E. Inc on $(-\infty, -2)$, $(0, \infty)$;
dec on $(-2, 0)$
F. Loc max $f(-2) = 4$;
loc min $f(0) = 0$
G. CU on $(-1, \infty)$; CD on $(-\infty, -1)$;
IP $(-1, 2)$
H. See graph at right.

3. A. $\mathbb{R}$ B. y-int 0; x-int 0, $\sqrt[3]{4}$
C. None   D. None
E. Inc on $(1, \infty)$; dec on $(-\infty, 1)$
F. Loc min $f(1) = -3$
G. CU on $(-\infty, \infty)$
H. See graph at right.

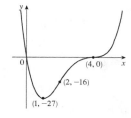

5. A. $\mathbb{R}$ B. y-int 0; x-int 0, 4
C. None   D. None
E. Inc on $(1, \infty)$; dec on $(-\infty, 1)$
F. Loc min $f(1) = -27$
G. CU on $(-\infty, 2)$, $(4, \infty)$;
CD on $(2, 4)$;
IPs $(2, -16)$, $(4, 0)$
H. See graph at right.

7. A. $\mathbb{R}$ B. y-int 0; x-int 0
C. About $(0, 0)$ D. None
E. Inc on $(-\infty, \infty)$
F. None
G. CU on $(-2, 0)$, $(2, \infty)$;
CD on $(-\infty, -2)$, $(0, 2)$;
IPs $\left(-2, -\frac{256}{15}\right)$, $(0, 0)$, $\left(2, \frac{256}{15}\right)$
H. See graph at right.

9. A. $\{x \mid x \neq 1\}$ B. y-int 0; x-int 0
C. None   D. VA $x = 1$, HA $y = 1$
E. Dec on $(-\infty, 1)$, $(1, \infty)$
F. None
G. CU on $(1, \infty)$; CD on $(-\infty, 1)$
H. See graph at right.

11. A. $(-\infty, 1) \cup (1, 2) \cup (2, \infty)$
B. y-int 0; x-int 0   C. None
D. HA $y = -1$; VA $x = 2$
E. Inc on $(-\infty, 1)$, $(1, 2)$, $(2, \infty)$
F. None
G. CU on $(-\infty, 1)$, $(1, 2)$;
CD on $(2, \infty)$
H. See graph at right.

13. A. $(-\infty, -2) \cup (-2, 2) \cup (2, \infty)$ **B.** y-int 0; x-int 0
C. About $(0, 0)$ **D.** VA $x = \pm 2$; HA $y = 0$
E. Dec on $(-\infty, -2)$, $(-2, 2)$, $(2, \infty)$
F. No local extrema
G. CU on $(-2, 0)$, $(2, \infty)$;
CD on $(-\infty, -2)$, $(0, 2)$; IP $(0, 0)$
H. See graph at right.

15. A. $\mathbb{R}$ **B.** y-int 0; x-int 0
C. About y-axis   **D.** HA $y = 1$
E. Inc on $(0, \infty)$; dec on $(-\infty, 0)$
F. Loc min $f(0) = 0$
G. CU on $(-1, 1)$;
CD on $(-\infty, -1)$, $(1, \infty)$; IPs $\left(\pm 1, \frac{1}{4}\right)$
H. See graph at right.

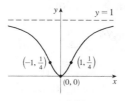

17. A. $(-\infty, 0) \cup (0, \infty)$ **B.** x-int 1
C. None   **D.** HA $y = 0$; VA $x = 0$
E. Inc on $(0, 2)$;
dec on $(-\infty, 0)$, $(2, \infty)$
F. Loc max $f(2) = \frac{1}{4}$
G. CU on $(3, \infty)$;
CD on $(-\infty, 0)$, $(0, 3)$; IP $\left(3, \frac{2}{9}\right)$
H. See graph at right.

19. A. $(-\infty, -1) \cup (-1, \infty)$
B. y-int 0; x-int 0   **C.** None
D. VA $x = -1$; HA $y = 1$
E. Inc on $(-\infty, -1)$, $(-1, \infty)$;
F. None
G. CU on $(-\infty, -1)$, $\left(0, \sqrt[3]{\frac{1}{2}}\right)$;
CD on $(-1, 0)$, $\left(\sqrt[3]{\frac{1}{2}}, \infty\right)$;
IPs $(0, 0)$, $\left(\sqrt[3]{\frac{1}{2}}, \frac{1}{3}\right)$
H. See graph at right.

21. A. $[0, \infty)$ **B.** y-int 0; x-int 0, 3
C. None   **D.** None
E. Inc on $(1, \infty)$; dec on $(0, 1)$
F. Loc min $f(1) = -2$
G. CU on $(0, \infty)$
H. See graph at right.

23. A. $(-\infty, -2] \cup [1, \infty)$
B. x-int -2, 1   **C.** None
D. None
E. Inc on $(1, \infty)$; dec on $(-\infty, -2)$
F. None
G. CD on $(-\infty, -2)$, $(1, \infty)$
H. See graph at right.

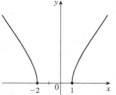

25. A. $\mathbb{R}$ **B.** y-int 0; x-int 0
C. About $(0, 0)$
D. HA $y = \pm 1$
E. Inc on $(-\infty, \infty)$ **F.** None
G. CU on $(-\infty, 0)$;
CD on $(0, \infty)$; IP $(0, 0)$
H. See graph at right.

27. A. $[-1, 0) \cup (0, 1]$ **B.** x-int ± 1 **C.** About $(0, 0)$
D. VA $x = 0$
E. Dec on $(-1, 0)$, $(0, 1)$
F. None
G. CU on $\left(-1, -\sqrt{2/3}\right)$,
$\left(0, \sqrt{2/3}\right)$;
CD on $\left(-\sqrt{2/3}, 0\right)$, $\left(\sqrt{2/3}, 1\right)$;
IPs $\left(\pm\sqrt{2/3}, \pm 1/\sqrt{2}\right)$
H. See graph at right.

29. A. $\mathbb{R}$ **B.** y-int 0; x-int $\pm 3\sqrt{3}$, 0   **C.** About $(0, 0)$
D. None   **E.** Inc on $(-\infty, -1)$, $(1, \infty)$; dec on $(-1, 1)$
F. Loc max $f(-1) = 2$;
loc min $f(1) = -2$;
G. CU on $(0, \infty)$;
CD on $(-\infty, 0)$; IP $(0, 0)$
H. See graph at right.

31. A. $\mathbb{R}$ **B.** y-int -1; x-int ± 1
C. About the y-axis   **D.** None
E. Inc on $(0, \infty)$; dec on $(-\infty, 0)$
F. Loc min $f(0) = -1$
G. CU on $(-1, 1)$;
CD on $(-\infty, -1)$, $(1, \infty)$; IPs $(\pm 1, 0)$
H. See graph at right.

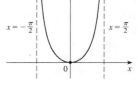

33. A. $\mathbb{R}$ **B.** y-int 0; x-int $n\pi$ (n an integer)
C. About $(0, 0)$, period 2π **D.** None
E–G answers for $0 \le x \le \pi$:
E. Inc on $(0, \pi/2)$; dec on $(\pi/2, \pi)$ **F.** Loc max $f(\pi/2) = 1$
G. Let $\alpha = \sin^{-1}\sqrt{2/3}$; CU on $(0, \alpha)$, $(\pi - \alpha, \pi)$;
CD on $(\alpha, \pi - \alpha)$; IPs at $x = 0, \pi, \alpha, \pi - \alpha$
H.

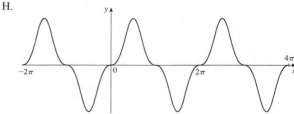

35. A. $(-\pi/2, \pi/2)$ **B.** y-int 0; x-int 0   **C.** About y-axis
D. VA $x = \pm\pi/2$
E. Inc on $(0, \pi/2)$;
dec on $(-\pi/2, 0)$
F. Loc min $f(0) = 0$
G. CU on $(-\pi/2, \pi/2)$
H. See graph at right.

37. A. $[-2\pi, 2\pi]$ **B.** y-int $\sqrt{3}$; x-int $-4\pi/3$, $-\pi/3$, $2\pi/3$, $5\pi/3$ **C.** Period 2π **D.** None
E. Inc on $(-2\pi, -11\pi/6)$, $(-5\pi/6, \pi/6)$, $(7\pi/6, 2\pi)$; dec on $(-11\pi/6, -5\pi/6)$, $(\pi/6, 7\pi/6)$
F. Loc max $f(-11\pi/6) = f(\pi/6) = 2$; loc min $f(-5\pi/6) = f(7\pi/6) = -2$
G. CU on $(-4\pi/3, -\pi/3)$, $(2\pi/3, 5\pi/3)$;
CD on $(-2\pi, -4\pi/3)$, $(-\pi/3, 2\pi/3)$, $(5\pi/3, 2\pi)$;
IPs $(-4\pi/3, 0)$, $(-\pi/3, 0)$, $(2\pi/3, 0)$, $(5\pi/3, 0)$
H. See graph at right.

39. A. All reals except $(2n + 1)\pi$ (n an integer)
B. y-int 0; x-int $2n\pi$ **C.** About the origin, period 2π
D. VA $x = (2n + 1)\pi$ **E.** Inc on $((2n - 1)\pi, (2n + 1)\pi)$
F. None   **G.** CU on $(2n\pi, (2n + 1)\pi)$;
CD on $((2n - 1)\pi, 2n\pi)$; IPs $(2n\pi, 0)$
H.

41. A. $\mathbb{R}$ **B.** y-int $\pi/4$
C. None
D. HA $y = 0$, $y = \pi/2$
E. Inc on $(-\infty, \infty)$ **F.** None
G. CU on $(-\infty, 0)$;
CD on $(0, \infty)$; IP $(0, \pi/4)$
H. See graph at right.

43. A. $\mathbb{R}$ **B.** y-int $\frac{1}{2}$ **C.** None
D. HA $y = 0$, $y = 1$
E. Inc on $\mathbb{R}$ **F.** None
G. CU on $(-\infty, 0)$;
CD on $(0, \infty)$;   IP $\left(0, \frac{1}{2}\right)$
H. See graph at right.

45. A. $(0, \infty)$ **B.** None
C. None   **D.** VA $x = 0$
E. Inc on $(1, \infty)$; dec on $(0, 1)$
F. Loc min $f(1) = 1$
G. CU on $(0, 2)$; CD on $(2, \infty)$;
IP $\left(2, \frac{1}{2} + \ln 2\right)$
H. See graph at right.

47. A. $\mathbb{R}$ **B.** y-int $\frac{1}{4}$
C. None
D. HA $y = 0$, $y = 1$
E. Dec on $\mathbb{R}$ **F.** None
G. CU on $\left(\ln \frac{1}{2}, \infty\right)$;
CD on $\left(-\infty, \ln \frac{1}{2}\right)$; IP $\left(\ln \frac{1}{2}, \frac{4}{9}\right)$
H. See graph at right.

49. A. All x in $(2n\pi, (2n + 1)\pi)$ (n an integer)
B. x-int $\pi/2 + 2n\pi$ **C.** Period 2π **D.** VA $x = n\pi$
E. Inc on $(2n\pi, \pi/2 + 2n\pi)$; dec on $(\pi/2 + 2n\pi, (2n + 1)\pi)$
F. Loc max $f(\pi/2 + 2n\pi) = 0$ **G.** CD on $(2n\pi, (2n + 1)\pi)$
H.

51. A. $(-\infty, 0) \cup (0, \infty)$
B. None   **C.** None
D. VA $x = 0$
E. Inc on $(-\infty, -1)$, $(0, \infty)$;
dec on $(-1, 0)$
F. Loc max $f(-1) = -e$
G. CU on $(0, \infty)$; CD on $(-\infty, 0)$
H. See graph at right.

53. A. $\mathbb{R}$ **B.** y-int 1
C. None   **D.** HA $y = e^{\pm \pi/2}$
E. Inc on $\mathbb{R}$ **F.** None
G. CU on $\left(-\infty, \frac{1}{2}\right)$; CD on $\left(\frac{1}{2}, \infty\right)$;
IP $\left(\frac{1}{2}, e^{\arctan(1/2)}\right)$
H. See graph at right.

55.

57. (a) When $t = (\ln a)/k$ **(b)** When $t = (\ln a)/k$
(c)

59.

61. $y = x - 1$ **63.** $y = 2x - 3$

65. A. $(-\infty, 1) \cup (1, \infty)$
B. y-int 0; x-int 0   C. None
D. VA $x = 1$; SA $y = x + 1$
E. Inc on $(-\infty, 0)$, $(2, \infty)$;
dec on $(0, 1)$, $(1, 2)$
F. Loc max $f(0) = 0$;
loc min $f(2) = 4$
G. CU on $(1, \infty)$; CD on $(-\infty, 1)$
H. See graph at right.

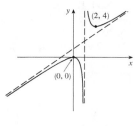

67. A. $(-\infty, 0) \cup (0, \infty)$
B. x-int $-\sqrt[3]{4}$ C. None
D. VA $x = 0$; SA $y = x$
E. Inc on $(-\infty, 0)$, $(2, \infty)$;
dec on $(0, 2)$
F. Loc min $f(2) = 3$
G. CU on $(-\infty, 0)$, $(0, \infty)$
H. See graph at right.

69. A. $\mathbb{R}$ B. y-int 2
C. None
D. SA $y = 1 + \frac{1}{2}x$
E. Inc on $(\ln 2, \infty)$;
dec on $(-\infty, \ln 2)$
F. Loc min $f(\ln 2) = \frac{3}{2} + \frac{1}{2}\ln 2$
G. CU on $(-\infty, \infty)$
H. See graph at right.

71.

75. VA $x = 0$, asymptotic to $y = x^3$

EXERCISES 4.6 ■ PAGE 329

1. Inc on $(-\infty, -1.50)$, $(0.04, 2.62)$, $(2.84, \infty)$; dec on
$(-1.50, 0.04)$, $(2.62, 2.84)$; loc max $f(-1.50) \approx 36.47$,
$f(2.62) \approx 56.83$; loc min $f(0.04) \approx -0.04$, $f(2.84) \approx 56.73$;
CU on $(-0.89, 1.15)$, $(2.74, \infty)$; CD on $(-\infty, -0.89)$, $(1.15, 2.74)$;
IPs $(-0.89, 20.90)$, $(1.15, 26.57)$, $(2.74, 56.78)$

3. Inc on $(-1.31, -0.84)$, $(1.06, 2.50)$, $(2.75, \infty)$; dec on
$(-\infty, -1.31)$, $(-0.84, 1.06)$, $(2.50, 2.75)$; loc max
$f(-0.84) \approx 23.71, f(2.50) \approx -11.02$; loc min
$f(-1.31) \approx 20.72, f(1.06) \approx -33.12, f(2.75) \approx -11.33$; CU
on $(-\infty. -1.10)$, $(0.08, 1.72)$, $(2.64, \infty)$; CD on $(-1.10, 0.08)$,
$(1.72, 2.64)$; IPs $(-1.10, 22.09)$, $(0.08, -3.88)$, $(1.72, -22.53)$,
$(2.64, -11.18)$

5. Inc on $(-\infty, -1.47)$, $(-1.47, 0.66)$; dec on $(0.66, \infty)$;
loc max $f(0.66) \approx 0.38$; CU on $(-\infty, -1.47)$, $(-0.49, 0)$,
$(1.10, \infty)$; CD on $(-1.47, -0.49)$, $(0, 1.10)$;
IPs $(-0.49, -0.44)$, $(1.10, 0.31)$, $(0, 0)$

7. Inc on $(-1.40, -0.44)$, $(0.44, 1.40)$; dec on $(-\pi, -1.40)$,
$(-0.44, 0)$, $(0, 0.44)$, $(1.40, \pi)$; loc max $f(-0.44) \approx -4.68$,
$f(1.40) \approx 6.09$; loc min $f(-1.40) \approx -6.09, f(0.44) \approx 4.68$;
CU on $(-\pi, -0.77)$, $(0, 0.77)$; CD on $(-0.77, 0)$, $(0.77, \pi)$;
IPs $(-0.77, -5.22)$, $(0.77, 5.22)$

9. Inc on $\left(-8 - \sqrt{61}, -8 + \sqrt{61}\right)$; dec on $\left(-\infty, -8 - \sqrt{61}\right)$,
$\left(-8 + \sqrt{61}, 0\right)$, $(0, \infty)$; CU on $\left(-12 - \sqrt{138}, -12 + \sqrt{138}\right)$,
$(0, \infty)$; CD on $\left(-\infty, -12 - \sqrt{138}\right)$, $\left(-12 + \sqrt{138}, 0\right)$

11. (a)

(b) $\lim_{x \to 0^+} f(x) = 0$ (c) Loc min $f(1/\sqrt{e}) = -1/(2e)$;
CD on $(0, e^{-3/2})$; CU on $(e^{-3/2}, \infty)$

13. Loc max $f(-5.6) \approx 0.018$, $f(0.82) \approx -281.5$,
$f(5.2) \approx 0.0145$; loc min $f(3) = 0$

15. $f'(x) = -\dfrac{x(x+1)^2(x^3 + 18x^2 - 44x - 16)}{(x-2)^3(x-4)^5}$

$f''(x) = 2\dfrac{(x+1)(x^6 + 36x^5 + 6x^4 - 628x^3 + 684x^2 + 672x + 64)}{(x-2)^4(x-4)^6}$

CU on $(-35.3, -5.0)$, $(-1, -0.5)$, $(-0.1, 2)$, $(2, 4)$, $(4, \infty)$;
CD on $(-\infty, -35.3)$, $(-5.0, -1)$, $(-0.5, -0.1)$;
IPs $(-35.3, -0.015)$, $(-5.0, -0.005)$, $(-1, 0)$, $(-0.5, 0.00001)$,
$(-0.1, 0.0000066)$

17. Inc on $(-9.41, -1.29)$, $(0, 1.05)$;
dec on $(-\infty, -9.41)$, $(-1.29, 0)$, $(1.05, \infty)$;
loc max $f(-1.29) \approx 7.49$, $f(1.05) \approx 2.35$;
loc min $f(-9.41) \approx -0.056$, $f(0) = 0.5$;
CU on $(-13.81, -1.55)$, $(-1.03, 0.60)$, $(1.48, \infty)$;
CD on $(-\infty, -13.81)$, $(-1.55, -1.03)$, $(0.60, 1.48)$;
IPs $(-13.81, -0.05)$, $(-1.55, 5.64)$, $(-1.03, 5.39)$, $(0.60, 1.52)$,
$(1.48, 1.93)$

19. Inc on $(-4.91, -4.51)$, $(0, 1.77)$, $(4.91, 8.06)$, $(10.79, 14.34)$,
$(17.08, 20)$;
dec on $(-4.51, -4.10)$, $(1.77, 4.10)$, $(8.06, 10.79)$, $(14.34, 17.08)$;
loc max $f(-4.51) \approx 0.62$, $f(1.77) \approx 2.58$, $f(8.06) \approx 3.60$,
$f(14.34) \approx 4.39$;
loc min $f(10.79) \approx 2.43$, $f(17.08) \approx 3.49$;

CU on $(9.60, 12.25)$, $(15.81, 18.65)$;
CD on $(-4.91, -4.10)$, $(0, 4.10)$, $(4.91, 9.60)$, $(12.25, 15.81)$,
$(18.65, 20)$;
IPs $(9.60, 2.95)$, $(12.25, 3.27)$, $(15.81, 3.91)$, $(18.65, 4.20)$

21. Inc on $(-\infty, 0)$, $(0, \infty)$;
CU on $(-\infty, -0.42)$, $(0, 0.42)$;
CD on $(-0.42, 0)$, $(0.42, \infty)$;
IPs $(\mp0.42, \pm0.83)$

23.

25. (a)

(b) $\lim_{x \to 0^+} x^{1/x} = 0$, $\lim_{x \to \infty} x^{1/x} = 1$
(c) Loc max $f(e) = e^{1/e}$ (d) IPs at $x \approx 0.58, 4.37$

27. Max $f(0.59) \approx 1$, $f(0.68) \approx 1$, $f(1.96) \approx 1$;
min $f(0.64) \approx 0.99996$, $f(1.46) \approx 0.49$, $f(2.73) \approx -0.51$;
IPs $(0.61, 0.99998)$, $(0.66, 0.99998)$, $(1.17, 0.72)$,
$(1.75, 0.77)$, $(2.28, 0.34)$

29. For $c < 0$, there is a loc min that moves toward $(-3, -9)$ as c increases. For $0 < c < 8$, there is a loc min that moves toward $(-3, -9)$ and a loc max that moves toward the origin as c decreases. For all $c > 0$, there is a first quadrant loc min that moves toward the origin as c decreases. $c = 0$ is a transitional value that gives the graph of a parabola. For all nonzero c, the y-axis is a VA and there is an IP that moves toward the origin as $|c| \to 0$.
$c \leq 0$:

$c \geq 0$:

31. For $c < 0$, there is no extreme point and one IP, which decreases along the x-axis. For $c > 0$, there is no IP, and one minimum point.

33. For $c > 0$, the maximum and minimum values are always $\pm\frac{1}{2}$, but the extreme points and IPs move closer to the y-axis as c increases. $c = 0$ is a transitional value: when c is replaced by $-c$, the curve is reflected in the x-axis.

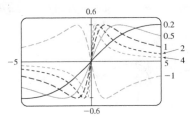

35. For $|c| < 1$, the graph has loc max and min values; for $|c| \geq 1$ it does not. The function increases for $c \geq 1$ and decreases for $c \leq -1$. As c changes, the IPs move vertically but not horizontally.

37.

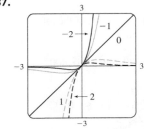

For $c > 0$, $\lim_{x \to \infty} f(x) = 0$ and $\lim_{x \to -\infty} f(x) = -\infty$.
For $c < 0$, $\lim_{x \to \infty} f(x) = \infty$ and $\lim_{x \to -\infty} f(x) = 0$.
As $|c|$ increases, the max and min points and the IPs get closer to the origin.

39. (a) Positive   (b)

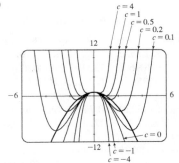

EXERCISES 4.7 ■ PAGE 336

1. (a) 11, 12   (b) 11.5, 11.5   **3.** 10, 10   **5.** $\frac{9}{4}$
7. 25 m by 25 m   **9.** $N = 1$
11. (a)

(b)

(c) $A = xy$ (d) $5x + 2y = 750$ (e) $A(x) = 375x - \frac{5}{2}x^2$
(f) $14{,}062.5 \text{ ft}^2$
13. 1000 ft by 1500 ft   **15.** 4000 cm^3 **17.** \$191.28
19. $20\sqrt{30}$ ft by $\frac{40}{3}\sqrt{30}$ ft   **21.** $\left(-\frac{6}{5}, \frac{3}{5}\right)$
23. $\left(-\frac{1}{3}, \pm\frac{4}{3}\sqrt{2}\right)$ **25.** Square, side $\sqrt{2}\,r$
27. $L/2, \sqrt{3}\,L/4$ **29.** Base $\sqrt{3}\,r$, height $3r/2$
31. $4\pi r^3/(3\sqrt{3})$ **33.** $\pi r^2(1 + \sqrt{5})$ **35.** 24 cm by 36 cm
37. (a) Use all of the wire for the square
(b) $40\sqrt{3}/(9 + 4\sqrt{3})$ m for the square
39. 16 in.   **41.** $V = 2\pi R^3/(9\sqrt{3})$ **45.** $E^2/(4r)$
47. (a) $\frac{3}{2}s^2\csc\theta\,(\csc\theta - \sqrt{3}\cot\theta)$ (b) $\cos^{-1}(1/\sqrt{3}) \approx 55°$
(c) $6s\left[h + s/(2\sqrt{2})\right]$
49. Row directly to B **51.** ≈ 4.85 km east of the refinery
53. $10\sqrt[3]{3}/(1 + \sqrt[3]{3})$ ft from the stronger source
55. $(a^{2/3} + b^{2/3})^{3/2}$ **57.** $2\sqrt{6}$
59. (b) (i) \$342,491; \$342/unit; \$390/unit   (ii) 400
(iii) \$320/unit
61. (a) $p(x) = 19 - \frac{1}{3000}x$ (b) \$9.50
63. (a) $p(x) = 500 - \frac{1}{8}x$ (b) \$250   (c) \$310
69. 9.35 m   **73.** $x = 6$ in.   **75.** $\pi/6$
77. At a distance $5 - 2\sqrt{5}$ from A **79.** $\frac{1}{2}(L + W)^2$
81. (a) About 5.1 km from B (b) C is close to B; C is close
to D; $W/L = \sqrt{25 + x^2}/x$, where $x = |BC|$
(c) ≈ 1.07; no such value   (d) $\sqrt{41}/4 \approx 1.6$

EXERCISES 4.8 ■ **PAGE 348**

1. (a) $x_2 \approx 7.3, x_3 \approx 6.8$ (b) Yes
3. $\frac{9}{2}$ **5.** a, b, c **7.** 1.5215   **9.** -1.25
11. 2.94283096   **13.** (b) 2.630020   **15.** -1.964636
17. $-3.637958, -1.862365, 0.889470$
19. $-1.257691, 0.653483$ **21.** $0, \pm0.902025$
23. $-1.69312029, -0.74466668, 1.26587094$
25. 0.76682579   **27.** 0.21916368, 1.08422462
29. (b) 31.622777
35. (a) $-1.293227, -0.441731, 0.507854$ (b) -2.0212
37. $(1.519855, 2.306964)$ **39.** $(0.410245, 0.347810)$
41. 0.76286%

EXERCISES 4.9 ■ **PAGE 355**

1. $F(x) = 2x^2 + 7x + C$ **3.** $F(x) = \frac{1}{2}x^4 - \frac{2}{9}x^3 + \frac{5}{2}x^2 + C$
5. $F(x) = 4x^3 + 4x^2 + C$ **7.** $F(x) = 5x^{7/5} + 40x^{1/5} + C$
9. $F(x) = \sqrt{2}x + C$ **11.** $F(x) = 2x^{3/2} - \frac{3}{2}x^{4/3} + C$
13. $F(x) = \begin{cases} \frac{1}{5}x - 2\ln|x| + C_1 & \text{if } x < 0 \\ \frac{1}{5}x - 2\ln|x| + C_2 & \text{if } x > 0 \end{cases}$
15. $G(t) = 2t^{1/2} + \frac{2}{3}t^{3/2} + \frac{2}{5}t^{5/2} + C$
17. $H(\theta) = -2\cos\theta - \tan\theta + C_n$ on $(n\pi - \pi/2, n\pi + \pi/2)$,
n an integer
19. $F(x) = 2^x/\ln 2 + 4\cosh x + C$
21. $F(x) = x^2 + 4x + 1/x + C, x > 0$
23. $F(x) = x^5 - \frac{1}{3}x^6 + 4$
25. $f(x) = x^5 - x^4 + x^3 + Cx + D$
27. $f(x) = \frac{1}{3}x^3 + 3e^x + Cx + D$

29. $f(t) = 2t^3 + \cos t + Ct^2 + Dt + E$
31. $f(x) = x + 2x^{3/2} + 5$ **33.** $f(t) = 4\arctan t - \pi$
35. $f(x) = 3x^{5/3} - 75$
37. $f(t) = \tan t + \sec t - 2 - \sqrt{2}$
39. $f(x) = -x^2 + 2x^3 - x^4 + 12x + 4$
41. $f(\theta) = -\sin\theta - \cos\theta + 5\theta + 4$
43. $f(x) = 2x^2 + x^3 + 2x^4 + 2x + 3$
45. $f(x) = e^x + 2\sin x - \dfrac{2}{\pi}(e^{\pi/2} + 4)x + 2$
47. $f(x) = -\ln x + (\ln 2)x - \ln 2$ **49.** 8   **51.** b
53.

55.

57.

59. $s(t) = 1 - \cos t - \sin t$ **61.** $s(t) = \frac{1}{3}t^3 + \frac{1}{2}t^2 - 2t + 3$
63. $s(t) = -10\sin t - 3\cos t + (6/\pi)t + 3$
65. (a) $s(t) = 450 - 4.9t^2$ (b) $\sqrt{450/4.9} \approx 9.58$ s
(c) $-9.8\sqrt{450/4.9} \approx -93.9$ m/s   (d) About 9.09 s
69. 225 ft   **71.** \$742.08   **73.** $\frac{130}{11} \approx 11.8$ s
75. $\frac{88}{15} \approx 5.87 \text{ ft/s}^2$ **77.** 62,500 km/h^2 $\approx 4.82 \text{ m/s}^2$
79. (a) 22.9125 mi   (b) 21.675 mi   (c) 30 min 33 s
(d) 55.425 mi

CHAPTER 4 REVIEW ■ **PAGE 358**
True-False Quiz
1. False   **3.** False   **5.** True   **7.** False   **9.** True
11. True   **13.** False   **15.** True   **17.** True
19. True   **21.** False

Exercises
1. Abs max $f(2) = f(5) = 18$, abs min $f(0) = -2$,
loc max $f(2) = 18$, loc min $f(4) = 14$
3. Abs max $f(2) = \frac{2}{5}$, abs and loc min $f(-\frac{1}{3}) = -\frac{9}{2}$
5. Abs and loc max $f(\pi/6) = \pi/6 + \sqrt{3}$,
abs min $f(-\pi) = -\pi - 2$, loc min $f(5\pi/6) = 5\pi/6 - \sqrt{3}$
7. 1   **9.** 4   **11.** 0   **13.** $\frac{1}{2}$
15.

17.

19. A. $\mathbb{R}$ B. y-int 2
C. None   D. None
E. Dec on $(-\infty, \infty)$ F. None
G. CU on $(-\infty, 0)$;
CD on $(0, \infty)$; IP $(0, 2)$
H. See graph at right.

21. A. $\mathbb{R}$ B. y-int 2
C. None   D. None
E. Inc on $(1, \infty)$; dec on $(-\infty, 1)$
F. Loc min $f(1) = 1$
G. CU on $(-\infty, 0)$, $\left(\frac{2}{3}, \infty\right)$;
CD on $\left(0, \frac{2}{3}\right)$; IPs $(0, 2)$, $\left(\frac{2}{3}, \frac{38}{27}\right)$
H. See graph at right.

23. A. $\{x \mid x \neq 0, 3\}$
B. None   C. None
D. HA $y = 0$; VA $x = 0$, $x = 3$
E. Inc on $(1, 3)$;
dec on $(-\infty, 0)$, $(0, 1)$, $(3, \infty)$
F. Loc min $f(1) = \frac{1}{4}$
G. CU on $(0, 3)$, $(3, \infty)$;
CD on $(-\infty, 0)$
H. See graph at right.

25. A. $(-\infty, 0) \cup (0, \infty)$
B. x-int 1   C. None
D. VA $x = 0$; SA $y = x - 3$
E. Inc on $(-\infty, -2)$, $(0, \infty)$;
dec on $(-2, 0)$
F. Loc max $f(-2) = -\frac{27}{4}$
G. CU on $(1, \infty)$; CD on $(-\infty, 0)$,
$(0, 1)$; IP $(1, 0)$
H. See graph at right.

27. A. $[-2, \infty)$
B. y-int 0; x-int $-2, 0$
C. None   D. None
E. Inc on $\left(-\frac{4}{3}, \infty\right)$, dec on $\left(-2, -\frac{4}{3}\right)$
F. Loc min $f\left(-\frac{4}{3}\right) = -\frac{4}{9}\sqrt{6}$
G. CU on $(-2, \infty)$
H. See graph at right.

29. A. $[-\pi, \pi]$ B. y-int 0; x-int $-\pi, 0, \pi$
C. None   D. None
E. Inc on $(-\pi/4, 3\pi/4)$; dec on $(-\pi, -\pi/4)$, $(3\pi/4, \pi)$
F. Loc max $f(3\pi/4) = \frac{1}{2}\sqrt{2}\,e^{3\pi/4}$,
loc min $f(-\pi/4) = -\frac{1}{2}\sqrt{2}\,e^{-\pi/4}$

G. CU on $(-\pi/2, \pi/2)$; CD on $(-\pi, -\pi/2)$, $(\pi/2, \pi)$;
IPs $(-\pi/2, -e^{-\pi/2})$, $(\pi/2, e^{\pi/2})$
H.

31. A. $\{x \mid |x| \geq 1\}$
B. None   C. About $(0, 0)$
D. HA $y = 0$
E. Dec on $(-\infty, -1)$, $(1, \infty)$
F. None
G. CU on $(1, \infty)$;
CD on $(-\infty, -1)$
H. See graph at right.

33. A. $\mathbb{R}$
B. y-int -2; x-int 2
C. None   D. HA $y = 0$
E. Inc on $(-\infty, 3)$; dec on $(3, \infty)$
F. Loc max $f(3) = e^{-3}$
G. CU on $(4, \infty)$;
CD on $(-\infty, 4)$;
IP $(4, 2e^{-4})$
H. See graph at right.

35. Inc on $\left(-\sqrt{3}, 0\right)$, $\left(0, \sqrt{3}\right)$;
dec on $\left(-\infty, -\sqrt{3}\right)$, $\left(\sqrt{3}, \infty\right)$;
loc max $f\left(\sqrt{3}\right) = \frac{2}{9}\sqrt{3}$,
loc min $f\left(-\sqrt{3}\right) = -\frac{2}{9}\sqrt{3}$;
CU on $\left(-\sqrt{6}, 0\right)$, $\left(\sqrt{6}, \infty\right)$;
CD on $\left(-\infty, -\sqrt{6}\right)$, $\left(0, \sqrt{6}\right)$;
IPs $\left(\sqrt{6}, \frac{5}{36}\sqrt{6}\right)$, $\left(-\sqrt{6}, -\frac{5}{36}\sqrt{6}\right)$

37. Inc on $(-0.23, 0)$, $(1.62, \infty)$; dec on $(-\infty, -0.23)$, $(0, 1.62)$;
loc max $f(0) = 2$; loc min $f(-0.23) \approx 1.96$, $f(1.62) \approx -19.2$;
CU on $(-\infty, -0.12)$, $(1.24, \infty)$;
CD on $(-0.12, 1.24)$; IPs $(-0.12, 1.98)$, $(1.24, -12.1)$

39.

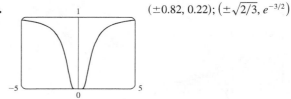

$(\pm 0.82, 0.22)$; $\left(\pm\sqrt{2/3},\, e^{-3/2}\right)$

41. $-2.96, -0.18, 3.01; -1.57, 1.57; -2.16, -0.75, 0.46, 2.21$
43. For $C > -1$, f is periodic with period 2π and has local maxima at $2n\pi + \pi/2$, n an integer. For $C \leq -1$, f has no graph. For $-1 < C \leq 1$, f has vertical asymptotes. For $C > 1$, f is continuous on $\mathbb{R}$. As C increases, f moves upward and its oscillations become less pronounced.
49. (a) 0   (b) CU on $\mathbb{R}$ **53.** $3\sqrt{3}\,r^2$
55. $4/\sqrt{3}$ cm from D **57.** $L = C$ **59.** \$11.50
61. 1.297383   **63.** 1.16718557
65. $F(x) = \frac{8}{3}x^{3/2} - 2x^3 + 3x + C$
67. $F(t) = -2\cos t - 3e^t + C$
69. $f(t) = t^2 + 3\cos t + 2$
71. $f(x) = \frac{1}{2}x^2 - x^3 + 4x^4 + 2x + 1$
73. $s(t) = t^2 - \tan^{-1}t + 1$
75. (b) $0.1e^x - \cos x + 0.9$ (c)

77. No
79. (b) About 8.5 in. by 2 in.   (c) $20/\sqrt{3}$ in. by $20\sqrt{2/3}$ in.
85. (a) $20\sqrt{2} \approx 28$ ft
(b) $\dfrac{dI}{dt} = \dfrac{-480k(h-4)}{[(h-4)^2 + 1600]^{5/2}}$, where k is the constant of proportionality

PROBLEMS PLUS ■ PAGE 363

3. Abs max $f(-5) = e^{45}$, no abs min   **7.** 24
9. $(-2, 4), (2, -4)$ **13.** $\left(1 + \sqrt{5}\,\right)/2$ **15.** $(m/2, m^2/4)$
17. $a \leq e^{1/e}$
21. (a) $T_1 = D/c_1$, $T_2 = (2h\sec\theta)/c_1 + (D - 2h\tan\theta)/c_2$, $T_3 = \sqrt{4h^2 + D^2}/c_1$
(c) $c_1 \approx 3.85$ km/s, $c_2 \approx 7.66$ km/s, $h \approx 0.42$ km
25. $3/(\sqrt[3]{2} - 1) \approx 11\frac{1}{2}$ h

CHAPTER 5

EXERCISES 5.1 ■ PAGE 375

1. (a) $R_5 \approx 12$, $L_5 \approx 22$

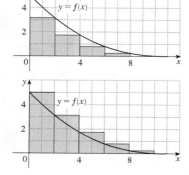

(b) $R_{10} \approx 14.4$, $L_{10} \approx 19.4$

3. (a) 0.6345, underestimate   (b) 0.7595, overestimate

5. (a) 8, 6.875   (b) 5, 5.375

(c) 5.75, 5.9375

(d) M_6

7. $n = 2$: upper $= 3\pi \approx 9.42$, lower $= 2\pi \approx 6.28$

$n = 4$: upper $= \left(10 + \sqrt{2}\right)(\pi/4) \approx 8.96$,

$\quad\quad$ lower $= \left(8 + \sqrt{2}\right)(\pi/4) \approx 7.39$

$n = 8$: upper ≈ 8.65, lower ≈ 7.86

9. 0.2533, 0.2170, 0.2101, 0.2050; 0.2
11. (a) Left: 0.8100, 0.7937, 0.7904;
right: 0.7600, 0.7770, 0.7804
13. 34.7 ft, 44.8 ft   **15.** 63.2 L, 70 L   **17.** 155 ft
19. 7840   **21.** $\lim\limits_{n\to\infty} \sum\limits_{i=1}^{n} \dfrac{2(1 + 2i/n)}{(1 + 2i/n)^2 + 1} \cdot \dfrac{2}{n}$
23. $\lim\limits_{n\to\infty} \sum\limits_{i=1}^{n} \sqrt{\sin(\pi i/n)} \cdot \dfrac{\pi}{n}$
25. The region under the graph of $y = \tan x$ from 0 to $\pi/4$
27. (a) $L_n < A < R_n$
29. (a) $\lim\limits_{n\to\infty} \dfrac{64}{n^6} \sum\limits_{i=1}^{n} i^5$ (b) $\dfrac{n^2(n + 1)^2(2n^2 + 2n - 1)}{12}$
(c) $\dfrac{32}{3}$
31. $\sin b$, 1

EXERCISES 5.2 ■ **PAGE 388**

1. -10
The Riemann sum represents
the sum of the areas of the two
rectangles above the x-axis minus
the sum of the areas of the three
rectangles below the x-axis; that is,
the *net area* of the rectangles with
respect to the x-axis.

3. $-\dfrac{49}{16}$
The Riemann sum represents the
sum of the areas of the two rectangles above the x-axis minus the
sum of the areas of the four
rectangles below the x-axis.

5. (a) 6   (b) 4   (c) 2
7. Lower, $L_5 = -64$; upper, $R_5 = 16$
9. 6.1820   **11.** 0.9071   **13.** 0.9029, 0.9018
15.

n	R_n
5	1.933766
10	1.983524
50	1.999342
100	1.999836

The values of R_n appear to be approaching 2.

17. $\int_0^1 \dfrac{e^x}{1 + x}\, dx$ **19.** $\int_2^7 (5x^3 - 4x)\, dx$
21. -9 **23.** $\dfrac{2}{3}$ **25.** $-\dfrac{3}{4}$
29. $\lim\limits_{n\to\infty} \sum\limits_{i=1}^{n} \sqrt{4 + (1 + 2i/n)} \cdot \dfrac{2}{n}$
31. $\lim\limits_{n\to\infty} \sum\limits_{i=1}^{n} \left(\sin \dfrac{5\pi i}{n}\right) \dfrac{\pi}{n} = \dfrac{2}{5}$
33. (a) 4   (b) 10   (c) -3 (d) 2
35. $\dfrac{3}{2}$ **37.** $3 + \dfrac{9}{4}\pi$ **39.** $\dfrac{25}{4}$ **41.** 0   **43.** 3
45. $e^5 - e^3$ **47.** $\int_{-1}^5 f(x)\, dx$ **49.** 122
51. $B < E < A < D < C$ **53.** 15
59. $0 \leq \int_0^1 x^3\, dx \leq 1$ **61.** $\dfrac{\pi}{12} \leq \int_{\pi/4}^{\pi/3} \tan x\, dx \leq \dfrac{\pi}{12}\sqrt{3}$
63. $0 \leq \int_0^2 xe^{-x}\, dx \leq 2/e$ **67.** $\int_1^2 \arctan x\, dx$
73. $\int_0^1 x^4\, dx$ **75.** $\dfrac{1}{2}$

EXERCISES 5.3 ■ **PAGE 399**

1. One process undoes what the other one does. See the Fundamental Theorem of Calculus, page 398.
3. (a) 0, 2, 5, 7, 3   (d)
(b) (0, 3)
(c) $x = 3$

5.

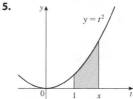

(a), (b) x^2

7. $g'(x) = \sqrt{x + x^3}$ **9.** $g'(s) = (s - s^2)^8$

11. $F'(x) = -\sqrt{1 + \sec x}$ **13.** $h'(x) = xe^x$

15. $y' = \dfrac{3(3x + 2)}{1 + (3x + 2)^3}$

17. $y' = -\frac{1}{2}\tan\sqrt{x}$ **19.** $\frac{26}{3}$ **21.** 2   **23.** $\frac{52}{3}$

25. $1 + \sqrt{3}/2$ **27.** $-\frac{37}{6}$ **29.** $\frac{82}{5}$ **31.** 1   **33.** $\frac{15}{4}$

35. $\ln 2 + 7$ **37.** $\dfrac{1}{e + 1} + e - 1$ **39.** $4\pi/3$

41. $\dfrac{15}{\ln 2}$ **43.** 0   **45.** $\frac{16}{3}$ **47.** $\frac{32}{3}$

49. $\frac{243}{4}$ **51.** 2

53. 3.75

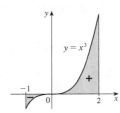

55. The function $f(x) = x^{-4}$ is not continuous on the interval $[-2, 1]$, so FTC2 cannot be applied.

57. The function $f(\theta) = \sec\theta \tan\theta$ is not continuous on the interval $[\pi/3, \pi]$, so FTC2 cannot be applied.

59. $g'(x) = \dfrac{-2(4x^2 - 1)}{4x^2 + 1} + \dfrac{3(9x^2 - 1)}{9x^2 + 1}$

61. $F'(x) = 2xe^{x^4} - e^{x^2}$

63. $y' = \sin x \ln(1 + 2\cos x) + \cos x \ln(1 + 2\sin x)$

65. $(-4, 0)$ **67.** $y = e^4x - 2e^4$ **69.** 29

71. (a) $-2\sqrt{n}$, $\sqrt{4n - 2}$, n an integer > 0

(b) $(0, 1)$, $\left(-\sqrt{4n - 1}, -\sqrt{4n - 3}\right)$, and $\left(\sqrt{4n - 1}, \sqrt{4n + 1}\right)$, n an integer > 0 (c) 0.74

73. (a) Loc max at 1 and 5; loc min at 3 and 7

(b) $x = 9$

(c) $\left(\frac{1}{2}, 2\right)$, $(4, 6)$, $(8, 9)$

(d) See graph at right.

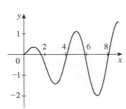

75. $\frac{7}{10}$ **83.** $f(x) = x^{3/2}$, $a = 9$

85. (b) Average expenditure over $[0, t]$; minimize average expenditure

EXERCISES 5.4 ■ **PAGE 408**

5. $\frac{1}{2.3}x^{2.3} + 2x^{3.5} + C$ **7.** $5x + \frac{2}{9}x^3 + \frac{3}{16}x^4 + C$

9. $\frac{2}{3}u^3 + \frac{9}{2}u^2 + 4u + C$ **11.** $\ln|x| + 2\sqrt{x} + x + C$

13. $-\cos x + \cosh x + C$ **15.** $\theta + \tan\theta + C$

17. $\dfrac{2^t}{\ln 2} + \dfrac{10^t}{\ln 10} + C$

19. $\sin x + \frac{1}{4}x^2 + C$

21. $-\frac{10}{3}$ **23.** $\frac{21}{5}$ **25.** -2 **27.** $5e^\pi + 1$ **29.** 36

31. $\frac{55}{63}$ **33.** $\frac{3}{4} - 2\ln 2$ **35.** $\dfrac{1}{11} + \dfrac{9}{\ln 10}$

37. $1 + \pi/4$ **39.** $\frac{69}{4}$ **41.** $\pi/3$ **43.** $\pi/6$ **45.** -3.5

47. ≈ 1.36 **49.** $\frac{4}{3}$

51. The increase in the child's weight (in pounds) between the ages of 5 and 10

53. Number of gallons of oil leaked in the first 2 hours

55. Increase in revenue when production is increased from 1000 to 5000 units

57. Newton-meters (or joules)   **59.** (a) $-\frac{3}{2}$ m   (b) $\frac{41}{6}$ m

61. (a) $v(t) = \frac{1}{2}t^2 + 4t + 5$ m/s   (b) $416\frac{2}{3}$ m

63. $46\frac{2}{3}$ kg   **65.** 1.4 mi   **67.** \$58,000   **69.** 39.8 ft/s

71. 5443 bacteria   **73.** 4.75×10^5 megawatt-hours

EXERCISES 5.5 ■ **PAGE 418**

1. $\frac{1}{2}\sin 2x + C$ **3.** $\frac{2}{9}(x^3 + 1)^{3/2} + C$

5. $\frac{1}{4}\ln|x^4 - 5| + C$ **7.** $-\frac{1}{3}(1 - x^2)^{3/2} + C$

9. $-\frac{1}{20}(1 - 2x)^{10} + C$ **11.** $(2/\pi)\sin(\pi t/2) + C$

13. $-\frac{1}{3}\ln|5 - 3x| + C$

15. $-\frac{1}{4}\cos^4\theta + C$ **17.** $\dfrac{1}{1 - e^u} + C$

19. $\frac{2}{3}\sqrt{3ax + bx^3} + C$ **21.** $\frac{1}{3}(\ln x)^3 + C$ **23.** $\frac{1}{4}\tan^4\theta + C$

25. $\frac{2}{3}(1 + e^x)^{3/2} + C$ **27.** $\frac{1}{15}(x^3 + 3x)^5 + C$

29. $-\dfrac{1}{\ln 5}\cos(5^t) + C$ **31.** $\frac{1}{3}(\arctan x)^3 + C$

33. $\frac{1}{5}\sin(1 + 5t) + C$ **35.** $-\frac{2}{3}(\cot x)^{3/2} + C$

37. $\frac{1}{3}\sinh^3 x + C$ **39.** $-\ln(1 + \cos^2 x) + C$

41. $\ln|\sin x| + C$ **43.** $\ln|\sin^{-1}x| + C$

45. $\tan^{-1}x + \frac{1}{2}\ln(1 + x^2) + C$

47. $\frac{1}{40}(2x + 5)^{10} - \frac{5}{36}(2x + 5)^9 + C$

49. $\frac{1}{8}(x^2 - 1)^4 + C$ **51.** $-e^{\cos x} + C$

53. $2/\pi$ **55.** $\frac{45}{28}$ **57.** $2/\sqrt{3} - 1$ **59.** $e - \sqrt{e}$

61. 0   **63.** 3   **65.** $\frac{1}{3}(2\sqrt{2} - 1)a^3$ **67.** $\frac{16}{15}$ **69.** 2

71. $\ln(e + 1)$ **73.** $\frac{1}{6}$ **75.** $\sqrt{3} - \frac{1}{3}$ **77.** 6π

79. All three areas are equal.   **81.** ≈ 4512 L

83. $\dfrac{5}{4\pi}\left(1 - \cos\dfrac{2\pi t}{5}\right)$ L

85. $C_0(1 - e^{-30r/V})$; the total amount of urea removed from the blood in the first 30 minutes of dialysis treatment
87. 5   **93.** $\pi^2/4$

CHAPTER 5 REVIEW ▪ PAGE 421

True-False Quiz
1. True   **3.** True   **5.** False   **7.** True   **9.** True
11. False   **13.** True   **15.** False   **17.** False

Exercises
1. (a) 8   (b) 5.7

      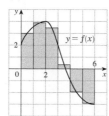

3. $\frac{1}{2} + \pi/4$ **5.** 3   **7.** f is c, f' is b, $\int_0^x f(t)\,dt$ is a
9. 3, 0   **11.** 37   **13.** $\frac{9}{10}$ **15.** -76 **17.** $\frac{21}{4}$
19. Does not exist   **21.** $\frac{1}{3}\sin 1$ **23.** 0
25. $-(1/x) - 2\ln|x| + x + C$
27. $\sqrt{x^2 + 4x} + C$ **29.** $[1/(2\pi)]\sin^2\pi t + C$
31. $2e^{\sqrt{x}} + C$ **33.** $-\frac{1}{2}[\ln(\cos x)]^2 + C$
35. $\frac{1}{4}\ln(1 + x^4) + C$ **37.** $\ln|1 + \sec\theta| + C$ **39.** $\frac{23}{3}$
41. $2\sqrt{1 + \sin x} + C$ **43.** $\frac{64}{5}$ **45.** $F'(x) = x^2/(1 + x^3)$
47. $g'(x) = 4x^3\cos(x^8)$ **49.** $y' = (2e^x - e^{\sqrt{x}})/(2x)$
51. $4 \leqslant \int_1^3 \sqrt{x^2 + 3}\,dx \leqslant 4\sqrt{3}$ **57.** 0.280981
59. Number of barrels of oil consumed from Jan. 1, 2000, through Jan. 1, 2008
61. 72,400   **63.** 3   **65.** $c \approx 1.62$
67. $f(x) = e^{2x}(2x - 1)/(1 - e^{-x})$ **73.** $\frac{2}{3}$

PROBLEMS PLUS ▪ PAGE 425

1. $\pi/2$ **3.** $2k$ **5.** -1 **7.** e^{-2} **9.** $[-1, 2]$
11. (a) $\frac{1}{2}(n - 1)n$
(b) $\frac{1}{2}[\![b]\!](2b - [\![b]\!] - 1) - \frac{1}{2}[\![a]\!](2a - [\![a]\!] - 1)$
17. $y = -\dfrac{2b}{a^2}x^2 + \dfrac{3b}{a}x$ **19.** $2(\sqrt{2} - 1)$

21. $y = 6x - 15$ **23.** $2x - 3y + 19 = 0$

25. $5x + y = 11$ **27.** $y = 3x - 2$ **29.** $y = 3x - 3$

31. $y = 5$ **33.** $x + 2y + 11 = 0$ **35.** $5x - 2y + 1 = 0$

37. $m = -\frac{1}{3},$ **39.** $m = 0,$ **41.** $m = \frac{3}{4},$
   $b = 0$ $b = -2$ $b = -3$

APPENDIXES

EXERCISES A ■ PAGE A9

1. 18   **3.** π **5.** $5 - \sqrt{5}$ **7.** $2 - x$

9. $|x + 1| = \begin{cases} x + 1 & \text{for } x \geq -1 \\ -x - 1 & \text{for } x < -1 \end{cases}$ **11.** $x^2 + 1$

13. $(-2, \infty)$ **15.** $[-1, \infty)$

17. $(3, \infty)$ **19.** $(2, 6)$

21. $(0, 1]$ **23.** $\left[-1, \frac{1}{2}\right)$

25. $(-\infty, 1) \cup (2, \infty)$ **27.** $\left[-1, \frac{1}{2}\right]$

29. $(-\infty, \infty)$ **31.** $\left(-\sqrt{3}, \sqrt{3}\right)$

33. $(-\infty, 1]$ **35.** $(-1, 0) \cup (1, \infty)$

37. $(-\infty, 0) \cup \left(\frac{1}{4}, \infty\right)$

39. $10 \leq C \leq 35$ **41.** (a) $T = 20 - 10h, 0 \leq h \leq 12$
(b) $-30°C \leq T \leq 20°C$ **43.** $\pm\frac{3}{2}$ **45.** $2, -\frac{4}{3}$
47. $(-3, 3)$ **49.** $(3, 5)$ **51.** $(-\infty, -7] \cup [-3, \infty)$
53. $[1.3, 1.7]$ **55.** $[-4, -1] \cup [1, 4]$
57. $x \geq (a + b)c/(ab)$ **59.** $x > (c - b)/a$

EXERCISES B ■ PAGE A15

1. 5   **3.** $\sqrt{74}$ **5.** $2\sqrt{37}$ **7.** 2   **9.** $-\frac{9}{2}$

17.   **19.**

43.   **45.**

47.   **49.**

51.

53. $(0, -4)$ **55.** (a) $(4, 9)$ (b) $(3.5, -3)$ **57.** $(1, -2)$

59. $y = x - 3$ **61.** (b) $4x - 3y - 24 = 0$

EXERCISES C ■ PAGE A23

1. $(x - 3)^2 + (y + 1)^2 = 25$ **3.** $x^2 + y^2 = 65$

5. $(2, -5), 4$ **7.** $\left(-\frac{1}{2}, 0\right), \frac{1}{2}$ **9.** $\left(\frac{1}{4}, -\frac{1}{4}\right), \sqrt{10}/4$

11. Parabola   **13.** Ellipse

15. Hyperbola

17. Ellipse

19. Parabola

21. Hyperbola

23. Hyperbola

25. Ellipse

27. Parabola

29. Parabola

31. Ellipse

33.

35. $y = x^2 - 2x$

37.

39.

1. $7\pi/6$ **3.** $\pi/20$ **5.** 5π **7.** $720°$ **9.** $75°$
11. $-67.5°$ **13.** 3π cm   **15.** $\frac{2}{3}$ rad $= (120/\pi)°$
17.

19.

21.

23. $\sin(3\pi/4) = 1/\sqrt{2}$, $\cos(3\pi/4) = -1/\sqrt{2}$, $\tan(3\pi/4) = -1$,
$\csc(3\pi/4) = \sqrt{2}$, $\sec(3\pi/4) = -\sqrt{2}$, $\cot(3\pi/4) = -1$
25. $\sin(9\pi/2) = 1$, $\cos(9\pi/2) = 0$, $\csc(9\pi/2) = 1$, $\cot(9\pi/2) = 0$
, $\tan(9\pi/2)$ and $\sec(9\pi/2)$ undefined
27. $\sin(5\pi/6) = \frac{1}{2}$, $\cos(5\pi/6) = -\sqrt{3}/2$, $\tan(5\pi/6) = -1/\sqrt{3}$,
$\csc(5\pi/6) = 2$, $\sec(5\pi/6) = -2/\sqrt{3}$, $\cot(5\pi/6) = -\sqrt{3}$
29. $\cos\theta = \frac{4}{5}$, $\tan\theta = \frac{3}{4}$, $\csc\theta = \frac{5}{3}$, $\sec\theta = \frac{5}{4}$, $\cot\theta = \frac{4}{3}$
31. $\sin\phi = \sqrt{5}/3$, $\cos\phi = -\frac{2}{3}$, $\tan\phi = -\sqrt{5}/2$, $\csc\phi = 3/\sqrt{5}$,
$\cot\phi = -2/\sqrt{5}$
33. $\sin\beta = -1/\sqrt{10}$, $\cos\beta = -3/\sqrt{10}$, $\tan\beta = \frac{1}{3}$,
$\csc\beta = -\sqrt{10}$, $\sec\beta = -\sqrt{10}/3$
35. 5.73576 cm   **37.** 24.62147 cm   **59.** $\frac{1}{15}(4 + 6\sqrt{2})$
61. $\frac{1}{15}(3 + 8\sqrt{2})$ **63.** $\frac{24}{25}$ **65.** $\pi/3, 5\pi/3$
67. $\pi/4, 3\pi/4, 5\pi/4, 7\pi/4$ **69.** $\pi/6, \pi/2, 5\pi/6, 3\pi/2$
71. $0, \pi, 2\pi$ **73.** $0 \leq x \leq \pi/6$ and $5\pi/6 \leq x \leq 2\pi$
75. $0 \leq x < \pi/4, 3\pi/4 < x < 5\pi/4, 7\pi/4 < x \leq 2\pi$
77.

79.

81.

89. 14.34457 cm^2

EXERCISES E ▪ PAGE A38

1. $\sqrt{1} + \sqrt{2} + \sqrt{3} + \sqrt{4} + \sqrt{5}$ **3.** $3^4 + 3^5 + 3^6$

5. $-1 + \frac{1}{3} + \frac{3}{5} + \frac{5}{7} + \frac{7}{9}$ **7.** $1^{10} + 2^{10} + 3^{10} + \cdots + n^{10}$

9. $1 - 1 + 1 - 1 + \cdots + (-1)^{n-1}$ **11.** $\displaystyle\sum_{i=1}^{10} i$

13. $\displaystyle\sum_{i=1}^{19} \frac{i}{i+1}$ **15.** $\displaystyle\sum_{i=1}^{n} 2i$ **17.** $\displaystyle\sum_{i=0}^{5} 2^i$ **19.** $\displaystyle\sum_{i=1}^{n} x^i$

21. 80 **23.** 3276 **25.** 0 **27.** 61 **29.** $n(n+1)$

31. $n(n^2 + 6n + 17)/3$ **33.** $n(n^2 + 6n + 11)/3$

35. $n(n^3 + 2n^2 - n - 10)/4$

41. (a) n^4 (b) $5^{100} - 1$ (c) $\frac{97}{300}$ (d) $a_n - a_0$

43. $\frac{1}{3}$ **45.** 14 **49.** $2^{n+1} + n^2 + n - 2$

EXERCISES G ▪ PAGE A56

1. (b) 0.405

EXERCISES H ▪ PAGE A64

1. $8 - 4i$ **3.** $13 + 18i$ **5.** $12 - 7i$ **7.** $\frac{11}{13} + \frac{10}{13}i$

9. $\frac{1}{2} - \frac{1}{2}i$ **11.** $-i$ **13.** $5i$ **15.** $12 + 5i,\ 13$

17. $4i, 4$ **19.** $\pm\frac{3}{2}i$ **21.** $-1 \pm 2i$

23. $-\frac{1}{2} \pm (\sqrt{7}/2)i$ **25.** $3\sqrt{2}\,[\cos(3\pi/4) + i\sin(3\pi/4)]$

27. $5\{\cos[\tan^{-1}(\frac{4}{3})] + i\sin[\tan^{-1}(\frac{4}{3})]\}$

29. $4[\cos(\pi/2) + i\sin(\pi/2)],\ \cos(-\pi/6) + i\sin(-\pi/6),$
$\frac{1}{2}[\cos(-\pi/6) + i\sin(-\pi/6)]$

31. $4\sqrt{2}\,[\cos(7\pi/12) + i\sin(7\pi/12)],$
$(2\sqrt{2})[\cos(13\pi/12) + i\sin(13\pi/12)], \frac{1}{4}[\cos(\pi/6) + i\sin(\pi/6)]$

33. -1024 **35.** $-512\sqrt{3} + 512i$

37. $\pm1, \pm i, (1/\sqrt{2})(\pm1 \pm i)$ **39.** $\pm(\sqrt{3}/2) + \frac{1}{2}i, -i$

41. i **43.** $\frac{1}{2} + (\sqrt{3}/2)i$ **45.** $-e^2$

47. $\cos 3\theta = \cos^3\theta - 3\cos\theta\sin^2\theta,$
$\sin 3\theta = 3\cos^2\theta\sin\theta - \sin^3\theta$

Index

RP denotes Reference Page numbers.

ALGEBRA

Arithmetic Operations

$$a(b + c) = ab + ac$$

$$\frac{a + c}{b} = \frac{a}{b} + \frac{c}{b}$$

$$\frac{a}{b} + \frac{c}{d} = \frac{ad + bc}{bd}$$

$$\frac{\frac{a}{b}}{\frac{c}{d}} = \frac{a}{b} \times \frac{d}{c} = \frac{ad}{bc}$$

Exponents and Radicals

$$x^m x^n = x^{m+n}$$

$$(x^m)^n = x^{mn}$$

$$(xy)^n = x^n y^n$$

$$x^{1/n} = \sqrt[n]{x}$$

$$\sqrt[n]{xy} = \sqrt[n]{x}\,\sqrt[n]{y}$$

$$\frac{x^m}{x^n} = x^{m-n}$$

$$x^{-n} = \frac{1}{x^n}$$

$$\left(\frac{x}{y}\right)^n = \frac{x^n}{y^n}$$

$$x^{m/n} = \sqrt[n]{x^m} = \left(\sqrt[n]{x}\right)^m$$

$$\sqrt[n]{\frac{x}{y}} = \frac{\sqrt[n]{x}}{\sqrt[n]{y}}$$

Factoring Special Polynomials

$$x^2 - y^2 = (x + y)(x - y)$$
$$x^3 + y^3 = (x + y)(x^2 - xy + y^2)$$
$$x^3 - y^3 = (x - y)(x^2 + xy + y^2)$$

Binomial Theorem

$$(x + y)^2 = x^2 + 2xy + y^2 \qquad (x - y)^2 = x^2 - 2xy + y^2$$
$$(x + y)^3 = x^3 + 3x^2y + 3xy^2 + y^3$$
$$(x - y)^3 = x^3 - 3x^2y + 3xy^2 - y^3$$
$$(x + y)^n = x^n + nx^{n-1}y + \frac{n(n - 1)}{2}x^{n-2}y^2$$
$$+ \cdots + \binom{n}{k}x^{n-k}y^k + \cdots + nxy^{n-1} + y^n$$

$$\text{where } \binom{n}{k} = \frac{n(n - 1) \cdots (n - k + 1)}{1 \cdot 2 \cdot 3 \cdot \cdots \cdot k}$$

Quadratic Formula

If $ax^2 + bx + c = 0$, then $x = \dfrac{-b \pm \sqrt{b^2 - 4ac}}{2a}$.

Inequalities and Absolute Value

If $a < b$ and $b < c$, then $a < c$.

If $a < b$, then $a + c < b + c$.

If $a < b$ and $c > 0$, then $ca < cb$.

If $a < b$ and $c < 0$, then $ca > cb$.

If $a > 0$, then

$$|x| = a \quad \text{means} \quad x = a \quad \text{or} \quad x = -a$$
$$|x| < a \quad \text{means} \quad -a < x < a$$
$$|x| > a \quad \text{means} \quad x > a \quad \text{or} \quad x < -a$$

GEOMETRY

Geometric Formulas

Formulas for area A, circumference C, and volume V:

Triangle	Circle	Sector of Circle
$A = \frac{1}{2}bh$	$A = \pi r^2$	$A = \frac{1}{2}r^2\theta$
$\quad = \frac{1}{2}ab\sin\theta$	$C = 2\pi r$	$s = r\theta$ (θ in radians)

Sphere	Cylinder	Cone
$V = \frac{4}{3}\pi r^3$	$V = \pi r^2 h$	$V = \frac{1}{3}\pi r^2 h$
$A = 4\pi r^2$		$A = \pi r\sqrt{r^2 + h^2}$

Distance and Midpoint Formulas

Distance between $P_1(x_1, y_1)$ and $P_2(x_2, y_2)$:

$$d = \sqrt{(x_2 - x_1)^2 + (y_2 - y_1)^2}$$

Midpoint of $\overline{P_1P_2}$: $\left(\dfrac{x_1 + x_2}{2}, \dfrac{y_1 + y_2}{2}\right)$

Lines

Slope of line through $P_1(x_1, y_1)$ and $P_2(x_2, y_2)$:

$$m = \frac{y_2 - y_1}{x_2 - x_1}$$

Point-slope equation of line through $P_1(x_1, y_1)$ with slope m:

$$y - y_1 = m(x - x_1)$$

Slope-intercept equation of line with slope m and y-intercept b:

$$y = mx + b$$

Circles

Equation of the circle with center (h, k) and radius r:

$$(x - h)^2 + (y - k)^2 = r^2$$

TRIGONOMETRY

Angle Measurement

π radians $= 180°$

$1° = \dfrac{\pi}{180}$ rad   1 rad $= \dfrac{180°}{\pi}$

$s = r\theta$

(θ in radians)

Right Angle Trigonometry

$\sin\theta = \dfrac{\text{opp}}{\text{hyp}}$ $\csc\theta = \dfrac{\text{hyp}}{\text{opp}}$

$\cos\theta = \dfrac{\text{adj}}{\text{hyp}}$ $\sec\theta = \dfrac{\text{hyp}}{\text{adj}}$

$\tan\theta = \dfrac{\text{opp}}{\text{adj}}$ $\cot\theta = \dfrac{\text{adj}}{\text{opp}}$

Trigonometric Functions

$\sin\theta = \dfrac{y}{r}$ $\csc\theta = \dfrac{r}{y}$

$\cos\theta = \dfrac{x}{r}$ $\sec\theta = \dfrac{r}{x}$

$\tan\theta = \dfrac{y}{x}$ $\cot\theta = \dfrac{x}{y}$

Graphs of Trigonometric Functions

Trigonometric Functions of Important Angles

θ	radians	$\sin\theta$	$\cos\theta$	$\tan\theta$
0°	0	0	1	0
30°	$\pi/6$	$1/2$	$\sqrt{3}/2$	$\sqrt{3}/3$
45°	$\pi/4$	$\sqrt{2}/2$	$\sqrt{2}/2$	1
60°	$\pi/3$	$\sqrt{3}/2$	$1/2$	$\sqrt{3}$
90°	$\pi/2$	1	0	—

Fundamental Identities

$\csc\theta = \dfrac{1}{\sin\theta}$ $\sec\theta = \dfrac{1}{\cos\theta}$

$\tan\theta = \dfrac{\sin\theta}{\cos\theta}$ $\cot\theta = \dfrac{\cos\theta}{\sin\theta}$

$\cot\theta = \dfrac{1}{\tan\theta}$ $\sin^2\theta + \cos^2\theta = 1$

$1 + \tan^2\theta = \sec^2\theta$ $1 + \cot^2\theta = \csc^2\theta$

$\sin(-\theta) = -\sin\theta$ $\cos(-\theta) = \cos\theta$

$\tan(-\theta) = -\tan\theta$ $\sin\left(\dfrac{\pi}{2} - \theta\right) = \cos\theta$

$\cos\left(\dfrac{\pi}{2} - \theta\right) = \sin\theta$ $\tan\left(\dfrac{\pi}{2} - \theta\right) = \cot\theta$

The Law of Sines

$$\dfrac{\sin A}{a} = \dfrac{\sin B}{b} = \dfrac{\sin C}{c}$$

The Law of Cosines

$a^2 = b^2 + c^2 - 2bc\cos A$

$b^2 = a^2 + c^2 - 2ac\cos B$

$c^2 = a^2 + b^2 - 2ab\cos C$

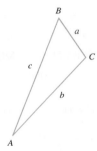

Addition and Subtraction Formulas

$\sin(x + y) = \sin x \cos y + \cos x \sin y$

$\sin(x - y) = \sin x \cos y - \cos x \sin y$

$\cos(x + y) = \cos x \cos y - \sin x \sin y$

$\cos(x - y) = \cos x \cos y + \sin x \sin y$

$\tan(x + y) = \dfrac{\tan x + \tan y}{1 - \tan x \tan y}$

$\tan(x - y) = \dfrac{\tan x - \tan y}{1 + \tan x \tan y}$

Double-Angle Formulas

$\sin 2x = 2\sin x \cos x$

$\cos 2x = \cos^2 x - \sin^2 x = 2\cos^2 x - 1 = 1 - 2\sin^2 x$

$\tan 2x = \dfrac{2\tan x}{1 - \tan^2 x}$

Half-Angle Formulas

$\sin^2 x = \dfrac{1 - \cos 2x}{2}$ $\cos^2 x = \dfrac{1 + \cos 2x}{2}$

Cut here and keep for reference ✂

SPECIAL FUNCTIONS

Power Functions $f(x) = x^a$

(i) $f(x) = x^n$, n a positive integer

n even

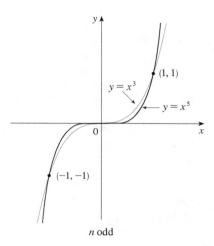

n odd

(ii) $f(x) = x^{1/n} = \sqrt[n]{x}$, n a positive integer

$f(x) = \sqrt{x}$

$f(x) = \sqrt[3]{x}$

(iii) $f(x) = x^{-1} = \dfrac{1}{x}$

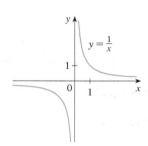

Inverse Trigonometric Functions

$$\arcsin x = \sin^{-1}x = y \iff \sin y = x \quad \text{and} \quad -\frac{\pi}{2} \leqslant y \leqslant \frac{\pi}{2}$$

$$\arccos x = \cos^{-1}x = y \iff \cos y = x \quad \text{and} \quad 0 \leqslant y \leqslant \pi$$

$$\arctan x = \tan^{-1}x = y \iff \tan y = x \quad \text{and} \quad -\frac{\pi}{2} < y < \frac{\pi}{2}$$

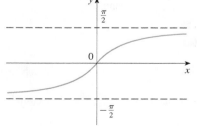

$y = \tan^{-1}x = \arctan x$

$$\lim_{x \to -\infty} \tan^{-1}x = -\frac{\pi}{2}$$

$$\lim_{x \to \infty} \tan^{-1}x = \frac{\pi}{2}$$

SPECIAL FUNCTIONS

Exponential and Logarithmic Functions

$$\log_b x = y \iff b^y = x$$

$$\ln x = \log_e x, \quad \text{where} \quad \ln e = 1$$

$$\ln x = y \iff e^y = x$$

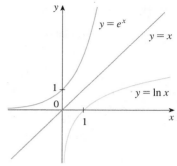

Cancellation Equations

$$\log_b(b^x) = x \qquad b^{\log_b x} = x$$

$$\ln(e^x) = x \qquad e^{\ln x} = x$$

Laws of Logarithms

1. $\log_b(xy) = \log_b x + \log_b y$

2. $\log_b\left(\dfrac{x}{y}\right) = \log_b x - \log_b y$

3. $\log_b(x^r) = r\log_b x$

$$\lim_{x \to -\infty} e^x = 0 \qquad\qquad \lim_{x \to \infty} e^x = \infty$$

$$\lim_{x \to 0^+} \ln x = -\infty \qquad \lim_{x \to \infty} \ln x = \infty$$

Exponential functions

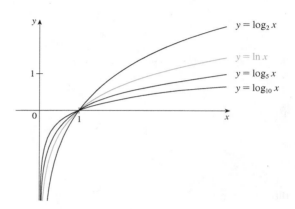

Logarithmic functions

Hyperbolic Functions

$$\sinh x = \frac{e^x - e^{-x}}{2} \qquad\qquad \operatorname{csch} x = \frac{1}{\sinh x}$$

$$\cosh x = \frac{e^x + e^{-x}}{2} \qquad\qquad \operatorname{sech} x = \frac{1}{\cosh x}$$

$$\tanh x = \frac{\sinh x}{\cosh x} \qquad\qquad \coth x = \frac{\cosh x}{\sinh x}$$

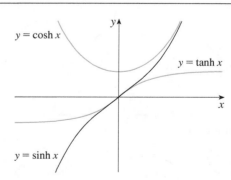

Inverse Hyperbolic Functions

$$y = \sinh^{-1}x \iff \sinh y = x$$

$$y = \cosh^{-1}x \iff \cosh y = x \quad \text{and} \quad y \geqslant 0$$

$$y = \tanh^{-1}x \iff \tanh y = x$$

$$\sinh^{-1}x = \ln\left(x + \sqrt{x^2 + 1}\right)$$

$$\cosh^{-1}x = \ln\left(x + \sqrt{x^2 - 1}\right)$$

$$\tanh^{-1}x = \tfrac{1}{2}\ln\left(\frac{1 + x}{1 - x}\right)$$

DIFFERENTIATION RULES

General Formulas

1. $\dfrac{d}{dx}(c) = 0$

2. $\dfrac{d}{dx}[cf(x)] = cf'(x)$

3. $\dfrac{d}{dx}[f(x) + g(x)] = f'(x) + g'(x)$

4. $\dfrac{d}{dx}[f(x) - g(x)] = f'(x) - g'(x)$

5. $\dfrac{d}{dx}[f(x)g(x)] = f(x)g'(x) + g(x)f'(x)$ (Product Rule)

6. $\dfrac{d}{dx}\left[\dfrac{f(x)}{g(x)}\right] = \dfrac{g(x)f'(x) - f(x)g'(x)}{[g(x)]^2}$ (Quotient Rule)

7. $\dfrac{d}{dx}f(g(x)) = f'(g(x))g'(x)$ (Chain Rule)

8. $\dfrac{d}{dx}(x^n) = nx^{n-1}$ (Power Rule)

Exponential and Logarithmic Functions

9. $\dfrac{d}{dx}(e^x) = e^x$

10. $\dfrac{d}{dx}(b^x) = b^x \ln b$

11. $\dfrac{d}{dx}\ln|x| = \dfrac{1}{x}$

12. $\dfrac{d}{dx}(\log_b x) = \dfrac{1}{x \ln b}$

Trigonometric Functions

13. $\dfrac{d}{dx}(\sin x) = \cos x$

14. $\dfrac{d}{dx}(\cos x) = -\sin x$

15. $\dfrac{d}{dx}(\tan x) = \sec^2 x$

16. $\dfrac{d}{dx}(\csc x) = -\csc x \cot x$

17. $\dfrac{d}{dx}(\sec x) = \sec x \tan x$

18. $\dfrac{d}{dx}(\cot x) = -\csc^2 x$

Inverse Trigonometric Functions

19. $\dfrac{d}{dx}(\sin^{-1}x) = \dfrac{1}{\sqrt{1 - x^2}}$

20. $\dfrac{d}{dx}(\cos^{-1}x) = -\dfrac{1}{\sqrt{1 - x^2}}$

21. $\dfrac{d}{dx}(\tan^{-1}x) = \dfrac{1}{1 + x^2}$

22. $\dfrac{d}{dx}(\csc^{-1}x) = -\dfrac{1}{x\sqrt{x^2 - 1}}$

23. $\dfrac{d}{dx}(\sec^{-1}x) = \dfrac{1}{x\sqrt{x^2 - 1}}$

24. $\dfrac{d}{dx}(\cot^{-1}x) = -\dfrac{1}{1 + x^2}$

Hyperbolic Functions

25. $\dfrac{d}{dx}(\sinh x) = \cosh x$

26. $\dfrac{d}{dx}(\cosh x) = \sinh x$

27. $\dfrac{d}{dx}(\tanh x) = \operatorname{sech}^2 x$

28. $\dfrac{d}{dx}(\operatorname{csch} x) = -\operatorname{csch} x \coth x$

29. $\dfrac{d}{dx}(\operatorname{sech} x) = -\operatorname{sech} x \tanh x$

30. $\dfrac{d}{dx}(\coth x) = -\operatorname{csch}^2 x$

Inverse Hyperbolic Functions

31. $\dfrac{d}{dx}(\sinh^{-1}x) = \dfrac{1}{\sqrt{1 + x^2}}$

32. $\dfrac{d}{dx}(\cosh^{-1}x) = \dfrac{1}{\sqrt{x^2 - 1}}$

33. $\dfrac{d}{dx}(\tanh^{-1}x) = \dfrac{1}{1 - x^2}$

34. $\dfrac{d}{dx}(\operatorname{csch}^{-1}x) = -\dfrac{1}{|x|\sqrt{x^2 + 1}}$

35. $\dfrac{d}{dx}(\operatorname{sech}^{-1}x) = -\dfrac{1}{x\sqrt{1 - x^2}}$

36. $\dfrac{d}{dx}(\coth^{-1}x) = \dfrac{1}{1 - x^2}$

TABLE OF INTEGRALS

Basic Forms

1. $\displaystyle\int u\, dv = uv - \int v\, du$

2. $\displaystyle\int u^n\, du = \frac{u^{n+1}}{n+1} + C, \quad n \neq -1$

3. $\displaystyle\int \frac{du}{u} = \ln|u| + C$

4. $\displaystyle\int e^u\, du = e^u + C$

5. $\displaystyle\int b^u\, du = \frac{b^u}{\ln b} + C$

6. $\displaystyle\int \sin u\, du = -\cos u + C$

7. $\displaystyle\int \cos u\, du = \sin u + C$

8. $\displaystyle\int \sec^2 u\, du = \tan u + C$

9. $\displaystyle\int \csc^2 u\, du = -\cot u + C$

10. $\displaystyle\int \sec u \tan u\, du = \sec u + C$

11. $\displaystyle\int \csc u \cot u\, du = -\csc u + C$

12. $\displaystyle\int \tan u\, du = \ln|\sec u| + C$

13. $\displaystyle\int \cot u\, du = \ln|\sin u| + C$

14. $\displaystyle\int \sec u\, du = \ln|\sec u + \tan u| + C$

15. $\displaystyle\int \csc u\, du = \ln|\csc u - \cot u| + C$

16. $\displaystyle\int \frac{du}{\sqrt{a^2 - u^2}} = \sin^{-1}\frac{u}{a} + C, \quad a > 0$

17. $\displaystyle\int \frac{du}{a^2 + u^2} = \frac{1}{a}\tan^{-1}\frac{u}{a} + C$

18. $\displaystyle\int \frac{du}{u\sqrt{u^2 - a^2}} = \frac{1}{a}\sec^{-1}\frac{u}{a} + C$

19. $\displaystyle\int \frac{du}{a^2 - u^2} = \frac{1}{2a}\ln\left|\frac{u+a}{u-a}\right| + C$

20. $\displaystyle\int \frac{du}{u^2 - a^2} = \frac{1}{2a}\ln\left|\frac{u-a}{u+a}\right| + C$

Forms Involving $\sqrt{a^2 + u^2},\ a > 0$

21. $\displaystyle\int \sqrt{a^2 + u^2}\, du = \frac{u}{2}\sqrt{a^2 + u^2} + \frac{a^2}{2}\ln\left(u + \sqrt{a^2 + u^2}\right) + C$

22. $\displaystyle\int u^2\sqrt{a^2 + u^2}\, du = \frac{u}{8}(a^2 + 2u^2)\sqrt{a^2 + u^2} - \frac{a^4}{8}\ln\left(u + \sqrt{a^2 + u^2}\right) + C$

23. $\displaystyle\int \frac{\sqrt{a^2 + u^2}}{u}\, du = \sqrt{a^2 + u^2} - a\ln\left|\frac{a + \sqrt{a^2 + u^2}}{u}\right| + C$

24. $\displaystyle\int \frac{\sqrt{a^2 + u^2}}{u^2}\, du = -\frac{\sqrt{a^2 + u^2}}{u} + \ln\left(u + \sqrt{a^2 + u^2}\right) + C$

25. $\displaystyle\int \frac{du}{\sqrt{a^2 + u^2}} = \ln\left(u + \sqrt{a^2 + u^2}\right) + C$

26. $\displaystyle\int \frac{u^2\, du}{\sqrt{a^2 + u^2}} = \frac{u}{2}\sqrt{a^2 + u^2} - \frac{a^2}{2}\ln\left(u + \sqrt{a^2 + u^2}\right) + C$

27. $\displaystyle\int \frac{du}{u\sqrt{a^2 + u^2}} = -\frac{1}{a}\ln\left|\frac{\sqrt{a^2 + u^2} + a}{u}\right| + C$

28. $\displaystyle\int \frac{du}{u^2\sqrt{a^2 + u^2}} = -\frac{\sqrt{a^2 + u^2}}{a^2 u} + C$

29. $\displaystyle\int \frac{du}{(a^2 + u^2)^{3/2}} = \frac{u}{a^2\sqrt{a^2 + u^2}} + C$

TABLE OF INTEGRALS

Forms Involving $\sqrt{a^2 - u^2}$, $a > 0$

30. $\displaystyle\int \sqrt{a^2 - u^2} \, du = \frac{u}{2} \sqrt{a^2 - u^2} + \frac{a^2}{2} \sin^{-1} \frac{u}{a} + C$

31. $\displaystyle\int u^2 \sqrt{a^2 - u^2} \, du = \frac{u}{8} (2u^2 - a^2) \sqrt{a^2 - u^2} + \frac{a^4}{8} \sin^{-1} \frac{u}{a} + C$

32. $\displaystyle\int \frac{\sqrt{a^2 - u^2}}{u} \, du = \sqrt{a^2 - u^2} - a \ln \left| \frac{a + \sqrt{a^2 - u^2}}{u} \right| + C$

33. $\displaystyle\int \frac{\sqrt{a^2 - u^2}}{u^2} \, du = -\frac{1}{u} \sqrt{a^2 - u^2} - \sin^{-1} \frac{u}{a} + C$

34. $\displaystyle\int \frac{u^2 \, du}{\sqrt{a^2 - u^2}} = -\frac{u}{2} \sqrt{a^2 - u^2} + \frac{a^2}{2} \sin^{-1} \frac{u}{a} + C$

35. $\displaystyle\int \frac{du}{u\sqrt{a^2 - u^2}} = -\frac{1}{a} \ln \left| \frac{a + \sqrt{a^2 - u^2}}{u} \right| + C$

36. $\displaystyle\int \frac{du}{u^2 \sqrt{a^2 - u^2}} = -\frac{1}{a^2 u} \sqrt{a^2 - u^2} + C$

37. $\displaystyle\int (a^2 - u^2)^{3/2} \, du = -\frac{u}{8} (2u^2 - 5a^2) \sqrt{a^2 - u^2} + \frac{3a^4}{8} \sin^{-1} \frac{u}{a} + C$

38. $\displaystyle\int \frac{du}{(a^2 - u^2)^{3/2}} = \frac{u}{a^2 \sqrt{a^2 - u^2}} + C$

Forms Involving $\sqrt{u^2 - a^2}$, $a > 0$

39. $\displaystyle\int \sqrt{u^2 - a^2} \, du = \frac{u}{2} \sqrt{u^2 - a^2} - \frac{a^2}{2} \ln \left| u + \sqrt{u^2 - a^2} \right| + C$

40. $\displaystyle\int u^2 \sqrt{u^2 - a^2} \, du = \frac{u}{8} (2u^2 - a^2) \sqrt{u^2 - a^2} - \frac{a^4}{8} \ln \left| u + \sqrt{u^2 - a^2} \right| + C$

41. $\displaystyle\int \frac{\sqrt{u^2 - a^2}}{u} \, du = \sqrt{u^2 - a^2} - a \cos^{-1} \frac{a}{|u|} + C$

42. $\displaystyle\int \frac{\sqrt{u^2 - a^2}}{u^2} \, du = -\frac{\sqrt{u^2 - a^2}}{u} + \ln \left| u + \sqrt{u^2 - a^2} \right| + C$

43. $\displaystyle\int \frac{du}{\sqrt{u^2 - a^2}} = \ln \left| u + \sqrt{u^2 - a^2} \right| + C$

44. $\displaystyle\int \frac{u^2 \, du}{\sqrt{u^2 - a^2}} = \frac{u}{2} \sqrt{u^2 - a^2} + \frac{a^2}{2} \ln \left| u + \sqrt{u^2 - a^2} \right| + C$

45. $\displaystyle\int \frac{du}{u^2 \sqrt{u^2 - a^2}} = \frac{\sqrt{u^2 - a^2}}{a^2 u} + C$

46. $\displaystyle\int \frac{du}{(u^2 - a^2)^{3/2}} = -\frac{u}{a^2 \sqrt{u^2 - a^2}} + C$

(continued)

TABLE OF INTEGRALS

Forms Involving $a + bu$

47. $\displaystyle\int \frac{u\,du}{a+bu} = \frac{1}{b^2}\left(a+bu - a\ln|a+bu|\right) + C$

48. $\displaystyle\int \frac{u^2\,du}{a+bu} = \frac{1}{2b^3}\left[(a+bu)^2 - 4a(a+bu) + 2a^2\ln|a+bu|\right] + C$

49. $\displaystyle\int \frac{du}{u(a+bu)} = \frac{1}{a}\ln\left|\frac{u}{a+bu}\right| + C$

50. $\displaystyle\int \frac{du}{u^2(a+bu)} = -\frac{1}{au} + \frac{b}{a^2}\ln\left|\frac{a+bu}{u}\right| + C$

51. $\displaystyle\int \frac{u\,du}{(a+bu)^2} = \frac{a}{b^2(a+bu)} + \frac{1}{b^2}\ln|a+bu| + C$

52. $\displaystyle\int \frac{du}{u(a+bu)^2} = \frac{1}{a(a+bu)} - \frac{1}{a^2}\ln\left|\frac{a+bu}{u}\right| + C$

53. $\displaystyle\int \frac{u^2\,du}{(a+bu)^2} = \frac{1}{b^3}\left(a+bu - \frac{a^2}{a+bu} - 2a\ln|a+bu|\right) + C$

54. $\displaystyle\int u\sqrt{a+bu}\,du = \frac{2}{15b^2}(3bu - 2a)(a+bu)^{3/2} + C$

55. $\displaystyle\int \frac{u\,du}{\sqrt{a+bu}} = \frac{2}{3b^2}(bu - 2a)\sqrt{a+bu} + C$

56. $\displaystyle\int \frac{u^2\,du}{\sqrt{a+bu}} = \frac{2}{15b^3}(8a^2 + 3b^2u^2 - 4abu)\sqrt{a+bu} + C$

57. $\displaystyle\int \frac{du}{u\sqrt{a+bu}} = \frac{1}{\sqrt{a}}\ln\left|\frac{\sqrt{a+bu}-\sqrt{a}}{\sqrt{a+bu}+\sqrt{a}}\right| + C, \quad \text{if } a > 0$

$\displaystyle\qquad\qquad = \frac{2}{\sqrt{-a}}\tan^{-1}\sqrt{\frac{a+bu}{-a}} + C, \qquad \text{if } a < 0$

58. $\displaystyle\int \frac{\sqrt{a+bu}}{u}\,du = 2\sqrt{a+bu} + a\int \frac{du}{u\sqrt{a+bu}}$

59. $\displaystyle\int \frac{\sqrt{a+bu}}{u^2}\,du = -\frac{\sqrt{a+bu}}{u} + \frac{b}{2}\int \frac{du}{u\sqrt{a+bu}}$

60. $\displaystyle\int u^n\sqrt{a+bu}\,du = \frac{2}{b(2n+3)}\left[u^n(a+bu)^{3/2} - na\int u^{n-1}\sqrt{a+bu}\,du\right]$

61. $\displaystyle\int \frac{u^n\,du}{\sqrt{a+bu}} = \frac{2u^n\sqrt{a+bu}}{b(2n+1)} - \frac{2na}{b(2n+1)}\int \frac{u^{n-1}\,du}{\sqrt{a+bu}}$

62. $\displaystyle\int \frac{du}{u^n\sqrt{a+bu}} = -\frac{\sqrt{a+bu}}{a(n-1)u^{n-1}} - \frac{b(2n-3)}{2a(n-1)}\int \frac{du}{u^{n-1}\sqrt{a+bu}}$

TABLE OF INTEGRALS

Trigonometric Forms

63. $\int \sin^2 u \, du = \frac{1}{2}u - \frac{1}{4}\sin 2u + C$

64. $\int \cos^2 u \, du = \frac{1}{2}u + \frac{1}{4}\sin 2u + C$

65. $\int \tan^2 u \, du = \tan u - u + C$

66. $\int \cot^2 u \, du = -\cot u - u + C$

67. $\int \sin^3 u \, du = -\frac{1}{3}(2 + \sin^2 u)\cos u + C$

68. $\int \cos^3 u \, du = \frac{1}{3}(2 + \cos^2 u)\sin u + C$

69. $\int \tan^3 u \, du = \frac{1}{2}\tan^2 u + \ln|\cos u| + C$

70. $\int \cot^3 u \, du = -\frac{1}{2}\cot^2 u - \ln|\sin u| + C$

71. $\int \sec^3 u \, du = \frac{1}{2}\sec u \tan u + \frac{1}{2}\ln|\sec u + \tan u| + C$

72. $\int \csc^3 u \, du = -\frac{1}{2}\csc u \cot u + \frac{1}{2}\ln|\csc u - \cot u| + C$

73. $\int \sin^n u \, du = -\frac{1}{n}\sin^{n-1}u \cos u + \frac{n-1}{n}\int \sin^{n-2}u \, du$

74. $\int \cos^n u \, du = \frac{1}{n}\cos^{n-1}u \sin u + \frac{n-1}{n}\int \cos^{n-2}u \, du$

75. $\int \tan^n u \, du = \frac{1}{n-1}\tan^{n-1}u - \int \tan^{n-2}u \, du$

76. $\int \cot^n u \, du = \frac{-1}{n-1}\cot^{n-1}u - \int \cot^{n-2}u \, du$

77. $\int \sec^n u \, du = \frac{1}{n-1}\tan u \sec^{n-2}u + \frac{n-2}{n-1}\int \sec^{n-2}u \, du$

78. $\int \csc^n u \, du = \frac{-1}{n-1}\cot u \csc^{n-2}u + \frac{n-2}{n-1}\int \csc^{n-2}u \, du$

79. $\int \sin au \sin bu \, du = \frac{\sin(a-b)u}{2(a-b)} - \frac{\sin(a+b)u}{2(a+b)} + C$

80. $\int \cos au \cos bu \, du = \frac{\sin(a-b)u}{2(a-b)} + \frac{\sin(a+b)u}{2(a+b)} + C$

81. $\int \sin au \cos bu \, du = -\frac{\cos(a-b)u}{2(a-b)} - \frac{\cos(a+b)u}{2(a+b)} + C$

82. $\int u \sin u \, du = \sin u - u \cos u + C$

83. $\int u \cos u \, du = \cos u + u \sin u + C$

84. $\int u^n \sin u \, du = -u^n \cos u + n \int u^{n-1}\cos u \, du$

85. $\int u^n \cos u \, du = u^n \sin u - n \int u^{n-1}\sin u \, du$

86. $\int \sin^n u \cos^m u \, du = -\frac{\sin^{n-1}u \cos^{m+1}u}{n+m} + \frac{n-1}{n+m}\int \sin^{n-2}u \cos^m u \, du$
$= \frac{\sin^{n+1}u \cos^{m-1}u}{n+m} + \frac{m-1}{n+m}\int \sin^n u \cos^{m-2}u \, du$

Inverse Trigonometric Forms

87. $\int \sin^{-1}u \, du = u \sin^{-1}u + \sqrt{1-u^2} + C$

88. $\int \cos^{-1}u \, du = u \cos^{-1}u - \sqrt{1-u^2} + C$

89. $\int \tan^{-1}u \, du = u \tan^{-1}u - \frac{1}{2}\ln(1+u^2) + C$

90. $\int u \sin^{-1}u \, du = \frac{2u^2-1}{4}\sin^{-1}u + \frac{u\sqrt{1-u^2}}{4} + C$

91. $\int u \cos^{-1}u \, du = \frac{2u^2-1}{4}\cos^{-1}u - \frac{u\sqrt{1-u^2}}{4} + C$

92. $\int u \tan^{-1}u \, du = \frac{u^2+1}{2}\tan^{-1}u - \frac{u}{2} + C$

93. $\int u^n \sin^{-1}u \, du = \frac{1}{n+1}\left[u^{n+1}\sin^{-1}u - \int \frac{u^{n+1}\,du}{\sqrt{1-u^2}}\right], \quad n \neq -1$

94. $\int u^n \cos^{-1}u \, du = \frac{1}{n+1}\left[u^{n+1}\cos^{-1}u + \int \frac{u^{n+1}\,du}{\sqrt{1-u^2}}\right], \quad n \neq -1$

95. $\int u^n \tan^{-1}u \, du = \frac{1}{n+1}\left[u^{n+1}\tan^{-1}u - \int \frac{u^{n+1}\,du}{1+u^2}\right], \quad n \neq -1$

(continued)

TABLE OF INTEGRALS

Exponential and Logarithmic Forms

96. $\int u e^{au} \, du = \dfrac{1}{a^2}(au - 1)e^{au} + C$

97. $\int u^n e^{au} \, du = \dfrac{1}{a} u^n e^{au} - \dfrac{n}{a}\int u^{n-1} e^{au} \, du$

98. $\int e^{au} \sin bu \, du = \dfrac{e^{au}}{a^2 + b^2}(a \sin bu - b \cos bu) + C$

99. $\int e^{au} \cos bu \, du = \dfrac{e^{au}}{a^2 + b^2}(a \cos bu + b \sin bu) + C$

100. $\int \ln u \, du = u \ln u - u + C$

101. $\int u^n \ln u \, du = \dfrac{u^{n+1}}{(n+1)^2}[(n+1)\ln u - 1] + C$

102. $\int \dfrac{1}{u \ln u} \, du = \ln|\ln u| + C$

Hyperbolic Forms

103. $\int \sinh u \, du = \cosh u + C$

104. $\int \cosh u \, du = \sinh u + C$

105. $\int \tanh u \, du = \ln \cosh u + C$

106. $\int \coth u \, du = \ln|\sinh u| + C$

107. $\int \operatorname{sech} u \, du = \tan^{-1}|\sinh u| + C$

108. $\int \operatorname{csch} u \, du = \ln\left|\tanh \tfrac{1}{2} u\right| + C$

109. $\int \operatorname{sech}^2 u \, du = \tanh u + C$

110. $\int \operatorname{csch}^2 u \, du = -\coth u + C$

111. $\int \operatorname{sech} u \tanh u \, du = -\operatorname{sech} u + C$

112. $\int \operatorname{csch} u \coth u \, du = -\operatorname{csch} u + C$

Forms Involving $\sqrt{2au - u^2}$, $a > 0$

113. $\int \sqrt{2au - u^2} \, du = \dfrac{u - a}{2}\sqrt{2au - u^2} + \dfrac{a^2}{2}\cos^{-1}\left(\dfrac{a - u}{a}\right) + C$

114. $\int u\sqrt{2au - u^2} \, du = \dfrac{2u^2 - au - 3a^2}{6}\sqrt{2au - u^2} + \dfrac{a^3}{2}\cos^{-1}\left(\dfrac{a - u}{a}\right) + C$

115. $\int \dfrac{\sqrt{2au - u^2}}{u} \, du = \sqrt{2au - u^2} + a\cos^{-1}\left(\dfrac{a - u}{a}\right) + C$

116. $\int \dfrac{\sqrt{2au - u^2}}{u^2} \, du = -\dfrac{2\sqrt{2au - u^2}}{u} - \cos^{-1}\left(\dfrac{a - u}{a}\right) + C$

117. $\int \dfrac{du}{\sqrt{2au - u^2}} = \cos^{-1}\left(\dfrac{a - u}{a}\right) + C$

118. $\int \dfrac{u \, du}{\sqrt{2au - u^2}} = -\sqrt{2au - u^2} + a\cos^{-1}\left(\dfrac{a - u}{a}\right) + C$

119. $\int \dfrac{u^2 \, du}{\sqrt{2au - u^2}} = -\dfrac{(u + 3a)}{2}\sqrt{2au - u^2} + \dfrac{3a^2}{2}\cos^{-1}\left(\dfrac{a - u}{a}\right) + C$

120. $\int \dfrac{du}{u\sqrt{2au - u^2}} = -\dfrac{\sqrt{2au - u^2}}{au} + C$

CHAPTER 1 CONCEPT CHECK ANSWERS

1. (a) What is a function? What are its domain and range?

A function f is a rule that assigns to each element x in a set D exactly one element, called $f(x)$, in a set E. The domain is the set D and the range is the set of all possible values of $f(x)$ as x varies throughout the domain.

(b) What is the graph of a function?

The graph of a function f consists of all points (x, y) such that $y = f(x)$ and x is in the domain of f.

(c) How can you tell whether a given curve is the graph of a function?

Use the Vertical Line Test: a curve in the xy-plane is the graph of a function of x if and only if no vertical line intersects the curve more than once.

2. Discuss four ways of representing a function. Illustrate your discussion with examples.

A function can be represented verbally, numerically, visually, or algebraically. An example of each is given below.

Verbally: An assignment of students to chairs in a classroom (a description in words)

Numerically: A tax table that assigns an amount of tax to an income (a table of values)

Visually: A graphical history of the Dow Jones average (a graph)

Algebraically: A relationship between the area A and side length s of a square: $A = s^2$ (an explicit formula)

3. (a) What is an even function? How can you tell if a function is even by looking at its graph? Give three examples of an even function.

A function f is even if it satisfies $f(-x) = f(x)$ for every number x in its domain. If the graph of a function is symmetric with respect to the y-axis, then f is even. Examples are $f(x) = x^2$, $f(x) = \cos x$, $f(x) = |x|$.

(b) What is an odd function? How can you tell if a function is odd by looking at its graph? Give three examples of an odd function.

A function f is odd if it satisfies $f(-x) = -f(x)$ for every number x in its domain. If the graph of a function is symmetric with respect to the origin, then f is odd. Examples are $f(x) = x^3$, $f(x) = \sin x$, $f(x) = 1/x$.

4. What is an increasing function?

A function f is increasing on an interval I if $f(x_1) < f(x_2)$ whenever $x_1 < x_2$ in I.

5. What is a mathematical model?

A mathematical model is a mathematical description (often by means of a function or an equation) of a real-world phenomenon. (See the discussion on pages 23–24.)

6. Give an example of each type of function.

(a) Linear function: $f(x) = 2x + 1$, $f(x) = ax + b$

(b) Power function: $f(x) = x^2$, $f(x) = x^n$

(c) Exponential function: $f(x) = 2^x$, $f(x) = b^x$

(d) Quadratic function: $f(x) = x^2 + x + 1$, $f(x) = ax^2 + bx + c$

(e) Polynomial of degree 5: $f(x) = x^5 + 2x^4 - 3x^2 + 7$

(f) Rational function: $f(x) = \dfrac{x}{x + 2}$, $f(x) = \dfrac{P(x)}{Q(x)}$

where $P(x)$ and $Q(x)$ are polynomials

7. Sketch by hand, on the same axes, the graphs of the following functions.

(a) $f(x) = x$ **(b)** $g(x) = x^2$

(c) $h(x) = x^3$ **(d)** $j(x) = x^4$

8. Draw, by hand, a rough sketch of the graph of each function.

(a) $y = \sin x$

(b) $y = \tan x$

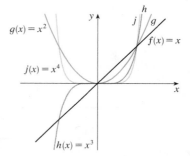

(c) $y = e^x$ **(d)** $y = \ln x$

(continued)

(e) $y = 1/x$

(f) $y = |x|$

(g) $y = \sqrt{x}$

(h) $y = \tan^{-1}x$

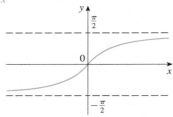

9. Suppose that f has domain A and g has domain B.

(a) What is the domain of $f + g$?

The domain of $f + g$ is the intersection of the domain of f and the domain of g; that is, $A \cap B$.

(b) What is the domain of fg?

The domain of fg is also $A \cap B$.

(c) What is the domain of f/g?

The domain of f/g must exclude values of x that make g equal to 0; that is, $\{x \in A \cap B \mid g(x) \neq 0\}$.

10. How is the composite function $f \circ g$ defined? What is its domain?

The composition of f and g is defined by $(f \circ g)(x) = f(g(x))$. The domain is the set of all x in the domain of g such that $g(x)$ is in the domain of f.

11. Suppose the graph of f is given. Write an equation for each of the graphs that are obtained from the graph of f as follows.

(a) Shift 2 units upward: $y = f(x) + 2$

(b) Shift 2 units downward: $y = f(x) - 2$

(c) Shift 2 units to the right: $y = f(x - 2)$

(d) Shift 2 units to the left: $y = f(x + 2)$

(e) Reflect about the x-axis: $y = -f(x)$

(f) Reflect about the y-axis: $y = f(-x)$

(g) Stretch vertically by a factor of 2: $y = 2f(x)$

(h) Shrink vertically by a factor of 2: $y = \frac{1}{2}f(x)$

(i) Stretch horizontally by a factor of 2: $y = f(\frac{1}{2}x)$

(j) Shrink horizontally by a factor of 2 $y = f(2x)$

12. (a) What is a one-to-one function? How can you tell if a function is one-to-one by looking at its graph?

A function f is one-to-one if it never takes on the same value twice; that is $f(x_1) \neq f(x_2)$ whenever $x_1 \neq x_2$. When looking at a graph, use the Horizontal Line Test: a function is one-to-one if and only if no horizontal line intersects its graph more than once.

(b) If f is a one-to-one function, how is its inverse function f^{-1} defined? How do you obtain the graph of f^{-1} from the graph of f?

If f is a one-to-one function with domain A and range B, then its inverse function f^{-1} has domain B and range A and is defined by

$$f^{-1}(y) = x \iff f(x) = y$$

for any y in B.

The graph of f^{-1} is obtained by reflecting the graph of f about the line $y = x$.

13. (a) How is the inverse sine function $f(x) = \sin^{-1}x$ defined? What are its domain and range?

The inverse sine function $f(x) = \sin^{-1}x$ is defined as

$$\sin^{-1}x = y \iff \sin y = x$$

and $\qquad -\pi/2 \leq y \leq \pi/2$

Its domain is $-1 \leq x \leq 1$ and its range is $-\pi/2 \leq y \leq \pi/2$.

(b) How is the inverse cosine function $f(x) = \cos^{-1}x$ defined? What are its domain and range?

The inverse cosine function $f(x) = \cos^{-1}x$ is defined as

$$\cos^{-1}x = y \iff \cos y = x$$

and $\qquad 0 \leq y \leq \pi$

Its domain is $-1 \leq x \leq 1$ and its range is $0 \leq y \leq \pi$.

(c) How is the inverse tangent function $f(x) = \tan^{-1}x$ defined? What are its domain and range?

The inverse tangent function $f(x) = \tan^{-1}x$ is defined as

$$\tan^{-1}x = y \iff \tan y = x$$

and $\qquad -\pi/2 < y < \pi/2$

Its domain is $\mathbb{R}$ and its range is $-\pi/2 < y < \pi/2$.

1. Explain what each of the following means and illustrate with a sketch.

   (a) $\lim_{x \to a} f(x) = L$ means that the values of $f(x)$ approach L as the values of x approach a (but $x \neq a$).

   (b) $\lim_{x \to a^+} f(x) = L$ means that the values of $f(x)$ approach L as the values of x approach a through values greater than a.

   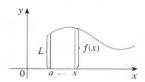

   (c) $\lim_{x \to a^-} f(x) = L$ means that the values of $f(x)$ approach L as the values of x approach a through values less than a.

   (d) $\lim_{x \to a} f(x) = \infty$ means that the values of $f(x)$ can be made arbitrarily large by taking x sufficiently close to a (but not equal to a).

   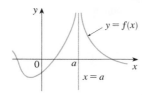

   (e) $\lim_{x \to \infty} f(x) = L$ means that the values of $f(x)$ can be made arbitrarily close to L by requiring x to be sufficiently large.

   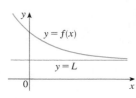

2. Describe several ways in which a limit can fail to exist. Illustrate with sketches.

   In general, the limit of a function fails to exist when the function values do not approach a fixed number. For each of the following functions, the limit fails to exist at $x = 2$.

    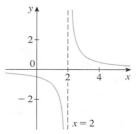

   The left and right limits are not equal.

   There is an infinite discontinuity.

   The function values oscillate between 1 and -1 infinitely often.

3. State the following Limit Laws.

   (a) Sum Law

   The limit of a sum is the sum of the limits:
   $$\lim_{x \to a} [f(x) + g(x)] = \lim_{x \to a} f(x) + \lim_{x \to a} g(x)$$

   (b) Difference Law

   The limit of a difference is the difference of the limits:
   $$\lim_{x \to a} [f(x) - g(x)] = \lim_{x \to a} f(x) - \lim_{x \to a} g(x)$$

   (c) Constant Multiple Law

   The limit of a constant times a function is the constant times the limit of the function: $\lim_{x \to a} [cf(x)] = c \lim_{x \to a} f(x)$

   (d) Product Law

   The limit of a product is the product of the limits:
   $$\lim_{x \to a} [f(x)g(x)] = \lim_{x \to a} f(x) \cdot \lim_{x \to a} g(x)$$

   (e) Quotient Law

   The limit of a quotient is the quotient of the limits, provided that the limit of the denominator is not 0:
   $$\lim_{x \to a} \frac{f(x)}{g(x)} = \frac{\lim\limits_{x \to a} f(x)}{\lim\limits_{x \to a} g(x)} \quad \text{if } \lim_{x \to a} g(x) \neq 0$$

(continued)

(f) Power Law

The limit of a power is the power of the limit:
$$\lim_{x \to a} [f(x)]^n = \left[\lim_{x \to a} f(x)\right]^n \quad \text{(for } n \text{ a positive integer)}$$

(g) Root Law

The limit of a root is the root of the limit:
$$\lim_{x \to a} \sqrt[n]{f(x)} = \sqrt[n]{\lim_{x \to a} f(x)} \quad \text{(for } n \text{ a positive integer)}$$

4. What does the Squeeze Theorem say?

If $f(x) \leq g(x) \leq h(x)$ when x is near a (except possibly at a) and $\lim_{x \to a} f(x) = \lim_{x \to a} h(x) = L$, then $\lim_{x \to a} g(x) = L$. In other words, if $g(x)$ is squeezed between $f(x)$ and $h(x)$ near a, and if f and h have the same limit L at a, then g is forced to have the same limit L at a.

5. (a) What does it mean to say that the line $x = a$ is a vertical asymptote of the curve $y = f(x)$? Draw curves to illustrate the various possibilities.

It means that the limit of $f(x)$ as x approaches a from one or both sides is positive or negative infinity.

(b) What does it mean to say that the line $y = L$ is a horizontal asymptote of the curve $y = f(x)$? Draw curves to illustrate the various possibilities.

It means that $\lim_{x \to \infty} f(x) = L$ or $\lim_{x \to -\infty} f(x) = L$.

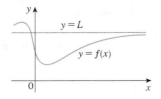

6. Which of the following curves have vertical asymptotes? Which have horizontal asymptotes?

(a) $y = x^4$: No asymptotes

(b) $y = \sin x$: No asymptotes

(c) $y = \tan x$: Vertical asymptotes $x = \pi/2 + \pi n$, n an integer

(d) $y = \tan^{-1} x$: Horizontal asymptotes $y = \pm\pi/2$

(e) $y = e^x$: Horizontal asymptote $y = 0$ $\left(\lim_{x \to -\infty} e^x = 0\right)$

(f) $y = \ln x$: Vertical asymptote $x = 0$ $\left(\lim_{x \to 0^+} \ln x = -\infty\right)$

(g) $y = 1/x$: Vertical asymptote $x = 0$, horizontal asymptote $y = 0$

(h) $y = \sqrt{x}$: No asymptotes

7. (a) What does it mean for f to be continuous at a?

A function f is continuous at a number a if the value of the function at $x = a$ is the same as the limit when x approaches a; that is, $\lim_{x \to a} f(x) = f(a)$.

(b) What does it mean for f to be continuous on the interval $(-\infty, \infty)$? What can you say about the graph of such a function?

A function f is continuous on the interval $(-\infty, \infty)$ if it is continuous at every real number a.

The graph of such a function has no hole or break in it.

8. (a) Give examples of functions that are continuous on $[-1, 1]$.

$f(x) = x^3 - x$, $g(x) = 2^x$, $y = \sin x$, $y = \tan x$, $y = 1/(x - 3)$, and $h(x) = |x|$ are all continuous on $[-1, 1]$.

(b) Give an example of a function that is not continuous on $[0, 1]$.

$$f(x) = \frac{1}{x - \frac{1}{2}} \quad \left(f(x) \text{ is not defined at } x = \tfrac{1}{2}\right)$$

9. What does the Intermediate Value Theorem say?

If f is continuous on $[a, b]$ and N is any number between $f(a)$ and $f(b)$ $[f(a) \neq f(b)]$, Then there exists a number c in (a, b) such that $f(c) = N$. In other words, a continuous function takes on every intermediate value between the function values $f(a)$ and $f(b)$.

10. Write an expression for the slope of the tangent line to the curve $y = f(x)$ at the point $(a, f(a))$.

The slope of the tangent line is given by

$$\lim_{x \to a} \frac{f(x) - f(a)}{x - a} \quad \text{or} \quad \lim_{h \to 0} \frac{f(a + h) - f(a)}{h}$$

(continued)

11. Suppose an object moves along a straight line with position $f(t)$ at time t. Write an expression for the instantaneous velocity of the object at time $t = a$. How can you interpret this velocity in terms of the graph of f?

The instantaneous velocity at time $t = a$ is

$$v(a) = \lim_{h \to 0} \frac{f(a + h) - f(a)}{h}$$

It is equal to the slope of the tangent line to the graph of f at the point $P(a, f(a))$.

12. If $y = f(x)$ and x changes from x_1 to x_2, write expressions for the following.

(a) The average rate of change of y with respect to x over the interval $[x_1, x_2]$:

$$\frac{\Delta y}{\Delta x} = \frac{f(x_2) - f(x_1)}{x_2 - x_1}$$

(b) The instantaneous rate of change of y with respect to x at $x = x_1$:

$$\lim_{\Delta x \to 0} \frac{\Delta y}{\Delta x} = \lim_{x_2 \to x_1} \frac{f(x_2) - f(x_1)}{x_2 - x_1}$$

13. Define the derivative $f'(a)$. Discuss two ways of interpreting this number.

$$f'(a) = \lim_{h \to 0} \frac{f(a + h) - f(a)}{h}$$

or, equivalently,

$$f'(a) = \lim_{x \to a} \frac{f(x) - f(a)}{x - a}$$

The derivative $f'(a)$ is the instantaneous rate of change of $y = f(x)$ (with respect to x) when $x = a$ and also represents the slope of the tangent line to the graph of f at the point $P(a, f(a))$.

14. Define the second derivative of f. If $f(t)$ is the position function of a particle, how can you interpret the second derivative?

The second derivative of f, f'', is the derivative of f'.

If $f(t)$ is the position function of a particle, then the first derivative is velocity and the second derivative is the instantaneous rate of change of velocity, namely, acceleration.

15. (a) What does it mean for f to be differentiable at a?

f is differentiable at a if the derivative $f'(a)$ exists.

(b) What is the relation between the differentiability and continuity of a function?

If f is differentiable at a, then f is continuous at a.

(c) Sketch the graph of a function that is continuous but not differentiable at $a = 2$.

The graph of f changes direction abruptly at $x = 2$, so f has no tangent line there.

16. Describe several ways in which a function can fail to be differentiable. Illustrate with sketches.

A function is not differentiable at any value where the graph has a "corner," where the graph has a discontinuity, or where it has a vertical tangent line.

A corner   A discontinuity

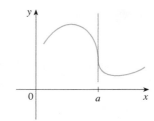

A vertical tangent

1. State each differentiation rule both in symbols and in words.

(a) The Power Rule

If n is any real number, then $\dfrac{d}{dx}(x^n) = nx^{n-1}$.

To find the derivative of a variable raised to a constant power, we multiply the expression by the exponent and then subtract one from the exponent.

(b) The Constant Multiple Rule

If c is a constant and f is a differentiable function, then

$$\frac{d}{dx}[cf(x)] = c\,\frac{d}{dx}f(x)$$

The derivative of a constant times a function is the constant times the derivative of the function.

(c) The Sum Rule

If f and g are both differentiable, then

$$\frac{d}{dx}[f(x) + g(x)] = \frac{d}{dx}f(x) + \frac{d}{dx}g(x)$$

The derivative of a sum of functions is the sum of the derivatives.

(d) The Difference Rule

If f and g are both differentiable, then

$$\frac{d}{dx}[f(x) - g(x)] = \frac{d}{dx}f(x) - \frac{d}{dx}g(x)$$

The derivative of a difference of functions is the difference of the derivatives.

(e) The Product Rule

If f and g are both differentiable, then

$$\frac{d}{dx}[f(x)g(x)] = f(x)\,\frac{d}{dx}[g(x)] + g(x)\,\frac{d}{dx}[f(x)]$$

The derivative of a product of two functions is the first function times the derivative of the second function plus the second function times the derivative of the first function.

(f) The Quotient Rule

If f and g are both differentiable, then

$$\frac{d}{dx}\left[\frac{f(x)}{g(x)}\right] = \frac{g(x)\,\dfrac{d}{dx}[f(x)] - f(x)\,\dfrac{d}{dx}[g(x)]}{[g(x)]^2}$$

The derivative of a quotient of functions is the denominator times the derivative of the numerator minus the numerator times the derivative of the denominator, all divided by the square of the denominator.

(g) The Chain Rule

If g is differentiable at x and f is differentiable at $g(x)$, then the composite function defined by $F(x) = f(g(x))$ is differentiable at x and F' is given by the product

$$F'(x) = f'(g(x))g'(x)$$

The derivative of a composite function is the derivative of the outer function evaluated at the inner function times the derivative of the inner function.

2. State the derivative of each function.

(a) $y = x^n$: $y' = nx^{n-1}$

(b) $y = e^x$: $y' = e^x$

(c) $y = b^x$: $y' = b^x \ln b$

(d) $y = \ln x$: $y' = 1/x$

(e) $y = \log_b x$: $y' = 1/(x \ln b)$

(f) $y = \sin x$: $y' = \cos x$

(g) $y = \cos x$: $y' = -\sin x$

(h) $y = \tan x$: $y' = \sec^2 x$

(i) $y = \csc x$: $y' = -\csc x \cot x$

(j) $y = \sec x$: $y' = \sec x \tan x$

(k) $y = \cot x$: $y' = -\csc^2 x$

(l) $y = \sin^{-1} x$: $y' = 1/\sqrt{1 - x^2}$

(m) $y = \cos^{-1} x$: $y' = -1/\sqrt{1 - x^2}$

(n) $y = \tan^{-1} x$: $y' = 1/(1 + x^2)$

(o) $y = \sinh x$: $y' = \cosh x$

(p) $y = \cosh x$: $y' = \sinh x$

(q) $y = \tanh x$: $y' = \operatorname{sech}^2 x$

(r) $y = \sinh^{-1} x$: $y' = 1/\sqrt{1 + x^2}$

(s) $y = \cosh^{-1} x$: $y' = 1/\sqrt{x^2 - 1}$

(t) $y = \tanh^{-1} x$: $y' = 1/(1 - x^2)$

3. (a) How is the number e defined?

e is the number such that $\displaystyle\lim_{h \to 0} \frac{e^h - 1}{h} = 1$.

(b) Express e as a limit.

$$e = \lim_{x \to 0} (1 + x)^{1/x}$$

(c) Why is the natural exponential function $y = e^x$ used more often in calculus than the other exponential functions $y = b^x$?

The differentiation formula for $y = b^x$ $[dy/dx = b^x \ln b]$ is simplest when $b = e$ because $\ln e = 1$.

(continued)

(d) Why is the natural logarithmic function $y = \ln x$ used more often in calculus than the other logarithmic functions $y = \log_b x$?

The differentiation formula for $y = \log_b x$ $[dy/dx = 1/(x \ln b)]$ is simplest when $b = e$ because $\ln e = 1$.

4. (a) Explain how implicit differentiation works.

Implicit differentiation consists of differentiating both sides of an equation with respect to x, treating y as a function of x. Then we solve the resulting equation for y'.

(b) Explain how logarithmic differentiation works.

Logarithmic differentiation consists of taking natural logarithms of both sides of an equation $y = f(x)$, simplifying using the Laws of Logarithms, differentiating implicitly with respect to x, and then solving the resulting equation for y'.

5. Give several examples of how the derivative can be interpreted as a rate of change in physics, chemistry, biology, economics, or other sciences.

In physics, interpretations of the derivative include velocity, linear density, electrical current, power (the rate of change of work), and the rate of radioactive decay. Chemists can use derivatives to measure reaction rates and the compressibility of a substance under pressure. In biology the derivative measures rates of population growth and blood flow. In economics, the derivative measures marginal cost (the rate of change of cost as more items are produced) and marginal profit. Other examples include the rate of heat flow in geology, the rate of performance improvement in psychology, and the rate at which a rumor spreads in sociology.

6. (a) Write a differential equation that expresses the law of natural growth.

If $y(t)$ is the value of a quantity y at time t, then

$$\frac{dy}{dt} = ky \qquad \text{where } k > 0 \text{ is a constant}$$

(b) Under what circumstances is this an appropriate model for population growth?

The equation in part (a) is an appropriate model when there is enough room and nutrition to support growth.

(c) What are the solutions of this equation?

If $y(0) = y_0$, then the solutions are $y(t) = y_0 e^{kt}$.

7. (a) Write an expression for the linearization of f at a.

$$L(x) = f(a) + f'(a)(x - a)$$

(b) If $y = f(x)$, write an expression for the differential dy.

$$dy = f'(x)\, dx$$

(c) If $dx = \Delta x$, draw a picture showing the geometric meanings of Δy and dy.

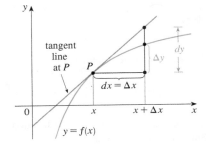

1. Explain the difference between an absolute maximum and a local maximum. Illustrate with a sketch.

   The function value $f(c)$ is the absolute maximum value of f if $f(c)$ is the largest function value on the entire domain of f, whereas $f(c)$ is a local maximum value if it is the largest function value when x is near c.

2. (a) What does the Extreme Value Theorem say?

   If f is a continuous function on a closed interval $[a, b]$, then it always attains an absolute maximum and an absolute minimum value on that interval.

   (b) Explain how the Closed Interval Method works.

   To find the absolute maximum and minimum values of a continuous function f on a closed interval $[a, b]$, we follow these three steps:

   - Find the critical numbers of f in the interval (a, b) and compute the values of f at these numbers.
   - Find the values of f at the endpoints of the interval.
   - The largest of the values from the previous two steps is the absolute maximum value; the smallest of these values is the absolute minimum value.

3. (a) State Fermat's Theorem.

   If f has a local maximum or minimum at c, and if $f'(c)$ exists, then $f'(c) = 0$.

   (b) Define a critical number of f.

   A critical number of a function f is a number c in the domain of f such that either $f'(c) = 0$ or $f'(c)$ does not exist.

4. (a) State Rolle's Theorem.

   Let f be a function that satisfies the following three hypotheses:

   - f is continuous on the closed interval $[a, b]$.
   - f is differentiable on the open interval (a, b).
   - $f(a) = f(b)$

   Then there is a number c in (a, b) such that $f'(c) = 0$.

   (b) State the Mean Value Theorem and give a geometric interpretation.

   If f is continuous on the interval $[a, b]$ and differentiable on (a, b), then there exists a number c between a and b such that

   $$f'(c) = \frac{f(b) - f(a)}{b - a}$$

   Geometrically, the theorem says that there is a point $P(c, f(c))$, where $a < c < b$, on the graph of f where the tangent line is parallel to the secant line that connects $(a, f(a))$ and $(b, f(b))$.

   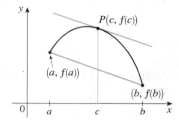

5. (a) State the Increasing/Decreasing Test.

   If $f'(x) > 0$ on an interval, then f is increasing on that interval.

   If $f'(x) < 0$ on an interval, then f is decreasing on that interval.

   (b) What does it mean to say that f is concave upward on an interval I?

   f is concave upward on an interval if the graph of f lies above all of its tangents on that interval.

   (c) State the Concavity Test.

   If $f''(x) > 0$ on an interval, then the graph of f is concave upward on that interval.

   If $f''(x) < 0$ on an interval, then the graph of f is concave downward on that interval.

   (d) What are inflection points? How do you find them?

   Inflection points on the graph of a continuous function f are points where the curve changes from concave upward to concave downward or from concave downward to concave upward. They can be found by determining the values at which the second derivative changes sign.

6. (a) State the First Derivative Test.

   Suppose that c is a critical number of a continuous function f.

   - If f' changes from positive to negative at c, then f has a local maximum at c.
   - If f' changes from negative to positive at c, then f has a local minimum at c.
   - If f' is positive to the left and right of c, or negative to the left and right of c, then f has no local maximum or minimum at c.

   (b) State the Second Derivative Test.

   Suppose f'' is continuous near c.

   - If $f'(c) = 0$ and $f''(c) > 0$, then f has a local minimum at c.
   - If $f'(c) = 0$ and $f''(c) < 0$, then f has a local maximum at c.

(continued)

(c) What are the relative advantages and disadvantages of these tests?

The Second Derivative Test is sometimes easier to use, but it is inconclusive when $f''(c) = 0$ and fails if $f''(c)$ does not exist. In either case the First Derivative Test must be used.

7. (a) What does l'Hospital's Rule say?

L'Hospital's Rule says that *if* the limit of a quotient of functions is an indeterminate form of type $0/0$ or ∞/∞, then the limit is equal to the limit of the quotient of their derivatives:

$$\lim_{x \to a} \frac{f(x)}{g(x)} = \lim_{x \to a} \frac{f'(x)}{g'(x)}$$

(b) How can you use l'Hospital's Rule if you have a product $f(x)g(x)$ where $f(x) \to 0$ and $g(x) \to \infty$ as $x \to a$?

Write fg as $\dfrac{f}{1/g}$ or $\dfrac{g}{1/f}$ so that the limit becomes an indeterminate form of type $0/0$ or ∞/∞.

(c) How can you use l'Hospital's Rule if you have a difference $f(x) - g(x)$ where $f(x) \to \infty$ and $g(x) \to \infty$ as $x \to a$?

Convert the difference into a quotient by using a common denominator, rationalizing, factoring, or by some other method.

(d) How can you use l'Hospital's Rule if you have a power $[f(x)]^{g(x)}$ where $f(x) \to 0$ and $g(x) \to 0$ as $x \to a$?

By taking the natural logarithm of both sides of $y = [f(x)]^{g(x)}$, we can convert the power to a product by writing $\ln y = g(x) \ln f(x)$. Alternatively, we can write the function as an exponential: $[f(x)]^{g(x)} = e^{g(x) \ln f(x)}$.

8. State whether each of the following limit forms is indeterminate. Where possible, state the limit.

(a) $\dfrac{0}{0}$: indeterminate. L'Hospital's Rule can be applied to this form. Note that every derivative is a limit of this form.

(b) $\dfrac{\infty}{\infty}$: indeterminate. L'Hospital's Rule can be applied to this form.

(c) $\dfrac{0}{\infty}$: not indeterminate. A limit of this form has value 0.

(d) $\dfrac{\infty}{0}$: not indeterminate. A limit of this form could equal ∞, $-\infty$, or may not exist (but it cannot equal a finite value).

(e) $\infty + \infty$: not indeterminate. A limit of this form is equal to ∞.

(f) $\infty - \infty$: indeterminate

(g) $\infty \cdot \infty$: not indeterminate. A limit of this form is equal to ∞.

(h) $\infty \cdot 0$: indeterminate

(i) 0^0: indeterminate

(j) 0^∞: not indeterminate. A limit of this form has value 0.

(k) ∞^0: indeterminate

(l) 1^∞: indeterminate

9. If you have a graphing calculator or computer, why do you need calculus to graph a function?

Calculus reveals all the important aspects of a graph, such as local extreme values and inflection points, that can be missed when relying solely on technology. In many cases we can find exact locations of these key points rather than approximations. Using derivatives to identify the behavior of the graph also helps us choose an appropriate viewing window and alerts us to where we may wish to zoom in on a graph.

10. (a) Given an initial approximation x_1 to a root of the equation $f(x) = 0$, explain geometrically, with a diagram, how the second approximation x_2 in Newton's method is obtained.

We find the tangent line L to the graph of $y = f(x)$ at the point $(x_1, f(x_1))$. Then x_2 is the x-intercept of L.

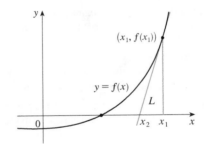

(b) Write an expression for x_2 in terms of x_1, $f(x_1)$, and $f'(x_1)$.

$$x_2 = x_1 - \frac{f(x_1)}{f'(x_1)}$$

(c) Write an expression for x_{n+1} in terms of x_n, $f(x_n)$, and $f'(x_n)$.

$$x_{n+1} = x_n - \frac{f(x_n)}{f'(x_n)}$$

(d) Under what circumstances is Newton's method likely to fail or to work very slowly?

Newton's method is likely to fail or to work very slowly when $f'(x_1)$ is close to 0. It also fails when $f'(x_i)$ is undefined.

11. (a) What is an antiderivative of a function f?

A function F is an antiderivative of f if $F'(x) = f(x)$.

(b) Suppose F_1 and F_2 are both antiderivatives of f on an interval I. How are F_1 and F_2 related?

They are identical or they differ by a constant.

1. (a) Write an expression for a Riemann sum of a function f. Explain the meaning of the notation that you use.

A Riemann sum of f is

$$\sum_{i=1}^{n} f(x_i^*)\, \Delta x$$

where x_i^* is a point in the ith subinterval $[x_{i-1}, x_i]$ and Δx is the length of the subintervals.

(b) If $f(x) \geqslant 0$, what is the geometric interpretation of a Riemann sum? Illustrate with a diagram.

If f is positive, then a Riemann sum can be interpreted as the sum of areas of approximating rectangles, as shown in the figure.

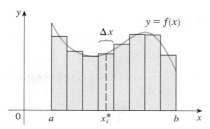

(c) If $f(x)$ takes on both positive and negative values, what is the geometric interpretation of a Riemann sum? Illustrate with a diagram.

If f takes on both positive and negative values then the Riemann sum is the sum of the areas of the rectangles that lie above the x-axis and the negatives of the areas of the rectangles that lie below the x-axis (the areas of the blue rectangles minus the areas of the gray rectangles).

2. (a) Write the definition of the definite integral of a continuous function from a to b.

If f is a continuous function on the interval $[a, b]$, then we divide $[a, b]$ into n subintervals of equal width $\Delta x = (b - a)/n$. We let $x_0\,(= a)$, $x_1, x_2, \ldots$, $x_n\,(= b)$ be the endpoints of these subintervals. Then

$$\int_a^b f(x)\, dx = \lim_{n \to \infty} \sum_{i=1}^{n} f(x_i^*)\, \Delta x$$

where x_i^* is any sample point in the ith subinterval $[x_{i-1}, x_i]$.

(b) What is the geometric interpretation of $\int_a^b f(x)\, dx$ if $f(x) \geqslant 0$?

If f is positive, then $\int_a^b f(x)\, dx$ can be interpreted as the area under the graph of $y = f(x)$ and above the x-axis for $a \leqslant x \leqslant b$.

(c) What is the geometric interpretation of $\int_a^b f(x)\, dx$ if $f(x)$ takes on both positive and negative values? Illustrate with a diagram.

In this case $\int_a^b f(x)\, dx$ can be interpreted as a "net area," that is, the area of the region above the x-axis and below the graph of f (labeled "+" in the figure) minus the area of the region below the x-axis and above the graph of f (labeled "−").

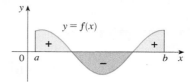

3. State the Midpoint Rule.

If f is a continuous function on the interval $[a, b]$ and we divide $[a, b]$ into n subintervals of equal width $\Delta x = (b - a)/n$, then

$$\int_a^b f(x)\, dx \approx \sum_{i=1}^{n} f(\bar{x}_i)\, \Delta x$$

where $\bar{x}_i = $ midpoint of $[x_{i-1}, x_i] = \frac{1}{2}(x_{i-1} + x_i)$.

4. State both parts of the Fundamental Theorem of Calculus.

Suppose f is continuous on $[a, b]$.

Part 1. If $g(x) = \int_a^x f(t)\, dt$, then $g'(x) = f(x)$.

Part 2. $\int_a^b f(x)\, dx = F(b) - F(a)$, where F is any antiderivative of f, that is, $F' = f$.

5. (a) State the Net Change Theorem.

The integral of a rate of change is the net change:

$$\int_a^b F'(x)\, dx = F(b) - F(a)$$

(b) If $r(t)$ is the rate at which water flows into a reservoir, what does $\int_{t_1}^{t_2} r(t)\, dt$ represent?

$\int_{t_1}^{t_2} r(t)\, dt$ represents the change in the amount of water in the reservoir between time t_1 and time t_2.

6. Suppose a particle moves back and forth along a straight line with velocity $v(t)$, measured in feet per second, and acceleration $a(t)$.

(a) What is the meaning of $\int_{60}^{120} v(t)\, dt$?

$\int_{60}^{120} v(t)\, dt$ represents the net change in position (the displacement) of the particle from $t = 60$ to $t = 120$ seconds, in other words, in the second minute.

(b) What is the meaning of $\int_{60}^{120} |v(t)|\, dt$?

$\int_{60}^{120} |v(t)|\, dt$ represents the total distance traveled by the particle in the second minute.

(continued)

(c) What is the meaning of $\int_{60}^{120} a(t)\,dt$?

$\int_{60}^{120} a(t)\,dt$ represents the change in velocity of the particle in the second minute.

7. (a) Explain the meaning of the indefinite integral $\int f(x)\,dx$.

The indefinite integral $\int f(x)\,dx$ is another name for an antiderivative of f, so $\int f(x)\,dx = F(x)$ means that $F'(x) = f(x)$.

(b) What is the connection between the definite integral $\int_a^b f(x)\,dx$ and the indefinite integral $\int f(x)\,dx$?

The connection is given by Part 2 of the Fundamental Theorem:

$$\int_a^b f(x)\,dx = \int f(x)\,dx \Big]_a^b$$

if f is continuous on $[a, b]$.

8. Explain exactly what is meant by the statement that "differentiation and integration are inverse processes."

Part 1 of the Fundamental Theorem of Calculus can be rewritten as

$$\frac{d}{dx} \int_a^x f(t)\,dt = f(x)$$

which says that if f is integrated and then the result is differentiated, we arrive back at the original function f.

Since $F'(x) = f(x)$, Part 2 of the theorem (or, equivalently, the Net Change Theorem) states that

$$\int_a^b F'(x)\,dx = F(b) - F(a)$$

This says that if we take a function F, first differentiate it, and then integrate the result, we arrive back at the original function, but in the form $F(b) - F(a)$.

Also, the indefinite integral $\int f(x)\,dx$ represents an antiderivative of f, so

$$\frac{d}{dx} \int f(x)\,dx = f(x)$$

9. State the Substitution Rule. In practice, how do you use it?

If $u = g(x)$ is a differentiable function and f is continuous on the range of g, then

$$\int f(g(x))\,g'(x)\,dx = \int f(u)\,du$$

In practice, we make the substitutions $u = g(x)$ and $du = g'(x)\,dx$ in the integrand to make the integral simpler to evaluate.

WebAssign Student Quick Start Guide

You can use WebAssign to access your homework, quizzes, and tests — whatever your instructor chooses — at any time of day or night, from any computer with a connection to the Internet and a Web browser. Your instructor creates your assignments, schedules them, and decides how many submissions you get. Your instructor also determines if you can have an extension, if you can save your work without submitting it at the time, and how much feedback you get after you submit an assignment.

The WebAssign support staff cannot change your username or password, give extensions, change your score, give you extra submissions, or help you with the content of your assignments.

Logging In
You can log in to WebAssign using any Web browser connected to the Internet. There are two different ways to log in to WebAssign. Each requires information from your teacher. If you are unsure about how to log in, please check with your teacher or another student in your class.

Go to the login page at http://webassign.net/login.html or the web address provided by your teacher. The way you log in depends on how your instructor set up the class:

- If your teacher created a WebAssign account for you, they will provide you with a **Username**, an **Institution** code and a **Password**. Simply enter this information in the boxes provided and click the **Log In** button.

WebAssign.

- If your teacher wants you to **Self-Enroll** in the WebAssign course they will provide you with a **Class Key**. You will create your own username and password. It is important that you remember this information so you can log in for the remainder of the class. In this case, just click the **I have a Class Key** button. You don't need to enter any other information on this page.

  Then, enter the **Class Key** your instructor provided and click **Submit**. Verify you are enrolling in the correct class on the next page.

Class Key

Enter the Class Key that you received from your instructor. You will only need to complete this once. After you have created your account, you can log in on the main page.

Class Key

[   ] [   ] [   ]

Class Keys generally start with an institution code, followed by two sets of four digits.

[Submit]

- Enter your preferred Login and Student information.

- Click the **Create My Account** button to complete the enrollment process.

- A review screen will display, showing your username, institution code, and password. **Retain a copy of this information.** You will need it to log into WebAssign.

Log In Information

Required fields are marked with an asterisk (*).

Preferred Username	* [ ] [Check Availability]
	Your username may contain letters, numbers, and the following characters: underscore (_), hyphen (-), period (.)
Institution Code	**webassign**
Password	* [ ]
Re-Enter Password	* [ ]
	Passwords are case-sensitive.

Student Information

Required fields are marked with an asterisk (*).

First Name	* [ ]
Last Name	* [ ]
Email Address	* [ ]
Student ID Number	[ ]

[Create My Account]

Access Codes

Once you log in, you may see a WebAssign Notice about entering an access code for your class. You can get an Access Code from any of the following places if you need to use one:

- A new textbook you purchased for the class.
- Your bookstore, which may sell Access Code cards.
- Online, where you can purchase an access code with a credit card.

You have a 14 day grace period to use WebAssign, starting with the WebAssign class start date. During this time you can work on and view your WebAssign assignments without registering a code.

After the grace period is over you will only see the code registration message until you submit or purchase a code.

There are two types of WebAssign access code cards. The small card requires you to scratch off the silver surface in order to reveal the complete access code.

WebAssign.

The larger security envelope card requires you to open the card to reveal the access code number.

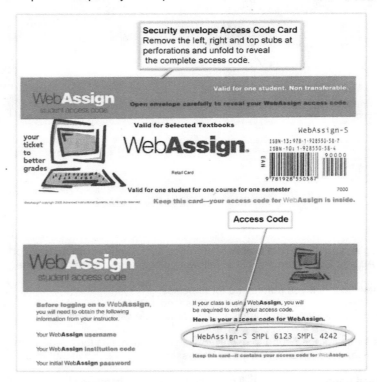

If you would like to purchase an access code directly from WebAssign online, you may do so with a credit card. Your code will be automatically registered to your WebAssign account as soon as the transaction is complete. You will receive an email confirmation. Please keep a copy for your records.

WebAssign.

Your WebAssign Home Page

Once you have successfully logged in you will see your WebAssign homepage. If you are taking more than one WebAssign class, you will need to select which class you wish to view first.

The upper right corner features links to a complete Student **Guide**, as well as a link to WebAssign Technical Support under **Help**. If you want to change your password or add or update your email address, simply click **My Options** in the upper right hand corner.

You will see your assignments and due dates listed, as well as any Communications, Grades, and Announcements posted by your teacher.

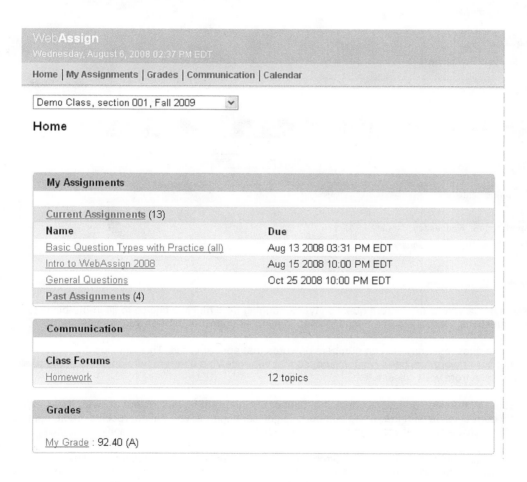

WebAssign
Wednesday, August 6, 2008 02:37 PM EDT

Home | My Assignments | Grades | Communication | Calendar

Demo Class, section 001, Fall 2009

Home

My Assignments	
Current Assignments (13)	
Name	**Due**
Basic Question Types with Practice (all)	Aug 13 2008 03:31 PM EDT
Intro to WebAssign 2008	Aug 15 2008 10:00 PM EDT
General Questions	Oct 25 2008 10:00 PM EDT
Past Assignments (4)	

Communication	
Class Forums	
Homework	12 topics

Grades
My Grade : 92.40 (A)

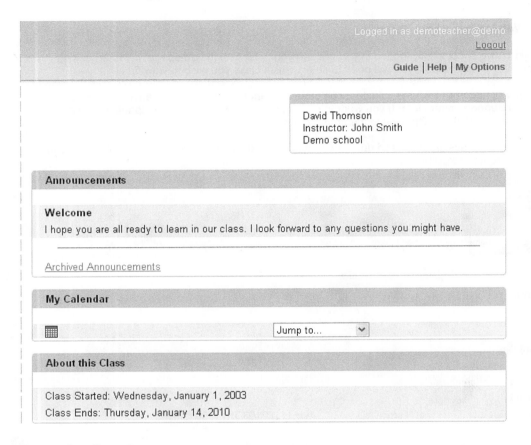

Guide | Help | My Options

David Thomson
Instructor: John Smith
Demo school

Announcements

Welcome

I hope you are all ready to learn in our class. I look forward to any questions you might have.

Archived Announcements

My Calendar

Jump to...

About this Class

Class Started: Wednesday, January 1, 2003
Class Ends: Thursday, January 14, 2010

Answering Questions

WebAssign has a variety of different question types, ranging from multiple choice to fill-in-the-blank to symbolic questions. Here are some things to keep in mind as you work through your assignments:

- Some questions may include numbers or words that appear in red. This signifies that the number or word has been randomized, so that you receive a different version of the same basic question from your classmates.

- Some WebAssign questions check the number of significant figures in your answer. If you enter the correct value with the wrong number of significant figures, you will not receive credit, but you will receive a hint that your number does not have the correct number of significant figures.

- Some questions require entering symbolic notation. Answer symbolic questions by using calculator notation. You must use the exact variables specified in the questions. The order is not important as long as it is mathematically correct. Clicking on the eye button previews the expression you enter in proper mathematical notation. Clicking on the symbolic formatting help button provides tips for using the correct keystrokes.

WebAssign.

- When you click on some WebAssign chemistry or math questions an input palette will open. These palettes, called chemPad and mathPad, will help you enter your answer in proper notation.

- Some questions may require the use of an Active Figure simulation. Active Figures require the free Macromedia Flash Player plug-in, downloadable from www.macromedia.com.

- If your instructor allows it, you can save your work without grading by selecting the Save Work button at the end of the question. After you save your work, it will be available to you the next time you click the assignment.

- Please note that WebAssign will **not** automatically submit your answers for scoring if you only **Save** your work. Your teacher will not be able to see your submissions. Please be sure to **Submit** prior to the due date and time.

- If your instructor allows it, you can submit answers by question part or for the entire assignment. To submit an individual question answer for grading, click the **Submit New Answers to Question __** button at the bottom of each question. To submit the entire assignment for grading, click the **Submit All New Answers** button at the end of the assignment.

Technical Support

If you are having difficulty logging in, please be sure to check with your teacher and verify whether an account has been created for you or whether you need to self-enroll. In either case your teacher needs to provide the appropriate information (username, institution code and password OR Class Key).

To email WebAssign Support go to http://www.webassign.net/info/support/report.html. This page also lists answers to **Common Problems**, and provides links to the **Student Guide**.

August 7, 2008